MATHEMATICAL CONNECTIONS: A MODELING APPROACH TO BUSINESS CALCULUS

MATHEMATICAL CONNECTIONS: A MODELING APPROACH TO BUSINESS CALCULUS

Preliminary Edition

Bruce Pollack-Johnson &
Audrey Fredrick Borchardt

The Villanova Project

PRENTICE HALL, Upper Saddle River, New Jersey 07458

Library of Congress Cataloging-in-Publication Data

Pollack-Johnson, Bruce,
Mathematical connections : a modeling approach to business calculus / by Bruce Pollack-Johnson & Audrey Fredrick Borchardt : the Villanova Project. — Preliminary ed.
p. cm.
Includes bibliographical references and index.
ISBN 0-13-576398-3
1. Calculus. I. Borchardt, Audrey Fredrick. II. Villanova Project. III. Title.
QA303.P73 1998
515—dc21 97-34502
CIP

Senior Acquisitions Editor: **SALLY DENLOW**
Marketing Manager: **PATRICE LUMUMBA JONES**
Editorial Assistant: **APRIL THROWER**
Assistant Vice-President of Production and Manufacturing: **DAVID W. RICCARDI**
Editorial/Production Supervision: **RICHARD DeLORENZO and BARBARA MACK**
Managing Editor: **LINDA MIHATOV BEHRENS**
Executive Managing Editor: **KATHLEEN SCHIAPARELLI**
Manufacturing Buyer: **ALAN FISCHER**
Manufacturing Manager: **TRUDY PISCIOTTI**
Marketing Assistant: **PATRICK MURPHY**
Creative Director: **PAULA MAYLAHN**
Art Director: **JAYNE CONTE**
Cover Designer: **BRUCE KENSELAAR**

Simon & Schuster / A Viacom Company
Upper Saddle River, NJ 07458

Printed in the United States of America

10 9 8 7 6 5 4 3 2 1

ISBN 0-13-576398-3

Prentice Hall International (UK) Limited, London
Prentice Hall of Australia Pty. Limited, Sydney
Prentice Hall Canada Inc., Toronto
Prentice Hall Hispanoamericana, S.A., Mexico
Prentice Hall of India Private Limited, New Delhi
Prentice Hall of Japan, Inc., Tokyo
Simon & Schuster Asia Pte. Ltd., Singapore
Editora Prentice-Hall do Brasil, Ltda., Rio de Janeiro

To our spouses
Linda and Hank
and our children
Amy, Paul, and Suzanne

Contents

PREFACE ix

1 PROBLEM SOLVING, FUNCTIONS, AND MODELS 1

Introduction 1
1.0 The Process of Problem Solving 2
1.1 Functions 14
1.2 Mathematical Models and Formulation from Verbal Descriptions 42
1.3 Linear Functions and Models 70
1.4 Functions with One Concavity: Quadratic, Exponential, Power 105
1.5 Functions with Changing Concavity: Cubic, Quartic, Logistic 131
Summary 153

2 RATES OF CHANGE 155

Introduction 155
2.1 Average and Percent Rate of Change Over an Interval 157
2.2 Instantaneous Rate of Change at a Point 176
2.3 Derivative Notation and Interpretation, Marginal Analysis 200
2.4 The Algebraic Definition of Derivative and Basic Derivative Rules 218
2.5 Composite Functions and the Chain Rule 244
2.6 The Product Rule 263
Summary 277

3 SINGLE-VARIABLE OPTIMIZATION AND ANALYSIS **280**

Introduction 280
3.1 Analysis of Graphs and Slope Graphs 282
3.2 Optimization—Algebraic Determination of Maxima and Minima 303
3.3 Testing of Critical Points, Concavity, and Points of Inflection 322
3.4 Post-Optimality Analysis 349
3.5* Percent Rate of Change at a Point, Elasticity, Average Cost 387
Summary 415

4 CONTINUOUS PROBABILITY AND INTEGRATION **418**

Introduction 418
4.1 Continuous Probability Distributions 421
4.2 Approximating Area under Curves (Subinterval Methods) 439
4.3 Finding Exact Areas Using Limits of Sums 465
4.4 Recovering Functions from their Derivatives 472
4.5 The Fundamental Theorem of Calculus 493
4.6 Variable Limits of Integration and Medians, Improper Integrals 511
4.7* Consumer and Producer Surplus 523
Summary 536

APPENDIX A MATHEMATICAL DETAILS **541**

APPENDIX B STUDENT-GENERATED PROJECTS **547**

APPENDIX C ANSWERS TO SELECTED EXERCISES **A-1**

INDEX **I-1**

*These sections are optional; later material does not depend on them.

Preface

AN OVERVIEW OF THIS BOOK

Problems! Problems! Problems! Life has many facets, including family, work, relationships, activities, and so much more. One dimension of life is the existence of problems and our attempts to solve them. A **problem** involves the awareness of a **challenge** or an **opportunity**, a belief that a situation could be improved. **Solving** a problem usually means finding some way to improve the situation, to the greatest extent possible. Frequently the best solutions to a problem are the most creative. Russ Ackoff tells a story of a company that was having a serious problem with employees complaining about long waits for elevators.[1] They solved it by installing mirrors! The idea of problem solving is closely related to the basic ideas of **decision making** and working toward **goals**. Some people might feel that the word *problem* has too many negative connotations, but we will use it because it is so deeply rooted in our vocabulary. We don't mean to imply anything negative, we are using it in the broadest, most inclusive sense.

In this book, we will focus on problems that can be represented or **modeled** in a **continuous** way; that is, the quantities involved are continuous by nature, *or close enough*. This book may not help you with *all* of your problems, but it is intended to help you to solve many problems that you are likely to encounter in your other college courses, in your career, and in your personal life.

Traditional math texts and courses are somewhat analogous to teaching someone to fly at 5000 feet (do the straight math needed to solve a problem); our emphasis in this book is to take time to teach you how to take off (set up a problem) and land (make sure it makes sense in the real world) as well.

In order to get a quick feel for the kinds of problems we will be studying in the two volumes of this book, here are some typical examples:

[1]Russell Ackoff, *The Art of Problem Solving*, (New York: John Wiley, 1978), p. 53.

- Your soccer team has decided to sell T-shirts to raise money to be able to attend a special all-day skills clinic in the state capital. You have been chosen to organize the entire effort. You want to be sure that you make money on this fundraiser—the last one was a disaster! How many T-shirts should you order? How much should you charge? Where do you begin?
- You play a sport or a musical instrument and want to figure out the ideal (**optimal**) amount of warm-up or practice before a match or performance. Or you want to find the optimal angle or distance for shots in a sport you play regularly.
- You are working for a small start-up firm with a radical new concept for cold-weather gloves. The firm has the initial capital to produce only one size glove. Fortunately, the design is such that, whatever actual size is produced, it should also fit people with a glove size that is up to 2 more or 2 less than the actual size produced. What single size, possibly fractional, should be produced to capture the largest possible market?
- We all have to learn how to manage our time. This can be very difficult if you are a first-year college student and this always had been done for you before now. You have to balance your need for sleep, exercise, studying, social life, extracurricular activities, and so forth. How can you get the maximum satisfaction with the way you divide your time? What is the best time of the day to do your math homework to get the highest grades? How many hours of exercise a week is best for you? How much sleep should you get every night to feel your best? How much should you study?
- You have some money you would like to invest. Your friend Rupert owns a newspaper delivery business. He is planning to go on a year-long walkabout in the Australian bush with an Aboriginal guide. He will not be in communication with the States, so he is looking for someone to take temporary ownership of the business (and keep the profits) while he is away. He is asking for an initial investment. You also have another investment opportunity to help another friend, Bonnie, who is a musician and wants to make a first CD. You and Bonnie have great confidence in the success of the album, and she agrees to give you a guaranteed amount one year from now. Which investment is better for you? (You don't have enough cash for both.)
- You are planning to mail a package and want to be sure that the post office will accept it. You called and they gave you information concerning the restrictions on the size of a package that they will accept. What dimension box should you get to stay within the regulations and still hold the greatest amount?
- You have decided to start a physical fitness/health/diet routine tomorrow. After consulting nutrition books and with your doctor, you are now working out guidelines for yourself on what to eat for breakfast. You decide that you want to get from your breakfast, on average, a certain limited amount of fat and calories, at least a fixed amount of fiber, and a specific amount of protein. You like to have just cereal for breakfast, and you have two favorite cereals. What should you eat for breakfast?

The first three examples above are typical of Volume 1 (Chapters 1–4), which focuses on situations where you want to find the optimal value and there is *one* major quantity over which you have control. The last three examples fall squarely in Volume 2 (Chapters 5–9), which focuses on situations where you want to optimize and there are *two or more* quantities over which you have control. The fourth example, about time management, could fall into either category, depending on the number of quantities over which you have control.

For some of these problems above, the **connection** to mathematics is easy to see. For others the connection is not so obvious, but it is definitely there. The power of mathematics can help us find solutions to these and many other problems, whether they arise in business, social science, or our personal lives. After you have worked with this book, we hope you will have substantially widened the categories of problems you are able to analyze quantitatively, including many problems you might never have guessed could be treated in that way.

The problems we will study in this book have a number of common characteristics. To distinguish this course from other quantitative courses, we will describe a number of these characteristics now.

Numerical quantities can be divided into two categories: **discrete** and **continuous**. **Discrete** means "composed of distinct parts." For example, you can *order* one hot dog or two hot dogs, but you normally can't order half of a hot dog. You can buy one T-shirt or two or three, but you usually can't buy one third of a T-shirt. A quantity being **continuous** (we will also use the term **divisible** as a synonym[2]) means exactly what it says: The set of possible values is an unbroken stream. You could *eat* one and a half hot dogs, or you can walk 1.235 miles or lose 5.3 pounds while on a diet. Some quantities are technically discrete, such as the price to charge for a T-shirt, which must be in whole cents, but are close enough to being continuous that we can *think* of them as continuous. The Census Bureau reports that the average family now has 1.20 children.[3] Of course, this sounds ridiculous, but we understand what they are trying to tell us. In fact, **we will assume throughout the book that *all variables are divisible* unless we specify otherwise. We will also assume that *all graphs and relationships are smooth* unless we specify otherwise.**

Look back over the problems listed above. Convince yourself that they do indeed involve quantities that can usefully be represented as divisible, and that the graphs related to them would be smooth. Note also that some of them involve concepts of **probability and statistics** (especially the problem about making the gloves, which involved knowing the relative probability or proportion of people with different glove sizes). In this book, we will talk about aspects of probability and statistics that are **continuous** in nature, for example, probabilities related to bell curves, such as for the distribution of SAT scores, and fitting curves to data. However, we will *not* discuss discrete aspects of

---

[2]Continuity has another meaning in mathematics; when the graph of a curve or a surface has no holes, gaps, or jumps, we say the *graph* is *continuous*. The difference is that the first kind of continuity mentioned above, which we will normally call *divisibility* from here on to avoid this confusion, refers to a *quantity* or variable, while the second kind of continuity refers to a *graph* or relationship between two or more variables. A property of graphs closely related to continuity is what we will call **smoothness**. A smooth curve is continuous **and has no angles or corners** (more precisely, if you kept magnifying it at any point, it would look more and more like a line there). In fact, **we will assume throughout the book that *all variables are divisible* (continuous in the first sense) unless we specify otherwise. We will also assume that *all graphs and relationships are smooth* (continuous in the second sense) unless we specify otherwise.**

[3]*Statistical Abstract of the United States, 1995*. U.S. Bureau of the Census.

probability, such as calculating probabilities of poker hands (sorry about that, poker players!). Most of the problems we will discuss will be making an assumption of **certainty**. This means **we assume that the relationships and equations hold *exactly* as they are written, without any fluctuations or uncertainty**. For instance, for the problem about choosing breakfast cereals, we can reasonably assume that the nutritional values and costs are stable and known. They may change slightly from time to time, but not so much that the uncertainty would be a major factor in solving the problem.

TO THE STUDENT

Most developments in mathematics, like that of our numeration system, have come about because people wanted to solve a problem, either theoretical or practical. Powerful mathematical tools have been developed from the time of the ancient Greeks through the present to help us solve many types of problems. In this book, we want to show you how to use these powerful tools to help you solve the problems that you face, both now and in the future. In order to use these mathematical tools, we must be able to express a problem in mathematical terms, we must have a **mathematical model** of the problem.

Traditionally, students in a math class are *given* equations that represent problems and are asked to solve them: Find the optimal solutions or to forecast from them. Occasionally, they are asked to formulate models from verbal descriptions, the dreaded *word problems.* In the real world, there will rarely be anyone to hand you these models on a silver platter, and if they did, how much confidence could you place in them? If your job depended on the analysis that you did on some model, you would probably want to know where that model came from. The most that you will probably have, particularly in this age of computers, is a lot of **data** and information about different quantities (variables). One of the skills that you will learn in this course is how to take data values and turn them into a mathematical expression of the relationship between variables. For example, you will learn how to analyze market research (the relationship between the selling price of something and the quantity that consumers will buy) to estimate a demand function. Some models can be constructed from definitions and basic principles, such as the fact that the revenue from selling a product equals the selling price times the quantity sold, and profit equals revenue minus cost. You will develop skills in recognizing and formulating this kind of model as well. You will start to learn how to solve **realistic problems** based on **realistic information.**

In this book we will explain how the power of mathematics can be used to take some of the guesswork out of decision making. We will study some of the mathematical concepts and techniques that have been developed to solve such problems. Most traditional mathematics courses proceed deductively. They introduce the theory first, then discuss its applications. In this book we **start with problems** that other students, business and social science faculty, and people in the nonacademic work world have found interesting and useful and then develop the theory needed to solve them. This way you can use the examples to build your intuition and understanding during the development of the various topics.

This book emphasizes the **understanding of the underlying concepts** involved in these mathematical techniques. When you really understand where a formula comes from or what a process means, you will know how and when to use it effectively. You will learn how to find something called a "derivative" in this course, but it is not enough to know "how to *take* the derivative." You have to know the *meaning* of the derivative that you have found in order to use it to solve problems. Knowing the proper way to use a tool (like a power saw or a sewing machine) is good, but we also want to know how to use that tool to actually *do* something. Our goal is for you to learn about the mathematical concepts involved so that you can solve simplistic problems by hand. However, working out solutions to *real-world* problems by hand can be a waste of time when we have the technology to "crunch the numbers," so we want you to be able to solve larger and messier problems by using this technology.

When you have finished this course, you should be able to solve many types of problems. When presented with a problem, you should be able to express it clearly and concisely, set it up mathematically, solve it mathematically, validate your model and assumptions, and finally reach conclusions and implement your solution. Each of these steps is equally important. For example, it is virtually worthless to be able to set up the problem and solve it mathematically if you do not know how valid your assumptions are. Computer people are fond of saying "Garbage In, Garbage Out"; a solution is only as good as the information on which it is based.

Many traditional textbooks in mathematics focus on only one or two of these five stages (solving the problem mathematically, and maybe setting it up). These are certainly very important, but not enough in themselves to solve real-world problems. This book will discuss each stage, and will help you practice them individually and then all together. This will help you **solve problems in future courses, in your career, and in your life in general.**

This is a course in *applied* mathematics. It is designed to prepare you to work with *real* data from the *real* world. In order to deal with real data, we will use all the modern **technology** available to us as tools. We will use the technology to draw pictures—graphs—of the numerical data. It is often very difficult to see a trend in numerical data. A graph of the data makes it much easier to see a trend, if there is one. We will use the technology and our common sense to help us determine a mathematical expression to represent these data for the intended purpose and context—to find a mathematical model. Once we have the mathematical model, we may use the mathematical technique of differentiation to study how our data are changing or to find an optimal solution. Or we might use an optimization program in a spreadsheet to determine prices to charge to maximize profit or a product mix to meet certain conditions and minimize costs. Using technology frees us from much of the tedious calculation commonly associated with doing mathematics and allows us to concentrate on the formulation of the problem and the interpretation of the results.

Many students don't see the usefulness of the mathematics they are studying, or any connection to other courses they are taking or will be taking in the future. We have spent a great deal of time and study with business and social science faculty and students to determine exactly what **mathematical tools you will need to succeed in your business and social science courses, careers, and personal lives.** The topics presented in this book are the results of those studies. You *will* be *using* this math and studying topics that are based on it, so it is important that you understand it thoroughly.

When you decide to learn to play a new game or a sport like tennis or basketball, it is important to spend some time learning the rules and the vocabulary of the game. And if you want to excel, you typically *must* spend quite a lot of time practicing the techniques involved in playing the game: for instance, forehand shots and backhand shots and serves for tennis; ball handling, shooting baskets, and rebounding for basketball. You *cannot* learn to play a sport simply by watching good players. If this were true we could all be pros! *Mathematics is like that—it is not a spectator sport.* If you want to be able to do mathematics, to use the power of mathematics to help solve problems, you must first learn the rules and the terminology and then **practice, practice, practice!** Shooting hoops for several hours may be more fun than doing practice problems in mathematics, but the sense of achievement when the skills are mastered can be quite similar. We wish you well in your efforts!

BASIC ORGANIZATION OF THE BOOK AND GENERAL TOPICS

Volume 1 (Chapters 1–4) focuses on problems with a **single decision variable** (one thing to be decided, like the amount of exercise to get each week), while Volume 2 (Chapters 5–9) focuses on problems with **two or more decision variables** (such as choosing between two breakfast cereals). In each part, the initial chapters (Chapters 1, 5, and 6) focus on the **process of problem solving and modeling** with appropriate **categories of functions** including how to **formulate models both from data and from verbal descriptions**. The other chapters in each part then analyze the models. Chapters 2, 3, and 8 explore **rates of change** of one variable with respect to another (such as velocity or marginal profit), and how these can be used to **optimize** the types of functions being studied. This is done both by hand and using technology. The major mathematical subjects we will explore are **single and multivariable calculus** (Chapters 2–4 and 8), which you can think of as the **language of change** (for continuous quantities), **linear (matrix) algebra** (Chapter 7), which studies how to **solve systems of linear equations** (useful for multivariable optimization), and **linear and nonlinear programming** (Chapter 9), which involve multivariable optimization with constraints. We will also discuss **continuous probability** and **descriptive statistics** (in Chapters 4 and 6), **compound interest** and **time value of money** (in Chapter 5), and the method of **least squares regression** (in Chapters 1, 5, 6, 7, and 8).

CORRECTIONS, COMMENTS, AND SUGGESTIONS

If you have any corrections, comments, or suggestions related to this material, we would greatly appreciate hearing them! You can mail hard copy to us at the Department of Mathematical Sciences, Villanova University, Villanova PA 19085, send us E-mail at brucepj@ucis.vill.edu, call us at (610) 519-6926, or fax us at (610) 519-6928. Thanks!

TO THE INSTRUCTOR

In recent decades, the first-year course in mathematics for business and social science undergraduates has normally consisted of some combination of calculus and "finite math." The finite math usually includes matrices, linear programming, some topics in probability and statistics, and possibly specific business topics such as compound interest, present value, break-even analysis, and so on. Faculty and students on both sides (math and business/social science) have become increasingly dissatisfied with the course for two reasons: It has been perceived as having no real connection to the students' lives, curriculum, or careers, and it has seemed to be a random jumble of unrelated topics. Students tend to see the course as a hazing to be endured and forgotten as soon as possible after the trauma—a torture that is an initiation ritual into the "club" of being a business or social science graduate. Faculty and administrators have tended to see the course as an academic filter, to weed out weaker students from academic programs.

This does not need to be the case! As a former chief financial officer at Dupont has said, "***Mathematics is the language of business***." Whether it is profit and loss, risk, net present value, internal rate of return, forecasts, probabilities, expected values, market research, hypothesis testing, maximizing profit or efficiency, or minimizing cost, business is all about numbers, estimates, calculations, and optimization. **Calculus** can be thought of as the **language of continuous change**. If there is anywhere in the world that is continuously changing, it is the business world, especially now with the advent of global competition. In fact, most of the topics in most traditional versions of this course *do* have relevance to business and social science, but the connections are not communicated to the students (and often not to the math faculty teaching the course).

The Villanova Project text takes a very different approach to this course. Our goal is to address the causes of the dissatisfaction. Our goal is to show the **connections between the mathematical topics** in the course **and the student's world** (present, and future) in as many ways as possible, and to communicate the **connections between the topics themselves**. Our primary focus is on the **process of problem solving in the real world**. In the introduction to Chapter 1, we give both brief and detailed outlines of this process and suggest the metaphor that most math texts teach students how to fly an airplane at a cruising altitude of 5000 feet (solve *mathematical* problems) but do not really teach them how to take off (formulate a *real-world* problem) or how to land (validate their model and solution). Our purpose in this book is to teach ***all* of the steps** needed to use mathematics **to solve real problems in the real world**. This includes clarifying a problem, making ballpark estimates, collecting data, defining variables clearly and unambiguously (including units), fitting models to data, understanding concepts in order to select appropriate methods and technologies, finding solutions, verifying (double-checking) calculations, validating assumptions of the models, performing sensitivity analysis (seeing the effect on the solution of plausible changes in the data), estimating a rough margin of error, and writing up conclusions.

To give students experience with this entire process from beginning to end, we strongly recommend the use of **student-generated projects**. The idea is to have students select topics in which they have a strong personal interest and apply the problem-solving process to the topic. Examples of successful topics from our experience include deciding the quantity to order and the price to charge for T-shirts being sold to raise money for a student group; determining the optimal number of hours to study (or sleep, or exercise, or the optimal *combination* of two or more of these) in an average week to maximize personal satisfaction; determining the optimal speed in a car to maximize fuel efficiency; finding the optimal combination of breakfast foods to eat to meet nutritional goals at minimum cost; and determining the optimal angle and/or distance for a shot in basketball (or hockey or water polo or . . .) to maximize the scoring percentage. When the students see how math can really help *them*, it makes all the concepts of the course come alive. After internalizing the process of problem solving with a topic they care about, they can then apply it to all kinds of problems. In fact, in their future careers, problems from work will have the same intensity and interest to them as personal topics do *now*, as beginning college students.

Because of our emphasis on solving real-world problems, we focus heavily on **mathematical modeling** with real data, usually by fitting curves. Thus, instead of *giving* students a function to be optimized (they typically *cannot imagine* how this would be given to them in the real world, and they are right!), we teach them how to take realistic data (which they *can imagine* being given to them, or getting themselves) and fit a model to those data points. We explain the *idea* of least squares from the beginning and teach that choosing a model depends on **how the model will be used** as well as on the data (e.g., one model may be great for interpolation but terrible for extrapolation). We explain that the Sum of the Squared Errors (SSE) or coefficient of determination (R^2) should **not** be the main criteria in choosing a model, but only used to choose *between* models which both reflect the shape of the data.

Given that our world is **technology** oriented, we seek to make the most of this reality. This is the career environment in which these students will find themselves. Although we expect students to be able to **solve simplistic problems by hand** to show mastery of the concepts, it does not make sense for them to try to solve realistic problems by hand. **Technology** now makes it possible to solve real problems relatively quickly.

We focus on two technologies that are commonly used by almost everyone in business: calculators and spreadsheets. (Other technologies, such as computer algebra systems, can also be used.) The calculator we recommend is the TI-83 (an old TI-82 will work fine for this course as well) or an equivalent calculator, for its curve-fitting ability and general level (not too elementary and not needlessly advanced or expensive). The TI-83 includes business functions (such as present value) and statistical programs, so this *can* be the only calculator the student will need for many years. The graphing capabilities make many calculus topics much less important (such as curve sketching and the First and Second Derivative Tests). On the other hand, spreadsheets are quite interchangeable these days, but we recommend one with some kind of optimization function (both linear and nonlinear), such as Quattro Pro or Excel. Many calculators and spreadsheets have similar features, so either technology could work alone, but we believe the combination is most powerful: graphing calculators for the single-variable calculus and matrices and spreadsheets for the multivariable regression and constrained optimization.

Technology reduces the need for some mathematical and algorithmic skills but *increases* the need for many other skills. Near the top of this list is the **ability to translate between words** (or the real world) **and algebraic symbols**, as in **problem formulation**. Students typically have a lot of trouble with this (hence their common aversion to "story problems"). We distinguish two major types of formulation: The first is to take **verbal descriptions** and put them together into relevant equations and functions such as formulating a linear program, and the second is to take or gather **data** and find a model by fitting curves. The latter type comes up with a modeling approach and focuses more on unambiguous definitions of variables and functions, as well as least-squares regression models. In both cases, translating back from symbols (and computer/calculator output) to the problem or real-world situation is also necessary. We have segments in the book to focus on all of these processes.

Other skills essential to problem solving include the ability to make ballpark estimates by hand or in your head (to know if a solution makes sense), verifying solutions (double-checking, for example, by a different technique or technology), sensitivity analysis, and validation, as discussed above. Again, we have sections of the book devoted to these topics, and the students can gain hands-on experience with them on their projects.

The topics of the course and book have been worked out as a result of exhaustive discussions among faculty in mathematics, business, and the social sciences. The idea of the course is to teach the fundamentals of problem solving and modeling listed above and to cover the mathematical topics which are the foundation of quantitative concepts that arise in these disciplines. The unifying mathematical theme is the analysis of quantities that can usefully be modeled as being **continuous** (as opposed to discrete). Unlike most traditional finite math courses, we do not try to teach **probability and statistics** in themselves (we leave this for a true statistics course), but we do use problems in these areas **to motivate and apply the basic concepts of the single-variable calculus material**. For example, people in business and social science use hypothesis testing, means, medians, modes, and variances all the time. To understand these ideas mathematically, one must have some understanding of the concept of an integral. We start by calculating the mode of a distribution as an example of single-variable optimization. We then use the idea of calculating probabilities for continuous random variables as the main (but not only) motivation for finding the area under a curve, after which we use the calculation of expected values and medians as applications of integration.

We do cover a small amount of **multivariate calculus** (partial derivatives and optimization) which is often not covered in finite math. For all of our calculus material, we focus more on **intuitive understanding** of the concepts graphically, numerically, algebraically, and verbally (sometimes called the "**rule of four**"), rather than on theoretical details that are less relevant for these disciplines. For example we do not get into calculating limits in general or focus on discontinuous functions. We also do not cover implicit differentiation, related rates, most techniques of integration, or Taylor series, because our business/social science colleagues felt these were not a high priority for their first-year students. We plan to develop a more advanced third semester course for students in majors such as economics to cover these and other topics. One of the unifying and culminating topics for *this* full two-semester course is the mathematical derivation of the **method of least squares**, including the use of matrices to calculate the regression parameters. Since both semesters use the idea of fitting models to data, this is a perfect capstone to the calculus material and is one of the reasons why we consider it so important to include the material on partial derivatives.

The sequence of the course is designed to maximize the **connections and interactions among the topics**. We strongly recommend the full-year sequence: The first semester focuses on single-variable analysis and optimization, and then the second semester focuses on multivariate analysis and optimization. This means that students with AP Calculus credit can jump into the second semester. Within the second semester, we use an original sequence by covering matrices *before* multivariate optimization (so matrices can be used to solve a system of linear equations when optimizing a quadratic objective). We then cover linear programming *after* multivariate calculus so that the idea of a partial derivative can be used to better understand the simplex method and the idea of a shadow (dual) price.

We describe the development in the book as **problem driven** (sometimes called the "*way of Archimedes*"). The book as a whole, each chapter, and each section start off with a number of typical problem types to motivate and give students a feel for the usefulness of what follows. It is our experience that students have quite extraordinary intuition with real problems involving real numbers. Thus we use these realistic problems to develop that intuition and then generalize it to understand the major concepts of the unit. This way the theory comes out of the applications, and the applications are always there to make the theoretical abstractions concrete and comprehensible.

A number of sections, and some topics within sections, are not essential to any later topics and so can be considered **optional**. These have been indicated with asterisks. For optional topics within a section, the exercises at the end of the section that relate to them are also marked with asterisks. This material is purely at the discretion of those who determine your syllabus.

The book has been used in a course spanning two 14-week semesters. Each section is designed for one 75-minute Tuesday/Thursday class session. For 50-minute Monday/Wednesday/Friday classes, covering two sections in three class sessions works about right. The Instructor's Manual indicates which sections are likely to need two such sessions and which can be done in one. Shorter semesters can cut out some of the optional sections and topics. Quarters and trimester systems could break up the chapters differently, such as Chapters 1–3, then 4–6 and finally 7–9. Or Chapter 7 and 9 could be covered more thoroughly for a short quarter course on matrices and linear programming. If you have the luxury of a fourth hour per week, it would be a great opportunity to go into the technology in more depth, and to spend more time on group learning and discovery-oriented laboratories.

Technically, the material in Chapter 5 (multivariable functions and models, including compound interest and present value and future value) and Chapters 7 and 9 (matrices and linear programming) could be covered independently at any time. As discussed above, however, we *strongly* recommend the sequence as presented here to optimize motivation and interconnectedness of topics.

An additional feature of the Villanova Project is a collection of brief **videos of faculty and nonacademics from business and social science** showing how the math of this course underlies what they do in the concurrent and subsequent courses they teach and in their other work. We recommend that each school actually film its own faculty (we can provide scripts to help make this easier), so the students will be watching familiar faces and names they know they will encounter later in their degree programs. We will also have videotapes available for general use. These should be available by mid-1998. This is one more way to keep students from believing "I'll never see or need this stuff again."

In summary, this book is a model of a truly collaborative service course. It has been designed to supply the concepts and skills that colleagues in business and social science want for their students, while maintaining mathematical integrity with an innovative selection and sequence of topics. It emphasizes the interrelationships among the mathematical topics, as well as the connections to students' curriculum, careers, and personal lives via realistic problems, videos, and student-generated projects. It is problem driven, uses technology to minimize drudgery, and makes it possible to solve real-world problems. It adopts a modeling approach and emphasizes **critical thinking** and understanding of concepts. It focuses on ***all* of the steps in the process of problem solving** (including translating between the real world, mathematical symbols, and computer/calculator output), reinforced by student-generated projects.

ACKNOWLEDGMENTS

Work on this project was made possible by funding from the United States Department of Education's Fund for the Improvement of Post-Secondary Education (FIPSE), Grant

P116B51374, and from the National Science Foundation, Grant DUE-9552464. We would like to thank our colleagues at Villanova's College of Commerce and Finance for their valuable contributions, and our colleagues in the Department of Mathematical Sciences of the College of Liberal Arts and Sciences for their enthusiasm, support, and extensive input. We want to acknowledge the groundbreaking work by the Clemson University FIPSE project spearheaded by Don LaTorre, John Kenelly, Iris Fetta, and others, which helped give us a starting point for our own work in this area, and to appreciate the support of Tami Wederbrand, Sally Denlow, Jerome Grant, Alan MacDonell, David Chelton, Barbara Mack, Rick DeLorenzo, John Tweeddale, and everyone else at Prentice Hall. We would also like to thank graduate students Cathryn Matuzza, Chad Cimo, Warren Towns, Laura Parese, and Greg Liano for their help with creating and writing up solutions to exercises and other comments, suggestions, and general assistance. Finally, we would like to thank all of our students for their creative projects and suggestions and for putting up with the frustrations of being guinea pigs, and to thank our families for being so patient, understanding, and supportive and coping with long working hours and hogged computers. Without them this book could not have been written.

Bruce Pollack-Johnson
Audrey Fredrick Borchardt

MATHEMATICAL CONNECTIONS: A MODELING APPROACH TO BUSINESS CALCULUS

CHAPTER 1

Problem Solving, Functions, and Models

INTRODUCTION

In the Preface, we list some examples of problems that this book will help you to solve. One goal of the book is for you to come away with a general strategy for the entire **process of problem solving**. We will start this discussion in Section 1.0, where we present the five major stages of problem solving and our 12-Step Program for Plugaholics. The types of problems we will be studying are those for which it is helpful to have a mathematical representation of the relationships between the relevant quantities involved. *The linear equation for the total cost of an item as its given price times the number of units bought is an example* of a special kind of relationship between variables called a **function**, which we will define and explore in Section 1.1. Throughout Volume 1, our functions will always be in terms of just one variable. This mathematical representation of a real-world situation is called a **mathematical model** and will be discussed in detail in Section 1.2, including how to specify a verbal definition as well as a formula for the function and stating any assumptions being made that are significant. In the rest of the chapter, we then discuss specific **categories of functions** and models: **linear functions** in Section 1.3, functions with a single type of curvature or **concavity** (such as quadratic, exponential, and power functions) in Section 1.4, and functions with changing concavity (such as cubic, quartic, and logistic functions) in Section 1.5.

Here are some examples of the kinds of problems this chapter will show you how to solve:

- You help run a local coffeehouse. You have past data about how much you have charged for concerts by a particular group, and how many people bought tickets. You are considering changing the price for an upcoming concert. You want to predict paid attendance, as well as your ticket revenue, for several different prices.
- You take weekly quizzes in one of your classes. You have been keeping track of how long you studied for a quiz, and the grade you ended up with on that quiz. You want to use this data to predict how long you should study to get at least an 80 on the next quiz.
- You have a favorite computer game on your PC. You have noticed that when you play for a while, your scores at first get better and then they start to get worse as you get tired and lose concentration. You gather data to see what your score is after different amounts of practice. You want to find a formula to predict what your score would be given a specified amount of practice time.

By the end of this chapter, in addition to being able to solve problems like those above, you should

- Know the five major stages of the problem-solving process.
- Understand how to apply the first 2 stages of the 12-step problem-solving process.
- Understand what is meant by a mathematical model and how to fully define one.
- Understand what a single-variable function is.
- Know when a situation corresponds to one quantity being a function of another, from a verbal description, table, formula, or a graph.
- Understand what is meant by the sum, difference, and product of two functions and by a composite function.
- Know what the basic shapes of linear, quadratic, exponential, power, cubic, and logistic functions look like and be able to look at a data graph to pick which of these function types could be part of a reasonable model for the data.
- Understand what interpolation and extrapolation mean and how the choice of one of these and the context of a given situation affect the selection of a model, in addition to the shape of the data.
- Understand the concepts of slope, vertical intercept, concavity, and points of inflection.

1.0 THE PROCESS OF PROBLEM SOLVING

There are two major categories of problems that we will consider in this book. The first category is **optimization** problems, which involve finding the best way to do something in order to maximize or minimize some quantity. The second category is **forecasting** problems, which involve making projections or predictions about what will occur under circumstances *different* from the values in your data, often involving projections into the future. If you look at the example problems listed in the beginning of the Preface, all of them involve optimization, such as maximizing profit, minimizing cost, maximizing personal happiness, or maximizing volume (cost-effectiveness). In addition, the T-shirt and glove examples both involve making forecasts of sales in some fashion.

In business, the main goal is usually optimization (especially maximizing profit or shareholders' wealth), but often forecasts are needed to be able predict the value being optimized. In social science, the goal of research is often to understand and quantify a relationship between different variables (such as time and knowledge, in a learning curve), which means being able to forecast an outcome knowing the values of certain variables. In our personal lives, we are usually trying to optimize, whether personal happiness, income, costs, justice, quality of life, or whatever. We may try to build forecasts, such as a forecast of our financial position in a few years, to *identify* potential problems. For example, if our forecasted debt payments five years from now are too high a percentage of our forecasted income five years from now, we may try to take action soon, such as to get an extra part-time job, to change that.

Sample Problem 1: Suppose you are Illinois Jones, on a trip in a strange land, trying to find an ancient sacred temple in a jungle. You are told by your guide that the only way to get there is through a huge maze built to protect the entrance to the temple. If you make a wrong turn, you could encounter a deadly beast. You have just landed at the tiny local airport and notice that there are some small airplanes available for rent (and pilots for hire). What can you do?

Solution: One strategy would be to rent one of the planes and hire a pilot and tell the pilot to fly low over the maze. While flying low over the maze, you could look down on it to see the layout and the location of the beast, draw a sketch (or take a Polaroid™ photograph) of it, and note the location of the beast's lair on your picture. You could then tell the pilot to climb to a safer and higher cruising altitude. Here you could take some time to study your picture and visually find the best path through the maze, check that it indeed seems to be the best, and draw it in on your picture. On your way back down to land, you could then have the pilot fly low over the maze again so you could check between the maze and your drawn-in solution path on the picture, to make sure you have drawn it as accurately as possible. If necessary you could then make corrections or improvements. Back on the ground, you could then follow the solution path you drew on the picture you made, using it as a map through the maze, and find the temple! □

What does this example tell us about the **process of problem solving**? In it, we can see the **five major stages in solving real problems**:

A. **Clarify the problem**
B. **Formulate a model**
C. **Solve/Analyze the model**
D. **Validate the model and solution**
E. **Draw conclusions and implement a solution**

This example is an analogy for how we propose to solve optimization and forecasting problems. In this analogy, the real world is like the ground, and the world of math is "up in the clouds." Stage A, clarifying the problem, happens on the ground (talking to the guide, noticing the airplane). Stage B, formulating a model (the picture of the maze), is the connection (or translation) from the real world to the symbolic world, including gathering data (flying over the maze, getting the picture). Stage C, solving the model, occurs in the world of math (up in the clouds, visually finding the most direct path through the maze and drawing it in). Stage D, validating the model, consists of flying lower again to check and make sure the model (the picture) accurately represents the real problem (the maze), and that the solution makes sense, making corrections and improvements where necessary. And stage E, drawing conclusions and implementing a solution, corresponds to coming back to the real world (the ground) and using the abstract solution (the map/picture with the solution path drawn on) to solve the real problem (negotiate the physical maze).

Many math books focus on stage C, but spend little or no time on the other four stages. This is like teaching someone to fly at an altitude of 5000 feet (which is relatively easy), but never teaching them to take off or land (the really challenging parts)! No wonder many students have trouble with "story problems," where more of the stages than just stage C are involved! In reality, pilot training consists of many more hours of takeoff and landing practice than straight level flying. In this book, besides some of the fairly standard mathematical material (stage C), we will also focus on the other stages to give you the tools you need to really make *use* of the mathematical knowledge.

Let's now look at an example of a project actually done by a student like you, to illustrate how this same process can be applied to a problem that is more conducive to quantitative analysis, such as those we will emphasize in this book. This is a good *model* (pun intended) of the process of solving a real-world problem.

EXAMPLE 1: A student named Meg wanted to know what was the best amount of exercise for her to get. How could she figure this out?

Stage A: Clarify the Problem

The first step in solving this problem was to clarify exactly what the problem was. Was Meg trying to get into shape? Was she trying to lose weight? Was she trying to find the right balance between exercise and the rest of her life? Was she trying to find the level of exercise that would give her the most energy for her other activities? The initial description of this problem could have meant many different things, not only to different people, but even to Meg herself at different points in time.

Meg decided that she was primarily interested in the effect of exercise (running, in her case) on how she felt. But what did she really mean by "how she felt"? At first, she thought that perhaps this meant "happiness." Then she realized that, although her running did indeed affect her happiness, so did many other things that were much harder to control (relationships, school, etc.). The running might not have been a major factor in her happiness much of the time. She realized that what she was really interested in was how she felt *physically*, which she also thought of as her *energy level*. When she exercised too little, she felt lazy and lethargic, but if she exercised too much, she could feel exhausted, which would also correspond to a low energy level for other activities. She realized that her energy level might also be affected by her amount of sleep, but at the time, she was very careful to get eight hours of sleep every night, so that was not really a factor to be considered. Finally, she realized that she didn't always have time to run every day, so decided she would focus on determining the optimal amount of running each *week*.

Clarified Statement of the Problem: Meg decided that her problem was to find the amount of running each week that would maximize her overall energy level for the week.

Stage B: Formulate a Model

Now the question was how to gather data to help find the optimal amount of running. She decided to keep a log of her running, with the date, time of day when she started, and length of time (in minutes) running. But how should she record her energy level? She decided to use a subjective 0–100 scale, where 0 meant no energy (comatose) and 100 meant a virtually infinite energy level. She made up a scale as follows:

100: Infinite energy
90: *Extremely* high energy level
80: Very high energy level
70: High energy level
60: Pretty high level of energy
50: Medium energy level
40: Moderately low energy level
30: Low energy level
20: Very low energy level
10: *Extremely* low energy level
0: No energy

The next question was: How often should she record her energy level? She could record it every day (say, just before going to bed at night), and average the values over a week. She could record it just before running every time, and again take some kind of average. Or she could record it at the end of each week, evaluating for the week as a whole. In some ways, the weekly assessment seemed closest to the idea of her problem, as long as she felt her memory was good enough to properly reflect the week overall. Meg felt she could do this, so that was how she gathered her data. She could have decided to gather data in several ways, do different analyses, and compare the results, for even more confidence in her conclusions.

The following table gives the results of her data collection, with the number of minutes of running and the corresponding energy level for each of ten weeks:

Running (min)	380	465	420	240	275	390	730	355	440	605
Energy (0–100)	68	90	80	30	40	70	45	60	75	85

The table is interesting and gives some clues about the optimal solution. Looking for the highest energy levels, they occur at 465, 420, and 605 minutes, so it would seem that the optimal level should be somewhere in that vicinity numerically. But it is hard to get a good feel for data just from a table. The old cliché says "a picture is worth a thousand words," and that is as true in math as anywhere else. For data values, a good way to make a picture out of them is to draw a graph. In this case, we can draw a graph, with running time in a week (in minutes) on the horizontal axis and energy level (using the 0–100 scale) on the vertical axis. Each data observation (week) can then be represented as a data **point** on the graph. For instance, the fourth data point can be written as the **ordered pair** (240,30) and plotted as the point that lines up with 240 on the horizontal axis and 30 on the vertical axis. On most graphing calculators and spreadsheets, such data can easily be entered and plotted on a graph. You are probably used to calling the first value in the ordered pair x and the second value y. If we use that convention here, then x is

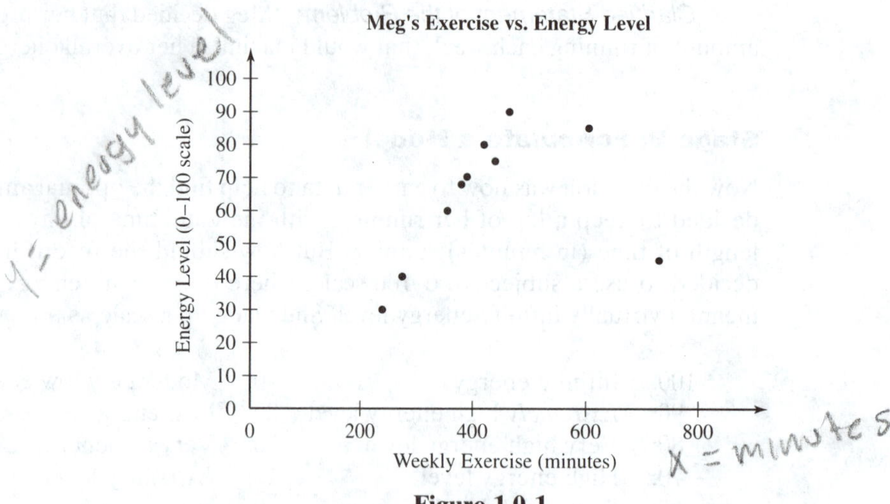

Figure 1.0-1

defined to be the amount of exercise in a week (in minutes) and y is the energy level (using the 0–100 scale). The resulting graph is given in Figure 1.0-1.

This graph gives a much better feel for the data, and we can already get a better estimate of what seems to be optimal: between 465 and 605 minutes (the highest two points), and probably closer to 465 than to 605. To make this more precise, we could try to find a mathematical equation whose graph would come as close as possible to the data points, sometimes called a **best-fit curve**. From looking at the plot of the data points, what general category of graph would seem to fit reasonably closely? To help refresh your memory, a **parabola** can be either the shape of a "U" or an upside-down "U," and its general equation can be written $y = ax^2 + bx + c$. Graphing calculators and spreadsheet programs use a technique called **least-squares regression** (discussed briefly in Sections 1.3–1.5 and in great detail in Chapters 5, 6, 7, and 8) to find the best-fit parabola for this set of data points. You will learn how to do this in Sections 1.3–1.5. The equation and graph that result for these data are given in Figure 1.0–2 (the numbers in the equation have been rounded).

We call this quadratic **function**, $y = -0.000786x^2 + 0.809x - 123$, along with the definitions of the variables and the assumptions we have made, a **mathematical model** of the relationship between exercise and energy level for Meg (analogous to the sketch of the maze). The useful x values for this model (the model equation is sometimes called our **objective function**) would seem to be values between about 200 and 800, since that is the approximate interval spanned by the data. Notice that the numbers in the model have been **rounded off**. We will discuss rounding in more detail later in this section, in Section 1.2, and in Section 3.4.

Stage C: Solve/Analyze the Model

The y value in the model is expressed in terms of the x value, and we could define it as y = the energy level (on the 0–100 scale) *on average* that results from running x minutes in a week. The concepts of functions and models will be further explained in Sections 1.1 and 1.2. We can now use our model to find the optimal average level of exercise, at least according to the model. You will learn a number of ways to do this in Section 3.2, both

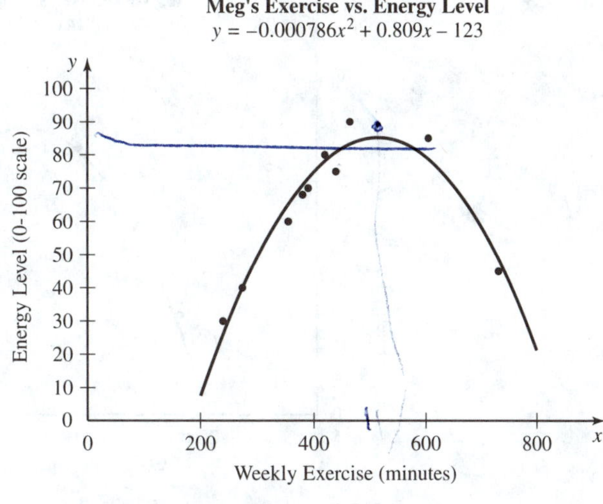

Figure 1.0-2

by hand (using calculus concepts directly) and using technology. The answer comes out to approximately 515 minutes. To **verify** that this calculation is correct, we can use multiple methods (for example, do it *both* using technology and by hand). In this example, both methods give the same answer, 515 minutes.

Stage D: Validate the Model and Solution

How good is this answer? Visually, we can see that the curve is pretty close to the data overall, although it is not perfect. You may have noticed that one of the data points [if you look carefully at the graph and the table, you can see it is the point (440,75)] seems to be a little out of line compared to the others. Perhaps some other factor that had nothing to do with the exercise she got affected Meg's energy level that week. How might the curve and the optimal solution change if we removed that data point? When checking her records and her memory, Meg decided that this unusual data point was *not* representative (she had been sick that week, which lowered her energy level, independent of her running), so she decided it was best to remove it. (She could also have tried to estimate what her energy level would have been if she *hadn't* been sick, and use that *y* value instead of the 75 to go with the *x* value of 440.)

We should emphasize that **just because data do not form a nice pattern is *not* a reason to throw them out!** We only do so in this case because it indicated an *unrepresentative* answer. There could well have been nothing special about the circumstances of that data point, in which case we would *not* remove that data point. This would simply mean that the relationship between the variables had more **random fluctuation** or **uncertainty**, and so we should not expect the model to fit as closely. This would also suggest less precision and a larger **margin of error** in the solution.

After removing that data point, the resulting data plot and best-fit curve turned out as given in Figure 1.0-3. You can see that the model has not changed much, and the new optimum comes out to be about 514 minutes. This process is an example of what is called **sensitivity analysis**, a process in which you try to determine which of your data values are most questionable, think of what might be more representative values (or remove those

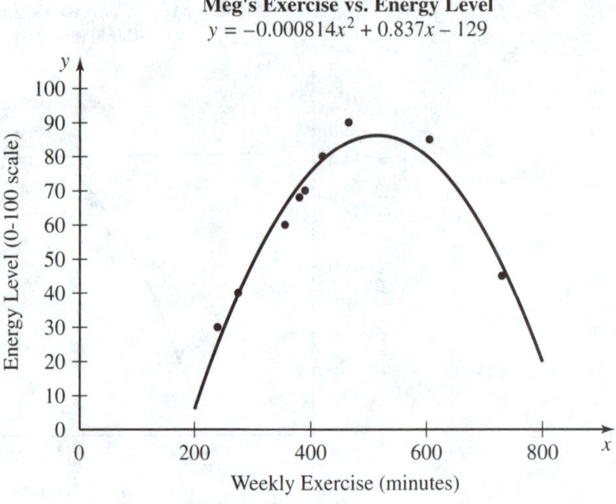

Figure 1.0-3

data points), and see what effect the change has on your solution. In other words, you explore how "sensitive" your model and solution are to changes in your data that seem possible or plausible.

In looking at this second graph (Figure 1.0-3), the fit is not bad, but it also seems that we might be able to find a smooth curve that would fit even better. Later on, we will look at more complicated functions we could try that would fit these data points better, but we will try a simpler strategy for now. Since what Meg is *really* interested in is her *optimal* amount of exercise, we could focus on trying to model the relationship closely *only near the optimal point*, rather than everywhere. For example, if you look at the graph, you can probably see that if we could choose three or four data points near the optimum and fit a parabola to them, it would seem to represent what happens in that vicinity much better than the parabola fit to all the data points in Figure 1.0-3.

If we choose just the three highest data points, we can actually use algebra to find the parabola that best fits them, because there is exactly *one* parabola that will go through three different points that are not all on the same line. In Chapter 7, you will learn some concepts and tools to be able to do this quickly and easily on a graphing calculator or spreadsheet. You can also use the same best-fit method we used earlier to find this parabola. Either way, the result is shown in Figure 1.0-4. The optimum this time comes out to be about 522 minutes.

This process of checking how our models and solutions fit the real-world situation is one component of what is called **validation**. One method of validation, as we just saw, is to visually evaluate the fit of a model to the data. When you see that the fit of a model is not great, you can try *different* models, and see which seem(s) to make the most sense for the context of the problem, as we did above. Validation also involves checking the validity of the assumptions you have made in your model, whether intentionally or not. In this example, and in most of the examples in this book, the main assumptions we will be making are the assumption of **divisibility** of the variables (so they can take on *any* values in the interval of interest, including fractions and decimals) and the assumption of **certainty** (that the mathematical equation gives the *exact* relationship between the variables).

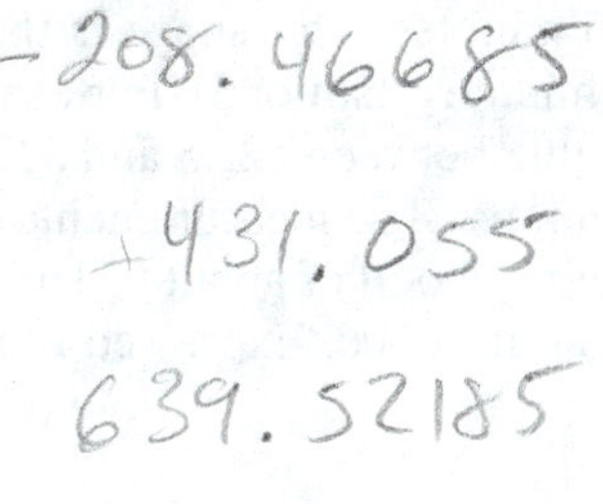

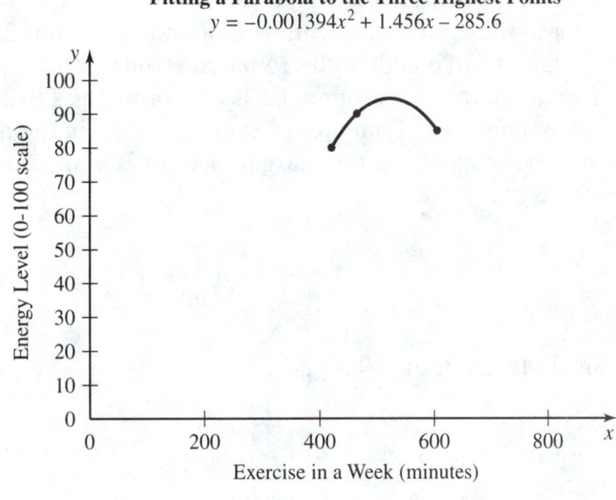

Figure 1.0-4

For this example, the assumption of divisibility for the variables seems very reasonable. She *could* have run for 341.5 minutes or 408.24 minutes. She *could* have recorded her energy level to any degree of precision, 67.3 for example. It would be hard to estimate that precisely, but she *could* have done it. You could even say that the assumption of divisibility in this situation holds *perfectly*, since neither variable is in any way discrete by nature.

Evaluating the fit is really testing the certainty assumption. In Meg's case, her data did not seem to have a lot of random fluctuations (especially after removing the one unusual data point), so the better fitting model seems justified. Another method of validation is to gather *new* data *after* fitting your model, to see how the model prediction for the dependent variable (y here) compares to the actual value. Meg tried running 515 minutes and had an energy level of 93 for that week. The first quadratic model—including the unusual data point—predicts an energy level of $y \approx 85$ when you plug in 515 for x in the model: $y = -0.000786(515)^2 + 0.809(515) - 123 \approx 85$. The second quadratic model—using all of the data except the unusual data point—predicts an energy level of approximately 86 for $x = 515$: $y = -0.000814(515)^2 + 0.837(515) - 129 \approx 86$. The third quadratic model—using only the three highest data points—predicts about 96: $y = -0.00139(515)^2 + 1.46(515) - 286 \approx 96$. This gives further support to the use of the third model, since it gives the best prediction. If it's not possible to gather new data values like these (such as if annual data won't be released for a few months), you can also validate your model by *not* including the most recent data point when fitting your model and then using that withheld data point to check how the model prediction compares to the actual value. It is also possible to withhold *several* data points for the same purpose.

Stage E: Draw Conclusions and Implement a Solution

From all of this analysis, it is important to try to assess how far off your answers could be, largely because of the lack of precision or accuracy in your data, or because of uncertainty about what model is best. This is called the **margin of error** in the results. From looking at the data, the exercise times seem quite accurate, probably to within a few minutes, since they are rounded off to the nearest multiple of 5 minutes. This suggests that the error in these measurements is about plus or minus 2.5 minutes (half of 5). For example, a value of 525 could be the rounded estimate for any value between 522.5 and 527.5, so the margin of error around 525 is plus or minus 2 to 3 minutes. The measurements of the optimal number of minutes of exercise are all in the neighborhood of about 500 minutes, so roughly speaking, the margin of error is about 2 or 3 out of 500. Expressed as a percent, 2 out of 500 is

$$\frac{2}{500} = 0.004 = 0.4\%$$

Similarly, 3 out of 500 is

$$\frac{3}{500} = 0.006 = 0.6\%$$

Since we are simply trying to get a *rough* idea of the margin of error of this measurement, we can say it is *below 1%*.

In general, **significant figures** (sometimes called **significant digits**) means the number of meaningful digits after and including the first nonzero digit of a numerical quantity. From the perspective of **significant figures** of accuracy in the exercise time measurements, 3 significant figures would mean that the exercise times were correct to the nearest minute (a 3-digit number, like 482 minutes), and 2 significant figures would mean that the exercise times were correct to the nearest 10 minutes (if you round off to the nearest multiple of 10, a value of 480 minutes would have 2 digits, or 2 significant figures, of accuracy).

For Meg's energy level, her estimates seem only accurate to plus or minus 2 energy level points or so (many are rounded off to the nearest multiple of 5, but other values like 68 are also given). Since the magnitudes of the energy level values near the optimum are mostly around 90, this suggests a margin of error of plus or minus 2 energy level points out of 90, which works out to a margin of error of about 2 to 3% and a precision of only 1 or 2 significant figures (1 would mean accurate to the nearest multiple of 10, such as rounding to 60, and 2 would mean accurate to the nearest unit, such as rounding to 63).

In general, **the accuracy of your final result is determined by the *least* accurate values in the calculation.** This would suggest that the most accurate final answer we can expect is a precision of about 1 or 2 significant figures, and an error of about 2 to 3%. Notice that the different optimal solutions from the three models we tried, ranging from 514 to 522, reflect a similar margin of error. For instance, 522 minus 514 is 8, and 8 out of 500 (the neighborhood of the time values) is a 1.6% error. Calculating the margin of error and most of the last two stages will be discussed more fully in Section 3.4.

So what should Meg do? What **conclusions** should she come to? The best guess at an answer would seem to be for her to run about 522 minutes (about 8.75 hours) each week, give or take 10–15 minutes (roughly 2–3% of 522, expressed in a way corresponding to how people tend to think), averaging about an hour and a quarter (8.75 / 7 = 1.25) per day. With that level of exercise, she should be able to achieve an energy level of about 90, give or take 4–6 points or so, based on the sensitivity analysis and the different model results. Now it's up to Meg to **implement** this solution! □

With the above example under our belts, let's now expand upon the **problem-solving process** to see more of the details. This list may seem overwhelming, but as you go along and start to actually apply the steps to problems, the meaning should become increasingly clear, and you will probably appreciate having a list like this to refer to when you get stuck!

Twelve Steps for Problem Solving

We call the process our **12-Step Program for Plugaholics**. By "plugaholic," we mean someone who is addicted to blindly following "plug-and-chug" or "cookbook" solution methods to standard problems: Plug in the numbers and grind out the answer with no thinking involved. The following 12-step process is a way to "de-tox" from such an addiction by inducing you to think about problems critically: To understand the concepts and assumptions behind methods so you can use technology intelligently; to make ballpark estimates to see if your results make sense; to understand how to work with appropriate levels of accuracy and precision; to translate between the real world and symbols (in both directions); and to check your model, solution, and assumptions. We have tried to give you a "model" to follow by pointing out the aspects of Meg's problem that correspond to each step.

Stage A: Clarify the Problem

1. **Identify, bound, and define** the problem. In the example, identifying the problem corresponded to the initial basic idea of finding the optimal amount of exercise. Bounding would include specifying that she was focusing on conscious specific exercising and not including just walking around campus. Defining the problem involved deciding that her objective was to maximize her average overall energy level.
2. **Understand the problem** and **restate** it clearly, concisely, and unambiguously, including any **assumptions** being made, usually related to how the problem was bounded and defined. In some cases, sketches or organizing the given information into tables, graphs, and so forth can help in understanding the problem. In Meg's example, the problem was to find the amount of time for her to spend per week on conscious exercise (running) to maximize her subjective evaluation of her average overall energy level for the week. In this example, there were no sketches or tables that would particularly help clarify this any more. The main assumption was that she was only focusing on *conscious* exercise. At this point in the process, it is a good idea to make an initial ballpark guess at what you think the solution might be. For example, Meg felt from the beginning that her optimal solution would probably be at least an hour a day (420 minutes per week) of exercise.

Stage B: Formulate a Model

3. **Define all relevant variables clearly and unambiguously (including units)**, with the eventual analysis in mind. To figure out what you should use for your decision variable(s), determine what questions need to be answered, and what

actions or choices need to be decided. In the example, the only decision variable was the amount of exercise (x), since this is the decision Meg had to make to optimize her energy level. Remember that we were careful to specify the units (minutes). The energy level was then the variable, y, which represented Meg's objective function, using her 0–100 scale.

4. **Plan for and gather data** as necessary. In general, **it is best to collect as much data as possible** (such as writing dates and time of day for Meg), because you might think of new analyses as you get into the problem. If you collect unnecessary information (such as Meg recording energy level both daily and weekly), you can always disregard it. If you *missed* something important, you would have to go back and do it all again. Once you have data, you can again make a ballpark guess at an optimal solution to help verify and validate your solution. In Meg's problem, eyeballing the graph suggested a ballpark solution between 465 and 605 minutes of exercise per week.
5. **Formulate a mathematical model** for the problem in symbols, including unambiguously defining all expressions, including their variables and domains (possible values for the variables), and specifying all modeling assumptions. We have gone through an example of formulating a mathematical model from a set of data. (You may want to consider withholding some of your data for validation; see step 9). For models based on data, you can use technology to find a best-fit model.

   If you are formulating a mathematical model from a verbal description of the problem, it is often helpful to assign a set of specific numerical values to the variables ("**plug in some numbers**") and do the calculations of the relevant quantities to see if the conditions of the problem are satisfied. If necessary, assign another set of numerical values to the variables, and perhaps even a third or fourth set. Finally, look for patterns so you can generalize symbolically with variables and parameters. **Parameters** are values that stay constant for a *specific* model from a category but may change from one data set to another in that same category. For example, when looking at models of the form

$$y = mx + b$$

   the parameters are m and b.

Stage C: Solve/Analyze the Model

6. Based on a good **understanding** of the **concepts** involved, **select an appropriate method of analysis and an appropriate technology**, whether paper and pencil or a supercomputer. From having taken this course, Meg knew which methods of analysis were appropriate, and knew that finding the model and maximizing it could be done both using technology and by hand.
7. Perform the **analysis**; that is, **solve** the mathematical problem. This is the focus of most math books, and probably what you associate with math.

Stage D: Validate the Model and Solution (Post-Optimality Analysis)

8. **Verify** (double-check) your calculations (for example, solve with a different technology). Meg did her calculations using technology and by hand for verification. **If you do not get the same answer when verifying, redo all of your cal-**

culations (perhaps try a third method) until you understand why the answers were different and have full confidence in the answer.

9. **Validate your model**, that is, see that it makes sense back in the real-world situation. The simplest validation of a model is looking at a graph of the model *with* the data points to visually **see how good the fit is**, if this is possible. Where this is not possible, there are statistical measures (such as R^2, which we will discuss in Chapter 6) to help assess how good the fit is. Besides applying your common sense, you can **gather new data and see how closely the model prediction matches what really happened** (or, you can **withhold data** when formulating your initial model to use similarly). Do whatever you can to **check your modeling assumptions**. As with verification, if the validity of a model is questionable, you could try using a different model or method of analysis to see if you get similar results. Meg validated her model by getting a new data point, and trying different models. Her actual new observation did match her models reasonably well, although the fit was not perfect. Her only real assumption (other than divisibility, which makes sense intuitively) was the certainty assumption, which says that the model tells the *exact* relationship between exercise and average overall energy level. The fact that the data did not vary much from a smooth curve and the good fit of the model validated that assumption nicely.

10. Perform **sensitivity analysis** to **see the effects of plausible changes** in the data on the final solution given by the model(s). In particular, think of which data values you have least confidence in and try other reasonable values for them. For example, you may want to adjust for dubious extreme values, biases, or unusual circumstances or conditions. In Meg's example, the main sensitivity analysis was removing the point that seemed out of line, *since she felt it was not representative*. If she had not felt that way, it should have been included when fitting the model. The fit would not have been as good, but it would have been important to include that point to accurately reflect the real normal situation.

Stage E: Draw Conclusions and Implement a Solution

11. Evaluate the overall **accuracy and precision (margin of error)** of your solution, based on your validation and sensitivity analysis, the reliability of the data, any biases, and so forth. Remember that the **accuracy and precision of your answer are determined by the *least* accurate data values**. Technology often gives you answers to many significant figures of precision, but for most realistic problems, three to four significant figures is the best you can hope for. It is a good idea to carry several extra digits in your calculations (*as many as possible* in your intermediate calculations, but at least two extra digits wherever possible), and then round off only for your final results and conclusions.

12. **Synthesize and communicate a summary of your conclusions** in clear and correct language that is **consistent with the original statement of the problem**, including areas for further exploration. Of course, it is then the responsibility of the decisionmaker (you or the person for whom you are doing the analysis) to **implement the solution** in the real world! In Meg's case, if her level of sleep started to vary more, she could do an analysis involving **two** independent variables (exercise and sleep) to see how they affected her energy level.

In this book, we will focus on applying this process of problem solving to optimization and forecasting problems involving divisible variables. The problems have been identified by students like you, as well as faculty and alumni, as being most useful and interesting for someone at your stage in your academic career. We hope the first example helps give you a feel for how mathematical models can be useful for both optimization and forecasting. Meg used her model to help her identify the number of minutes of exercise she should get each week to maximize her energy level. Meg could also use her model to predict her energy level if she runs for 600 minutes in a week by plugging in $x = 600$ into her model.

Assumptions

As discussed in the Preface, in this book we will focus on problems involving **variables** that are **divisible**, which means that they can take on *any* value in an interval, including *any* fraction, decimal, or irrational number like π, $\sqrt{2}$, or e. Even when a situation by its nature is discrete, such as involving money or the number of T-shirts ordered, we will normally **assume** divisibility throughout the mathematical analysis, then round off as needed in the final conclusion. We will also make the assumption of **certainty** throughout our mathematical analyses. This simply means that we assume that our equations and expressions give the *exact* relationships between variables, even though in the real situation there is likely to be some uncertainty or margin of error. Once again, this margin of error can be brought in during the conclusions. We will also assume that the expressions and graphs we work with are **smooth**, which means there are no holes, gaps, jumps, or angles in the graphs. **Throughout this book, if not otherwise stated, you can take for granted the assumptions of divisibility of variables, certainty of equations and expressions, and smoothness of expressions and graphs.**

1.1 *FUNCTIONS*

The purpose of this course is to help you learn how to solve problems; in particular, to use the power of mathematics to solve problems. To do this, we must first build mathematical models, as we discussed in Section 1.0. A central part of the model for Meg's problem was the mathematical expression, the equation that was fit to the data. This is probably the part of the model with which you are most familiar—you have worked with equations in algebra. In this section, we will explore the notion of a particular kind of mathematical object (the kind we will focus on in this book) that *typically* corresponds to an equation, called a **function**.

Here are the kinds of problems that the material in this section will help you solve:

- You are a big hockey fan and really keep up on all the team standings, determined by points. A friend bets you that you can't identify the team if you are told the number of league points. Should you take the bet?
- You have a syllabus for this course that lists the point range corresponding to each letter grade. If you know your final average, can you figure out what letter

grade you should get? If a friend of yours in the class receives her final letter grade, can she figure out what her average was?

- You have a fixed amount to spend for T-shirts for your fund-raiser. You have gotten a price list that includes the order price (fixed cost) and price per shirt. How many shirts can you order?

By the end of this section you should be able to solve these and similar problems and

- Have a clear idea of what a function is.
- Have a good understanding of the concepts of ordered pair, relation, inverse relation, and the Vertical Line Test.
- Have a firm grasp of functional notation.
- Feel comfortable with the terms input, output, independent variable, and dependent variable.
- Be able to identify the input (independent variable) and the output (dependent variable) in a real-world situation.
- Be able to define a function for a simple word problem, including a specification of the domain.
- Be able to find the output of a function for any specified input value, given the definition of the function, even for symbolic input values such as $(a + h)$.
- Be able to find one or more input values that yield a given output value for a function, either by hand (algebraically) or using technology.

In Section 1.0, we worked through an example to help Meg determine her optimal amount of exercise per week to maximize her average overall energy level. We fit a quadratic equation to the data given: $y = -0.000786x^2 + 0.809x - 123$ where y = the energy level (on a scale of 0–100) on average that results from running x minutes a week.

Notice that the effect of this mathematical equation is that, if we choose a value for x (the number of minutes of exercise Meg gets in a week), it gives us a formula to find the value of y, and that this formula will give us *exactly one* answer for this value of x. In fact, that is why we used x as part of our definition of y. The mathematical expression or equation gives a method for finding *the* value of y that goes with any given value of x within the set of possible values. This is exactly the basic idea behind what is called a **function** in mathematics. Such a function can then be used to find an x value that goes with the *best* y value (an **optimal solution**). Let's look at some other examples of functions now.

Examples

Here are some other examples of functions to give you a feel for what we will be doing in this section. See if you can see what they have in common with the equation for Meg's energy level above.

EXAMPLE 1: The following table appeared in the sports section of a newspaper.

NHL Eastern Conference Atlantic [Division]	Pts.
N.Y. Rangers	88
[Philadelphia] Flyers	87
Florida	85
New Jersey	77
Tampa Bay	77
Washington	76
N.Y. Islanders	48

Source: Philadelphia *Inquirer*, March 26, 1996.

EXAMPLE 2: The following table might appear on a class syllabus:

Semester Average	Letter Grade
90–100	A
80–89	B
70–79	C
60–69	D
59 or below	F

EXAMPLE 3: You want to order some T-shirts. The basic order cost is $50 and the shirts cost $7 each. The mathematical expression that describes the cost of ordering the T-shirts is

$$C = 7x + 50, \text{ for } x \geq 0$$

where x is the number of T-shirts ordered, and C is the total cost of the order.[1]

EXAMPLE 4: You help run a local coffeehouse and have had the same musical group perform three years in a row. You have charged a different admission price each year, and attendance has varied. The first year you charged $7 per ticket and 100 people attended. The second year 80 tickets were sold at $8 per ticket. Last year you charged $9 and sold 70 tickets. Using the data, you were able to determine that the number of tickets you can expect to sell for a concert by the same group this year can be expressed by

$$n = -15p + 203, \text{ for } 0 \leq p \leq 13.50$$

where n is the number of tickets you can sell and p is the price, in dollars, of a ticket.[2] The graph is shown in Figure 1.1-1.

[1]We will discuss the formulation of this cost equation (and general formulation *from a verbal description* like this) in more detail in Section 1.2.

[2]We will discuss the formulation of this demand equation (and general formulation *from data* like this) in more detail in Section 1.2. For simplicity here, the numbers have been rounded.

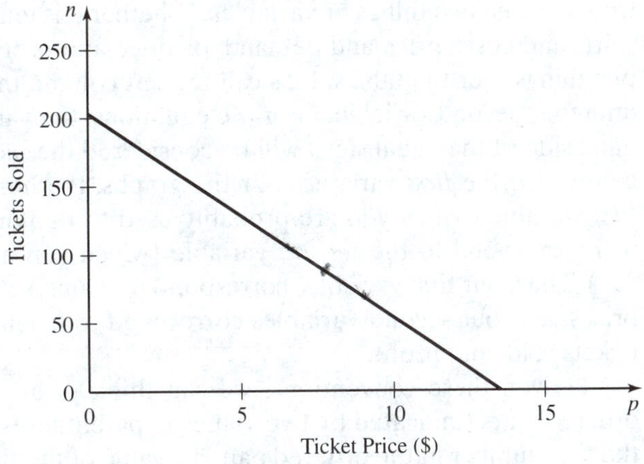

Figure 1.1-1

EXAMPLE 5: Knowing that the cost of the musical group is a fixed \$300, the profit from the sale of tickets can be expressed as

$$P = -15p^2 + 203p - 300, \text{ for } 0 \le p \le 13.50$$

where P is the profit and p is the ticket price.[3] The graph of this model is shown in Figure 1.1-2.

Take a minute now to look back over the information given in Examples 1–5. Just focus on the tables, graphs, and equations. As you do, try to see what the given relationships have in common.

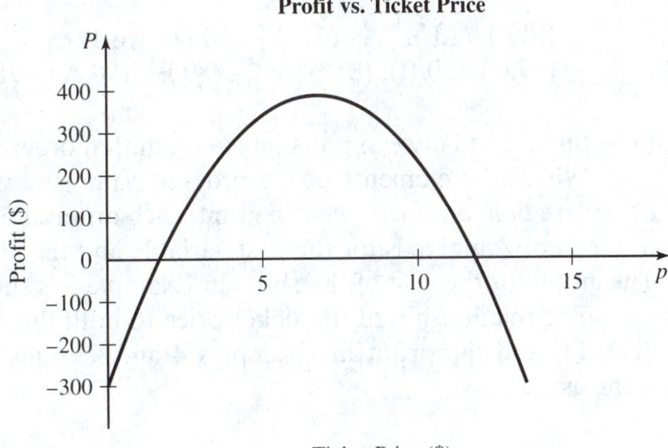

Figure 1.1-2

[3]We will discuss the formulation of this profit equation (and general formulation *from both a verbal description and data values* like this) in more detail in Section 1.2. For simplicity here, the numbers have been rounded.

One common characteristic of all of these problems is that they involve a *relationship* between two things or variables. Whether it is teams and points, averages and grades, shirts and costs, price and demand, or price and profit, all depict a relationship between two things. For the tables, let's call the left column the "first variable" and the right column the "second variable." For the equations, the variable that is solved *for* (on the left-hand side of the equal sign) will be considered the *second* variable, and the other will be considered the *first* variable. For the graphs, the horizontal axis will correspond to the *first* variable (which you are probably used to thinking of as "x"), and the vertical axis will correspond to the *second* variable (which you are probably used to thinking of as "y"). Thus, our first variables correspond to teams, semester averages, T-shirts, and ticket prices, and our second variables correspond to league points, letter grades, order costs, tickets sold, and profit.

Given these conventions, we can think of any of these relationships as a set of ordered pairs (indicated by two values in parentheses, with a comma in between), where the first number in the ordered pair is a value of the first variable and the second number is a value of the second variable that is paired with the value of the first. The mathematical term for such a set of ordered pairs is a **relation**. For Example 1, the relationship could be written in this form as the relation

{(N.Y. Rangers, 88), (Philadelphia Flyers, 87), (Florida, 85), (New Jersey, 77), (Tampa Bay, 77), (Washington, 76), (N.Y. Islanders, 48)}

(You may recall from high school that squiggly braces, {}, are used to indicate a set.) For Example 2, we have to clarify an ambiguity. When we consider values for semester averages, will they be rounded off or in "raw" form? For now, let's take the simpler choice and assume that our semester averages are already rounded off. In that case, we do not want to waste space and enumerate all of the ordered pairs, but we can represent them as follows:

{(0,F), (1,F), . . . , (59, F), (60,D), (61,D), . . . , (69,D), (70,C), (71,C), . . . , (79,C), (80,B), (81,B), . . . , (89,B), (90,A), (91,A), . . . , (100,A)}

(The three dots between pairs indicate omitted ordered pairs that follow the pattern.)

Where the elements of the ordered pairs (the values of the first and second variables) are *both numbers*, we can **graph** each ordered pair as a point. The convention is to use the horizontal axis for the first variable and the vertical axis for the second variable. The graph for the cost of the shirts in Example 3 is shown in Figure 1.1-3.

The relationship of the ticket price to both the demand (expected number of tickets sold) and the profit in Examples 4 and 5 could be expressed in a similar general form, as

$$\{(p, -15p + 203), \text{ for } 0 \le p \le 13.50\} \text{ and}$$
$$\{(p, -15p^2 + 203p - 300), \text{ for } 0 \le p \le 13.50\}$$

respectively, with graphs as shown in Figure 1.1-4.

As mentioned earlier, any set of ordered pairs is called a **relation** in math. Thus, all of our sample problems are relations. What else do they have in common?

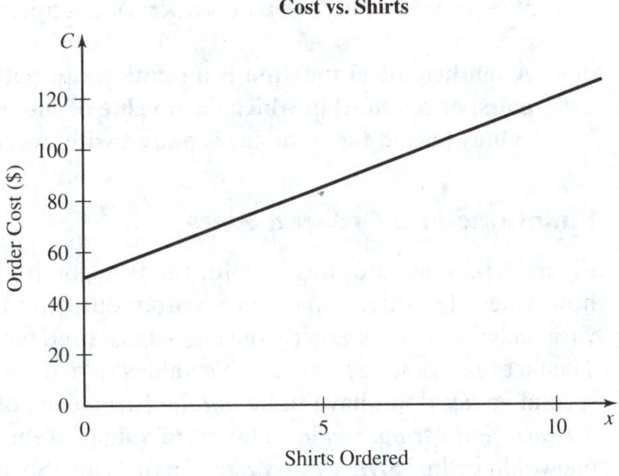

Figure 1.1-3

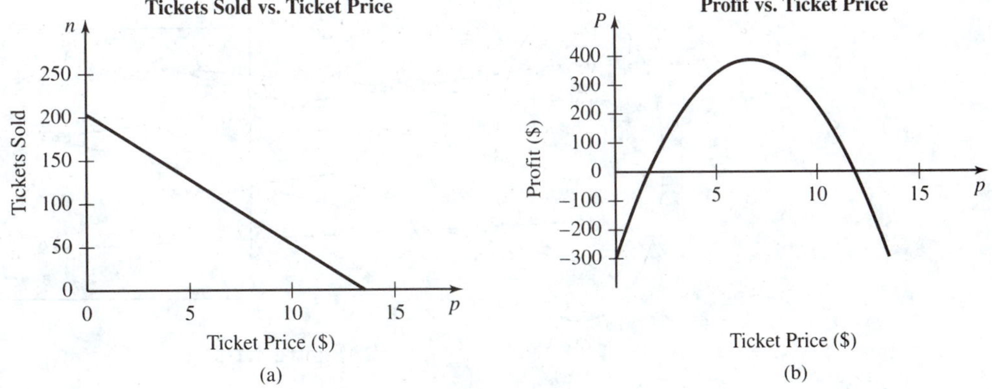

(a)

(b)

Figure 1.1-4

Definition of a Function

You might have noticed that in *all* of these examples, once an allowable value of the first variable is specified, there is *exactly one* value of the second variable that is paired with it. For example, in the hockey problem, once a team is specified, there is exactly one point total for that team. Notice that the point values are not all *different* (two teams have 77 points), but there is always exactly one point value for each team. Thinking of the relationships as ordered pairs, this means that there are no two ordered pairs with the *same* first variable value and *different* second variable values (check this with the hockey problem). This type of relationship is exactly what is meant by a **function** in mathematics. Notice that we specified that we have to start with a value of the first variable that is *allowable*. **The set of all allowable values of the first variable is called the *domain* of a function.**

We are now ready for a definition of a mathematical function:

A mathematical function is a relationship between two variables (a set of ordered pairs, or relation) in which each value of the first variable from the set of allowable values (called the domain) is paired with *exactly* one value of the second variable.

Functions and Ordered Pairs

Figure 1.1-5 illustrates this idea for the function {(0,7),(1,6),(2,8),(3,6),(4,9),(5,7)}. Notice how, when the ordered pairs are written out, with no duplicate ordered pairs, each first variable value occurs exactly once (so it is paired with exactly one second variable value). The fact that some second variable values (like 6 and 7) occur twice is not a problem. The second values don't have to be *unique* (from each other); there just has to be *exactly one for each first variable value*. **The set of values of the second variable**, which in this example would be {6,7,8,9}, **is the *range* of a function**. Specifying the range can get complicated, but we will normally not need to do so.

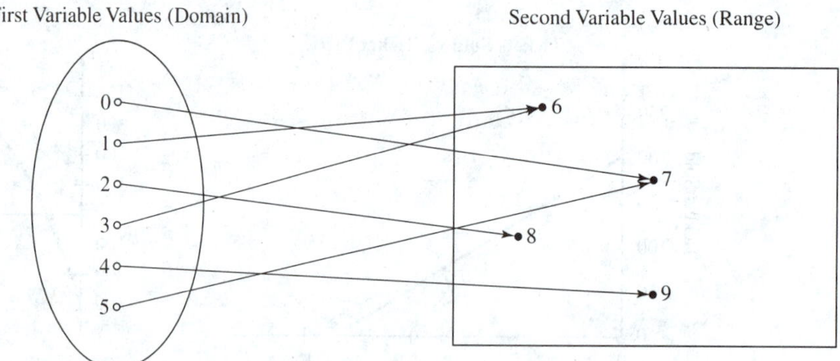

Figure 1.1-5

The main situation in which you would have to specify the range occurs when you are graphing a function based solely on its equation. Using technology, you can make a table that spans the allowable first values, the domain, and then plug these values into the equation to determine the function values, which helps determine the range, at least approximately. The concert ticket problem stated that the number of tickets sold could be expressed by

$$n = -15p + 203, \text{ for } 0 \leq p \leq 13.50$$

This means that the domain was prices ranging from 0 to 13.50. Table 1 shows the calculations for the range, the possible numbers of tickets sold. From the table we can see the reason for limiting the domain to values of p less than or equal to 13.50—the number of tickets sold would have become negative for values higher than 13.53, and we rounded this down to 13.50. For graphing purposes, you can hedge a bit (round to more extreme values at both ends), since you will not have checked *every* value in the domain. For this example, you could use an interval like 0 to 250 on the vertical axis to cover the full range.

TABLE 1

Price	Tickets Sold
0	203
5	128
10	53
11	38
12	23
13	8
13.5	0.5
13.53	0.05
13.54	−0.1

Thus the essence of a function is its set of possible first variable values (its **domain**) and a **rule** (or set of directions, analogous to the arrows in Figure 1.1-5) that describes how to determine the second value for *each* possible first value. Given these two pieces, the full relationship can be constructed; that is, the entire set of ordered pairs can be defined. The question to ask yourself when determining if a relationship is a function is: **given any possible value of the first variable in the domain, can you determine "beyond the shadow of a doubt" the (one) value of the second variable that is paired with it?**

We can represent this way of specifying a function visually with the diagram in Figure 1.1-6.

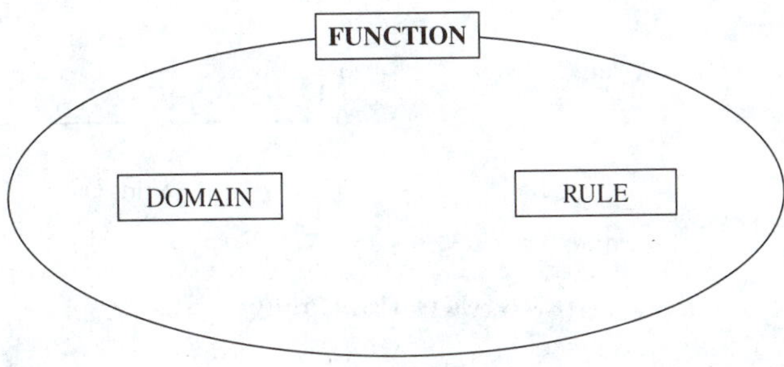

Figure 1.1-6

Functions and Their Graphs; The Vertical Line Test

We mentioned above that if both variables in a relation (set of ordered pairs) are numbers, then we can graph the relation as a set of points, using the horizontal axis for the first variable and the vertical axis for the second. We also said that if a relation is a function, then no two different ordered pairs can have the same first value and different second values. Graphically, two ordered pairs with the same first value and different second values would line up vertically, so they would lie on the same vertical line. This leads us to the **Vertical Line Test** for determining from a graph whether a relation is a function or not:

The graph of a relation (a set of points) where both variables are real numbers represents a function if, and *only* if, there are *no* vertical lines that intersect the graph in two or more points.

Put differently, saying that a graph represents a function is **equivalent to saying that *every* vertical line intersects the graph in *at most* one point**. It's OK if some vertical lines don't intersect the graph at all; that just means those corresponding first values are not in the domain. Consequently, if you can find *any* vertical line that intersects a graph in two or more points (even if there is only *one* such vertical line), then the graph does *not* represent a function. Look at the graphs for Examples 3, 4, and 5 (Figure 1.1-7) to confirm that they all pass the Vertical Line Test, and so indeed correspond to functions.

Figure 1.1-8 shows the graph of the equation $y^2 = x$. This graph does not pass the Vertical Line Test: The vertical line at $x = 25$ hits the graph at $y = 5$ and at $y = -5$ (at the points (25,5) and (25,−5)). Thus the equation $y^2 = x$ represents a *relation*, but does *not* represent a *function*.

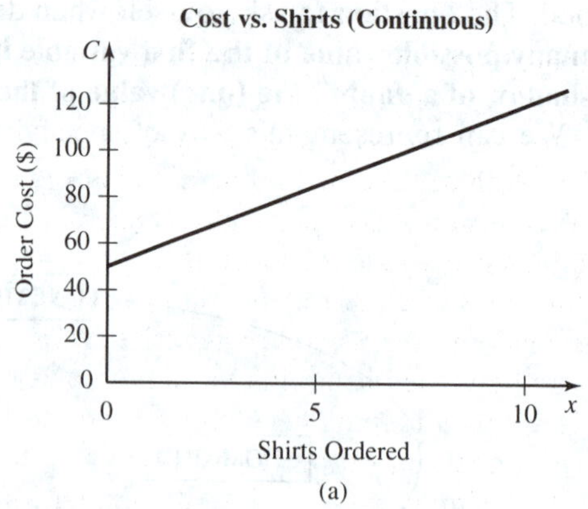

(a)

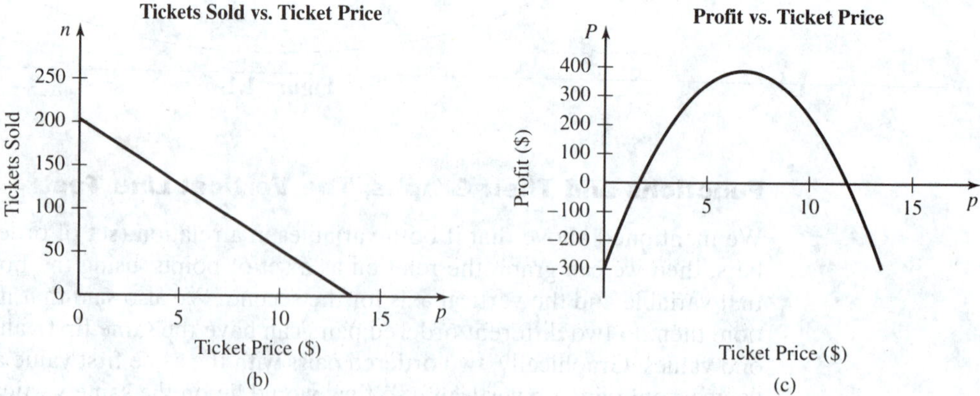

(b)

(c)

Figure 1.1-7

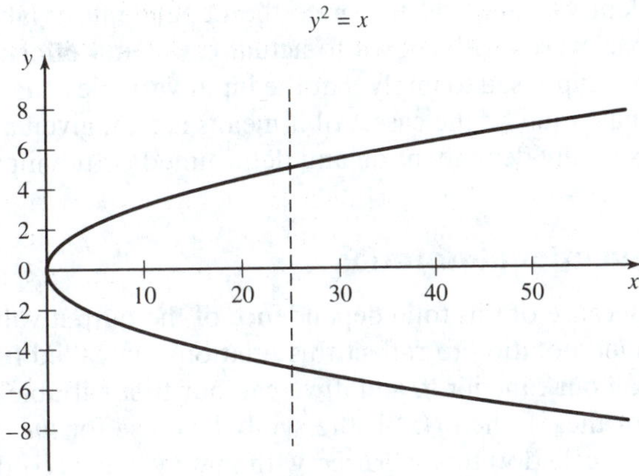

Figure 1.1-8

Function Terminology

Since the nature of a function is that there is exactly one value of the second variable that is paired with each value of the first variable in the domain, we can think of a function as an input-output machine, as illustrated in Figure 1.1-9. If you put a value of the first variable (say, x) from the domain *into* the machine, it will spit *out* the one value of the second variable (say, y) that gets paired with it. For this reason, when we have a function, we will often call the first variable the **input** and the second variable the **output**. Another common terminology, especially in statistics, is to call the first variable the **independent variable**, since you can choose it freely (independently) out of all the possible values in the domain. Once the independent (first) variable value is chosen, however, the value of the second variable is completely determined by the function's rule, so is totally *dependent* on the value of the first variable and is called the **dependent variable**. Each expression is useful in helping us to understand the concept: Input is what we "put into" the formula or function, output is what "comes out" when we *plug in* the input. The term "variable" simply means a letter standing for a quantity that can change within a given problem, such as the price of something you want to sell.

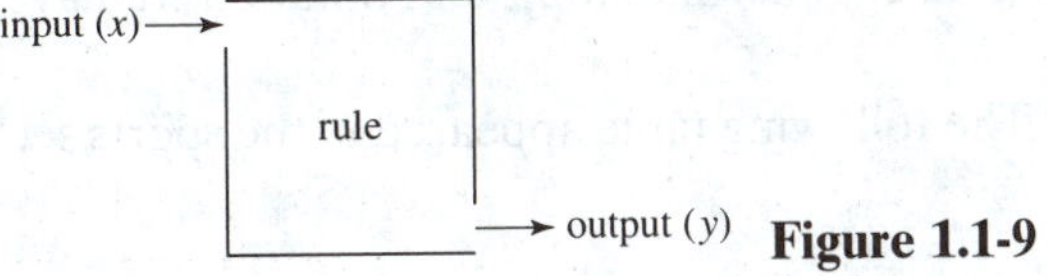

Figure 1.1-9

In the demand function of the concert ticket example (Example 4), the price is the independent variable (input) and the number of tickets sold is the dependent variable (output). This makes sense, because we are free to choose the ticket price independently, and the number of tickets sold will follow from that. We should point out, however, that sometimes economists think of quantity (like tickets sold) as the first variable and selling price as the second variable. In such a case, we would call the quantity sold the independent variable and the price the dependent variable, even though intuitively this seems backwards with regard to actual **cause and effect**. "Independent" and "dependent" are not supposed to imply that the input variable *causes* the value of the output, although this may often be the case. All it means is that, given a possible input value, the output value is totally dependent on and determined by the input value (as defined by the rule).

Functional Notation

Because of this total dependence of the output value on the input value, there is a shorthand notation to reflect this relationship, called **functional (or function) notation**. If we call our function f, and if we call our first variable x (the standard symbols used in math for these), then $f(x)$ is the symbol we use for *the one* value of the second variable (usually called y) that is paired with the x value, according to the rule of the function. This is read "eff of ex" and can be thought of in words as "the output value that is paired with the input value x." Note that the notation *looks* as if means f multiplied by x, *but it doesn't*! Try not to interpret it that way by accident! There is really no way to know the difference other than the context of the problem, since f multiplied by x **could** be written in exactly the same way. In this book, we will try to avoid such a notation; when multiplying a single letter by something, we will try to use $(f)(x)$ for such multiplication. Normally, we will use a single letter to name our functions. Often we will use letters for our variables and our functions that help you remember what they stand for (like n for number of tickets sold and p for price).

Functions Expressed in Tables; Inverse Relations

We now know all we need to know in order to be able to fully define any function. For a mathematical or symbolic definition of a function, we only need to specify the domain and the rule. Our convention from this point on will be to do this using functional notation.

"OK," you might ask (or maybe you wouldn't, but you really should), "now we know how to use the Vertical Line Test to recognize when a graph does or does not represent a function, but how can we tell whether a table or algebraic expression represents a function or not?" What do you think? Let's look again at some examples to explore this question and apply the concepts, terminology, and notation we have just discussed.

Sample Problem 1: The following table appeared in the sports section of a newspaper:

NHL Eastern Conference Atlantic [Division]	Pts.
N.Y. Rangers	88
[Philadelphia] Flyers	87
Florida	85
New Jersey	77
Tampa Bay	77
Washington	76
N.Y. Islanders	48

Source: Philadelphia *Inquirer*, March 26, 1996.

For the Atlantic Division of the Eastern Conference of the National Hockey League on the morning of March 26, 1996, was the *number of league points* a **function** of the team? If so, use functional notation to define it. Was the *team* a function of the *number of points*? If so, use functional notation to define it.

Solution: Given a team listed in the table, there is exactly one point value that is paired with it for that day, so the *number of points* is a function of the *team*. The domain is the set of teams in the Atlantic Division of the Eastern Conference of the NHL, or {N.Y. Rangers, Philadelphia Flyers, Florida, New Jersey, Tampa Bay, Washington, N.Y. Islanders}. Recall that with sets, order is not important. The rule in this case could be stated in words as "for a given input value (team), the output value is the number of league points for that team on the morning of March 26, 1996, as listed in the Philadelphia *Inquirer* that day." Or, given the table, the rule could be described as "for a given team (input value), look in the table for the number of points that corresponds to that team (the value in line with the team in the table) to get the output value."

Tables can be lined up in rows or columns. The usual convention is to put the input (independent variable) in the **first** row or column, and the output (dependent variable) in the **second** row or column. Functions in which one or both of the variables are discrete are often defined using tables (such as the hockey team and letter grade problems).

Notice that, in the hockey team example, we could not graph the function as we described earlier, since the input values are not numbers. We could, however, make a bar chart, with the team names as labels for vertical bars representing the number of points for each team, as shown in Figure 1.1-10.

Even though this problem does not have numbers for its input values, we can still use functional notation. For example, if we define

$p(t)$ = the number of league points for team t as recorded in the March 26, 1996, Philadelphia *Inquirer*, for t = any of the teams in the Atlantic Division of the Eastern Conference of the NHL

then we have specified both the domain (Atlantic Division teams) and the rule (points for that team at that particular time), so we have defined the function. Thus, if someone

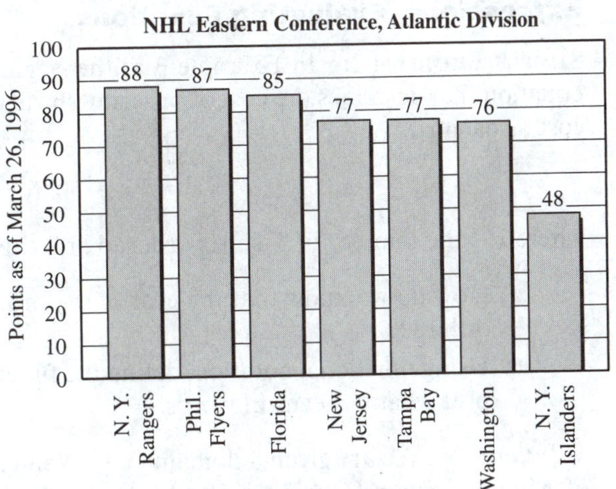

Figure 1.1-10

wanted to know how many points the N.Y. Islanders had that day, they would really be looking for the value of p(N.Y. Islanders). From the table, we see that the answer is p(N.Y. Islanders) = 48. Translated back into words, this means that the Islanders had 48 points that day. What if someone said "I know *one* of the Atlantic Division teams had 87 points. Which one was it?" Answering this question corresponds to "solving the equation" $p(t) = 87$. Using the table, we see that there is a unique answer of t = Philadelphia Flyers. So of course, the answer to the question is the Philadelphia Flyers. Notice, however, that the equation

$$p(t) = 77$$

does **not** have a unique solution; both t = New Jersey and t = Tampa Bay are correct solutions.

Solving equations like this is similar to reversing the role of the first variable and the second variable, because we *start* with a value of the *second* variable, and look for a value of the first variable that is paired with it. If we consider our *first* variable to be the league points that day and our *second* variable to be the team, we would have the set of ordered pairs

{(88, N.Y. Rangers), (87, Philadelphia Flyers), (85, Florida), (77, New Jersey), (77, Tampa Bay), (76, Washington), (48, N.Y. Islanders)}

This set of ordered pairs (relation) obtained by **switching the first and second variables** of a given relation is called the **inverse** of the original given relation. Notice that *this* inverse relation is *not* a function, since the first variable value 77 is paired with *two* different second variable values, New Jersey and Tampa Bay. In the above example, the original relation *was* a function, and the inverse listed out just above is *not* a function. Any combination of function and relation is possible. For instance, the graph of a circle is not a function, and neither is its inverse. For this example, then, the *number of points* *is* a function of the *team*, but the *team is not* a function of the *number of points*. □

Functions Defined Using Equations or Algebraic Expressions; Evaluating Functions

Sample Problem 2: In Example 3 at the beginning of this section we stated that the equation that describes the cost of ordering shirts when the fixed cost is \$50 and the shirts cost \$7 each is

$$C = 7x + 50, \text{ for } x \geq 0$$

where x is the number of T-shirts ordered and C is the total cost of the order, in dollars.

a. How do we know that the cost of the order is a function of the number of shirts ordered?

b. Using functional notation, define the function as $C(x)$, then find $C(80)$, and interpret what it means in words.

Solution: **a.** We are given a domain ($x \geq 0$) and a rule for finding C for any given value of x in the domain ($C = 7x + 50$). Since this equation results in exactly one output (C) for any given input value (x), this defines a function. Note also that the graph shown in

Figure 1.1-3 passes the vertical line test, since no vertical line can hit the graph of the cost function in 2 or more points.

b. $C(x) = 7x + 50$, for $x \geq 0$, where $C(x)$ is the total cost, in dollars, of ordering x T-shirts.

To find $C(80)$, we simply plug in 80 for x in the mathematical definition of $C(x)$:

$$C(80) = 7(80) + 50 = 560 + 50 = 610.$$

In other words, if 80 shirts are ordered, the total cost of the order will be \$610. □

Sample Problem 3: If your group has \$400 to spend for the initial order of shirts from Sample Problem 2, how many shirts can you order?

Solution: This means we are looking for the value of x for which $C(x) = 400$. In other words, we need to solve the equation $C(x) = 400$.

$$C(x) = 400$$
$$7x + 50 = 400$$
$$7x = 400 - 50 = 350$$
$$x = 350/7 = 50$$

If the group has \$400 to spend, it could order up to 50 shirts. □

Suppose we want to generalize this and take a value of the total cost, C, and find out what order size x would correspond to it. This would boil down to solving the equation $C(x) = C$, analogous to what we did with the \$400:

$$C(x) = C$$
$$7x + 50 = C$$
$$7x = C - 50$$
$$x = (C - 50)/7$$

Given a specific value of C, this expression for x clearly gives exactly one answer. If we specify a domain of C values, then we have a function. To correspond to the C values of our original cost function, we can specify $C \geq 50$. This is because, when $x = 0$,

$$C(0) = (7)(0) + 50 = 50$$

The resulting set of ordered pairs is the **inverse** of our original cost function, since we have simply reversed the roles of the number of shirts and total cost with regard to input versus output. The inverse is a function, and could be written

$$x(C) = \text{the number of shirts ordered for } C \text{ dollars}$$

$$x(C) = \frac{C - 50}{7}, \text{ for } C \geq 50.$$

The number of shirts ordered *is* a function of the total cost, in addition to the total cost being a function of the number of shirts. So, in this case, the inverse of the function is also a function.

Sample Problem 4: In the relation defined by $y^2 = x$, where x is the first variable and y is the second variable (as usual), is y a function of x?

Solution: If we solve the equation for y in terms of x, we get

$$y^2 = x$$

$$y = \pm\sqrt{x}$$

We assume here, as we do throughout the book, that we are restricting ourselves to *real numbers*, which means that even roots of negative numbers (imaginary numbers) are not defined. Because of this, our equation only makes sense if $x \geq 0$. For any strictly positive value of x, like 25, there will be *two* values of y whose squares equal that value of x, the positive and negative square roots of x (such as 5 and -5, whose squares both equal 25). Remember that the symbol $\sqrt{x}$ means the *nonnegative* square root of x. For example, $\sqrt{25} = 5$, but $\sqrt{25} \neq -5$.

Our expression for y, the second variable value, is given by $y = \pm\sqrt{x}$, so for some values of the first variable, x (any positive value, in fact), there will be *two* values of the second variable paired with them. This means that this relation is not a function. In general, **if an expression for the second variable in terms of the first can have more than one value for a single value of the first variable, then the relation is not a function. One clue to look for is a ± sign in the formula. But make sure you *actually can get* two answers for a *particular* value of the first variable before coming to a final conclusion.** □

Functions Expressed Graphically

We have already seen how functions can be defined by tables or mathematical expressions (both defined from verbal descriptions, like the shirt cost function, and fit to a set of data, like the ticket sales function). We have also seen that when both variables have numerical values, functions (and relations in general) can be represented graphically. Sometimes, this is the *only* information we have about a function, such as when you see a graph in the newspaper with no table or formula accompanying it. This is not ideal, because it is very hard to estimate values on a graph accurately. As a result, the definition of a function in this form is very approximate, but can still be useful. The problem below gives a feel for how you can work with such a function.

Sample Problem 5: A smooth graph was drawn representing the horizontal distance an arrow traveled when the arrow was initially aimed at different angles. Figure 1.1-11 shows the distance in inches and the initial angle of the arrow in degrees (horizontal ↔ 0 degrees).[4] Is the distance traveled a function of the arrow angle? Estimate the distance traveled in inches for arrow angles of 0 and 20 degrees (assume the pull, height, wind, and terrain stay constant).

Solution: If you draw a vertical line up from any point on the x (degree) axis, you will find the one distance that is associated with it. No vertical line would hit the graph in 2 or more points, so this graph passes the Vertical Line Test and does represent a function. Clearly, the distance that the arrow traveled was a function of the angle at which the arrow was aimed.

---

[4]Data adapted from a student project.

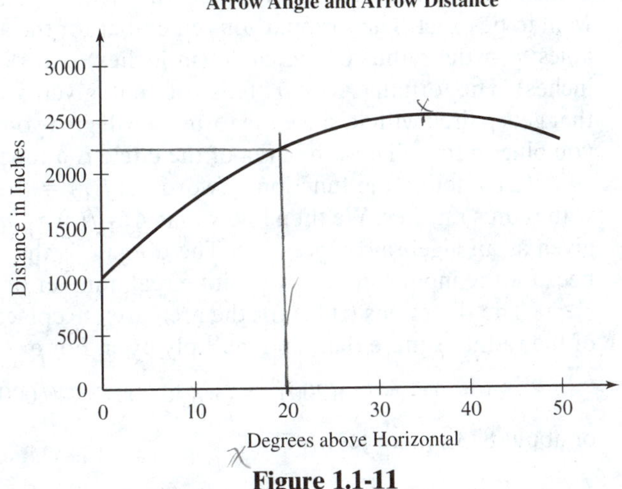

Figure 1.1-11

Let's define x to be the arrow angle (in degrees from the horizontal). The implied domain of this function is $0 \leq x \leq 50$. The rule for the function is: "Look along the horizontal axis until you find the angle at which the arrow is initially held (x), then go straight up until you hit the point on the line; the height of that point, in the units of the vertical axis, represents the predicted distance." We will define the function verbally as $f(x)$ = the distance (in inches) traveled on average when the initial arrow angle is x degrees above the horizontal. To find the distance traveled when the arrow angle is 0 degrees, we look on the graph directly over 0 on the x-axis. This point will lie on the vertical axis and be the vertical intercept, the place on the vertical axis where the graph hits. The symbol for the value of the function f when x is equal to 0 is $f(0)$. Visually estimating the intercept, it seems to be about at 1050, so $f(0) \approx 1050$, or on the average the arrow traveled approximately 1050 inches (about 90 feet, or 30 yards) when the arrow was initially held level. Similarly, $f(20) \approx 2200$, or on the average, the arrow traveled approximately 2200 inches (about 180 feet, or 60 yards) when the arrow was initially aimed 20 degrees above horizontal. □

In general, **for any function defined by a graph** the directions (rule) are: **"Look along the horizontal axis until you reach the input or independent variable value you're interested in, and then look straight up until you hit the graph; its height, in the units of the vertical axis, represents the value of the output or dependent variable."** As we have said earlier, if you look along this vertical line and it crosses the graph in more than one place, then there is not *exactly one* output associated with that input, and so the graph does not represent a function, according to the Vertical Line Test.

Sample Problem 6: You are planning a fair to celebrate Earth Day. You want to paint a large circular area on the floor of the gym so it can be used for some activities that you and your committee have planned. Obviously you had better use some specialized paint that will last for the day of the fair, but can then be cleaned up easily afterwards. This paint is quite expensive so you need to know the exact area of the circle. You don't remember a lot from your grammar school math, but you do remember the formula for the area of a circle: $A = \pi r^2$. Is the area a function of the radius? What is the area if the radius is 200 inches?

Solution: You measured the radius in inches—this stuff is really expensive, and you want to be exact. The formula you remember for the area has implicitly defined your variables: r = the radius of the circle (in inches) and A = the area of the circle (in square inches). The formula $A = \pi r^2$ tells you that, given a value for r, there is exactly one area that gets paired with it, since the formula will give you exactly one answer, no matter what you plug in for r. Thus, the area of the circle *is* a function of its radius.

Let's define the function in words as $f(r)$ = the area (in square inches) of a circle with radius r inches. We then know that $A = f(r) = \pi r^2$. In this example, the function was given as an algebraic expression. The domain is the set of all nonnegative real numbers because the input can be any positive real number or zero. We can write this in symbols: $r \geq 0$. The directions (rule) for the area are: "Replace the letter r with the measurement of the radius, square that, and multiply by π. For example, if $r = 200$ inches,

$$A = f(r) = (\pi)(200)^2 = (\pi)(40{,}000) = 40{,}000\pi \approx 125{,}664 \text{ square inches}$$

or about 873 square feet. □

The general directions **to evaluate an algebraic expression** are: **"Replace the symbol representing the known variable (the input or independent variable) with the input value (whatever is inside the parentheses in function notation) and carry out the indicated calculations."** It's a good idea to put the input value in *parentheses* in place of the input variable in the definition, wherever it occurs.

Notice that in this example, we used two different letters: A for the output variable and f for the name of the function. It is normally not necessary to do this; you can use the same letter for your output variable and your function. The time when it is most meaningful to use two different letters is when graphing. If your function is named $f(x)$, technically the equation defining your graph would be $y = f(x)$, since this says that the value of the second variable y is given by the function f at each possible value of x. Otherwise, a *single* letter symbol is fine. After all, everyone is in such a rush today that we love to use "shorthand," abbreviations, or symbols that immediately convey the idea and save a little time or space. Sports announcers refer to RBI's, and very rarely say "runs batted in." You have probably heard references to your GPA. In the interest of brevity and to save a little writing, we could have just named our function $A(r)$.

Sample Problem 7: In the ASCII code used in some computers,[5] each capital letter of the alphabet is assigned a binary number:

A → 01000001
B → 01000010
C → 01000011

and so on. Is the ASCII code a function of the capital letter? Is the capital letter a function of the ASCII code?

Solution: For each capital letter exactly one binary number is assigned. When you type a capital "A" the computer changes this to its binary representation: 01000001. Thus, the

[5]American Standard Code for Information Interchange. Larry Long, *Introduction to Computers and Information Processing* (Englewood Cliffs, NJ: Prentice Hall, 198

binary number (the output) is a **function** of the letter of the alphabet (the input). When you type "A" the number 01000001 is put into memory. The domain in this case is the set of all capital letters in English, {A,B,C, . . . , Z}. The rule is the assignment of the code number to each capital letter (A is 65 written in 8-digit binary notation, B is 66, and so on, up to Z, which is 90). If we define c = capital letter, then we could define the function $A(c)$ = the ASCII code for the capital letter c, for any c in the set of all capital letters in English. Thus, $A(\text{A}) = 01000001$, $A(\text{B}) = 01000010$, and so on. The relation (set of ordered pairs) in this case could be written {(A,01000001),(B,01000010), . . . , (Z,01011010)}. If we reverse the order of the values in these pairs, we get {(01000001,A),(01000010,B), . . . ,(01011010,Z)}, which is the **inverse** of the original relation. In this example, the letter is a function of the binary number, since the code numbers for the capital letters are **unique** (there is not only exactly one output value for each input value, but also **the output values are all different from each other** as well). Such a function is sometimes called a **one-to-one correspondence**. When this is the situation, the inverse will always be a function as well. For example, if the binary number 01000001 is sent to your printer, it knows to print "A". □

Sample Problem 8: You are taking flying lessons. On your first solo cross-country flight you take off, climb to cruising altitude, cruise at that altitude until you reach your check-point, turn around and fly back to the airport at the same altitude, and then come in for a landing. The graph in Figure 1.1-12 shows your time into the flight and your altitude.

Is your altitude a function of your time into flight? Is your time into flight a function of your altitude? If someone accidentally heard you say on your radio what your altitude was at some point during your trip, would they be able to tell if you were on your way out or on your way back in, given the graph in Figure 1.1-12?

Solution: Since no vertical line hits the graph in two or more points (the graph is steep, but never directly vertical), we can see that the altitude is a function of the time into flight. We could define $a(t)$ = the altitude (in feet) t minutes into the flight. From the graph, the domain would seem to be about $0 \le t \le 120$. Also estimating from the graph,

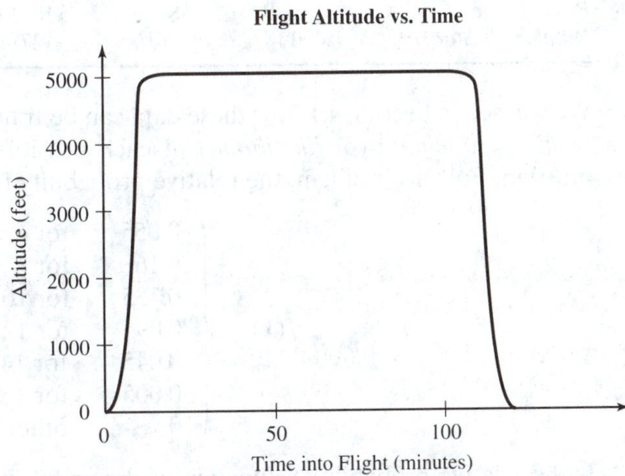

Figure 1.1-12

it would appear that $a(50) \approx 5000$. This means that 50 minutes into the flight, the altitude was approximately 5000 feet. If we reverse the roles of the two variables to see if time into the flight is a function of the altitude, we also reverse the roles of the horizontal and vertical axes. What we are really doing is forming the inverse of the original relation. Since the roles of the axes are reversed to check to see if the inverse of the original relation is a function, we can use the **Horizontal Line Test**.

Horizontal Line Test: If any horizontal line intersects a graph in two or more points, the *inverse* of the relationship shown in the graph is *not* a function.

The original relation has points like (50,5000) and (100,5000) in it, so the inverse has (5000,50) and (5000,100) in it, which means the inverse is *not* a function. This corresponds to two points being in line **horizontally** in the graph of the original relation. Thus, the time is *not* a function of the altitude because you flew quite a while at your cruising altitude of 5000 feet, and you were also at the same heights (between 0 and 5000) on *both* landing and takeoff. □

Notice that graphs like this can tell a story. In this case, the graph is saying that the plane climbed to 5000 feet in the first 10 minutes or so, stayed at that altitude until about 110 minutes into the flight, and then came down to land in about 10 minutes.

If the graph of a function passes the Horizontal Line Test (or, in other words, if a graph passes *both* the Vertical and Horizontal Line Tests), then it represents a one-to-one correspondence, and its inverse is also a function. If you think about it, this would be true for *any* slanted line (as in Figures 1.1-3 and 1.1-4(a), and *not* the case for any parabola opening up or down (as in Figures 1.1-4(b) and 1.1-11).

Can you think of any other sets of data in which one variable is a function of the other, but not the other way?

Sample Problem 9: A market survey asks 100 students at random what is the most they would pay for a particular T-shirt. The results are as follows:

Price Range	\$6–8	\$8–10	\$10–12	\$12–14	\$14–16	\$16–20
Number of Students	11	32	27	19	9	2

We will see in Section 4.1 how these data can be translated into a function expressing the *relative probability* (or *likelihood*) of each possible price, called a **probability density function**. For this problem, the relative probability function is given by

$$f(x) = \begin{cases} 0.055 & \text{for } 6 \le x < 8 \\ 0.16 & \text{for } 8 \le x < 10 \\ 0.135 & \text{for } 10 \le x < 12 \\ 0.095 & \text{for } 12 \le x < 14 \\ 0.045 & \text{for } 14 \le x < 16 \\ 0.005 & \text{for } 16 \le x < 20 \\ 0 & \text{otherwise} \end{cases}$$

Is the relative probability a function of the price? Is the price a function of the relative probability? What are $f(8)$, $f(11)$, and $f(20.3)$?

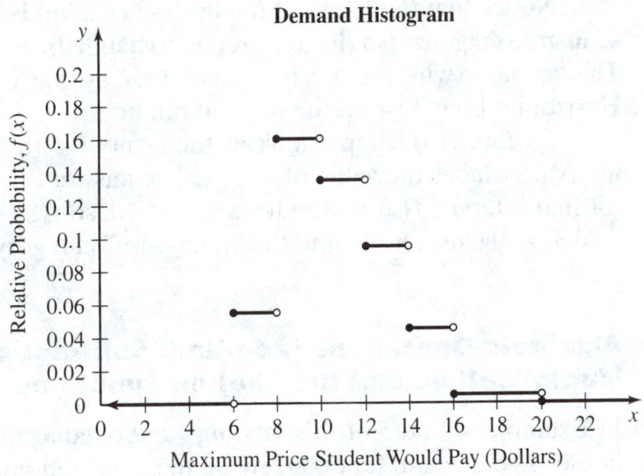

Figure 1.1-13

Solution: Graphically, the function, sometimes called a **histogram**, is shown in Figure 1.1-13. (*Note*: The graph is along the horizontal axis for values of x less than 6 or greater than or equal to 20.)

Each price is paired with exactly one relative probability, and the graph passes the Vertical Line Test, so the relative probability *is* a function of the price.

What about the inverse? Is the price a function of the relative probability? The graph of the relative probability and the prices paired with them is shown in Figure 1.1-14.

It is obvious that this graph does not pass the Vertical Line Test. Given a particular relative probability, there are many prices that get paired with it. This is equivalent to saying that the graph of the original function (Figure 1.1-13) does not pass the *Horizontal* Line Test. For example, for the relative probability value 0.16, prices of \$8 and \$9, or anything in between, get paired with it. Thus, the price is *not* a function of the relative probability.

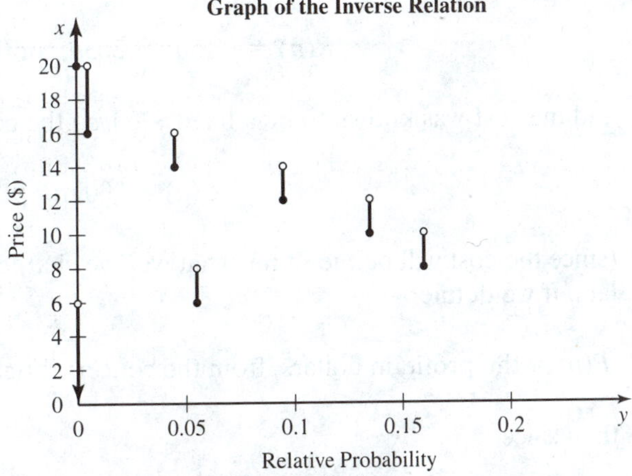

Figure 1.1-14

Notice that the graph of the inverse relation is like flipping the original graph on a 45 degree diagonal (so the axes get interchanged), as long the units also get interchanged. This explains why the Vertical Line Test on the inverse relation is equivalent to the Horizontal Line Test on the original relation.

To find $f(8)$, simply look at the symbol definition of the function and see which interval includes the value of 8 for x. The answer is $8 \leq x < 10$, so you use the rule given for that interval: $f(x) = 0.16$ for $8 \leq x < 10$, so $f(8) = 0.16$. Similarly $f(11) = 0.135$, and $f(20.3) = 0$ (since it fits into the "otherwise" category). □

Algebraic Operations (Addition, Subtraction, Multiplication, and the like) on Functions

In Examples 4 and 5 of this section, we looked at the finances of a concert at a coffee-house. The demand function, giving the expected number of tickets sold (n) in terms of the price of the tickets (p) was given to be

$$n = -15p + 203, \text{ for } 0 \leq p \leq 13.50$$

and the cost of the concert was fixed at $300.

As manager of the coffeehouse, let's assume that you want to maximize your profit for the concert (we will discuss other possibilities later in Section 1.2). What is profit? Intuitively, you probably already know that

Profit = Revenue minus Cost $\quad \boldsymbol{P = R - C}$

(or revenue minus expenses). We already know the cost, but how can we find the revenue expected from the concert? To find this, we need to know another basic relationship in business: When you are selling something at a fixed price, your revenue is given by

Revenue = (selling price) times (quantity sold) $\quad \boldsymbol{R = pq}$

In our example, the revenue function was given by

$$R(p) = -15p^2 + 203p, \text{ for } 0 \leq p \leq 13.50$$

and the cost was known to be a fixed $300, so the cost function is given by

$$C(p) = 300, \text{ for } p \geq 0$$

(since the cost will be the same, whatever we charge) where p, $R(p)$, and $C(p)$ are in dollars. If we define

$P(p)$ = the profit, in dollars, from the concert if tickets are sold for p dollars per ticket,

then since

Profit = Revenue minus Cost $\quad \boldsymbol{P = R - C}$

(or revenue minus expenses), we can define the profit function to be

$$P(p) = R(p) - C(p) = -15p^2 + 203p - 300, \text{ for } 0 \le p \le 13.50$$

This is an example of a **difference** of two functions. A **sum** of two functions would be constructed analogously.

This process of forming a new function from an arithmetic operation on two functions is common in many situations. In general, you just do the desired operation to the formulas of the functions being combined. The only subtlety to keep in mind is that the domain of the final (result) function must be the overlap, or **intersection**, of the domains of the functions being combined by arithmetic operations. In the profit example above, the revenue function domain ($0 \le p \le 13.50$) was more restrictive than that of the cost function ($p \ge 0$), so the more restrictive domain, which was the intersection of the two, was used ($0 \le p \le 13.50$).

In the same example, suppose we now used x to denote the selling price, in dollars, of the tickets. We can verbally define the functions as follows:

$p(x)$ = the selling price, in dollars, of the tickets if the selling price is x dollars
$q(x)$ = the quantity of tickets that can be expected to be sold if the price is x dollars

The definition of $p(x)$ sounds odd, and is done simply for notational generality, and means that $p(x) = x$. Since revenue is selling price times quantity sold, the revenue function, in dollars, is given by

$$R(x) = p(x)q(x) = (x)(-15x + 203) = -15x^2 + 203\,x, \text{ for } 0 \le x \le 13.50$$

This is an example of a product of two functions; the domain derivation is as above.

Composite Functions

Sample Problem 10: The following table might appear on a class syllabus:

Semester Average (Rounded Off)	Letter Grade
90–100	A
80–89	B
70–79	C
60–69	D
59 or below	F

Is the letter grade a function of the *raw* semester average (as opposed to the *rounded* semester average)?

Solution: Let's define

a = raw semester average, $0 \le a \le 100$

r = rounded semester average

l = letter grade

We discussed earlier that l is a function of r (as given by the table). Let's define

$$l(r) = \text{the letter grade that goes with a rounded semester average of } r \text{ (from the table)}$$

So, for example, $l(80) = \text{B}$.

By the usual rules of rounding (round up any decimal of 0.5 or more, and down otherwise), r is a function of a, since for any given raw semester average, the rules of rounding give exactly one rounded value. Thus we can define

$$r(a) = \text{the rounded semester average (using the standard rule) for a raw semester average of } a,\ 0 \le a \le 100$$

For example, $r(79.5) = 80$, by the usual rules.

How would we find the letter grade for a raw semester average of 79.5 ? We would first round the 79.5 to 80, then use the table to find the letter grade for 80, which is a B. In symbols, this means first finding $r(a)$, then plugging *that answer* into the letter grade function, which in symbols would be written $l(r(a))$. In our example, we had

$$l(r(79.5)) = l(80) = \text{B}$$

This situation of **a function *of* a function is called a composite function.** To evaluate such a function, you **first evaluate the inner function and then plug that answer into the outer function**. To work right, **the *value* of the inner function must lie within the domain of the outer function**. In this example, you can probably see that by doing these two functions sequentially (rounding, then using the table), we can start with a raw semester average, and end with exactly one letter grade. Thus the letter grade *is* a function of the raw semester average. If we want to define it as a separate function, we should use a different letter than l, because we already defined a function named l that is not the same (its domain was *rounded* semester averages rather than *raw* semester averages). Let's call the new function

$$g(a) = \text{the letter grade that goes with a raw semester average of } a$$

The mathematical definition of this function is then

$$g(a) = l(r(a)), \text{ for } 0 \le a \le 100$$

So $g(79.5) = l(r(79.5)) = l(80) = \text{B}$, as before.

We could also define $g(a)$ directly:

$$g(a) = \begin{cases} \text{A}, & \text{for } 89.5 \le a \le 100 \\ \text{B}, & \text{for } 79.5 \le a < 89.5 \\ \text{C}, & \text{for } 69.5 \le a < 79.5 \\ \text{D}, & \text{for } 59.5 \le a < 69.5 \\ \text{F}, & \text{for } 0 \le a < 59.5 \end{cases}$$

Either way, we have seen that, yes, the letter grade *is* a function of the raw average. □

Notice, however, that the raw average is not a function of the letter grade (the inverse relation is not a function).

In some math books, the operation of *composition* of functions is denoted as follows:

$$(f \circ g)(x) = f(g(x)) \text{ (Read "eff circle gee of } x\text{" or "eff of gee of } x\text{.")}$$

In our above example, we could have written

$$g(a) = (l \circ r)(a) = l(r(a))$$

If you really understand the concept of a function, you should be able to explain it to a friend who doesn't understand. Or, if you can't find a friend who wants or needs your help, try writing out the explanation in your own words. How many times do we think we really understand something until we try to explain it to someone else? Certainly, as teachers, we have had this happen more than once, and it can be a bit embarrassing! The only thing we can do then is admit that we need to work on the concept a little more and that we will get back to you. *Now* is the time for you to be certain that you have a very good grasp of the concept of a function.

Section Summary

Before you start the exercises for this section, be sure that you

- Know that a function is a particular type of relationship between two variables in which the value of one of them (which we call the **output**, or **dependent variable**) is completely determined by some **rule** (usually a mathematical formula) once you have specified **any** value of the other variable (which we call the **input**, or **independent variable**) that is within a set of possible values for the input variable (called the **domain**). If the function is well defined, you should be willing to bet the farm that exactly one output value exists for *any* input value in the domain.
- Know that any function can be thought of as a special kind of set of ordered pairs of symbols (any set of ordered pairs in math is called a **relation**), in which each value of the first variable is paired with *exactly one* value of the second variable. Thus, if a relation has two ordered pairs with the *same* value for the first variable and *different* values for the second variable, then that relation is *not* a function. The set of second values is called the **range**.
- Recognize the special **functional notation** that makes it easy to refer to the output value (such as $f(x)$) that is paired with a specific given input value (such as x).
- Know that when the elements of ordered pairs are numbers, relations and functions can be graphed, with the input value on the horizontal axis and the output value on the vertical axis.
- Know that a graph can be tested to see if it represents a function by the **Vertical Line Test**: The graph represents a function if *no* vertical line hits the graph in *two or more* points (that is, if *every* vertical line hits the graph in *at most one* point).

If a single vertical line intersects the graph in two or more points, then the graph does not represent a function.

- Recognize whether relations are functions or not when they are presented as tables (see if each first variable value is paired with exactly one second value), algebraic expressions (see if the formula gives exactly one value of the second variable for any individual value of the first variable), and/or graphs (using the Vertical Line Test).
- Understand that we can reverse the roles of the two variables in a relation to define the **inverse relation** (just switch the order of the two values in *all* of the ordered pairs), which is useful when you want to solve for the original *input* variable in terms of the original *output* variable. If a relation passes the Horizontal Line Test (no *horizontal* line hits the graph in two or more points), then the inverse relation is a function. If the original relation was also a function, this relationship is called a **one-to-one correspondence**.
- Understand the idea of sums, differences, and products of functions, and of composite functions (functions of functions), and know how to evaluate them all.

EXERCISES FOR SECTION 1.1

Warm Up

1. Given the following table:

x	2	4	8	9	10	11	10	15
y	4	5	7	9	11	13	15	17

(a) Is y a function of x? Explain your answer.
(b) Is x a function of y? Explain your answer.

2. Given the following table:

x	1.22	1.43	1.08	1.99	1.51	1.13	1.51	1.72
y	4	5	7	9	10	11	9	15

(a) Is y a function of x? Explain your answer.
(b) Is x a function of y? Explain your answer.

3. Given the following table:

Name	Smith	Jones	White	Brown	Black	Eeny	Meeny	Moe
Grade	85	89	93	87	85	78	75	91

(a) Is the grade a function of the name? Explain your answer.
(b) Is the name a function of the grade? Explain your answer.

4. Given the following table:

Day	Mon.	Tues.	Wed.	Thurs.	Fri.	Sat.	Sun.	Mon.
Hours Worked	2.5	1.7	3.8	2.6	4.1	3.0	1.8	2.9

(a) Is the number of hours worked a function of the day? Explain your answer.
(b) Is the day a function of the number of hours worked? Explain your answer.

5. Which of the graphs in Figure 1.1-15 are functions and which are not? Explain your answers.

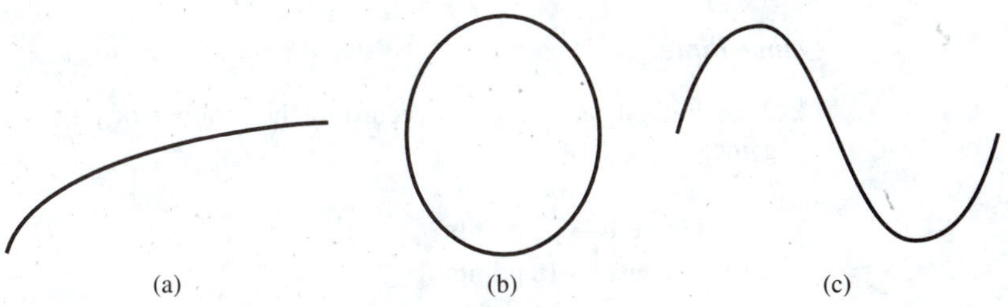

Figure 1.1-15

6. Which of the graphs in Figure 1.1-16 are functions and which are not? Explain your answers.

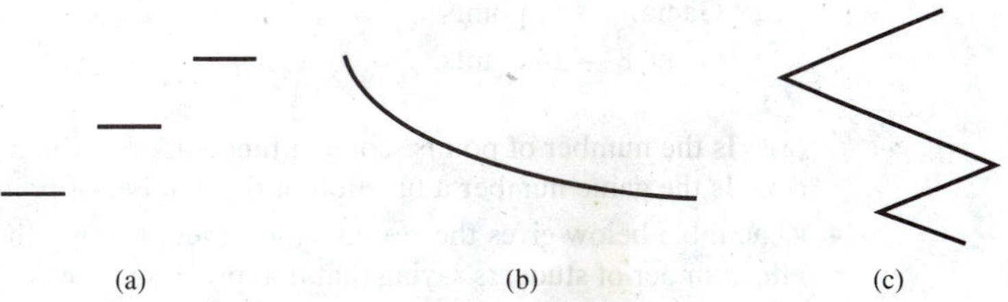

Figure 1.1-16

7. If $f(x) = 3x^2 - 5x + 12$ for $0 \le x \le 10$, evaluate the following:
(a) $f(3)$ **(b)** $f(4.5)$ **(c)** $f(8)$ **(d)** $f(12)$

8. If $f(x) = -2x^2 + 7x - 5$ for $-2 \le x \le 2$, evaluate the following:
(a) $f(1)$ **(b)** $f(0.5)$ **(c)** $f(-1)$ **(d)** $f(3)$

9. If $f(x) = 4x^3 + 12$ for $0 \le x \le 4$, evaluate the following:
(a) $f(0)$ **(b)** $f(4.5)$ **(c)** $f(1)$ **(d)** $f(12)$

10. If $f(x) = x^3 + 0.03$ for $0 \le x \le 3$, evaluate the following:
(a) $f(3)$ **(b)** $f(4.5)$ **(c)** $f(0)$ **(d)** $f(-1)$

11. If

$$f(x) = \begin{cases} 5 & \text{for } 0 \le x < 3 \\ 10 & \text{for } 3 \le x < 4 \\ 15 & \text{for } 4 \le x < 8 \\ 19 & \text{for } 8 \le x < 13 \\ 9 & \text{for } 13 \le x < 16 \\ 2 & \text{for } 16 \le x < 20 \end{cases}$$

Evaluate
(a) $f(2)$ **(b)** $f(8)$ **(c)** $f(0)$ **(d)** $f(22)$

12. If

$$f(x) = \begin{cases} -3 & \text{for } -1 \leq x < 0 \\ 0 & \text{for } 0 \leq x < 1 \\ 2 & \text{for } 1 \leq x < 5 \\ 7 & \text{for } 5 \leq x < 9 \\ 9 & \text{for } 9 \leq x < 12 \\ 12 & \text{for } 12 \leq x \end{cases}$$

Evaluate:
(a) $f(2)$ **(b)** $f(-1)$ **(c)** $f(0)$ **(d)** $f(15)$

Game Time

13. A basketball team keeps a record of the number of points scored by the star player in each game:

Game 1 → 23 points
Game 2 → 18 points
Game 3 → 21 points
Game 4 → 18 points
Game 5 → 19 points
Game 6 → 22 points
Game 7 → 17 points
Game 8 → 24 points

(a) Is the number of points scored a function of the game number? Why or why not?
(b) Is the game number a function of the number of points scored? Why or why not?

14. The table below gives the results of a survey showing the price charged for a sweatshirt and the number of students saying that that price is the most they would pay, to the nearest $2.50.

Price ($)	0	2.50	5.00	7.50	10	12.50	15	17.50	20	22.50	25
Number of Students	9	3	2	12	15	18	17	12	8	3	1

(a) Is the number of students a function of the price? Why or why not?
(b) Is the price a function of the number of students? Why or why not?

15. After you paid your toll at the toll booth, you increased your speed from 0 to 55 mph at a steady rate over 3 minutes and then maintained that speed for the next 15 minutes. Draw a graph of speed as a function of time, when time varies from 0 to 18 minutes. Define the relationship in *words* (*not* the *formula*) using functional notation. Is the inverse relation a function? Sketch its graph.

16. You leave your room to go to your first class in the morning. It is one-fourth of a mile to your classroom. You walk at a steady pace and arrive in 5 minutes. Draw a graph of distance from your room as a function of time. Define the relationship *in words* (*not* the *formula*) and *then* give the formula, using functional notation for both. Is the inverse relation a function? At what time are you 1/10 of a mile from your room?

17. You leave your room to go to your first class, walking at a steady pace. Halfway there you remember that you have to hand in a paper. You rush back to your room at twice the pace, get the paper and rush back to class. Draw a graph of distance from your room as a function of time. Define the relationship *in words* (*not* the *formula*) using functional notation.

18. The *Statistical Abstract of the United States* has a table showing average prices of aluminum each year from 1980 to 1992. Do you think the average prices of aluminum are a function of the year? If so, use functional notation to verbally define this relationship.

19. You go to the store to buy some oranges, which are selling for \$0.35 each. Will the total cost of the oranges be a function of the quantity that you buy? Use functional notation to write the rule for this function, including a definition in words of what everything stands for.

20. The graphs in Figure 1.1-17 show the relationship between time elapsed and distance from the starting point for various trips.

(a) Write a possible verbal description of a trip that would explain each graph.

(b) For which graphs is distance a function of time?

(c) For which graphs would time be a function of distance?

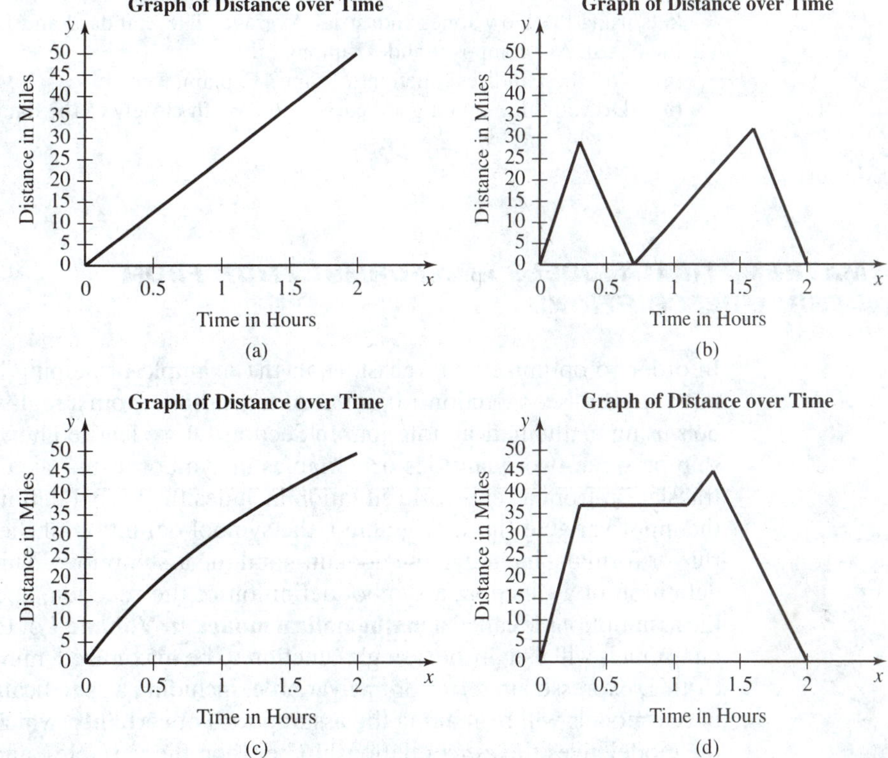

Figure 1.1-17

21. Mathematically speaking, is a U.S. Social Security number a function of a person's legal name? Is legal name a function of U.S. Social Security number? Explain your answers. If you answered yes for either, use functional notation to express these relationships.

22. Mathematically speaking, is a person's name a function of an SAT math score? Could an SAT math score be a function of a person's name? If so for either, use functional notation to define this relationship in words, and be sure to specify the domain and the rule carefully.

23. The 1995 individual tax rates are shown in the following table.

Taxable Income	Marginal Tax Rate (Single Rate)
$0 to $23,350	15%
$23,351 to $56,550	28%
$56,551 to $117,950	31%
$117,951 to $256,500	36%
More than $256,500	39.6%

Source: *World Almanac Book of Facts 1996*, p. 731.

Is marginal tax rate a function of income? Is income a function of marginal tax rate? Explain your answers. For any that *are* functions (of the two), define the relationships in words using functional notation.

24. Newspapers and radio and television newscasts regularly report the activities in the major markets using the Dow Jones Industrial Average, The Standard and Poor's 500 stock index, and the NASDAQ composite index, among others.

(a) Are these indices functions of time? Explain.

(b) Do you think that a good curve could be fit closely to the data from these indexes? Why or why not?

1.2 MATHEMATICAL MODELS AND FORMULATION FROM VERBAL DESCRIPTIONS

In order to optimize or forecast, as in the example of helping Meg decide her optimal level of exercise in Section 1.0, we need to translate from a real-world situation into symbols using mathematical notation. In Section 1.1 we learned how to express the relationship between two quantities or variables in symbol form—as a function. The complete translation from a real-world situation includes the verbal definition of the function and the input variable (including units), the symbol definition of the function (the algebraic rule or formula, including the domain), and the assumptions. This triumvirate of a verbal definition of a function, a symbol definition of the function, and verbal specification of the assumptions is called a **mathematical model**. In **Volume 1** of this book (Chapters 1–4), our models will usually be a single function to be maximized, minimized, or used for forecasting, expressed in terms of **one variable**, including a specification of the domain. Most of our models will be making the assumptions of **certainty**, which means that we assume the model gives the *exact* relationship between the variables, and **divisibility** of the variables, which means that we assume *all* real number values in an interval are possible, including all fractions and decimals. There are two major categories of models: those derived from verbal descriptions (such as traditional "story problems") and those that are derived from raw data (for example, the exercise model used to solve Meg's problem). Sometimes a model is a combination of both types.

Here are the kinds of problems that the material in this section will help you solve:

- Your soccer team has decided to sell T-shirts to raise money for attending a special all-day skills clinic in the state capital. You have been chosen to organize the entire effort. You know how much it costs to place an order to buy the shirts (covering design, making the silk screens, and shipping and handling), and the cost of each shirt. Formulate a model for the cost.

- Define a model to help Meg determine her optimal amount of exercise to maximize her energy level.
- You help run a local coffeehouse and have had the same musical group perform for three years in a row. You have been raising your ticket prices, and the attendance has varied. Define a model of the attendance in terms of the ticket price.
- You have a special cheese that you particularly like. You have to travel a considerable distance to buy it, so you don't want to make a lot of trips. On the other hand, it costs you money to store the cheese. Formulate a model for the cost of keeping yourself supplied with the cheese.

When you have finished this section, you should

- Understand what is meant by the term "mathematical model."
- Have a feel for how to model real-world problems by defining a function, including what its input variable means and all units involved, writing out the symbol definition of the function (the algebraic formula, including the domain), and specifying the assumptions and their implications.
- Realize that there are two major ways to formulate models: from verbal descriptions and from data, and that some models involve both.
- Understand why models are useful in solving real-world problems.
- Know how to translate from verbal descriptions of problems to mathematical models of these problems.
- Be able to translate mathematical statements involving a model into well-constructed statements about a real-world situation.

Sample Problem 1: You know that it costs \$50 just to place an order to buy T-shirts (covering design, making the silk screens, and shipping and handling), and then each shirt costs \$7. Formulate and define the function for the cost.

Solution: If you order 10 shirts, the total cost will be (\$7)(10) = \$70 for the direct cost of the shirts, plus \$50 for the order, for a total cost of

$$\text{Total Cost} = (7)(10) + 50 = 70 + 50 = 120 \text{ dollars}$$

If you order 20 shirts, the total cost will be

$$\text{Total Cost} = (7)(20) + 50 = 140 + 50 = 190 \text{ dollars}$$

At this point, the pattern of the numbers is visible, so that if we define

x = the number of shirts ordered and C = the total cost (in dollars)

we can see that

$$C = (7)(x) + 50 = 7x + 50$$

The process we just used to formulate this equation is a useful one that you may want to try when formulating problems. The idea is to choose specific numerical values for your **decision variables** (the variables you have some control over and want to find optimal values for). For these specific numerical values, try to choose *different* values from the other numbers involved in the problem. Plug these specific numerical values in for your variables, and calculate the quantities of interest in the problem (see if they represent a solution). **If you choose two or three different sets of sample values (numbers to plug in) and write out the actual calculations, you can usually see the patterns.** Sometimes, just trying *one* set of numerical values is sufficient to see the patterns. Next, you can **generalize the patterns and replace the numerical values of a quantity with the *variable* that represents the quantity** in general. This way you can obtain the general mathematical expression (the function or equation) for the problem.

We can now define x and C in a more complete way:

$$C(x) = \text{the total cost, in dollars, if } x \text{ T-shirts are ordered}$$

and specify (since negative values of x do not make sense here):

$$C(x) = 7x + 50, \text{ for } x \geq 0 \quad \square$$

Definition of a Mathematical Model

When we use the function we formulated for the cost above to predict the cost if we order 80 T-shirts, what **assumptions** were we making? For one thing, we were assuming that we would not get any discounts or wait for a sale at the T-shirt distributor. The way we defined our function, we were also assuming that we could order a fractional number of T-shirts, since we did not say that x had to be a whole number (a nonnegative **integer**).

By now, we have given a full specification (definition) of a **mathematical model** for the cost of the T-shirts in Sample Problem 1. What exactly do we mean by a mathematical model? We are all familiar with model railroads, fashion models, and role models, but not necessarily with mathematical models.

A mathematical model is a specification of a verbal definition of a function and its variables (including units), a symbol definition (algebraic formula) for the function (including the domain), and the assumptions being made, that is used to represent a real-world problem or situation mathematically (Figure 1.2-1).

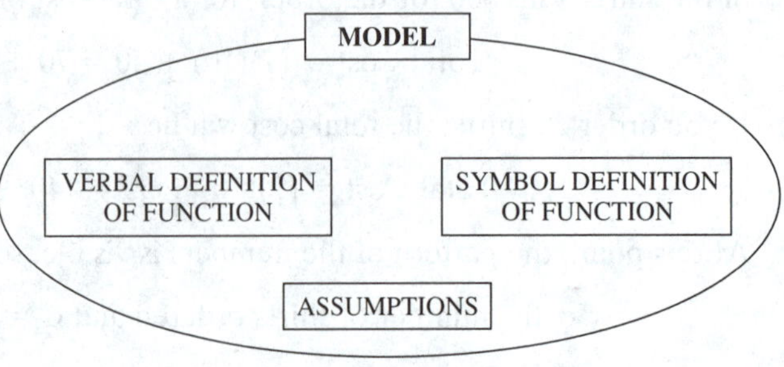

Figure 1.2-1

Our **verbal definition** was the definition (in **words**) of the expression for the total cost (C) in terms of x:

Verbal Definition: $C(x)$ = the total cost, in dollars, if x T-shirts are ordered

Notice that in our verbal definition we were careful to specify the **units** for all quantities (both the value of the function and the input variable, in this example). For example, the cost could have been in cents, or hundreds of dollars, but we took the simplest and most obvious choice of selecting dollars for our units. The units for x were T-shirts. These **verbal definitions are what *connect* the mathematical expressions to the real world**—what give them context and meaning.

Our **symbol definition** was the specification of the **algebraic (formula) definition** of the cost as a **function** of the number of shirts, and the **domain** of that function:

Symbol Definition: $C(x) = 7x + 50$, for $x \geq 0$

When solving a real-world problem, carefully identifying what your input and output values mean and the domain of the function is just as important as having the right function or formula that describes the relationship between the input and the output variables. NO NAKED NUMBERS OR SYMBOLS! In this situation, a function with no verbal definition is as useless as a message on your voice mail that just says: "Your friend told me to tell give you this message: Sounds great. See you there." No identification of the caller or the friend, no idea as to *what* sounds great or where "there" is. The message is worse than useless: You could end up in big trouble for not showing up at the appointed place, wherever that is.

Our major assumptions are that the relationship between x and C is given *exactly* by the equation (the **certainty** assumption) and that the variables can take on *any* real-number value, including *any* fraction or decimal, in an interval (the assumption of **divisibility** of the variables). If the price is exactly as described (with no quantity discounts or sales, for example), then we could say that the certainty assumption holds *perfectly*. If the shirt manufacturer has quantity discounts or periodic sales, then the assumption of certainty might be reasonable, but not perfect. In the real world, neither x nor C is truly continuous, since fractions of T-shirts do not make sense in this context, and costs must be in cents, not fractions of cents. But for the given situation, the assumption of divisibility for both quantities makes good sense. We can always round off our answers if necessary and still have a result that is meaningful and useful.

You should be aware that there is a difference between *specifying* assumptions and their implications and *commenting* on them and how well they hold. Commenting on assumptions is called **validation** and is step 9 in our 12-step process for problem solving. Clearly, you want to choose a model that is as valid as possible, so this needs to be thought about when choosing your model. But in *specifying* a model, you only *need* to say what the assumptions *are*, and their implications. In the above discussion, we have done a little of both.

The verbal definition, the symbol definition for the cost in terms of the number of shirts ordered (the functional representation of the relationship between the cost and the number of shirts ordered), and the specification of the assumptions are the specification of the mathematical model for this problem. This model could then be used to *project* the cost of different shirt orders (a **forecasting** problem), or could be used to build a model of the profit from selling the shirts, which could then be *maximized* (an **optimization** problem).

SAMPLE PROBLEM 2: You now have a mathematical expression that defines the relationship between the cost of ordering shirts when the fixed cost is \$50 and the price per shirt is \$7. *Fully define* a **model** for the cost of ordering the shirts, and find the cost of ordering 80 T-shirts.

Solution: We will adopt the following **format throughout the text to fully define a mathematical model:**

Verbal Definition: $C(x)$ = the cost, in dollars, if we order x T-shirts.
Symbol Definition: $C(x) = 7x + 50$, for $x \geq 0$.
Assumptions: Certainty and divisibility. Certainty implies no sales or discounts, and stability of the order cost. Divisibility implies we can order any fraction of a T-shirt.

Given this model, we now have a concise way to calculate the cost of ordering any number (x) of shirts, $C(x) = 7x + 50$. For example, the cost of ordering 80 can be calculated by evaluating $C(80)$ (plugging in 80 in place of x in the function rule):

$$C(80) = 7(80) + 50 = 560 + 50 = 610$$

Translation: The cost of ordering 80 T-shirts is \$610. □

Assumptions

In Sample Problem 2, recall that our **assumptions** were that the cost equation gives the *exact* relationship between the number of shirts ordered and the order cost (the assumption of **certainty**) and that the variables can take on *any* value in an interval, including any fraction or decimal, (the assumption of **divisibility** of the variables). Since these may not hold exactly in the real world situation, our mathematical model is *not a perfect representation* of the real situation but is close enough to be very useful. This is similar to the fact that a wind tunnel is not a perfect representation of actual flying conditions but is useful in testing plane designs. Assumptions also provide a *connection* between the mathematical expressions and the real-world situation.

Virtually all of the models we will work with in this book will assume both certainty and divisibility. The main challenge in working with a model, then, is to clarify what these two assumptions *imply* in the real-world situation. In this example, the implications include no volume discounts or sales, and stability of the costs. Sometimes we make some of these assumptions *early* in the process, as we are deciding what types of models to consider. For example, in Sample Problem 1, we could have first thought about the fact that we would assume there are no discounts or sales involved, which could then help us choose the simple linear model. Once we have chosen the model, such assumptions will clearly hold, since the model was chosen to incorporate them. Even if we hadn't thought about the implications for discounts and sales, those assumptions are still *implicit* in our model. **It is always best to try to smoke out, and be aware of and *explicit* about, all such practical implications of any model that is used.** The model may be used by other people who would not be aware of the assumptions that were made when the model was developed.

Sample Problem 3: Formulate (define) a model to help Meg determine her optimal amount of exercise to maximize her energy level.

Solution: We have already done most of the work for this in Section 1.0. If necessary, look back at that section to review what we did.

We started by clarifying the problem and defining the variables of interest:

x = the number of minutes of exercise Meg gets in a week

y = Meg's overall energy level in a week (on the 0–100 scale, subjectively determined by her)

Given these definitions, Meg then gathered data over 10 weeks, resulting in the following table:

Running (min)	380	465	420	240	275	390	730	355	440	605
Energy (0–100)	68	90	80	30	40	70	45	60	75	85

These data values were plotted graphically and are shown again in Figure 1.2-2.

At this point, we wanted to find a curve that would represent the relationship **on average** (or in general) between Meg's energy level and the number of minutes she exercised. We can define a *function* to represent this idea:

$E(x)$ = Meg's expected energy level *on average* in a week, using her subjective 0–100 scale, if she exercises x minutes in that week

This looks very similar to the definition of y we made earlier. The difference is that it would be possible for Meg to have gotten exactly the same number of minutes of exercise during two different weeks. Suppose she had obtained data points (400,80) and (400,85). We would know from this data that the energy level as recorded by Meg would

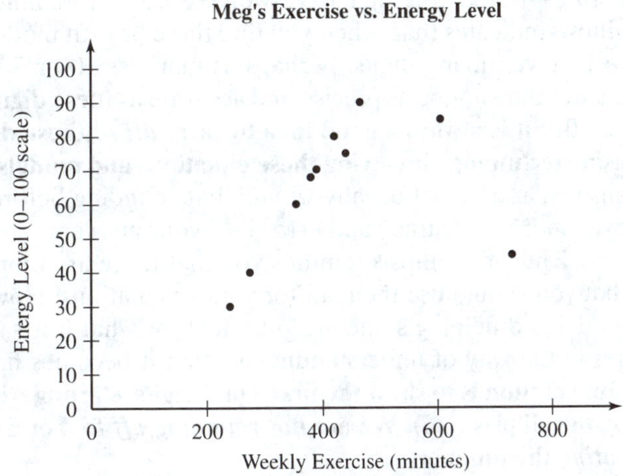

Figure 1.2-2

not be a function of the number of minutes of exercise in a week, since there would be two different second variable values (80 and 85) paired with the same first variable value (400). We could use our definition of y to say that, for one week, $y = 80$, and for the other week, $y = 85$. This does *not* mean that we should not fit a model to the data that will give a functional relationship. When we are making a model, we want the model to just give us *one* predicted energy level, on average, for a given x value. For example, a reasonable model might have $E(400) \approx 83$ (the rounded average).

Looking at the data, we visually saw that the data seemed as if it would fit a parabolic graph quite well, and used technology to find the best-fit parabola (a **quadratic** model, of the form $ax^2 + bx + c$). Then after removing the questionable data point and trying a few more types of curves, we ended up concluding that the best model *for Meg's problem* was the quadratic function (based on fitting the top three points):

$$E(x) = -0.001394\ldots x^2 + 1.456\ldots x - 285.6\ldots$$

Notice that in the last sentence we said that "the best model . . . was the quadratic function." This usage of "model" is a useful shorthand, since the heart of a model is the symbol definition of the function. However, you should keep in mind that the *full* meaning of "model" always includes the verbal definition and the specification of the assumptions, which are sometimes implied or assumed in a discussion. **By definition, a model *has* to be connected to the real world in some way—it *must* have a *context*.** The exact same quadratic function above could also be part of many other models, with different contexts and definitions, such as total revenues in thousands of dollars from the sale of x tickets, as an arbitrary example.

Before continuing with our discussion of this model, let's take a minute to talk about how to *present* models to other people.

Standard Numerical Format of a Model: Significant Digits and Rounding

In the above expression for $E(x)$, we have written each numerical constant (parameter) with a number of digits followed by three periods (you probably would read this as "dot dot dot"; it is also referred to as an **ellipsis**). You have probably never seen this notation before—frankly, neither have we! We use it to communicate a number of things. The ellipsis indicates that, when you find these best-fit models using technology, you are likely to be given many digits, perhaps as many as 10 or 20 digits, for each parameter. Your answer may not be as precise and accurate as these digits suggest, as we will see in Section 8.3. But **it is always a good idea to carry *all* of these digits in any calculations you make using technology involving these equations and models**. You can always round your final answer, and in fact usually *should*, but rounding before that point is likely only to inject even more inaccuracy and error into your answers.

Thus, the ellipsis reminds you that there are more digits than we are showing, and that you should use them all for your calculations. However, for the sake of presenting a model and helping someone get a feel for what it is saying, showing *all* of the digits can get in the way of understanding, so that it becomes hard to see the forest for the trees. **Our solution is to show the first 4 or 5 digits, starting with the first nonzero digit, followed by the ellipsis (. . .). We are *not rounding off* to 4 or 5 significant digits, but simply *truncating* the number.**

For example, if your calculator gave an answer of −0.00013946825465, we could write this as "−0.0001394 . . . ." With this system, **the number can easily be rounded to one less significant digit or fewer**. For example, the quadratic coefficient in our model above, −0.0001394 . . . , could be rounded to 3 significant digits as −0.000139 or to 2 significant digits as −0.00014. However, if the digit after the 4 in the −0.0001394 . . . had been, for example, an 8 (−0.00013948 . . .), then the coefficient rounded off to 4 significant digits would have been −0.0001395. On the other hand, if the next digit after the 4 had been, for example, a 2 (−0.00013942 . . .), then the coefficient rounded off to 4 significant digits would have been −0.0001394. Remember that **rounding off numbers always rounds *up* (adds one to the last digit) when the next digit is a 5 or more, and rounds *down* (keeps the last digit the same) when the next digit is 4 or less**. Since most real-world examples are not likely to have more than 3 or 4 significant digits of accuracy in their data, truncating after 4 or 5 digits seems like a reasonable compromise between completeness and visual comprehensibility. **We will adopt this convention from this point on in the book.**

We should also mention here that in some models our parameters (or some of them) can be *exact*, such as in our model for the cost of T-shirts. For such cases the ellipsis notation may not be necessary, and may not be used. On the other hand, fractions such as 1/3 can be written exactly, but in decimal form cannot be denoted without a bar to denote a repeated decimal, such as $0.\overline{3}$, in which case the ellipsis notation, 0.3 . . . , can also be useful and convenient.

Our assumptions, as usual, were **divisibility** of the variables and **certainty**. This time, divisibility is not a problem; Meg could theoretically run for *any positive number* of minutes. With regard to certainty, recall again that we obtained our model by fitting a parabola to the top three points from Meg's data. In fact, there is only one parabola that goes through those three points, so you might be inclined to think that the relationship *is* exact, as certainty implies. If you think about it more, though, you realize that if Meg gathered more data in the same vicinity of x values, they would probably not fit *exactly* to the model, although they may be very close. Judging from the way most of the data seems very close to a smooth curve (except the one unusual data point), certainty would seem to be very close to being satisfied, but there is still a small amount of error between the model predictions and the actual values. If Meg gathered more data, it is possible that there would be *more* fluctuation than the original data set might suggest.

Some other **implications of the certainty assumption are that the conditions, patterns, and forces under which the data were gathered will continue into the future** (such as Meg's level of sleep, nutrition, health, etc.), that there are no cumulative *training* effects, and that the relationship of exercise to energy is only *immediate*. The idea of *immediacy* is that we only are looking at the effect on energy *the same week*, rather than possible lag or momentum effects that might be more complicated.

A model is used for a particular **purpose**, which could also have a major effect on what model is selected. In Meg's problem, she wanted to find the *optimal amount* of exercise. The best model for that purpose may not be the best model for *predicting* her energy level given any possible amount of exercise.

Recall that this symbol definition that we defined for Sample Problem 3 came from fitting the highest three data points from Meg's data given in Section 1.0, which we felt gave us the best representation of the relationship near the maximum. Because of this, the interval of x values for which this model makes sense is somewhat limited. As suggested by the graph in Figure 1.0-4, this model would seem to make sense mainly for values of x roughly between 400 and 600 minutes, which we can express mathematically as $400 \le x \le 600$.

Here, then, is **our model** for Meg's problem.

The verbal definition for our model is

x = the number of minutes of exercise in a week

$E(x)$ = Meg's expected overall energy level on average in a week, using her 0–100 scale, if she gets x minutes of intentional exercise that week

The symbol definition of the function for this model is

$$E(x) = -0.001394\ldots x^2 + 1.456\ldots x - 285.6\ldots, \text{ for } 400 \le x \le 600$$

Our assumptions are:

Divisibility of the variables, and certainty that the model gives the exact relationship between Meg's exercise and her average energy level.

Using our standard format, we can now write our model for Meg's average energy level:

Verbal Definition: $E(x)$ = Meg's average overall expected energy level, on her 0–100 scale, in a week if she consciously exercises for x minutes that week

Symbol Definition: $E(x) = -0.0001394\ldots x^2 + 1.456\ldots x - 285.6\ldots,$ for $400 \le x \le 600$

Assumptions: Certainty and divisibility. Certainty implies that other factors influencing Meg's energy level (including sleep, diet, etc.) remain relatively constant.

The assumption of divisibility of the variables holds perfectly. The assumption of certainty seems to hold quite well but implies that the conditions, patterns, and forces from when the data were gathered will continue into the future (including Meg's level of sleep, health, and nutrition). □

Notice that we have been careful to say that this was the best model we could find *for Meg's problem.* If, for example, Meg had decided that the unusual data point *was* legitimate and should be included, we probably would have chosen a different model as being best. Or, we might have chosen a different model *if* our main purpose had been to **forecast** Meg's energy level if she runs for 800 minutes in a week. In other words, **the choice of a model depends on the *context* of the problem and the *purpose* for the model, as well as on the data!** For example, if you are **forecasting for a value of a variable that lies *within* the extremes of the input values of the given data** (a process called **interpolation**), you may choose a very different model than if you are **forecasting for a value *outside* the extremes of the input values of the given data** (a process called **extrapolation**). And if you are optimizing, you may choose an entirely *different* model.

Sample Problem 4: You help run a local coffeehouse, and have had the same musical group perform three years in a row. You have charged different admission prices each year, and attendance has varied. The first year you charged $7 per ticket and 100 people bought tickets. The second year 80 tickets were sold at $8 per ticket. Last year you charged $9 and sold 70 tickets. You have agreed to guarantee to pay the band $300 this year. You now need to decide what price to charge for tickets. Describe the process of formulating a model to represent the profit from the concert, so that you can then determine the price to charge to maximize this profit.

Solution: *Clarifying the Problem.* Before we go any further, let's clarify our problem (Stage A of our process of problem solving).

First, we need to identify, bound, and define our problem (Step 1 of the 12-step program). The problem has been identified for us by the question itself (unlike a student-generated project, where *you* have to come up with the question to be answered.). We are asked to maximize profit, and only ticket sales and the cost of the band are referred to explicitly. Suppose that, in reality, other revenues (such as income from the sale of refreshments or recordings) and costs (such as space rental, copying and postage for publicity, etc.) do exist. Should we include these factors in the definition of our problem?

When solving problems and formulating models, the specification of assumptions is somewhat relative. For example, you could make simplifying assumptions in the **definition of your problem**, such as our saying we will ignore the other revenues and costs in defining our problem. These assumptions will then normally be carried directly into the assumptions of your **model**. Or, you can define your problem to be broader (*not* ignoring the other revenues and costs), but choose to make those simplifying assumptions for your **model**, at least initially. Either approach can be justified. **Our advice in general is to make the definition of the problem as close to reality as possible, and make your simplifying assumptions as you formulate your model.** In the present problem, if you cannot easily get information about the other revenues and costs, then it makes sense to define your problem to limit your focus to the ticket revenue and cost of the band. Let's suppose that is indeed the situation.

The above discussion is one form of **bounding** the problem. In this problem, bounding could also mean that we are only considering the problem of what price to charge for this concert and not worrying about the longer-term implications. For example, charging a low price might be a good strategic move to start to build a regular faithful audience in the future, but we are *not* considering such a larger perspective here. That is an important problem for the club to consider but is not the focus now. Bounding usually reflects the extent of time, or the scope and span in other dimensions, such as the level of the organization, the people involved, and so on.

We now want to be sure we understand the problem (step 2) and make a ballpark guess at the solution. The main work to do in this problem is to try to forecast what our ticket revenue will be for different ticket prices. Put together with the cost of the band, this will give us a model for the profit we want to maximize. Conceptually, the more we charge, the fewer tickets we will sell. If we charge too much, almost no one will come, and we will probably not make much in revenue. If we charge too little, we could get a lot of people but still might not make much in revenue. The revenue will be the number of tickets sold times the price of tickets. At this point, we need to know whether all tickets are sold at the same price (whether or not there are discounts for students, seniors, etc., for example). Let's suppose that they are. Even if they were not in reality all the same price, if the effect on the revenue were small, we could still make this assumption for the sake of formulating a model. If the effect on the revenue were not small, we could formulate a more complicated model.

Clarified Statement of the Problem: We are trying to find the price to charge for tickets to maximize our profit for the concert, considering only ticket revenue and the \$300 cost for the band. We assume there is only one ticket price, which all people attending pay.

Since you have been involved in the club for several years, you have some intuition about what you think is the best price to charge. Let's suppose your gut feeling is that somewhere around \$8 would be best.

Formulating the Model: Now we are ready to define relevant variables. This is step 3, the first step in formulating our model (stage B). Let

p = price charged for the concert (in dollars)
n = number of people attending the concert, paying full price
R = revenue (in dollars) from tickets
C = cost (in dollars) of the act
P = profit (in dollars)

We are *assuming* that the profit in the model includes only ticket sales and the cost of the act.

Notice that we have specified units for all of these definitions and have tried to make them as clear, precise, and unambiguous as possible. Our basic decision variable is the price to charge, p, and the variable representing our objective (the quantity we are primarily interested in) is the profit, P. The other variables are needed for intermediate calculations.

Next, to really understand the relationship between the price and the profit, we need some data, or some other form of information. Since the band has played at the club three times before, we can get data about how much was charged and how many people came in the past. Fortunately, we have been careful about keeping such records. Here are the data, as specified in the description of the problem above:

Price (dollars)	7	8	9
Paid Attendance	100	80	70

Now we have the basis for another ballpark estimate. Since the revenue is simply the ticket price times the number of tickets (attendance), $R = pn$, we can do a quick rough calculation. When the price was \$7, 100 tickets were sold, for a revenue of (\$7)(100) = \$700. When the price was \$8, 80 tickets were sold, for a revenue of \$640, and when the price was \$9, 70 tickets were sold, for a revenue of \$630. Since the cost is fixed at \$300 this year, the highest revenue will give the highest profit, which suggests that the best price might be around \$7, or possibly less.

Where do we go from here? We can try to formulate a model of the demand for tickets—the number of tickets sold in terms of the ticket price. From that, we can then derive an expression for the revenue and finally get a model of the profit by subtracting the cost of $300 from the revenue expression.

How do we find a model for the demand? As we said earlier, if there is a way to **visualize** an aspect of a problem, it is usually a good thing to do, to help us get a deeper understanding. In this case, let's look at a **graph** of the data in Figure 1.2-3. From looking at this plot, we see that we could fit a straight line, even though the data do look a bit curved. What kind of a model should we try?

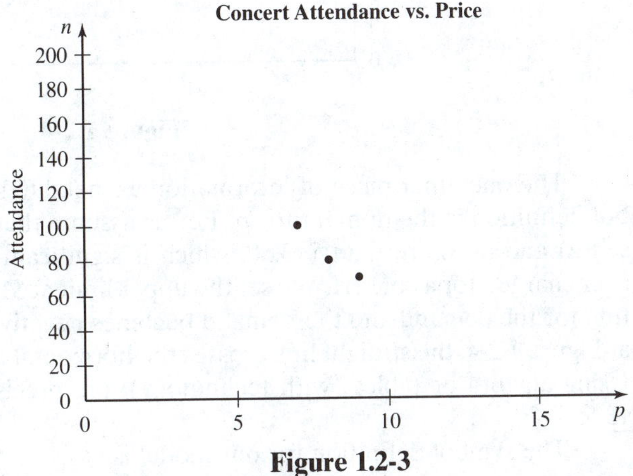

Figure 1.2-3

One of the principles of modeling is **simplicity**. In a given situation, if you are choosing between several options, **always give preference to the simpler model unless you have compelling reasons to do otherwise**. You can always go back and do a more complicated analysis later if you feel it is justified. Some people would call this the principle of **Keep It Simple, Stupid! (KISS!)** In more esoteric modeling terminology, it is often called the principle of **parsimony**, which means *stinginess*, as in being stingy with the number of parameters, or level of complexity, that you allow yourself in your model.

If we apply the KISS! principle, it would suggest trying a linear model first. After doing this, if we feel that we have oversimplified, we can always try a more complicated model.

You will learn in Section 1.3 how to use technology to find a best-fit linear model for given data in this kind of situation. When we do this, the linear equation we get is given by

$$y = -15x + 203.33\ldots$$

and its graph is shown in Figure 1.2-4.

Thus our expression for the number of tickets sold in terms of the ticket price is given by

$$n(p) = -15p + 203.33\ldots$$

This relationship between the price of something and the quantity *demanded* by consumers, is called a **demand** function in economics.

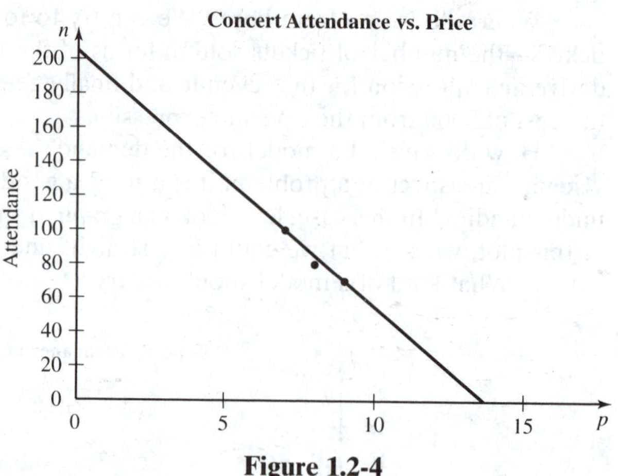

Figure 1.2-4

The one other piece of information we need to specify before writing out our symbol definition is the domain for p. Let us assume that p must be nonnegative (common sense) and no more than \$13.50, which is significantly more than the coffeehouse had ever charged for a concert. We set the upper limit at \$13.50 because we used a linear function for the demand and the demand becomes negative at prices over \$13.55. If you look at Figure 1.2-4, the straight line crosses the horizontal axis somewhere between 13 and 14. Using algebra or tables (with technology), the precise crossing point is approximately 13.55.

The symbol definition for our model is

$$n(p) = -15p + 203.33\ldots, \text{ for } 0 \leq p \leq 13.50$$

As we build a model for the profit in this problem, we will need to make some assumptions. For example, we will assume certainty for the demand equation (the curve fit above). This implies that the future will be comparable to the past; that is, the relationship between price and attendance represented by the *past* data would hold *this* year as well. For example, it assumes that there is no "staleness" effect (people getting tired of seeing the act every year), and that inflation is not too high. We will, as usual, be assuming divisibility of the variables, while in reality they are all discrete. The number of tickets sold will have to be a nonnegative integer. The other variables are all monetary, so must be multiples of .01 (that is, in no more detail than whole numbers of cents). In fact, the admission price is more likely to be restricted to multiples of \$0.25, if not whole dollars.

As discussed earlier, we will also be assuming that tickets are the only source of revenue, and that the act is the only cost (this is actually an additional consequence of the certainty assumption). These are really assumptions we made in defining our problem, but we need to keep them in mind.

Now that we have the demand equation, the rest of our model comes from the given information and the following basic business definitions:

Revenue = (selling price)(quantity sold)	$R = pq$
Profit = Revenue − Cost	$P = R - C$

For example, when the selling price of the tickets was \$7, and 100 were sold, the revenue was (\$7)(100) = \$700, and the profit to the coffeehouse was \$700 − \$300 = \$400.

In our problem, we are using n for the quantity sold instead of q, so we have $R = pn$. Furthermore, the cost will be \$300. And, for completeness, let us just repeat the demand function found above:

$$n(p) = -15p + 203.33\ldots \text{ for } 0 \le p \le 13.50\ldots$$

We now have everything we need to find a model for profit in terms of the concert price. Let's define a variable

$$P = \text{the profit, in dollars, from the concert}$$

All we need to do now is a sequence of substitutions:

$$P = R - C = (pn) - C = p(-15p + 203.33\ldots) - 300 = -15p^2 + 203.33\ldots p - 300$$

What we did above is a very general strategy for finding the main expression for a model. We wanted P in terms of p. We had several equations involving several variables, including p. **To eliminate the variables we didn't want in the final expression, we solved for them in the additional equations, then substituted back.** For example, the demand equation had n solved in terms of p

$$n = -15p + 203.33\ldots$$

so we substituted for n in the revenue equation $R = pn$, and got R in terms of p

$$R = (p)(-15p + 203.33\ldots)$$

Then, since the cost was fixed at \$300, we knew that we could replace C with 300. Finally, since P was a function of R and C ($P = R - C$), we substituted for each of them to get P in terms of p:

$$P = [p(-15p + 203.33\ldots)] - 300$$

We are finally able to write out our full **model** for the profit in this problem!

Verbal Definition: $P(p)$ = profit, in dollars, from the concert if tickets are sold for p dollars per ticket.

Symbol Definition: $P(p) = -15p^2 + 203.33\ldots p - 300$, for $0 \le p \le 13.50$.

Assumptions: Certainty and divisibility. Certainty implies that we are considering only the ticket sales and cost of the act in our measurement of profit (that we are ignoring refreshment sales revenue and other costs), that all people attending pay the full price, and that the sales will follow the demand function exactly. Divisibility assumes that fractions of tickets can be sold.

The above verbal definition, the symbol definition, and the statement of the assumptions and their implications are thus our **model** of the profit for this problem. The club wants to maximize the profit function of this model over its domain. Notice that, along the way, we could also have fully defined models of the demand, revenue, and cost, but we focused only on our profit. □

Sample Problem 5: In most states a sales tax is added when items are purchased. If the state sales tax is 5%, is the amount of sales tax a function of the purchase total?

Formulate a model of the final bill (cost of the purchase plus the sales tax) in terms of the purchase total.

Solution: Remember that a model involves a verbal and symbol definition of a function, and assumptions. In this case, we could start by defining some variables:

c = the total cost of the purchases, in dollars

t = the total amount of sales tax, in dollars

$F(c)$ = the final bill, in dollars, if the total cost of the purchases is c dollars

What is the mathematical function that describes the relationship between the total amount of the sales tax and the total cost of the purchase? Let's plug in some simple numbers so we can get a picture of what is going on. Let's say your purchase was $1.00. The sales tax of five percent will be calculated on $1.00: 0.05 times 1.00 = .05, so the sales tax will be $0.05. If the purchase were $2.00, the sales tax would be 0.05 times 2.00 = 0.10, so the sales tax would be $0.10. The pattern here is easy to see: Multiply the total cost of the purchase time five percent, or 0.05 times c. We can write this as

$$t = 0.05c$$

We now have a mathematical expression for the sales tax in terms of the cost of the purchases. What would the final bill be? For our $1.00 purchase the tax was $0.05, so the final bill would be the cost of the purchase plus the tax, $1.00 + $0.05, or $1.05. For our $2.00 purchase, the tax was $0.10, so the final bill would be the cost of the purchase plus the tax, $2.00 + $0.10 = $2.10. Substituting our variable c for the cost we have

$$F(c) = c + t = c + 0.05c = 1.05c$$

Our model is

Verbal Definitions: $F(c)$ = the final bill, in dollars, on a total purchase of c dollars.

Symbol Definitions: $F(c) = 1.05c$, for $c \geq 0$.

Assumptions: Certainty and divisibility. Certainty implies that the sales tax follows the mathematical formula exactly. Divisibility implies that fractions of cents are possible for both the original purchase cost and the final bill amount.

In reality, both variables are *not* divisible, but discrete (they must be in cents, or units of $.01), but are close enough for this to be a reasonable assumption. The certainty assumption would not be quite accurate in some states where sales tax is computed according to tables that may not correspond exactly to the formula. □

Sample Problem 5, like Sample Problem 1, was expressed *directly* in functional notation. We did not start with a table of data and find a model to describe the relationship. We went directly from the words describing the relationship to the mathematical expression describing the relationship. Such situations are commonly referred to as the dreaded "word problems." As we begin to find mathematical models to describe relationships, we will work with three types of problems: problems that can be formulated from data that are presented in tables or graphs, problems that can be formulated directly from **verbal descriptions** of relationships, and problems that are a combination of the two. We need to be able to work with "word problems." When you get out of school and are working in the "real world" (a rather strange expression, because school is quite real, but you know what we mean), you would not expect to have your boss present you with a mathematical equation to solve. It is more likely that you will be presented with a much broader and less well defined problem: "How can we increase our profits on this item?" "How can we reduce our shipping costs?" "How much should we put into advertising?" You will be expected to gather the necessary data, develop the models, use mathematical techniques to analyze the models and validate and interpret the results! Big order? Yes! Can you do it? We think you can, and this course is designed to help you do it.

Aligning Data

Sample Problem 6: Enrollments at Alma Mater University in recent years have been as follows:[6]

Year	84–85	85–86	86–87	87–88	88–89	89–90	90–91	91–92	92–93	93–94	94–95
Enroll-ment	6020	6327	6221	6189	6289	6411	6313	6460	6289	6103	6113

where the enrollments are as of the *beginning* of each academic year (for example, "84–85" means "1984–1985." Give the verbal definition for a model of the enrollment as a function of time.

Solution: If you are modeling the enrollment in your school over a period of years, starting with the academic year 1984–1985, you might want to use $E = f(y)$ = some expression or formula involving y, where $E = f(y)$ is the enrollment in your school as of the beginning of the academic year that starts in year y.

Notice that our definition of y really only makes sense if y is an integer. Partly to make the divisibility assumption possible, when fitting a model to a **time series (set of values over time)**, as in this example, we usually **align** the data. This means that we set the initial time period equal to 0 and count subsequent time periods from that point. There are several reasons for this. In some cases no data existed before the initial entry. For example, data relating to commercial aviation would begin in 1911, or space travel with the first orbital flight of Yuri Gagarin on the twelfth of April 1961. In other cases, data may exist prior to the first year of the series but are not included in the study. Since we have no data prior to these years, we would set the initial year equal to zero and count years after that time. If we do not align our data, our models can become quite cumbersome and the real relationship of the variables can be difficult to see. When we actually fit a model to a time series, you will be able to see why this is important.

In this example, let us align our data by defining a different input variable. Let

$$t = \text{the number of years after September 1, 1984}$$

By this definition, $t = 0$ corresponds to September 1, 1984 (the beginning of the academic year 1984–1985), $t = 1$ is one year later, so corresponds to September 1, 1985 (the

[6]Adapted from a student project.

beginning of the academic year 1985–1986), and so forth. In the following table our years are shown with 1984–85 aligned to 0 in this way:

Year	0	1	2	3	4	5	6	7	8	9	10
Enroll-ment	6020	6327	6221	6189	6289	6411	6313	6460	6289	6103	6113

By aligning our data in this way, we at least have a mechanism for talking about other points in time during the academic year other than the beginning. For example, $t = 0.5$ means one half of a year after $t = 0$, or 6 months after September 1, 1984, which we could interpret as March 1, 1985. For business and economics problems, this kind of precision can sometimes be very important.

It is very important that this ***aligning*** **of the data** be **included in the definition of the variables.** Our verbal definition now would be

$$E(t) = \text{enrollment in your school } t \text{ years after September 1, 1984}$$

Writing the data in this form makes it more convenient to fit a model to it. □

This problem is a reminder in general, and especially when working with time, to **define all functions and variables as clearly and precisely as possible, including units for everything.**

Average Cost

Sample Problem 7: Suppose you and some friends decide to order T-shirts from the same company described in Sample Problem 1. It will cost you $50 to place the order and each shirt costs $7. Find a model for the *average* cost of the shirts (the cost per shirt each of you would have to pay, incorporating the fixed cost by splitting it evenly) as a function of the number of shirts ordered.

Solution: You need to find a mathematical expression for the *average* cost as a function of the number of shirts ordered. Let's try plugging in some numbers. Say you start out with 10 friends wanting the shirts. We already have a model for the cost of the shirts, from Sample Problem 2:

Verbal Definition: $C(x)$ = the cost, in dollars, if we order x T-shirts.
Symbol Definition: $C(x) = 7x + 50$, for $x \geq 0$.
Assumptions: Certainty and divisibility. Certainty implies no sales or discounts, and stability of the order cost. Divisibility implies fractions of shirts can be ordered.

Let's **plug in some numbers, chosen arbitrarily to make calculations relatively easy.** Suppose there are 10 of you interested in ordering the shirts. The cost would then be $C(10) = 7(10) + 50 = 70 + 50 = 120$. The total bill would be $120. Since there are 10 of you ordering, each of you would have to pay $120/10, or $12 for your shirt. The average cost per shirt when there are 10 shirts ordered is $12 per shirt. What if you could get 20 people to order the shirts? $C(20) = 7(20) + 50 = 140 + 50$. The total bill would be $190. Since there were 20 of you ordering, each of you would pay $190/20 or $9.50 for your shirt. The average cost per shirt when 20 shirts are ordered is \$9.50. Suppose there were 25 of you. The calculations would be $C(25) = 7(25) + 50 = 175 + 50 = 225$, and you each would pay \$225/25 = \$9.00. Let's recap what we did. **If we write out the calculations for each case, without simplifying, we can see the patterns best:**

10 shirts: Average cost $= \dfrac{7(10) + 50}{10}$

20 shirts: Average cost $= \dfrac{7(20) + 50}{20}$

25 shirts: Average cost $= \dfrac{7(25) + 50}{25}$

For each calculation, we multiplied 7 times the number of shirts ordered, added 50 to that, and then divided the total by the number of shirts ordered. Using x for the number of shirts ordered, the pattern would suggest

x shirts: Average cost $= \dfrac{\text{total cost of shirts}}{\text{number of shirts}} = \dfrac{7(x) + 50}{(x)}$

We can now fully define our model:

Verbal Definition: $A(x) =$ the average cost, in dollars per shirt, if we order x T-shirts.

Symbol Definition: $A(x) = \dfrac{7x + 50}{x}$, for $x > 0$.

Assumptions: Certainty and divisibility. Certainty implies no sales or discounts, and stability of the order cost. Divisibility implies fractions of shirts can be ordered.

Notice that, if $x = 0$, the average cost would not be defined by the formula, so we have excluded this value from the domain that we used for the total cost function in Sample Problem 2. □

The average is usually denoted by placing a bar over the letter, so we could have called our average cost $\overline{C}$ or $\overline{c}$ instead of A. If $C(x) =$ the total cost (in dollars) for x items the general formula for the average cost is $\overline{C}(x) =$ the average cost (in dollars) for x items:

$$\overline{C}(x) = \frac{C(x)}{x}.$$

Sample Problem 7 above is a good example of what we described in step 5 (formulating a model) of our 12-step process for problem solving: Pick some sample numerical values for the decision variables; then do whatever calculations you need to see if those values represent a solution (numbers are easier than x's). If necessary, use two or three sets of numerical values. The important thing to concentrate on is to write out the calculations *without simplifying*, to see what those basic calculations were, and to make the patterns most clear. When deciding what numerical values to choose, the best criteria are

- Easy to calculate with (such as 10, 20, 100),
- Different from each other (if several variables are involved), and

- Different from the given constants in the problem (like the 7 and the 50 in Sample Problem 7 above).

Sample Problem 8: Suppose you have a favorite kind of cheese that you like to snack on almost every day. Unfortunately, this cheese is not very common, so you can't just get it at the corner grocery store—you have to go downtown. You estimate that such a trip costs you about $5.00, taking into consideration transportation costs and the value of your time. The cheese itself costs about $4.00 per 8 ounces. If you buy a lot of cheese, you figure there is also a cost for storing it, which includes the cost of electricity to keep it cold in your refrigerator, the interest you are losing by tying up your money in cheese rather than keeping it in your bank account, and possibly a cost in deterioration of the cheese over time (in taste, or in the chances it will go bad). You lump these all together, and estimate the storage cost to be about 20 cents per ounce, per month. For example, if you stored 2 ounces of cheese for 3 months, the storage cost would be $2(20) = 40$ cents per month, so for 3 months, the storage cost would be $3(40) = 120$ cents, or $1.20. Finally, you estimate that you eat about 2 pounds of the cheese every month. The problem is that it's a hassle to go to this special cheese store downtown, so you want to figure out how often you should go, and how much cheese you should buy each time. Formulate a model.

Solution: Let's apply our first five steps for problem solving to formulate the model for this problem.

Stage A: Clarify the Problem

Step 1. Identify, Bound, and Define the Problem. We know that the problem is to decide how often to go to the store, and how much to buy each time; it has already been identified for us in the description of the problem.

We can bound the problem by limiting the scope of what we consider. In the real situation, perhaps you sometimes go downtown for other reasons, so the fixed cost of $5 could be shared with other errands. Since we are not given any of this information, we will assume that this cheese excursion is the only reason you go downtown, or at least that the $5 is the exact fixed cost for each trip. Similarly, we will assume away other realistic complicating factors and just work with the information given in the simplest possible way. For example, we can assume that you essentially eat the cheese continuously (as if taking it intravenously!), so your consumption is steady at all times.

There is one major part of the problem we have not defined yet: we need to say what the criteria are for deciding which answer is best. Clearly, it has something to do with minimizing cost, but exactly *what* cost? We want a measure of the *total* cost, including the fixed cost, the cost of the cheese itself, and the storage cost, but if the problem is ongoing, the total cost is essentially *unlimited*! There are two major possibilities then: the total cost in between trips to the store (which we'll call the cost *per cycle*), or the total cost *per month*.

A little thought can help decide which of these two is better. In one cycle, you incur the fixed cost, the cost of the cheese, and the storage cost. How could you minimize this total cost? Well, the fixed cost is always the same, but the other two will be smaller if you order less. This means the cost per cycle will be minimized if you order the *least* cheese possible, as close to 0 as you can get. But then you'd have to go to the store *constantly* to keep up with your consumption (demand), and this would mean you're *constantly* incurring the fixed cost of $5! Besides, you wouldn't have time for a life! So minimizing the cost per cycle can't be your objective. Cost per month makes a lot of sense, though, as anyone on a budget would understand.

Step 2. Understand and Restate the Problem. Now we know that we're trying to determine how often you should go to the cheese store downtown, and how much cheese you should buy each time, to minimize your total cost per month. As we said above, we assume all the given information is exact, and that there are no other complicating factors we will consider. One way to help understand a problem is with some kind of a picture or graph or table. In this problem, a very helpful graph is the graph of the amount of cheese you have on hand versus time. In the simplest case, the graph would look something like Figure 1.2-5.

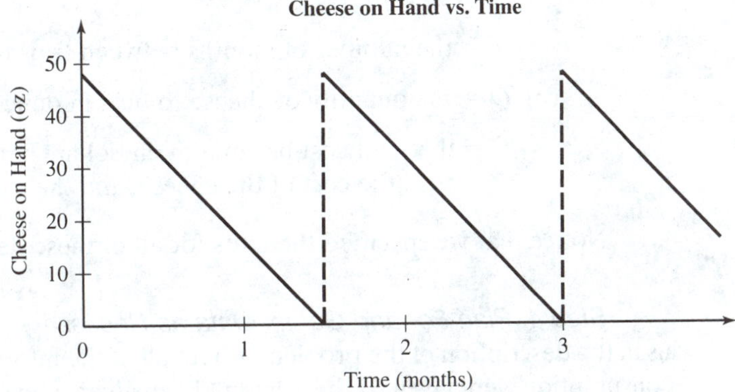

Figure 1.2-5

For obvious reasons, this graph is sometimes called the **sawtooth graph** of a simple inventory problem. The graph helps point out some assumptions we are making. For example, we are assuming that we will wait until we run out of cheese before we go to the store, but that we will go *immediately* as soon as we run out. Another way of saying the same thing is that we assume that we are not allowing any *shortages* of cheese. The graph also emphasizes that we will order the *same* quantity of cheese every cycle. At this stage, we do not need to try to make the graph too accurate, it is really just to give us a visual image of the problem. We could have left off the units on the axes and still accomplished this.

Stage B: Formulate a Model

Step 3. Define All Variables Clearly, Including Units. As a guide for what your decision variable(s) should be, always look at the question you are trying to answer or the problem you are trying to solve. What quantity (or quantities) that you have some control over are you trying to find the *best* value(s) of? In our problem, there are *two* such quantities: how often to go to the store, and how much cheese to buy each time. For now, let's define a variable for each. Sometimes we can eliminate variables because they are directly linked to each other, and by doing so, we can simplify the problem and the solution. But you don't need to do this at the very beginning, if it's not obvious to you. You can also define a variable for your objective, the criteria you will use to decide what solutions are best.

Before defining variables, it is a good idea to look at the units involved in your problem and, if possible, pick *one* unit for each kind of quantity (time, money, etc.) and *stick* with that unit throughout the problem. Especially when you are relatively inexperienced in applied mathematics, this is a good idea, because otherwise it is too easy to mix up your units and get answers that are meaningless. For this example, let's use "majority rule" to decide which units to choose for each type of quantity. For money, two of our costs are given in dollars, and one in cents, so let's choose dollars (either would be fine; *consistency* is what matters). For time, let's use months (a no-brainer). For cheese, let's use ounces rather than pounds, since they were mentioned more.

Now let's define our variables. Let

T = the number of months between trips to the cheese store

Q = the quantity of cheese to buy, in ounces, each trip

C = the total cost per month, in dollars per month (including the fixed cost, the cost of the cheese, and the storage cost)

Notice that we specified the units for all of these definitions.

Step 4. Plan For and Gather Data as Necessary. Fortunately, this was done for us in the description of the problem, where all of the information about the costs and your consumption were given. If these hadn't been given, you would have had to keep a record of your consumption, the cost of the cheese, the travel costs, and the time to make the trip (and an estimate of the value of your time), and the costs related to storing the cheese.

Step 5. Formulate a Model. We can start by specifying the **parameters** of the problem (the values that stay fixed or constant throughout *this particular* problem, although they could be different for other examples of the *same kind* of problem), using the units we chose in step 3. In terms of these units, we know that your fixed cost is \$5 per trip (cycle), the cost of the cheese is \$0.50 per ounce (\$4/8 ounces), your consumption is 32 ounces per month (since there are 16 ounces in a pound, and you consume 2 pounds per month), and the storage cost is (\$0.20 per ounce) per month.

Once we have the parameters, we want to formulate the **symbol definition** of our **objective function**, the quantity we are trying to optimize, including its domain. This is always the hardest part, although doing the first four steps well can help a lot. We know that we are trying to formulate an expression for the total cost per month, C. Since there are three components to this cost, let's break this task down and find each component individually.

The cost of the cheese per month is probably the easiest component to formulate. Since we know that you consume 32 ounces of cheese a month, and that the cheese costs \$0.50 per ounce, your cost of cheese will always be

$$\left(\frac{\$0.50}{\text{ounce}}\right)\left(\frac{32\text{ ounces}}{\text{month}}\right) = \frac{\$16}{\text{month}} = \$16\text{ per month}$$

Notice that writing out the units helps us to see that a calculation makes sense (for example, in the above calculation, we see that the ounces cancel). This is sometimes called **dimensional analysis** in the sciences and is a useful tool.

Next, let's look at *the fixed cost.* We know that this will be \$5 every cycle. But to calculate this cost per month, we need to know how long a cycle is. We have already defined this to be T, so our fixed cost per month will be

$$\frac{\$5}{T\text{ months}} = \$\left(\frac{5}{T}\right)\text{ per month}$$

Now, we get to *the storage cost.* If we go back to our graph, we can now label it using our variables, as in Figure 1.2-6. From the graph, we see that the inventory of cheese starts at Q, the amount we buy each time, and decreases uniformly down to 0. Since it decreases uniformly, what is the *average* inventory over each cycle? It will just be the *average* of Q and 0:

$$\text{average inventory of cheese} = \frac{Q+0}{2}\text{ ounces} = \frac{Q}{2}\text{ ounces}$$

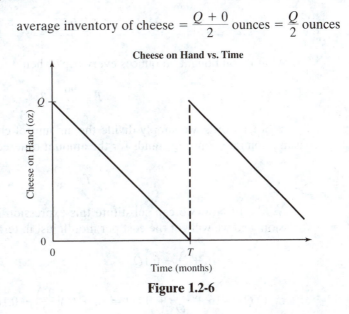

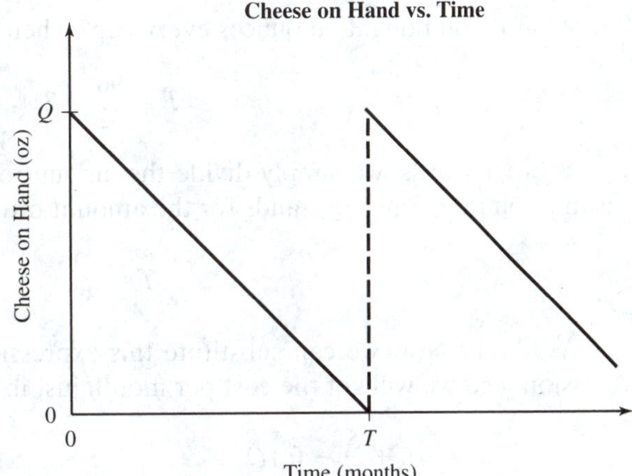

Figure 1.2-6

We already know that the storage cost is (\$0.20 per ounce) per month. Thus, if we have an average of $\frac{Q}{2}$ ounces on hand overall, we can simply multiply this by (\$0.20 per ounce) to get the average storage cost per month,

$$\left(\frac{\$0.20}{\text{ounce}}\right)\left(\frac{Q}{2}\text{ ounces}\right) = \$0.10Q$$

We now have expressions for all three components of the cost, so let's put them together:

$$C = (\text{cost of cheese per month}) + (\text{fixed cost per month}) + (\text{storage cost per month})$$

$$C = 16 + \frac{5}{T} + 0.1Q$$

(Remember, we defined the units of C to be dollars per month, so we don't need to put them into our expression.) This looks fine, but we have said several times that in Volume 1 of this text we will only focus on functions of *one* variable. Can we get an expression for the cost in terms of just one variable? We saw this in Sample Problem 4 already. It required finding another equation relating the variables in question, solving for one of them, and substituting back to eliminate that variable.

Can we find an equation relating T (the time between trips) and Q (the quantity of cheese purchased each trip)? Is either one a *function* of the other? If we knew the value of Q, for example, could we figure out a corresponding value of T, and would there be exactly one such value? Here is where trying out specific numerical values can help. Suppose you bought 64 ounces of cheese every time. How long would it last?

Well, since you eat 32 ounces of cheese each month, 64 ounces would last 2 months. What calculation would give us this answer? We just divided 32 into 64 (or, equivalently, divided 64 by 32):

$$T = \frac{64}{32} = 2$$

What if you bought 96 ounces every trip? Then it would last 3 months, or

$$T = \frac{96}{32} = 3$$

In other words, we simply divide the amount of cheese ordered each time by the consumption rate. Since Q stands for the amount ordered, the pattern suggests that

$$T = \frac{Q}{32}$$

We did it! Now we can substitute this expression in for T in our cost per month expression, and we will get the cost per month just in terms of Q:

$$C = 16 + \frac{5}{T} + 0.1Q$$

$$C(Q) = 16 + \frac{5}{\frac{Q}{32}} + 0.1Q = 16 + (5)\left(\frac{32}{Q}\right) + 0.1Q = 16 + \frac{160}{Q} + 0.1Q$$

(Remember that dividing by a fraction is the same as multiplying by the reciprocal of the fraction, so dividing 5 by $Q/32$ is the same as multiplying 5 by $32/Q$.)

Finally, we can write out our model for the cost per month:

Verbal Definition: $C(Q)$ = the total cost per month, in dollars per month, if you order Q ounces of cheese every trip.

Symbol Definition: $C(Q) = 16 + \frac{160}{Q} + 0.10Q$, for $Q > 0$.

Assumptions: Certainty and divisibility. Certainty implies that no shortages are allowed, that we go to the store as soon as we run out of cheese, that our consumption is continuous and uniform, that there are no discounts or sales and that all of the costs remain exactly the same, that the storage cost is directly proportional to the amount of cheese and the time it is held, and that the cost estimates are exactly right. Divisibility implies that we can buy any quantity of cheese (not just packages of a fixed size).

It was a long haul, but we did it! □

Getting a good mathematical model to describe a problem is essential to solving that problem and it involves a great deal more than just fitting a function to data or a verbal description. Mary Poppins was right on the mark when she told the children: "Well begun is half done."[7] **If the model is not correct, it can be worse than useless.** If the definitions are vague or missing altogether, it will be impossible to interpret the solution when it is found. If the assumptions are not plainly laid out so that the limitations on the solution are understood, the model could be used inappropriately and the "solution" might not apply to the actual problem. If the mathematical definition of the function does not correctly express the relationships of the things involved, then the solution most likely will not be correct. ***No* answer may well be better than an incorrect answer.** Decisions based on an inappropriate "solution" to a problem could be disastrous.

To avoid such errors, always be sure to double-check, or **verify**, your calculations, **validate** your solution and model (check them with your intuition and common sense and against the data, see how well the model predicts reality, and think carefully about the assumptions you have made), and perform **sensitivity analysis** (see how sensitive your solution is to possible changes or corrections to the data to adjust for measurement errors, possible biases, etc.). The example we worked out of finding the optimum level of exercise for Meg is a good model for how to put these concepts into practice. We will talk more about these processes over the course of this book, especially in Section 3.4.

Section Summary

Before you begin the exercises, be sure that you

- Know that a **mathematical model** is a collection of a verbal definition and a symbol definition of a function and the specification of the assumptions being made that are used to represent a real-world problem or situation mathematically.
- Know how to define a single-variable function and its variable in words (give its **verbal definition**) clearly and precisely, including all units involved.
- Know how to specify the **symbol definition** of a function, including the domain, to model a real-world situation.
- Know how to specify **assumptions**, which will normally include **certainty** that the relationship is exactly as specified in the symbol definition of the function and **divisibility** of the function and its variable (allowing *all* decimal values), and how to write out the *implied assumptions* related to these as well.

---

[7]Robert Stevenson, *Mary Poppins* (Walt Disney, 1964).

- Understand *why* models are useful in solving real-world problems.
- Realize that there are two major ways to formulate models: directly from **verbal descriptions**, and indirectly from **data**.
- Know how to *translate* from verbal descriptions of problems to symbol definitions of functions for mathematical models of those problems, if necessary by plugging in sample numerical values and looking for patterns.
- Know when you should **align data** (for example, when the input variable stands for *time*), and know that this alignment must be reflected in the model definitions.

EXERCISES FOR SECTION 1.2

Warm Up

1. Suppose you use technology to fit a straight line equation to some data and get the result $y = 13.46582674645x + 0.003756193675$.
(a) Express your equation, with each parameter *rounded off* to 5 places after the decimal point (5 decimal places).
(b) Express your equation, with each parameter *rounded off* to 5 significant digits (5 significant figures).
(c) Express your equation, with each parameter *truncated* to 5 significant digits, using the ellipsis (. . .) notation.

2. Suppose you use technology to fit a straight line equation to some data and get the result $y = 0.015679835x + 200.0036987$.
(a) Express your equation, with each parameter *rounded off* to 5 places after the decimal point (5 decimal places).
(b) Express your equation, with each parameter *rounded off* to 5 significant digits (5 significant figures).
(c) Express your equation, with each parameter *truncated* to 5 significant digits, using the ellipsis (. . .) notation.

3. Suppose you use technology to fit a quadratic equation to some data and get the result $y = -0.00238577287347x^2 - 4985.7285934x + 8236573.939067$.
(a) Express your equation, with each parameter *rounded off* to 5 places after the decimal point (5 decimal places).
(b) Express your equation, with each parameter *rounded off* to 5 significant digits (5 significant figures).
(c) Express your equation, with each parameter *truncated* to 5 significant digits, using the ellipsis (. . .) notation.

4. Suppose you use technology to fit a quadratic equation to some data and get the result $y = 10.05783259347x^2 - 0.0004049855934x + 92.4585939067$.
(a) Express your equation, with each parameter *rounded off* to 5 places after the decimal point (5 decimal places).
(b) Express your equation, with each parameter *rounded off* to 5 significant digits (5 significant figures).
(c) Express your equation, with each parameter *truncated* to 5 significant digits, using the ellipsis (. . .) notation.

Game Time

5. You help run a local coffeehouse, and have had the same musical group perform three years in a row. You have charged different admission prices each year, and attendance has varied. The first year you charged \$7 per ticket and 100 people bought tickets. The second year 80 tickets were sold at \$8 per ticket. Last year you charged \$9 and sold 70 tickets. You have agreed to guarantee to pay the band \$300 this year. In Sample Problem 4, we implicitly formulated a model for the number of tickets sold in terms of the price of each ticket:

Verbal Definition: $n(p)$ = the number of tickets sold if they cost p dollars per ticket.
Symbol Definition: $n(p) = -15p + 203.33\ldots$, for $0 \le p \le 13.50$.
Assumptions: Certainty and divisibility. Certainty implies that the sales will follow the demand function exactly, no discounts, and all pay full price. Divisibility implies that fractions of tickets can be sold.

You eventually need to decide what price to charge for tickets. Suppose you believe that you will receive an additional \$2 in revenue for food and drinks on average from every person attending the concert. Assume that the food and drink have been donated at no cost.

(a) Formulate and *fully define* a model for the revenue from food and drink in terms of p.
(b) Now find an expression for the *total* revenue (from ticket sales *and* food and drink) for the concert, in terms of p.
(c) Now *fully define* a model for the profit (including *both* categories of revenue).

6. You help run a local coffeehouse, and have had the same musical group perform three years in a row. You have charged different admission prices each year, and attendance has varied. The first year you charged \$7 per ticket and 100 people bought tickets. The second year 80 tickets were sold at \$8 per ticket. Last year you charged \$9 and sold 70 tickets. You have agreed to guarantee to pay the band \$400 this year. In Sample Problem 4, we implicitly formulated a model for the number of tickets sold in terms of the price of each ticket:

Verbal Definition: $n(p)$ = the number of tickets sold if they cost p dollars per ticket.
Symbol Definition: $n(p) = -15p + 203.33\ldots$, for $0 \le p \le 13.50$.
Assumptions: Certainty and divisibility. Certainty implies that the sales will follow the demand function exactly, no discounts, and all pay full price. Divisibility implies that fractions of tickets can be sold.

You eventually need to decide what price to charge for tickets. Suppose you believe that you will receive an additional \$3 in revenue for food and drinks on average from every person attending the concert. Assume the food and drink have been donated at no cost.

(a) Formulate and *fully define* a model for the revenue from food and drink in terms of p.
(b) Now find an expression for the *total* revenue (from ticket sales *and* food and drink) for the concert, in terms of p.
(c) Now *fully define* a model for the profit (including *both* categories of revenue).

7. You know that it costs \$50 just to place an order to buy T-shirts (covering design, making the silk screens, and shipping and handling), and then each shirt costs \$7 (as in Sample Problem 1). Suppose there is a 6% sales tax on the T-shirts, but not on the fixed cost for an order. *Fully define* a model for the total cost in terms of x.

8. You know that it costs \$50 just to place an order to buy T-shirts (covering design, making the silk screens, and shipping and handling), and then each shirt costs \$7 (as in Sample Problem 1). Suppose there is a 5% sales tax on the T-shirts, and a 3% tax on the fixed cost for an order. *Fully define* a model for the total cost in terms of x.

9. The U.S. Postal Service lists the following rates for insured mail and third- and fourth-class matter:

Dollar Value		
Low	High	Cost
$ 0.01	$ 50.00	$ 0.75
$ 50.01	$ 100.00	$ 1.60
$ 100.01	$ 200.00	$ 2.50
$ 200.01	$ 300.00	$ 3.40
$ 300.01	$ 400.00	$ 4.30
$ 400.01	$ 500.00	$ 5.20
$ 500.01	$ 600.00	$ 6.10
Liability for insured mail is limited to $600		

Formulate and fully define a model for the rates for insured mail in the United States as a function of the dollar value. Think carefully about how you can define this model to satisfy the assumption of divisibility, and have the domain include all values between 0 and 600.

10. A survey of 100 people was taken of the highest prices people would pay to have a pizza delivered.

Price Range	$4.00–5.00	$5.00–5.50	$5.50–6.00	$6.00–$6.50	$6.50–$10.00
Relative Probability	0.19	0.42	0.60	0.32	0.04

Formulate and fully define a model for the relative probability as a function of the cost of delivery.

11. You are thinking of ordering some baseball caps to sell as a fund-raiser. There is a basic order cost of $20 and each cap costs $3.50.

(a) Formulate and fully define a model for the cost of ordering the caps.

(b) Formulate and fully define a model for the average cost of ordering the caps.

12. You are thinking of ordering some baseball caps to sell as a fund-raiser. There is a basic order cost of $25 and each cap costs $3.25 each.

(a) Formulate and fully define a model for the cost of ordering the caps.

(b) Formulate and fully define a model for the average cost of ordering the caps.

13. The cost of ordering your group's own *Strawberry Sensations* 200-page cookbook is as follows: The initial set up cost is $100.00, regardless of how many books you order. The first hundred books are $3.95 each. For every book over the initial 100 books there is a 5% discount on the book price. Formulate and fully define a model for the cost of the books as a function of the number of books ordered.

14. The cost of ordering your class yearbook is as follows: The initial setup cost is $500.00, regardless of how many books you order. The first 50 books are $15.00 each. On every book over the initial 50 books there is a 8% discount on the book price. Formulate and fully define a model for the cost of the books as a function of the number of books ordered.

15. Suppose you like to keep a supply of soda in your dorm room. You buy them at a convenience store near campus, but it takes a while to walk there, and your time is valuable, so you estimate the fixed cost to go to the store each time to be about $2. The soda costs about $1.00 for a 2-liter bottle, and you consume an average of half a liter per day. Your dorm room is small, so storage causes some inconvenience, electricity for your refrigerator costs money, and your tied-up cash all combine into a storage cost you roughly estimate to be about 8 cents per 2-liter bottle per day. Formulate and fully define a model for the average cost per day.

16. You are trying to figure out how long to wait between sessions of doing laundry. For simplicity, you assume that you accumulate laundry at a rate of about 1 load per week. Since the laundry facility you use has lots of machines, the time to do laundry is about the same, however many loads you do at a time, and for the value of your time, you estimate this cost to be about \$4. Each load you do costs about \$2 (including the cost of soap, washing, and drying). As laundry accumulates and takes up more space, your choices of what you can wear (or the number of people who are willing to get close to you) gets more limited. You estimate all this together as an accumulated laundry "storage" cost of (80 cents per load) per week. Formulate and fully define a model for the average cost per week.

17. In the concert ticket price problem (Sample Problem 4), we found that the demand (n = number of tickets sold) could be modeled using the equation

$$n = -15p + 203.33\ldots, \text{ for } 0 \leq p \leq 13.50$$

where p is the price (in dollars) of a ticket.

(a) Can the price be thought of as a function of the demand (number of people buying tickets)? If so, formulate and fully define this model.

(b) If you were able to define the model in (a), formulate and fully define a model for the revenue as a function of demand (number of tickets sold).

18. In the ticket price problem (Sample Problem 4), we derived an expression for the revenue from ticket sales (R) in terms of the ticket price (p), both in dollars, to be

$$R = -15p^2 + 203.33\ldots p, \text{ for } 0 \leq p \leq 13.50$$

Suppose you have agreed to pay the band a straight 60% of the ticket revenues.

(a) Show how the cost of the band can be expressed as a composite function, and define the three different functions involved.

(b) Formulate and fully define a model for the cost as a function of the price charged for the tickets.

If you are going to be doing a student-generated project in your class, Exercises 19 and 20 can help you think out a possible topic.

19. Think of a problem (a question) that you would like solved that involves optimizing something. Try to think of something that you really care about. This could involve a sport or recreation that you particularly enjoy, the best use of your time for study and other activities, or a fund-raiser for your group. The possibilities are unlimited.

(a) Try to state the problem in clear and concise terms.

(b) What kind of information would you need to solve this problem?

(c) How could you gather this information?

(d) What would be the best way to organize the data?

(e) What kinds of assumptions might you make?

(f) What variables would you need to define? What do you expect to be the verbal definition for the model you end up using?

(g) How might you validate your model and solution?

20. Think of a problem involving optimizing that a small business might have.

(a) Try to state the problem in clear and concise terms.

(b) What kind of information would you need to solve this problem?

(c) How could you gather this information?

(d) What would be the best way to organize the data?

(e) What kinds of assumptions might you make?

(f) What variables would you need to define? What do you expect to be the verbal definition for the model you end up using?
(g) How might you validate your model and solution?

21. Pick a data series over time of interest to you (such as inflation rate, Dow Jones average, etc.) and define it using functional notation, aligning your data.

1.3 LINEAR FUNCTIONS AND MODELS

One of the simplest types of models is the linear model. The name "linear" comes from the fact that the graph of a linear model is a straight line. This will occur if, *whenever* the input changes by a constant fixed amount, the output also changes by a constant fixed amount, not necessarily the *same* fixed amount as the change of the input. In this section, we will study a number of situations in which linear models are useful, derived both directly from given information and from data. Here are the types of problems that the material in this section will help you solve:

- You want to start a CD club on campus. How many CDs would you have to sell to break even (start making a profit)?
- You are asked to predict company sales for the next three years.
- You really want to make the basketball team. The team needs a good three-point shooter. About how many hours would you have to practice to make approximately 80 percent of your shots?

When you have finished this section you should be able to solve these and similar problems and

- Recognize when a linear model is appropriate from a plot of the data.
- Verify that the data is linear or nearly linear by checking that the first differences are relatively constant.
- Recognize the general form of a linear model in its different variations: $y = mx + b$, $y = ax + b$, and $y = a + bx$.
- Express a linear model using correct functional notation.
- Interpret the slope and intercept of a linear model.
- Understand the concepts of marginal cost, marginal revenue, and marginal analysis.
- Know how to formulate linear models from verbal descriptions.
- Know how to use technology to find the best-fit least-squares linear regression line for a given set of data.
- Find the intersection of two linear functions algebraically and graphically.

Sample Problem 1: You are 10 miles from your home and you start to walk home. (Obviously you are big on walking, or maybe you haven't any money to do anything else. Walking is cheap.) You have walked this way a lot of times and on several occasions have checked your watch at the mile markers along the way:

Time (hours)	0	1	2	3	4	5
Distance (miles)	10	8	6	4	2	0

Your normal walking speed is 2 miles per hour (your knee has been acting up a bit lately, so you walk at a leisurely pace). How far will you be from home in one hour? How far will you be from home in an hour and a half? In fact, how far will you be from home at any given time? If it is now exactly 10:00 A.M. and you are expecting a very important phone call at 3:30, will you make it home in time or should you try to speed up a bit?

Solution: Since you started out 10 miles from home and are walking at the rate of 2 miles per hour, your distance from home at any given time will be 10 miles **minus** 2 miles for each hour of walking. If you have walked for 3 hours, you will have covered 6 miles (since your rate is 2 miles per hour), and so your distance from home will be

$$10 - 2(3) = 10 - 6 = 4 \text{ miles}$$

In general, your distance can be expressed as

$$D(t) = \text{the distance from home (in miles) } t \text{ hours since you started walking}$$

From the pattern of the example, we can express the relation in general as

$$D(t) = 10 - 2t$$

As usual, the major assumptions are divisibility of the variables and certainty. Some implications of certainty include

- You can continue the pace.
- You won't be run over by a car or eaten by a bear.
- You started *exactly* 10 miles from home, and are walking at *exactly* 2 mph.

Can you think of any more?

Figure 1.3-1 shows a plot of the data and the same plot with the line added.

In this case the domain is from 0 to 5 hours. We can write this as $t \geq 0$ and $t \leq 5$, or more concisely as $0 \leq t \leq 5$. After 5 hours of walking, you will be home. (If t continued to increase, the distance would go negative; can you think of how this could be inter-

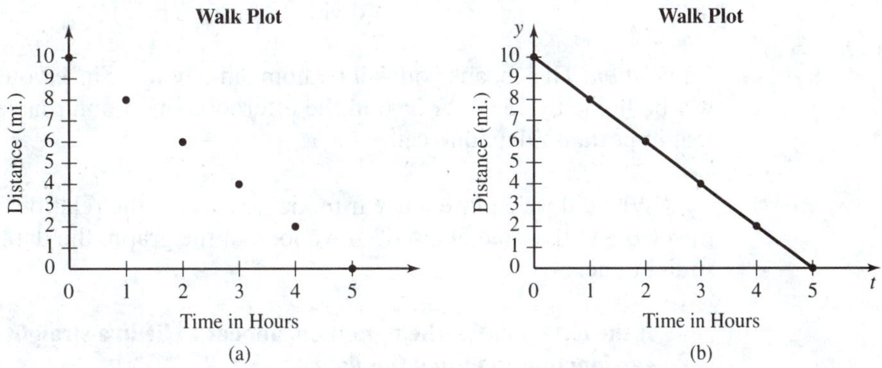

Figure 1.3-1

preted?) The outputs would range from 10 to 0 miles. There is no problem with fitting a continuous model (no breaks, holes, or jumps in the graph) to the data: Both time and distance are divisible by their nature. As we learned earlier, we should include the domain for our input in our model:

Verbal Definition: $D(t)$ = the distance from home (in miles) t hours since you started walking.

Symbol Definition: $D(t) = 10 - 2t$, for $0 \leq t \leq 5$.

Assumptions: Certainty and divisibility. Certainty implies that your pace is constant.

Remember, the model must always include the units of the input and output, and the domain of the input, as well as the rule and the assumptions.

Now that we have a model for the distance as a function of time we can answer the questions posed in the problem:

How far will you be from home in one hour? To find this, we can simply plug in 1 for t in the function for the distance from home:

$$D(1) = 10 - 2(1) = 10 - 2 = 8.$$

Translation: In one hour, you will be 8 miles from home.

How far from home will you be in one and one half hours? This time, plug in 1.5 for t:

$$D(1.5) = 10 - 2(1.5) = 10 - 3 = 7.$$

Translation: In one and one half hours, you will be 7 miles from home.

Will you be home by 3:30? We can answer this question by finding out what time you will be home; that is, what time the distance from home is 0. Thus, we want to know when

$$D(t) = 10 - 2t = 0$$

Solving this equation we have

$$\begin{aligned} 10 - 2t &= 0 \\ 10 &= 2t \quad \text{(add } 2t \text{ to both sides of the equation)} \\ 5 &= t \quad \text{(divide both sides by 2)} \end{aligned}$$

Translation: This means you will be home in 5 hours. Since you started out at 10:00, you will be home by 15:00, or 3:00 in the afternoon. You don't have to worry about missing that important telephone call. □

Why did we choose a linear model to express the relationship between the distance traveled and the time elapsed? If we look at the graph, the data points appear to lie in a straight line.

If the data points, when graphed, appear to lie in a straight line, we should use a linear function to model the data.

TABLE 2

Time (hours)	0	1	2	3	4	5
Distance (miles)	10	8	6	4	2	0
Change in Miles		−2	−2	−2	−2	−2

Could we have determined that the relationship was linear without looking at the graph? Let's look at the table of the data in Table 2.

For each change of one unit in the input, the change in the distance—the change in the output—was always exactly −2. These changes are often called the **first differences** because they are calculated by finding the differences between each consecutive pair of outputs when the inputs are in increasing order and equally spaced throughout. The distance was always changing by the same amount. You were walking at a steady rate of speed.

If the change in the output is exactly constant for equally spaced input values, then a linear function will fit the data perfectly. If it is nearly constant, then a linear model is likely to be reasonable.

Linear Functions: Slopes and Intercepts

The general form of a linear function is

$$y = f(x) = mx + b$$

where

y is the output, or dependent variable,
x is the input, or independent variable,
m is the slope of the line, and
b is the intercept with the y axis (the value of y when $x = 0$)

This is often referred to as the "slope-intercept form" of a linear function.

Sometimes the "m" is replaced by an "a" and the general form of the linear equation is written as

$$y = f(x) = ax + b$$

Don't let the change in letters from "m" to "a" confuse you. Whether we use "a" or "m" as the coefficient of x, they both refer to the slope of the line. Some calculators also allow the general form $y = f(x) = a + bx$. Be very careful. In this form, the coefficient of the x value is b, so b would be the slope of the line.

In Sample Problem 1, the coefficient of the input variable was negative (−2). This corresponded to the fact that our distance from home was **decreasing** as we walked. In general, if the "m" is negative, then the slope is negative, and the function will be a **decreasing function** (falls from left to right). If the "m" is positive, then the slope is positive, and the function will be an **increasing function** (rises from left to right). This would be the case if we were describing the distance traveled from the starting point as a function of time:

Verbal Definition: $S(t)$ = the distance from the starting point (in miles) t hours since you started walking.
Symbol Definition: $S(t) = 2t,\ t \geq 0$.
Assumptions: Certainty and divisibility. Certainty implies a constant pace.

Since the concept of the slope of a line is so important in our study of calculus, let's take some time and review this idea. Many of you will be familiar with the definition of the slope of a line as the "rise over run." We can also write this as

$$\textbf{slope} = \frac{\text{rise}}{\text{run}} = \frac{\text{change in output}}{\text{change in input}} = \frac{\text{change in } y}{\text{change in } x} = \frac{y_2 - y_1}{x_2 - x_1}$$

It is sometimes convenient to use symbols to abbreviate expressions. The "change in y" can be abbreviated Δy and the "change in x" abbreviated Δx. The symbol Δ ("delta") is the uppercase Greek letter equivalent to "D," for "difference." Think of it as meaning "change in." The letters used don't have to be x and y, but whatever your corresponding input and output variables are, respectively. So we can write (Figure 1.3-2)

$$\text{slope} = m = \frac{\text{change in } y}{\text{change in } x} = \frac{y_2 - y_1}{x_2 - x_1} = \frac{\Delta y}{\Delta x}$$

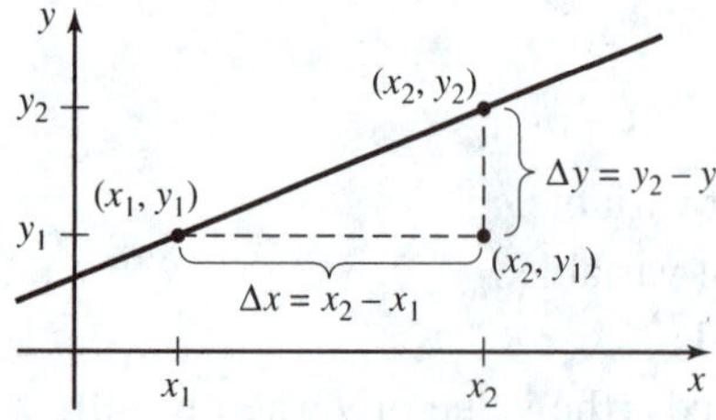

Figure 1.3-2

These are all just different ways of describing the slope. You might ask "Why can't we just settle on one way to express this idea?" That is a fair question. The expressions using words, rise over run, change in output over change in input, and the like, help us to understand the concept. The symbolic expressions $\frac{y_2 - y_1}{x_2 - x_1}$ and $\frac{\Delta y}{\Delta x}$ allow us to work mathematically with the concept. The "m" in the general form of the linear function refers to this slope,

$$m = \frac{y_2 - y_1}{x_2 - x_1}$$

The interpretation of "b" is even easier to see. In our example, the constant term in our function $D(t) = 10 - 2t$ was 10. This corresponded to our initial distance from home.

In general, ***b* is the output value that is paired with the input value 0**. You can see this, since if $y = mx + b$ and $x = 0$, then

$$y = (m)(0) + b$$

$$y = 0 + b$$

$$y = b$$

The value b is called the ***y* intercept** (or, more generally, the **vertical intercept**), because it is where the graph of the linear function crosses the vertical axis. Since everything on the vertical axis corresponds to points with an input value of 0, we have come back full circle.

How does the definition of slope relate to our general expression for a linear function, $y = mx + b$? We said that the b represents the y intercept, which is the value of y when $x = 0$, or the point $(0,b)$. Any *other* point on the line could be represented by (x,y), as in Figure 1.3-3.

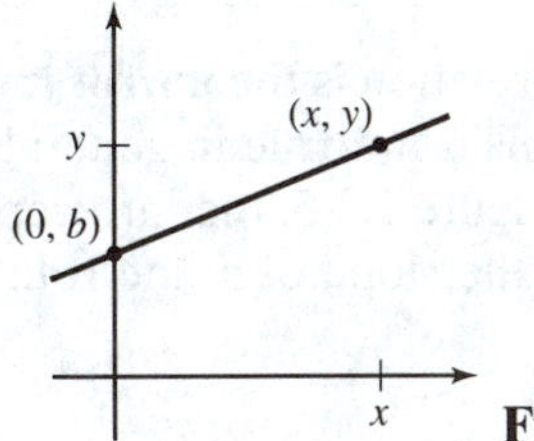

Figure 1.3-3

The slope is the change in the output divided by the change in the input: $\dfrac{y_2 - y_1}{x_2 - x_1}$ so

$$m = \frac{y - b}{x - 0}$$

Clearing the fraction (multiplying both sides by x, which is fine as long as $x \neq 0$, which must be true since our second point was *different* from $(0,b)$) gives

$$m(x - 0) = y - b, \text{ or } mx = y - b$$

Solving for y gives

$$y = mx + b$$

In the four lines shown on the graph in Figure 1.3-4, the slopes are all the same, $m = 1$, but the intercepts, the b values, are different. Changing the constant moves the line up or down. Notice that these lines do not form a 45-degree angle with the horizontal axis because the scales on the two axes are not the same. If they were the same, any line with a slope of 1 would form a 45-degree angle. But when you are working with real problems, it is rarely convenient to have the scales the same on both axes, so some properties you might take for granted may no longer hold. For example, if the axes have the

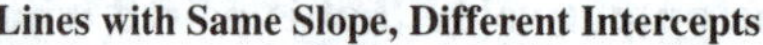

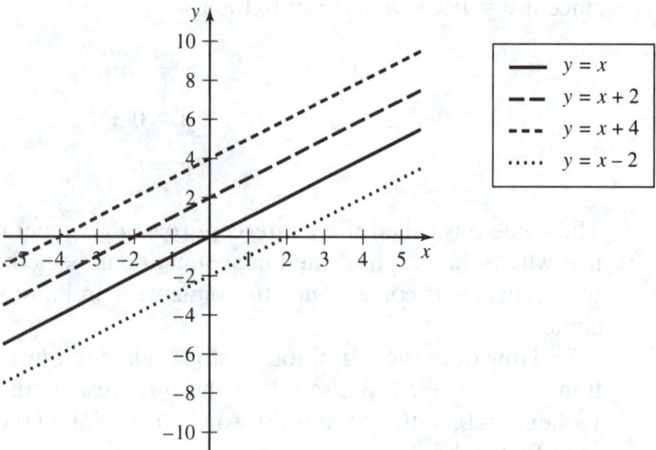

Figure 1.3-4

same scales, then the graph of the inverse of a relation is the **mirror image** (about the line $y = x$) of the graph of the original relation. This is not true in general, however.

In the four lines shown on the graph in Figure 1.3-5, the intercepts are all the same, $b = 2$, but the slopes are different. Changing the slope of a line rotates the line around the y-intercept.

Lines with Same Intercept , Different Slopes

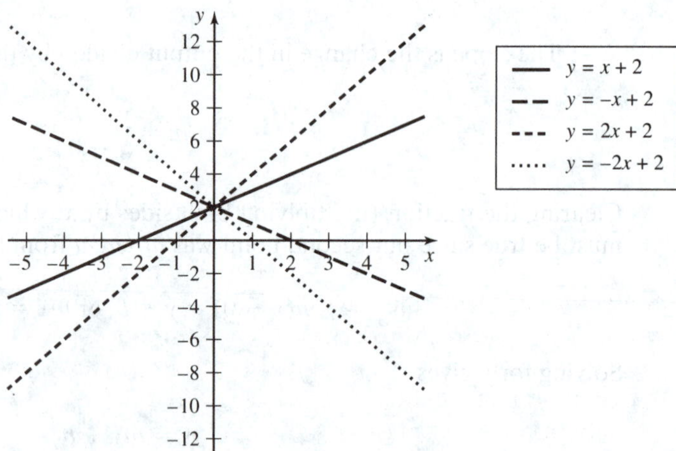

Figure 1.3-5

Sample Problem 2: Another example of data in which the change in the outputs remains constant would be a paycheck receipt that shows the total accumulated pay throughout the year. If you are paid a flat monthly amount and have not received a raise, the total will increase by the same amount each pay period, as shown in Table 3. Figure 1.3-6 is a plot of the data. Find a model to represent your accumulated income at any time during the year.

TABLE 3

Month	Total Pay $
1	1600
2	3200
3	4800
4	6400
5	8000
6	9600
7	11200
8	12800
9	14400
10	16000
11	17600
12	19200

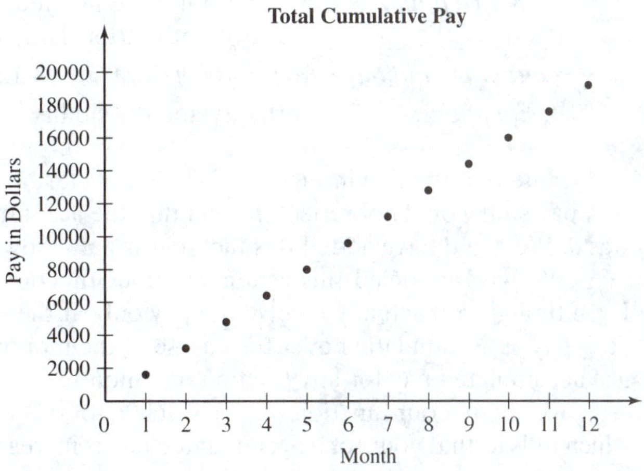

Figure 1.3-6

Solution: Examine Table 3. If you are paid monthly, the first pay day is at the end of January, and the last pay day of the year would be at the end of December. The input values range from 1 to 12 (domain) and the output values from 1600 to 19200. Since your total accumulated pay was 0 on the first day of January and increased by $1600 each month, your accumulated pay can be expressed as $A(t) = 1600t$ = total accumulated pay for the year (in dollars) t months after the end of December of last year. For example, the end of January would correspond to $t = 1$. Look at the data plot in Figure 1.3-6: The data lie in a straight line. Since we can interpret the points in between the months as days, or even hours or minutes that we have worked, and can interpret the output as the amount *earned* up to that point (even if we haven't *received* it yet!), the continuous model makes sense. Figure 1.3-7 is the data plot with the line drawn joining the points. However, we might wish to start our input and output at 0. The domain of the function would then be 0 to 12, because we are interested in the amount accumulated throughout the year, even during the first month. So our model would be

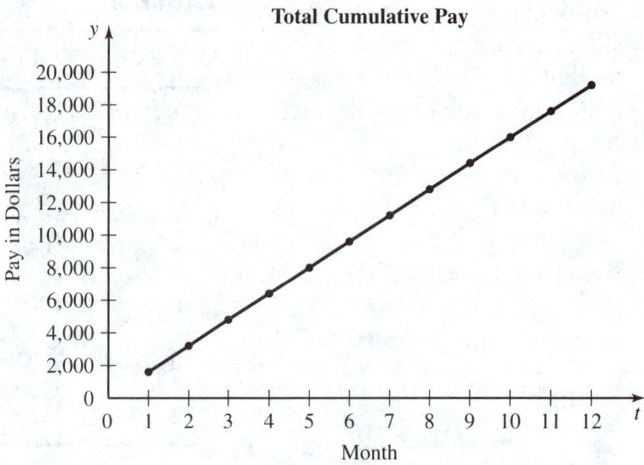

Figure 1.3-7

Verbal Definition: $A(t)$ = total accumulated pay for the year (in dollars) t months after the end of December of last year.

Symbol Definition: $A(t) = 1600t$, for $0 \le t \le 12$.

Assumptions: Certainty and divisibility.

Note that the line in Figure 1.3-7 did not start at 0 because the data did not start at 0. A pay stub would not reflect the fact that the accumulated pay at the start of the period was 0. We could have added this fact to our data, however (see Figure 1.3-8).

We have modeled this as a continuous function (with no holes, breaks, or gaps). Even though you actually receive the pay only at the end of the month, you can *think* of your pay as accumulating over the course of the month and we can calculate the theoretical accumulated pay for any fraction of a month.

Recall that our function is $A(t) = 1600t$, for $0 \le t \le 12$. In this case the slope is 1600, which tells us that your total accumulated pay is increasing by \$1600 per month. This does

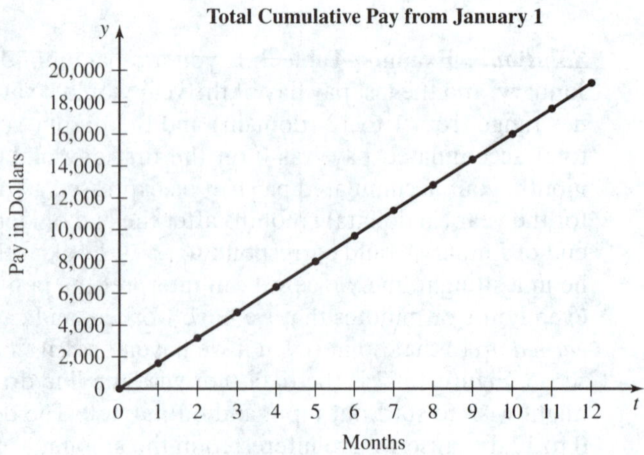

Figure 1.3-8

not change from month to month; the change in output per unit change in input, the slope, is constant, as it is for all linear models. There is no "b" in the model; that is, the intercept is 0. This tells you that when you began the year, at $t = 0$, your accumulated pay was 0. On the graph of the data you can see that the line would go through the point (0,0). □

Sample Problem 3: You have decided to start a CD club on campus. You write to wholesalers to find out what they would charge to ship CDs to you. It costs \$25.00 to place an order for CDs, and \$8.00 per CD, so the total cost of the order will increase by \$8.00 for each CD ordered. The minimum order is 5 CDs, to get this discounted price. You plan to charge a flat \$10.00 for each CD. Find a model for the total cost of an order, and a model for your revenue function.

Solution: We know that 5 CDs would cost \$8 each, so together they would cost \$8(5) = \$40, plus the \$25 order cost, for a grand total of

$$\$8(5) + \$25 = \$40 + \$25 = \$65$$

To generalize this pattern, if you order x CDs, the total cost in dollars will be

$$8(x) + 25 = 8x + 25$$

If we define $C(x)$ = the total cost (in dollars) of an order of x CDs, then in algebraic form this could be expressed as

$$C(x) = 25 + 8x, \text{ for } x \geq 5$$

A table and the scatter plot for selected values of x are shown below and in Figure 1.3-9.

Number Ordered	5	6	7	8	9	10	11	12	13
Cost (\$)	65	73	81	89	97	105	113	121	129

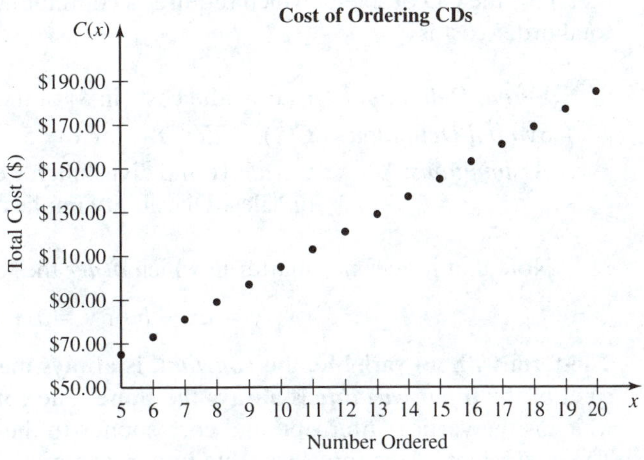

Figure 1.3-9

Since you must order at least 5 CDs, the input values would start at 5 and could go as high as you like. However, we should note that the input values would always be whole numbers, or integers. You cannot order half of a CD. The output values would start at \$65 and increase in increments of \$8.00 for each CD ordered. If we graph the function

$$C(x) = 25 + 8x, \text{ for } x \geq 5$$

we will see a continuous line. We have fit a continuous model to data that, by its nature, is discrete, not continuous. It can still be a useful model, but we should be careful in interpreting it. A line fit to the data is shown by Figure 1.3-10.

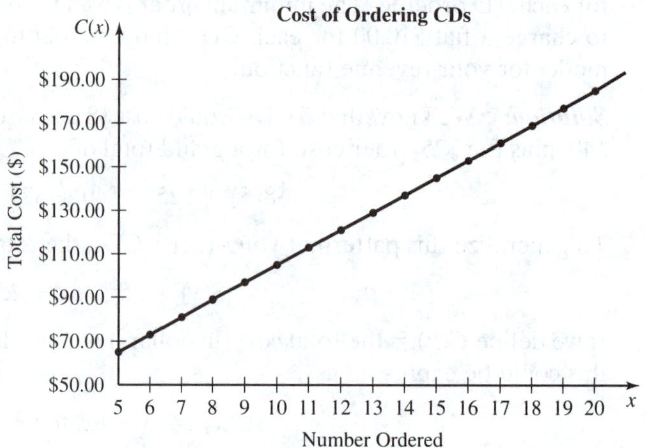

Figure 1.3-10

In fact, in some ways, the real situation is most accurately modeled by the original graph showing the discrete points. However, it is sometimes convenient to fit a continuous model to data that are not, by their nature, continuous. For example, we will later do a calculation of a break-even point (the number of items you need to sell to cover your costs) for the CD problem, which requires a continuous model. So our final model for the total order cost is

Verbal Definition:	$C(x)$ = the cost (in \$) of the order when x CDs are ordered.
Symbol Definition:	$C(x) = 25 + 8x$, for $x \geq 5$.
Assumptions:	Certainty and divisibility. Certainty implies no other discounts or sales. Divisibility implies fractions of CDs can be ordered.

Note that it does not matter in which *order* the terms of a linear model appear:

$$y = ax + b \text{ or } y = a + bx$$

The term with no variable, the *constant*, is always the vertical intercept and the coefficient of the *input variable* is always the slope. The constant in our cost model is the 25, which is the vertical intercept, and corresponds to the fixed price of \$25 for each order. The coefficient of the input variable in our cost model is 8, which is the slope, and tells you that your cost is increasing \$8 for each CD bought (in other words, this is the cost of each CD).

All costs that depend upon the quantity (like the \$8 per CD) are called **variable costs**, because they vary with the quantity. The intercept in this case is 25, which tells you that you must pay $25 to order *any* CDs. This is called the **fixed cost** because it does not change with the quantity, and would be the C intercept if the x's started at 0.

Since you are charging $10 per CD, your revenue will simply be $10 times the number of CDs sold, since

Revenue = selling price times quantity sold $\quad$ $R = pq$

so our revenue can be modeled by $r = f(x) = 10x$ = the revenue in dollars from the sale of x CDs. Since x is the number of CDs sold, x could be an integer greater than or equal to zero. The graph of the revenue is shown in Figure 1.3-11.

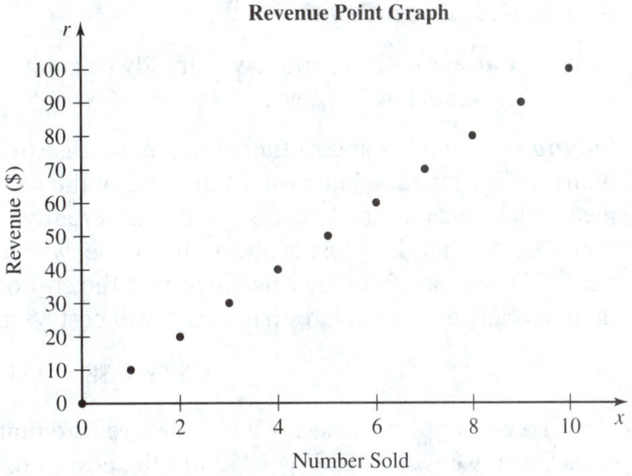

Figure 1.3-11

Again, we will graph the data as a continuous line (Figure 1.3-12), but the data values are, by their nature, discrete. As mentioned earlier, assuming divisibility in this way makes possible certain calculations that we could not otherwise do, such as finding the break-even point, to be discussed later in this section.

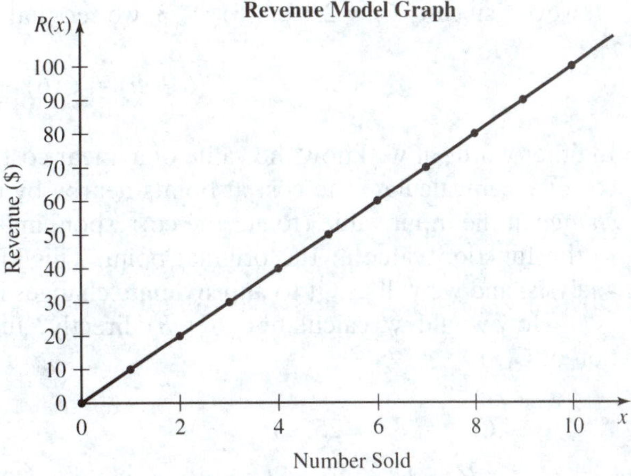

Figure 1.3-12

The model for these data values is thus

Verbal Definition: $R(x)$ = the revenue in dollars when x CDs are sold.
Symbol Definition: $R(x) = 10x$, for $x \geq 0$.
Assumptions: Certainty and divisibility. Certainty implies no discounts or sales (such as to friends or family). Divisibility implies fractions of CDs can be sold. □

Marginal Cost, Marginal Revenue, and Break-Even Points

Sample Problem 4: If you have already calculated that 6 CDs cost \$73, how much will 7 cost? How much will 9 cost?

Solution: For a linear cost function, the increase in cost when a quantity increases by one item is called the **marginal cost**. In the case of the CDs, since the cost of each CD is \$8, this means that our marginal cost is \$8. Put differently, since the cost is modeled by a linear function, the marginal cost is always the slope, which in the cost model was 8, meaning \$8 per CD. If we have already calculated that the cost of 6 CDs is \$73, and each CD costs \$8, then we can easily calculate that 7 CDs will cost \$8 more than 6, for a total of

$$\$73 + \$8 = \$81$$

To calculate the cost for 9 CDs, we realize that this is 3 more than 6, and each one costs \$8, so we will add \$8(3) = \$24 to the cost of 6, for a total cost of

$$\$73 + \$8(3) = \$73 + \$24 = \$97 \quad \square$$

In general, if we use the variable h to stand for the number of CDs **more** than 6 we are ordering, then the total cost will be $73 + 8(h)$. Put differently,

$$C(6 + h) = 73 + 8h$$

However, since $C(6) = 25 + 8(6) = 73$, we see that

$$C(6 + h) = C(6) + 8h$$

In other words, if we know the value of a linear cost function at a point and the marginal cost, we can calculate the cost at points nearby by multiplying the marginal cost by the *change* in the input value (to get the corresponding change in the output), and adding it to the function value at the original point. This kind of calculation is called **marginal analysis**, and we will use it to **approximate** changes in nonlinear functions later.

How would we calculate $C(6 + h)$ directly? Just plug in $(6 + h)$ for x in the definition of $C(x)$:

$$C(x) = 8x + 25, \text{ so}$$

$$C(6 + h) = 8(6 + h) + 25 = 48 + 8h + 25 = 73 + 8h = C(6) + 8h$$

Let's take another look at the revenue function from Sample Problem 3:

$$R(x) = 10x$$

The intercept here is 0—if you don't sell any CDs, you don't take in any money. The slope is 10, which tells you that your revenue is increasing at the rate of \$10 per CD sold: $R(5) = 50$, $R(6) = 60$. In economics this rate has a special name: **marginal revenue**, the revenue from the sale of one more item for a linear revenue function. We will come back to these very important ideas later and use calculus to find the marginal cost and marginal revenue functions for more complicated cost and revenue functions.

Another concept that is of great interest to all business people is the **break-even point**. The break-even point is the point at which the revenue, the money received from sales, equals the cost. Once the break-even point has been reached, the seller is beginning to make a profit. (You may have heard the day after Thanksgiving referred to as "Black Friday." Curious name. The day after Thanksgiving is a major shopping day, and, for many merchants, is the day on which they finally get "in the black;" that is, start to make a profit for the *year*.) Always remember the primary purpose in business is to make a profit or maximize shareholder wealth. The bottom line on a financial spreadsheet shows the profit or loss, hence the expression "the bottom line."

Sample Problem 5: Find the break-even point and the profit function for the sale of the CDs.

Solution: The break-even point is the point at which the total cost equals the revenue. Algebraically, we set the two functions equal to each other and solve:

$$\begin{aligned} C(x) &= R(x) \\ 25 + 8x &= 10x \\ 25 &= 10x - 8x \\ 25 &= 2x \\ x &= \frac{25}{2} = 12.5 \end{aligned}$$

Since it is impossible to sell 12.5 CDs, you would have to sell 13 CDs before you are no longer losing money (no longer in the red).

Let's look at this important point as shown graphically in Figure 1.3-13. Profit equals revenue minus cost, the amount that you take in minus the amount that you had to spend. Your profit function in dollars would be

$$P(x) \text{ or } \pi(x) = 10x - (25 + 8x) = 10x - 25 - 8x = 2x - 25$$

Note that we have indicated profit by two different symbols: P and π. You should be comfortable with both notations, as they are both used in business and economics. It is also a common practice to use "q" for the number of units manufactured or the number of units sold. Our profit function might have been written $P(q) = 2q - 25$. □

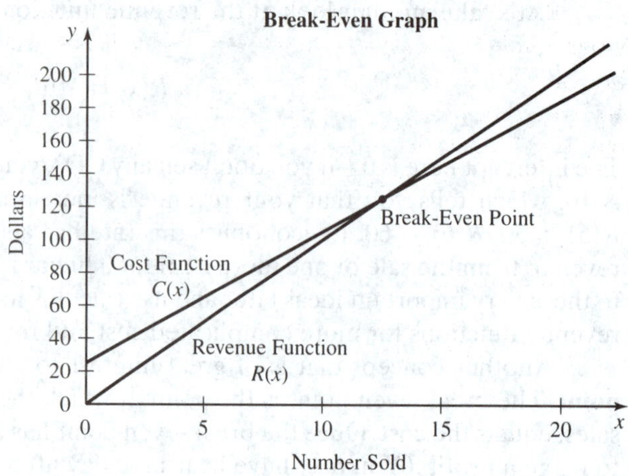

Figure 1.3-13

Costs That Vary with Quantity

Sample Problem 6: You want to sell reusable mugs as part of a conservation project. The supplier charges \$2.25 each for the first 100 mugs and \$2.00 for each mug after that. Can you fit a linear model to this cost function?

Solution: Recall the problem concerning the insurance rates for packages. The rate was a flat amount for different intervals. This problem is similar to that in some ways. The mathematical expression for the first part is fairly easy: The cost in dollars to order x mugs is $C(x) = 2.25x$, $0 \le x \le 100$. So we would have

Order Size	Cost	
99	(99)(\$2.25) or \$222.75	
100	(100)(\$2.25) or 225.00	exactly \$2.25 more

The second part is not so easy. We might be tempted to say it is just $C(x) = 2x$, $x > 100$. Let's look at what happens:

Order Size	Cost
101	(101)(\$2.00) = \$202.00!!!

That's \$23 less than the cost of ordering 100 mugs! This can't be right. We know we have to pay \$2.00 for the next mug instead of \$2.25. What actually does happen? Well, we pay \$2.25 for the first 100 mugs, *then* we pay \$2 for the next one, so we get

Order Size	Cost
101	(100)(\$2.25) + (1)(\$2.00) = \$227.00

That's more like it. How about the next mug? We have to pay \$2.00 for that mug, so the cost should be \$227.00 + \$2.00.

Order Size	Cost
102	(100)(\$2.25) + (1)(\$2.00) + (1)(\$2.00) = (100)(\$2.25) + (2)(\$2.00)

If we order 103 mugs the next mug will also cost \$2.00, so we add \$2.00 to our previous total. In fact every mug after 100 will cost \$2.00. How can we express this algebraically? If we let x equal the number of mugs bought, the number of mugs **after** the first 100 would be $x - 100$. We pay \$2.25 for the first 100 mugs, for a cost of \$225.00, and then pay \$2.00 for each mug after that: $(\$2.00)(x - 100)$. Thus our total cost in dollars when ordering more than 100 mugs will be

$$225 + 2(x - 100) = 225 + 2x - 200 = 25 + 2x$$

We are now ready to write our cost model:

Verbal Definition: $C(x)$ = the cost in dollars to order x mugs.

Symbol Definition: $C(x) = \begin{cases} 2.25x & \text{for } 0 \le x \le 100 \\ 225 + 2(x - 100) = 25 + 2x & \text{for } x > 100 \end{cases}$

Assumptions: Certainty and divisibility. Certainty implies no other discounts or sales. Divisibility implies fractions of mugs can be ordered.

Note: Once again, we are assuming divisibility for variables that are by nature not continuous, to enable calculations like the break-even point.

Let's look at a graph of the function and some sample points (Figure 1.3-14).

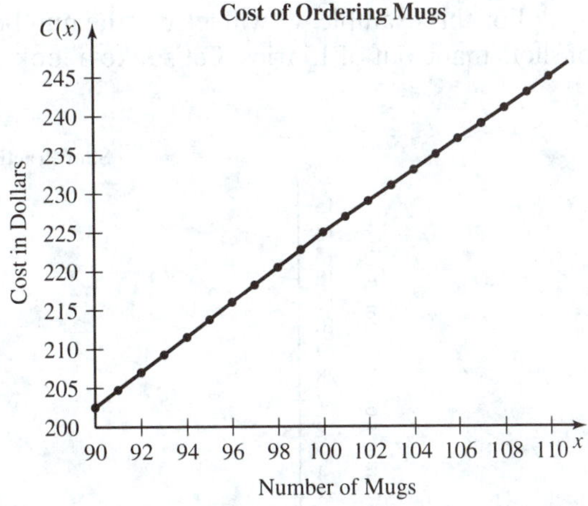

Figure 1.3-14

It is somewhat difficult to see on this graph, but there are two separate lines: The slope of the line from 90 to 100 is 2.25 and the slope of the line from 100 to 110 is 2. This kind of function is called **piecewise linear**, since the individual pieces are linear (from 0 to 100, then from 100 on). This graph is **continuous (no holes, jumps, or breaks)**, but it is **not smooth**, since it contains an angle. We repeat that **throughout this book we assume the functions we are working with are continuous and smooth (continuous, with no angles), unless otherwise indicated.**

How can you figure the cost of 125 mugs? The order is for more than 100 mugs, so we will use the second part of the model function: $C(125) = 225 + 2(125 - 100) = 225 + 2(25) = 275 = 25 + 2(125)$. So it would cost \$275 to order 125 mugs. □

In all of the examples above the data were perfectly linear, that is; the differences between the outputs were constant when the differences between the inputs were constant. When we plotted the points and drew the line through them, the line went exactly through each point. In fact, we were able to determine the equation for the line, the model, by simply studying the data. This is not always the case. In fact, this is probably the exception rather than the rule. If the plot of the data is approximately linear, but not perfectly linear, how do we determine the slope and intercept of the line that will give the best fit?

Best-Fit Line, Sum of the Squares of the Errors (SSE)

Suppose you have kept a record of how many basketball free throws you made out of 15 tries after various amounts of practice. After 1 hour of practice you could only make 3 shots out of 15 tries. After two hours, you made 8. One more hour of practice and you made 11 of the shots, but after the fourth hour of practice you were down to 10 shots.

Hours	Shots made
1	3
2	8
3	11
4	10

For this example we will let x = the number of hours practiced and y = the number of shots made out of 15 tries. Let's take a look at the graph of this data (Figure 1.3-15).

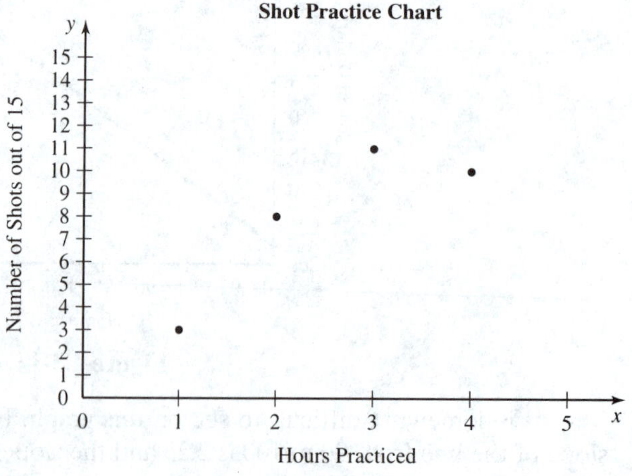

Figure 1.3-15

Suppose, for example, we had guessed that the best linear model to describe this data was $y = 2x + 3$ (since, from the graph, it looks as if the line with a vertical intercept of about 3, and a slope of about 2, would fit pretty well), where x = the number hours practiced and y = the number of shots made. Figure 1.3-16 shows the graph of this line along with the data points, so you can judge the fit.

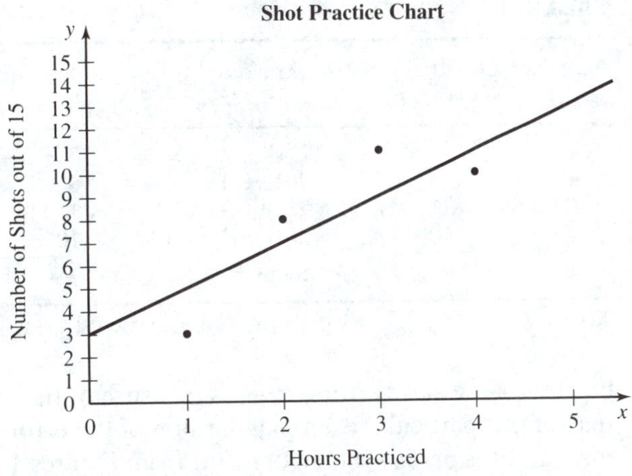

Figure 1.3-16

We can get a measure of the fit between the data and the model by looking at the vertical distance from the model to each data point, which we call the **errors** (or, in statistics, the **residuals**). These are shown graphically in Figure 1.3-17.

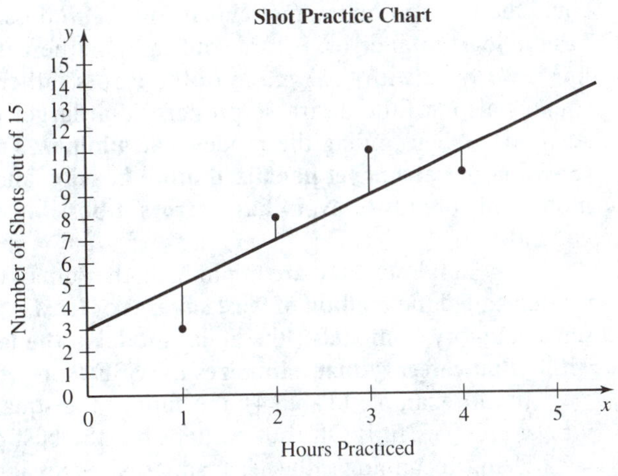

Figure 1.3-17

We have defined these errors so that **when the data point is *above* the model, we call the error *positive***, and when the data point is *below*, we call it *negative*. Since these are vertical distances, **the errors correspond to the *differences in the y values* between the two points.** To make the signs come out right, we calculate: **the *y* value of the data point minus the *y* value of the corresponding point that lines up with it vertically on the graph of the model.** Since our models are always functions, there will always be exactly one such point, as long as the corresponding input value is in the domain.

Table 4 shows the calculations of the errors for this model.

How can we get a measure of how good or bad the fit is between the data and our model? Since we have calculated an error for each data point, an obvious choice would

TABLE 4

Actual x (x_{data})	Actual y (y_{data})	Predicted y $y_{\text{model}} = 2x_{\text{data}} + 3$	Errors $y_{\text{data}} - y_{\text{model}}$	Errors Squared $(y_{\text{data}} - y_{\text{model}})^2$
1	3	$2(1) + 3 = 5$	$3 - 5 = -2$	$(-2)^2 = 4$
2	8	$2(2) + 3 = 7$	$8 - 7 = +1$	$(1)^2 = 1$
3	11	$2(3) + 3 = 9$	$11 - 9 = +2$	$(2)^2 = 4$
4	10	$2(4) + 3 = 11$	$10 - 11 = -1$	$(-1)^2 = 1$
		Sums ⇒	$-2 + 1 + 2 + -1 = 0$	$4 + 1 + 4 + 1 = 10$

be the *sum* of these errors, to get a measure of the *total* error. But look at Table 4 to see that, in this particular example, the sum of the errors is zero! This would *seem* to suggest that the fit is *perfect* (no error)! But from Figures 1.3-16 and 1.3-17, we can see that the fit is *far* from perfect. *None* of the data points lies *exactly* on the line. What happened?

You can probably see that the problem is that the positive and negative errors canceled each other out. We don't really want that to happen; we want to get a useful measure of the *total* error. One possibility would be to simply add the **absolute values** of the errors (make them all positive and add). Another would be to *square* the errors first (since squaring also has the effect of always resulting in positive answers or 0), and then add up the squares of the errors. Whichever method we use, the idea would be that we could then try to find the model (say, out of *all* possible linear functions in this case) that has the **least total error**. It turns out that, mathematically, it is easier to find this **best-fit line** if we work with the *squares* of the errors rather than the absolute values. Using the squares also has the desirable property that larger errors get magnified (when they get squared). When finding the model that **minimizes the sum of the squares of the errors (SSE), larger errors get penalized more heavily**, and so the best model adjusts for them more, which **tends to avoid huge errors if possible**. Intuitively, this usually seems like a good idea.

In general, since we are minimizing the sum of the squares of the errors, this method is often called the **method of least squares** or **least-squares regression**. Once we have chosen a category of models (like linear models), **the least-squares method finds the model within that category that minimizes the SSE (sum of the squared errors).**

In our example (Table 4), the sum of the squares of the errors is 10 for our model $y = 2x + 3$. The fit is OK, but perhaps not the best possible fit.

Using technology, the least-squares regression line for this data turns out to be $y = 2.4x + 2$, shown in Figure 1.3-18. See your technology supplement to find out how to do this for yourself.

The calculations for the sum of the squares of the errors of this best-fit least-squares linear model are shown in Table 5.

The sum of the squares went *down*, from 10 to 9.2. This suggests that we have found a line with a better fit. In fact, it is the *best* fitting line (from the SSE perspective, anyway), because technology gives us the linear model with the SSE that is the *smallest* possible.

Advanced calculus techniques are used to minimize the sums of the squares of the errors and determine the line with the best fit. We will come back to this discussion later. In Chapters 5, 6, and 8, you can learn the calculus necessary to do the derivation yourself.

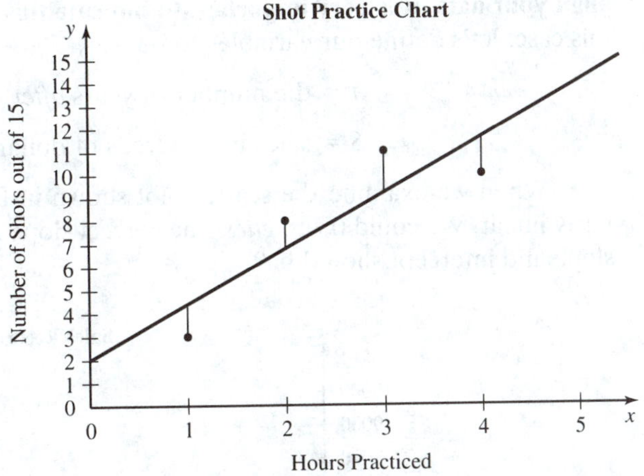

Figure 1.3-18

In the meantime we will use technology, such as a graphing calculator or a spreadsheet program, to find the best-fit models for our data.

TABLE 5

Actual x (x_{data})	Actual y (y_{data})	Predicted y $y_{model} = 2.4x_{data} + 2$	Errors $y_{data} - y_{model}$	Errors Squared $(y_{data} - y_{model})^2$
1	3	4.4	−1.4	1.96
2	8	6.8	+1.2	1.44
3	11	9.2	+1.8	3.24
4	10	11.6	−1.6	2.56
		Sums	0	9.20

Sample Problem 7: Let's examine the following table which shows the total sales in hundreds of dollars for each year from 1990 to 1995 (calculated as of the last business day of the year):

Year	1990		1991		1992		1993		1994		1995
Year's Sales ($100s)	5085		6003		6945		7923		8835		9815
Change in Sales ($100s)		918		942		978		912		980	

Fit a linear model to this data, and then project the sales for 1996 and 1997.

Solution: The change from year to year is almost, but not exactly, constant. If the first differences are constant or fairly constant, a linear model can be used.

As we mentioned in the last section, for problems like this over time, it is a good idea to **align** your input values (make the first value 0). We ordinarily align the data when we are dealing with years as inputs because it simplifies the model fit to the data. **If you**

align your data, it is very important to indicate this when you identify your variables. In this case, let's define our variables to be

t = the number of years *after* the *end* of 1990

S = sales in hundreds of dollars

When we examine the scatter plot shown in Figure 1.3-19, the data appears to be fairly linear. We could try to *guess* the correct slope and intercept. What do you think the slope and intercept should be?

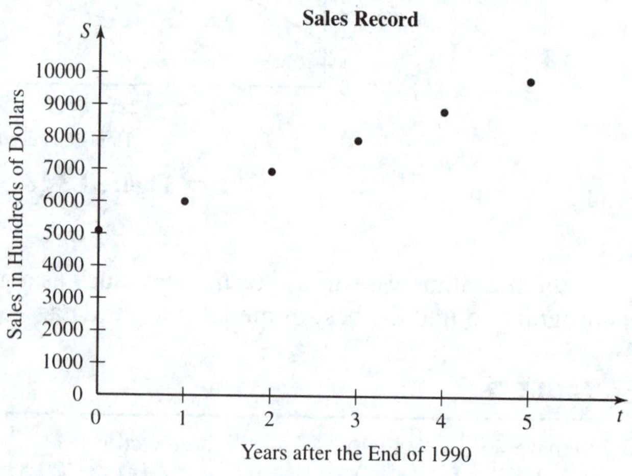

Figure 1.3-19

A guess is all right, but if the model we are trying to find is important, we would *not* want to rely on a guess. It would not give us the best fit. We need an exact method of finding the best fit. The best fit model for this is

Verbal Definition: $S(t)$ = sales in hundreds of dollars for the *year preceding* the point in time that is t years after the end of 1990.
Symbol Definition: $S(t) = 946.4\,t + 5068.3\ldots$, for $t \geq 0$.
Assumptions: Certainty and divisibility.

From our definition, when $t = 0$, $S(0)$ means total sales for the year preceding the end of 1990; in other words, total sales for calendar year 1990. $S(0.5)$ would mean sales in the year preceding the middle of 1991 (mid-1990 to mid-1991). This definition is necessary to give meaning to $S(t)$ for noninteger values, because of the divisibility assumption. □

How close was your guess? Would you have been willing to make an estimate of next year's sales based on your guess?

Trendlines and Extrapolation

Technology tools, such as a graphing calculator or a computer spreadsheet, will give you the best fitting least-squares line or **trendline**: the linear function that describes the relationship between the independent and dependent variables. The term used to describe this process is **regression analysis**, and the line or curve fit is called the regression model. "Trendline" is the word most commonly used to describe the relationship of output to input when the input is periods of time, as in Sample Problem 7 relating to the Sales Record for the years after 1990. The line shows the **trend** that the output is following over the period. We speak of "upward trends" or "downward trends."

Let's look again at the Sales Record graph in Figure 1.3-19. Note that the plot of the data shows the t-axis, or year values, as 0, 1, 2, 3, 4, 5 not 1990, 1991, 1992, 1993, 1994, and 1995. We changed the first input value to 0 and then calculated the other input values as the number of years *after* 1990; we **aligned the data**. The model for this data is given by

Verbal Definition: $S(t)$ = sales in hundreds of dollars for the 12 months prior to t years after the end of 1990 (so $S(0.5)$ means mid-1990 to mid-1991).

Symbol Definition: $S(t) = 946.4\ldots t + 5068.3\ldots$, for $t \geq 0$.

Assumptions: Certainty and divisibility. These imply sales can be calculated for *any* 12-month span of time, and correspond *exactly* to the given linear function.

If we were trying to use this data to predict sales for 1996, we would use technology to find $S(6)$, since $1996 - 1990 = 6$. The predicted sales would be approximately 10750 hundred dollars, or \$1,075,000. If we failed to note the aligning of 1990 to zero and calculated $S(1996)$, the predicted sales would be about 1,900,000 hundred dollars, or *\$190,000,000*!!! This is a difference of *over 189 million dollars*! This is not a little error. **Defining your variables correctly and continuing to work with those same units consistently is very important.**

We *could* have fit the data using the years as shown (in other words, *not* aligning the data), and the resulting model would have been

Verbal Definition: $Y(x)$ = sales in hundred of dollars for the 12 months prior to the end of year x ($Y(1990.5)$ would mean mid-1990 to mid-1991).

Symbol Definition: $Y(x) = 946.4\ldots x - 1878268.6\ldots$, for $x \geq 0$.

Assumptions: Certainty and divisibility. These imply sales can be calculated for *any* 12-month span of time, and correspond *exactly* to the given linear function.

In this case $Y(1996) \approx 10800$ hundred dollars again. The total sales predictions are the same at this level of precision, but the big difference lies in the complexity of the function. **Not aligning data *can* result in serious roundoff errors because of the large numbers involved and will cause errors when trying to fit some categories of models for the same reason. In general, when in doubt, it is usually a good idea to align your data, especially when working with calendar years or other large numbers for inputs.** This is even more true for nonlinear models than for linear models.

In the example above a model was fit to determine a **trendline**, and this trend line was used to predict future results: to **extrapolate** beyond the existing inputs. This is a very important application of mathematical models, but great caution must be used when

extrapolating or going beyond the known inputs to predict future outputs. When we extrapolate, we are assuming that whatever forces have caused the data to act in a certain way in the past will continue to do so in the future. Obviously, this is not always the case. The demand for a certain product may virtually disappear and sometimes this happens rather quickly. There aren't many LP records being sold today. Pop stars come and go. There are even fads in automobiles, such as the current popularity of sport utility vehicles. Population trends change: in recent years, away from the cities into the suburbs. **We must be aware of, and alert to, the inherent dangers in extrapolating, particularly if we extrapolate very far beyond the data points that we have.**

Just a reminder here. When you use a technology tool to fit a trendline to data, the result of the regression analysis will, in general, be given as an equation. For a line, the calculator might give the following: $y = ax + b$, $a = 2.4$, $b = 2$ for our data relating to the shots made after x hours of practice. *This is not a model*, although you *could* say it is the *heart* of a model. **To fully define a model, you must verbally define your function and your variables (including units) clearly, write out the symbol definition (mathematical formula, including the domain), and verbally indicate all assumptions made.**

Linear Demand Models

Sample Problem 8: You belong to a group that is selling sweatshirts to raise money. The sweatshirts will cost \$9 each, with a fixed order cost of \$25 . You did a simple survey of people in your dorm and determined that the number of sweatshirts you could sell at several different prices is given in Table 6.

(a) Use technology to find and *fully define* a linear model for the number of sweatshirts you could sell as a function of the selling price (quantity as a function of price).

(b) Interpret the slope and vertical intercept of your model.

(c) Use technology to find *and fully define* a linear model for the price you would have to charge in order to sell a given number of sweatshirts (price as a function of quantity).

(d) Interpret the slope and intercept of your model.

Solution: Let's define some variables first:

p = selling price, in dollars
q = quantity sold (number of sweatshirts)

(a) To find quantity as a function of price, we will be thinking of p as the independent variable and q as the dependent variable. This is probably the way that most of us would think of this situation, since we can decide the price, then see how many sweatshirts we sell. Let's check the graph to make sure a linear model seems reasonable. Figure 1.3-20 shows this graph.

TABLE 6

Selling Price	\$10	\$15	\$20	\$25
Quantity Sold	40	25	13	5

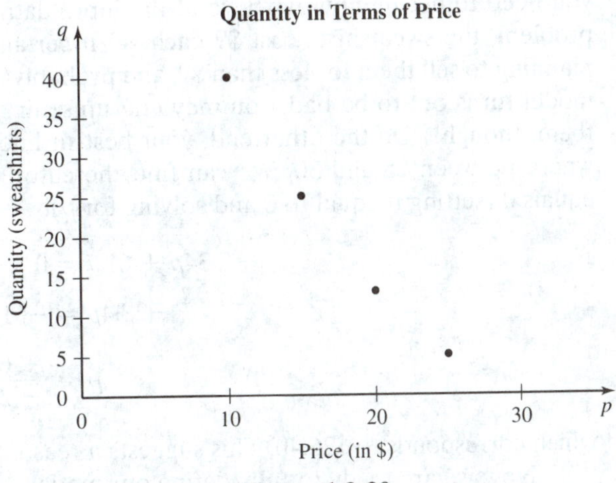

Figure 1.3-20

The data points do seem to be fairly close to being linear, although they have a slight curve to them. But, based on the KISS! principle discussed in Section 1.2, it makes sense to formulate a simple model first, which we can complicate later if necessary.

Using technology, the least-squares best-fit line is given by

$$q = -2.34p + 61.7$$

The parameters (-2.34 and 61.7) are not rounded but are actually exact. This is primarily because of the small number of data points. The graph is shown in Figure 1.3-21.

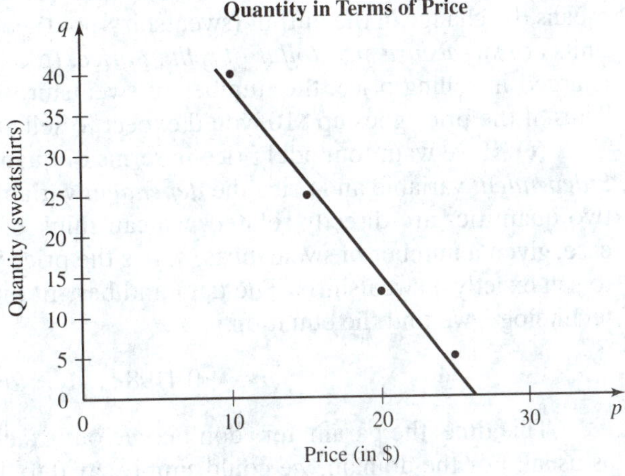

Figure 1.3-21

The last step before being able to fully define a model is to decide what domain is appropriate. **It is always safe to restrict your domain to the interval between the extremes of your input data values.** This would then limit use of your model to interpolation. If your intended use of the model definitely or possibly could involve extrapolation, then you need to decide how far beyond the input data values is reasonable. In our present problem, the sweatshirts cost \$9 each, so it certainly wouldn't make sense in advance planning to sell them for less than \$9, and probably not for less than \$10. (If your demand model turns out to be bad, you may end up selling some at cut rates later to get rid of them, though!) On the other end, your best-fit line looks as if it will go negative somewhere between 25 and 30. We can find the cutoff point by finding when the equation equals 0 (setting it equal to 0 and solving for p):

$$-2.34p + 61.7 = 0$$

$$-2.34p = -61.7$$

$$p = \frac{-61.7}{-2.34} \approx 26.4$$

which corresponds to \$26.40. This suggests a reasonable domain would be $9 \leq p \leq 26.4$.

Now we are ready to fully define our model:

Verbal Definition: $D(p)$ = the number of sweatshirts you expect to sell (demand) if you charge p dollars each for the selling price.

Symbol Definition: $D(p) = -2.34p + 61.7$, for $9 \leq p \leq 26.4$.

Assumptions: Certainty and divisibility. Certainty implies no sales or discounts. Divisibility implies fractions of sweatshirts can be sold.

(b) The vertical intercept is the constant term, 61.7. Since this is a value of q, the units are sweatshirts. The vertical intercept is the output value when the input is 0, which here means the number of sweatshirts sold if the price is \$0 (so the number you could give away for free is about 62 sweatshirts—note that a price of \$0 is not in the domain). The slope is given by the coefficient of the input variable, which here is −2.34. Since this means the change in the output (sweatshirts sold) per unit change in the input (price), the units are *sweatshirts per dollar of selling price*. In other words, for each additional dollar charged in selling price, the number of sweatshirts sold goes *down* by 2.34 sweatshirts. Thus, if the price goes up \$10, you'd expect to sell about 23 fewer.

(c) If we want to model price in terms of quantity, we now want quantity to be the *independent* variable and price the *dependent* variable. This is less intuitive, but since the two quantities are directly related, we can think of the relationship either way. In this case, given a number of sweatshirts, q, p is the price we would have to charge (in dollars) to sell exactly q sweatshirts. The data and best-fit line are shown in Figure 1.3-22. Using technology, we find the equation is

$$p = -0.4198\ldots q + 26.21\ldots$$

This time, the parameters don't come out exact, so we are using the "..." notation as usual. For the domain, we could simply say 0 to 40 (since the price to sell 40 is about \$10, which barely makes a profit, since they cost \$9 each), or we could hedge a little bit and say 0 to 45. If we wanted, we could calculate the q value that corresponds to $p = 9$, but it is not really worth the effort here. Let's play it safe and use the 0 to 45 domain. So this time, our model is given by

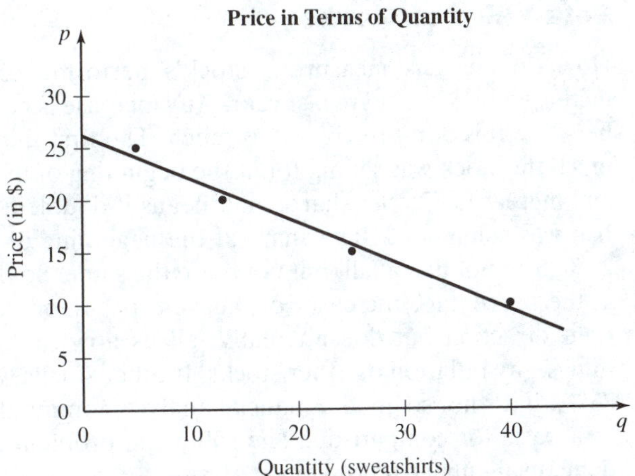

Figure 1.3-22

Verbal Definition: $d(q)$ = the price, in dollars, you'd have to charge to sell exactly q sweatshirts.

Symbol Definition: $d(q) = -0.4198\ldots q + 26.21\ldots$, for $0 \leq q \leq 45$.

Assumptions: Certainty and divisibility. Certainty implies no sales or discounts. Divisibility implies fractions of sweatshirts can be sold.

(d) This time the intercept is about 26.2 and is a value of p, so the units are dollars. The value is the price that goes with a quantity demand of 0. In other words, if you charge \$26.20 or more, you won't sell any sweatshirts. This is close to the value of \$26.40 we obtained as the endpoint of the domain for our first model, in part (a), but not exactly the same. This means that our two models are very similar, but not the same. In fact, one way to formulate the second model would have been to solve the first model for p:

$$q = -2.34p + 61.7$$

$$-2.34p = q - 61.7$$

$$p = \frac{q - 61.7}{-2.34} \approx -0.4273\ldots q + 26.36\ldots$$

This linear function is the **exact inverse** of the first function (since the graphs are slanted lines, they pass both the vertical and horizontal line tests). Notice that our second model is *very close* to the inverse of the first, but not exactly the same. **You should be aware that your choice of which variable is input and which is output *does* affect your model and so affects any solutions based on it.** The differences are not great here, but for nonlinear functions they could be significant, and one choice of roles for the variables may not correspond to a function. Economists tend to work with price as a function of quantity, which has advantages for certain kinds of analysis. Some people and texts use x as the variable for quantity. Try to be comfortable with either form.

The slope of our second model, $d(p)$, is about -0.43, and the units are dollars per sweatshirt. This means that for every additional sweatshirt you want to sell, you need to lower the price by about \$0.43. □

Beta Value of Stocks*

How do analysts measure a stock's performance? Suppose the value of the stock increased by $25 a share last year. Any increase is good, but was it really great? Raw numbers like this don't really tell us much. The first thing you would want to know was how much the stock was selling for at the beginning of the year: If it was selling for $25 a share and increased $25 per share, that means it doubled. The percentage increase was 100%! If it was selling at $100 a share at the beginning of the year, a $25 per share increase is 25%. Still not bad at all, but not quite the same. So the first thing we want to do is to look at the percentage increase (or, heaven forbid, decrease). This is definitely a step in the right direction but doesn't really tell us how the stock performed compared to other things, say inflation or other stocks. If other stocks went up 35% while this stock went up 25%, that's not so good. Financial analysts commonly use the Standard and Poor's 500™ as a basis for comparison. The following problem gives a feel for how this is typically done, using linear regression analysis.

Sample Problem 9: The Intel Corporation shows information about its relative stock price compared to the Standard & Poor's (S&P) 500 Stock Index in an annual report for the benefit of its stockholders. Table 7 gives the stock and index values (assuming you started with $100 and reinvested all dividends), then expresses them as percent change values (in decimal form) for each one-year interval. (To calculate the 1991–92 percent change for Intel, for example, the calculation was

$$\frac{234 - 123}{123} = \frac{111}{123} \approx 0.90$$

Note that this really means 90%; that is, the percent change is expressed in pure decimal form, not as a number of percentage points.)

Find a linear function for the percent increase in Intel stock value as a function of the percent increase in the S&P 500 Index value.

TABLE 7

Year (as of the end of Dec.)	1990	1991	1992	1993	1994	1995
Intel Stock Value	$100	$123	$234	$324	$335	$596
S&P 500 Index Value	$100	$130	$140	$155	$157	$215

Year (end Dec.-end Dec.)	1990–91	1991–92	1992–93	1993–94	1994–95
Intel Stock (% chg.)	0.23 = 23%	0.90 = 90%	0.38 = 38%	0.03 = 3%	0.78 = 78%
S&P 500 (% chg.)	0.30 = 30%	0.08 = 8%	0.11 = 11%	0.01 = 1%	0.37 = 37%

*This topic may be omitted without affecting the flow of the development. No later topics depend on it. Homework exercises relating to it are marked with the same symbol.

Solution: The graph in Figure 1.3-23 shows the data points and the trendline. Using technology tools to find the best-fitting least-squares line gives

$$y = 0.7226\ldots x + 0.3382\ldots$$ [8]

The slope coefficient of this equation, approximately 0.72, is called the stock's **beta** value (written β in symbols and pronounced "BAY-tuh"). It gives a measure of how sensitive the stock is to changes in the market. $\beta > 1$ means the stock changes proportionately *more* than the market, while $\beta < 1$ means it changes proportionately less, on average. This is sometimes referred to as a measure of the stock's **undiversifiable risk** (the risk that is associated with the changes in the market). Thus Intel's beta value indicates it changed proportionately *less* than the overall market over the period being studied. □

Before you begin the exercises be sure that you

- Recognize that a linear model is appropriate when a plot of the data appears to lie in a straight line.
- Can verify that the data is linear or nearly linear by checking that the first differences are relatively constant.
- Recognize the general form of a linear model in its different variations:

$$y = mx + b, y = ax + b, \text{and } y = a + bx$$

- Can correctly write out a model for a linear function, including carefully and precisely defining your variables (with units); writing the function using correct functional notation (with the domain); and listing your assumptions.

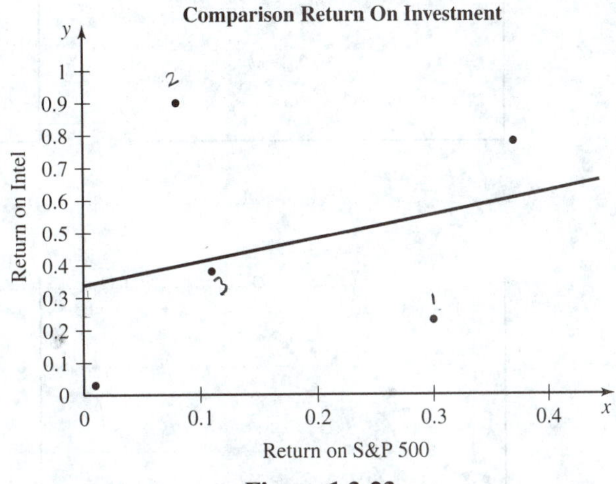

Figure 1.3-23

[8] *Note*: Our calculator gave the slope value as 0.7226944326 and the intercept as 0.3382511687. We recorded the numbers using . . . for visual purposes, as we discussed in Section 1.2. When using a model for calculations, use all of the decimal places. When writing a model four or five significant places are generally sufficient for visual purposes, and can then be rounded to three or four significant digits. When writing a numerical answer to a problem using a model, we will usually round off the numbers. This will be discussed further in Section 3.4.

- Know that the slope of a linear model is the change in output divided by the change in input, or $m = \dfrac{y_2 - y_1}{x_2 - x_1}$.
- Know that the vertical intercept of a linear model is the point at which the line meets the vertical axis (the output value when the input is 0).
- Understand that, for linear cost functions, marginal cost is the increase in cost when the quantity increases by one item.
- Understand that, for linear revenue functions, marginal revenue is the increase in revenue when the quantity increases by one item.
- Understand that marginal analysis involves the calculation of marginal cost and marginal revenue, and the use of these to project changes in the output.
- Can find the intersection of two linear functions algebraically by setting them equal to each other.

EXERCISES FOR SECTION 1.3

Warm Up

1. Which of the following plots could be reasonably modeled by a linear function?

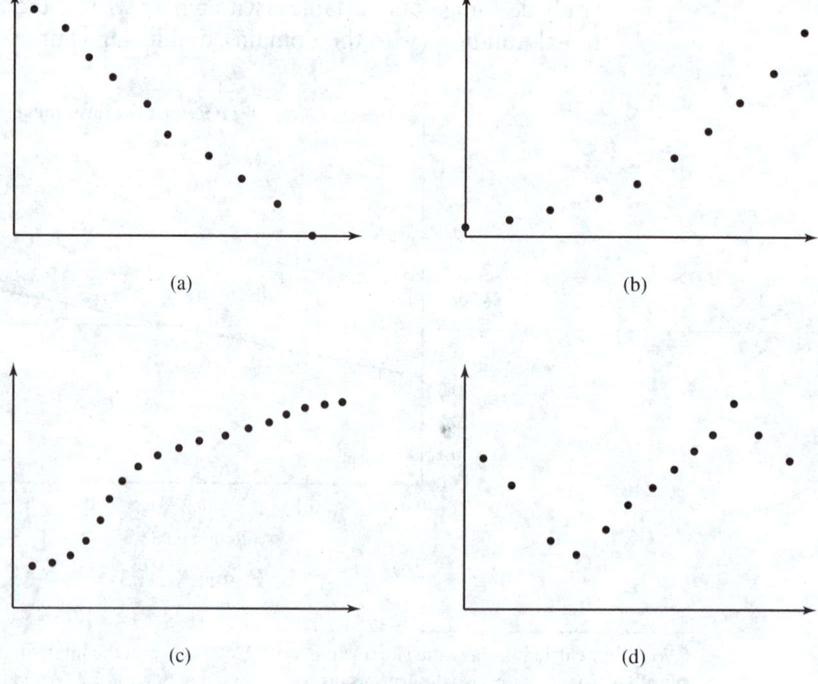

Figure 1.3-24

2. Which of the following plots could be reasonably modeled by a linear function?

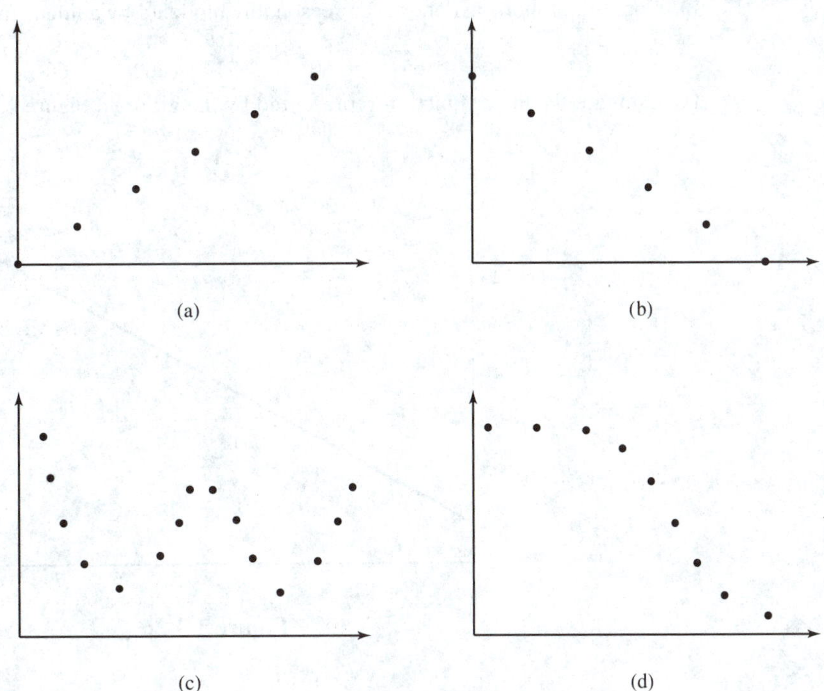

Figure 1.3-25

3. Could the following data set be reasonably modeled by a linear function?

0	1	2	3	4	5	6	7
1.5	2.4	3.6	4.5	5.6	6.7	7.4	8.2

4. Could the following data set be reasonably modeled by a linear function?

0	1	2	3	4	5	6	7
15	14	12	8	6	5	3	2

5. Could the following data set be reasonably modeled by a linear function?

10	14	18	22	25	30	41	49
20	21	22	23	24	25	26	27

6. Could the following data set be reasonably modeled by a linear function?

0	1	2	3	4	5	6	7
22	24	26	28	27	25	23	21

7. Could the following data set be reasonably modeled by a linear function?

0	1	2	3	4	5	6	7
5.1	6.2	7.1	8.3	9.2	10.0	11.5	13

8. Could the following data set be reasonably modeled by a linear function?

0	1	2	3	4	5	6	7
28	25	20	18	15	12	8	5

9. Could the following data set be reasonably modeled by a linear function?

10	14	18	22	25	30	41	49
25	24	23	22	21	20	19	18

10. Could the following data set be reasonably modeled by a linear function?

0	1	2	3	4	5	6	7
78	65	60	54	60	66	72	80

11. Estimate the linear function represented by the graph in Figure 1.3-26.

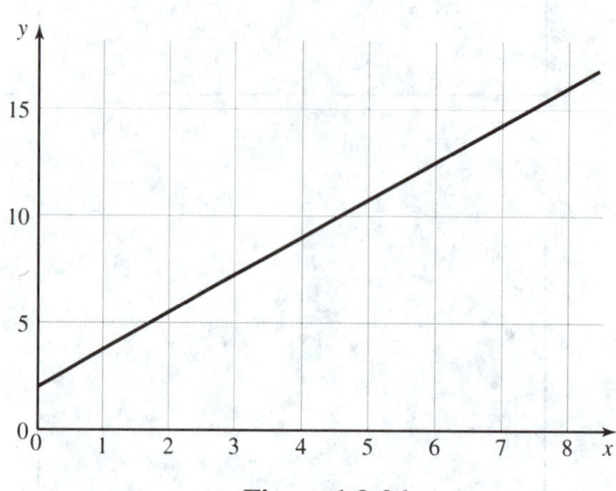

Figure 1.3-26

12. Estimate the linear function represented by the graph in Figure 1.3-27.

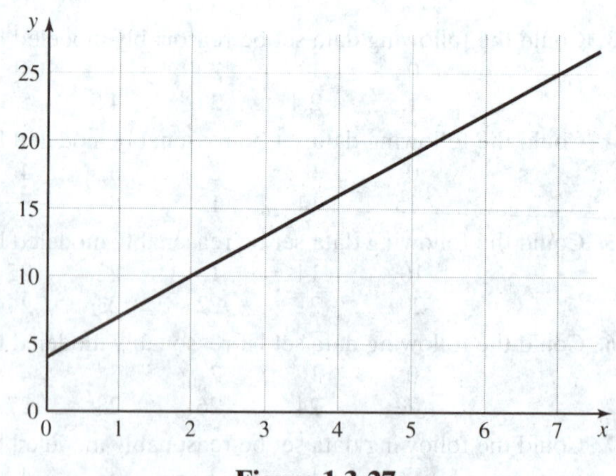

Figure 1.3-27

13. Estimate the linear function represented by the graph in Figure 1.3-28.

14. Estimate the linear function represented by the graph in Figure 1.3-29.

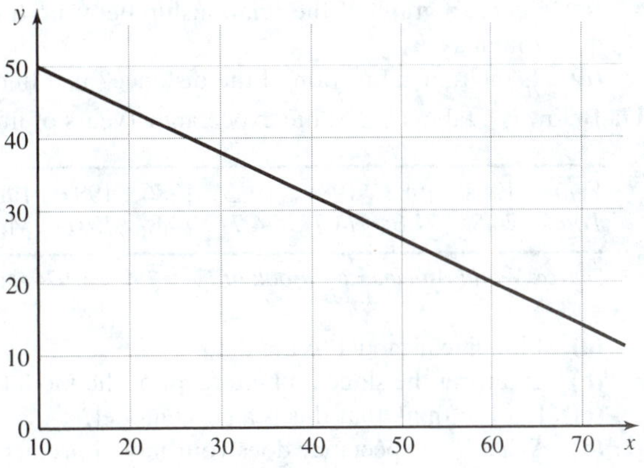

Figure 1.3-28

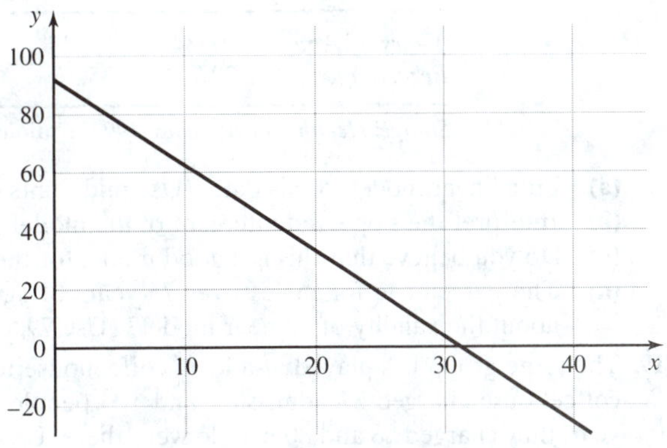

Figure 1.3-29

Game Time

15. You have been training to run in your first marathon. You certainly don't expect to win, but you would like to make a decent showing, hopefully finishing in 4 hours. The race is usually 26 miles, 385 yards in length. (There are 1760 yards in a mile.) Your strategy is to set your pace and stick with it throughout the race.

(a) Determine what this pace must be in minutes per mile if you are to finish in 4 hours.

(b) Write the distance you should have covered as a function of the time elapsed.

(c) Sketch a graph of this function.

16. On a regular class day, you start out from your room for class, which starts in 10 minutes. The classroom is approximately a quarter of a mile away. You start out at a leisurely pace and arrive at class early. Suddenly you remember that you left your assignment in your room. You double your speed, go back to your room, grab the paper and go to class, arriving just in time.

(a) Sketch a graph of the relationship between time and distance from your room, with time on the horizontal axis.

(b) Is the distance a function of the time? Is it linear?

(c) Sketch a graph of the relationship between time and distance, with distance on the horizontal axis.

(d) Is the time a function of the distance? Is it linear?

17. Below is a table of U.S. life expectancy (years of life expected at birth).

Year	1982	1983	1984	1985	1986	1987	1988	1989	1990	1991	1992	1993
Total	74.5	74.6	74.7	74.7	74.8	75.0	74.9	75.1	75.4	75.5	75.5	75.5

Source: *World Almanac and Book of Facts 1996*, p. 974 (National Center for Health Statistics).

(a) Fit a linear model to the data.

(b) Interpret the slope and intercept of the model.

(c) Do you think that this is a good model?

(d) What life expectancy does your model predict for 1994? 1995? 1996? 2000?

(e) Explain why you would or would not have confidence in these figures.

18. Below is a table showing the percentage of overweight males in the United States for the years 1988–91:

Years of Age	20–34	35–44	45–54	55–64	65–74
Percentages	22.2	35.3	35.6	40.1	42.9

Source: *Health United States, 1994* (National Center for Health Statistics).

(a) Fit a linear model to this data. (Use midpoints of intervals.)

(b) Interpret the slope and intercept of the model.

(c) Do you believe that this is a good model for the data?

(d) The percentage for males over 75 years of age is 26.4. Does this support your opinion about the validity of a linear model? (Use 79.5 years.)

19. The same group has played at a local coffeehouse for three years running.[9] The first year the coffeehouse charged $7 admission and 100 people attended. Due to rising costs the second year, they charged $8 and 80 people were there. Last year they raised the price to $9, and only 70 people came.

(a) Fit a linear model to this data.

(b) Interpret the slope and intercept of the model.

(c) Do you think their strategy of raising the prices each year was a good one?

(d) What other things besides the increased ticket prices may have influenced attendance?

20. A group of students decided to earn some money selling "scrunchies." They estimated that they could sell 310 scrunchies at $2 each, 200 at $3.50 each and 100 at $5 each.[10]

(a) Find and *fully define* the best-fit linear demand model of quantity as a function of price (the quantity they can sell at a given price).

(b) Interpret the slope and intercept of your demand model.

(c) Now find the best fitting linear demand model, expressing the *price* as a function of the *quantity* (the price they would have to charge to sell that quantity).

(d) Interpret the slope and intercept of the demand model from part (c).

21. Students took a poll and found the number of team T-shirts that they could expect to sell at four different prices:[11]

---

[9] Data from the Folk Factory Coffeehouse in Philadelphia, PA.

[10] Data adapted from a student project.

[11] Data adapted from student projects.

Price	\$8	\$10	\$12	\$15
Number Sold	302	231	175	123

(a) Find and *fully define* the best-fit linear model for the demand for T-shirts as a function of the price.
(b) Interpret the slope and intercept of the model.
(c) Do you think this is a good model? Why or why not?
(d) From your model, at what price would they be unable to sell any T-shirts?
(e) Now find the best-fit linear demand model, expressing the *price* as a function of the *quantity* (the price they would have to charge to sell that quantity).
(f) Interpret the slope and intercept of the demand model from part (e).

22. Some students making and selling hair scrunchies researched their costs and found that it would cost them \$4.95 to set up (cost of scissors and needles, etc.) and \$0.27 to make each scrunchie.
(a) Write a model for the cost in terms of the quantity of scrunchies.
(b) If they sell the scrunchies for \$8 each, what is the revenue function?
(c) Find the break-even point, and explain what it means.
(d) Find a model for the average cost per scrunchie (the total cost divided by the number of scrunchies).

23. Some students selling T-shirts researched their costs and found that it would cost them \$5.55 per T-shirt plus a fixed cost of \$45.
(a) Write a model for the cost in terms of the quantity of T-shirts.
(b) If they sell the T-shirts for \$7 each, what is the revenue function?
(c) Find the break-even point, and explain what it means.
(d) Find a model for the average cost per T-shirt (the total cost divided by the number of T-shirts).

24. The following table shows the number of hours spent studying the night before a quiz and the percent grade received on the quiz:

Hours Studied	1	2	3
Grade Received	45%	73%	91%

(a) Fit a linear model to these data values.
(b) Interpret the slope and intercept.
(c) What grade does your model predict for 4 hours studied?
(d) Do you think this is a good model? Why or why not?

25. The following table shows total personal income (in billions of dollars) in the United States from 1980–1994:

Year	Total Income (\$)	Year	Total Income (\$)
1980	2165.3	1988	4070.8
1981	2429.5	1989	4384.3
1982	2584.6	1990	4679.8
1983	2838.6	1991	4828.4
1984	3108.7	1992	5058.1
1985	3325.3	1993	5375.1
1986	3526.2	1994	5701.7
1987	3776.6		

Source: *The World Almanac and Book of Facts*, p. 127 (*source*: Bureau of Economic Analysis, U.S. Dept of Commerce).

(a) Fit a linear model to the data.
(b) Is the model a good fit for the data?
(c) What is the total personal income (in billions of dollars) that the model predicts for 1995? 1996? 2000?
(d) What is the annual rate of growth of personal income as calculated from your model?

26. The disposable personal income (in billions of dollars) for the United States from 1980–1994 is shown in the following table:

Year	Disposable Income ($)	Year	Disposable Income ($)
1980	1828.9	1988	3479.2
1981	2041.7	1989	3725.5
1982	2180.5	1990	4058.8
1983	2428.1	1991	4209.6
1984	2668.6	1992	4430.6
1985	2838.7	1993	4688.7
1986	3013.3	1994	4959.6
1987	3205.9		

Source: *The World Almanac and Book of Facts*, p. 127 (*source*: Bureau of Economic Analysis, U.S. Dept of Commerce).

(a) Fit a linear model to the data.
(b) Is the model a good fit for the data?
(c) What is the total disposable personal income (in billions) that the model predicts for 1995? 1996? 2000?
(d) What is the annual rate of growth of disposable personal income as calculated from your model?
(e) Disposable personal income is total personal income (see Exercise 25) minus personal taxes. Have taxes been growing at a faster or slower rate than total personal income?

27. The local pizza place will deliver pizza to your room up until 1 A.M. A small regular pizza costs $6, a medium regular pizza costs $8, and a large regular pizza (cheese and sauce) is $10.00. Each extra topping costs $1.00.
(a) Express the cost of each size pizza as a function of the number of toppings ordered.
(b) On weekends they have a special: small pizza with five toppings for $9, a medium pizza with five toppings for $11 and a large with five toppings for $13. If the small is half the size of the large and the medium is three-fourths the size of the large, it's cheaper to buy large pizzas than the smaller size. If six of you get together on a weekend and you each want to get medium pizzas with five toppings, how many of each size pizza should you order to get the same amount of pizza at the least cost?
(c) How would you divide the pizzas so that everyone got a fair share? (If you are ever stuck in this situation with one other person try this: offer to let the other person divide, but then you get to choose (or reverse the roles). It does make for a greater effort to divide evenly.)

Overtime

28.* Public Storage, Inc., shows information about its relative stock price compared to the Standard and Poor's 500 Index for the benefit of its stockholders. The following table gives the stock and index values, assuming that the value invested in the Company's common stock and the S&P 500 was $100 each on Dec. 31, 1990 and that all dividends were reinvested.

*This problem involves the concept of a beta value of a stock, an optional topic at the end of the section.

	12/31/90	12/31/91	12/31/92	12/31/93	12/31/94	12/31/95
Public Storage	$100.00	$137.38	$161.85	$277.38	$296.25	$411.42
S&P 500	$100.00	$130.47	$140.41	$154.56	$156.60	$215.45

(a) Find the percent change for Public Storage, Inc. and the S&P 500 for 1990–1991, 1991–1992, 1992–1993, 1993–1994, and 1994–1995.

(b) Find a linear function for the percent increase in Public Storage, Inc. stock value as a function of the percent increase in the S&P 500 Index value.

(c) What is the beta value of Public Storage, Inc. stock for this period?

(d) What does the beta value mean?

(e) Do you think that a company would include this kind of information in its annual report if the beta value were negative? Why or why not?

29.* International Paper shows information about its relative stock price compared to the Standard and Poor's 500 Index for the benefit of its stockholders. The following table gives the stock and index values, assuming that the value invested in the Company's common stock and the S&P 500 was $100 each on Dec. 31, 1990 and that all dividends were reinvested.

	12/31/90	12/31/91	12/31/92	12/31/93	12/31/94	12/31/95
Int'l. Paper	$100	$136	$131	$137	$156	$160
S&P 500	$100	$131	$141	$155	$157	$216

(a) Find the percent change for International Paper and the S&P 500 for 1990–1991, 1991–1992, 1992–1993, 1993–1994, and 1994–1995.

(b) Find a linear function for the percent increase in International Paper stock value as a function of the percent increase in the S&P 500 Index value.

(c) What is the beta value of International Paper stock for this period?

(d) What does the beta value mean?

(e) Do you think that a company would include this kind of information in its annual report if the beta value were negative? Why or why not?

(f) International Paper also gave a comparison of ten-year total returns in the same annual report. Why do you think they did this?

1.4 FUNCTIONS WITH ONE CONCAVITY: QUADRATIC, EXPONENTIAL, POWER

Not all phenomena can be described by linear models. In a way this is unfortunate, because linear models are fairly easy to understand and to work with. On the other hand, in many ways they are not very interesting. Linear models are always increasing or always decreasing. They have no distinctive points, no peaks (maximums), or troughs (minimums). They never change their shape. In the section on functions, we saw graphs of many different shapes, some representing functions, and some that did not represent functions. **We will normally confine ourselves in this text to data that can usefully be**

---

*This problem involves the concept of a beta value of a stock, an optional topic at the end of the section.

modeled by a *function.* We are now going to look at some types of functions that show similar characteristics based on their curvature, a property called **concavity**.

Here are some kinds of problems that the material in this section will help you solve:

- You have a juggling act. You want to add a new twist to your act: Juggle three raw eggs, toss one of the eggs high in the air, turn your back and continue to juggle the other two. At the last minute, you'll turn around and catch the third egg. If you don't time it just right, you could end up with egg on your face in more ways than one. Because you are a juggler, your tosses are very consistent, but you still need to get your timing down pat.
- You want to sell reusable mugs as part of an environmental project. You hope to achieve two things: promote the idea of reusing resources to as many people as possible and raise money for the environmental effort. You're not sure how to price the mugs. Is there any way to determine the profit as a function of the price charged?
- What would be the best price to charge for tickets to a performance at a local coffeehouse? The same musical group has been there before and you have always made money, but lately profits seem to be dropping.
- You and a friend like to play computer games. In one of the games the object is to remove as many tiles as you can from a board before the game is ended. You think that you are pretty good at the game and decide on a little wager. The loser has to go out and get the pizza. You have some time to "warm up" before the actual contest. How long should you play in your warm-up?
- You are interested in the relationship between ticket prices and the number of tickets sold at a local sporting event. Is it possible to determine approximately how many tickets can be sold at a variety of different prices?

When you have finished this section you should be able to solve problems like those above and

- Know the shapes of quadratic, exponential, and power functions.
- Know that quadratic functions are characterized by equal second differences for equally spaced inputs.
- Know that exponential functions are characterized by equal percentage change (ratio) for equally spaced inputs.
- Choose models with context, purpose, and expected behavior in mind.
- Fit quadratic, exponential, and power models using technology.
- Solve equations involving quadratic, exponential, and power functions.

Quadratic Models

Sample Problem 1: You have a juggling act. You want to add a new twist to your act: Juggle three raw eggs, toss one of the eggs high in the air, turn your back, and continue to juggle the other two. At the last minute, you'll turn around and catch the third egg. If you don't time it just right, you could end up with egg on your face in more ways than one. Because you are a juggler, your tosses are very consistent, but you still need to get your timing down pat. You decide to practice with a ball and learn to time how long it takes the ball to come back to you. (You remember from high school physics that Galileo concluded that a ball and an egg should be equivalent.) After you have practiced long enough with the ball, you'll give the egg a try. If you throw a ball straight up into the air and catch it as it comes down, the height of the ball above the ground will increase rapidly when thrown and then gradually slow until the ball stops rising and begins to fall back down again. As a function of time, the height is both **increasing** (goes *up* from left to right) and **decreasing** (goes *down* from left to right). The table below shows the height of the ball (above the ground) at different given times:

Time/Sec	0	0.25	0.5	0.75	1	1.25	1.5	1.75	2	2.25
Height/Ft	5	16	25	32	37	40	41	40	37	32

Find a model to represent this situation, and use it to determine how long after you throw the egg it will take for the egg to come back down to where you can catch it.

Solution: The data is plotted in the graph in Figure 1.4-1. Obviously, we should not use a linear model to describe these data points. What kind of a model should we use?

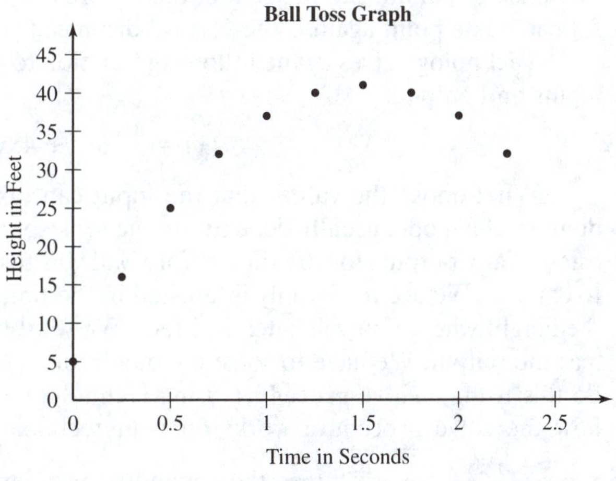

Figure 1.4-1

You probably recognize from previous math courses that these points seem to look something like a **parabola**. The general equation for a parabola opening up or down is

$$y = ax^2 + bx + c$$

Expressed as a *function*, the equation becomes

$$f(x) = ax^2 + bx + c$$

This is called a **quadratic function.**

The curvature of this scatter plot and the fact that it is increasing over part of the input and decreasing over another are what distinguish it from linear models. **Concavity** is the word used to describe the curvature of a graph. The curvature shown in this graph is described as **concave down** (rhymes with frown). The three graphs in Figure 1.4-2 are all concave down.

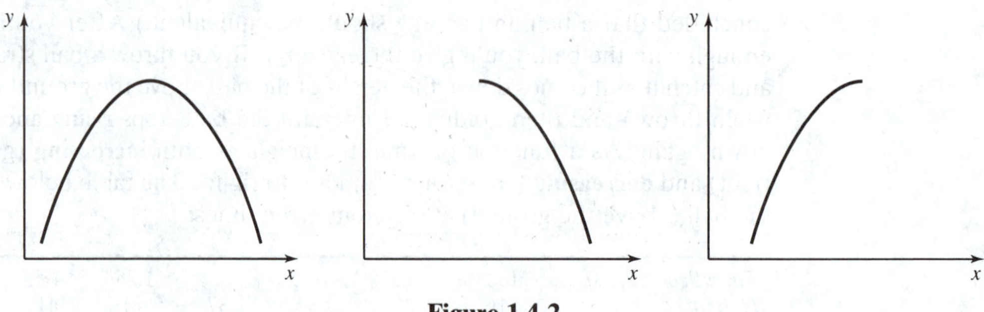

Figure 1.4-2

The model for the ball height data will give the height of the ball H, in feet, t seconds after the toss. Once we have decided on the type of function that would best describe the relationship between the height of the ball (egg) and the time since the toss, we can use technology to find that function. We have said that the graph looks as if it could be a parabola. Technology will find the best-fit parabola (quadratic function), in the sense of minimizing the sum of the squares of the errors (SSE), as before. The error for each data point again is the vertical distance from the model curve to the data point.

Technology gives us the following function to describe the relationship between the inputs and outputs:

$$f(x) = -16x^2 + 48x + 5$$

What about the values that the input can take, the domain of the function? The domain of a model usually depends on the sense of the model. We are certainly not interested in any outputs for the time before ball was tossed, so t will be greater than or equal to 0: $t \geq 0$. We are really only interested in the time it takes for the ball to come back to the height where you will catch it, 5 feet. We want to determine what input x will give us 5 as the output. We have to solve the quadratic equation: $5 = -16x^2 + 48x + 5$. We can do this using a **solve procedure** using technology (see your technology supplement for how the solve procedure works on your technology) or using the quadratic formula: $x = \dfrac{-b \pm \sqrt{b^2 - 4ac}}{2a}$ (see the appendix for a derivation of the quadratic formula). In most cases, we must express the equation in a form that has 0 on one side. In this problem, we can subtract 5 from both sides to get

$$0 = -16x^2 + 48x$$

Using either of these methods we find that the output will equal 5 when the input is 0 or 3.

Translation: The ball will be at a height of 5 feet (in our hand) initially (after 0 seconds) and after 3 seconds. We already knew the 0 value; it is the 3 that interests us. Our input values, times, should be at least 0 and less than or equal to 3 (seconds). We are now ready to write our model:

Verbal Definition: $H(t)$ = the height of the ball (in feet), t seconds after the toss.
Symbol Definition: $H(t) = -16t^2 + 48t + 5$, for $0 \le t \le 3$.
Assumptions: Certainty and divisibility.

Let's see how well this model fits. The following table has the time (input value), the actual height (data output value paired with that input value), and the predicted height (output value predicted by the model for each given input):

Input	0	0.25	0.5	0.75	1	1.25	1.5	1.75	2	2.25
Predicted	5	16	25	32	37	40	41	40	37	32
Actual	5	16	25	32	37	40	41	40	37	32

They are identical! This means that the quadratic model is a *perfect* fit! Thus you can be quite sure that you will have about 3 seconds to do the full turning around routine before getting egg on your face. □

Quadratic Functions and Second Differences

Let's take another look at the data.

Time/Sec	0	0.25	0.5	0.75	1	1.25	1.5	1.75	2	2.25
Height/Ft	5	16	25	32	37	40	41	40	37	32

When we studied linear models we looked at first differences in the tabular data:

First differences

Time/Sec	0		0.25		0.5		0.75		1		1.25		1.5		1.75		2		2.25
Height/Ft	5		16		25		32		37		40		41		40		37		32
First Differences		+11		+9		+7		+5		+3		+1		−1		−3		−5	

The first differences here are not at all constant; in fact, they change from being positive to being negative.

Obviously, the first differences are not constant, they themselves are changing. How are they changing? There does seem to be a nice pattern here, but what is it? Let's find the differences *for the first differences* listed above. These are called the **second differences**.

Second differences

Time/Sec	0		0.25		0.5		0.75		1		1.25		1.5		1.75		2		2.25
Height/Ft	5		16		25		32		37		40		41		40		37		32
First Differences		+11		+9		+7		+5		+3		+1		−1		−3		−5	
Second Differences			−2		−2		−2		−2		−2		−2		−2		−2		

The second differences are exactly equal! **If the second differences are exactly equal for equally spaced inputs, the data can be modeled exactly by a quadratic function.** If the second differences are relatively close to each other for equally spaced inputs, then a quadratic model should fit quite well.

The General Form of Quadratic Functions and Concavity

Recall that the general form of a quadratic function is

$$f(x) = ax^2 + bx + c$$

The "a" is called the **leading coefficient: the (nonzero) coefficient of the largest power of x.** The sign of the leading coefficient of a quadratic function tells us whether the curve is concave up or concave down. A positive coefficient tells us that the curve is concave up. ("Up" starts with a "u" and concave up is shaped like any part of a "u;" think positively; think a smile; the cup is up.) The three curves in Figure 1.4-3 are concave up.

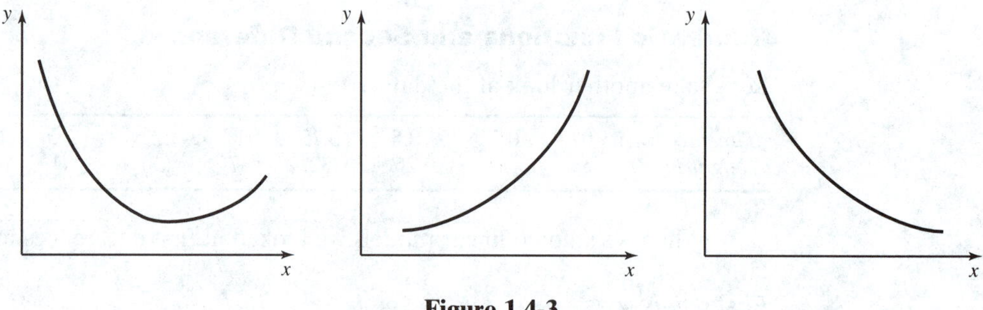

Figure 1.4-3

A negative leading coefficient tells us that the curve is concave down (looks like a frown; shaped like any part of an upside-down "u"; think negatively; the cup is turned down). The three graphs in Figure 1.4-2 are all concave down.

The "c" in the general quadratic function is the constant, and it tells us the vertical intercept, just as it did for a linear model. This is true because the value of $f(0)$ (which is equal to $0 + 0 + c = c$) is the output value that goes with an input value of 0. The graph of this point $(0,c)$ is thus on the vertical axis. In Example Problem 1,

$$H(t) = -16t^2 + 48t + 5$$

so

$$c = H(0) = 0 + 0 + 5 = 5$$

which tells us that the ball was five feet above the ground at the time that it was tossed (after 0 seconds; when $t = 0$).

Sample Problem 2: You like to play a computer game that involves removing as many tiles from a board as possible before the game is ended. There are 1000 tiles to start. You know that you get better with practice, but you also tend to get tired after a while and lose your touch. You have kept the following record:[12]

Minutes Practiced	10	20	30	40	50	60
Tiles Left	540	260	80	15	75	380

How long should you practice to be at your best?

Solution: The graph of the data is shown in Figure 1.4-4. These data points certainly do not look linear. In fact, they look very much like the first example shown above for concave up functions. Let's look at the second differences.

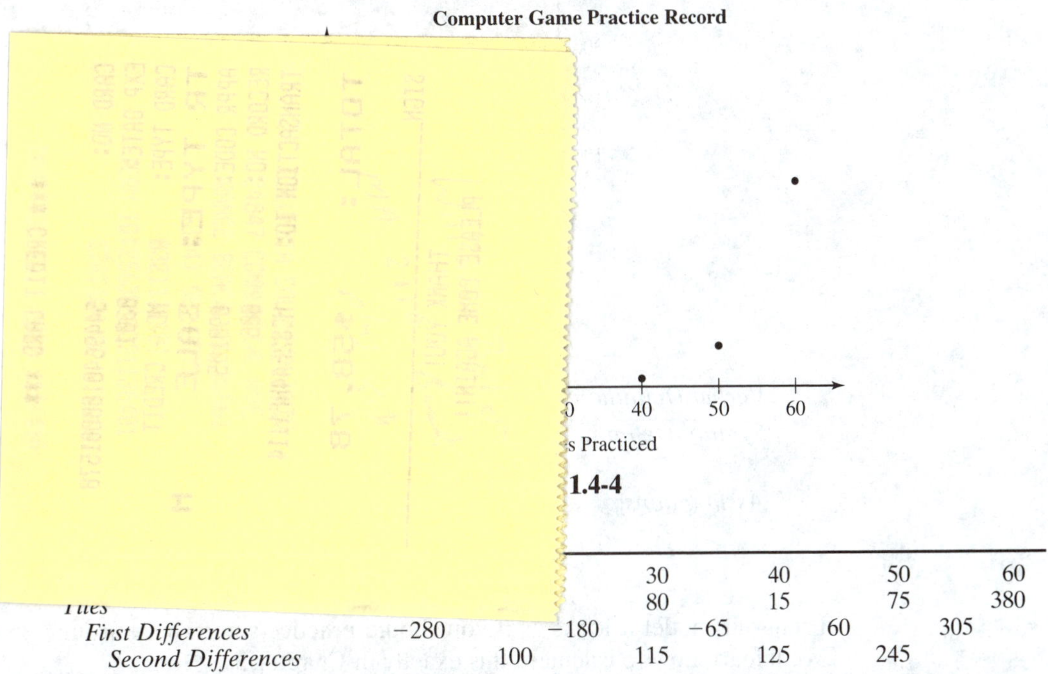

1.4-4

[illegible]	[illegible]		[illegible]		30		40		50		60
Tiles	[illegible]		[illegible]		80		15		75		380
First Differences		−280		−180		−65		60		305	
Second Differences			100		115		125		245		

The second differences are not constant, but the relative range is less than that of the first differences, and the magnitudes are roughly comparable.

Should we fit a quadratic model to data for which the second differences are not constant? If the shape of the data is clearly close enough to quadratic, the answer is yes. The first example involved a physical law, the law of falling bodies. We rarely, if ever, find data in real-world situations that are perfectly quadratic. The analysis of differences in outputs when inputs are evenly spaced is helpful, but not conclusive. If the analysis of differences (first or second) shows a strong consistency, that should influence us to choose a particular model (linear or quadratic, respectively). If the analysis does not show a strong consistency, that should not exclude that particular model. The second differences

[12]Data adapted from a student project.

for this data were not perfectly consistent, but the shape certainly appears to be quadratic, so it is reasonable to fit a quadratic model to the data. The graph in Figure 1.4-5 shows the best-fit quadratic model for the data, as given by technology. This seems to be a very good fit. Thus, a reasonable model for this data would be

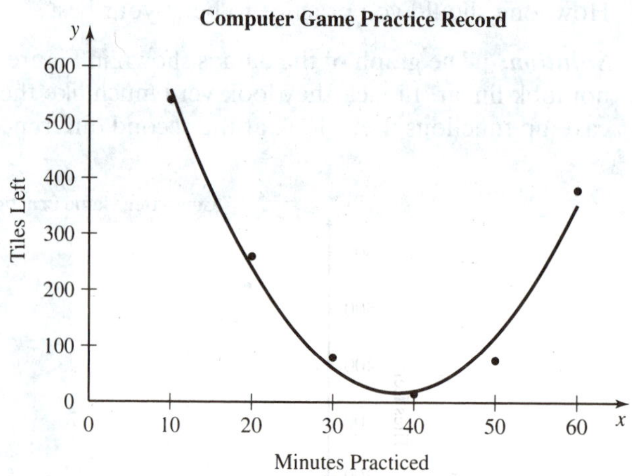

Figure 1.4-5

Verbal Definition: $T(x)$ = the number of tiles left after playing x minutes.
Symbol Definition: $T(x) = 0.693\ldots x^2 - 52.61\ldots x + 1014.5\ldots$, for $10 \leq x \leq 60$.
Assumptions: Certainty and divisibility. Certainty implies the result will always be exactly the same. Divisibility implies fractions of tiles can be left.

From our model it looks as if you should practice somewhere around 35 to 40 minutes. We'll learn how to calculate this exactly in Chapter 3. □

Exponential Models

Sample Problem 3: Ten years ago you deposited \$1000 in an account that pays 12% interested compounded annually. If you continue to leave the account untouched, how much will you have in the account 5 years from now? (Incidentally, if you can find an account that pays 12%, even if it is only compounded annually, please let us know!)

You checked your statements for the past 10 years and then graphed the data as shown in Figure 1.4-6.

Here are the data values and the first and the second differences:

Years	0		1		2		3		4		5		6		7
Total in \$	1000.00		1120.00		1254.40		1404.93		1573.52		1762.34		1973.82		2210.68
First Differences		120.00		134.40		150.53		168.59		188.82		211.48		236.86	
Second Differences			14.40		16.13		18.06		20.23		22.66		25.38		

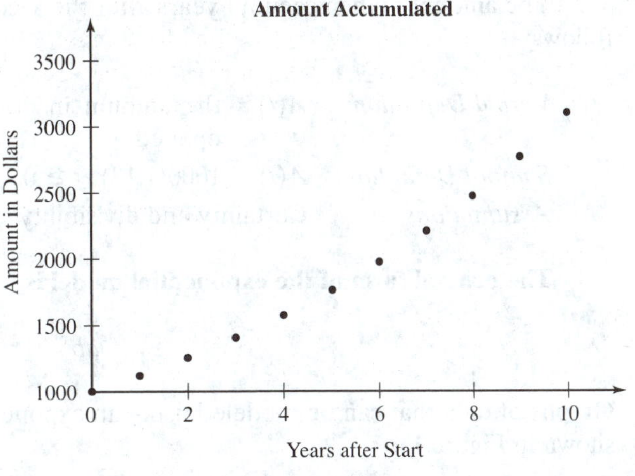

Figure 1.4-6

These data points are concave up and increasing. Certainly it will never change to decreasing as time goes on unless you take your money out of the account. Should we fit a quadratic model?

Solution: The second differences are *not* constant. In fact, they are always increasing, and by increasing amounts. Is there something about the way the output is changing that is more consistent and can guide us in the choice of a model for this data?

Let's check the **ratios** of each new output to the previous output:

$1120.00/1000.00 = 1.12$

$1254.40/1120.00 = 1.12$

$1404.93/1254.40 \approx 1.12$

$1573.52/1404.93 \approx 1.12$

$1762.34/1573.52 \approx 1.12$

$1973.82/1762.34 \approx 1.12$

$2210.68/1973.82 \approx 1.12$

The ratios are not only *more* consistent, they are also *exactly* constant. How can we relate this, as we did in the linear models (first differences are constant) and quadratic models (second differences are constant)? Well, if the ratios are constant, then the percent change should also be constant (if the ratio were 1.5 always, it would mean a 50% increase always). So let's check the **percent change** expressed as a pure decimal (rather than a number of percentage points). Divide the first difference by the previous output number, for example 120/1000 and 134.40/1120, and so forth, we get:

Years	0	1	2	3	4	5	6	7
% Change		0.12	0.12	0.12	0.12	0.12	0.12	0.12

The percent change is constant.

Data that are concave up *or* concave down (not both), always increasing *or* always decreasing, and characterized by an *exactly constant* percent change (or ratio) for equally spaced inputs can be modeled *exactly* using exponential models.

The amount in the account t years after the account was opened can be modeled as follows:

Verbal Definition: $A(t)$ = the amount in dollars t years after the account was opened.
Symbol Definition: $A(t) = 1000(1.12)^t$, $t \geq 0$.
Assumptions: Certainty and divisibility. □

The general form of the exponential model is

$$y = ab^x$$

Graphs of data that can be modeled using an exponential model have the general shape shown in Figure 1.4-7.

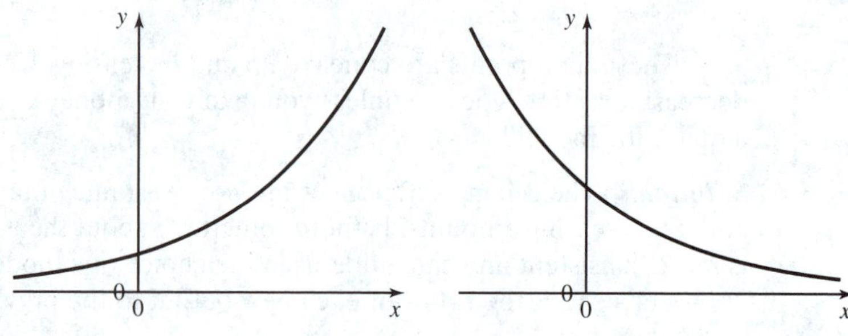

Figure 1.4-7

Note that some technologies use the form

$$y = ce^{kx}$$

where c and k are constants and e is Euler's number, approximately 2.71828. See the appendix for details on exponents and logarithms, including how to convert between these two different forms of the exponential function.

Also, it is possible for the a (or the c) in an exponential function to be negative, in which case the graph would be concave down and the *mirror image* of those in the above graphs, on the other side of (below) the horizontal axis. This is illustrated in Figure 1.4-8 (which shows the general shape for values of b that are strictly positive) and in the next sample problem.

Sample Problem 4: A market survey was taken to estimate the number of tickets that could be sold to a local sporting event when different prices were charged. The table below shows the prices charged per ticket and the number of tickets sold. Find a model for the number of tickets sold as a function of the price per ticket (the demand function).

Price ($)	10	15	20	25	30	35	40
Number of Tickets	1880	1450	1120	860	660	500	400

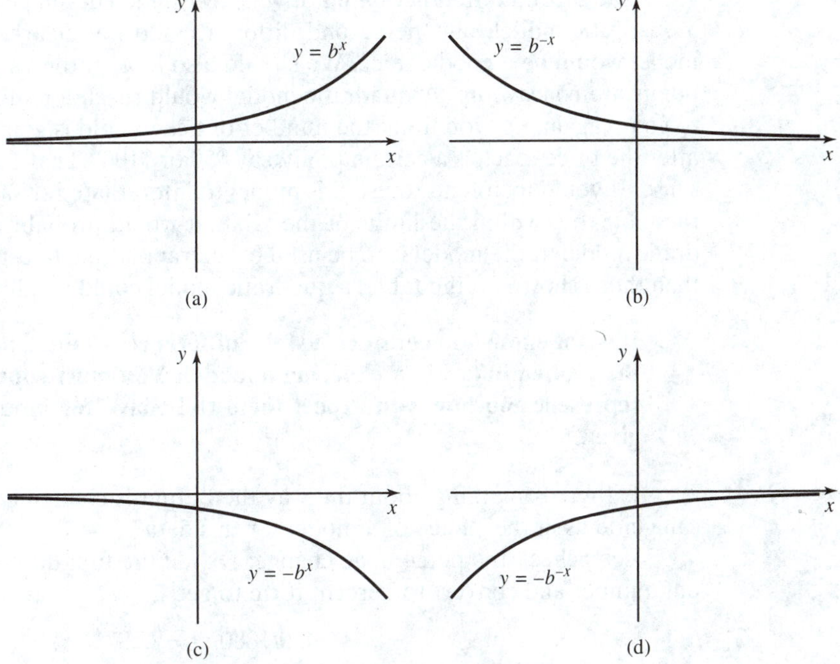

Figure 1.4-8

Solution: First we graph the data (Figure 1.4-9). This curve is not linear, it is basically concave up. Should we fit a quadratic or an exponential model? Since the inputs are evenly spaced, we can look at the second differences.

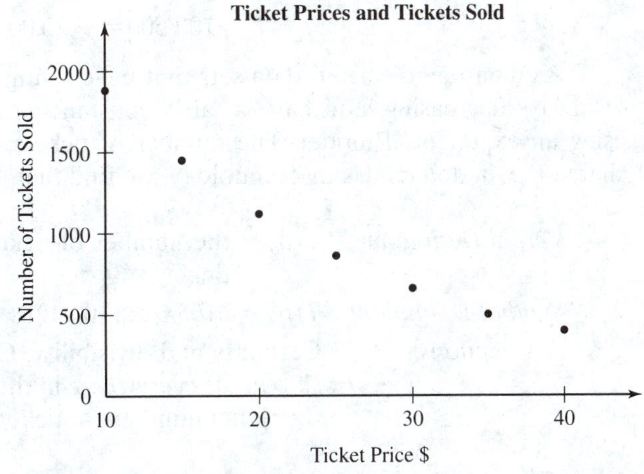

Figure 1.4-9

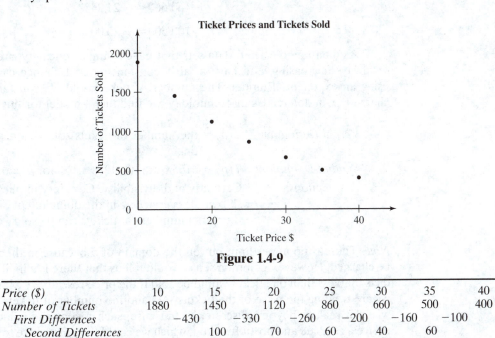

Price ($)	10		15		20		25		30		35		40
Number of Tickets	1880		1450		1120		860		660		500		400
First Differences		−430		−330		−260		−200		−160		−100	
Second Differences			100		70		60		40		60		

The second differences are not very constant. The shape appears to be all right for a quadratic model, but the second differences do not clearly indicate that a quadratic model would be a good choice. We should also look at the *sense* of the data. These data points are concave up. A quadratic model would reach a minimum value and then start to go up again. Do you think the number of tickets sold is going to start increasing again after the price reaches a certain point, say \$50 or \$100? That doesn't seem to make much sense. If you only intend to use this model to interpolate for sales at prices not shown on the table, but within the limits of the table, it would probably be all right to use a quadratic model. If the model is to be used for extrapolation, to estimate sale at prices higher than those shown on the table, a quadratic model could be all wrong.

It is not enough to consider just the differences of the tabular data and the shape of the plotted data when choosing a model. You must consider what the data points represent and how you expect them to behave for input values other than those given.

Is there something about the way the output is changing that is more constant and can guide us in the choice of a model for this data?

Let's check the percentage change: Divide the first difference by the previous output number and convert to percent form to get:

$$-430/1880 \approx -22.87\%$$

$$-330/1450 \approx -22.76\%$$

$$-260/1120 \approx -23.21\%$$

$$-200/860 \approx -23.26\%$$

$$-160/660 \approx -24.24\%$$

$$-100/500 = -20.00\%$$

As mentioned earlier, data sets that have a single concavity, are always increasing or always decreasing, and have a fairly constant percent change can be modeled well using an exponential model. The number of tickets sold, T, is a function of the price charged, p, in dollars. Using technology, we find that a model for this data is

Verbal Definition: $T(p)$ = the number of tickets sold at a price of p dollars per ticket.

Symbol Definition: $T(p) = 3165.6\ldots(0.94918\ldots)^p$, for $p \geq 0$.

Assumptions: Certainty and divisibility. Certainty implies that this relationship holds *everywhere* in the domain. Divisibility assumes that fractional numbers of tickets can be sold.

Note: There is no upper limit put on the domain of p because in theory any price could be charged. However, common sense would tell us that there is a limit, as our model predicts that less than one ticket would be sold if the price rose above \$150. For convenience we are treating the price of the tickets sold and the number of tickets sold as continuous when in reality they are discrete. In this case, such a continuous model could help us derive a revenue and profit function, which we could then maximize to find the best ticket price. □

Power Models

Sample Problem 5. The following appeared in *SAIL* magazine in April 1995:

> For many years conventional wisdom held that the best arrangement for "house" batteries on a boat is to have two separate battery banks that alternated in daily use . . . However, evolving technology is changing the way in which we manage and monitor battery banks, eliminating the advantages of having two house-battery banks. This arrangement no longer represents the best use of batteries.

The article included the following table which shows the number of cycles to failure (that is, the number of recharges before the battery dies) versus the depth of discharge at each cycle (how far down the battery is run before each recharge).

% Discharge	2	3	4	5	10	20	40
Number of Cycles	12000	8000	6000	4400	2000	800	380

Can you find a good model for this relationship?

Solution: The graph of the data points is shown in Figure 1.4-10. This curve is most certainly not linear; it looks as if it could be quadratic or exponential. Since the first four inputs are evenly spaced, we can look at the second differences.

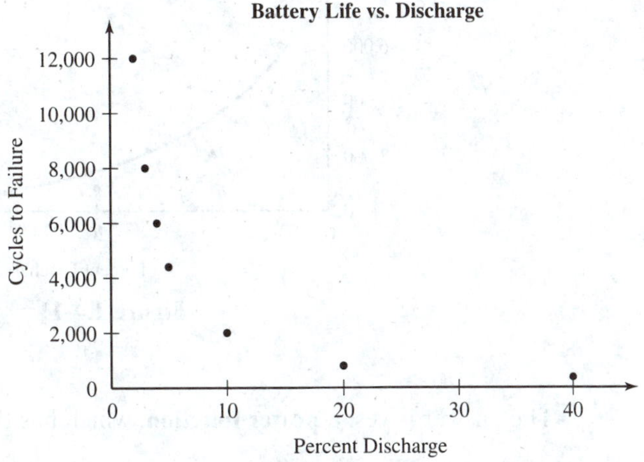

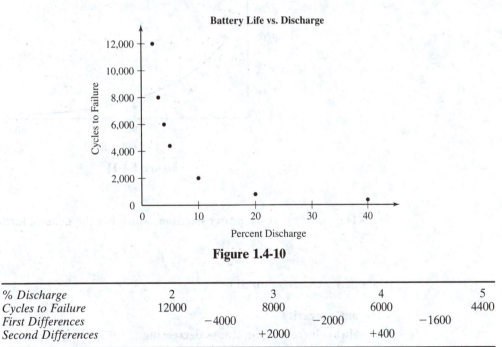

Figure 1.4-10

% Discharge	2		3		4		5
Cycles to Failure	12000		8000		6000		4400
First Differences		−4000		−2000		−1600	
Second Differences			+2000		+400		

(*Note*: We did not include the last three data points given in the table because the changes in the input—the percent discharge—were not constant between these points. **If the changes in the inputs are not constant, we cannot use the differences or percent change guidelines directly to help us choose a model.**)

Unfortunately, there are not a lot of data points to work with, but it certainly does not seem that the second differences are constant. Therefore, we need to look for a different type of model. Could these data points be fit with an exponential model?

Let's look at the percent change expressed as a pure decimal:

$$-4000/12000 = -0.333$$
$$-2000/8000 = -0.25$$
$$-1600/6000 = -0.267$$

The percent change is not very constant either, However, the data points are characterized by one concavity, up, and we would certainly not expect the number of cycles to failure to increase again after there is a 40% discharge of the batteries. For these reasons, we will try an exponential model (Figure 1.4-11). This is not a particularly good fit. Is there another type of model that has the same general characteristics as the exponential model?

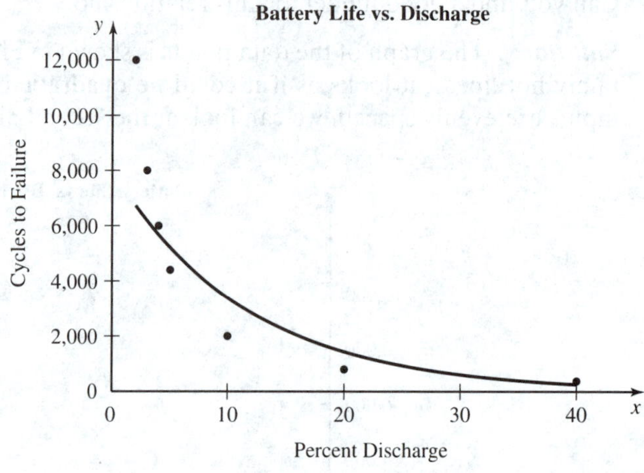

Figure 1.4-11

The answer is yes: a **power function**, which has the general form

$$y = ax^b$$

Data that are characterized by

one concavity
always increasing or always decreasing
domain $x \geq 0$
ratio of the percent change in y **to the percent change in** x **is exactly constant**
can be modeled exactly using a **power model**.

Note the difference between the exponential model and the power model: In the **exponential** model **a constant "b" is raised to a variable "x" power**. In the **power** model **the variable "x" is raised to a constant "b" power.**[13]

Let's see how closely our data come to fitting the ratio condition:

% Discharge	2	3	4	5	10	20	40
Number of Cycles	12000	8000	6000	4400	2000	800	380
% Change in y (decimal)	−0.333	−0.25	−0.267	−0.545	−0.6	−0.525	
% Change in x (decimal)	0.5	0.333	0.25	1	1	1	
Ratio (% chg. y/% chg. x)	−0.667	−0.75	−1.07	−0.545	−0.6	−0.525	

We can see that the ratios are fairly stable, although they still vary quite a bit. Let's try looking at a graph of a fit power model using technology. Figure 1.4-12 shows the result. This is an excellent fit! So a power function is a good choice to model the number of cycles to failure as a function of the percentage discharge. A model for these data points is:

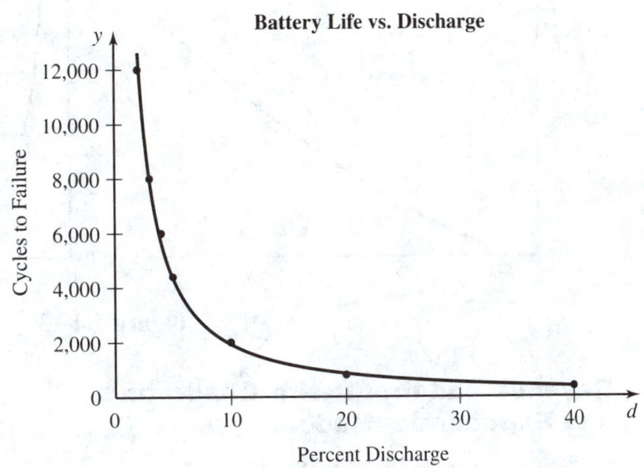

Figure 1.4-12

Verbal Definition: $C(d)$ = the cycles to failure (battery life) corresponding to a d percent discharge during each cycle

Symbol Definition: $C(d) = 28906.2\ldots d^{-1.176\ldots}$, for $2 \leq d \leq 40$.

Assumptions: Certainty and divisibility. □

[13]This is related to our linear function, where the variable x is raised to the first power: $y = ax + b$, and the quadratic function, where the variable x is raised to the second power as well as the first: $y = ax^2 + bx + c$. Linear and quadratic functions are part of a group of functions called **polynomial functions**. In polynomial functions, each of the *terms* has the form ax^b, but all of the exponents must be positive integers. In power functions the exponents can be any real number.

Other shapes that a power function can take are shown in Figure 1.4-13. Note that changing the sign of the coefficient of the x value from positive to negative would flip the curve around the x-axis (upside down).

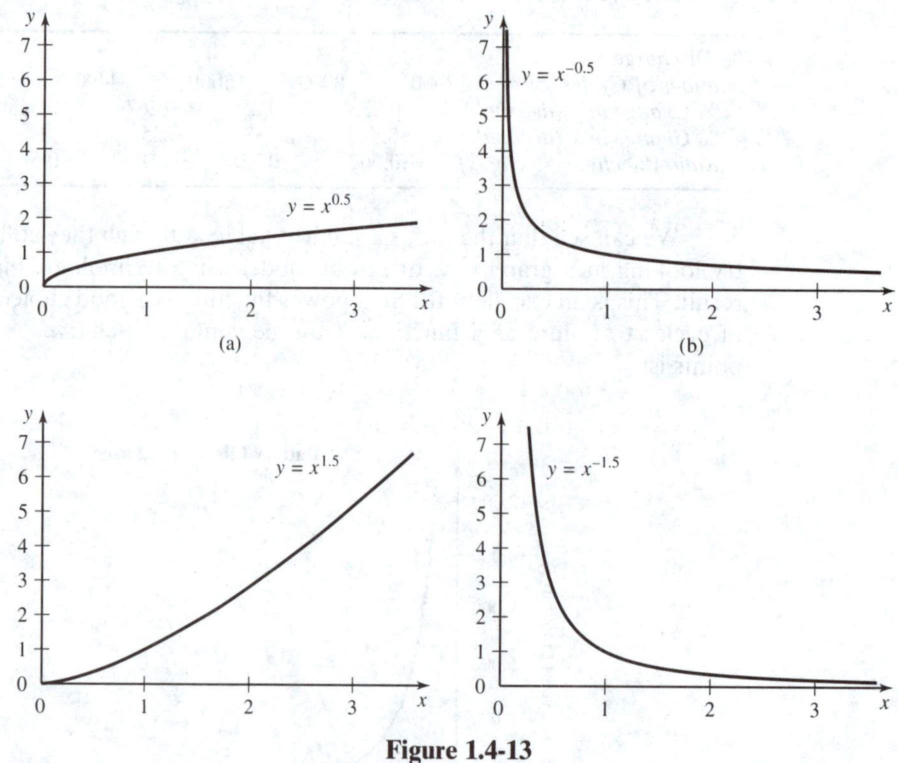

Figure 1.4-13

Revenue and Profit with Quadratic and Exponential Models

In Sections 1.2 and 1.3 we discussed the concept of revenue. Revenue from items sold at a single price is equal to the selling price charged per item times the number (quantity) of items sold, or price times quantity, for short,

Revenue = selling price times quantity sold $\qquad$ $\boldsymbol{R = pq}$

In Sample Problem 3 of Section 1.3 we sold CDs for \$10 each. We let x equal the number of CDs sold, and the revenue then equaled $10x$. In this case we made the assumption that the selling price was fixed.

Sample Problem 6: Let's return to Sample Problem 8 of Section 1.3. A group you belong to is raising money by selling sweatshirts, which cost \$9 each, with a fixed order cost of \$25. You did a survey in your dorm to estimate the demand. In Section 1.3, we used these data to derive the following two demand functions:

Verbal Definition: $D(p)$ = the number of sweatshirts you expect to sell (demand) if you charge \$$p$ each for the selling price.

Symbol Definition: $D(p) = -2.34p + 61.7$, for $9 \le p \le 26.4$.
Assumptions: Certainty and divisibility. Certainty implies no sales or discounts. Divisibility implies fractions of sweatshirts can be sold.
Verbal Definition: $d(q)$ = the price, in dollars, you'd have to charge to sell exactly q sweatshirts.
Symbol Definition: $d(q) = -0.4198\ldots q + 26.21\ldots$, for $0 \le q \le 45$.
Assumptions: Certainty and divisibility. Certainty implies no sales or discounts. Divisibility implies fractions of sweatshirts can be sold.

(a) Formulate and fully define a model for the cost of the sweatshirts in terms of the number of sweatshirts ordered.
(b) Formulate and fully define a model for the revenue from the sweatshirts in terms of the number of sweatshirts ordered.
(c) Formulate and fully define a model for the profit from the sweatshirts in terms of the number of sweatshirts ordered.
(d) Formulate and fully define a model for the cost of the sweatshirts in terms of the price you charge for the sweatshirts.
(e) Formulate and fully define a model for the revenue from the sweatshirts in terms of the price you charge for the sweatshirts.
(f) Formulate and fully define a model for the profit from the sweatshirts in terms of the price you charge for the sweatshirts.

Solution: **(a)** The cost function is similar to several we have seen, such as Sample Problem 1 in Section 1.2. Since the fixed cost is \$25 and the variable cost is \$9, the total cost will be given by the following model:

Verbal Definition: $c(q)$ = the cost, in dollars, if q sweatshirts are ordered.
Symbol Definition: $c(q) = 9q + 25$, for $q \ge 0$.
Assumptions: Certainty and divisibility. Certainty implies no sales or discounts. Divisibility implies fractions of sweatshirts can be ordered.

(b) Since **Revenue is selling price times quantity**, and the second demand function gives the selling price in terms of the quantity ($p = d(q)$), the revenue is given by

$$\boldsymbol{R = pq}$$

$$\begin{aligned} r(q) &= [d(q)]q \\ &= [-0.4198\ldots q + 26.21\ldots]q \\ &= -0.4198\ldots q^2 + 26.21\ldots q \end{aligned}$$

Thus our revenue model is given by

Verbal Definition: $r(q)$ = the revenue, in dollars, if you charge the price at which you should sell exactly q sweatshirts (as determined by the demand function).

Symbol Definition: $r(q) = -0.4198\ldots q^2 + 26.21\ldots q$, for $0 \le q \le 45$.
Assumptions: Certainty and divisibility. Certainty implies that the demand function is exactly right (if you charge $p = d(q)$, you will sell exactly q sweatshirts). Divisibility implies that you can sell fractions of sweatshirts.

(c) Since **profit is revenue minus cost**, we simply take our revenue function minus our cost function:

$$\boldsymbol{P = R - C}$$

$$\begin{aligned}
\pi(q) &= r(q) - c(q) \\
&= [-0.4198\ldots q^2 + 26.21\ldots q] - [9q + 25] \\
&= -0.4198\ldots q^2 + 26.21\ldots q - 9q - 25 \\
&= -0.4198\ldots q^2 + 17.21\ldots q - 25
\end{aligned}$$

Recall that this kind of function is the **difference of two functions**, as discussed in Section 1.1, and its domain is the *intersection* (overlap) of the domains of the functions being operated upon. The intersection of $q \ge 0$ and $0 \le q \le 45$ is $0 \le q \le 45$, since the second is completely contained within the first. Thus our profit model is given by

Verbal Definition: $\pi(q)$ = the profit, in dollars, if you charge the price at which you sell exactly q sweatshirts (as determined by the demand function $p = d(q)$), and order exactly q sweatshirts
Symbol Definition: $\pi(q) = -0.4198\ldots q^2 + 17.21\ldots q - 25$, for $0 \le q \le 45$
Assumptions: Certainty and divisibility. Certainty implies that the demand function is *exactly* right (if you charge $p = d(q)$, you will sell *exactly* q sweatshirts). Divisibility implies that you can order and sell fractions of sweatshirts.

(d) This is similar to before, but now we want the answers in terms of p instead of in terms of q. Thus wherever we had a q before, we can now substitute the function that expresses q in terms of p: $q = D(p) = -2.34p + 61.7$:

$$\begin{aligned}
c(q) &= 9q + 25 \\
C(p) = c(D(p)) &= 9[D(p)] + 25 \\
&= 9[-2.34p + 61.7] + 25 \\
&= -21.06p + 555.3 + 25 \\
&= -21.06p + 580.3
\end{aligned}$$

This is a prime example of a **composite function**, since $c(D(p))$ is a function of a function, as discussed in Section 1.1. Notice that we used a lowercase c for the cost function in terms of q (as we used a lowercase d for the demand function), and used uppercase letters for the same functions when they were in terms of p. This is necessary because they are different functions mathematically, and different models, so should have different names within the same problem.

To find the domain of a composite function, you need to find input values in the domain of the inner function ($D(p)$ here) whose corresponding output values are in the domain of the outer function ($c(q)$ here). The domain for $D(p)$ was $9 \le p \le 26.4$, as determined in Sample Problem 8 of Section 1.3. Part of the determination of this domain (see Figure 1.3-21) included ensuring that $q = D(p)$ stays nonnegative. In other words, *all* of the domain values for $D(p)$ have corresponding output values ($q = D(p)$) that are in the domain of $c(q)$. Thus, the domain of $C(p)$ is also $9 \le p \le 26.4$. As a result, the cost model is given by

Verbal Definition: $C(p) =$ the cost, in dollars, if sweatshirts are sold for p dollars, and $D(p)$ are ordered (and sold).

Symbol Definition: $C(p) = -21.06p + 580.3$, for $9 \le p \le 26.4$.

Assumptions: Certainty and divisibility. Certainty implies no sales or discounts (buying or selling), and that the demand function $D(p)$ gives exactly how many sweatshirts we will order (and sell) at a selling price of p dollars. Divisibility implies fractions of sweatshirts can be ordered.

(e) Our formulation of the revenue function is very similar to before, except that this time we substitute $q = D(p)$ in for q in the basic revenue formula:

$$
\begin{aligned}
R &= pq \\
R(p) &= p[D(p)] \\
&= p[-2.34p + 61.7] \\
&= -2.34p^2 + 61.7p
\end{aligned}
$$

The revenue formula only requires that p and q both be nonnegative, and so we can use the same domain as that of $D(p)$: $9 \le p \le 26.4$. Thus our revenue model is given by

Verbal Definition: $R(p) =$ the revenue, in dollars, if you charge p dollars for each sweatshirt, and order and sell *exactly* $q = D(p)$ sweatshirts (the demand function).

Symbol Definition: $R(p) = -2.34p^2 + 61.7p$, for $9 \le p \le 26.4$

Assumptions: Certainty and divisibility. Certainty implies that the demand function is *exactly* right (if you charge p, you will sell *exactly* $q = D(p)$ sweatshirts). Divisibility implies that you can sell fractions of sweatshirts.

(f) Again, the profit function is simply the revenue function minus the cost function, or

$$P = R - C$$

$$\begin{aligned}\Pi(p) &= R(p) - C(p)\\ &= [-2.34p^2 + 61.7p] - [-21.06p + 580.3]\\ &= -2.34p^2 + 61.7p + 21.06p - 580.3\\ &= -2.34p^2 + 82.76p - 580.3\end{aligned}$$

Since both the revenue and the cost functions had the same domains, that is also the domain of the profit function, so the profit model is given by

Verbal Definition: $\Pi(p)$ = the profit, in dollars, if you charge p dollars per sweatshirt and order and sell *exactly* q sweatshirts (as determined by the demand function $q = D(p)$).

Symbol Definition: $\Pi(q) = -2.34p^2 + 82.76p - 580.3$, for $9 \le p \le 26.4$

Assumptions: Certainty and divisibility. Certainty implies that the demand function is *exactly* right (if you charge p, you will sell *exactly* $q = D(p)$ sweatshirts). Divisibility implies that you can order and sell fractions of sweatshirts. □

If you look at Figure 1.3-20, you can see that the data seem to have a little curve to them. It would be very appropriate to try fitting a quadratic demand function to capture this curvature. However, you would also have to realize that the graph of this model would go down at first but then go back up again. This would mean that, at first, the more you charge, the fewer sweatshirts people would buy (as you would expect). But then, beyond some price value, if you started to charge more, people would actually buy *more*! There may be situations where this makes sense (a high price can make people *think* something is better, even if it isn't, or you might have a special marketing strategy to appeal to status and ego), but normally it does not seem likely. This means you need to be sure to restrict your domain to capture only the part of the parabola that has the quantity sold decrease as the price increases. Try this yourself!

Note that you could also try fitting an exponential or power function to the sweatshirt demand data. The procedure for finding the cost, revenue, and profit would be exactly analogous to above.

Before you start the exercises for this section, be sure that you

- Recognize plots that are concave up (shaped like the letter U) and concave down (look like a frown).
- Know that data whose plots have exactly one concavity can be modeled using quadratic, exponential, or power functions.
- Know that quadratic functions are characterized by equal second differences for equally spaced inputs.
- Know that exponential functions are characterized by equal percentage change (ratio) for equally spaced inputs, are always increasing or decreasing, and have a lower limit or upper limit.

- Know that power functions are characterized by equal ratios of (% change in y) to (% change in x) for equally-spaced inputs.
- Choose models with the context of the problem, your purpose (optimization versus forecasting, interpolation versus extrapolation, etc.), and expected behavior (does the shape of a function correspond to what you expect in the real situation?) in mind.
- Know how to fit quadratic, exponential, and power models, and can solve equations involving them, using a graphing calculator or a computer.

EXERCISES FOR SECTION 1.4

Warm Up

1. For each of the graphs shown in Figure 1.4-14, indicate what type of function or functions you would consider for a possible model. Explain your reasons for choosing each type of function.

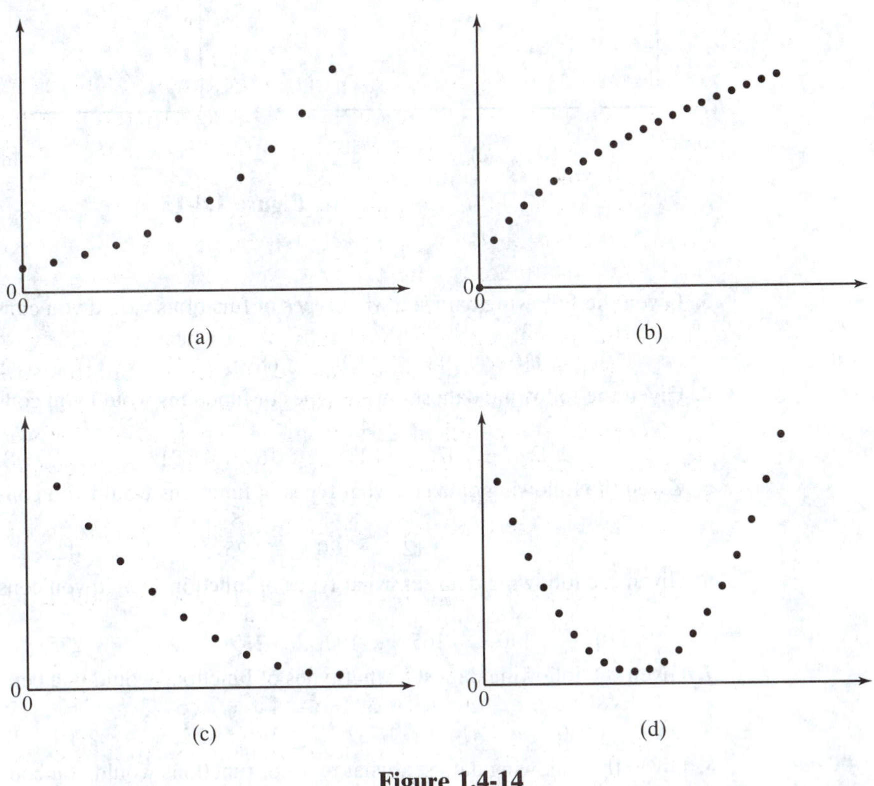

Figure 1.4-14

2. For each of the graphs shown in Figure 1.4-15, indicate what type of function or functions you would consider for a possible model. Explain your reasons for choosing each type of function.

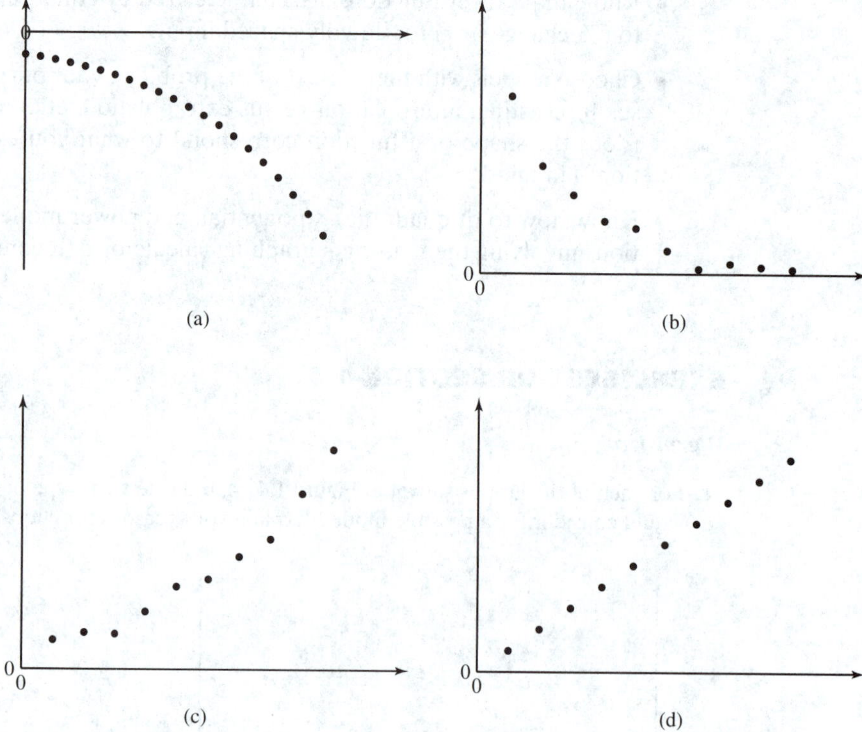

Figure 1.4-15

3. Given the following data set, what types of functions would you consider for a model?

1	2	3	4	5	6	7	8	9	10
119	110	98	80	68	45	37	25	12	3

4. Given the following data set, what types of functions would you consider for a model?

1	2	3	4	5	6	7	8	9	10
1	2.02	3.07	4.11	5.16	6.21	7.28	8.34	9.41	10.47

5. Given the following data set, what types of functions would you consider for a model?

1	2	3	4	5	6	7	8	9	10
3	8	12	20	25	33	42	56	65	78

6. Given the following data set, what types of functions would you consider for a model?

1	2	3	4	5	6	7	8	9	10
120	140	165	190	215	245	275	310	345	380

7. Given the following data set, what types of functions would you consider for a model?

1	2	3	4	5	6	7	8
1.000	1.414	1.732	2	2.236	2.450	2.646	2.828

8. Given the following data set, what types of functions would you consider for a model?

1	2	3	4	5	6	7	8
1.00	0.574	0.415	0.323	0.276	0.238	0.211	0.189

9. Given the following data set, what types of functions would you consider for a model?

1	2	3	4	5	6	7	8
1.00	−2.83	−5.20	−8.00	−11.2	−14.7	−18.5	−22.6

10. Given the following data set, what types of functions would you consider for a model?

1	2	3	4	5	6	7	8
3.00	1.85	1.39	1.14	0.972	0.856	0.768	0.700

Game Time

11. Shown in Figure 1.4-16 is a graph of the revenue realized from the sale of T-shirts:

(a) What type of model would you fit to these data points?

(b) Explain why you chose this type of model.

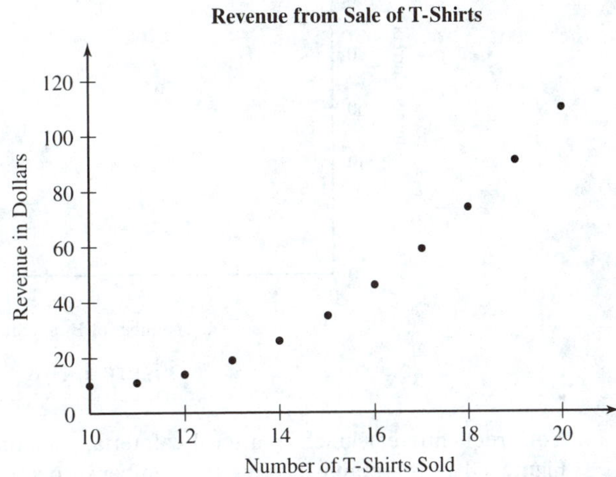

Figure 1.4-16

12. Shown in Figure 1.4-17 is a graph of the prices charged for souvenir programs and the number of programs sold:

(a) What type of model would you fit to these data points?

(b) Explain why you chose this type of model.

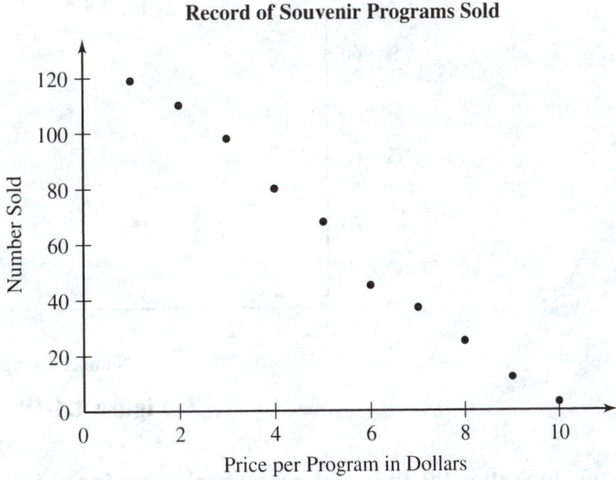

Figure 1.4-17

13. Shown in Figure 1.4-18 is a graph of the number of hours spent studying for a test in the 24-hour period prior to the test and the number of correct answers (out of 250) on a test.
(a) What type of model would you fit to these data points?
(b) Explain why you chose this type of model.

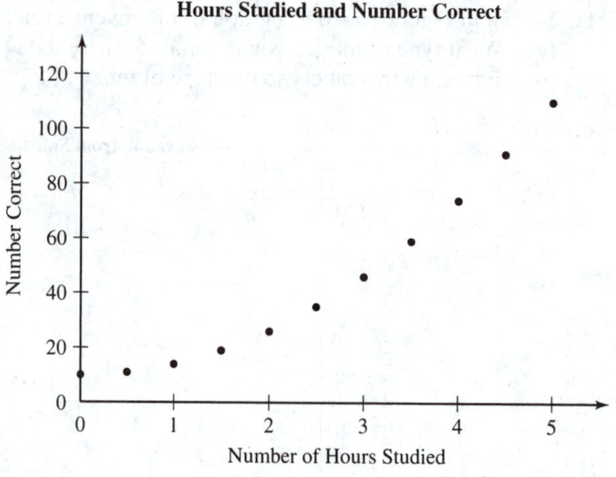

Figure 1.4-18

14. You frequently eat lunch in a local cafeteria, sometimes alone and sometimes with friends. Figure 1.4-19 is a graph showing the number of people eating and the total lunch bill for the group in the cafeteria.
(a) What type of model would you fit to these data points?
(b) Explain why you chose this type of model.

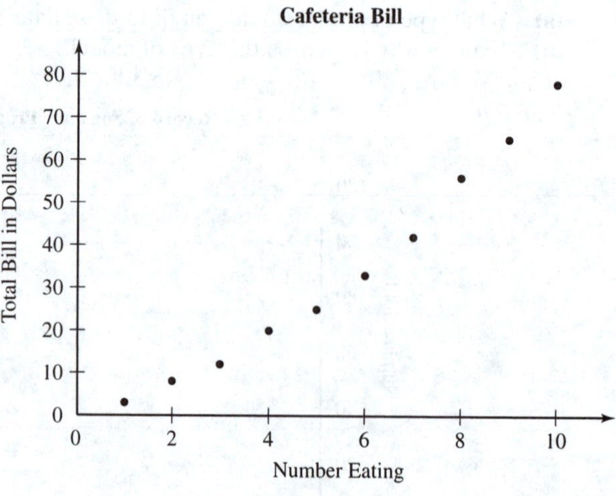

Figure 1.4-19

15. In Section 1.0, three different models were fit to Meg's exercise problem. The first model used all of the data. Verify the model given by fitting a quadratic model yourself.

16. In Section 1.0, three different models were fit to Meg's exercise problem. In the second model, one data point, (440,75), was dropped. Verify the model given by fitting a quadratic model yourself.

17. In Section 1.0, three different models were fit to Meg's exercise problem. In the third model, only three data points were used. Verify the model given by fitting a quadratic model yourself.

18. In Exercise 21 of Section 1.3 you fit a linear model to the following data gathered in a poll about the sale of T-shirts:

Price	\$8	\$10	\$12	\$15
Number Sold	302	231	175	123

(a) From what you have learned in this section, do you think that this is the best model for the data?
(b) Find the best model you can for the data.
(c) Explain why you chose this model.
(d) How many T shirts would you expect to sell if you charged \$18?
(e) How does this answer differ from that predicted by your linear model?

19. The following table shows the total number deaths each year from AIDS in the United States, 1988–1994:

Yr. Ending Dec. 31	1988	1989	1990	1991	1992	1993	1994
Total Deaths That Year	21,019	27,653	31,339	36,246	40,072	42,572	44,052

Source: Health United States 1994, (National Center for Health Statistics).

(a) Find a model for the total deaths due to AIDS.
(b) What does your model predict for the year ending December 31, 1996?
(c) If you were an AIDS activist helping to raise money for research and AIDS hostels, how could you use these data values?

20. The following table shows the estimated total population of the world:

Year	1650	1750	1850	1900	1950	1980	1995
Population Estimate (millions)	550	725	1,175	1,600	2,556	4,458	5,734

Source: Bureau of the Census, U.S. Dept of Commerce; prior to 1950, Rand McNally & Co.

(a) Find a model for total world population. Explain why you chose this model.
(b) What total world population does your model predict for 2000?
(c) What world population does your model predict for 2020?
(d) The land area of the world is estimated to be 57,900,000 square miles. From your model, what would be the world population density in 2020?

21. The following table gives the number of bank failures (closed or assisted) from 1978 to 1987:

Year	1978	1979	1980	1981	1982	1983	1984	1985	1986	1987
Number	7	10	11	10	42	48	80	120	145	203

Source: *The World Almanac and Book of Facts 1996*, p. 119, (Federal Deposit Insurance Corporation).

(a) Find a model for the number of bank failures (assisted or closed) from 1978 to 1987.
(b) Why did you choose this type of model?

(c) How many bank failures does your model predict for 1988?
(d) How many bank failures does your model predict for 1989?

22. The following table gives the number of bank failures (closed or assisted) from 1988 to 1994:

Year	1988	1989	1990	1991	1992	1993	1994
Number	221	207	169	127	122	41	13

Source: *The World Almanac and Book of Facts 1996*, p. 119, (Federal Deposit Insurance Corporation).

(a) Find a model for the number of bank failures (assisted or closed) from 1988 to 1994.
(b) Why did you choose this type of model?
(c) How many bank failures does your model predict for 1987?
(d) How many bank failures does your model predict for 1986?
(e) How did the model predictions from this problem and the previous problem compare with the actual data?
(f) Can you find any explanation for this?

23. For the sweatshirt demand data given in Sample Problem 8 of Section 1.3, find a quadratic model for the quantity as a function of the price. What is an appropriate domain? Then find models for the revenue, cost and profit as functions of the price.

24. For the sweatshirt demand data given in Sample Problem 8 of Section 1.3, find exponential and power functions for the quantity in terms of the price. Then derive a model for the profit in each case.

25. You own a landscaping business. You measure your productitvity by the number of lawns that you maintain for customers. You have been faced with the choice of whether to spend money to hire more workers or purchase new equipment. You currently have $15,000 worth of equipment; this has been relatively constant in recent years, and you don't expect to change it significantly. You have kept records concerning the number of lawns, on average, that you were able to maintain as you hired more full-time workers.

Number of Full-Time Workers	1	2	3	4	5
Avg. Number Lawns Maintained	15	22	27	32	36

(a) Find a model for the productivity as a function of the number of full-time workers.
(b) On average, how many lawns could you expect to maintain if you had eight full-time employees but didn't invest in any more equipment?

26. You own a landscaping business. You measure your productitvity by the number of lawns that you maintain for customers. You have been faced with the choice of whether to spend money to hire more workers or purchase new equipment. You currently have $15,000 worth of equipment. You have been renting some extra equipment and have kept records concerning the number of lawns, on average, that you were able to maintain with three full-time workers and varying amounts of equipment. Find a model for the productivity as a function of the amount of money invested in equipment.

Equipment ($1000)	15	16	17	18	19
Avg. Number Lawns Maintained	34.5	35.5	36.0	36.6	37.1

(a) Find a model for the productivity as a function of the dollar amount of equipment.
(b) On average, how many lawns could you maintain with three full-time workers and $20,000 worth of equipment?

27. You regularly compete with another local company for business, particularly the revenues realized from new products. You have been keeping track of how long it takes for the competitor to come out with a rival product after you come out with a new product. You have recorded this as the percentage of rival products that come out within a given period of time after yours.

Number of Months	0	1	2	3	4	5
% New Products	20.0	16.4	13.4	11.0	8.99	7.36

(a) Find a model for the percentage of new competing products that will come out as a function of the time.

(b) According to your model, what percentage of the new competing products will come out 7 months after you introduce your new product?

28. You spend a lot of time waiting for buses. To help pass the time, you have kept a record of how long you spend waiting for the bus. You recorded the percent of the time that you waited for different numbers of minutes.

Number of Minutes	0	1	2	3	4	5	6
% of Times	10	9.0	8.2	7.4	6.7	6.1	5.5

(a) Find a model for the percentage of times that you waited as function of the number of minutes waited.

(b) According to your model, what percent of the time will you have to wait 8 minutes?

1.5 FUNCTIONS WITH CHANGING CONCAVITY: CUBIC, QUARTIC, LOGISTIC

In the preceding section we learned about data whose plots are either concave up or concave down (not both) and data that can be modeled by quadratic, exponential, or power functions. These types of functions allowed us to fit good models to a variety of data, from demand functions to learning curves on computer games to energy levels. Do you think that all data will behave in one of these ways: no concavity or just concave up or just concave down? Any of you who play sports are familiar with "the slump." You keep improving as you practice; then you go into a slump. Your batting average falls or your golf score rises. Generally, this is a temporary condition: Hang in there, keep practicing, and you will come out of the slump and start to improve again. If you drew a graph of your performance over time, you would see that the data had been increasing and concave down, then decreasing and still concave down (Figure 1.5-1). Then it began to change and became concave up.

We will now look at functions that can be used to model data points that suggest changing concavity.

Here are the kinds of problems that the material in this section will help you solve:

- You have decided to grow plants commercially and just put in a large number of seedlings. Suddenly a cold front comes through. You are watching the temperature very carefully and must decide if you need to cover the plants. You have a special material for this, but it is quite expensive and you can't reuse it. It will take you at least an hour to cover your plants. You don't have much working cap-

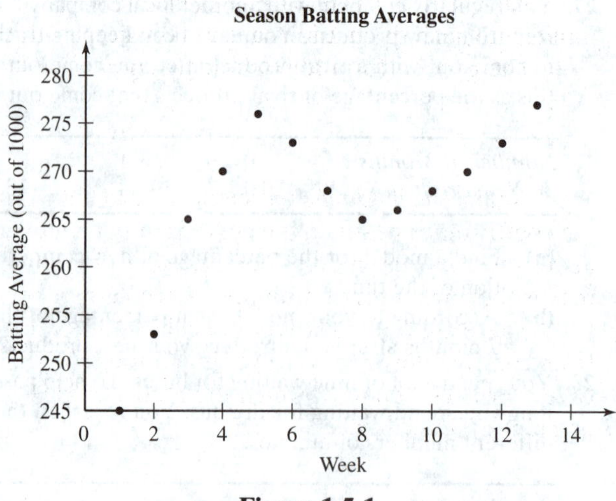

Figure 1.5-1

ital and can't afford to waste the material, but losing all of your plants would be a financial disaster. Is there anything you can do to help with your decision?

- Recently a concert was held to raise money for AIDS research. This concert was broadcast across the United States and raised a great deal of money. You are in charge of the sales of the CD recorded at the concert. You initially ordered a large number of the CDs but saw immediately that sales were greatly exceeding all expectations. Thrilled with the results, you ordered more. Sales continued very briskly for a while, but then began tapering off, as you expected they would. The initial enthusiasm is dieing out as the memory of the concert fades. You don't have many CDs left. You don't want to run out, but you really expect the market to dry up in a while and you don't want to be stuck with the CDs. Should you reorder or just stick with what you have?

By the end of this section you should be able to solve problems like those above and

- Know the shapes of cubic, logistic, and quartic functions.
- Recognize a cubic function by its single inflection point and behavior at the extremes.
- Recognize a logistic function by its single inflection point and finite limits at both extremes.
- Select an appropriate model based on the expected end behavior.
- Fit cubic, quartic, and logistic models using technology.
- Solve equations involving cubic, quartic, and logistic functions.

What kind of models can be used to describe data like the batting average graph in Figure 1.5-1?

Data whose plots **change concavity exactly once (are *both* concave up and concave down)** can be modeled using **cubic** models.

An exact characterization is given by equal third differences for equally spaced inputs, but such calculations are usually not worth the effort.

The general form for a cubic function is

$$y = f(x) = ax^3 + bx^2 + cx + d$$

Again, the leading coefficient "a" can tell us something about the curve: If the leading coefficient is positive, the curve is initially increasing; if the leading coefficient is negative, the curve is initially decreasing. Cubic functions are polynomial functions. This means they are sums of multiples of whole-number powers of the variable.

There are four basic types of cubic models (Figure 1.5-2). Graphs (a) and (c) change from concave down to concave up. Graphs (b) and (d) change from concave up to concave down. The point at which the model changes concavity is called an **inflection point** or **point of inflection.**

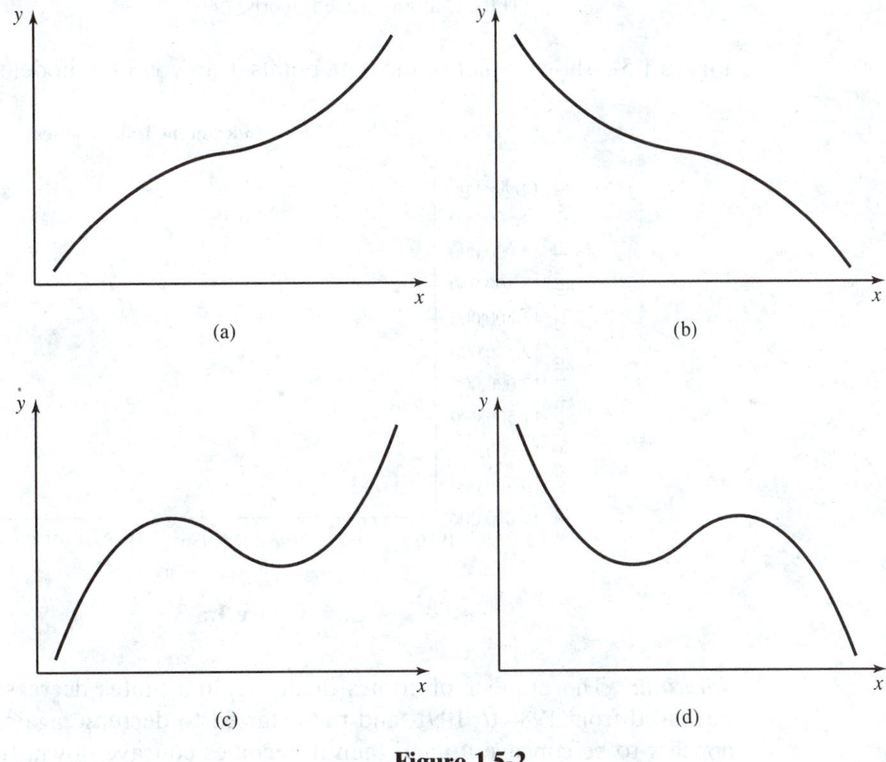

Figure 1.5-2

Look at the graphs in Figure 1.5-2 and try to identify the places where the concavity changes (the points of inflection) before reading on.

Inflection points can be very important, telling us when the *acceleration* of an increasing or decreasing function changes direction (the function changes from speeding up to slowing down, or vice versa). In Chapter 3 you will learn how to calculate the exact location of inflection points using calculus.

Sample Problem 1: You are running a reelection campaign for a friend. Part of the campaign platform for the first election was a tough stance on crime. You want to analyze the crime statistics to see if you can use them in the reelection campaign. The table below shows the reports of crime for the years 1981 through 1993. (The source does not state it, but you assume that the numbers given represent total crimes reported in the U.S. through December 31 of each year.)

Year	Crime Index	Year	Crime Index
1981	13,432,800	1988	13,923,100
1982	12,974,400	1989	14,251,400
1983	12,108,600	1990	14,475,600
1984	11,881,800	1991	14,872,900
1985	12,431,400	1992	14,438,200
1986	13,211,900	1993	14,141,000
1987	13,508,700		

Source: *The World Almanac and Book of Facts 1996*, p. 957 (FBI, Uniform Crime Reports, 1993).

Figure 1.5-3 shows a plot of the data points. Can you fit a model to them?

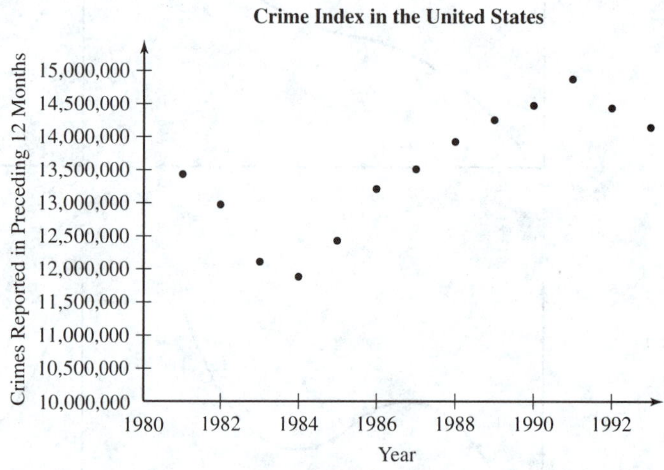

Figure 1.5-3

Solution: The number of crimes in the United States decreased from 1981 to 1984, increased from 1984 to 1991, and then started to decrease again. From 1981 to 1986 it appears to be concave up and then it becomes concave down. It is certainly not linear because it is curved (has concavity), nor can it be quadratic or exponential because it *changes concavity*. These data points certainly cannot be modeled well using a linear model, nor will quadratic, exponential, or power models be appropriate. Obviously, we need a different kind of model to fit to these data points.

If we **align** the data, (that is, treat 1981 as zero and count the years after 1981 as 1982 = 1, 1983 = 2, etc.) and fit a cubic model, we get the graph in Figure 1.5-4.

The model for these data values is

Verbal Definition: $C(x)$ = crimes committed in the preceding 12 months in the United States x years after December 31, 1981.

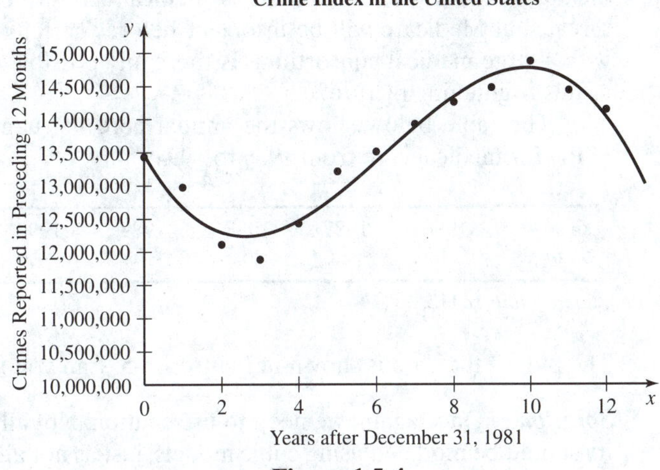

Figure 1.5-4

Symbol Definition:	$C(x) = -13{,}150x^3 + 247{,}506x^2 - 1{,}030{,}143x + 13{,}471{,}438, x \geq 0.$
Assumptions:	Certainty and divisibility. Certainty implies the relationship is exact. Divisibility implies that fractional values of x make sense.

This is not a simple model, but it could reasonably be used to interpolate and extrapolate for short periods. Since the coefficients all have at least 5 significant digits, we have simply rounded them to the nearest whole number here and later in the book when we use this example. □

The importance of aligning data in time series, particularly time series involving years, can be seen from the model we would have gotten if we had not aligned 1981 to zero:

Verbal Definition:	$C(x)$ = crimes committed in the previous 12 months in the United States in year x (1987.5 means mid-1987).
Symbol Definition:	$C(x) = -13{,}150x^3 + 78{,}399{,}756x^2 - 1.5580\text{E}11x + 1.0320\text{E}14, x \geq 0.$
Assumptions:	Certainty and divisibility. Certainty implies the relationship is exact. Divisibility implies that fractional values of x make sense.

What in the world do E11 and E14 mean? They are scientific notation, and mean "times 10 raised to the 11th power" and "times 10 raised to the 14th power," respectively. We should move the decimal point 11 places to the right for the coefficient of x to the first power and 14 places to the right for the constant. That means the constant for this model would be 103,200,000,000,000! This model is more difficult to work with than the model of the aligned data. And it doesn't make much sense to have a model from the year 0 because there was no U.S. prior to 1776!

SAMPLE PROBLEM 2: The costs of medical care are constantly in the news. There is a threat that Medicare will be insolvent before too long. You have a public relations job with a large medical consortium. Is there any possible way you can show a positive side to this frightening picture?

The table below shows the annual percent change in the Consumer Price Index (CPI) for medical care from 1986 to 1994:

Year	1986	1987	1988	1989	1990	1991	1992	1993	1994
Change	7.5	6.6	6.5	7.7	9.0	8.7	7.4	5.9	4.8

Source: *Ibid*, p. 112.

The plot of the data is shown in Figure 1.5-5. Can you fit a model?

Solution: Once again, we need to use caution. Not all scatter plots that change concavity should be modeled using cubic models, just as not all scatter plots that are always concave up should be modeled using quadratic models. We saw in some instances that such data should be modeled using exponential or power models. This was determined by the end behavior. If the data would naturally approach a limiting value at an endpoint, it should not be modeled using a quadratic model. Expected end behavior must also be considered in data points that change concavity.

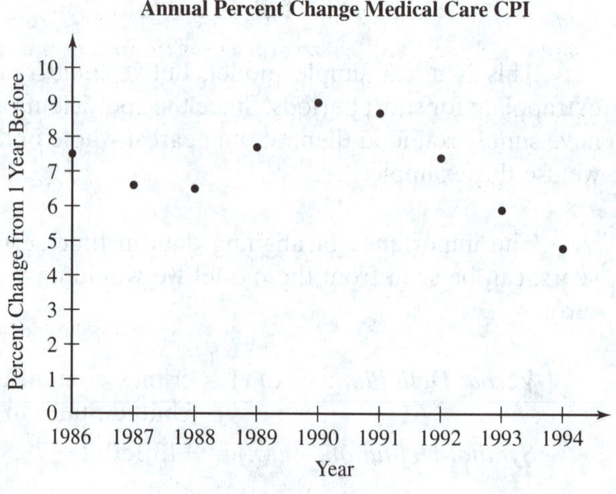

Figure 1.5-5

What kind of model should be used for these data points?
The fitted cubic model is shown in Figure 1.5-6.
The model for these data points (rounded to 3 significant digits) is

Verbal Definition: $I(x)$ = the percent change (in %) in the CPI for medical care x years after December 31, 1986, from 1 year before.
Symbol Definition: $I(x) = -0.0451\,x^3 + 0.392\,x^2 - 0.606\,x + 7.16,\ x \geq 0.$
Assumptions: Certainty and divisibility.

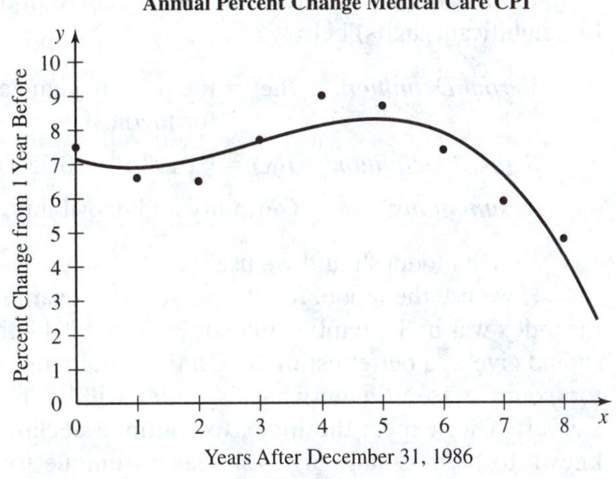

Figure 1.5-6

This is obviously not as good a fit as the last example. It looks as if there is a second point of inflection around 1992 or 1993 (look carefully at the data plot: It appears to change from concave down to concave up when x is about 6 or 7). We might conclude that somehow we could find a better model to describe our data. If we expect the index to start rising again in the near future, we might wish to fit a **quartic** model to the data points, showing that they will change concavity again and actually have two inflection points. **The general form of the quartic function is** $y = ax^4 + bx^3 + cx^2 + dx + e$.

It can be shown that for a polynomial function, **the highest power of the variable that has a nonzero coefficient** (the **degree** of the polynomial) determines the highest number of points of inflection possible. **If the degree is n, then the graph can have at most $n - 2$ points of inflection.** Thus, quadratic functions ($n = 2$) have no points of inflection, cubic functions ($n = 3$) have at most one, quartics have at most two, and so on.

The quartic fit is shown in Figure 1.5-7.

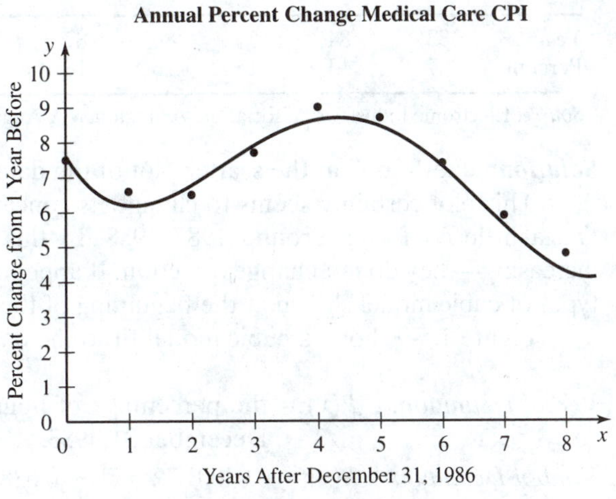

Figure 1.5-7

In this case the quartic model does seem to fit the data better. We define it (rounded to 3 significant digits) below:

Verbal Definition: $I(x)$ = the percent change (in %) in the Consumer Price Index for medical care x years after, from 1 year before.

Symbol Definition: $I(x) = 0.0193\,x^4 - 0.355\,x^3 + 1.93\,x^2 - 3.02\,x + 7.62, x \geq 0.$

Assumptions: Certainty and divisibility.

Which model should we use?

If we use the model to interpolate—for example, to determine approximately what the index was in September 1990 or in June 1994—the quartic is certainly a better fit and should give us a better estimate. On the other hand, if we use the model to extrapolate—for example, to estimate what the index will be in 1998—the question is not as easy to answer. If we expect the index to continue declining (for example if year 8, 1994, was known to be unusually high for reasons unique to that year), we should use the cubic model. If, on the other hand, we expect the index to start rising again, we should use the quartic model. We might want to include both predictions, along with the reasons that support the choice of each model. Here is an example of possible manipulation of data. If you wanted the picture for future medical costs to look good, you might use the cubic. **The higher the degree of the function**, fourth power versus third power, third power versus second power, and so forth, **the better the fit will be**. The curve will follow more changes in the data, **but the best fit is not always the best model**. Remember KISS!: **Keep It Simple, Stupid!** □

SAMPLE PROBLEM 3: In the first two examples we have seen data that were decreasing, then increasing, then decreasing again. In both of these cases we could reasonably hope that the downward trend will continue. We would not expect that the crime index would ever become negative, but it is certainly possible for the percent change in the medical care Consumer Price Index to be negative. Now consider the following data, showing the percentage of households with computers. Find a model for the percentage of households with computers.

Year	83	84	85	86	87	88	89	90	91	92	93
Percent	7	13	15	16	18	20	21	22	25	27	30

Source: Electronic Industries Association, Arlington, VA August 1995 quoted in 1996 World Almanac p. 171.

Solution: Let's look at the scatter plot of the data (Figure 1.5-8).

This plot certainly seems to change its concavity, first concave down and then up, and a little confusion around 1987–1988. In this case, however, the data are always increasing—they do not change direction. It appears to be an example of one of the other types of cubic model shown at the beginning of this section (Figure 1.5-2).

Figure 1.5-9 shows a cubic model fit to the data. The model is

Verbal Definition: $P(x)$ = the percentage of households with computers x years after December 31, 1983.

Symbol Definition: $P(x) = 0.04001\ldots x^3 - 0.6282\ldots x^2 + 4.532\ldots x + 7.776\ldots,$ $0 \leq x \leq 17.$

Assumptions: Certainty and divisibility.

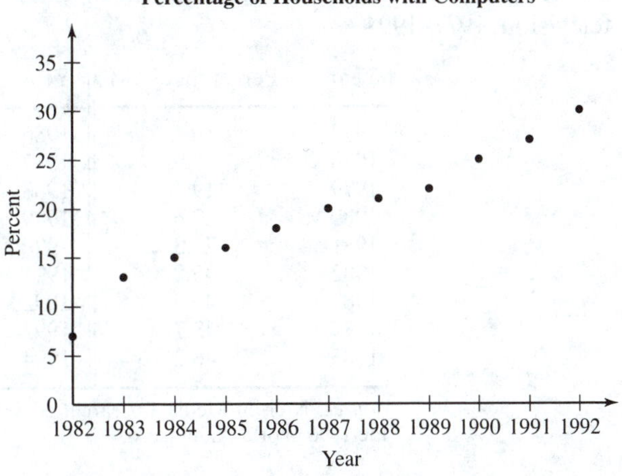

Figure 1.5-8

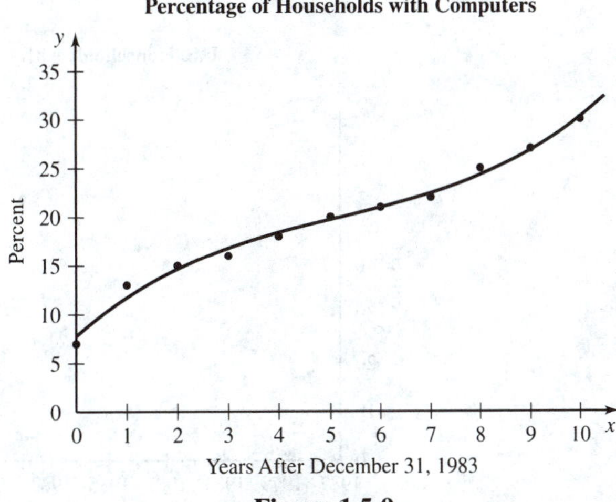

Figure 1.5-9

We used a graphing calculator to find the input year x for which the percentage was 100. That occurs when x is about 17, so we restricted the domain to values of x less than or equal to 17. That only takes us to the end of the year 2000.

The cubic model fits the data very well, but we must be cautious using it for extrapolation. The percentage of households with computers will probably continue to increase for quite a few years, but there is a natural upper limit—the percentage cannot exceed 100 percent. Expected end behavior is an important consideration when fitting a model to data. This is probably the best model to use for short-term (very short-term, such as 3 or 4 years) extrapolation but would not be appropriate for long-term extrapolation. If we decided to use this model, we should probably restrict the domain to $0 \leq x \leq 13$ or $0 \leq x \leq 14$. □

Sample Problem 4: A company that makes satellite dishes for TVs has asked you to analyze the market. Their biggest competitors are the cable companies. What does this market look like? The following table shows the number of U.S. households with cable television, 1977–1994.

Year	Percentage (%)	Year	Percentage (%)
1977	16.6	1986	48.1
1978	17.9	1987	50.5
1979	19.4	1988	53.8
1980	22.6	1989	57.1
1981	28.3	1990	59.0
1982	35.0	1991	60.6
1983	40.5	1992	61.5
1984	43.7	1993	62.5
1985	46.2	1994	63.4

Source: Nielson Media Research, New York, NY, quoted in the 1996 World Almanac.

Find a model for the number of households with cable television.

Solution: First let's look at a plot of the data (Figure 1.5-10).

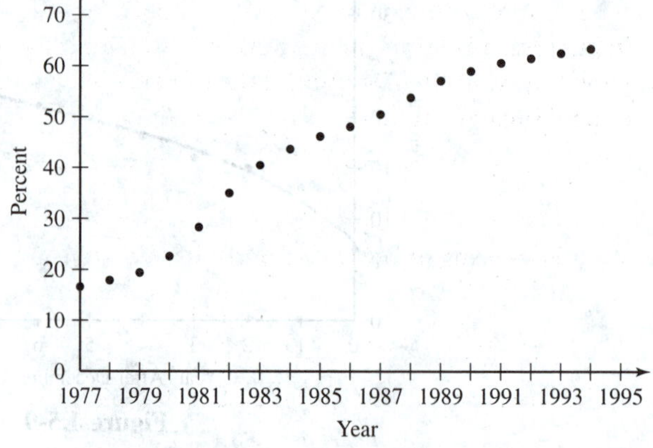

Figure 1.5-10

What kind of a model should we use for these data? Like the last example the concavity changes from up to down, and the data are always increasing. However, in this case the increase in the years since 1991 is quite small. The great explosion in cable television seems to have leveled off.

The data appear to be concave up at first, and then change to concave down. We could fit a cubic model to the data as shown in Figure 1.5-11.

The cubic model (rounded off) is

Verbal Definition: $P(x)$ = the percentage of U.S. households with cable television x years after the end of 1975.

Symbol Definition: $P(x) = -0.0083\,x^3 + 0.0948\,x^2 + 3.6412\,x + 14.194,\ x \geq 0.$

Assumptions: Certainty and divisibility.

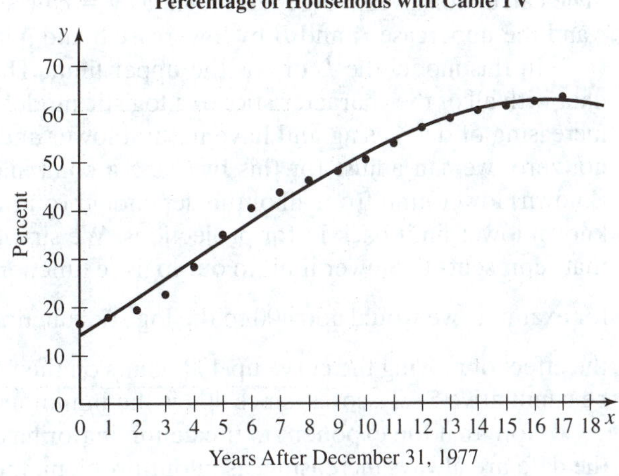

Figure 1.5-11

The cubic model is a good fit, but there is a problem. In fact, it is a very serious problem: The model predicts that the percentage of households with cable television will start **declining** in 1994 and actually be negative by 2006! Obviously, we need a different type of model to extrapolate for this data, or at least to extrapolate that far out. In fact, **in general, it is better to be cautious about your domain and restrict it to the interval from your lowest to highest input data values, unless you have a good reason to believe it can extend further**. In this situation, the actual percentage of households with cable would be expected to continue to rise, but very slowly. There is a natural upper limit.

Data that are **concave up *and* concave down, have one inflection point, are always increasing or always decreasing, and are characterized by *both* upper and lower limits** can be modeled using a **logistic function.**

The two characteristic shapes of logistic curves are shown in Figure 1.5-12. The general form of the logistic function is

$$f(x) = \frac{L}{1 + Ae^{-Bx}}$$

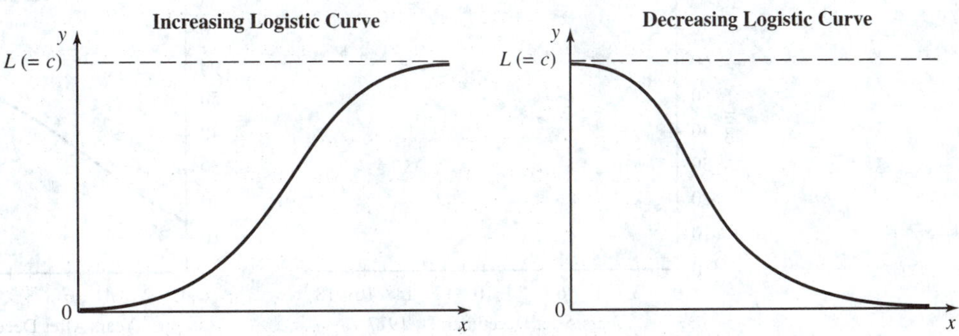

Figure 1.5-12

(The calculator gives the general form as $y = c/(1 + ae\wedge(-bx))$). The L is replaced by a c and the uppercase A and B by lowercase a and b, respectively.)

In this model the L or c is the upper limit. The lower limit is always 0. If we have data with all of the characteristics of a logistic model—both concave up and down, always increasing or decreasing and have natural lower and upper limits—but the lower limit is not zero, we can adjust for this by using a combination of functions. First subtract the (known) lower limit from all of the dependent (output) values, fit the curve, then add the known lower limit back in for projections. We simply add the function $y =$ the constant that represents the lower limit to our logistic function. If the natural lower limit were 400, for example, we would add 400 to the logistic function: $g(x) = \dfrac{L}{1 + Ae^{-Bx}} + 400$. This has the effect of moving the curve up 400 points on the y-axis. (A good example for this could be cumulative SAT scores, since 400 is the minimum score.)

Note that the exponent in the denominator has a *negative* sign before the Bx. When the data are always increasing, as in our problem, the B value that is shown for the logistic model will be positive. The negative sign before the coefficient means that the second term in the denominator is getting smaller and smaller as the variable gets larger. As the second term in the denominator gets smaller, the denominator gets closer to just 1, that is, it decreases to 1. As the denominator gets smaller, the whole expression gets larger, approaching the upper limit of the numerator in Figure 1.5-12(a). If the value for B is negative the sign of the exponent becomes positive and the second term in the denominator will increase as the value of x increases. This means that the function will always be decreasing (Figure 1.5-12(b)).

A logistic model for the percentage of U.S. households with cable TV is

Verbal Definition: $P(x) =$ the percent (in %) of U.S. households with cable TV x years after December 31, 1977.

Symbol Definition: $P(x) = \dfrac{65.65\ldots}{1 + 3.527\ldots e^{-0.2621\ldots x}}, x \geq 0.$

Assumptions: Certainty and divisibility.

The data points and the curve are shown in Figure 1.5-13. This model predicts that the

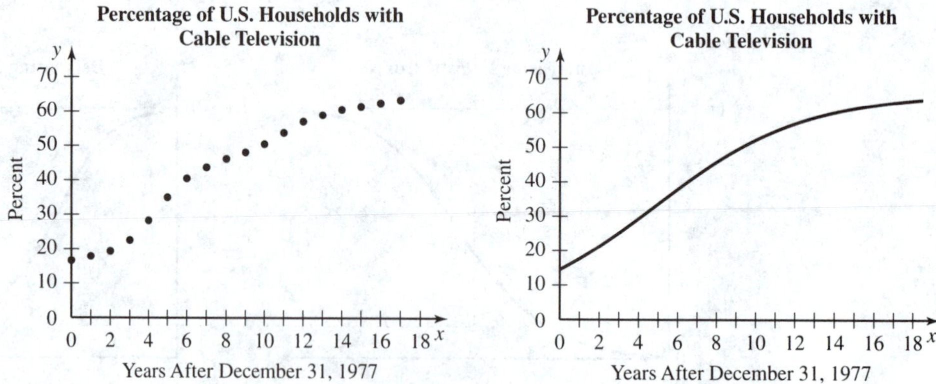

Figure 1.5-13

percent of U.S. households with cable television will level off at approximately 66 percent. Do you think that this is a good model? □

Sample Problem 5: A group of clever and enterprising students had followed their basketball team's success throughout the season. When the announcement came that the team had received a bid to the NCAA tournament, they were ready to send in the order for T-shirts that read "All the way in the NCAA—Go Dunkers." There were about 24,000 students on campus. They had no way to advertise their shirts but relied on other students seeing the shirts and spreading the word. They were very optimistic about sales because basketball is very big on campus and so the initial order was 6500 T-shirts. The accumulated total sales of the T-shirts from the day they went on sale are shown in the following table:

Days	1	2	3	4	5	6	7
Total Sales	158	317	793	1823	3170	5310	5944

Should they order more shirts, or will this be enough to satisfy demand? How many more shirts should they order?

Solution: A graph of the data is shown in Figure 1.5-14. This scatter plot resembles the previous example. It is always increasing and changes concavity from concave up to concave down. It also has a natural upper limit: there are only 24,000 students on campus and not all of them will buy the shirt. These facts indicate that a logistic model would be appropriate. The model is

Verbal Definition: $N(t)$ = the total number of T-shirts sold t days after the beginning of the sale.

Symbol Definition: $N(t) = \dfrac{6802.8\ldots}{1 + 190.48\ldots e^{-1.0501\ldots t}}, 0 \le t.$

Assumptions: Certainty and divisibility.

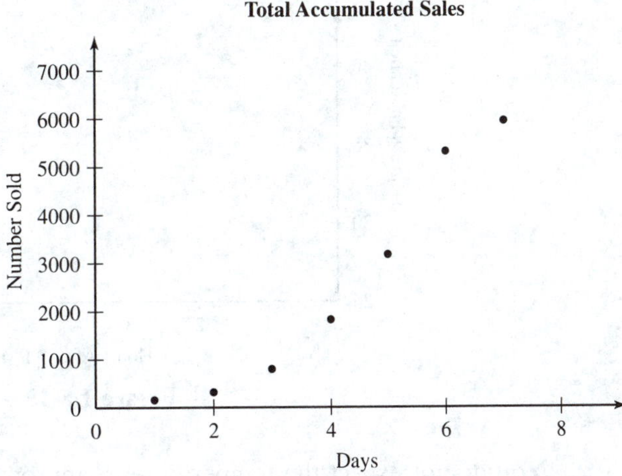

Figure 1.5-14

This model indicates that the maximum number of T-shirts that they could sell would be approximately 6800, or 300 more T-shirts than they have on hand. The decision to reorder would depend on the cost to reorder, minimum size of a reorder, and the time for delivery. □

Sample Problem 6: You have decided to grow plants commercially and just put in a large number of seedlings. They are very susceptible to cold temperatures and will not survive in temperatures below freezing (32 degrees Fahrenheit) for more than a few minutes. Suddenly a cold front comes through. You are watching the temperature very carefully and must decide if you need to cover the plants. You have a special material for this, but it is quite expensive and you can't reuse it. It will take you at least an hour to cover your plants. You don't have much working capital and can't afford to waste the material, but losing all of your plants would be a financial disaster. You have been watching the weather very closely. If there is no sudden change in the weather pattern, the lowest temperatures are usually recorded around 5 A.M. in your area at this time of year. The cold front has already come through and the weather service is not predicting any further sudden changes in the weather. You recorded the following times and temperatures (in degrees Fahrenheit), the last reading being taken at 4 A.M., just a few moments ago:

Time	21:00	22:00	23:00	24:00	01:00	02:00	03:00	04:00
Temp	43	43	42	41	39	36	34	33

What should you do?

Solution: First we get a plot of the data, shown in Figure 1.5-15. The graph definitely shows an inflection point:, it is concave down then concave up. The points seem to flatten out a little bit at the last reading. Which kind of model do you think is appropriate?

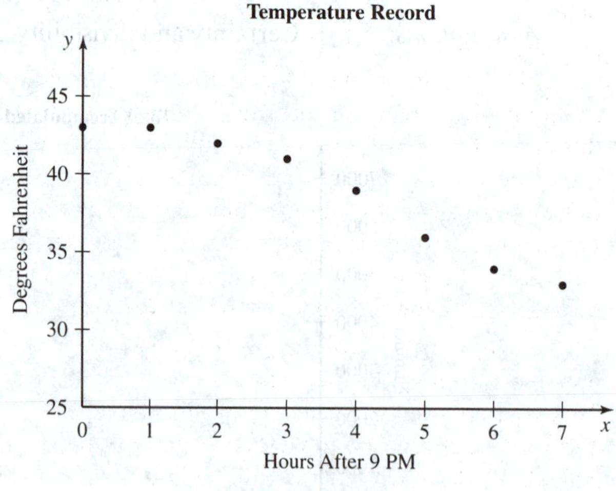

Figure 1.5-15

You do not expect the temperature to approach a limit, say 31 or 32 degrees and stay there. On the contrary, you expect the lowest temperature to occur around 5 A.M., and then the temperatures will gradually begin to rise. This would indicate that you should fit a cubic model to the data. Figure 1.5-16 shows a cubic model.

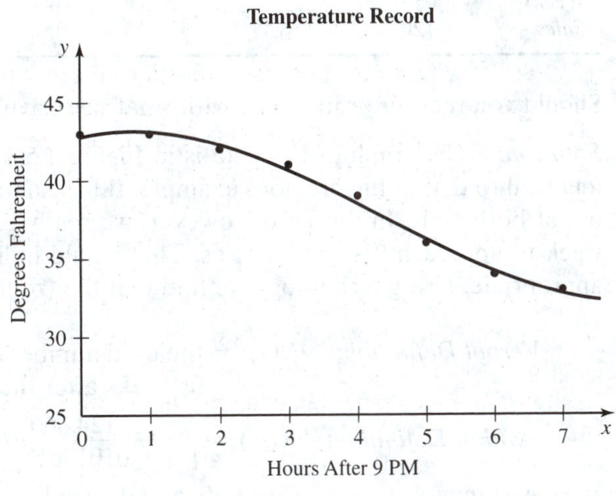

Figure 1.5-16

The cubic model (rounded to 3 significant digits) is

Verbal Definition: $T(x)$ = temperature (degrees Fahrenheit) x hours after 21:00 (9:00 P.M.).
Symbol Definition: $T(x) = 0.0581\,x^3 - 0.759\,x^2 + 1.03\,x + 42.8, x \geq 0.$
Assumptions: Certainty and divisibility.

You are particularly interested in the temperature at 5 A.M., traditionally the time of the lowest temperature. 5 A.M. will be exactly 8 hours after 9:00 P.M. the previous evening: $T(8) = 32.29$. Very, very close. Is this actually the minimum temperature that the model predicts? We'll learn how to determine that in Chapter 3. Should you spend the money to cover the plants? The model assumption of certainty is very important in this case. If you had time, you would be able to take some steps to validate your model, such as waiting an hour or even a half hour to see if the actual temperature agrees with the model prediction. You could also check some temperature history from the past. Unfortunately, time is of the essence in this case. The final decision remains with you: we have used the power of mathematics to give you as much information as possible, but mathematics cannot give you "the answer." □

Sample Problem 7: Recently a concert was held to raise money for AIDS research. This concert was broadcast across the United States and raised a great deal of money. You are in charge of the sales of the CD taped at the concert. You initially ordered 500,000 of the CDs but saw immediately that sales were greatly exceeding all expectations. Thrilled with the results, you ordered 150,000 more. Sales continued very briskly for a while but then began tapering off, as you expected they would. The initial enthusiasm is dieing out as the memory of the concert fades. You don't have many CDs left. You don't want to run out, but you really expect the market to dry up in a while and you don't want to be stuck with the CDs. You have the record of sales (in thousands of CDs) up to this time, for each week after the concert:

Weeks	1	2	3	4	5	6	7	8
Sales	120	120	110	100	80	50	30	20

Should you reorder or just stick with what you have?

Solution: The graph of the data is in Figure 1.5-17. The shape of the graph is almost identical to that of the previous example. It has an inflection point and appears to flatten out at both ends. In this case, however, we really expect the number of CDs sold each week to approach 0 as time passes. That would indicate that a logistic model would be appropriate. Using technology we find that the (rounded) logistic model is

Verbal Definition: $N(w)$ = the total number of CDs sold per week (in thousands) w weeks after the concert.

Symbol Definition: $N(w) = \dfrac{124.360}{1 + 0.01019e^{0.81197821\,w}},\ 0 \le w.$

Assumptions: Certainty and divisibility.

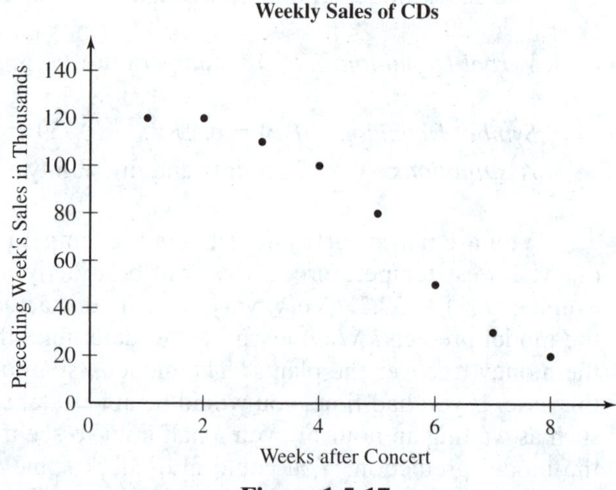

Figure 1.5-17

$N(9) \approx 7.675$ and $N(10) \approx 3.528$. In the next two weeks you might expect to sell approximately 11,203 CDs, and by the 11th week sales would be just over 100 CDs. You have ordered a total of 650,000 CDs, sold 630,000 so far and can expect to sell fewer than 15,000 in the next three weeks. The data do not seem to warrant a reorder, but it could be pretty close. □

Before you begin the exercises for this section be sure that you

- Know that data characterized by an inflection point—a point where the concavity changes—can be modeled using a cubic, quartic, or logistic function.

- Know that when the data also change direction, from increasing to decreasing and then increasing again, or decreasing, increasing, decreasing, a cubic or quartic function will probably be better than a logistic model.
- Recognize that even if the data do not change direction, that is, continue to increase or decrease, a cubic model may still be best. For example, data that have only one concavity but are not symmetrical (are skewed) may be fit with a cubic model.
- Know that when the data are characterized by an inflection point and by finite natural output limits at both extremes, a logistic model is indicated.
- Know that, when considering the choice of a function for a model, you should consider the sense of the data and the purpose for which the model will be used.
- Can use technology to solve equations involving cubic, quartic, and logistic functions.
- Know that the maximum number of inflection points a polynomial can have is *two less* than its degree.

EXERCISES FOR SECTION 1.5

Warm Up

1. For each of the graphs shown in Figure 1.5-18, indicate what type of model or models you would fit for interpolating. Explain your reasons for choosing each type of model.

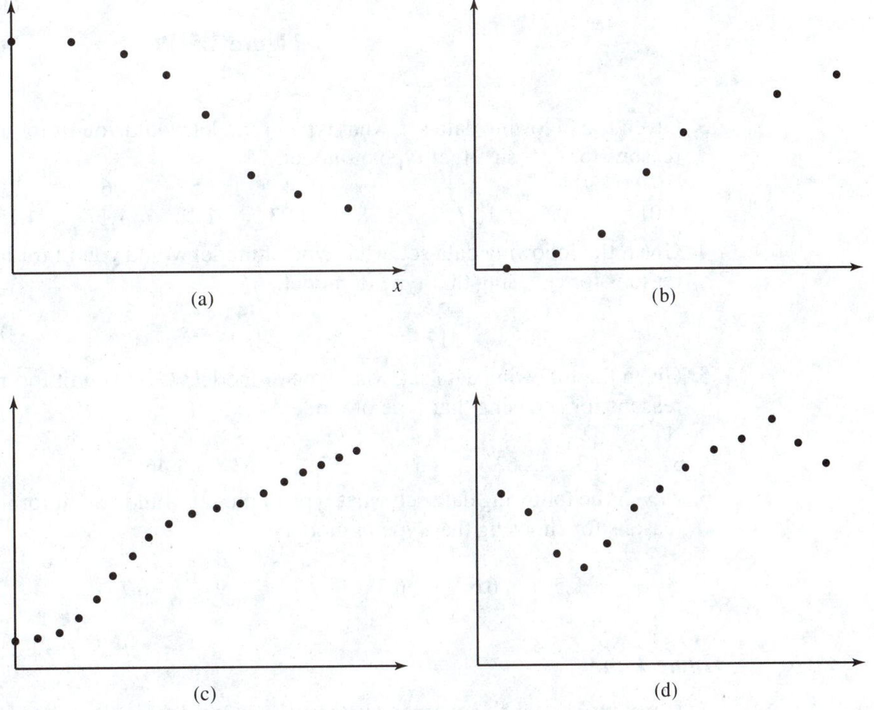

Figure 1.5-18

2. For each of the graphs shown in Figure 1.5-19, indicate what type of model or models you would fit for interpolating. Explain your reasons for choosing each type of model.

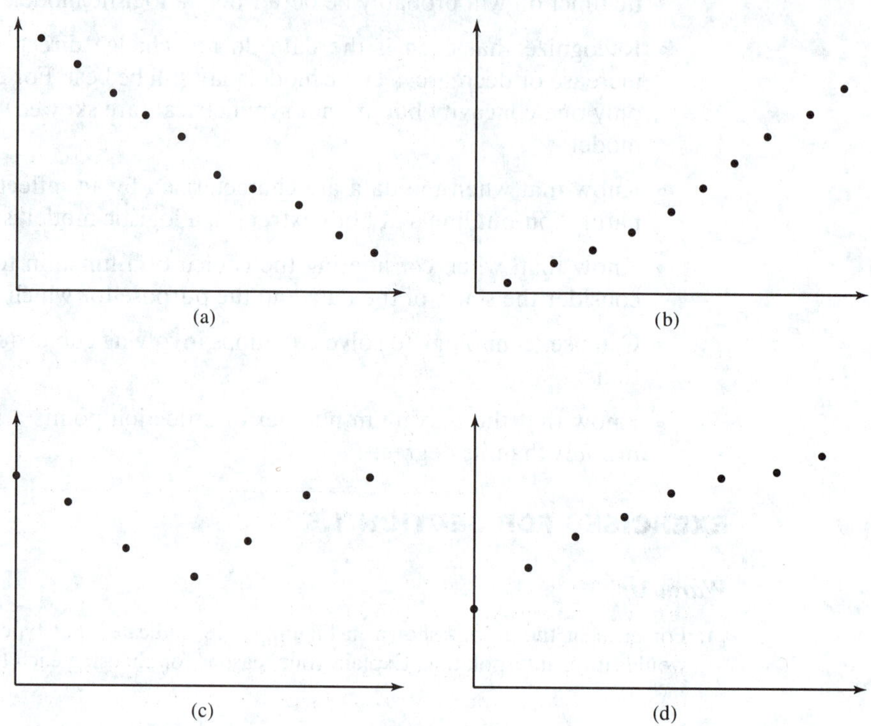

Figure 1.5-19

3. Given the following data set, what type of model would you fit for interpolation? Explain your reasons for choosing that type of model.

0	1	2	3	4	5	6	7	8	9	10
1.01	1.02	1.17	1.28	1.23	1.18	1.17	1.15	1.21	1.29	1.37

4. Given the following data set, what type of model would you fit for interpolation? Explain your reasons for choosing that type of model.

1	2	3	4	5	6	7
158	317	793	1823	3170	5310	5944

5. Given the following data set, what type of model would you fit for interpolation? Explain your reasons for choosing that type of model.

1	2	3	4	5	6	7	8	9	10	11	12
64	63	62	57	53	53	46	43	39	38	38	36

6. Given the following data set, what type of model would you fit for interpolation? Explain your reasons for choosing that type of model.

0	1	2	3	4	5	6	7	8
7.5	6.6	6.5	7.7	9	8.7	7.4	5.9	4.8

Game Time

7. Everyone should be concerned with our growing debt in the United States. The following table shows the per capita public debt of the United States (in dollars):

Fiscal Year	Per capita ($)
1975	2475
1976	2852
1977	3170
1978	3463
1979	3669
1980	3985
1981	4338
1982	4913
1983	5870
1984	6640
1985	7598
1986	8774
1987	9615
1988	10534
1989	11545
1990	13000
1991	14436
1992	15846
1993	16871
1994	18026

Source: *World Almanac and Book of Facts* 1996, p. 112 (Bureau of Public Debt, U.S. Dept. of the Treasury).

(a) Fit a model to the data.
(b) Explain why you chose this model.
(c) What does your model predict for the per capita debt in 1995? 1996?
(d) Do you think these are reasonable predictions? Explain your answer.
(e) If possible, locate the actual figures for 1995 and 1996. Are the predictions from your model close to the actual figures?

8. Shown below is the percent of federal outlays for the interest paid on public debt in the United States:

Fiscal Year	% of Federal Outlays
1975	9.8
1976	10.0
1977	10.2
1978	10.6
1979	11.9
1980	12.7
1981	14.1
1982	15.7
1983	15.9
1984	18.1
1985	18.9
1986	19.2
1987	19.5
1988	20.1
1989	21.0
1990	21.1
1991	21.5
1992	21.1
1993	20.8
1994	20.3

Source: *World Almanac and Book of Facts* 1996, p. 112 (Bureau of Public Debt, U.S. Dept. of Treasury).

(a) Fit a model to the data.
(b) Explain why you chose this model.
(c) What does your model predict for the percent expenditures in 1995? 1996?
(d) Do you think these are reasonable predictions? Explain your answer.
(e) If possible, locate the actual figures for 1995 and 1996. Are the predictions from your model close to the actual figures?
(f) Does this model present a different picture from that given in the previous problem?

9. The data for unemployment in the U.S. civilian population is shown in the following table:

Year Ending Dec. 31	Number Unemployed (thousands)
1983	10717
1984	8539
1985	8312
1986	8237
1987	7425
1988	6704
1989	6528
1990	6874
1991	8426
1992	9384
1993	8734
1994	7996

Source: *World Almanac and Book of Facts* 1996, p. 146 (Bureau of Labor Statistics, U.S. Dept. of Labor).

(a) Find a model for the total number unemployed.
(b) Explain why you chose this model.
(c) What does your model predict for total unemployment in the United States for 1995? 1996?
(d) If possible, locate the actual figures for 1995 and 1996. Are the predictions from your model close to the actual figures?

10. The following table gives the number of votes cast for president by Republican popular vote for 1972 to 1992:

Year	1972	1976	1980	1984	1988	1992
Number (1000s)	47170	39148	43904	54455	48886	39104

(a) Find a model for the Republican popular vote.
(b) What does your model predict for the Republican popular vote in 1996?
(c) How does this compare to the actual vote?

11. The table below shows foreign exchange rates, 1970 to 1994 (National currency units per dollar; data are annual averages).

Year	Canada (dollars)	Germany (marks)[a]	Japan (yen)
1970	1.0103	3.6480	357.60
1975	1.0175	2.4613	296.78
1980	1.1693	1.8175	226.63
1985	1.3655	2.9440	238.54
1988	1.2307	1.7562	128.15
1989	1.1840	1.8800	137.96
1990	1.1668	1.6157	144.79
1991	1.1457	1.6595	134.71
1992	1.2087	1.5617	136.65
1993	1.2901	1.6533	111.20
1994	1.3656	1.6228	102.21

[a] West Germany prior to 1991.
Source: *World Almanac and Book of Facts* 1996, p. 125 (International Monetary Fund).

(a) Find a model for the foreign exchange rate for each of the three countries.
(b) If you were traveling in each of the countries, in which ones, if any, was the exchange rate improving for U.S. citizens?
(c) Find models for 1988 through 1994. How do these models differ from the ones found in part a)?

12. The following table shows drug use among America's high school seniors (percent ever used by graduating class). (*Note*: No data for 1989.)

	1975	1980	1986	1987	1988	1990	1991	1992	1993	1994
Pot[a]	47.3	60.3	50.9	50.2	47.2	40.7	36.7	32.6	35.3	38.2
Stimulants	22.3	26.4	23.4	21.6	19.8	17.5	15.4	13.9	15.1	15.7
Cigarettes	73.6	71.0	67.6	67.2	66.4	64.4	63.1	61.78	61.9	62.0

[a] Marijuana/hashish.
Source: *World Almanac and Book of Facts* 1996, p. 965 (University of Michigan Institute for Social Research).

(a) Fit models for each of the drugs.
(b) Do the data and the models present an optimistic picture?
(c) What conclusions, if any, can you draw from your models?

13. The following table shows the average kilowatt hours (kWh) electric use pattern per day and average hundred cubic feet (CCF) gas use pattern per day by month for one year for a private residence in the Northeastern United States:

Avg.	Jan.	Feb.	Mar.	Apr.	May	June	July	Aug.	Sept.	Oct.	Nov.	Dec.
kWh	32	27	27	23	24	62	77	63	37	25	28	30
CCF	8.8	7.7	5.5	2.5	1.7	0.5	0.5	0.5	1.5	2.5	4	6.5

Note: Since these figures are monthly averages, use 0.5 for January, 1.5 for February, and so on.

(a) Find models for the average electric and gas use patterns.
(b) If this is the usage for a house in the Northeast corridor, what gas and electric appliances are probably in use?
(c) What conditions would cause this pattern of usage to vary from one year to the next?

14. The following table shows the average kilowatt hours (kWh) electric use pattern per day and average hundred cubic feet (CCF) gas use pattern per day by month for one year:

Avg.	Jan.	Feb.	Mar.	Apr.	May	June	July	Aug.	Sept.	Oct.	Nov.	Dec.
kWh	33	30	29	28	26	34	65	60	53	29	30	33
CCF	7.2	8.1	4	3.8	2	0.5	0.4	0.4	0.6	1.1	3.9	5.2

(a) Find models for the average electric and gas use patterns.
(b) If this is the usage for a house in the Northeast corridor, what gas and electric appliances are probably in use?
(c) What conditions would cause this pattern of usage to vary from one year to the next?

15. The following table shows the percent of the resident U.S. population by age:

	Midpoint	1970	1980	1990
Under 5 years	2.5	8.4	7.2	7.5
5–9 years	7	9.8	7.4	7.3
10–14 years	12	10.2	8.1	6.9
15–19 years	17	9.4	9.3	7.2
20–24 years	22	8.1	9.4	7.7
25–29 years	27	6.6	8.6	8.6
30–34 years	32	5.6	7.8	8.8
35–39 years	37	5.5	6.2	8.0
40–44 years	42	5.9	5.2	7.1
45–49 years	47	6.0	4.9	5.5
50–54 years	52	5.5	5.2	4.5
55–59 years	57	4.9	5.1	4.2
60–64 years	62	4.2	4.5	4.3
65–74 years	69.5	6.1	6.9	7.3
75–84 years	79.5	3.0	3.4	4.0
85 years and over	90	0.7	1.0	1.2

Source: *Statistical Abstract of the United States, 1995*, Table No. 14, p. 15.

(a) Find a model for the percent of the population in each age group (use the midpoint of the age range for the independent variable, 90 for last interval) for each of the years.
(b) If you wanted to start a business that targeted a certain age group, under 10, 10 to 20, 20 to 30, 40 to 50, etc., for example, which do you think would be the best age group to target? Explain your answer.
(c) What implications, if any, do your models have regarding Social Security and Medicare?

16. Calculate the cumulative total percent of the population under 5 years, under 10 years, under 15 years, and so forth, for each of the years.
(a) Find models for the total populations.
(b) Why did you select these models?
(c) How do the three models differ?

CHAPTER 1 SUMMARY

The Twelve Steps of Problem Solving

In this chapter you have been introduced to a process for problem solving using the power of mathematics. We listed twelve steps necessary to do this:

A. Clarify the Problem.
   1. Identify, bound, and define the problem.
   2. Understand the problem (create a sketch or table).

B. Formulate a Model.
   3. Define all variables (including units).
   4. Gather data if necessary.
   5. Formulate and fully define a mathematical model.

C. Solve/Analyze the Model.
   6. Select an appropriate method of analysis.
   7. Solve the problem mathematically (optimize/forecast).

D. Validate the Model (Post-Optimality Analysis).
   8. Verify (double-check) the calculations.
   9. Make sure your model and solution make sense (validate).
   10. Perform sensitivity analysis.

E. Draw Conclusions and Implement the Solution.
   11. Evaluate overall accuracy and precision, rough margin of error.
   12. Summarize your conclusions and implement them.

Mathematical Models

Mathematical models consist of three parts:

1. Verbal Definition: A verbal definition, including units of the function and all variables.
2. Symbol Definition: A mathematical definition of the function, including the domain.
3. Assumptions: The usual assumptions are certainty (the model gives the exact relationship between the variables) and divisibility (the variables can be fractions or *any* value in an interval).

Functions

A mathematical **function** is a relationship between two variables, a set of ordered pairs (relation) in which each value of the first variable from the domain is paired with exactly one value of the second variable.

The **domain** of a function is the set of allowable values for the first variable.

The first variable is also called the **independent variable** or **input**.

The second variable is also called the **dependent variable** or **output**.

Functions can be represented with tables, algebraic expressions, graphs, or in words.

Functional notation allows us to express the relationship of the independent (i) and dependent (d) variables: $d(i)$ = some expression involving i.

Seven Basic Types of Models

Model Type	General Shape	Concavity	Inflection Point(s)	Constant	General Form
Linear		None	0	1st differences	$y = ax + b$
Quadratic		Up or Down	0	2nd differences	$y = ax^2 + bx + c$
Exponential		Up or Down	0	% change	$y = ab^x$ or ce^{kx}
Power		Up or Down	0	(% change y) ÷ (% change x)	$y = ax^b$
Cubic		Up and Down	1	3rd differences	$y = ax^3 + bx^2 + cx + d$
Logistic		Up and Down	1	NA	$y = \frac{L}{1 + Ae^{-Bx}}$ or $\frac{c}{1 + ae^{-bx}}$
Quartic		$\leq$ 2 Changes	≤ 2	4th differences	$y = ax^4 + bx^3 + \cdots + e$

When fitting a model, all of the above should be considered, but always consider the context of the problem, the expected behavior *beyond* the given data intervals, the sense of the data, and the purpose for which the model will be used, including interpolation (estimating output values *within* the interval of the given input data values) versus extrapolation (estimating output values *beyond* the interval of the given input data values) and optimization versus forecasting.

CHAPTER 2

Rates of Change

INTRODUCTION

In the first chapter we learned how to take the first steps toward solving real-life problems by clarifying problems and formulating mathematical models for them. One factor influencing our choice of a model from data depended on how the independent variable was changing as the dependent variable changed (first and second differences, percent change, etc.). We noted these changes numerically from the data values themselves and graphically from plots of the data. Understanding how a function is changing is essential to the analysis of the function. When we discussed linear functions in Section 1.3, we talked about the concept of slope, defined as the change in the output divided by the change in the input, or the change in the output *per unit change* in the input. This idea is also called a **rate of change**. When we calculated slope, it was always over an *interval* of input values. As in computing your average speed on a car trip, this is called an **average rate of change over an interval**. However, in the car example, we can also think of our speed at any *moment* in time (as given by the speedometer, for example), which is called an **instantaneous rate of change at a point**. In this chapter we will begin the step of analyzing and solving problems by learning how to find and interpret the rate of change of a model.

We start in Section 2.1 by showing how to calculate average rates of change from tables, from models, and from graphs, and what they correspond to graphically. Sometimes it is helpful to express a rate of change in *relative* (percentage) rather than *absolute* terms, and we show how to do that by dividing the average rate of change by the initial output value, and call it the **percent rate of change**. This can also be thought of as the *average percent change over the interval* (the percent change divided by the length of the interval). In Section 2.2 we then show how we can find the instantaneous rate of change (also called the **derivative**) at a point as the **limit** of average rates of change, and what this corresponds to graphically, which involves the idea of a **tangent** line. In Section 2.3 we focus on standard notation for derivatives and how to interpret them in real-world contexts. In Section 2.4 we define the derivative more formally in mathematical symbols and use this definition to derive rules for finding the derivative of basic functions such as polynomials, sums and differences of functions, and multiplication of a function by a constant. In Section 2.5 we develop a rule called the **Chain Rule** for finding the derivative of a composite function (function of a function), useful for derivatives of many complicated expressions and functions, including the exponential and logistic functions. Finally, in Section 2.6, we derive the rule called the **Product Rule** for finding the derivative of a product (or quotient) of two functions, often needed when working with revenue and profit functions.

Here are some types of problems that the material in this chapter will help you analyze:

- Your coach wants you to practice free throws for a couple of hours and report back on your progress. How should you express your rate of improvement?
- How should you express the rate of change in your GPA over the last few semesters to your advisor (or your parents)?
- You bought two stocks last year. In the past year the price of each stock rose at the rate of $1 per month. What form of rate of change makes sense to use to compare their performance?
- You work in the admissions office of a school. You need to know how many applicants to accept for the coming fall. You have information about the relationship between the number accepted and how many students send in a deposit, and also about the relationship between how many send in a deposit and how many actually enroll and appear in the fall. You want to find the number who enroll and appear in the fall as a function of the number accepted.
- You are selling mugs as part of an environmental project. You hope to achieve two things: communicate to as many people as possible the idea of reusing resources and raise money for the environmental effort. How would your revenue change as you change the price of the mugs?

These problems all refer to how things are changing. By the end of this chapter you should be able to answer these and similar questions and

- Know how to calculate and interpret the average rate of change over an interval and the percent rate of change over an interval from data, a graph, and a model and know what the average rate of change corresponds to graphically.
- Know that the instantaneous rate of change of a function at a point is the slope of the tangent to the curve at that point and that the slope of the tangent at a point is the limiting value of the slopes of secants drawn from that point to a second point on the curve, as the second point gets closer and closer to the first.

- Be able to recognize, use, and interpret derivative notation.
- Know the algebraic definition of the derivative and the basic derivative rules, including the Chain Rule and the Product Rule.

2.1 AVERAGE AND PERCENT RATE OF CHANGE OVER AN INTERVAL

In Chapter 1 we looked at various types of models and what distinguishes one from another. The simplest model was linear, distinguished by the fact that the first differences of the model were constant for equally spaced inputs. In the quadratic model the second differences were constant for equally spaced inputs. In an exponential model the percent change was constant for equally spaced inputs. The differences told us how the dependent variable was changing as the independent variable changed. For linear functions, we saw that the slope, which is the change in the output divided by the change in the input, was constant. The slope of a linear function is also the **rate of change** of that function. When we calculated slope, we always did it over an *interval* of input values. This corresponds to what is called the ***average* rate of change of a function over an interval** of input values. Sometimes, it is convenient to express the rate of change in *relative* (as a percent) rather than *absolute* terms. We can express the average rate of change as a *percentage* of the initial output value, which is called the **percent change of a function over an interval**. In this section we will learn how to find the average rate of change and percent rate of change of a function over an interval.

Here are the kinds of problems that the material in this section will help you solve:

- Your basketball coach wants you to practice your free throws (from the foul line) and report back on your progress. How should you express your rate of improvement?
- If you practice a computer game, at what rate is your game improving from the first game to the third? From the third to the fifth?
- You have just added some new electronic navigation gear to your boat that will increase the drain on your batteries. When your battery discharge goes from 4% to 10%, how fast are your cycles to failure (the number of recharges before the battery dies) changing? How fast are they changing when you go from 10% to 20%?
- You bought two stocks last year. In the past year the price of each stock rose at the rate of $1 per share per month. What form of rate of change makes sense to use to compare their performance?
- How should you express the rate of change in your GPA over the last few semesters to your advisor (or your parents)?

By the end of this section you should be able to solve problems like those above and should also

- Know how to calculate and interpret the average rate of change of a function over an interval from data, from a graph, or from a model.

- Know how to calculate and interpret the percent rate of change of a function over an interval from data, from a graph, or from a model.
- Recognize the average rate of change of a function over an interval as the slope of the secant line from the point on the curve at one endpoint of the interval through the point on the curve at the other endpoint.

Turn to the business section of the newspaper and you will see statements such as, "The December vehicle sales were equivalent to a seasonally adjusted annual selling rate of 15.9 million autos, one of the strongest sales rates of what turned out to be a disappointing 1995. Since September, sales had been running at rates between 14.5 million and 14.8 million units a year."[1] ". . . growing at the rate of 50,000 customers a month."[2] ". . . annual demand growth of 1.6% for the next twenty years." ". . . and the equivalent net change from 1985 to 1996 is 18 million barrels a day."[3] Other sections of the paper report such things as "The temperature dropped at the rate of 4 degrees per hour." "The snow fell at the rate of 2 inches per hour." "The population of Big Bear fell at an average rate of 46 persons per year from 1980 to 1990." What do these statements really mean? How are they calculated?

These statements all refer to the rate of change of a function over an interval. Sometimes the change was reported as a percent change, sometimes as an actual change. Because these terms appear so often, it is important that we have a good understanding of exactly what they mean and how they are calculated. Being able to understand and interpret such statements is one of the goals of this section.

Average Rate of Change

Sample Problem 1: If you drove on a car trip for 4 hours and traveled 240 miles, what was your average speed?

Solution: Since you traveled 240 miles in 4 hours, your average speed was

$$\frac{240 \text{ miles}}{4 \text{ hours}} = 60 \text{ miles per hour} \quad \square$$

If you traveled the whole 4 hours using cruise control at exactly 60 miles per hour, the distance traveled could be modeled by

Verbal Definition: $D(t)$ = the distance, in miles, traveled after t hours on the trip.
Symbol Definition: $D(t) = 60t$, for $t \geq 0$.
Assumptions: Certainty and divisibility. Certainty implies your speed was *exactly* 60 mph the whole way. Divisibility implies fractions of time and miles are possible.

Notice that this is a simple linear model, with a slope of 60 and a vertical intercept of 0.

[1]Nichole M. Christian and Robert L. Simison, *Wall Street Journal*, January 5, 1996, p. 4.
[2]Douglas McIntyre, *Financial World*, January 2, 1996, p. 15.
[3]Arthur Jones, *Financial World*, January 2, 1996, p. 38.

What did our calculation of the average speed correspond to with respect to this model? Well, $D(0) = 0$ means that we had traveled no distance as of the start, and $D(4) = 60(4) = 240$ means that we had traveled 240 miles after 4 hours. So our average speed calculation was really the change in the output (distance) divided by the change in the input (time), which is the same as calculating the slope of the line joining the points at $t = 0$ and at $t = 4$.

This idea of calculating the change in output over the change in input (over an interval of input values) is called the **average rate of change of a function over an interval**. In Sample Problem 1, we can think of our average speed of 60 mph as the average rate of change in our *distance traveled* over the first 4 hours of the trip (the interval [0,4] for the input variable t). We will at times use the notation $[a,b]$ to refer to the interval from a to b, *including* the endpoints. In this example, [0,4] corresponds to the inequality $0 \le t \le 4$.

Suppose that your entire car trip was on one highway, and it was marked off with mile markers every tenth of a mile or so. Suppose you started your trip at mile marker 70, and traveled in the direction in which the mile marker values increased. We could define a model for the mile marker location at any point in your trip as

Verbal Definition:	$L(t)$ = your mile marker location after t hours on the trip.
Symbol Definition:	$L(t) = 60t + 70$, for $0 \le t \le 4$.
Assumptions:	Certainty and divisibility. Certainty implies your speed was *exactly* 60 mph the whole way. Divisibility implies fractions of time and miles are possible.

To find the average rate of change this time, the change in the input is still 4 $(4 - 0)$ as before, but the change in output calculation is more complicated. The initial location is $L(0) = 70$ and the final location is $L(4) = 60(4) + 70 = 240 + 70 = 310$, so the change in the output is given by

$$L(4) - L(0) = 310 - 70 = 240$$

Not surprisingly, we still get an average speed of $240/4 = 60$ mph. This time, however, we can think of it as the rate of change of our *position* over the first 4 hours of the trip.

When using the position function, we can also use the term **average velocity** over the interval. In physics, velocity and speed are very similar, but velocity also includes the idea of the *direction* of motion. With our position function, the direction traveled on the highway would correspond to whether the mile markers were *increasing* or *decreasing*. If they were increasing, the change in output would be *positive*, so our average rate of change would be positive, and vice versa.

With these conventions, suppose you traveled from marker 70 to marker 190 in the first 2 hours, then turned around and come back to marker 70 after 2 more hours. Technically, the change in your position (change in the output) over the *four* hours would be 0 (you were at marker 70 at time 0 and at time 4), and so your average velocity would also be 0 miles per hour. That sounds odd, but if you think of it as averaging 60 mph for the first 2 hours, then -60 mph for the second two hours (negative because you were going in the opposite direction), it makes sense that the average *velocity* is 0 mph, since the velocity reflects the direction of the motion in its sign, so the signs cancel when averaging. On the other hand, the average *speed* in the same situation would be 60 mph, since speed only involves the *magnitude* of the velocity and ignores the direction, so the average *speed* for *both* 2-hour periods was 60 mph. To emphasize this difference, you could think of the average velocity as the average *net* change in your position over the interval (allowing opposite signs to cancel).

From these examples, we can see that the general formula for finding the **average rate of change of a function f from x_1 to x_2 (over the interval $[x_1, x_2]$)** is

$$\frac{\text{change in output}}{\text{change in input}} = \frac{y_2 - y_1}{x_2 - x_1} = \frac{f(x_2) - f(x_1)}{x_2 - x_1}$$

Sample Problem 2: The ball toss graph from Section 1.4 is shown in Figure 2.1-1. How fast was the ball rising, on the average (what was its average velocity), for the first second after it was thrown? From 1 second to 1.5 seconds after being tossed?

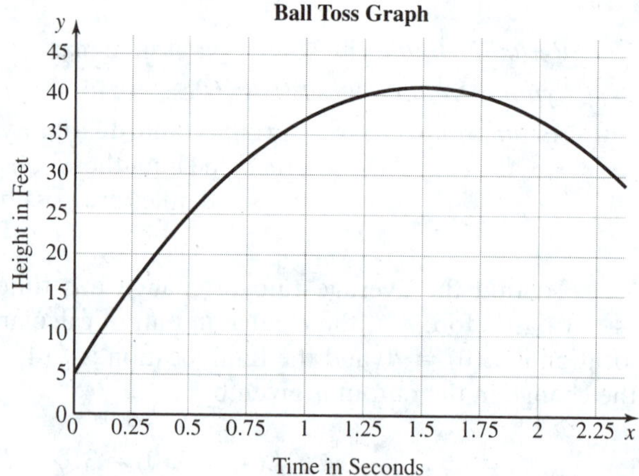

Figure 2.1-1

Solution: **The average rate of change of a function** over an interval is the change in the *output* over the interval divided by the change in the *input* over the interval. If we do not have the original data at hand we have to *estimate* the values for the output from the graph. Reading numbers from a graph is not easy. They are estimates at best, and two different people may get different readings. The height of the ball at time = 0 seconds appears to be about 5 feet. The height of the ball at time 1 (after 1 second) appears to be about 37 feet. The change in the height is 37 feet minus 5 feet, or 32 feet. The change in the time is 1 second minus 0 seconds, or 1 second. So the average change in height from time 0 to time 1 is **approximately**

$$\frac{(37 - 5)\text{ feet}}{(1 - 0)\text{ seconds}} = \frac{32\text{ feet}}{1\text{ second}} = 32\,\frac{\text{feet}}{\text{second}}$$

We usually write the final statement as 32 feet per second or 32 ft/sec. Estimating from the graph, from 0 to 1 second after the ball was tossed it was rising at an average rate of 32 feet per second. We could also say its *average (vertical) velocity* over the first second was 32 feet per second.

If you get in the habit of writing the units when you are calculating rates, identifying the units of the resulting rate will be easy. The units for the average rate of change are **(output units) per (input unit)**.

Let's look at the second part of the question: The average rate of change of the height of the ball from time 1 to time 1.5, since the height of the ball after 1.5 seconds seems to be about 42 feet, is **approximately**

$$\frac{(42 - 37)\text{ feet}}{(1.5 - 1)\text{ seconds}} = \frac{5\text{ feet}}{0.5\text{ seconds}} = 10\,\frac{\text{feet}}{\text{second}} = 10\text{ feet per second}$$

Again, estimating from the graph, from 1 to 1.5 seconds after the ball was tossed it was rising at the average rate of 10 feet per second.

The average rate of change of the curve is **positive** in both cases because, going from left to right, the curve is **increasing**, in both cases. However, the average rate of change of the curve is much larger from 0 seconds to 1 second than it is from 1 to 1.5 seconds.

Recall that the general formula for finding the **average rate of change of a function f from x_1 to x_2** is

$$\frac{\text{change in output}}{\text{change in input}} = \frac{y_2 - y_1}{x_2 - x_1} = \frac{f(x_2) - f(x_1)}{x_2 - x_1}$$

This is a little awkward to write out each time. Recall from Section 1.3 the Greek letter delta (Δ) used for "change in." We can shorten the formula to

Average rate of change of $f(x)$ over $[x_1,x_2]$ =

$$\frac{\text{change in output}}{\text{change in input}} = \frac{y_2 - y_1}{x_2 - x_1} = \frac{\Delta y}{\Delta x} = \frac{f(x_2) - f(x_1)}{x_2 - x_1}$$

This formula should look familiar: It is the formula **for the slope of a line**, discussed in Section 1.3. Many of you may be familiar with the expression "rise over run" to describe the slope of a line. A line drawn from one point on a curve through another point on the curve is called a **secant line** (Figure 2.1-2). When we calculate an **average**

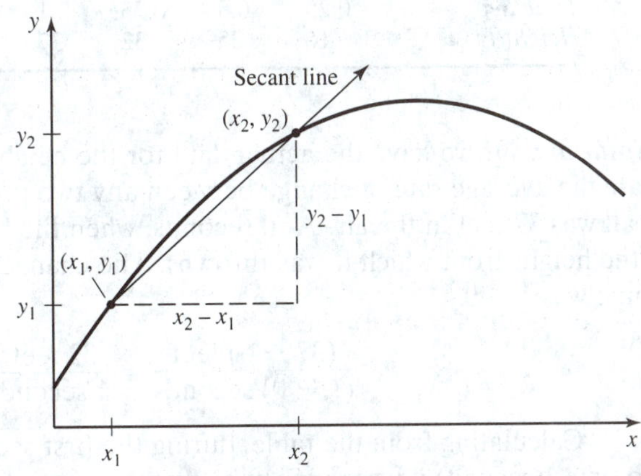

Figure 2.1-2

rate of change over an interval, we are calculating the slope of the **secant line** from one endpoint of the interval *through* the other (Figure 2.1-2). We could also simply think of it as the slope of the *line segment* joining the two points, but in Section 2.2 there is an advantage to thinking of the line as a secant, so it is a good idea for you to get used to that idea.

First we measure the **rise**, the change in the output, values: $y_2 - y_1$. Then the **run**, the change in the input values, $x_2 - x_1$. Finally, we divide the change in y by the change in x: $\frac{y_2 - y_1}{x_2 - x_1}$. Be careful that you do not switch the order of the subtraction. You may choose either point as the first point, (x_1,y_1). It doesn't matter; just be certain that your points are lined up in the formula $\frac{y_2 - y_1}{x_2 - x_1}$. If the numerator is the y value of the *second* point minus that of the *first*, then the denominator must be the x value of the *second* point minus that of the *first* as well. For our example we used the points (1.5,42) and (1,37). We did the following:

$$\frac{(42 - 37)\text{ feet}}{(1.5 - 1)\text{ seconds}} = \frac{5\text{ feet}}{0.5\text{ seconds}} = 10\,\frac{\text{feet}}{\text{second}} = 10\text{ feet per second}$$

This corresponded to thinking of (1,37) as the first point and (1.5,42) as the second point. Let's reverse the order (think of (1.5,42) as the first point and (1,37) as the second) now:

$$\frac{(37 - 42)\text{ feet}}{(1 - 1.5)\text{ seconds}} = \frac{-5\text{ feet}}{-0.5\text{ seconds}} = 10\,\frac{\text{feet}}{\text{second}} = 10\text{ feet per second}$$

The signs of *both* the numerator and the denominator switch (and so cancel each other out), so it doesn't matter how we set it up. Estimating from the graph, from 1 second to 1.5 seconds after the ball was tossed, the ball was rising at an average rate of 10 feet per second. □

Sample Problem 3: Let's look at the ball toss example again, only this time we have the actual data. How fast was the ball rising, *on the average*, for the first second after it was thrown? How fast, on the average, was it rising in the next half-second after that?

Time/Sec	0	0.25	0.5	0.75	1	1.25	1.5	1.75	2	2.25
Height/Ft	5	16	25	32	37	40	41	40	37	32

Solution: If we have the actual data for the height as a function of time, we can **calculate** the average rate of change between any two points on our data set. At 1 second, the ball was 37 feet in the air. At 0 seconds, when the ball was tossed, it was 5 feet in the air (the height from which it was thrown). The change in height was 37 − 5. The change in time was 1 − 0.

$$\frac{(37 - 5)\text{ feet}}{(1 - 0)\text{ seconds}} = \frac{32\text{ feet}}{1\text{ second}} = 32\,\frac{\text{feet}}{\text{second}}$$

Calculating from the table, during the first second after the ball was tossed (time 0 to time 1 second) it was rising at an average rate of 32 feet per second.

If we wanted to know the average rate of change of the height of the ball from one second after the toss to 1.5 seconds after the toss, time 1 to time 1.5, we would have

$$\frac{(41 - 37) \text{ feet}}{(1.5 - 1) \text{ seconds}} = \frac{4 \text{ feet}}{0.5 \text{ seconds}} = 8 \frac{\text{feet}}{\text{second}} = 8 \text{ feet per second} = 8 \text{ ft/sec}$$

Calculating from the table, from 1 second to 1.5 seconds after the ball was tossed it was rising at an average rate of 8 feet per second.

The average rates that we found for time 0 to 1 second from the graph and the data were exactly the same. This is not the usual case. When we estimated the average rate of change from 1 second to 1.5 seconds from the graph, we got an answer of 10 feet per second. When we calculated the average rate of change from the table we got 8 feet per second. This is an error of 10 − 8 = 2 ft/sec out of 8, or an error of 25%. This is not unusual for graphical estimation. □

Sample Problem 4: You have just added some new electronic navigation gear to your boat which will increase the drain on your batteries. You have a 20 amp-hour battery. Before you added the new gear you were discharging your batteries about 0.8 amp hours each use. The new gear will increase your amp-hour usage to about 2.8 amp hours each time. One component of the life of the battery is measured in cycles to failure, how many times you can use the battery and recharge it before it fails. When you increase your battery discharge from 4% to 14% (0.8/20 = 0.04 or 4%, 2.8/20 = 0.14 or 14%), at what rate is the number of cycles to failure of the battery changing? Figure 2.1-3 shows the graph of the battery's cycles to failure versus percent discharge.

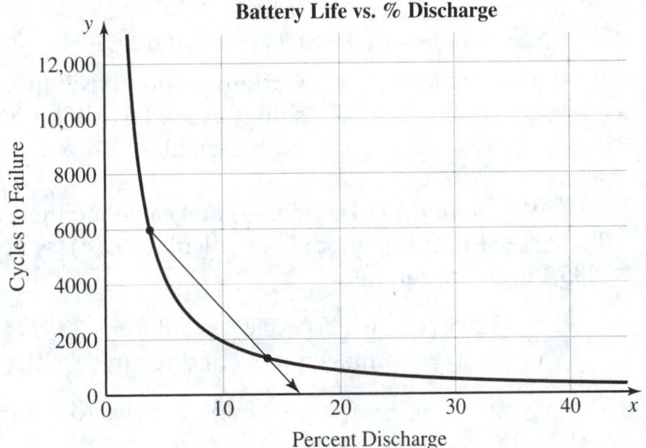

Figure 2.1-3

Source: Adapted from *Sail Magazine*, April 1995.

Solution: To calculate the average rate of change of the cycles to failure when the battery discharge changes from 4% to 14%, we begin by estimating the cycles to failure at 4% discharge and the cycles to failure at 14% discharge. Let's define

$C(x)$ = the number of cycles to failure if you discharge the battery by x percent each time

We estimate from Figure 2.1-3 that $C(4) \approx 6000$ and $C(14) \approx 1200$. So the average rate of change from 4 to 14% discharge is

$$\frac{(1200 - 6000) \text{ cycles to failure}}{(14 - 4) \text{ percentage points of discharge}} = \frac{-4800 \text{ cycles to failure}}{10 \text{ percentage points of discharge}}$$

$$= -480 \text{ cycles to failure per percentage point of discharge}$$

So, estimating from the graph, the number of cycles to failure was *decreasing* at an average rate of 480 cycles per percentage point of discharge when the percent discharge went from 4% to 14%.

The average rate of change is *negative* because the function is *decreasing*, that is, the number of cycles to failure decreases as the percent discharge increases. The more you discharge your battery each time that you use it, the fewer cycles it will last. The slope of the secant line is negative.

Sail Magazine also gave the following table:

% Discharge	2	3	4	5	10	20	40
Cycles	12000	8000	6000	4400	2000	800	380

In this case our table of data did not include the second input, 14% discharge, so we cannot calculate the rate of change from the table. However, we did fit a model for the data in Section 1.4:

Verbal Definition: $C(x)$ = cycles to failure if you discharge the battery by x percent each time.

Symbol Definition: $C(x) = 28906.\ldots x^{-1.176\ldots}$, for $0 \le x \le 100$.

Assumptions: Certainty and divisibility. Certainty implies that the relationship is *exact.* Divisibility implies that fractions are possible for both variables.

We can find $C(14)$ and $C(4)$ and calculate the average rate of change from 4 to 14% discharge. From our model we calculate $C(4) \approx 5656.5$ cycles to failure and $C(14) \approx 1295.2$ cycles to failure.

$$\frac{(1295.2 - 5656.5) \text{ cycles to failure}}{(14 - 4) \text{ percentage points of discharge}} = \frac{-4361.3 \text{ cycles to failure}}{10 \text{ percentage points of discharge}}$$

$$\approx -436.1 \text{ cycles to failure per percentage point of discharge}$$

The average rate of change of the cycles to failure is decreasing at a rate of about 436.1 cycles to failure per percentage point of discharge, when you increase the percent discharge from 4% to 14%.

In this case the error from the graph estimate was $480 - 436.1 = 43.9$ out of 436.1, an error of approximately 10%. □

Percent Rate of Change over an Interval

Sample Problem 5: As noted at the beginning of this section, rates of change over a period or an interval are often quoted as percentages rather than straight numbers. Often we can learn more from a percentage increase than we can from an average increase. For example, suppose you bought 100 shares of stock in the Calculus Company at $24 per share and 100 shares of stock in the Algebra Company at $120 per share. In the past 12 months the share price of each stock rose at an average rate of $1 per month. How should you compare these performance figures to each other?

Solution: The solution to this problem depends on how you interpret "performance." In the sense of actual dollars per month growth, the two stocks are growing at the same rate. The share price of both stocks rose by an average of $1 per month, or $12 over the 12 months. However, as an investor, you want to know which stock gave you a better **return on your investment**. Return on an investment is usually expressed as a percent, so you can compare the *relative* rates of return (per dollar invested rather than absolute return in dollars). Thus, to put the rise in stock prices in perspective we should find the **percent rate of change** of the stock prices over the 12 months.

Let's consider the Calculus Company first. We have already calculated that the rate of change of its share price was $1 per month in absolute dollars. What percent rate of change does this correspond to? We know that the Calculus Company stock was selling for $24 a share at the beginning of the 12 months. If we think of this as our starting base-line share price, what percent of this is represented by the $1 average increase every month? This just means finding what percent 1 is of 24. Expressed as a fraction, the answer would be 1/24 , which we can convert to a decimal or a percent, to give us an answer of

$$\frac{1}{24} = 0.04166\ldots = 4.166\ldots\%$$

per month. We could say that the percent rate of change is about 0.042 per month, or about 4.2% per month.

Now let's consider the Algebra stock, which was selling at $120 a share at the beginning of the 12 months. Since the average rate of change in absolute dollars was the same ($1 per month), we can use the same procedure to find the percentage rate of change over the 12 months:

$$\frac{\$1 \text{ per month}}{\$120} = 0.00833\ldots \text{ per month} = 0.833\ldots\% \text{ per month}$$

This tells quite a different story about how the stocks performed, and makes it clear that the Calculus Company was a much more lucrative investment. □

To generalize, the **percent rate of change of a function over an interval** is calculated by **dividing the average rate of change of the function over the interval by the output value at the beginning of the interval** (the *initial* output value):

$$\text{Percent rate of change over } [x_1, x_2] = \frac{\text{average rate of change of } f(x) \text{ over } [x_1, x_2]}{f(x_1)}$$

$$= \frac{\left(\dfrac{f(x_2) - f(x_1)}{x_2 - x_1}\right)}{f(x_1)}$$

In other words, the percent rate of change expresses the average rate of change of a function over an interval as a percentage of the function's initial value.

Recall from Section 1.4 the discussion on **percent change** over an interval. **The percent change over an interval is the change in the output of the function over that interval divided by the output of the function at the beginning of the interval**:

$$\text{Percent change over an interval} = \frac{f(x_2) - f(x_1)}{f(x_1)}$$

Don't confuse **percent change** over an interval and **percent *rate of* change** over an interval. These two expressions sound very much alike, but they tell us different things. For Sample Problem 5, the Calculus Company stock rose by \$12 over the 12 months and started at \$24 dollars per share, so the **percent change** over the year was

$$\frac{\$12}{\$24} = 0.5 = 50\%$$

Notice that **the units for percent change are just %**. When we calculated the **percent rate of change** we got

$$\frac{\$1 \text{ per month}}{\$24} = 0.04166\ldots \text{ per month, or } 4.166\ldots\% \text{ per month}$$

These two quantities are different, but closely related. **Another way to think of the percent rate of change is that it is the *average* percent change *over the interval*.** In our example, the percent change was 50% over the 12 months, so if we average this over the 12 months to see what percent it represents per month, we get

$$\frac{50\%}{12 \text{ months}} = \frac{50}{12}\% \text{ per month} = 4.166\ldots\% \text{ per month}$$

Either way of thinking of the percent rate of change is fine. Sometimes one way makes more sense in a particular situation. Use whatever is easiest and makes most sense to you. In either case, **the units for percent rate of change are *percent per (input unit)*.**

Let's go back to Sample Problem 2, the ball toss. The **average rate of change** from 1 to 1.5 seconds was

$$\frac{41 \text{ feet} - 37 \text{ feet}}{1.5 \text{ seconds} - 1 \text{ second}} = 8 \text{ feet per second}$$

The **percent rate of change** from 1 to 1.5 seconds was thus

$$\frac{8 \text{ feet per second}}{37 \text{ feet}} \approx 0.216 \text{ per second} = 21.6\% \text{ per second}$$

The **percent change** from 1 to 1.5 seconds was

$$\frac{41 \text{ feet} - 37 \text{ feet}}{37 \text{ feet}} = \frac{4}{37} \approx 0.108 = 10.8\%$$

Notice again that if we divide the percent change by the length of the interval, we get the percent rate of change:

$$\frac{10.8\%}{0.5 \text{ seconds}} = 21.6\% \text{ per second}$$

In Sample Problem 3 we looked at the number of cycles to failure for a battery as a function of the percent discharge of the battery each cycle. The **average rate of change** from 4% discharge to 14% was

$$-436.1 \text{ cycles to failure per percentage point of discharge}$$

The **percent rate of change** from 4% discharge to 14% is therefore

$$\begin{aligned} &\frac{-436.1 \text{ cycles to failure per percentage point of discharge}}{5656.5 \text{ cycles to failure}} \\ &\approx -0.077 \text{ per percentage point of discharge} \\ &= -7.7\% \text{ per percentage point of discharge} \end{aligned}$$

This answer is very hard to say with all the "percents" and "pers", but these are the correct units for this answer. Remember, the correct units are just as important as the correct numerical answer. Answers should always contain units. There should be **no naked numbers—only correctly dressed numbers!**

The percent change from 4% discharge to 14% discharge is

$$\frac{1295.2 \text{ cycles to failure} - 5656.5 \text{ cycles to failure}}{5656.5 \text{ cycles to failure}} \approx -0.7709 = -77.09\%$$

Since this percent change of about −77% is spread over an interval of 10 percentage points of battery discharge, the average percent change over the interval is

$$\frac{77\%}{10 \text{ percentage points of discharge}} = 7.7\% \text{ per percentage point of discharge}$$

which is the percent rate of change.

Sample Problem 6: You have increased your number of free throws made at the foul line from 1 out of 10 to 5 out of 10 over two hours practice time. After the first hour you were making 3 out of 10, and after the second hour you were making 5 out of 10. How should you report this to your coach? Should you tell the coach your average rate of change or your percent rate of change? The coach is no dummy and will get the picture very quickly, but you might as well present yourself in the best light possible.

Solution: You could report that you increased your number of free throws out of 10 tries at the average rate of 2 free throws out of 10 tries per hour:

$$\frac{5 \text{ made} - 1 \text{ made}}{2 \text{ hours} - 0 \text{ hours}} = \frac{4 \text{ shots made}}{2 \text{ hours}} = \text{an average rate of 2 shots per hour}$$

In other words, the number of shots out of 10 you got *increased* at a rate of 2 shots per hour over the two-hour time interval.

On the other hand, you could report your progress as a percent rate of change:

$$\frac{2 \text{ shots per hour}}{1 \text{ shot}} = 2 \text{ per hour} = 200\% \text{ per hour}$$

In this situation, your percent change is 4/1 = 400% and is spread over 2 hours, for an average percent change (percent rate of change) of 400/2 = 200% per hour.

Each of these answers is correct, although they each paint a very different picture. You actually improved at the rate of 2 shots per hour for the first two hours. This answer might be better if you wanted to convey to your coach a sense of steady progress. On the other hand, if you wanted to convey major *improvement*, you might use the 200% per hour (percent rate of change). □

Sample Problem 7: From the first semester to the second semester you and your best friend both raised your GPAs by one full point. Yours went from 1.5 to 2.5, and your friend's went from 2.5 to 3.5. How can each of you present your improvement in the very best light to your other friends and to your parents?

Solution: You started with a GPA of 1.5 for the first semester and so your average rate of change is

$$\frac{(2.5 - 1.5) \text{ points}}{(2 - 1) \text{ semester}} = \frac{1 \text{ point}}{1 \text{ semester}} = = 1 \text{ point per semester average rate of change}$$

If you calculated the **percent rate of change** it would be

$$\frac{1 \text{ point per semester}}{1.5 \text{ points}} = 0.6666\ldots \text{ or about } 66.7\% \text{ per semester.}$$

Your friend started with a GPA of 2.5 for the first semester, but the **average rate of change** would be the same:

$$\frac{(3.5 - 2.5) \text{ points}}{(2 - 1) \text{ semester}} = \frac{1 \text{ point}}{1 \text{ semester}} = 1 \text{ point per semester average rate of change}$$

The **percent rate of change** for your friend would be

$$\frac{1 \text{ point per semester}}{2.5 \text{ points}} = 0.4 \text{ per semester, or } 40\% \text{ per semester}$$

You would probably choose to present the increase to parents, dean, or whomever as a percent rate of change, since you started with the lower GPA compared to your friend, and so the percent increase is bigger. Of course, if your GPA *dropped* instead of *rising*, you might choose to present it the other way around. Similarly, if your friends were not overly academically inclined, you might also choose the other way around, to avoid showing off or looking too nerdy. □

As we have just seen, there are often different ways to present the same information. You should be aware of this and be able to interpret exactly what is being reported when rates of change and percent rates of change are used. The average rate of change tells us the *absolute* rate at which a function is changing, on average, over an interval. The percent rate of change tells us the *relative* rate at which the function is changing over the interval. Usually, if you see a reference to a rate of change over an interval and it does not specify whether it is an average rate of change or a percent rate of change, the best default assumption to make is that it refers to the *average* rate of change.

Section Summary

Before you begin the exercises for this section be sure that you can

- Understand that finding the average rate of change of a function over an interval is a simple slope calculation: Change in output divided by change in input.
- Calculate the average rate of change of a function $f(x)$ over an interval $[x_1, x_2]$ from a graph, data table, or model, using the formula $\dfrac{f(x_2) - f(x_1)}{x_2 - x_1}$.
- Correctly give the units of the average rate of change as (output units) per (input unit).
- Recognize the average rate of change over an interval as the slope of the secant line (or the line segment) from one endpoint of the interval through the other endpoint of the interval.
- Understand that the percent rate of change of a function over an interval is simply the average rate of change expressed as a percent (or fraction) of the *initial* output value.
- Calculate the percent rate of change over an interval from a graph, data table, or model using the formula $= \dfrac{\left(\dfrac{f(x_2) - f(x_1)}{x_2 - x_1}\right)}{f(x_1)}$.
- Correctly give the units of a percent rate of change over an interval as percent per input unit.
- Understand the difference and the relationship between percent change and percent rate of change. The percent rate of change is the percent change over the interval divided by the length of the interval (since the percent change is being spread and averaged over the number of units in the interval).

EXERCISES FOR SECTION 2.1

Warm Up

1. Consider the following table:

x	0	1	2	3	4	5
y	1.5	2.4	3.6	4.5	5.6	6.7

For each of the following intervals, find the average rate of change:
(a) [0,1] **(b)** [0,3] **(c)** [2,5] **(d)** [0,5]

2. Consider the following table:

x	1	2	3	4	5	6
y	24	26	28	27	25	23

For each of the following intervals, find the average rate of change:
(a) [1,3] **(b)** [2,4] **(c)** [2,6] **(d)** [1,5]

3. Consider the following table:

x	0	1	2	3	4	5
y	78	65	60	54	60	66

For each of the following intervals, find the average rate of change and the percent change. Then show *two* ways of calculating the percent rate of change.
(a) [0,1] **(b)** [4,5] **(c)** [2,4]

4. Consider the following table:

x	1	2	3	4	5	6
y	1	2.02	3.07	4.11	5.16	6.21

For each of the following intervals, find the average rate of change and the percent change. Then show *two* ways of calculating the percent rate of change.
(a) [1,2] **(b)** [2,5]

5. For Figure 2.1-4, estimate the average rate of change over each of the following intervals:
(a) [10,12] **(b)** [10,14] **(c)** [12,20] **(d)** [15,19]

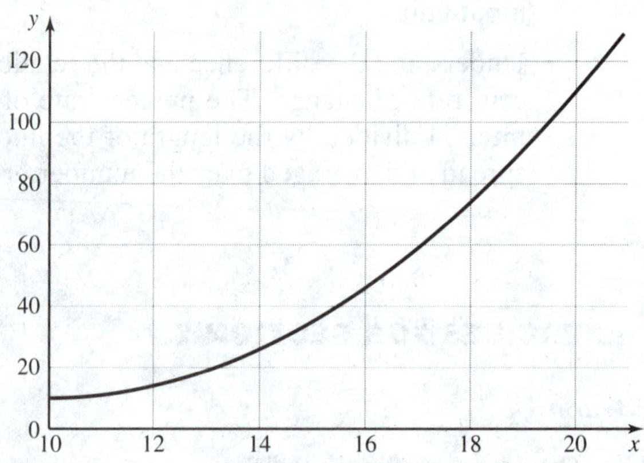

Figure 2.1-4

6. For Figure 2.1-5, estimate the average rate of change over each of the following intervals:
(a) [2,4] **(b)** [2,8] **(c)** [4,6] **(d)** [3,5]

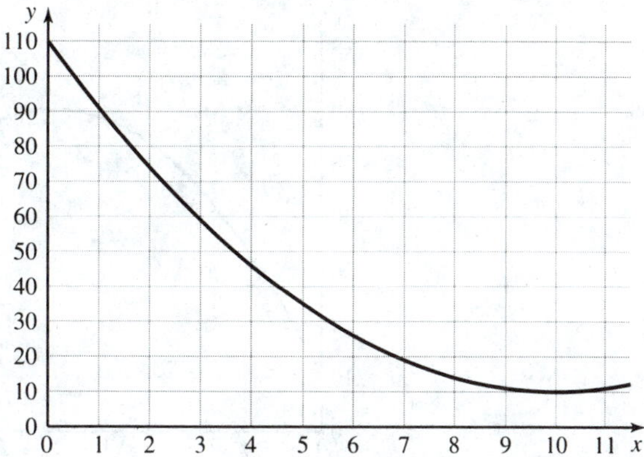

Figure 2.1-5

7. For Figure 2.1-6, estimate the average rate of change, the percent rate of change, and the percent change over each of the following intervals:
(a) [5,10] **(b)** [5,15] **(c)** [3,5]

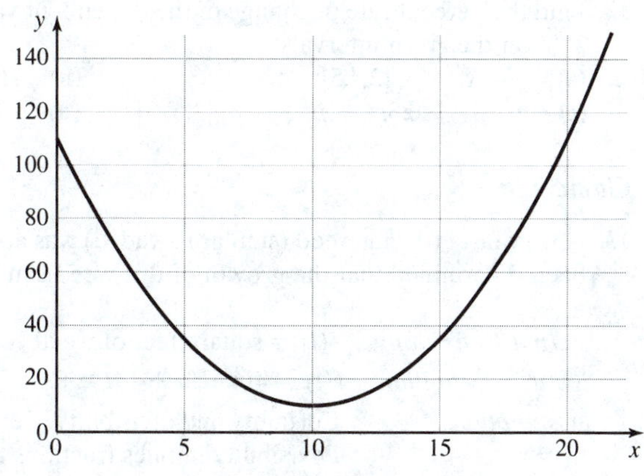

Figure 2.1-6

8. For Figure 2.1-7, estimate the average rate of change, the percent rate of change, and the percent change over each of the following intervals:
(a) [5,8] **(b)** [2,8] **(c)** [2,4]

9. Find the average rate of change of the dependent variable with respect to the independent variable on the given intervals:
(a) $y = x^2 + 2$; [3,3.5] **(b)** $y = 2t - t^2$; [1,2]
(c) $f(r) = r^3 - 3r^2 + 5r - 10$, [0,5] **(d)** $s(t) = -0.5t^3 + 1.4t^2 - 3.8t + 6$, [2.5,2.8]

10. Find the average rate of change of the dependent variable with respect to the independent variable on the given intervals:
(a) $y = x^2 - 5$, [2,4.5] **(b)** $f(x) = -3x^2 + 2x - 5$, [0,4]
(c) $q(p) = -0.3p^3 + p^2 - 4p$, [2.5, 3] **(d)** $d(w) = 4w^3 - 6w^2 + 9w - 34$, [10,11]

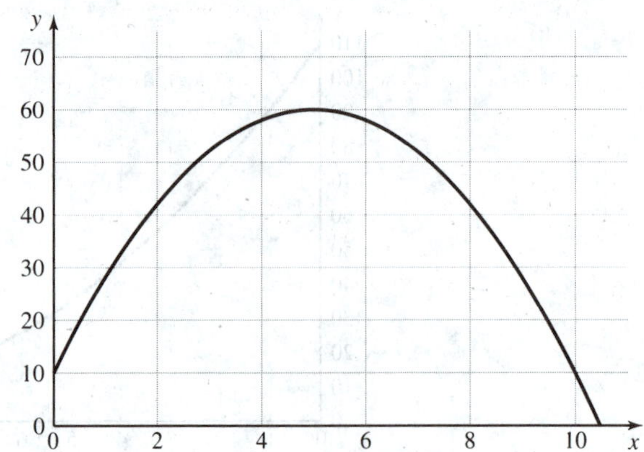

Figure 2.1-7

11. Find the percent rate of change of the dependent variable with respect to the independent variable on the given intervals:
(a) $y = x^2 + 2$; [3,3.5] **(b)** $y = 2t - t^2$; [1,2]
(c) $f(r) = r^3 - 3r^2 + 5r - 10$, [0,5] **(d)** $s(t) = -0.5t^3 + 1.4t^2 - 3.8t + 6$, [2.5,2.8]

12. Find the percent rate of change of the dependent variable with respect to the independent variable on the given intervals:
(a) $y = x^2 - 5$, [2,4.5] **(b)** $f(x) = -3x^2 + 2x - 5$, [0,4]
(c) $q(p) = -0.3p^3 + p^2 - 4p$, [2.5, 3] **(d)** $d(w) = 4w^3 - 6w^2 + 9w - 34$, [10,11]

Game Time

13. A very fast growing weed (similar to kudzu) was accidentally introduced into a local lake. It has been determined that the growth of this weed can be modeled by

Verbal Definition: $f(t)$ = square feet of weed coverage t years after introduction.
Symbol Definition: $f(t) = 4t^2 + 3t + 1, t \geq 0$.
Assumptions: Certainty and divisibility. Certainty implies the relationship holds *exactly*. Divisibility implies fractions are possible for both variables.

(a) For each of the following intervals, calculate the average rate of change: [0,0.5] [0.5,1.0]. Express your answers in complete sentences.
(b) What is the percent rate of change over these intervals? Express your answers in complete sentences.

14. A model for the percentage change in the Consumer Price Index for medical care is as follows:

Verbal Definition: $I(x)$ = the percentage change in the Consumer Price Index for medical care x years after Dec. 31, 1986 (for the preceding year).
Symbol Definition: $I(x) = 0.0193x^4 - 0.3547x^3 + 1.9314x^2 - 3.0161x + 7.6229, x \geq 0$.
Assumptions: Certainty and divisibility. Certainty implies the relationship holds *exactly*. Divisibility implies fractions are possible for both variables.

(a) For each of the following intervals, calculate the average rate of change: [0,1] [2,4]. Express your answers in complete sentences.

(b) What is the percent rate of change over these intervals? Express your answers in complete sentences.

15. The data for the unemployment in the U.S. civilian population is shown in the following table:

Year (Dec. 31)	Number Unemployed (thousands)
1983	10717
1984	8539
1985	8312
1986	8237
1987	7425
1988	6704
1989	6528
1990	6874
1991	8426
1992	9384
1993	8734
1994	7996

Calculate the average rate of change, the percent change, and the percent rate of change over each of the following intervals:

(a) 1986 to 1989
(b) 1989 to 1992
(c) 1986 to 1992
(d) Does the average rate of change from 1986 to 1992 fully reflect what occurred over this interval? Explain your answer.

16. Shown in the following table is the percent of federal outlays for the interest paid on public debt in the United States.

Fiscal Year	Percent of Federal Outlays
1984	18.1
1985	18.9
1986	19.2
1987	19.5
1988	20.1
1989	21.0
1990	21.1
1991	21.5
1992	21.1
1993	20.8
1994	20.3

Source: The World Almanac and Book of Facts 1996, p. 112 (Bureau of Public Debt, U.S. Dept. of the Treasury).

Calculate the average rate of change, the percent change, and the percent rate of change over each of the following intervals:

(a) 1986 to 1989
(b) 1990 to 1992
(c) 1988 to 1994
(d) Do the average rates of change over these intervals fully reflect what occurred over this interval? Explain your answer.

17. Figure 2.1-8 shows the number of tiles left in a computer game after playing the game for a period of time. From the graph, estimate the average rate of change and the percent rate of change over the following intervals. Write your answers in complete sentences.

(a) [5,10]
(b) [50,60]
(c) [30,40]

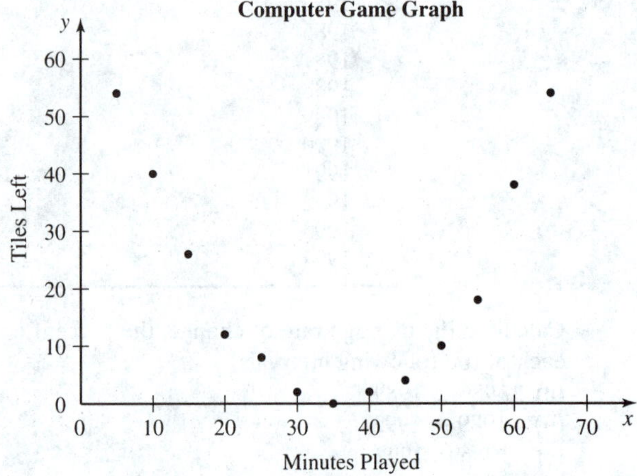

Figure 2.1-8

18. Figure 2.1-9 shows the annual percentage change in the medical Consumer Price Index (CPI). From Figure 2.1-9, estimate the average rate of change and the percent rate of change over the following intervals. Write your answers in complete sentences.

(a) [1986,1988]
(b) [1989,1992]
(c) [1990,1994]

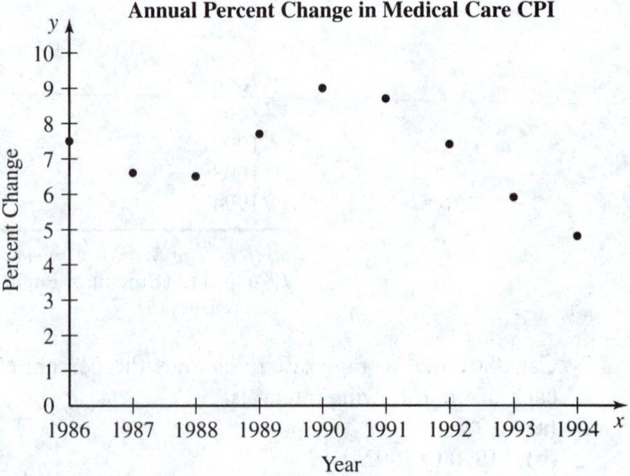

Figure 2.1-9

19. Figure 2.1-10 shows the annual percent change in the medical Consumer Price Index.

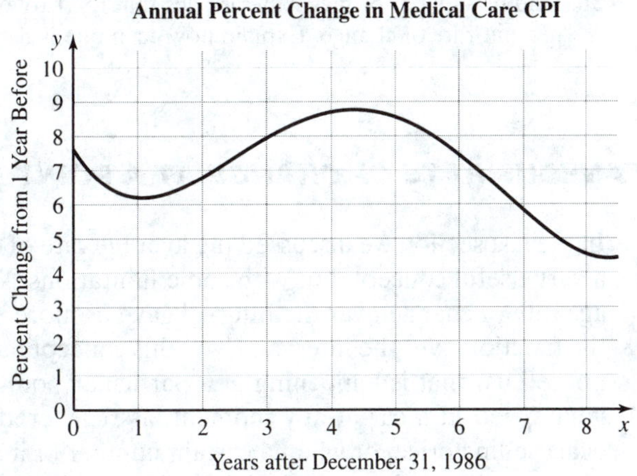

Figure 2.1-10

(a) For each of the following intervals estimate the average rate of change: [0,1] [2,4]. Express your answers in complete sentences.

(b) What is the percent rate of change over these intervals? Express your answers in complete sentences.

(c) Use the *model* given in Exercise 14 to perform the corresponding calculations, if you haven't already done so. How do your answers differ from those above? Why are the answers different?

20. Figure 2.1-11 shows the percentage of households with cable TV.

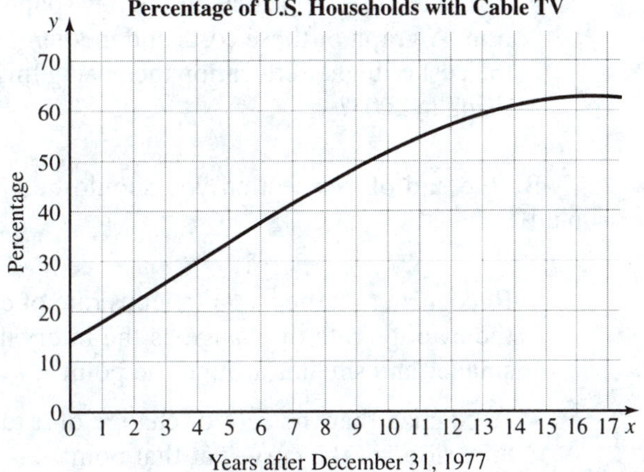

Figure 2.1-11

(a) For each of the following intervals, calculate the average rate of change: [0,5] [5,10] [10,17]. Express your answers in complete sentences.

(b) What is the percent rate of change over these intervals? Express your answers in complete sentences.

21. Find newspaper or magazine articles that refer to average rate of change, percent change, and percent rate of change. Explain how each one is used.

2.2 INSTANTANEOUS RATE OF CHANGE AT A POINT

In the last section we discussed the average rate of change of a function over an interval—a very useful concept, but with some limitations. We saw that in some cases the percentage rate of change over the interval gave us more useful information about the change in the function over the interval. Even this concept has some limitations. These measures do not tell us what is happening at a particular point, that is, at a given point in time (like your speed in a car at any moment, as registered by the speedometer), or at a certain exact selling price, or when a certain number of items are sold. In this section we will see how to estimate the rate of change of a function at a point by seeing what happens to the value of the average rate of change between that point and a second point, as the second point gets closer to the original point. If the value converges to a limiting value from both sides, we will call the result the instantaneous rate of change at that point. We will also discuss what this corresponds to graphically.

Here are the kinds of problems that the material in this section will help you solve:

- You are running a reelection campaign for a friend who took office a few years ago. Part of the campaign platform for the first election was a tough stance on crime. Should you use the crime rate in the United States as part of the reelection campaign? How could you use it?
- You have seen an article in the newspaper concerning the costs of higher education. A graph of these costs and median family income was shown. How fast was the cost of higher education increasing this year? How fast are the predicted costs rising in 2005?

By the end of this section you should be able to answer questions like these and should

- Recognize that the instantaneous rate of change at a point is the limiting value of the average rate of change as the interval over which that average is taken gets smaller and smaller around the point.
- Recognize that the rate of change of a function at a point can be thought of as "the slope of the curve" at that point.
- Know that "the slope of a curve at a point" really means the slope of the tangent line to the curve at that point.
- Know that the tangent line to a curve at a point can be thought of as a limit of secant lines (and its *slope* as the limit of the *slopes* of the secants) and is also the straight line through the point that most closely resembles the curve *near* the point.

- Be able to estimate the instantaneous rate of change of a function at a point from its graph by estimating the slope of the tangent line to the curve at that point.
- Be able estimate the rate of change of a function at a point from the symbol definition of a function or model using technology and the limit concept.
- Know that the instantaneous rate of change of a function can itself be thought of as a function of the independent variable (defined to be the slope of the tangent at that point).

Estimating Instantaneous Rate of Change at a Point from Models and Graphs

Sample Problem 1: In Section 2.1 we found the rate at which a ball was rising, on average, between time 1 and time 1.5 seconds (after it was thrown up in the air). How fast is the ball rising *exactly* 1 second after you tossed it?

Solution: As we did in the last section we'll start by looking at a table of data values (Table 1) and also the graph (Figure 2.2-1).

TABLE 1

Time/Sec	0	0.25	0.5	0.75	1	1.25	1.5	1.75	2	5
Height/Ft	5	16	25	32	37	40	41	40	37	32

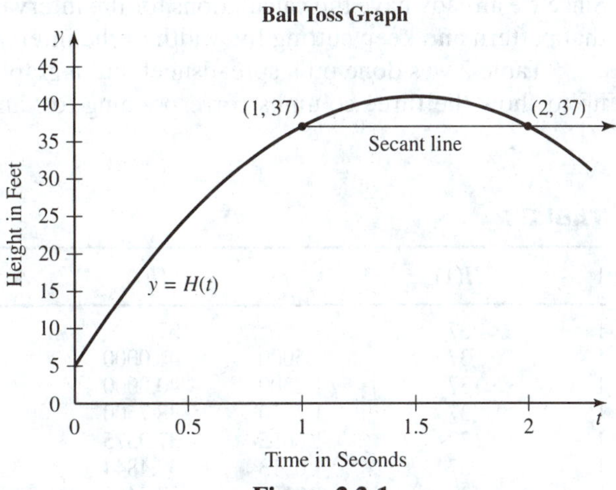

Figure 2.2-1

This time, let's also use the model we found for the height of the ball in Section 1.4:

Verbal Definition: $H(t)$ = the height of the ball in feet, t seconds after the toss.

Symbol Definition: $H(t) = -16t^2 + 48t + 5, 0 \leq t \leq 3$.

Assumptions: Certainty and divisibility. Certainty implies the relationship is exact, and divisibility implies that fractions are possible for both variables.

From the data, from the graph, or from the model, we are able to calculate the average rate of change of the height from 1 second to 2 seconds:

$$\frac{37 \text{ feet} - 37 \text{ feet}}{2 \text{ seconds} - 1 \text{ second}} = \frac{0 \text{ feet}}{1 \text{ second}} = 0 \text{ feet per second}$$

The average rate of change is *0*! This does not seem to represent what occurred at all. The ball did not maintain a constant height from 1 to 2 seconds; it went higher and then began to come back down again. Recall our discussion in Section 2.1 about velocity and average velocity. Since our output in this example is the height of the ball, the velocity can be positive (when the ball is going up) or negative (when it is going down). You should be able to verify that the average velocity on the interval [1,1.5] is

$$\frac{(41 - 37) \text{ feet}}{(1.5 - 1) \text{ seconds}} = \frac{4 \text{ feet}}{0.5 \text{ seconds}} = 8 \text{ feet per second}$$

and that on the interval [1.5,2] it is -8 feet per second, so the average velocity over [1,2] does turn out to be 0. If we really want to know what was happening to the ball at 1 second, however, we should pick a point closer to 1 second, say $1 + 0.25$ or 1.25 seconds. The average rate of change from 1 to 1.25 seconds is

$$\frac{(40 - 37) \text{ feet}}{(1.25 - 1) \text{ seconds}} = \frac{3 \text{ feet}}{0.25 \text{ seconds}} = 12 \text{ feet per second}$$

Intuitively, it makes sense that, the smaller the interval around 1 we use to find the average rate of change, the closer we would expect the answer to be to the "actual" rate of change at 1. Let's try using technology to construct a table to show these calculations. Since we already have the calculations for the intervals [1, 1.5] and [1, 1.25], let's continue that pattern and keep cutting the width of the interval in half.

Table 2 was done on a spreadsheet, but a graphing calculator can also be easily set up to show the three columns corresponding to t and $H(t)$ (both rounded to 4 decimal

TABLE 2

1	$H(1)$	t	$H(t)$	Avg. Rate on $[1,t] = [H(t) - H(1)]/(t - 1)$
1	37	1	37	#DIV/0!
1	37	1.5000	41.0000	8.00
1	37	1.2500	40.0000	12.00
1	37	1.1250	38.7500	14.00
1	37	1.0625	37.9375	15.00
1	37	1.0313	37.4844	15.50
1	37	1.0156	37.2461	15.75
1	37	1.0078	37.1240	15.88
1	37	1.0039	37.0623	15.94
1	37	1.0020	37.0312	15.97
1	37	1.0010	37.0156	15.98
1	37	1.0005	37.0078	15.99
1	37	1.0002	37.0039	16.00
1	37	1.0001	37.0020	16.00

places), and the average rate on $[1, t]$ (rounded to 2 decimal places), although the headings cannot be as specific.

In the first row of Table 2 we have shown what happens if you blindly try the average rate of change formula for the interval [1,1]: Since the width of the interval is 0, the formula would require division by 0, so the table indicates that this is not possible. In looking at the rest of Table 2, notice that the average rate of change seems to level off at the value 16.00 as the interval gets smaller and smaller (as t gets closer and closer to 1). This is a beautiful illustration of the mathematical concept called the **limit** of a function. In this case, the function we are taking the limit of is the expression for the average rate of change on $[1, t]$:

$$f(t) = \frac{H(t) - H(1)}{t - 1}, \text{ for } t \neq 1$$

You may have thought about the fact that we took intervals only to the *right* of 1 (had t approach 1 from the *right*), when we could also have done it from the left. In this case, the values of t would be *smaller* than 1, so the interval of interest would actually be $[t, 1]$ and the average rate of change calculation would be

$$f(t) = \frac{H(1) - H(t)}{1 - t}, \text{ for } t \neq 1$$

Notice that we have used the same function name as before, even though the subtractions in the numerator and denominator are reversed. This is because the two functions are identical: Reversing the subtraction is like changing the sign or multiplying by -1, so if we multiply the second function by -1 over -1 we get

$$f(t) = \frac{H(1) - H(t)}{1 - t} = \frac{H(1) - H(t)}{1 - t} \cdot \frac{-1}{-1} = \frac{-H(1) + H(t)}{-1 + t} = \frac{H(t) - H(1)}{t - 1}, \text{ for } t \neq 1$$

For this reason, we can put the results for the calculations from *both* sides of 1 into the *same* table, shown in Table 3.

From Table 3 you can see that as the value of t gets closer and closer to 1 from *below* (or from the *left*), the average rate of change again levels off at 16.00. This seems to be very strong evidence for saying that the rate of change of the function *at* $t = 1$ (which we will call the **instantaneous rate of change** of the height at $t = 1$) is approximately 16.00 feet per second. Keep in mind that if we showed more decimal places for the average rates of change, it would take longer for our values to converge (the input intervals would have to get even smaller). □

The Limit of a Function at a Point

Now that we have looked at what happened on *both* sides of 1, we are ready for a general definition of the limit of a function at a point:

> **The limit of a function $f(x)$ as x approaches a, written as $\lim_{x \to a} f(x)$, means the value that $f(x)$ approaches (levels off at, converges to) as x gets closer and closer to a from *both* sides, *ignoring* what happens when $x = a$. In order for "*the* limit" to exist, the values approached from the two sides must be the *same*.**

TABLE 3

t	$H(t)$	Avg. Rate between 1 and t = $[H(t) - H(1)]/(t - 1)$
1.5000	41.0000	8.00
1.2500	40.0000	12.00
1.1250	38.7500	14.00
1.0625	37.9375	15.00
1.0313	37.4844	15.50
1.0156	37.2461	15.75
1.0078	37.1240	15.88
1.0039	37.0623	15.94
1.0020	37.0312	15.97
1.0010	37.0156	15.98
1.0005	37.0078	15.99
1.0002	37.0039	16.00
1.0001	37.0020	16.00
1	37	#DIV/0!
0.9999	36.9980	16.00
0.9998	36.9961	16.00
0.9995	36.9922	16.01
0.9990	36.9844	16.02
0.9980	36.9687	16.03
0.9961	36.9373	16.06
0.9922	36.8740	16.13
0.9844	36.7461	16.25
0.9688	36.4844	16.50
0.9375	35.9375	17.00
0.8750	34.7500	18.00
0.7500	32.0000	20.00
0.5000	25.0000	24.00

Since the concept of a limit came initially from this idea of the instantaneous rate of change, we can see why we need to *ignore* what happens *at* the value the variable is approaching. For the rate of change calculation, the function is not defined at that point, since it involves division by 0. Note also that the definition involves approaching the value from *both* sides and requires that the values approached from each side must be the *same*.

In more formal math courses, the limit is defined more precisely, based on the idea that $\lim_{x \to a} f(x) = L$ means that $f(x)$ can be kept arbitrarily close (within some specified acceptable error magnitude) to the value L for *all* values of x that are chosen *close enough* to a. In our example, from Table 3, we can see that our L value is 16.00 and our a value is 1. If we want to keep our output values (the average rate of change values) within 0.1 of 16 (between 15.9 and 16.1), we could safely choose any value of x (or t in the example) between 0.997 and 1.003 . For this example, the acceptable error magnitude is the 0.1 and was chosen arbitrarily. If we chose an acceptable error magnitude of 0.01, then the required interval of x values would reduce to being between 0.9998 and 1.0002. From looking at the table, you can see that no matter how small the error tolerance, it appears that it would always be possible to find an interval on either side of 1 for which all the output values (average rates of change) would stay within that error tolerance of 16. As in the above examples, the required interval of input values typically will normally *depend* on the *size* of the acceptable error magnitude.

For general functions, this idea of a limit can get very complicated. In this course, as we have said before, we are focusing on models with functions that are *continuous* (no holes, gaps, or jumps) and *smooth* (no corners, angles, or cusps). **When a function is continuous at $x = a$, $\lim_{x \to a} f(x) = f(a)$.** And when a function is smooth, the limit corresponding to the instantaneous rate of change always exists (although the calculation can be complicated); and must be a finite number. For all of the limits we are likely to need in this course, using technology to approximate an answer should be sufficient.

In our example, the mathematical way to express the limit we found is

$$\lim_{t \to 1} f(t) = \lim_{t \to 1} \frac{H(t) - H(1)}{t - 1} \approx 16.00$$

Remember that our domain for $f(t)$ was $t \neq 1$, but the definition of limit says that we can ignore what happens when $t = 1$, so this does not create a problem.

Let's now consider what this limit process for finding the instantaneous rate of change at a point corresponds to graphically. Recall from Section 2.1 that we said that the average rate of change over an interval corresponds to the *slope* of a secant line drawn *from* one of the endpoints of the interval *through* the other. For the average rates of change calculated in Table 3, let's draw all of our secant lines starting at the point common to *all* of the calculations, the point (1,37). If we draw in all of these secant lines, we get the graph in Figure 2.2-2.

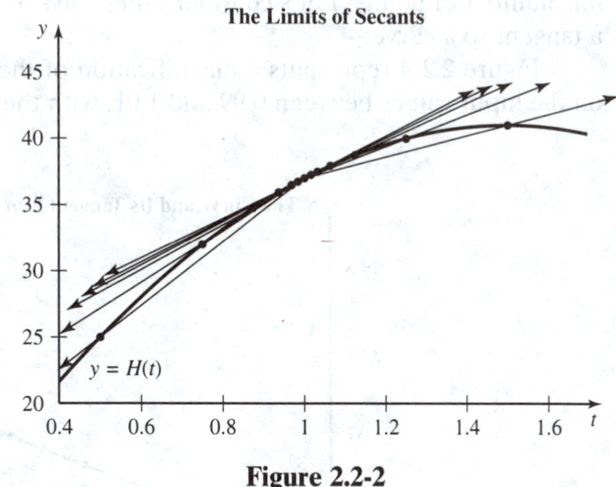

Figure 2.2-2

In Figure 2.2-2, notice that the secant lines drawn from the point at 1 through the points at 1.5, 1.25, 1.125, and so forth, gradually get closer and closer and seem to converge on a "limiting" secant line, and that the same thing happens with the secants drawn to the left. In either case, the closer the second point is to 1, the closer its secant line is to this limiting secant line. Notice also that the two limiting secant lines seem to be going in exact opposite directions, so they form a single line! This line that they form is called the **tangent** line to the curve at $t = 1$. Figure 2.2-3 shows the graph with just the tangent line drawn on.

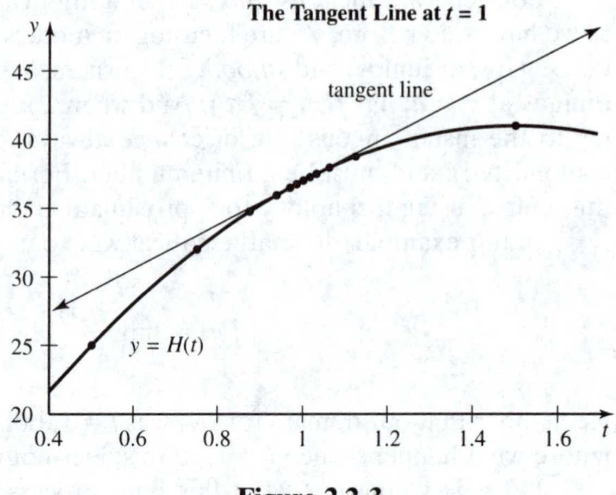

Figure 2.2-3

Notice that the tangent line just "kisses" the curve at $t = 1$, just touches it at that one point, just like the tangent lines to circles that you studied in geometry. This idea of just touching in one point is not the best way to think of a tangent line in calculus, however. For example, if our original function had been linear and we drew in secant lines to nearby points, *all* of the secant lines would be the same (the original line), and so the tangent line would also be the same as the original line and would touch the graph in an *infinite* number of points! Let's consider other ways to think about this more general idea of a tangent to a curve.

Figure 2.2-4 represents a magnification of the graph in Figure 2.2-3, "zooming in" on the input values between 0.99 and 1.01, with the tangent line again drawn in.

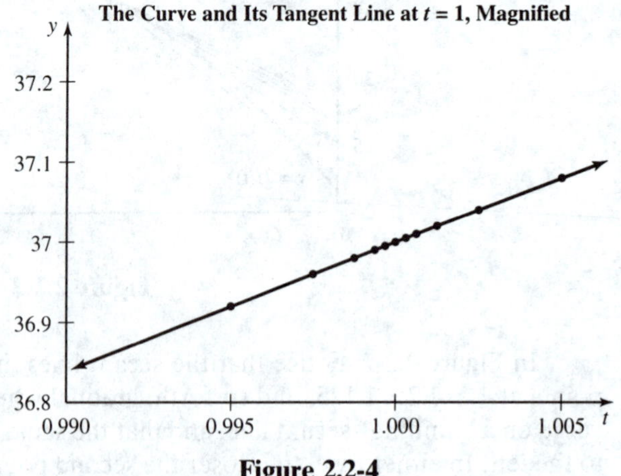

Figure 2.2-4

You probably can't even tell that the tangent line and the curve are different! They are so similar that, for all practical purposes, they are essentially identical in this interval.

This gives us a better way of thinking of the tangent line to a curve at a point:

> **The tangent line to a curve at a point is the straight line going through the point that most closely resembles the curve *near* the point.**

Another way of saying the same thing is that the tangent line is the **best linear approximation** of the function at and near the point. If you look back at Figure 2.2-3, you can see even there that the tangent is very similar to the curve between 0.9 and 1.1. There are a number of situations where we can use this similarity to *approximate* the value of the original function using the tangent line.

From yet another perspective, if you zoom in on a graph near a point (as you can with most graphing calculators and graphing software for PCs) and the graph looks more and more like a straight line, that line is the tangent line.

Now, recall that the average rate of change over an interval corresponded to the *slope* of the secant line between the endpoints of the interval. Thus, when we found the limiting value of these rates of change as the second point got closer and closer to the first, since the secant lines converged to the tangent line, our final answer was the *slope* of the tangent line.

> **The instantaneous rate of change of a function at a point is the *slope* of the tangent to the curve at that point. This is the *limit* of the *slopes* of the secant lines from the point to a second point, as the second point gets closer and closer to the original point.**

To emphasize this point, Figure 2.2-5 has an arrow showing the dynamic approach of the second point to the first from the right (the same would apply to the left as well). Picture *sliding* the point along the curve and watching the resulting secant lines.

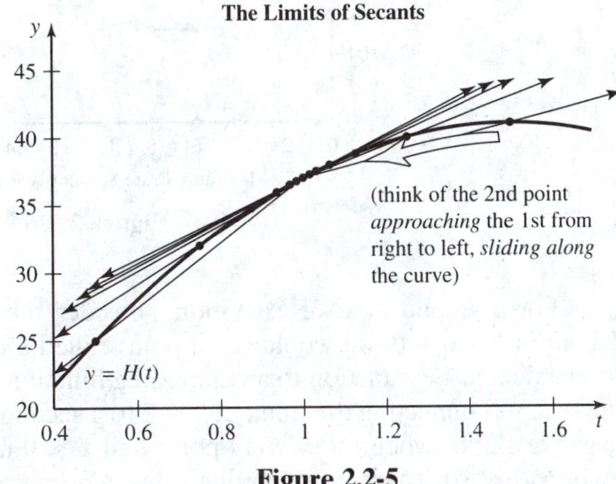

Figure 2.2-5

★ **Discovery Question:**

Can you think of some kind of graph where the graph would not look like a straight line as you zoomed in closer and closer at a point? (See if you can answer before reading ahead.)

Hint: The graph will not look like a straight line when there is a sudden abrupt change in the dependent variable. Recall some of the graphs from Section 1.1 on functions. Can you answer the question now? (Again, see if you can answer before reading ahead.)

Answer: Look back at the graph of the demand histogram in Figure 1.1-13, shown again here in Figure 2.2-6. If we "zoomed in" on this graph around the value of \$8 or \$10 or \$12, we would not see anything that looked like a single straight line; instead we would see *two* lines always, no matter how much we magnified, as long as our input interval included values on both sides of any of these endpoints. This is a discontinuous graph: It has holes, gaps, and jumps in it. On the other hand, notice that if we zoomed in on the graph at the input value of 9, it *would* look like a single straight line, so the tangent would be uniquely defined, and the instantaneous rate of change would also be uniquely defined. Since the tangent would be horizontal, the rate of change would be 0, since a horizontal line has a slope of 0.

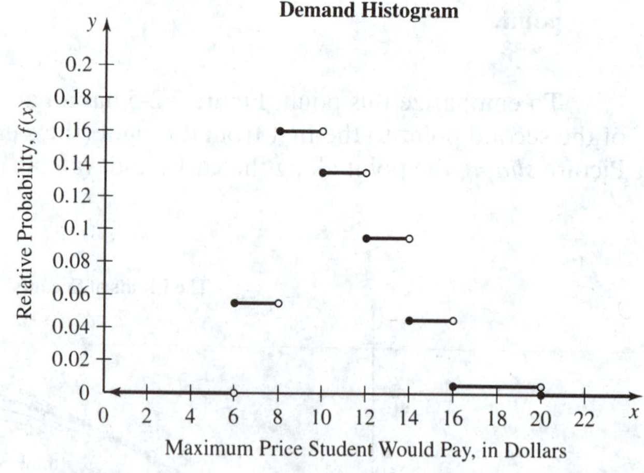

Figure 2.2-6

For a second kind of exception, consider the graph shown in Figure 2.2-7. If we "zoom in" at $x = 0$, we would see a point called a cusp. The function changes abruptly from a decreasing function to an increasing function. If we tried to calculate the slopes of the secants connecting the point at $x = 0$ to a second point, we would find that they were negative slopes when the second point had $x < 0$ and positive slopes when the second point had $x > 0$. There is no limiting value as x approaches 0. However, we *could* say that the secant lines would converge to a common secant line and could think of the extension of that (vertical) secant line as the tangent at $x = 0$. However, since the tangent is vertical, its slope is not defined, so the instantaneous rate of change is not defined. This is an example of a curve that is *not smooth*, since it has a cusp. Such a complication is not likely to arise in a realistic problem, since it would imply that something was changing at an *infinite* rate essentially! ★

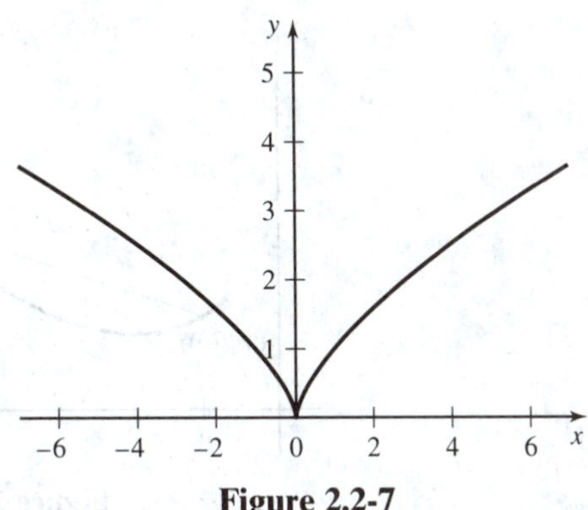

Figure 2.2-7

We now have a way of numerically estimating the instantaneous rate of change of a function at a particular point (if it exists), in addition to being able to calculate an average rate of change over an interval. **Sometimes we will not *specify* whether a rate of change is an *average* or *instantaneous* rate, and you will need to judge from the context. If the rate of change refers to an *interval*, it must be an *average* rate of change; if it refers to a *point*, it must be an *instantaneous* rate of change.**

The Symbol (Mathematical) Definition of the (Instantaneous) Rate of Change at a Point

The instantaneous rate of change of a function at a point is the limiting value of the slope of the secant line between that point and a second point on the curve as the second point approaches (but never actually reaches) the first point from both sides (the distance between the two points approaches zero but never equals zero). The values approached from the two sides must be the *same* for the rate of change to exist (be uniquely defined).

Figure 2.2-8 shows a generic graph of a curve corresponding to a function defined by $y = f(x)$ and the secant corresponding to the input interval $[a, x]$. We are eventually interested in the slope of the tangent at $x = a$ (the instantaneous rate of change at a), which means at the point $(a, f(a))$. To find it, we will be taking the slopes of secant lines from $(a, f(a))$ to a second point, which we will call $(x, f(x))$. In both cases, since the function is given by $y = f(x)$, to find the y value at a particular point, we can just substitute the x value into the function (put it within the parentheses). Using the slope formula, the slope of the secant is then

$$\text{slope of secant from } (a, f(a) \text{ to } (x, f(x)) = \frac{\Delta y}{\Delta x} = \frac{f(x) - f(a)}{x - a}$$

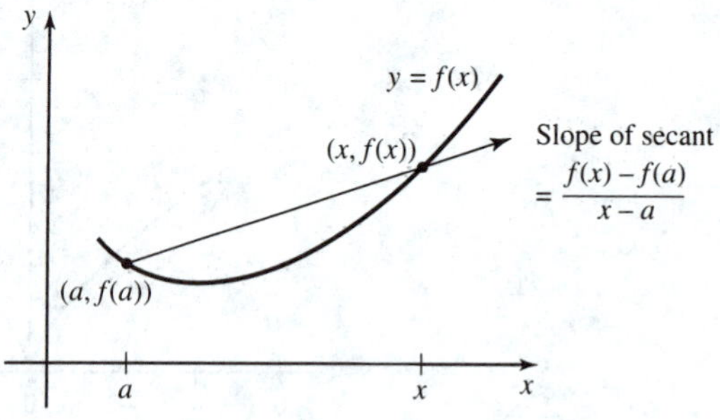

Figure 2.2-8

Since the slope of the tangent is the limit of the slopes of the secants, we can now write a mathematical definition for the instantaneous rate of change (slope of the tangent) at a:

The instantaneous rate of change of $f(x)$ at the point where $x = a$ is given by

$$\lim_{x \to a} \frac{f(x) - f(a)}{x - a}$$

Read this as "the limit as x approaches a of eff of x minus eff of a divided by x minus a." Recall that $\lim_{x \to a} g(x)$ means the value that $g(x)$ approaches as x gets closer and closer to a from both sides, ignoring what happens when $x = a$.

If we designate the difference between the input values of our given point at a and the other point at x as "h," then

$$h = x - a$$

and so we can write

$$x = a + h$$

As x approaches a (as the second point approaches the first), this difference between the input values will go to 0, so h will approach 0. Graphically, we can think of

h as the horizontal distance between the original point and the second point, so as they get closer, this distance will go to 0. We can then define the rate of change at a point as

$$\text{rate of change at } a = \lim_{h \to 0} \frac{f(a+h) - f(a)}{a+h-a} = \lim_{h \to 0} \frac{f(a+h) - f(a)}{h}$$

Read this as "the limit as h approaches zero of eff of a plus h minus eff of a divided by h." This last form is most commonly used in texts and it is this form that we will be using later as we look for a way to find the rate of change at a point algebraically.

Figure 2.2-9 illustrates this version of the definition.

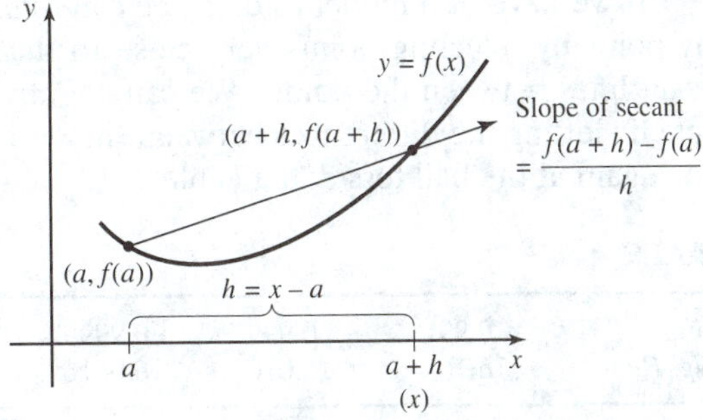

Figure 2.2-9

Terminology for Rate of Change at a Point

You won't hear the question, "What was the instantaneous rate of change of this or that at a particular point?" but you may hear, "How fast was this or that changing at a particular point?" How fast a function is changing at a point is called by several different names:

- **rate of change of a function at a point**
- **instantaneous rate of change at a point**
- **slope of the curve at a point**
- **the slope of the tangent to the curve at a point**
- **the derivative at a point**

The **derivative** is the mathematical name. You may have heard some reference to "derivatives" in the news in recent years. In the early 90s, Orange County, California invested its money in "derivatives" and lost almost all of it. Much was said in the news about the high risk in "derivatives." Don't be alarmed. The news stories were referring to investment derivatives, not mathematical derivatives. A mathematical derivative is a function "derived" from a second function, which gives the slope of that second function.

The derivative of a function at a point is the slope of the tangent line to the curve at that point and can be thought of as the "slope" of the graph of the function at that point.

Summarizing,

The slope of a curve at a point (the instantaneous rate of change at that point) is the slope of the tangent line to the curve at that point. The tangent line is the limiting line of the secant lines between the original point and a second point as the second point approaches the first, so the *slope* of the tangent is the limit of the *slopes* of the secant lines.

If we have fit a model to data, we can *estimate* the instantaneous rate of change at any point by selecting points very close to that point and calculating the slope of the secant lines between the points. We can get any desired degree of accuracy for the estimate by letting the differences between the points get smaller and smaller. For example, look again at the ball toss data (Table 4).

TABLE 4

Time	0.99	0.999	0.9999	1	1.0001	1.001	1.01
Avg. Rate	16.16	16.016	16.0016		15.9984	15.984	15.84

If we were satisfied with an estimate correct to the units' digit, we would have been satisfied with the values from 1 ± 0.01(0.99 and 1.01): 16.16 rounded to the nearest whole number is 16 and 15.84 rounded to the nearest whole number is 16. These numbers would not have been all right if we wanted the estimate to be correct to the first decimal place. If we wanted an answer that was correct to the first decimal place, 1 ± 0.001 would have been sufficient, since 16.016 rounded to the first decimal place is 16.0 and 15.984 rounded to the first decimal place is also 16.0. To be correct to the second decimal place we could use ± 0.0001, since 16.0016 rounded to the second decimal place is 16.00 and 15.9984 rounded to the second place is also 16.00.

Estimating the Rate of Change at a Point from a Graph

Sometimes we have the graph of a function but do not have the data from which the graph was drawn. How could this happen? Almost every graph that we see was created from a set of data, just as the graphs above were. (The only exceptions that immediately come to mind are a seismogram and an electrocardiogram.) However, we frequently see graphs in newspapers and other publications for which the data values are not given. If we have the graph of a function but do not have the data or a model for the function, we can **estimate** the slope of the curve at a point by drawing a **tangent line** at that point and estimating the slope of the tangent line. Alternatively, we could, of course, try to read the points from the graph and then fit a model, but this is equally open to error.

Drawing a tangent line to a curve at a point is not particularly easy. The tangent line at a point is the line that goes through the point and most closely resembles the curve *near*

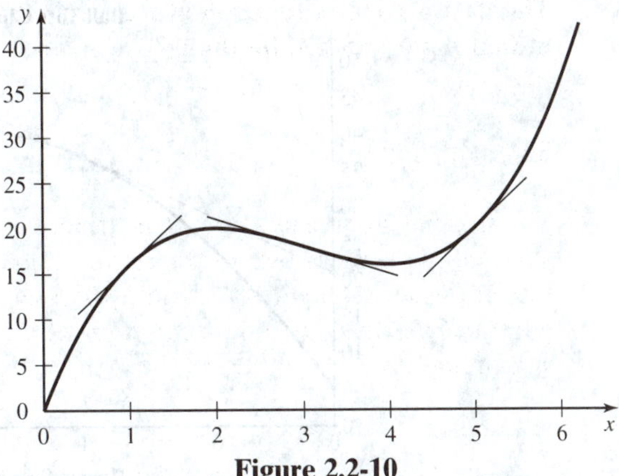

Figure 2.2-10

the point. If the curve is concave down at the point, the tangent line will be above the curve. If the curve is concave up at the point, the tangent line will be below the curve. If the curve changes concavity at the point (has an inflection point), the tangent line cuts through the curve at that point—above the curve where it is concave down and below the curve where it is concave up (Figure 2.2-10).

Recall that the tangent line is the line through the point that most closely resembles the curve *near* the point. It is *possible* for the tangent to hit the curve in *more* than one point. For example, a tangent drawn in at $x = 2$ on Figure 2.2-10 would hit the curve a second time somewhere around $x = 5$. A good way to draw the tangent is to put a ruler down on the curve (below the point if the curve is concave down, above the point if the curve is concave up) and slowly move the ruler toward the point until the point just shows and the curve around it looks somewhat like a line. Then draw in the tangent line. It is rough at best, but if we don't have the model so that we can calculate the limit of slopes of the secant lines, there is no other way.

Grid lines have been added to Figure 2.2-11 to make the reading of the graph a little easier. You can pick *any* two points on the tangent line to estimate the slope of the line, since the slope of a line is constant. It is a little easier if you select a point that is on either a grid line for the horizontal x axis or the vertical y axis. The farther apart the two points are, the lower the percent error, so you should always try to pick points as far apart as possible. Try reading the values for height when time is 0 seconds and time is 1 second. When time is 0 the tangent line appears to be at approximately 22; when the time is 1, the tangent line appears to be at about 37. These are *rough estimates* and you may have gotten different readings. That is to be expected. Using these figures we get a slope of

$$\frac{f(x_2) - f(x_1)}{x_2 - x_1} = \frac{(37 - 22)\text{ feet}}{(1 - 0)\text{ seconds}} = \frac{15\text{ feet}}{1\text{ second}} = 15\text{ feet per second}$$

We will probably get a different estimate if we use two different points. Let's try 0.5 and 1:

$$\frac{f(x_2) - f(x_1)}{x_2 - x_1} = \frac{(37 - 29)\text{ feet}}{(1 - 0.5)\text{ seconds}} = \frac{8\text{ feet}}{0.5\text{ second}} = 16\text{ feet per second}$$

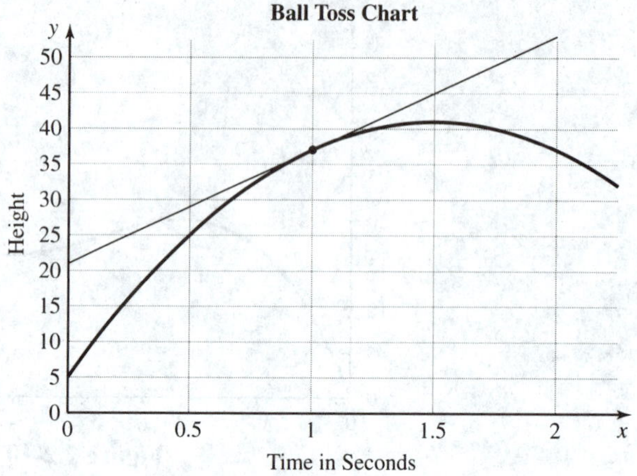

Figure 2.2-11

The difference between these answers is 1 foot per second, which out of 15 or 16 feet per second is about a 7% error. This is actually even a relatively *low* error for graphical estimates. With such a wide range possible in general, you certainly would not like to bet your life or your life savings on your graphical estimates! If you have to use a hand-drawn tangent line to estimate the rate of change of a function at a point, it is probably a good idea to take a couple of readings, and **be aware that the margin of error for graphical estimates can be quite high**.

Sample Problem 2: You are running a reelection campaign for a friend who took office in January 1991. Part of the campaign platform for the first election was a tough stance on crime. Should you use the crime rate in the United States (Figure 2.2-12) as part of the reelection campaign? How could you use it?

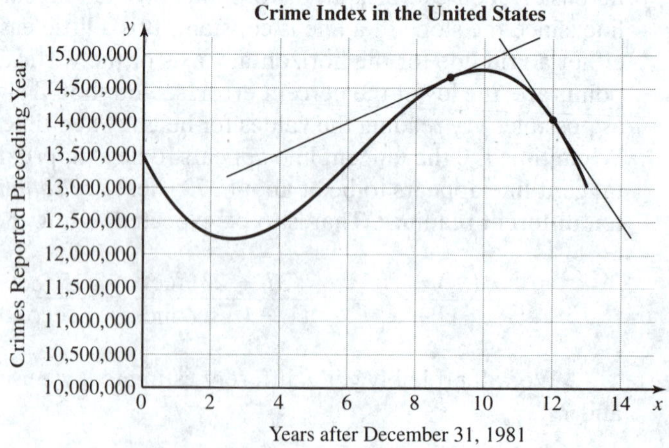

Figure 2.2-12

Solution: You now have a way of answering the problem stated at the beginning of this section. Should you use the crime rate in the United States as part of the campaign? You can estimate how the crime rate was changing before and after your candidate took office. If all we have is a graph of the data, we will have to estimate the rate of change of crime in the United States by drawing tangents to the curve at the desired points and estimating the slopes of the tangents. Since your friend took office in January 1991, you are interested in the figure for Dec. 31, 1990 (year 9) and also the last year for which you have the graph, year 12, or Dec. 31, 1993. In Figure 2.2-12 we have extended the curve beyond Dec. 31, 1993, based on the model that we developed in Section 1.5.

We have tried to draw a tangent line to the curve at $x = 9$, or Dec. 31, 1990, so that we can estimate the rate at which crime was changing at the end of 1990. In order to do this, we have to choose **any two points on the line** and estimate the values at those points. The output for 9 appears to be somewhere around 14,700,000. We could choose either 6 or 7 (or any other number where the tangent line appears) for our other input. It is easier to select an input that has a grid line on it. As we said earlier, it is also best to pick points as far from each other as possible. So let's use 6 first. The output value *on the tangent line* at 6 appears to be about 14,000,000. So our estimate of the rate of change at 9 is given by

$$\frac{(14{,}700{,}000 - 14{,}000{,}000) \text{ crimes}}{(9 - 6) \text{ years}} = \frac{700{,}000 \text{ crimes}}{3 \text{ years}} = 233{,}000 \text{ crimes per year}$$

Now let's try the calculation with 7. The line appears to be at an output value of 14,200,000 for 7. These are at best very **rough estimates**. The slope of the tangent line would then be

$$\frac{(14{,}700{,}000 - 14{,}200{,}000) \text{ crimes}}{(9 - 7) \text{ years}} = \frac{500{,}000 \text{ crimes}}{2 \text{ years}} = 250{,}000 \text{ crimes per year}$$

These two answers vary by approximately 17,000 out of 250,000, which is a margin of error of about 7%. Thus, we can say that crimes were increasing at the rate of at least approximately 250,000 crimes per year at the end of 1990 when your candidate first took office.

We have also drawn a tangent line to the curve at $x = 12$, or Dec. 31, 1993. Reading two values from this line we get

$$\frac{(14{,}000{,}000 - 14{,}800{,}000) \text{ crimes}}{(12 - 11) \text{ years}} = \frac{-800{,}000 \text{ crimes}}{1 \text{ year}} = -800{,}000 \text{ crimes per year}$$

Crimes were *decreasing* at the rate of approximately 800,000 crimes per year after your candidate had been in office three years.

Since we have fit a curve to this data, we can calculate the slope of secant lines drawn between 9 and points very close to 9 (Table 5).

TABLE 5

Time	8.99	8.999	8.9999	9	9.0001	9.001	9.01
Avg. Rate	230589	229623	229526		229504	229407	228438

Verbal Definition: $C(x)$ = crimes reported in the United States in the year preceding the point in time that is x years after Dec. 31, 1981.

Symbol Definition: $C(x) = -13150x^3 + 247506x^2 - 1030143x + 13471438$, for $0 \le x \le 12$.

Assumptions: Certainty and divisibility. Certainty implies the relationship is exact. Divisibility implies that fractions of crimes and years are possible.

The range here is between 229,526 crimes per year and 229,504 crimes per year. Our second estimate from the graph (250,000) was pretty bad (about 8% error), but the first estimate (233,000) was much better: It had approximately a 2% error.

Because the initial outputs are so large, the slopes of the secant lines do not *appear* to be **converging** as nicely as they did in the first example, but they are definitely converging, and we clearly have an answer to 4 significant digits already. We could feel comfortable with an estimation of a rate of 229,515 crimes per year (the average of 229,526 and 229,504) in 1990. If we needed a greater degree of accuracy, we would choose numbers even closer to 9, such 8.99999, which gives an average rate of 229,516, and 9.00001, which gives an average rate of 229,514. This confirms that an estimate of 229,515 crimes per year is reasonable.

Let's look at how the crime rate was changing at the end of 1993, year 12 (Table 6).

TABLE 6

Time	11.99	11.999	11.9999	12	12.0001	12.001	12.01
Avg. Rate	−768540	−770572	−770775		−770820	−771023	−773058

We could estimate that the crimes in the United States were declining at the rate of approximately 770,798 crimes per year at the end of 1993. Our estimate from the graph was not very far off for this year (about 4%)!

Whether we estimated from the graph or from the model, we could certainly use these data to show that the crime rate not only stopped increasing and started decreasing since your friend took office, but that the magnitude of the **rate of decrease** was almost three times as great as that of the previous rate of increase. □

Rate of Change as a Function

Think about what happens when you throw a ball in the air. The ball starts to go up rapidly, then slows, and finally starts to fall back down again. Just like the height of the ball, the slope of the graph that represents the height of the ball (the velocity of the ball) is always changing. In the beginning the height of the ball is increasing rather quickly and the slope of the graph is quite steep. The rate of change is fairly high and it is positive (the ball is going up at a fast speed). Then the change in the height of the ball (velocity) slows down; it is still increasing, but not as rapidly, and the slope of the graph is not so steep. The rate of change of the height of the ball is still positive, but it is much smaller. Finally, the height of the ball reaches its highest point and begins to decrease, at first slowly, and then more rapidly. The slope of the graph becomes a negative and gradually gets steeper and steeper. The slope of the graph changes over time (Figure 2.2-13).

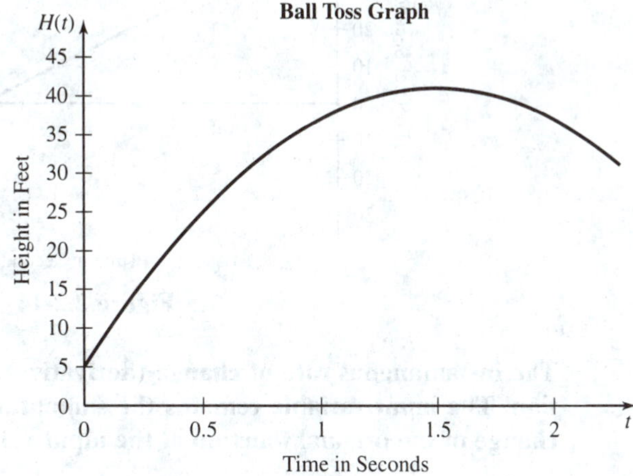

Figure 2.2-13

Table 7 shows the estimated rates of change (using technology and the limit concept) at various points of time.

TABLE 7

Time/Sec	0	0.25	0.5	0.75	1	1.25	1.5	1.75	2	2.25
Rate of Change	48	40	32	24	16	8	0	−8	−16	−24

We can fit a model to this data:

Verbal Definition: $R(t)$ = the rate of change of the height of the ball, in feet per second, t seconds after the toss.

Symbol Definition: $R(t) = -32t + 48, 0 \le t \le 3.$

Assumptions: Certainty and divisibility. Certainty implies the relationship is *exact.* Divisibility implies that fractions of seconds and feet are possible.

The rate of change of a smooth function at a point is a function of the input variable, just as the original function is. The function is defined as the slope of the tangent to the original function at that input value.

Let's look at the graph of this rate of change function (Figure 2.2-14).

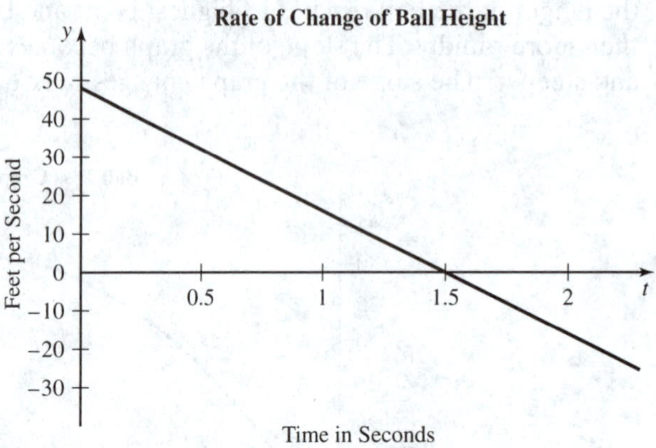

Figure 2.2-14

The instantaneous rate of change (derivative) of a smooth function is also a function. The input variable remains the same; the output variable is now the rate of change of the original function at the input value.

Section Summary

Before starting to do the exercises be sure that you

- Understand that the instantaneous rate of change of a function at a point can be found by taking the limit of the average rate of change between that point and a second point, as the second point gets closer and closer to the original point, from *both* sides.
- Know how to use technology to find the instantaneous rate of change at a point given the symbol definition of a function.
- Understand the general concept of the limit of a function as the variable approaches a given value: $\lim_{x \to a} f(x)$ means the value that $f(x)$ approaches as x gets closer and closer to a from *both* sides, *ignoring* what happens when $x = a$. For continuous functions, $\lim_{x \to a} f(x) = f(a)$. The values approached from each side must be the *same* in order for the limit to exist (be uniquely defined).
- Understand that the tangent line to a smooth curve at a point is the limit of the secant lines from that point to a second point, as the second point approaches the original point, from both sides.
- Understand that the tangent line to a smooth curve at a point is the straight line *through* that point which most closely resembles the curve *near* that point. If you keep zooming in (magnifying) the curve around that point, it will look more and more like a unique straight line, and that line is the tangent line.
- Know that the instantaneous rate of change (or derivative) of a function at a point can be thought of as the "slope of the curve" at that point, which is defined formally to be the slope of the tangent line to the curve at the point.

- Recognize that the rate of change of a function at a point is the slope of the tangent line at that point, which is the *limit* of the *slopes* of the secant lines from the point to a second point, as the second point approaches the original point from both sides.
- Recognize that the instantaneous rate of change (the derivative, or slope of the tangent) of $f(x)$ at a is defined as $\lim_{x\to a} \frac{f(x) - f(a)}{x - a} = \lim_{h\to 0} \frac{f(a + h) - f(a)}{h}$.
- Can estimate the rate of change of a function at a point from the graph of a function by drawing in the tangent line at that point and estimating its slope.
- Understand that the derivative of a function at a point is *not* defined if the graph is discontinuous (has a hole, jump, or gap) at the point, or if it has a corner, angle, or cusp there.
- Recognize that the derivative of a smooth function at a point is itself a function of the independent variable.

EXERCISES FOR SECTION 2.2

Warm Up

1. (a) Using technology and the *limit concept*, *estimate* the slope of the tangent line to the curve $y = f(x) = x^3$ at the point (2,8).

(b) Sketch a graph of the curve and the tangent line.

2. (a) Using technology and the *limit concept*, *estimate* the slope of the tangent line to the curve $y = f(x) = x^3 - 3x + 2$ at the point (1,0).

(b) Sketch a graph of the curve and the tangent line.

3. Using technology and the *limit concept*, *estimate* the slope of the tangent line to the curve $y = x^2 + 2$ at $x = 3$.

4. Using technology and the *limit concept*, *estimate* the slope of the tangent line to the curve $y = 2t - t^2$ at $t = 1$.

5. Using technology and the *limit concept*, *estimate* the slope of the tangent line to the curve $f(r) = r^3 - 3r^2 + 5r - 10$ at $r = 2.5$.

6. Using technology and the *limit concept*, *estimate* the slope of the tangent line to the curve $s(t) = -0.5t^3 + 1.4t^2 - 3.8t + 6$, at $t = 2.65$.

7. Using technology and the *limit concept*, *estimate* the slope of the tangent line to the curve $y = f(x) = 24(0.9)^x$ at $x = 5$.

8. Using technology and the *limit concept*, *estimate* the slope of the tangent line to the curve $y = f(x) = 3165(0.95)^x$ at $x = 8$.

9. (a) Find the average rate of change of the dependent variable with respect to the independent variable of $f(r) = r^3 - 3r^2 + 5r - 10$ over the interval [0,5].

(b) Using technology and the limit concept, estimate the slope of the tangent line to $f(r)$ at $r = 0$.

(c) Using technology and the limit concept, estimate the slope of the tangent line to $f(r)$ at $r = 5$.

(d) Using technology and the limit concept, estimate the slope of the tangent line to $f(r)$ at $r = 2.5$.

(e) Were the slopes better estimates of the average rate of change when the point was at an endpoint of the interval or in the middle of the interval? Why?

10. (a) Find the average rate of change of the dependent variable with respect to the independent variable of $f(x) = -3x^2 + 2x - 5$ over the interval [0,4].

(b) Using technology and the limit concept, estimate the slope of the tangent line to $f(x)$ at $x = 0$.

(c) Using technology and the limit concept, estimate the slope of the tangent line to $f(x)$ at $x = 4$.

(d) Using technology and the limit concept, estimate the slope of the tangent line to $f(x)$ at $x = 2$.

(e) Was the average rate of change a better estimate of the instantaneous rate of change when the point was at an endpoint of the interval or in the middle of the interval? Why?

Game Time

11. Figure 2.2-15 shows the annual percent change in the Medical Consumer Price Index.

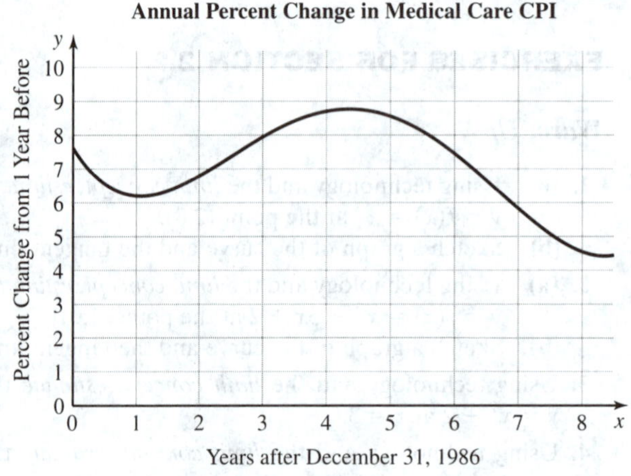

Figure 2.2-15

(a) Estimate the rate of change of the annual percent change in the medical CPI on Dec. 31, 1986. Express your answer in a complete sentence.

(b) Estimate the rate of change of the annual percent change in the medical CPI on Dec. 31, 1988. Express your answer in a complete sentence.

(c) Estimate the rate of change of the annual percent change in the medical CPI on Dec. 31, 1989. Express your answer in a complete sentence.

12. A model for the percent change in the Consumer Price Index for medical care is as follows:

Verbal Definition: $I(x)$ = the percent change in the Consumer Price Index for medical care x years after Dec. 31, 1986.

Symbol Definition: $I(x) = 0.0193x^4 - 0.3547x^3 + 1.9314x^2 - 3.0161x + 7.6229,\ x > 0.$

Assumptions: Certainty and divisibility. Certainty implies the relationship is exact. Divisibility implies fractions of years and percentage points are possible.

(a) Calculate the average rate of change from Dec. 31, 1988, to Dec. 31, 1990.
(b) From Figure 2.2-15, estimate the rate of change on Dec. 31, 1989.
(c) How do your answers compare?

13. Figure 2.2-16 shows the percentage of household with cable TV (0 corresponds to 12/31/77).

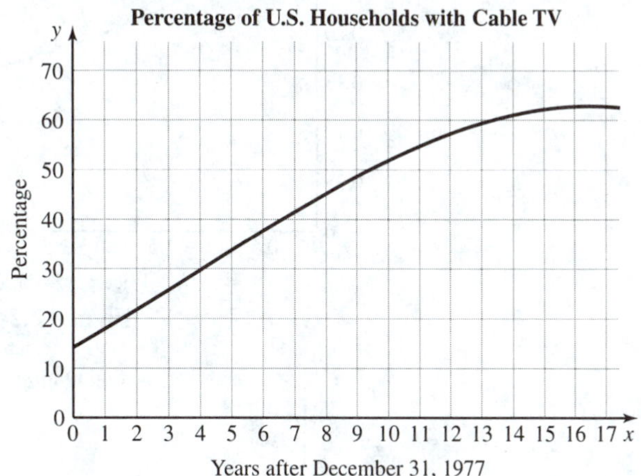

Figure 2.2-16

(a) Estimate the rate of change of the percentage of households with cable TV on Dec. 31, 1977. Express your answer in a complete sentence.
(b) Estimate the rate of change of the percentage of households with cable TV on Dec. 31, 1982. Express your answer in a complete sentence.
(c) Estimate the rate of change of the percentage of households with cable TV on Dec. 31, 1987. Express your answer in a complete sentence.

14. A model for the percentage of households with cable TV is

Verbal Definition: $P(x)$ = the percentage of U.S. households with cable TV x years after Dec. 31, 1977.

Symbol Definition: $P(x) = \dfrac{65.65\ldots}{1 + 3.527\ldots e^{-0.2621\ldots x}}$, for $0 \le x \le 17$.

Assumptions: Certainty and divisibility. Certainty implies the relationship is exact. Divisibility implies that fractions of years and percentage points are possible.

(a) Calculate the average rate of change from Dec. 31, 1980, to Dec. 31, 1982.
(b) From Figure 2.2-16, estimate the rate of change on Dec. 31, 1981.
(c) How do your answers compare?

15. You have been selling sodas at sporting events. Figure 2.2-17 shows the demand for the sodas.
(a) Sketch the tangents to the graph at \$1, \$1.50, and \$3.
(b) Estimate how the quantity of sodas sold changes with respect to the price charged at prices of \$1, \$1.50, and \$3.00

16. You have been selling sodas at sporting events. You get your sodas from several suppliers and have developed a model for the supply. The graph of the model is shown in Figure 2.2-18.

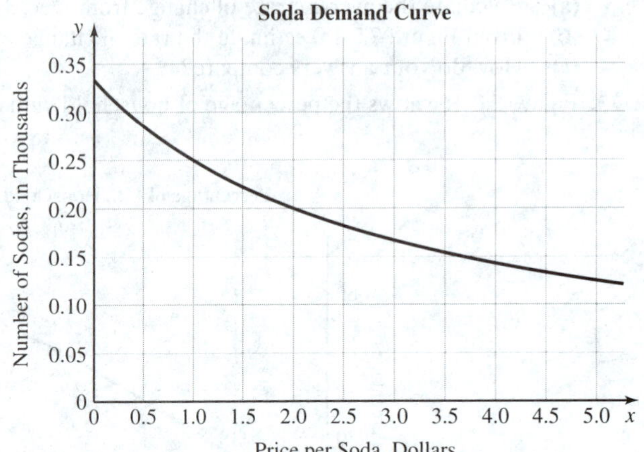

Figure 2.2-17

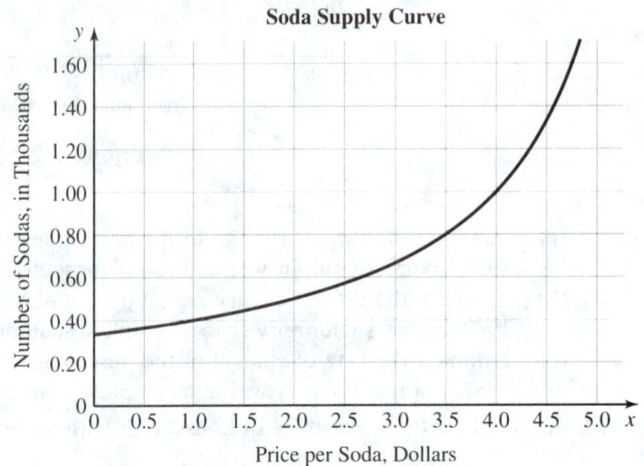

Figure 2.2-18

(a) Sketch the tangents to the graph at \$1, \$1.50, and \$3.

(b) Estimate how the quantity of sodas supplied changes with respect to the price charged at prices of \$1, \$1.50, and \$3.00

17. You have been selling sodas at sporting events and have developed the following model for the demand:

Verbal Definition: $q(p)$ = the quantity of sodas (in thousands) sold at p dollars per soda.

Symbol Definition: $q(p) = \dfrac{2}{p+3}$, $0.5 \le p \le 5$.

Assumptions: Certainty and divisibility. Certainty implies that the relationship is exact. Divisibility implies that the quantity of sodas and the price can be any fraction.

(a) Estimate the rate of change of the quantity sold (slope of the tangent line) when the price is \$1, \$1.50, and \$3.

(b) Find the average rate of change over the interval [1, 3].

(c) How did your answers to part (a) compare with your answer to part (b)?

18. You have been selling sodas at sporting events and have developed the following model for the supply:

Verbal Definition: $q(p)$ = the quantity of sodas (in thousands) provided at p dollars per soda.

Symbol Definition: $q(p) = \dfrac{1}{3 - 0.55p}$, $0.5 \le p \le 5$.

Assumptions: Certainty and divisibility. (Neither the independent variable, price, nor the dependent variable, quantity, is by nature divisible. We are assuming divisibility for the model.)

(a) Estimate the rate of change of the quantity (slope of the tangent line) when the price is \$1, \$1.50, and \$3 .
(b) Find the average rate of change over the interval [1, 3].
(c) How did your answers to part (a) compare with your answer to part (b)?

19. Figure 2.2-19 shows the accumulated sales of T-shirts in the weeks after a sale began.

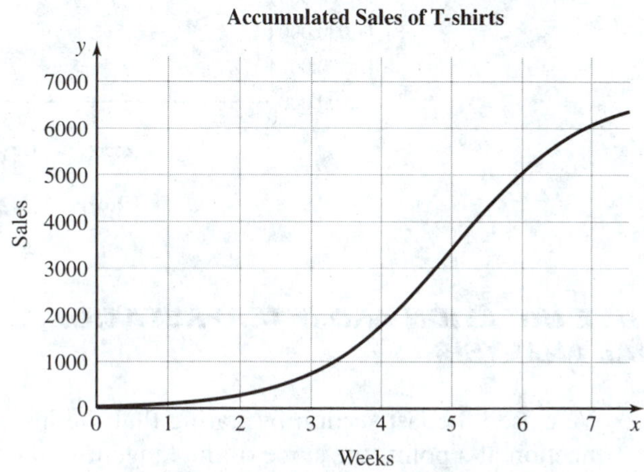

Figure 2.2-19

(a) Estimate how fast the accumulated sales of T-shirts was changing 3 weeks after the sale began. Be sure to write your answer in a complete sentence.
(b) Estimate how fast the accumulated sales of T-shirts was changing 5 weeks after the sale began. Be sure to write your answer in a complete sentence.

20. Figure 2.2-20 is the graph of a model of the crime index in the United States.
(a) Estimate how fast the crime index was changing Dec. 31, 1983.
(b) Estimate how fast the crime index was changing Dec. 31, 1987.
(c) Estimate how fast the crime index was changing Dec. 31, 1991.

21. We have fit a model to the accumulated sales of T-shirts in the weeks after a sale began:

Verbal Definition: $S(t)$ = total accumulated sales of T-shirts t weeks after the sale began.

Symbol Definition: $S(t) = \dfrac{6802}{1 + 190e^{-1.0502t}}$, $t \ge 0$.

Assumptions: Certainty and divisibility. Certainty implies the relationship is *exact.* Divisibility implies that fractions of weeks and T-shirts are possible.

(a) Estimate how fast the accumulated sales of T-shirts were changing 3 weeks after the sale began. Be sure to write your answer in a complete sentence.
(b) Estimate how fast the accumulated sales of T-shirts were changing 5 weeks after the sale began. Be sure to write your answer in a complete sentence.

(c) Find the average rate of change over the interval [1,3].
(d) How did the answer to part (c) compare with the answer to part (a)?

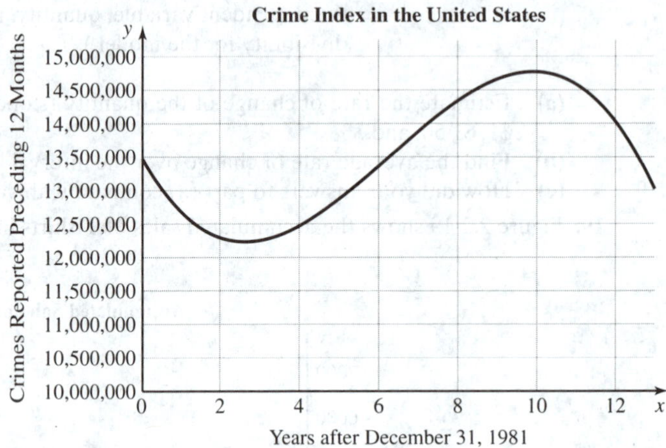

Figure 2.2-20

2.3 DERIVATIVE NOTATION AND INTERPRETATION, MARGINAL ANALYSIS

We closed the last section by stating that the instantaneous rate of change of a smooth function at a point (the slope of the tangent to the graph of the function at that point) is itself a function, called the **derivative**. The derivative is one of the two central themes of calculus. Derivatives have their own special symbols and it can take some practice to recognize and interpret expressions involving them. In Section 2.2 we saw that a tangent line can be thought of as a **linear approximation** of a curve. This makes it possible for us to make estimates based on the derivative that can be much simpler and easier than working with a complicated function. These estimates may be sufficient in situations where we don't need tremendous precision. Working with such estimates is called **marginal analysis** and it is used frequently in economics and business.

Here are the kinds of problems that the material in this section will help you solve:

- You have fit a model for the height of a tossed ball as a function of time. How can you interpret the value of its derivative at a point in time, and how can you communicate that to someone else? If you also know the height of the ball at that time, how could you project its height a short time afterward?
- You have fit a model for your expected grade on a certain test as a function of the number of hours you study for it. How can you interpret the derivative of this function at a point? If you have also calculated your expected grade at that point (level of studying), how could you quickly estimate the benefit from studying a couple of more hours?

- You are the director of a social service agency. You have past data about the number of families you serve, and the total cost of doing so, for different years. You have fit a model of total cost as a function of the number of families served. How can you interpret the derivative of this function at a point? If you have also calculated the estimated total cost at that point (number of families), how could you quickly estimate the total cost for serving 10 more families?
- From a market survey and past history, you have put together models for your total cost, demand, total revenue, and total profit from a computer game you developed as a function of the number of copies of the game you make this year. How can you interpret the derivatives of each of these functions at a point (number of copies), how do they relate to each other, and how can you use them and the values of the original cost, revenue, and profit functions at the same point, to estimate the value of each function if you make five more copies than you originally intended?

By the end of this section you should be able to solve problems like these and should

- Recognize and be able to use the various symbols used for the derivative, knowing which versions are better in which circumstances.
- Be able to interpret derivative notation related to a model in a real-world context and know how you can use it to understand the situation and make estimates using marginal analysis.
- Understand and be able to use the concepts of marginal cost, marginal revenue, and marginal profit in a business context.

Derivative Notation

Sample Problem 1: We wrote our model of the ball toss in Section 1.4 as

Verbal Definition: $H(t)$ = the height, in feet, t seconds after the ball was tossed.
Symbol Definition: $H(t) = -16t^2 + 48t + 5$, for $0 \le t \le 3$.
Assumptions: Certainty and divisibility. Certainty implies that the relationship is *exact.* Divisibility implies that any fraction of time in seconds or height in feet is possible.

How can we interpret the derivative of $H(t)$ at $t = 1$, and how can we use that value to estimate the height of the ball 1.1 seconds after it was tossed?

Solution: At the beginning of Section 2.2, we used the limit concept to calculate the instantaneous rate of change of the height of the ball 1 second after it was tossed, and we found it to be approximately 16.00 feet per second. We said that this is called the **derivative** of $H(t)$ at the point where $t = 1$, or simply at $t = 1$, and saw how this corresponds to the **slope of the tangent** to the curve $y = H(t)$ at the point where $t = 1$. We also pointed out that the *derivative* of a smooth function is *itself* a function, a function *derived* from the original function (hence the name *derivative*), with the same domain (except possibly the endpoints). To show the relationship to the original function, one notation for the derivative is to use the same function name, followed by a **prime** symbol ($'$), often used to denote something that is *changed* or modified in some way. In our case the derivative of $H(t)$ at $t = 1$ is denoted $H'(1)$ (read: "H prime of 1"), and in general

> **$H'(a)$ means the derivative (or instantaneous rate of change, or slope of the tangent line) of $H(t)$ at $t = a$.**

More generally, if $y = f(x)$, then **the derivative of $f(x)$ at $x = a$ is denoted $f'(a)$**. Due to the historical development of calculus and because of the many disciplines that make use of calculus for different kinds of purposes, there are in fact *many* different notations for this same idea of the derivative at a point. For example, since the derivative is a slope (change in y over change in x) and comes from the limit of the slopes of the secants, one common notation is

$$\frac{dy}{dx} \text{ at } x = a \text{ (read: "dee } y \text{ dee } x\text{," or sometimes "dee } y \text{ over dee } x\text{")}$$

which is sometimes also written as

$$\frac{df}{dx} \text{ at } x = a$$

to emphasize the *function* (rather than the dependent variable).

One way to see the origin of this notation even more clearly is shown below:

$$\frac{dy}{dx} = \lim_{\Delta x \to 0} \frac{\Delta y}{\Delta x}$$

In this equation, $\frac{\Delta y}{\Delta x}$ is the slope of the secant (change in y over change in x), and the Δx is the same as the h we used in Section 2.2 (the horizontal distance between the original point and the second point used to form the secant line), so we are again taking the limit of the slope of the secant as the distance from the original point to the second point goes to 0. Sometimes it is useful to think of the dy as a tiny little change in y and the dx as a tiny little change in x, measured along the *tangent* line rather than the original *curve*.

For our ball situation, then, we could say that

$$H'(1) \approx 16.00$$

or

$$\frac{dH}{dt} \approx 16.00 \text{ at } t = 1$$

or

$$\frac{dy}{dt} \approx 16.00 \text{ at } t = 1$$

How can we interpret this value? Well, we have already discussed this to some extent. Since this is the rate of change of the height with respect to time, it actually corresponds to the vertical **velocity** of the ball. In general, **velocity can be thought of as the rate of change of position with respect to time** (similar to what we usually think of as speed) **and incorporates the direction of motion**. We will only consider motion along a line, so the direction will be indicated by the *sign* of the velocity. **In our example, a positive velocity means the ball is moving *up*, and a negative sign means it is moving *down*.** This is closely related to the idea of **speed**, which **is simply the *magnitude* of the velocity (absolute value, in our case), or the rate of change of the *distance traveled* with respect to time**.

The units of the velocity of the ball help us understand how to interpret the derivative. We found the derivative after 1 second to be about 16 *feet per second*. We also calculated that the height of the ball at that time was 37 feet. *If* the ball continued at that velocity for *another* second, this means it would travel *another* 16 feet up, so would climb to a height of $37 + 16 = 53$ feet. *If* the ball continued at that velocity for *two* more seconds, it would travel a total of

$$\left(\frac{16 \text{ feet}}{\text{second}}\right)(2 \text{ seconds}) = 32 \text{ feet}$$

so at $t = 3$ (2 seconds after $t = 1$) it would be at a height of $37 + 32 = 69$ feet.

Unfortunately, there's a problem with these calculations. We know that the ball did *not* continue at a velocity of 16 feet per second, because we know it was slowing down, and even started going back *down* (a *negative* velocity) after time $t = 1.5$ seconds. So our estimates may be interesting, but not very useful. On the other hand, the question in Sample Problem 1 asks us to estimate the height of the ball at time $t = 1.1$ seconds. Now, we could substitute 1.1 for t in $H(t)$ and calculate $H(1.1)$ either by hand or with technology, but we could also *estimate* the answer just in our heads!

Since $t = 1.1$ is only 0.1 seconds after $t = 1$, we can estimate the change in the ball's height using the velocity (derivative) of 16 feet per second at $t = 1$, since the velocity won't have had time to change too much. At a rate of 16 feet per second, in one-tenth of a second, the ball would go one-tenth of 16 feet, or

$$\left(\frac{16 \text{ feet}}{\text{second}}\right)(0.1 \text{ seconds}) = 1.6 \text{ feet}$$

(which we could do in our heads by just moving the decimal point one place to the left), and so the height of the ball would be about $37 + 1.6 = 38.6$ feet. Let's see how good that estimate is by calculating $H(1.1)$:

$$H(1.1) = -16(1.1)^2 + 48(1.1) + 5 = -16(1.21) + 52.8 + 5 = -19.36 + 57.8$$
$$= 38.44$$

Not too bad! The estimate is off by 0.16 feet out of 38.44, which is less than half of one percent! □

What do these calculations correspond to graphically? Remember that the derivative at a point is the slope of the tangent line to the curve at that point. Remember also that we said that the tangent line to a curve (function) at a point is the straight line *through* the point that most closely resembles the curve *near* the point. In other words, **for input values *near* the point, the tangent is a very good *linear approximation* of the curve**. What we have done with our calculations is to **use the tangent line as a *substitute* for the curve**. Figure 2.3-1 shows what this looks like for the calculations in Sample Problem 1.

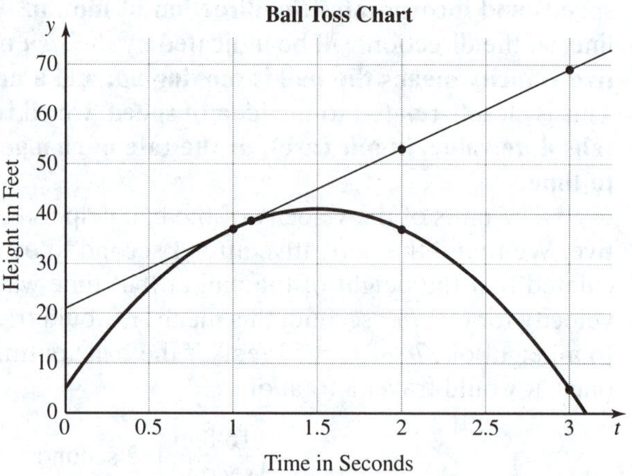

Figure 2.3-1

Notice how the tangent line is very close to the curve near the point at $t = 1$, and so the estimate of the y value (height) for $t = 1.1$ is very good, but for $t = 2$ and $t = 3$, the tangent is far from the curve, so the estimate is virtually useless.

Reading and Interpreting Derivative Notation

Let's use this situation and model to get more practice reading and interpreting derivative notation:

$H'(a)$ is read "H prime of a"

$\dfrac{dH}{dt}$ at $t = a$ is read "dee H dee t at $t = a$"

Both denote the **derivative of H with respect to t, or the rate of change of H with respect to t, at the point where $t = a$.**

It is not enough to be able to find the derivative (rate of change) and write it in correct notation—you must be able to interpret what it means. In many cases, modern technology can *find* the derivative, but it cannot *interpret* the derivative. If this course were simply teaching you to find the derivative, we could just show you how to use the technology and we would be finished in a few days. Problem formulation and interpretation of results are just as important as the actual mathematical solution of the problem, if not more so. It doesn't do you any good to be able to do mathematical manipulations if you don't know what they tell you when you are done.

Let's look at how to write and interpret these symbols.

$H(1) = 37$ (To find the output 37 we replaced the t in the original height function with 1.)
Read: "H of 1 is 37."
Interpret: "The height of the ball 1 second after it was thrown is 37 feet."

$H'(1) = 16$ (To find the output 16 we replaced the t in the rate of change function, the derivative, with 1.)

Read: "H prime of 1 is 16."
Interpret: "One second after the ball was tossed, the height of the ball was changing (or increasing) at the rate of 16 feet per second."

Note: In the second example the number in the parentheses tells you at what point the rate of change was calculated. Read the part before the equal sign: "When the time t was one second after the toss the height of the ball H was . . . ." The number following the equal sign is the rate of change at that point: "changing at the rate of 16 feet per second." Generically we read: **"When the [input variable] is [the number in the parentheses], the [output variable] is changing (increasing if the sign is positive or decreasing if the sign is negative) at the rate of [the number following the equals sign] [output units] per [input unit]."** This gives you a template for interpreting the rate of change function evaluated at a point.

$\frac{dH}{dt} = -16$ when $t = 2$

Read: "The derivative of H with respect to t is -16 when $t = 2$."
Interpret: "Two seconds after the ball was tossed, the height of the ball was changing at the rate of -16 feet per second (or decreasing at the rate of 16 feet per second)."

The actual grammatical structure of the interpretation can vary, but the sense of it cannot. For instance, we wrote: "Two seconds after the ball was tossed" rather than "When time was two seconds." It simply reads better and does not change the sense. However, be very careful if you try to change the form of the second part. It is very important that you say **"was increasing at the rate of"** or **"was decreasing at the rate of"**, followed by the **rate** (the number after the equals sign) and then **[output units] per [input unit]**. If you do not have these elements in your interpretation, the interpretation is incorrect or incomplete.

$H(0) = 5$

Read: "H of zero is five."
Interpret: "At time zero (the instant when the ball was thrown) it was 5 feet above the ground."

$H'(0) = 48$

Read: "H prime of zero is forty-eight."
Interpret: "At time zero (when the ball was thrown) the height of the ball was increasing at the rate of 48 feet per second."

$\frac{dH}{dt} = 48$ when $t = 0$

Read: "Dee H dee t equals 48 when $t = 0$. (The derivative of H with respect to t is 48 when $t = 0$.)"
Interpret: "At time zero (when the ball was thrown) the height of the ball was increasing at the rate of 48 feet per second."

(The physical interpretation of these last two is: "The ball was thrown up with an initial velocity of 48 feet per second.")

Why in the world do we have to have more than one symbol for the derivative? That is a fair question. Sometimes one of the symbols expresses the idea more appropriately and succinctly than the others. Mathematical symbols are created to *simplify* and *clarify* the writing of mathematical expressions, so we should use whatever symbols do the job best.

Let's look at the example relating to the crime index in the United States (Sample Problem 2 from Section 2.2). Our model was

Verbal Definition: $C(x)$ = the number of reported crimes in the United States in the 12 months prior to the point in time that is x years after the end of December 31, 1981.

Symbol Definition: $C(x) = -13150x^3 + 247506x^2 - 1030142x + 13471438$, for $0 \le x \le 14$.

Assumptions: Certainty and divisibility. Certainty implies that the relationship is *exact.* Divisibility implies that *any* fractions of years and crimes are possible.

$C(9) = 104661796$

Read: "C of 9 is 104661796."

Interpret: "The model estimates that, in 1990 (the 12 months prior to the point in time nine years after the end of 1981) 104,661,796 crimes (about 105 million crimes) were reported in the United States."

$C'(9) \approx 229516$ (Note the use of approximately here because we did not carry on our calculations of the slopes of the secants sufficiently to give us a good reading on the limit.)

Read: "C prime of 9 is approximately 229516."

Interpret: "According to the model, at the end of 1990 (nine years after the end of 1981) the number of reported crimes in the United States was increasing (or changing) at the rate of approximately 229,516 crimes per year (about 230,000 crimes per year)."

$\frac{dC}{dx} \approx 229516$ when $x = 9$

Read: "The derivative of C with respect to x is approximately 229,516 when $x = 9$."

Interpret: "According to the model, at the end of 1990, the number of reported crimes in the United States was increasing at the rate of approximately 229516 crimes per year (about 230,000 crimes/year)."

Sample Problem 2: Suppose that, based on a past test in your calculus class, you have estimated the grade you can expect on an upcoming test if you study different numbers of hours. You compiled some data and fit the following logistic model (rounded to 3 significant figures):

Verbal Definition: $G(t)$ = the grade (out of 100) you expect to get if you study for t hours for a calculus test.

Symbol Definition: $G(t) = 53 + \frac{41.1}{1 + 23.8e^{-0.948t}}$, for $0 \le t \le 8$.

Assumptions: Certainty and divisibility. Certainty implies that the relationship is *exact*. Divisibility implies that any fractions of hours and grade points are possible.

Using technology, you calculate that $G(4) \approx 79.8$ and, using a limit table as in Section 2.2, that $G'(4) \approx 8.85$. How can you interpret these values? If you have already studied for 4 hours, how could you estimate the approximate benefit of another half hour of study, and what could you expect roughly to get on the test if you stopped after that extra half hour?

Solution: $G(t)$ is defined in the verbal definition, so $G(4)$ is the grade you would expect on the test if you study for 4 hours. So $G(4) \approx 79.8$ is saying that if you study 4 hours, you can expect to get about an 80 on the test (a low B− according to the scale listed on the course syllabus). $G'(4)$ refers to the derivative, or instantaneous rate of change, or slope of the tangent, of $y = G(t)$ at $t = 4$. Since the input units for $G(t)$ are hours and the output units are grade points (out of 100), the units of $G'(4)$ (the slope units) will be grade points per hour. So $G'(4) \approx 8.85$ means that, after 4 hours of studying, the *additional* grade points you can expect on the test (*over* 79.8) will be earned at an instantaneous rate of about 9 points per hour (9 points per additional hour of study beyond 4 hours).

Thus, to see the additional effect of studying half an hour *more than* 4 hours, we can simply multiply the rate at $t = 4$ by one-half:

$$\left(\frac{8.85 \text{ grade points}}{\text{hour}}\right)(0.5 \text{ hours}) \approx 4.4 \text{ grade points}$$

So the extra half hour of studying should earn you approximately 4 or 5 points on the test, and you could then expect a total grade of approximately $79.8 + 4.4 = 84.2$ (about an 84), closer to a solid B according to the scale given on the syllabus.

For the sake of comparison, let's use technology to calculate the exact value from the model when $t = 4.5$: $G(4.5) \approx 83.8$. This is still approximately 84; the margin of error is about 0.4 out of 83.8, which is again about one-half of one percent. Not bad! As we saw in the ball toss problem, however, in general, **the accuracy of an estimate using the derivative (tangent line) at a point normally gets worse as the distance from the point increases, and is best when estimating quite close to the point**.

Note: The assumption of divisibility assumes a grade of 84.2 is possible, but in reality it probably isn't (most instructors round off to whole numbers), so the assumption does not hold perfectly, but we can easily round off and get meaningful results. □

Marginal Cost and Marginal Analysis

Sample Problem 3: Suppose you are the director of a social service agency, and from your historical records, you have fit a model to project the total cost of your operations as a function of the number of families you serve in a year. The model, rounded off to 3 significant digits, is given by

Verbal Definition: $C(x)$ = the total cost for one year, in thousands of dollars, if your agency serves x families that year.
Symbol Definition: $C(x) = 41.3 + 4.96x^{0.837}$, for $0 \le x \le 150$.

Assumptions: Certainty and divisibility. Certainty implies that the relationship is *exact.* Divisibility implies that *any* fractions of families and of money units (thousands of dollars) are possible.

You are currently planning to serve 100 families for the current year. Find the value of the cost function at this point, and use technology and the limit concept to find the derivative (instantaneous rate of change) at this point, and interpret each value. You are considering taking on an additional 10 families this year. Given the values of the cost function and its derivative at 100, use them to do a rough projection of what your incremental cost of serving 10 more families would be, and what your corresponding total cost would be. You have a chance to be awarded a government grant for \$20,000. Would this grant seem to be able to cover the incremental cost?

Solution: The model predicts that the cost of serving 100 families should be

$$C(100) = 41.3 + (4.96)(100)^{0.837} \approx 275$$

This means that the cost of serving 100 families this year should be about \$275,000. We can use technology to estimate the derivative, either by using the limit table as we did in Section 2.2, or by using a numerical derivative function. Either way, we find that

$$C'(100) \approx 1.96$$

This means that, at the point of serving 100 families, the instantaneous rate of change of the total cost is about 1.96 thousand dollars (\$1,960) per additional family served. Since this is the additional cost per additional family, this means that 10 additional families should cost approximately 10 times that value, or

$$\left(\frac{\$1{,}960}{\text{family}}\right)(10 \text{ families}) = \$19{,}600$$

or about \$20,000. This means that the grant should be just about right to cover these costs. The total cost for the agency for the year with those 110 families should then be about \$275,000 + \$20,000 = \$295,000.

Let's see how far off this projection is from the direct model prediction:

$$C(110) = 41.3 + 4.96(110)^{0.837} \approx 295$$

It was right-on, to 3 significant figures! Not bad! □

The term **marginal** is used frequently in economics. When used with cost as in **marginal cost**, it normally refers to the rate of change of cost with respect to the quantity produced (or serviced, as above). Since **marginal cost** is a rate of change, it is a **derivative**. We find the **marginal cost function** by finding the **derivative of the cost function**. Since the marginal cost is the rate of change of cost at a given point or quantity, the marginal cost can be interpreted as the **approximate cost of producing (or serving) one additional unit**.

We should point out that some texts *define* marginal cost as the cost of 1 additional unit [essentially the *average* cost over the interval $(x, x + 1)$]. As long as 1 unit is a relatively small proportion of the values of interest, the two interpretations should be very close to each other, as we have discussed above. Otherwise, you need to be clear in your own mind as to which definition you are using. **Throughout this text, the term "marginal"**

will always refer to the derivative, or instantaneous rate of change, so it will in general be *approximately* (but usually not *exactly*) the change in the output for 1 additional unit of the input.

There may be times when a business might be focusing on the best price to charge for an item they sell. From market survey data, they could express the demand (number of items they could sell) as a function of the selling price. Cost is normally thought of in terms of the quantity produced, but since the quantity can be expressed in terms of price, this can yield an expression for cost in terms of the selling price (a **composite function**, where we substitute the demand function expression for the quantity variable in the cost function), say $c(p)$, as we saw in Sample Problem 6 of Section 1.4. In such a case, we could still call the derivative of the cost function, say $c'(p)$, a marginal cost function, but this would not be the standard economics usage of the term.

In Sample Problem 3, then, we could say that the service agency's marginal cost at the point when they are serving 100 families is about $1,960 (per family). When we estimated the incremental and total cost of serving 10 more families, we were performing what is often called **marginal analysis**. ***Marginal analysis* is the process of using the derivative of a model at a point to make estimates of the change in (or the actual value of) the output variable value that result from small changes in the value of the input variable from that point. To estimate the change in the output value, multiply the *rate* at the original input value *times* the *change in the input.* To estimate the new output value, *add* this estimate of the change in the output value *to the value of the output* at the original input value:**

change in output ≈ [rate of change at original input][change in input]

new output ≈ [original output] + [estimated change in output]
= [original output] + [rate of change at original input][change in input]

Sample Problem 4: The exercises in Section 1.3 include a problem concerning the sale of "scrunchies." The cost function was modeled by

Verbal Definition: $C(n)$ = the cost C (in dollars) to make n scrunchies.
Symbol Definition: $C(n) = 4.95 + 0.27n,\ 0 \le n.$
Assumptions: Certainty and divisibility. Certainty implies that the relationship is *exact*. Divisibility implies that *any* fractions of scrunchies and dollars are possible.

Since this is a simple linear cost function, similar to what we saw in Section 1.3, we know that it corresponds to a **fixed cost** of $4.95 (for scissors, a thimble, etc.) and a **variable cost** of $0.27 per scrunchie (for fabric, thread, etc.). What is the marginal cost of producing scrunchies?

Solution: Since the graph will be a straight line, we know that it has the same slope everywhere, which is the 0.27. This means that the average rate of change over *any* interval is 0.27. Also, the tangent line to the graph will be the original line itself, and so the derivative or instantaneous rate of change will also be the same slope, 0.27. Therefore the marginal cost at *any* point is

$$\frac{dC}{dn} = 0.27 \text{ dollars per scrunchie.}$$

The cost of producing one more scrunchie is \$0.27. In this particular case, **the derivative gives us *exactly* the cost of one more unit, because the function is linear. With a linear cost function, the variable cost *is* the marginal cost.** □

With more complicated cost functions, the marginal cost will change as the quantity of items produced changes, as we saw in Sample Problem 3.

Marginal Revenue and Marginal Profit

The same kind of expression is used in connection with revenue: **Marginal revenue** is the **rate of change of revenue with respect to the quantity sold**, the **derivative of the revenue function**. Since the marginal revenue is the rate of change of revenue at a given point or quantity, the marginal revenue is interpreted as the **approximate revenue per additional item sold**. As we said earlier about marginal cost, it is possible to express revenue in terms of selling price (using the demand function), and so the term marginal revenue could also refer to the derivative of such a revenue function, but **the standard usage of the term *marginal revenue* in economics usually assumes that marginal revenue means the derivative of revenue with respect to *quantity***.

In Sample Problem 6 of Section 1.4, we derived cost, revenue, and profit functions for a small fund-raising effort selling sweatshirts. The cost model was given by

Verbal Definition: $c(q)$ = the cost, in dollars, if q sweatshirts are ordered.
Symbol Definition: $c(q) = 9q + 25$, for $q \geq 0$.
Assumptions: Certainty and divisibility. Certainty implies no sales or discounts. Divisibility implies fractions of sweatshirts can be ordered.

The revenue model, rounded to 3 significant digits, was given by

Verbal Definition: $r(q)$ = the revenue, in dollars, if you charge the price at which you sell exactly q sweatshirts (as determined by the demand function).
Symbol Definition: $r(q) = -0.420q^2 + 26.2q$, for $0 \leq q \leq 45$.
Assumptions: Certainty and divisibility. Certainty implies that the demand function is *exactly* right (if you charge $p = d(q)$, you will sell *exactly* q sweatshirts). Divisibility implies that you can sell fractions of sweatshirts.

The profit model (revenue minus cost), rounded to 3 significant digits, was given by

Verbal Definition: $\pi(q)$ = the profit, in dollars, if you charge the price at which you sell exactly q sweatshirts (as determined by the demand function $p = d(q)$), and order exactly q sweatshirts.
Symbol Definition: $\pi(q) = -0.420q^2 + 17.2q - 25$, for $0 \leq q \leq 45$.
Assumptions: Certainty and divisibility. Certainty implies that the demand function is *exactly* right (if you charge $p = d(q)$, you will sell

exactly q sweatshirts). Divisibility implies that you can order and sell fractions of sweatshirts.

TABLE 1

q	$r(q)$	$c(q)$	$\pi(q)$	$r'(q)$	$c'(q)$	$\pi'(q)$
10	220	115	105	17.8	9	8.8
20	356	205	151	9.4	9	0.4
30	408	295	113	1.0	9	−8.0
40	376	385	−9	−7.4	9	−16.4

Let's evaluate these functions at a few different points, and use technology to compute the derivatives at those points as well (Table 1).

Let's try to make sense of the numbers in Table 1, starting with $a = 10$. If we charge the price our demand function says we must in order to sell 10 sweatshirts, then our revenue will be \$220, our cost will be \$115, and so our profit will be \$105. The derivative of our revenue, the marginal revenue, will be \$17.80 per sweatshirt. This means that, when we are selling 10 sweatshirts, the rate of change of the revenue is about \$17.80 per sweatshirt. In other words, for each additional sweatshirt we sell (by reducing the price appropriately), we should make approximately \$17.80 more in revenue. Thus, if we sold 2 more sweatshirts (for a total of 12), our revenue should *increase* by approximately (17.80)(2) = \$35.60, for a total revenue of approximately

$$220 + 2(17.80) = 220 + 35.60 = \$255.60$$

From what we said earlier, since our cost function is linear, the marginal cost is simply the variable cost, which in this case is \$9 per sweatshirt. Since the cost of 10 is \$115, the cost of 2 more would be exactly 2(9) = \$18, for a total cost of \$133.

Notice that according to the table, our marginal profit at this same point (10 sweatshirts) is about \$8.80, which is the marginal revenue *minus* the marginal cost (17.80 − 9). This makes sense: Since profit is revenue minus cost, it makes sense that **the marginal profit is the marginal revenue minus the marginal cost** (intuitively, the profit from the *next* sweatshirt is the *revenue* from the next sweatshirt *minus* its *cost*). Notice that this pattern holds for all of the other values of q as well, and in fact is completely general.

The marginal profit of \$8.80 means that, when we sell 10 sweatshirts, the profit is *increasing* at a rate of \$8.80 per sweatshirt, so we can earn approximately an *extra* \$8.80 from *each* additional sweatshirt beyond 10 (at least for an increment of a small number of sweatshirts, where the tangent line should be quite close to the profit curve). Since our profit from 10 sweatshirts is about \$105, if we made 2 more sweatshirts, for a total of 12, we could expect a profit of approximately

$$105 + 2(8.80) = 105 + 17.60 = 122.60$$

or about \$123. We could have also made the estimate from our estimates of the total revenue and the total cost for 12 sweatshirts:

$$255.60 - 133 = \$122.60$$

Before we look at the big picture of what is going on in the table with respect to the profit, let's look at a graph (Figure 2.3-2) of the profit function to give a visual interpretation of the numbers.

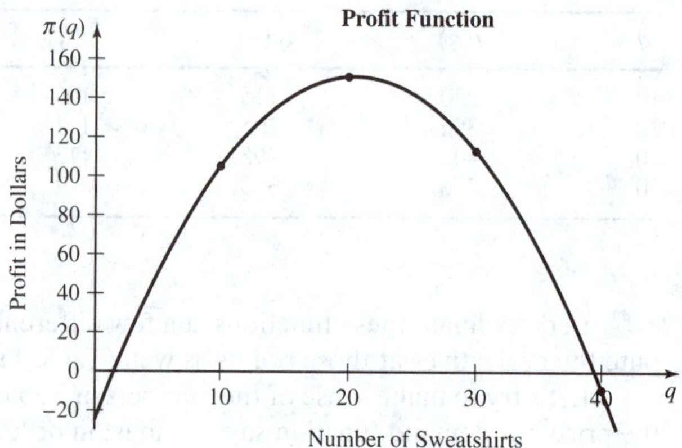

Figure 2.3-2

Notice from Table 1 and the graph in Figure 2.3-2 that the profit gets bigger when the number of sweatshirts goes from 10 to 20, but then gets smaller from 20 to 30. This suggests that the maximum profit occurs somewhere between 10 and 30, which the graph confirms. What additional information does the marginal profit (derivative) give us?

Notice from Table 1 and Figure 2.3-2 that the profit is *increasing* (the marginal profit is *positive*) at 10 sweatshirts and at 20 sweatshirts, indicating that we can make more profit by selling *more* sweatshirts in both cases, but at 30, the marginal profit is *negative*, so selling more would actually *decrease* our profit (because we'd have to lower the price so much to sell that many). This indicates that the best number to sell to maximize the profit must be somewhere between 20 and 30 sweatshirts. Notice that, even though the profit goes down from 20 to 30 (slope of the secant is negative), the marginal profit (slope of the tangent) is actually slightly positive at 20, suggesting that our maximum occurs to the *right* of 20, as we said above. Chapter 3 is largely about how to find that optimal point in a situation like this.

Since marginal profit equals marginal revenue minus marginal cost, the marginal profit is positive as long as the marginal revenue is larger than the marginal cost. In other words, as long as we are *taking in* more for the next unit than we are *spending* for it, our profit will keep increasing if we increase our quantity.

The Derivative as a *Function*

At the end of Section 2.2 we saw a table of the different rates of change of height with respect to time for the ball toss example and a graph of these data. From the graph and the table it appears that a linear model is appropriate: The plot appears linear and the first differences are exactly constant. These data follow a law of physics, so we can expect to get exact differences. Using technology we can fit a line to the data:

$$y = -32x + 48$$

This suggests that the value of the derivative at $t = x$ at can be thought of as a function of x, which we could write as

$$H'(x) = -32x + 48$$

or that the value of the derivative at $t = a$ can be thought of as a function of a:

$$H'(a) = -32a + 48$$

Or, for that matter, since our original input variable was t, we could express this same function using the variable t, since the choice of a symbol for the input variable is really optional, as long as we do not use the same symbol for two different quantities in the same problem.

The complete model for the derivative would then be

Verbal Definition: $H'(t)$ = rate of change of the height of the ball with respect to time (vertical velocity), in feet per second, t seconds after the toss.

Symbol Definition: $H'(t) = -32t + 48$, for $0 \leq t \leq 3$.

Assumptions: Certainty and divisibility. Certainty implies that the relationship is *exact.* Divisibility implies that *any* fractions of seconds or feet are possible.

Once we start to think of the derivative as a *function*, there are even *more* notations possible! If our original function is defined by

$$y = H(t) = -16t^2 + 48t + 5$$

then the derivative function can be written with any of the following notations:

$$H'(t) = -32t + 48$$

$$\frac{dH}{dt} = -32t + 48$$

$$\frac{dy}{dt} = -32t + 48$$

$$D_t[H(t)] = -32t + 48$$

$$D_t[-16t^2 + 48t + 5] = -32t + 48$$

$$\frac{d}{dt}[H(t)] = -32t + 48$$

$$\frac{d}{dt}[-16t^2 + 48t + 5] = -32t + 48$$

$$y' = -32t + 48$$

The two notations D_t and $\frac{d}{dt}$ can be interpreted to mean "the derivative of . . ." They are useful when giving rules for finding derivatives, or when showing the calculation of a derivative. We will make use of just about every form in this text to help you feel comfortable with all of them, since you're likely to see them in other courses.

This is a relatively short section, but a very important one. You have learned how to express rate of change in mathematical symbols and how to interpret those symbols. In the remaining sections in this chapter and in Chapter 3 you will be using and interpreting these symbols.

Before you start the exercises be sure that you can

- Recognize the various symbols used for the derivative.
- Interpret derivative symbols and be able to translate symbolic statements involving derivatives into words about the particular context being studied. In general $f'(a) = k$ means that, at $x = a$, the value of $f(x)$ is changing at a rate of k output units per unit increase in the input value (from a).
- Understand how to use the value of the derivative to approximate the change in the output of a function for a given small change in the input (by multiplying the derivative times the increment in the input), which can then be added to the initial output value to estimate the new output value. This process is called **marginal analysis** and corresponds to using the tangent line as a substitute for the original curve, so it is most accurate for relatively small increments in the input.
- Understand and know how to interpret marginal cost, marginal revenue, and marginal profit.
- Understand how a derivative of a function can be thought of as a function of the same input variable (defined as the value of the derivative at each value of the input variable).

EXERCISES FOR SECTION 2.3

Warm Up

1. If $y = f(x) = 5x^2 - 2x + 2, 0 \le x \le 3$ and $f'(x) = 10x - 2, 0 \le x \le 3$,
 (a) Find $f(1)$, $f(2.5)$, and $f(3)$.
 (b) Find $f'(1)$. Use this to estimate the change in y if x increases by 0.1 to 1.1, and then use this answer to *approximate* $f(1.1)$. Now calculate the actual *exact* value of $f(1.1)$. How far off is your approximation?
 (c) Find $\frac{df}{dx}$ when $x = 2.5$.
 (d) Find $\frac{dy}{dx}$ when $x = 3$.

2. If $y = g(x) = -2x^2 + 4x - 10, -10 \le x \le 10$, and $g'(x) = -4x + 4, -10 \le x \le 10$,
 (a) Find $g'(3)$, $g'(-10)$, and $g'(0)$.
 (b) Find $g(3)$, $g(-10)$, and $g(0)$.
 (c) Find $\frac{dg}{dx}$ when $x = -2$.

(d) Find $\frac{dy}{dx}$ when $x = 4$.

(e) Use your answers from parts (a) and (b) for $g(3)$ and $g'(3)$ to estimate the change in the value of $g(x)$ if the value of x increases from 3 to 3.1, and use this answer to approximate the value of $g(3.1)$. Now calculate the actual *exact* value of $g(3.1)$. How far off is your approximation?

3. If $y = g(x) = 347.0(0.9654)^x$, $x \geq 0$ and $\frac{dg}{dx} = -12.22(0.9654)^x$,

(a) Find $g'(3)$, $g'(0.2)$, and $g'(0)$.
(b) Find $g(3)$, $g(0.2)$, and $g(0)$.
(c) Find $\frac{dg}{dx}$ when $x = 0.5$.

(d) Find $\frac{dy}{dx}$ when $x = 1.2$

(e) Use your answers from parts (a) and (b) for $g(3)$ and $g'(3)$ to estimate the change in the value of $g(x)$ if the value of x increases from 3 to 3.1, and use this answer to approximate the value of $g(3.1)$. How far off is your approximation?

4. If $y = p(x) = 0.7520(0.8954)^x$, $x \geq 0$ and $\frac{dp}{dx} = -0.08308(0.8954)^x$;

(a) Find $p'(5)$, $p'(0.5)$, and $p'(0)$.
(b) Find $p(5)$, $p(0.5)$, and $p(0)$.
(c) Find $\frac{dp}{dx}$ when $x = 0.5$.

(d) Find $\frac{dy}{dx}$ when $x = 1.2$

(e) Use your answers from parts (a) and (b) for $p(5)$ and $p'(5)$ to estimate the change in the value of $p(x)$ if the value of x increases from 5 to 5.1, and use this answer to approximate the value of $p(5.1)$. How far off is your approximation?

Game Time

5. Rabbits are very prolific. You have been raising rabbits and determined that after buying one pair for breeding, you can model the number of rabbits that you have by

Verbal Definitions: $R(t)$ = the number of rabbits t years after the start of breeding.
Symbol Definition: $R(t) = 5t^2 - 2t + 2, 0 \leq t \leq 3$.
Assumptions: Certainty and divisibility. Certainty implies the relationship is *exact*. Divisibility implies that *any* fractions of rabbits and years are possible.

(a) How many rabbits will you have after one year? Write your answer in a complete sentence.
(b) How many rabbits will you have after two and a half years? Write your answer in a complete sentence.
(c) How many rabbits will you have after three years? Write your answer in a complete sentence.
(d) Use technology and the limit concept to estimate how fast the number of rabbits is changing after the one year. Write your answer in a complete sentence.
(e) Use technology and the limit concept to estimate how fast the number of rabbits is changing after two and a half years. Write your answer in a complete sentence.

(f) Use technology and the limit concept to estimate how fast the number of rabbits is changing after three years. Write your answer in a complete sentence. What is happening to the rate of change over time?

(g) Use your answers to parts (c) and (f) to *approximate* the number of additional rabbits you would expect between $t = 3$ and $t = 3.5$, and then use this answer to approximate the total number of rabbits 3.5 years after breeding. How far off is your approximation?

6. Recall the ball toss problem from the beginning of this section. The model was

Verbal Definition: $H(t)$ = the height in feet of the ball t seconds after it was tossed.
Symbol Definition: $H(t) = -16t^2 + 48t + 5, 0 \le t \le 3.$
Assumptions: Certainty and divisibility. Certainty implies the relationship is *exact*. Divisibility implies that *any* fractions of time and feet in height are possible.

(a) Use technology and the limit concept to estimate $H'(0.5)$. Write your answer in a complete sentence.

(b) Use technology and the limit concept to estimate $\frac{dH}{dt}$ when $t = 1.5$. Write your answer in a complete sentence.

(c) If we define $y = H(t)$, use technology and the limit concept to estimate $\frac{dy}{dt}$ when $t = 2$. Write your answer in a complete sentence.

7. Let $\pi = P(q)$ = the profit (in hundreds of dollars) from the sale of q thousand items. Interpret the following mathematical statements. Write your answers in complete sentences.

(a) $P(25) = 34758$
(b) $P'(25) = 23.58$
(c) $P(37) = 35289$
(d) $\frac{dP}{dq} = -1.45$ when $q = 37$
(e) $\frac{d\pi}{dq} = 0$ when $q = 32.5$
(f) Use your answers to parts (a) and (b) to *approximate* the values of $P(25.1)$, $P(26)$, and $P(28)$ using marginal analysis. Which *should* be most accurate?

8. Let $r = R(p)$ = the revenue (in hundreds of dollars) from the sale of items at p dollars per item. Interpret the following mathematical statements. Write your answers in complete sentences.

(a) $R(150) = 23$
(b) $R'(150) = 0.25$
(c) $\frac{dR}{dp} = -1.5$ when $p = 200$
(d) $\frac{dr}{dp} = 16$ when $p = 75$
(e) Use your answers to parts (a) and (b) to *approximate* the values of $R(151)$, $R(155)$, and $R(170)$ using marginal analysis. Which *should* be most accurate?

9. Let $c = I(x)$ = the annual percent change in the medical CPI x years after Dec. 31, 1986. Interpret the following mathematical statements. Write your answers in complete sentences

(a) $I(2.5) = 7.3656$
(b) $I(5.5) = 8.1067$
(c) $I'(4) = 0.3503$
(d) $\frac{dI}{dx} = -1.6651$ when $x = 6.5$
(e) $\frac{dc}{dx} = 0$ when $x = 1.07355$

10. Let $c = C(x)$ = crimes reported in the preceding year in the United States x years after Dec. 31, 1981. Interpret the following mathematical statements. Write your answers in complete sentences.

(a) $C(2.5) = 12{,}237{,}524$
(b) $C'(2.5) = -57{,}925$
(c) $C(6.5) = 13{,}621{,}318$
(d) $\frac{dC}{dx} = 99843$ when $x = 3$
(e) $\frac{dc}{dx} = 0$ when $x = 9.913889$

11. Let $P = \pi(p)$ = the profit in dollars at a price of p dollars per mug. Interpret the following mathematical statements. Write your answers in complete sentences.

(a) $\pi(5.5) = 4475$
(b) $\pi'(5.5) = -50$
(c) $\frac{d\pi}{dp} = 50$ when $p = 5$
(d) $\frac{dP}{dp} = 150$ when $p = 4.5$
(e) $\pi(7) = 4175$

12. Let $c = P(x)$ = the percentage of U.S. households with computers x years after Dec. 31, 1997. Interpret the following mathematical statements. Write your answers in complete sentences.

(a) $P(6) = 37.92$
(b) $P'(4.5) = 4.296$
(c) $P(10.5) = 53.602$
(d) $\frac{dP}{dx} = 0.65$ when $x = 17$
(e) $\frac{dc}{dx} = 3.227$ when $x = 9$

13. The marginal revenue function for a show at a coffeehouse can be modeled by

Verbal Definition: $R'(q)$ = the marginal revenue (dollars per ticket) from the sale of q tickets.
Symbol Definition: $R'(q) = 0.00510q^2 - 0.700q + 25.3,\ q \geq 0.$
Assumptions: Certainty and divisibility. Certainty implies the relationship is *exact*. Divisibility implies that *any* fractions of tickets and dollars are possible.

(a) Find and interpret the marginal revenue when 80 tickets are sold.
(b) Find and interpret the marginal revenue when 70 tickets are sold.
(c) Find and interpret the marginal revenue when 100 tickets are sold.

14. The marginal revenue function for a production at a local coffeehouse could also be modeled by

Verbal Definition: $R'(q)$ = the marginal revenue (dollars per ticket) from the sale of q tickets.
Symbol Definition: $R'(q) = 15.67(-0.00813q + 1)(0.9919)^q,\ q \geq 0.$
Assumptions: Certainty and divisibility.

(a) Find and interpret the marginal revenue when 80 tickets are sold.
(b) Find and interpret the marginal revenue when 70 tickets are sold.
(c) Find and interpret the marginal revenue when 100 tickets are sold.

15. The marginal profit (dollars per mug) from the sale of mugs is given by

$$\pi'(q) = -0.020q + 6.5,\ q \geq 100$$

Find and interpret the marginal profit when
(a) 100 mugs are sold.
(b) 250 mugs are sold.
(c) 500 mugs are sold.

16. The marginal profit (dollars per shirt) from the sale of q T-shirts is given by

$$\pi'(q) = 22.5(0.997)^q(-0.0783q + 1) - 5.55, \; q \geq 0$$

Find and interpret the marginal profit when
(a) 100 T-shirts are sold.
(b) 250 T-shirts are sold.
(c) 300 T-shirts are sold.

2.4 THE ALGEBRAIC DEFINITION OF THE DERIVATIVE AND BASIC DERIVATIVE RULES

We defined the derivative as the instantaneous rate of change of a function at a point as well as the slope of the tangent line to the graph at that point. We defined the slope of the tangent line as the limiting value of the slope of secant lines drawn from that point to a second point as the distance between the two points gets smaller and smaller, the second point approaching the original point. Finally, we wrote this symbolically: $\lim_{h \to 0} \frac{f(a+h) - f(a)}{h}$. In Section 2.2 we learned how to estimate the instantaneous rate of change of a function at a point, the derivative, graphically by drawing a tangent line to the graph and numerically by calculating slopes of secants from a model. Please note the word **estimate**. If we do not have or cannot get the numerical data from which a graph was drawn, we have to draw a tangent line to the graph at the point and then estimate the slope of the tangent line. Because both of these activities require judgments on our part, the graphical estimates are the least satisfactory. If we have a model or have numerical data and can fit a model, we can use technology to help us find the limit of the secant slopes to any degree of accuracy we choose by using the limit concept and a table, but the result is still an estimate of the instantaneous rate of change at a particular point. In this section we will investigate finding the rate of change of a function algebraically. (Note the use of "rate of change" instead of 'instantaneous rate of change.' When we use the term "rate of change," it assumes instantaneous rate at a point unless otherwise indicated. If an interval is specified, assume average rate of change.)

Here are the kinds of problems that the material in this section will help you solve:

- Section 2.2 showed a table of the rates of change of the height of a ball with respect to time after it was thrown (values of the derivative function). A model was fit to these data values. Is there some way to arrive at this derivative function without going through this rather long procedure?
- You have developed a model for cost as a function of quantity. Find a model for the marginal cost.
- You have developed a model for profit as a function of quantity. Find a model for the marginal profit function.

- How fast is the population of the world growing this year? What is the predicted rate of population growth in 2000? 2010?

By the end of this section you should be able to solve these problems and many other similar problems using the following skills:

- Use the algebraic definition of the derivative to find the rate of change of a function at a point.
- Use the algebraic definition of the derivative to find general rules for finding the derivative.
- Know the general rules for finding the derivative for polynomial functions, exponential functions, power functions, and natural logarithm functions.

The Mathematical Definition of the Derivative

In the definition of the rate of change, $\lim_{h \to 0} \frac{f(a + h) - f(a)}{h}$, we were always referring to a specific point, a. We drew a tangent line at a specific point or we calculated the slopes of secants drawn from a specific point. To generalize, we want to find the rate of change at *any* general point, x. **If we designate $y = f(x)$, then we can let $f'(x)$ be the derivative at x. The general algebraic definition of the derivative at x is then**

$$f'(x) = \lim_{h \to 0} \frac{f(x + h) - f(x)}{h} \text{ or } \lim_{\Delta x \to 0} \frac{\Delta y}{\Delta x}$$

When we are using this definition, we must always remember that **h or Δx approaches 0 as a limit but does not equal 0.**

The Derivative of a Constant Function

Sample Problem 1: Sometimes you like to have a salad for lunch. One local place has an "all you can eat" salad bar for \$4.95. What is the rate of change of the cost for these salads with respect to the amount (number of pounds) of salad taken?

Solution 1. This problem is almost too simple—you certainly don't need calculus to figure it out. On the other hand, it is always nice to start out with something that is easy. The model for the all you can eat salad cost C, in dollars, is given by: $C = 4.95$. There is no variable involved here: one salad, one price. If we want to make the model look like other models that we have fit, we can write the all you can eat model as

Verbal Definition:	$C(x)$ = the cost (in dollars) of x pounds of a salad.
Symbol Definition:	$C(x) = 4.95$ for $x \geq 0$.
Assumptions:	Certainty and divisibility. Certainty implies the relationship is *exact*. Divisibility implies that *any* fractions of pounds or dollars are possible.

The graph of the all you can eat model is in Figure 2.4-1.

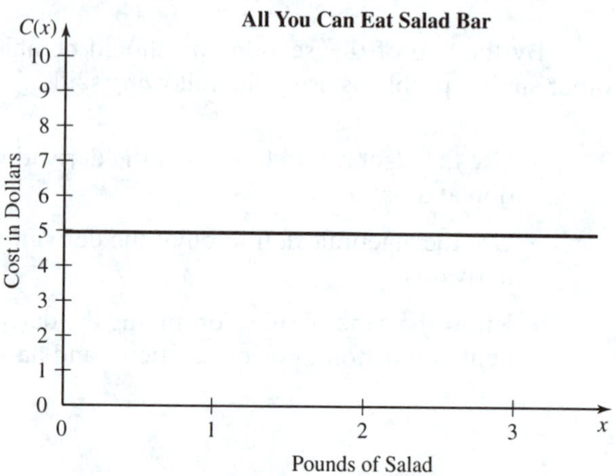

Figure 2.4-1

The graph is just a horizontal line because the price does not change with the quantity eaten: all you can eat for one price. The function is not changing with the pounds consumed. It neither increases nor decreases. The slope is 0. Graphically, since the tangent line to a function at a point means the line *through* the point that is most like the graph *near* the point, we can see here that the tangent line is the *same* as the original graph, which is a horizontal line, so its slope is 0.

Using the algebraic definition of the derivative we have

$$\lim_{h\to 0}\frac{C(x+h)-C(x)}{h}=\lim_{h\to 0}\frac{4.95-4.95}{h}=\lim_{h\to 0}\frac{0}{h}$$

We ignore $h = 0$ because the definition of a limit says to ignore what happens when the variable *equals* the value it is *approaching*. Since 0 divided by anything except 0, whether the value of h is large or small, is always 0, the limit is 0. If we had constructed a table of the average rates of change to calculate the limit, all of the entries would be 0 (except where $h = 0$, at the point itself, which would give an error as always), so the limit would clearly be 0.

The rate of change is 0 dollars per pound. The graph of the slope is shown in Figure 2.4-2.

Would it make any difference if the fixed price were $3.95 or $15.00? No, of course not. If the function never changes, the rate of change, the derivative, is always 0. □

The derivative of a constant is 0. If $f(x) = c$, then $f'(x) = 0$. In other words,[4]

$$D_x(c) = 0$$

[4]If $f(x) = c$, then (from the definition of the derivative) $f'(x) = \lim_{h\to 0}\frac{f(x+h)-f(x)}{h}$ so, plugging in the function, we get $\lim_{h\to 0}\frac{c-c}{h} = \lim_{h\to 0}\frac{0}{h} = 0$.

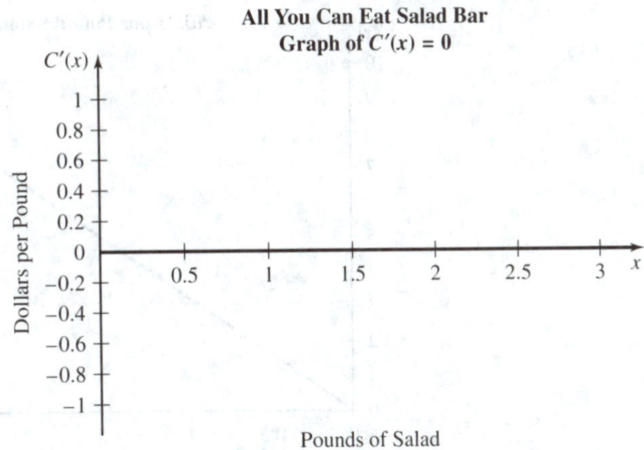

Figure 2.4-2

The first general rule for finding the derivative of a function is the rule for the derivative of a constant function. We will develop more general rules in this section and collect them into a reference table at the end of the section.

The Derivative of $f(x) = mx$

Sample Problem 2: Another local place has a salad bar for \$3.25 a pound. You fill your plate and it is weighed at the cashier's counter. The function for the cost of this salad is the cost $C(x)$, in dollars, for x pounds of salad: $C(x) = 3.25x$, $x \geq 0$. What is the rate of change (derivative) of this function?

Solution: Again, this is almost too easy. The cost is changing at the rate of \$3.25 per pound. That is even the way the price is expressed. The graph of the function is shown in Figure 2.4-3. We can see that the function is always increasing; the slope is positive. Not only that, but the function is always increasing at the same rate; the slope is constant. Since the tangent line is the line *through* the point most like the graph *near* the point, the tangent line is the *same* as the curve in this case.

Using the algebraic definition of the derivative, we have

$$\lim_{n\to 0}\frac{f(x+h)-f(x)}{h} = \lim_{h\to 0}\frac{3.25(x+h)-3.25x}{h}$$

Multiplying through and simplifying, we have

$$\lim_{h\to 0}\frac{3.25x+3.25h-3.25x}{h} = \lim_{h\to 0}\frac{3.25h}{h}$$

Let's stop for a minute to discuss finding limits. For any limit, you can first try plugging in the value the variable is approaching into the expression. If you get a real number as a result, that's the limit. In our case, this means substituting 0 in for h. In our algebraic definition this will always give $\frac{f(x+0)-f(x)}{0} = \frac{0}{0}$. This is not a real number, so to eval-

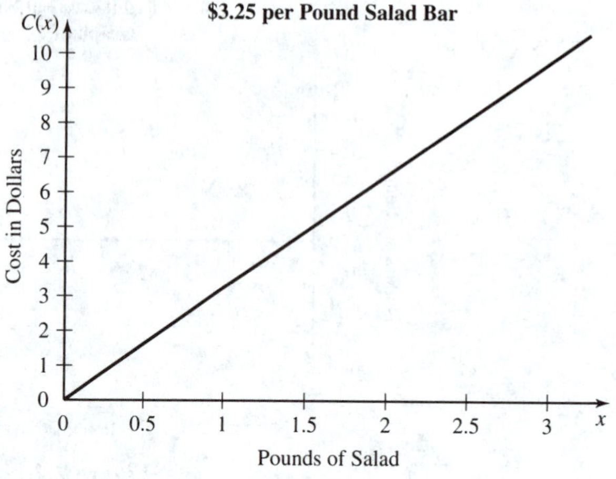

Figure 2.4-3

uate the limit we need to try to cancel out the h in the denominator. In general, **for evaluating a limit,**

1. **Substitute the limiting value into the expression. If you get a real number, then that is the limit. If you get $\frac{0}{0}$, then**

2. **Simplify the expression $\left(\text{for example, cancel terms causing the } \frac{0}{0}\right)$, and then substitute the limiting value into the simplified expression.**

When evaluating a limit to find a derivative, the main way to simplify the expression is to try to factor out an h in the numerator so that we can *cancel* it with the h in the denominator (divide both numerator and denominator by h) because ***h approaches* 0 but does not equal 0, according to the definition of a limit**. Division by 0 is undefined,[5] but we can *ignore* this complication in the case of evaluating this limit. Therefore, we get

$$\lim_{h\to 0} \frac{3.25h}{h} = \lim_{h\to 0} 3.25 = 3.25$$

Since there is no h term left after cancelling the h, letting h get smaller and smaller will have no effect. If we construct a table of the average rates of change to determine this limit, the values will *all* be 3.25 (except at $h = 0$, where it is undefined as usual, but can be ignored), so the limit will be 3.25.

The graph of the slope, the rate of change, of the per pound salad is shown in Figure 2.4-4. Since the original function is a line, the tangent will be the *same* line, as discussed earlier, so the *slope* of the tangent (the derivative) will be the *slope* of the original line, which was 3.25 here.

[5]Division is defined: $\frac{a}{b} = c$ if and only if $(c)(b) = a$. To divide $\frac{0}{0}$ there would have to be some number c that multiplied by zero gives zero. *Any* number satisfies this, so $\frac{0}{0}$ is not *uniquely* defined. To divide $\frac{a}{0}$ when $a \neq 0$, there would have to be some number c such that $(c)(0) = a \neq 0$. There is no such number, so this division is undefined. In *either* case, division by 0 is not (uniquely) defined.

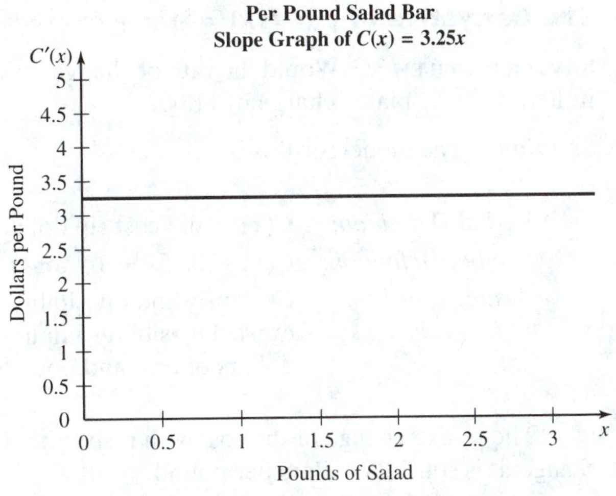

Figure 2.4-4

How would this change if the price were \$3.50 per pound? The rate of change would then be 3.50 dollars per pound. □

The derivative of a linear function is its slope. If $f(x) = mx$, then $f'(x) = m$. In other words,

$$D_x(mx) = m$$

If the derivative of $f(x) = mx$ is m, what is the derivative of $f(x) = x$? From the rule above, $m = 1$, so the derivative should be 1. Think about the line $y = f(x) = x$. The graph of this function is shown in Figure 2.4-5. For every unit change in x there is an equal unit change in y. This function has a rate of change of $\frac{1}{1} = 1$. The derivative of $f(x) = x$ is $f'(x) = 1$.

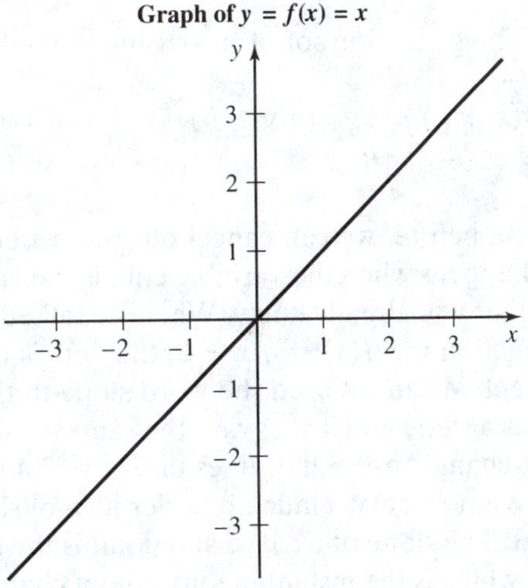

Figure 2.4-5

The Derivative of $y = f(x) = mx + b$

Sample Problem 3: Would the rate of change for the \$3.25 per pound salad bar change if there were a "plate" charge of \$1.00?

Solution: The model for this is

Verbal Definition:	$C(x)$ = the cost (in dollars) of x pounds of a salad.
Symbol Definition:	$C(x) = 3.25x + 1$ for $x \geq 0$.
Assumptions:	Certainty and divisibility. Certainty implies the relationship is *exact*. Divisibility implies that *any* fractions are possible for dollars of cost and pounds of salad.

The rate of change of the cost with respect to the number of pounds bought will not change; it is still 3.25 dollars per pound.

Using the definition of the derivative,

$$\lim_{h\to 0} \frac{3.25(x + h) + 1.00 - (3.25x + 1.00)}{h}$$

Multiplying through and then simplifying, we have

$$\lim_{h\to 0} \frac{3.25x + 3.25h + 1.00 - 3.25x - 1.00}{h} = \lim_{h\to 0} \frac{3.25h}{h} = 3.25 \text{ dollars per pound}$$

This is exactly what we got before. Remember that we can cancel the h's since the limit *ignores* what happens when $h = 0$. Notice that in this case, adding a constant to the cost function did not change the slope of the function. □

The derivative of a linear function is its slope. Thus, if $y = f(x) = mx + b$, then $f'(x) = m$. In other words,

$$D_x(mx + b) = m$$

The general proof of this, using the definition of the derivative, is

$$\lim_{h\to 0} \frac{m(x + h) + b - (mx + b)}{h} = \lim_{h\to 0} \frac{mx + mh + b - mx - b}{h} = \lim_{h\to 0} \frac{mh}{h} = \lim_{h\to 0}(m) = m$$

As before, we can cancel out the h's because, when finding a limit, we can ignore what happens when the variable equals the value it is approaching. Actually, this is something that you already knew. When we talked about linear functions we said that for a linear function $y = f(x) = mx + b$, the "m" is the slope of the line and the b is the vertical intercept. When we used the word slope in that context, its meaning was very simple: A line has a slope and it is always the same; it is the rise over run, or the change in y divided by the change in x. All of lines in Figure 2.4-6 have the same slope.

We now have a much broader idea of slope. We can think of the slope of a curve at a point. The slope of a curve at a point is the slope of the tangent line to the curve at that point, which is the instantaneous rate of change, or the rate of change, at the point. From

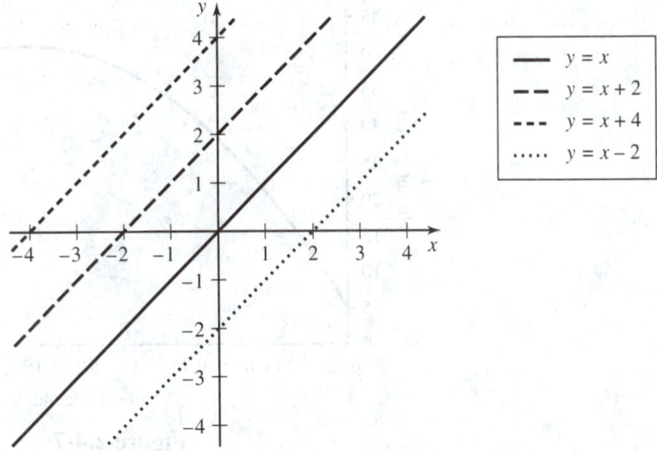

Figure 2.4-6

this concept came the definition of the derivative. Since the tangent to a line is the line itself, the slope of the tangent will simply be the slope of the original line. So it makes sense graphically that the derivative of a linear function is simply its slope. Notice that this rule covers the derivative of a constant function as well, since a constant function is a special case of a linear function where the slope is 0 ($m = 0$).

Sample Problem 4: Find the marginal cost function (the derivative) if the cost of producing n scrunchies is

$$C(n) = 4.95 + 0.27n, \quad \text{for } n \geq 0$$

Solution: In this case, we have $C(n)$ instead of $f(x)$. But, as $C(n)$ is a linear function, we know that its derivative is its slope, which in this case is 0.27. Thus, the rate of change of the cost, the marginal cost, is 0.27 dollars per scrunchie. Try checking this using the limit definition of the derivative (remember to use n instead of x). □

Derivatives of Polynomial Functions

Finding the derivatives of constant and linear functions is rather easy because the slopes never change. The functions are either constant (not changing), or increasing or decreasing at a constant rate. How can we find the derivative of a function that does not have a constant rate of change?

Let's look again at the graph of the ball toss model discussed in Sections 2.1–2.3, repeated here in Figure 2.4-7. The formula for the height of the ball was given by $H(t) = -16t^2 + 48t + 5$. As we discussed earlier, the ball is rising, rapidly at first and then more slowly, until it stops rising and starts to fall again. The height is increasing, sharply at first, then more gradually until it stops increasing and starts to decrease, gradually at first and then sharply. In Section 2.2, we numerically estimated the slope at several points (shown in Table 7 of Section 2.2) and then we graphed those points and fit a model (shown again here in Figure 2.4-8).

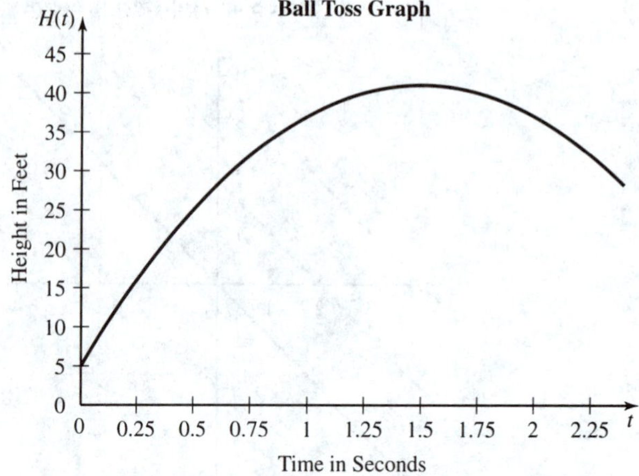

Figure 2.4-7

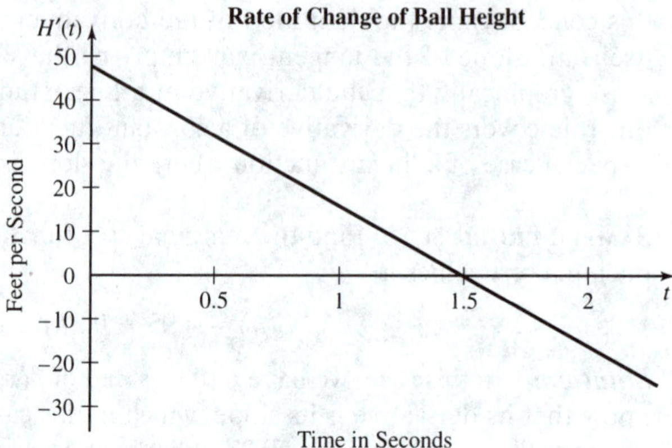

Figure 2.4-8

The graph in Figure 2.4-8 certainly describes the slope of the curve: positive and rather large at first, then gradually getting smaller until it is actually 0 at 1.5 seconds. Then it is negative (the ball starts to come down) and gradually gets larger (more negative). When we fit a model to these data we obtained $H'(t) = -32t + 48$ feet per second.

Will all quadratic functions have slope graphs that look like this? First let's look at the simplest quadratic function: $y = f(x) = x^2$. The graph of this function is shown in Figure 2.4-9. This graph is decreasing sharply, then more gradually until it stops decreasing and starts to increase, gradually at first and then sharply. The slope would be negative, then 0, and then positive.

Let's use the algebraic definition of the derivative:

$$\lim_{n\to 0}\frac{f(x+h)-f(x)}{h} = \lim_{h\to 0}\frac{(x+h)^2-x^2}{h}$$

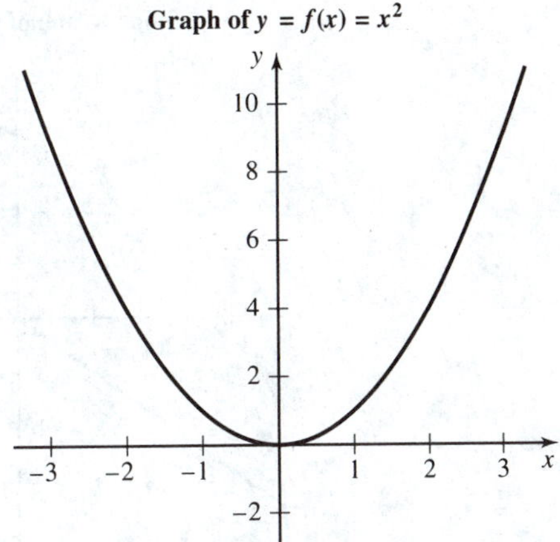

Figure 2.4-9

Expanding the first term in the numerator we have

$$\lim_{h \to 0} \frac{x^2 + 2xh + h^2 - x^2}{h}$$

Combining like terms we have

$$\lim_{h \to 0} \frac{2xh + h^2}{h}$$

Factoring h from the two terms in the numerator we have

$$\lim_{h \to 0} \frac{h(2x + h)}{h}$$

Once again, if we just substitute the limiting value of the variable, $h = 0$, we will get $\frac{0}{0}$, which is not a real number. But we can cancel out the h's (divide both the numerator and denominator by h) because h is approaching 0, and according to the definition of the limit, we ignore what happens when h equals 0, leaving us with

$$\lim_{h \to 0} (2x + h)$$

As h approaches 0, $2x + h$ approaches $2x$ as a limit. This follows from our principle of evaluating limits: If you substitute the value that the limit variable is approaching for the limit variable and get an answer that is defined as a real number, then that answer is the value of the limit.

Thus, the derivative of x^2 is $2x$. This is a line with a slope of 2 (Figure 2.4-10).

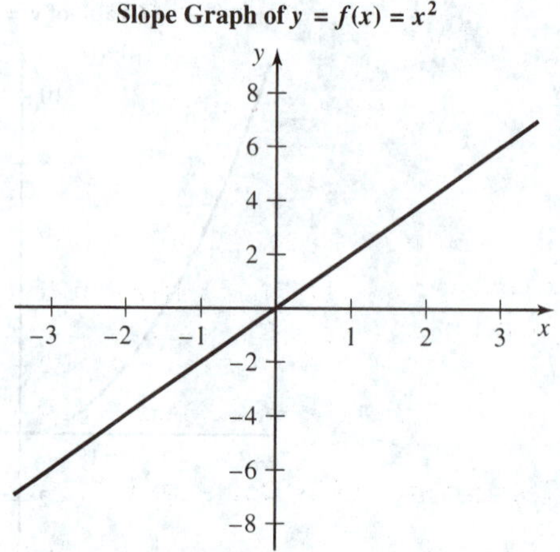

Figure 2.4-10

Sample Problem 5: What do you think will happen to the slope of x^2 if we add a constant term, say $f(x) = x^2 - 5$?

Solution: Let's use the algebraic definition of the derivative to find out.

$$\lim_{h \to 0} \frac{f(x+h) - f(x)}{h} = \lim_{h \to 0} \frac{[(x+h)^2 - 5] - (x^2 - 5)}{h}$$

Expanding the first term in the numerator and clearing the parentheses in the second term we have

$$\lim_{h \to 0} \frac{x^2 + 2xh + h^2 - 5 - x^2 + 5}{h}$$

Combining like terms, we have

$$\lim_{h \to 0} \frac{2xh + h^2}{h}$$

This is exactly what we had in Sample Problem 5 above. The limit is $2x$. □

Does adding a constant to *any* function have the same property? Well, if

$$g(x) = f(x) + c$$

then

$$g'(x) = \lim_{h \to 0} \frac{g(x+h) - g(x)}{h} \qquad \text{(definition of the derivative)}$$

$$= \lim_{h \to 0} \frac{[f(x+h) + c] - [f(x) + c]}{h} \qquad (g(x) = f(x) + c, \text{ so } g(x+h) = f(x+h) + c)$$

$$= \lim_{h\to 0} \frac{f(x+h) + c - f(x) - c}{h} \qquad \text{(removing the parentheses)}$$

$$= \lim_{h\to 0} \frac{f(x+h) - f(x)}{h} \qquad \text{(canceling the } c\text{'s)}$$

$$= f'(x) \qquad \text{(definition of the derivative)}$$

We have proved that **adding a constant to a function will not change its derivative, because the derivative of a constant is 0. In other words,**

$$\mathbf{D_x[f(x) + c] = f'(x) + 0 = f'(x)}$$

The Derivative of a Constant Times a Function

Sample Problem 6: Earlier we stated that the derivative of a constant times a variable x is the constant times the derivative of x, which is 1; for linear functions the derivative of $y = f(x) = x$ is $f'(x) = 1$. The derivative of $y = f(x) = mx$ is $f'(x) = m$. Is this also true for quadratic models? In other words, will the derivative of $y = f(x) = ax^2$ be $\frac{dy}{dx} = a(2x)$?

Solution: Let's look at two curves (Figure 2.4-11). Multiplying the x^2 by a negative three reversed the direction of the curve. The first curve is decreasing for all negative values of x, the second curve is increasing for the same values. Now look at the values on the y-axis: Instead of positive 10, they go to negative 10. The slope changes sign and is steeper.

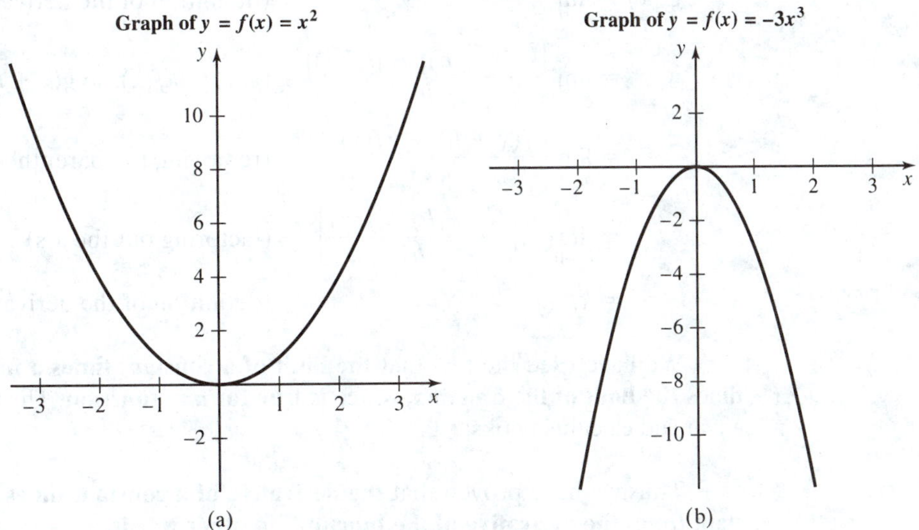

Figure 2.4-11

Let's use the algebraic definition of the derivative to see if we are correct. If this is true the derivative of $y = f(x) = -3x^2$ should be $f'(x) = -3(2x) = -6x$.

$$\lim_{h\to 0} \frac{f(x+h) - f(x)}{h} = \lim_{h\to 0} \frac{-3(x+h)^2 - 3x^2}{h}$$

Expanding the first term in the numerator we have

$$\lim_{h\to 0} \frac{-3(x^2 + 2xh + h^2) - (-3x^2)}{h}$$

Multiplying through and simplifying we have

$$\lim_{h\to 0} \frac{-3x^2 - 6xh - 3h^2 + 3x^2}{h} = \lim_{h\to 0} \frac{-6xh - 3h^2}{h}$$

Factoring h from the two terms in the numerator we have

$$\lim_{h\to 0} \frac{h(-6x - 3h)}{h}$$

We can cancel the h's, leaving

$$\lim_{h\to 0} (-6x - 3h)$$

As h approaches 0, $-6x - 3h$ approaches $-6x$ as a limit (plug in 0 for h).

Yes, it does work!

If we want to generalize this to see the effect of multiplying *any* function by a constant, the proof is quite simple. First we will define

$$g(x) = kf(x)$$

Then

$$g'(x) = \lim_{h\to 0} \frac{g(x+h) - g(x)}{h} \quad \text{(definition of the derivative)}$$

$$= \lim_{h\to 0} \frac{[kf(x+h)] - [kf(x)]}{h} \quad (g(x) = kf(x), \text{ so } g(x+h) = kf(x+h))$$

$$= \lim_{h\to 0} \frac{kf(x+h) - kf(x)}{h} \quad \text{(removing the parentheses)}$$

$$= \lim_{h\to 0} (k)\left[\frac{f(x+h) - f(x)}{h}\right] \quad \text{(factoring out the } k\text{'s)}$$

$$= kf'(x) \quad \text{(definition of the derivative)}$$

We have used the fact that **the limit of a constant times a function is the constant times the limit of the function, which is true for *any* function**. The proof can be found in advanced calculus books. □

Thus we have proved that **the derivative of a constant times a function is the constant times the derivative of the function. In other words,**

$$\mathbf{D_x[kf(x)] = kf'(x)}$$

The Derivative of the Sum (Difference) of Functions

Sample Problem 7: In the ball toss model we also had a term with a single power of the variable. The height, H, in feet, t seconds after the toss is $H(t) = -16t^2 + 48t + 5$,

$0 \le t \le 3$. What is the rate of change of the height of the ball with respect to the time after it was tossed?

Solution: We have seen that the derivative of $-16t^2$ is $-16(2t)$. We also know that the derivative of $48t$ is 48 and that adding a constant doesn't change anything. It would be really simple if we could just add these two and say the derivative of $-16t^2 + 48t + 5$ is $-32t + 48$. Back we go to the definition of the derivative to see if this is true. This time, instead of using the specific function $-16t^2 + 48t + 5$, we will work with the general quadratic $ax^2 + bx + c$.

$$\lim_{n\to 0} \frac{f(x+h) - f(x)}{h} = \lim_{h\to 0} \frac{a(x+h)^2 + b(x+h) + c - (ax^2 + bx + c)}{h}$$

Expanding we have

$$\lim_{h\to 0} \frac{ax^2 + 2axh + ah^2 + bx + bh + c - ax^2 - bx - c}{h}$$

Combining like terms,

$$\lim_{h\to 0} \frac{2axh + ah^2 + bh}{h}$$

Factoring an h out of the numerator, we have

$$\lim_{h\to 0} \frac{h(2ax + ah + b)}{h}$$

Canceling out the h's (dividing the numerator and denominator by h, which is possible since the limit *ignores* what happens when $h = 0$) we have

$$\lim_{h\to 0} (2ax + ah + b)$$

As h approaches 0, we can now simply substitute in 0 for h, so the first term is not changed, the second term goes to 0 and the third term is not changed. Thus, the limit is

$$2ax + b$$

The a in our ball toss example was -16, the b was 48, and the x was t. Substituting these values in the general form we have $2(-16)t + 48$ or $-32t + 48$. Just what we hoped: $H'(t) = -32t + 48$ feet per second. □

Let's show that, in general, the derivative of the sum of two functions is the sum of the derivatives. We can define

$$k(x) = f(x) + g(x)$$

Then

$$k'(x) = \lim_{h\to 0} \frac{k(x+h) - k(x)}{h}$$

$$= \lim_{h\to 0} \frac{[f(x+h) + g(x+h)] - [f(x) + g(x)]}{h}$$

$$= \lim_{h\to 0} \frac{f(x+h) + g(x+h) - f(x) - g(x)}{h}$$

$$= \lim_{h\to 0} \frac{f(x+h) - f(x) + g(x+h) - g(x)}{h}$$

$$= \lim_{h\to 0} \left\{\frac{f(x+h) - f(x)}{h} + \frac{g(x+h) - g(x)}{h}\right\}$$

$$= \lim_{h\to 0} \frac{f(x+h) - f(x)}{h} + \lim_{h\to 0} \frac{g(x+h) - g(x)}{h}$$

$$= f'(x) + g'(x)$$

Here we have used the fact that **the limit of the sum (or difference) of two functions is the sum (or difference) of their limits**, if the separate limits exist. The proof of this can be found in advanced calculus books. If we are subtracting a function instead of adding, we can simply think of this as adding (-1) times the function:

$$\begin{aligned} D_x[f(x) - g(x)] &= D_x\{f(x) + [-g(x)]\} = D_x[f(x)] + D_x[-g(x)] \\ &= f'(x) + D_x[(-1)g(x)] = f'(x) + (-1)D_x[g(x)] \\ &= f'(x) - g'(x) \end{aligned}$$

Thus we have shown that **the derivative of the sum (or difference) of two functions is the sum (or difference) of their derivatives.** In other words,

$$\boldsymbol{D_x[f(x) \pm g(x)] = f'(x) \pm g'(x)}$$

We can relate this to our concept of profit: profit = revenue minus cost,

$$P(x) = R(x) - C(x)$$

so the rate of change of profit is equal to the rate of change of revenue minus the rate of change of cost:

$$P'(x) = R'(x) - C'(x)$$

Intuitively, you can think of this as making sense, since the profit from the next unit equals the revenue from the next unit minus the cost of the next unit. Recall that we observed this when discussing Table 1 in Section 2.3.

Sample Problem 8: So far, we have discovered how to find the rate of change (take the derivative) of linear and quadratic functions. What about cubic functions? We estimated the rate of change of crime in the United States on Dec. 31, 1990, and Dec. 31, 1993, by finding the limit of the slopes of secants, a slow and laborious process. Is there a better way?

The model is

Verbal Definition: $C(x)$ = crimes reported in the preceding year in the United States x years after Dec. 31, 1981.

Symbol Definition: $C(x) = -13150x^3 + 247506x^2 - 1030143x + 13471438, x \geq 0.$

Assumptions: Certainty and divisibility. Certainty implies the relationship is *exact.* Divisibility implies that *any* fractions of crimes or years are possible.

Solution: Let's look again at the graph of our model of the crime index in the United States (Figure 2.4-12). This curve is decreasing (has a negative slope) up to about $x = 3$, then is increasing (has a positive slope) up to about $x = 10$, and then is decreasing (has a negative slope) again. Let's do what we did with the ball toss curve: Estimate the slope of the curve at a number of points, plot the points, and fit a model (Figure 2.4-13).

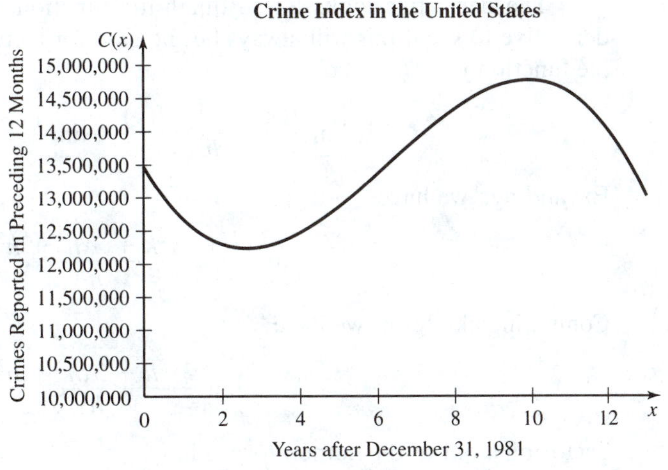

Figure 2.4-12

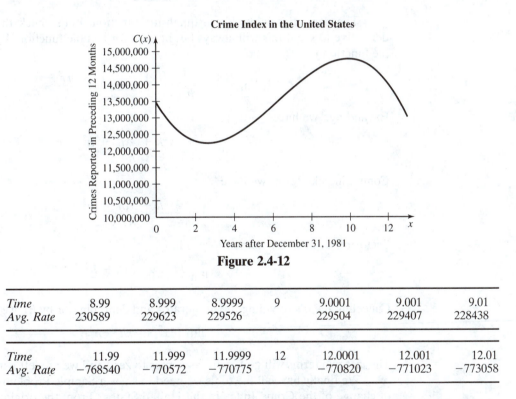

Time	8.99	8.999	8.9999	9	9.0001	9.001	9.01
Avg. Rate	230589	229623	229526		229504	229407	228438

Time	11.99	11.999	11.9999	12	12.0001	12.001	12.01
Avg. Rate	−768540	−770572	−770775		−770820	−771023	−773058

Was this what you expected? At first **the *derivative* is negative, so the *original function* is *decreasing*** sharply, then the derivative becomes less negative, then zero. Then **the *derivative* becomes *positive,* so the *original function* is *increasing*** slowly at first, then the derivative reaches a peak and starts to go down, so that the original function is still increasing, but more slowly again. Finally the derivative hits zero and becomes negative again, so the original function starts to decrease again, slowly at first and then more steeply.

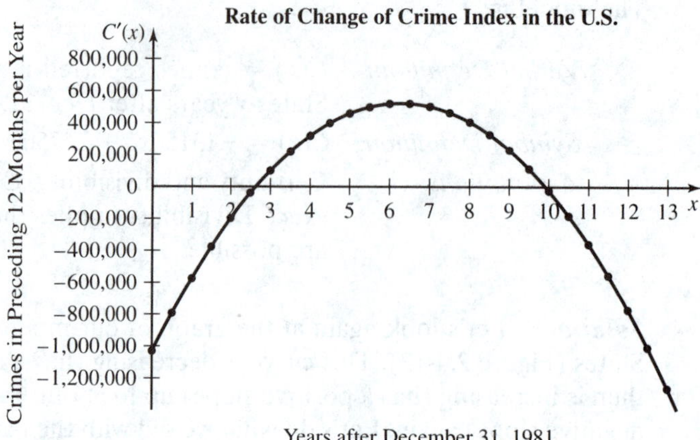

Figure 2.4-13

The derivative looks like a quadratic function. Let's check the definition of the derivative to see if this will always be the case for a cubic function. We will use the simple function $y = f(x) = x^3$:

$$\lim_{n \to 0} \frac{f(x+h) - f(x)}{h} = \lim_{h \to 0} \frac{(x+h)^3 - x^3}{h}$$

Expanding,[6] we have

$$\lim_{h \to 0} \frac{x^3 + 3x^2h + 3xh^2 + h^3 - x^3}{h}$$

Combining like terms we have

$$\lim_{h \to 0} \frac{3x^2h + 3xh^2 + h^3}{h}$$

Factoring h out of the numerator we have

$$\lim_{h \to 0} \frac{h(3x^2 + 3xh + h^2)}{h}$$

Canceling the h's (dividing the numerator and denominator by h), we have

$$\lim_{h \to 0} (3x^2 + 3xh + h^2)$$

The last two terms will go to zero as h goes to zero and we are left with the limit as $3x^2$.

We finally have all the tools we need to answer Sample Problem 9: What is the rate of change of the Crime Index in the United States? From the original model of crimes reported in the United States, we can now find the *rate of change* of crimes reported in the United States.

The formula from the original model was given as

$$C(x) = -13150x^3 + 247506x^2 - 1030143x + 13471438$$

[6]See Appendix for expansion of binomials.

so we can use our rules to find the derivative:

$$\begin{aligned} C'(x) &= D_x(-13150x^3 + 247506x^2 - 1030143x + 13471438) \\ &= D_x(-13150x^3) + D_x(247506x^2) - D_x(1030143x) + D_x(13471438) \\ &= (-13150)D_x(x^3) + (247506)D_x(x^2) - (1030143)D_x(x) + D_x(13471438) \\ &= (-13150)(3x^2) + (247506)(2x) - (1030143)(1) + 0 \end{aligned}$$

Simplifying, then, we get

$$C'(x) = -\,39450x^2 + 495012x - 1030143 \text{ (in crimes per year)} \quad \square$$

Basic Derivative Rules for Polynomials

So far we have the following basic derivative rules:

$f(x)$	$f'(x)$
x^1	$1(x^0)$
x^2	$2(x^1)$
x^3	$3(x^2)$

Recognizing patterns when they occur is a big part of problem solving. When we fit models to data, we were recognizing patterns of behavior of output with respect to input: The pattern of changes of the output relative to changes in the input determined what kind of a model we chose.

★ **DISCOVERY QUESTION:**

Do you see a pattern in the table above? What would you expect the next entry to be? What would the derivative of x^n be? (Try to answer this question before you read on.)

Hint: Look at the exponents and the coefficients. (Try again to answer the question before you read on.)

Answer: On the left side, showing the original functions, the exponents are increasing by one each time. On the right side, showing the derivatives, the coefficients are increasing by one each time—they match the *exponents* of the original functions. The exponents of the derivatives are always one less than the corresponding exponents of the original functions. This suggests that the derivative for x^4 would be $4x^3$. Using the same logic, the derivative for x^n would be nx^{n-1}, the exponent of the original function times the variable to the power of that original exponent minus one. ★

It can be proved[7] that for *any* real number *n*, the derivative of x^n is $n(x^{n-1})$. Using our rule concerning a constant times a function, **the derivative of** ax^n is anx^{n-1}.

[7] See the Appendix for a proof for any positive whole number value of n. For a general proof for any real number, see an advanced calculus book.

We can now find the derivative of any polynomial function directly using four rules:

Function	Derivative	
c	0	The derivative of a constant is 0.
$c[f(x)]$	$c[f'(x)]$	The derivative of a constant times a function is the constant times the derivative of the function.
$f(x) \pm g(x)$	$f'(x) \pm g'(x)$	The derivative of the sum or difference of two functions is the sum or difference of the derivatives of the functions.
x^n	$n(x^{n-1})$	The derivative of a constant power of x is the exponent times (the x raised to one less than the original power).

The Derivative of e^x

Sample Problem 9: Inflation continues to be a problem, particularly in some developing countries. In 1992 the inflation rate in Brazil was 100%! The interest paid by banks reflected this inflation rate. If you deposited 5000 cruzeiros at 100% interest compounded continuously, the amount in your account at any time t would be equal to $A(t) = 5000e^t$ (the derivation of this equation will be discussed in Chapter 5). How fast is your account growing? How fast is your account growing 1 year after the initial deposit?

Solution: This is an exponential function, not a polynomial function. Before we try to work with this function, let's start with the basic exponential function to see if we can discover some general rules for exponential functions as we did for polynomial functions.

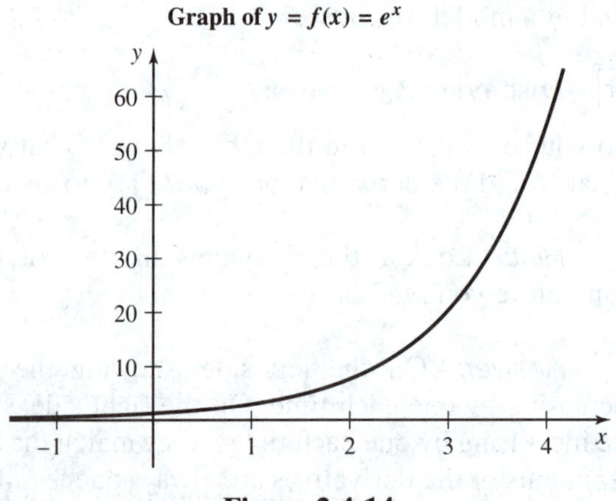

Figure 2.4-14

Figure 2.4-14 shows the graph of $y = f(x) = e^x$. We can see from the graph that the function is always increasing—the slope is always positive. The function is increasing very slowly at first, and then more and more rapidly. The slope curve will be positive and increasing. In Table 1, we have estimated the rate of change of the curve at a few points using technology and the limit of the slopes of secants from Section 2.2.

It looks as if the derivative is the same as the original function! Let's explore this further and see if we can obtain more evidence that this is indeed the relationship.

Figure 2.4-15 shows the original function $y = f(x) = e^x$ [in graph (a)] and the graph of the estimated rate of change function at the points calculated in Table 1 with an exponential curve fit to the points [in graph (b)].

TABLE 1

x	-1	0	1	2	3	4
$y = e^x$	.36788	1	2.7183	7.3891	20.086	54.598
Est. dy/dx	0.367	1	2.718	7.389	20.085	54.598

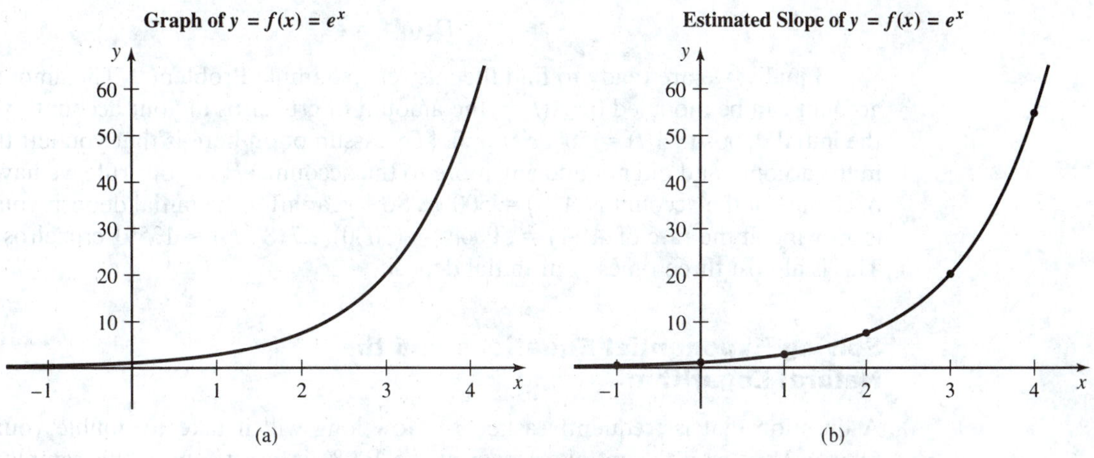

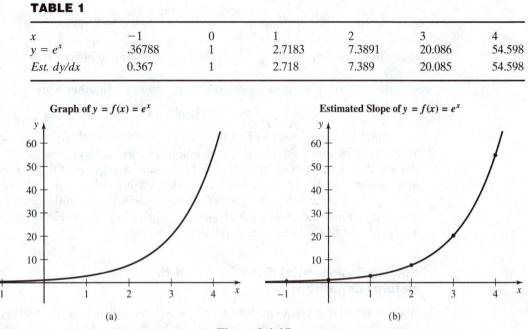

Figure 2.4-15

We put the graphs next to each other so that you could compare them. The curves look very much alike, but are they really? We can check this using the definition of the derivative and the concept of limits.

If $f(x) = e^x$, then

$$f'(x) = \lim_{h \to 0} \frac{f(x+h) - f(x)}{h}$$

$$= \lim_{h \to 0} \frac{e^{x+h} - e^x}{h}$$

Factoring out e^x, we have

$$\lim_{h \to 0} \frac{e^x(e^h - 1)}{h}$$

Since e^x does not involve the limit variable h, it acts like a constant and can be factored out, giving us

$$(e^x) \lim_{h \to 0} \frac{(e^h - 1)}{h}$$

We'll use technology to find $\lim_{h \to 0} \frac{e^h - 1}{h}$:

h	0.1	0.01	0.001	0.0001	0	-0.0001	-0.001	-0.01	-0.1
$\frac{e^h - 1}{h}$	1.0517	1.005	1.0005	1.0001	(Undefined)	0.99995	0.9995	0.99502	0.95163

As the values for h get closer and closer to 0, the values of $\frac{e^h - 1}{h}$ get closer and closer to 1.[8] So $(e^x) \lim_{h \to 0} \frac{(e^h - 1)}{h} = (e^x)(1) = e^x$. It can be proved that

If $y = f(x) = e^x$, then its derivative is $f'(x) = e^x$. In other words,

$$D_x(e^x) = e^x$$

Finally, we are ready to find the answer to Sample Problem 9. The amount in your account can be modeled by $A(t)$ = the amount of cruzeiros in your account t years after the initial deposit. $A(t) = 5000e^t$, $t \geq 0$. (The assumption here is that you left the money in the account and did not add any more to the account.) From our rule we have the rate of change of the account is $A'(t) = 5000e^t$. So 1 year after the initial deposit your account is growing at the rate of $A'(1) = 5000e^1 = (5000)(2.718\ldots) \approx 13591$ cruzeiros per year! This is almost three times your initial deposit. □

Solving Exponential Equations and the Natural Logarithm

A question that is frequently asked is: How long will it take to double your money? Obviously, it won't even take 1 year at the 100% interest rate in this sample problem, since the 5000 cruzeiros *more* than doubled in a year. How long would it take? To answer this we have to solve the equation

$$10{,}000 = 5000e^t$$

for t. To solve this type of equation we can first divide both sides by 5000, giving

$$2 = e^t$$

Now we want to solve this equation for t. We are looking for the exponent that the number e must be raised to, in order to get an answer of 2. As discussed in the Appendix, the natural logarithm of a number, $\ln x$, is the exponent to which e must be raised to equal that number, x. This means that the solution to our equation is

$$t = \ln 2$$

For a more general approach to solving equations like $2 = e^t$, **when the variable you want to solve for is located in an exponent, it is often a good idea to try taking the natural logarithm of both sides of the equation**. In our case, this would have resulted in the equation

$$\ln(2) = \ln(e^t)$$

In general, $\ln(x)$ means the exponent to which you have to raise e in order to get the answer x. So evaluating $\ln(e^t)$ means answering the question: To what exponent

[8]See an advanced calculus book for a proof that this limit is exactly 1, which then proves the derivative rule.

would you have to raise e to get e^t? One way to think of this is as the visual algebraic question

$$e^? = e^t$$

The answer, of course, is t. So $\ln e^t = t$. Thus our equation now becomes

$$\ln 2 = t$$

as we had determined already. Using our calculator we find that

$$t = \ln 2 = 0.69314\ldots.$$

So it would take approximately 0.69 of a year, or about eight and one-quarter months, to double your money.

Sample Problem 10: In Section 1.4 we discussed exponential functions. In the Appendix, we give the rule for changing an exponential function of the form $y = ax^b$ to the form $y = ce^{kx}$. This involves the use of $\ln x$, the natural logarithm function. We also used the natural log to answer the question about doubling time for continuously compounded interest in Sample Problem 9. Recall the discussion of inverse functions from Section 1.1:

The natural logarithm function, ln x, is the *inverse* of the exponential function, e^x:
If $y = f(x) = e^x$, then $x = g(y) = \ln y$

This means that e raised to the x power equals y (or x is the exponent to which e must be raised to equal y). This is the set of directions that take us from y as an input to x as an output.[9] The natural log is a type of function, just as polynomials and exponentials are types of functions. How can we find the rate of change—the derivative—of the natural log function? The graph of the natural log function is shown in Figure 2.4-16.

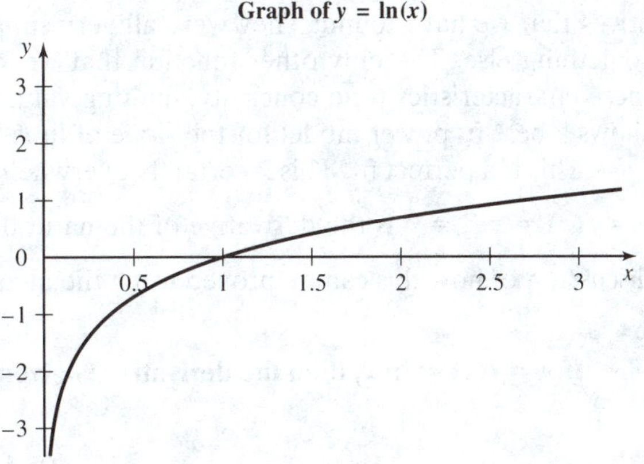

Figure 2.4-16

[9]See the Appendix for an expanded discussion of the natural logarithm.

Analyzing this curve, we can see that it is always increasing—the slope is everywhere positive—and that it is increasing sharply at first and then more and more gradually. We have estimated the slope of the tangent lines at several points using the limit of secant slopes and technology:

x	0.2	0.25	0.5	1	2	3	4	5
Est. dy/dx	5.00	4.00	2.00	1.00	0.50	0.33	0.25	0.20

★ **Discovery Question:**

Do you notice any pattern in the estimate of the derivative at x? See if you can infer a formula for $D_x(\ln x)$ intuitively from the table above. (See if you can answer this question before reading further.)

Hints: See if you can find a relationship between the value of x and the value of the estimated derivative of $\ln x$ right below it, and see if you can express the derivative in terms of the value of x. (Again, see if you can answer the question before reading further.)

Answer: If you looked carefully, you probably noticed that the derivative looks as if it is the *reciprocal* of the x value: when x is 2, the derivative is ½, when x is 4, the derivative is ¼, and so on. So the general pattern is that the derivative of $\ln x$ is $\frac{1}{x}$. Let's explore further to convince ourselves whether or not this is true in general. ★

Figure 2.4-17 is a graph of the estimated slopes at the various points. This plot is concave up. It appears to have a natural lower limit. Our initial reaction in trying to model this would probably be to try fitting an exponential function (Figure 2.4-18).

This is not a very good fit. And the model that was fit is nothing like the other derivatives that we have found: They were all very simple functions. Perhaps we need to try something else. The only other function that we know from this course so far that has these characteristics (one concavity, limiting value) is the power function. Figure 2.4-19 shows a best-fit power model for the slope of $\ln x$.

This is a perfect fit. This is certainly overwhelming evidence that the power function $y = f(x) = x^{-1} = \frac{1}{x}$ is the derivative of the natural log function. See the Appendix for a discussion of how this can be proved using the algebraic definition of the derivative.

If $y = f(x) = \ln x$, then the derivative is given by $f'(x) = \frac{1}{x}$. In other words,

$$D_x(\ln x) = \frac{1}{x}$$

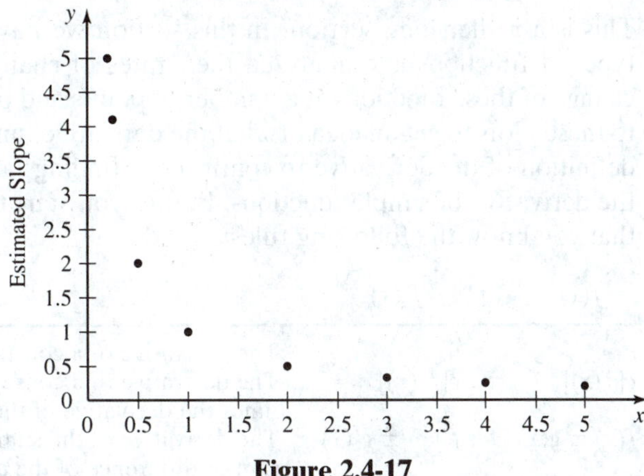

Figure 2.4-17

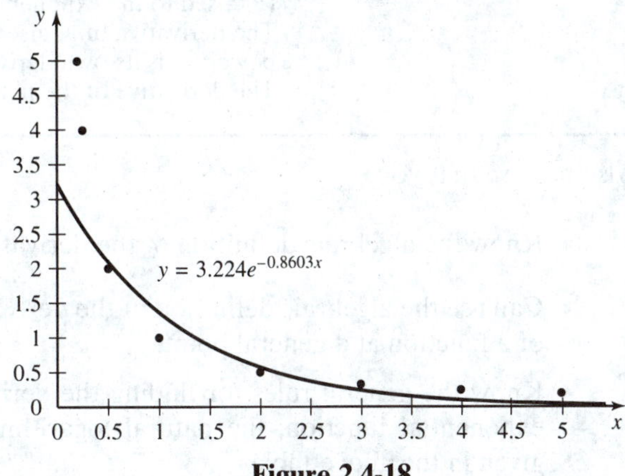

Figure 2.4-18

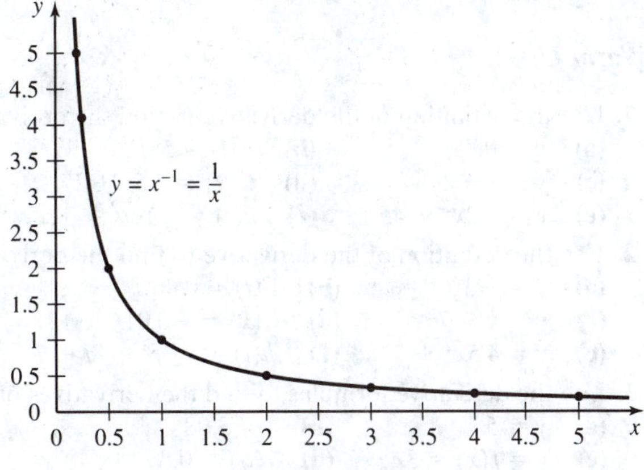

Figure 2.4-19

Section Summary

This is a rather long section. In this section we have looked at the graphs of different types of functions and analyzed their rates of change. We have estimated the rates of change of these functions at a number of points and plotted these estimates. We fit curves to these plots to get an idea of what the derivative function would be. Finally, we used the definition of the derivative to confirm our findings and develop general rules for finding the derivative of simple functions. Before you start the exercises for this section be sure that you know the following rules:

$f(x)$	$f'(x)$	
c	0	The derivative of a constant is 0.
$c[f(x)]$	$c[f'(x)]$	The derivative of a constant times a function is the constant times the derivative of the function.
$f(x) \pm g(x)$	$f'(x) \pm g'(x)$	The derivative of the sum or difference of two functions is the sum or difference of the derivatives of the functions.
$mx + b$	m	The derivative of a linear function is its slope.
ax^n	$an(x^{n-1})$	The derivative of a constant coefficient times x raised to a constant exponent is the coefficient times the exponent times (x raised to the exponent minus 1).
e^x	e^x	The derivative of e raised to the x power is e raised to the x power: e^x is its own derivative.
$\ln x$	$\frac{1}{x}$	The derivative of the natural log of x is one over x.

Also be sure that you

- Know the algebraic definition of the derivative: $f'(x) = \lim_{h \to 0} \frac{f(x+h) - f(x)}{h}$.
- Can use the algebraic definition of the derivative to try to find the rate of change of a function at a general point.
- Know the general rules for finding the derivative for polynomial functions, the exponential function, the natural logarithm function, and power functions as given in the above table.

EXERCISES FOR SECTION 2.4

Warm Up

1. Use the definition of the derivative to find the derivatives of the following functions:
(a) $y = 0.75$ **(b)** $V(t) = 55$
(c) $y = x + 5$ **(d)** $C(x) = 4.5x + 12$
(e) $y = -2x^2 + 45$ **(f)** $h(t) = -16t^2 + 12t$

2. Use the definition of the derivative to find the derivatives of the following functions:
(a) $y = -3$ **(b)** $V(t) = 0.45$
(c) $y = x - 7$ **(d)** $C(x) = -0.01x + 12$
(e) $y = 4.5x^2 + 5$ **(f)** $h(t) = -3t^2 + 8t + 7$

3. Use the derivative formulas to find the derivatives of the following functions:
(a) $y = 5$ **(b)** $t = 3.08$
(c) $y = f(x) = 3x$ **(d)** $r(q) = 0.8q$

4. Use the derivative formulas to find the derivatives of the following functions:
(a) $h(t) = 9t + 5$ **(b)** $y = -5x + 0.3$
(c) $y = x^2$ **(d)** $f(s) = 4s^2 - 5$

5. Use the derivative formulas to find the derivatives of the following functions:
(a) $c(x) = 0.09x^{0.5} - 0.5x^{0.1} + 7$ **(b)** $R(p) = -0.3p^{2.2} + 45p^{0.9} + 20$
(c) $y = f(x) = e^x$ **(d)** $A = f(r) = -0.5e^r$

6. Use the derivative formulas to find the derivatives of the following functions:
(a) $y = -0.5e^r$ **(b)** $A(r) = 100e^r$
(c) $d(x) = \ln x$ **(d)** $y = f(x) = -2.5 \ln(x)$

7. Use the derivative formulas to find the derivatives of the following functions:
(a) $m(r) = 3 \ln(r)$ **(b)** $w = -2 \ln(y)$
(c) $y = 3x^{2.5} - 2e^x$ **(d)** $g(t) = 0.0579t^2 + 0.0075t - 23.987$

8. Use the derivative formulas to find the derivatives of the following functions:
(a) $t = 0.45r^3 - 0.7r^2 + 25.89$ **(b)** $y = e^t$
(c) $s = 0.087t + e^t$ **(d)** $t = 45 + 23g - 21g^2$

9. Use the definition of the derivative to show *directly* that the derivative of the difference of two functions is the difference of the derivatives. (*Hint:* Define $k(x) = f(x) - g(x)$, similar to what we did for proving the analogous result for the sum.)

Game Time

10. The model for the profit from the sale of "Save the Environment" mugs is

Verbal Definition: $\pi(q)$ = profit in dollars from the sale of q mugs.
Symbol Definition: $\pi(q) = -0.01q^2 + 6.5q - 25,\ 0 \le q \le 850.$
Assumptions: Certainty and divisibility. Certainty implies that the relationship is *exact*. Divisibility implies that *any* fractions of mugs and dollars of profit are possible.

Find the value of the marginal profit (rate of change of profit with respect to quantity) function when
(a) 200 mugs are sold.
(b) 300 mugs are sold.
(c) 400 mugs are sold.
(d) What conclusions can be drawn about your goal for sales?

11. You have been selling sodas at sporting events and have developed the following model for the demand:

Verbal Definition: $q(p)$ = the quantity of sodas (in thousands) sold at p dollars per soda.
Symbol Definition: $q(p) = \frac{2}{p},\ 0.5 \le p \le 5.$
Assumptions: Certainty and divisibility. Certainty implies the relationship is *exact*. Divisibility implies that *any* fractions of sodas and dollars of selling price are possible.

Find the rate at which the quantity of sodas sold was changing when the price was
(a) \$1 per soda.
(b) \$2 per soda.
(c) \$3 per soda.
(d) Describe how the demand is changing as the price increases.

12. The following model was developed for the percentage change in the Consumer Price Index for Medical Care:

Verbal Definition: $I(x)$ = the percentage change in the Consumer Price Index for Medical Care x years after Dec. 31, 1986.

Symbol Definition: $I(x) = 0.0193x^4 - 0.3547x^3 + 1.9314x^2 - 3.0161x + 7.6229,\ x \geq 0.$

Assumptions: Certainty and divisibility. Certainty implies the relationship is *exact*. Divisibility implies that *any* fractions of the year and the percentage change are possible.

According to the model, determine whether the percentage change in the Consumer Price Index for Medical Care was increasing or decreasing at the end of

(a) 1990.
(b) 1991.
(c) 1994.
(d) In the years that the *percentage change* in the Consumer Price Index for Medical Care was decreasing, was the Consumer Price Index for Medical Care *itself* actually decreasing? Explain your answer.

13. The number of tickets sold to a production at a local coffeehouse can be modeled by

Verbal Definitions: $T(p)$ = the number of tickets sold at p dollars per ticket.

Symbol Definition: $T(p) = 5p^2 - 95p + 520,\ 0 \leq p \leq 9.50.$

Assumptions: Certainty and divisibility. Certainty implies the relationship is *exact*, and that there are no discounts or sales. Divisibility implies that *any* fractions of tickets and dollars of selling price are possible.

Find the rate of change of the number of tickets sold (per dollar of ticket price) when the price per ticket is

(a) \$8.00.
(b) \$9.00.
(c) \$9.50.
(d) When we fit a linear model to this data, the domain was $0 \leq p \leq 13.50$. Find the rate of change of the number of tickets per dollar ticket price when the price per ticket is \$10.00 if we had used the same domain for the quadratic model. Does your answer seem reasonable to you? Explain why it does or does not seem reasonable.

2.5 COMPOSITE FUNCTIONS AND THE CHAIN RULE

In some of the examples we worked out at the beginning of the book, we found times when we needed to take a function of a function. Such a function is called a **composite function**. In this section, we will discuss such functions, show how to evaluate them, and show how to take derivatives of them using what is called the **Chain Rule**.

After studying this section, you should be able to solve problems like the following:

- You are interested in teaching summer school. You are paid a certain percentage of your regular salary per credit hour taught in the summer. The city wage tax is a flat percentage of your summer salary income. You want to find the amount of city wage tax paid per credit hour taught.

- You are trying to decide what hair dryer you want to buy. More expensive ones tend to have more power, which reduces the time needed to dry your hair. You can put a dollar value on the time you would waste drying your hair over the life of the hair dryer. You want to find an expression for the dollar value of this wasted time as a function of the cost of the hair dryer. You can then *add* the direct cost of the hair dryer to the cost of the wasted time to get the *total cost* of any given hair dryer as a function of its selling price, in order to eventually find the hair dryer (price level) that will *minimize* that total cost over the life of the hair dryer. For now, you are considering a certain hair dryer and want to know if a more expensive one would be better.[10]
- You work in the admissions office of a school. You need to know how many applicants to accept for the coming fall. You have information about the relationship between the number accepted and how many students send in a deposit, and also about the relationship between how many send in a deposit and how many actually enroll and appear in the fall. You want to find the number who enroll and appear in the fall as a function of the number accepted.
- If SAT scores are *normalized* by subtracting the average and dividing by a constant, there is a particularly simple model that can represent the relative frequencies (how often different scores occur compared to each other). What would be the model for the relative frequencies of the *original* (as opposed to the *normalized*) scores?

In addition to being able to solve problems like those above, after studying this section, you should also

- Be able to evaluate composite functions.
- Understand the importance of using consistent letters for function and variable names when working with composite functions.
- Understand the conditions and restrictions on the domain of a composite function
- Understand the concept behind the Chain Rule.
- Know how to identify the inner and outer functions of a composite function and be able to apply the Chain Rule to find the derivative of a composite function or an expression involving one.
- Be able to write out, understand, and apply at least two different notational forms for the Chain Rule.
- Know when and how to apply the Power Rule and the rules for the derivatives of ce^{kx} and ab^x.

[10]Based on a student project.

Composite Functions

Sample Problem 1: You are interested in teaching summer school this summer. At your institution, the pay for this teaching is 2.5% of your school-year salary per credit hour taught. Your salary is \$34,000 for the current school year. The city wage tax is a straight 4.96% of your summer school income.

a. How much city wage tax would you pay if you taught 3 credit hours?
b. How much city wage tax would you be paying per credit hour taught?

Solution: **a.** From the information we are given, the summer income will be 2.5% of \$34,000 = 0.025(34000) = \$850 per credit hour taught. If you teach 3 credit hours, you would earn 850(3) = \$2550. The city tax on this would then be

$$0.0496(2550) = \$126.48$$

b. To generalize, let's define some variables and functions. Let

h = the number of summer school credit hours taught
$i = g(h)$ = the summer income (in dollars) if h credit hours are taught
$t = f(i)$ = the city wage tax (in dollars) on a summer income of i dollars

We know that your income from summer teaching is \$850 per credit hour, so

$$i = g(h) = 850h$$

We also know that the city wage tax is 4.96% of your summer income, so

$$t = f(i) = 0.0496i$$

Now, we want to find an expression for t as a function of h. We have t as a function of i, and i in terms of h, so we can simply substitute the expression for i in terms of h in the tax function. Since $i = g(h) = 850h$ and $t = f(i) = 0.0496i$, we can substitute to get

$$t = f(i) = f[g(h)] = f(850h) = 0.0496(850h) = 42.16h$$

Notice that $f(i) = 0.0496i$, so $f(850h) = 0.0496(850h)$, since when evaluating functions, you plug in whatever is in the parentheses ($850h$ here) for the variable of the function (i here) *wherever* it occurs in the symbolic definition of the function. This illustrates a general principle: **To evaluate a composite function, plug in the value of the input variable into the inner function, find the value of the inner function, and plug this answer into the outer function.**

Notice also that if $h = 3$ (if you teach 3 credit hours), $t = 42.16(3) = \$126.48$, as we found earlier (this is a *verification* of the answer). Put differently,

$$t = f[g(3)] = f[(850)(3)] = f(2550) = 0.0496(2550) = \$126.48$$

Recall from Section 1.1 that the form $t = f[g(h)]$, which indicates that we are **taking a function of a function**, is called a **composite function**. The function $g(h) = i$ is called the **inner function** (since it is nested, or plugged *in*, *within* the f function), and the function $f(i) = t$ is called the **outer function**, since the inner function is nested *within* it. Notice that, in order for such a function to make sense (be well defined), **the output of the inner function** (values of $g(h)$ here) **must be in the domain of the outer function. The domain of the composite function will therefore be all values in the domain of the inner function whose outputs are in the domain of the outer function**. In realistic problems, this relationship is normally very intuitive and not a problem. We saw an example of this in Sample Problem 6 of Section 1.4. In our current problem, the domain of the outer function is essentially any nonnegative number (wage tax can be calculated for any nonnegative income), and our inner function will always result in a nonnegative value (since the number of credit hours taught must be nonnegative, resulting in a nonnegative salary), so any nonnegative number of credit hours will be possible as the domain of the composite function. For general mathematical problems, finding the domain of a composite function can get *very* complicated.

Thus, if you teach h hours of summer school, you will pay \$42.16h in city wage tax (\$42.16 times the number of credit hours taught). This is equivalent to saying that you are paying \$42.16 *per credit hour*. In other words, the rate of change of the wage tax with respect to the number of credit hours taught (t with respect to h) is \$42.16 per credit hour. To understand this with units, we could write

$$\left(\frac{0.0496 \text{ tax dollars}}{\text{salary dollar}}\right) \cdot \left(\frac{850 \text{ salary dollars}}{\text{credit hour taught}}\right) = \frac{(0.0496)(850) \text{ tax dollars}}{\text{credit hour taught}}$$

$$= \frac{42.16 \text{ tax dollars}}{\text{credit hour taught}} = \$42.16 \text{ / credit hour}$$

(the salary dollar units cancel out). The graph of the City Tax [graph (a)] and the city tax rate [graph (b)] are shown in Figure 2.5-1.

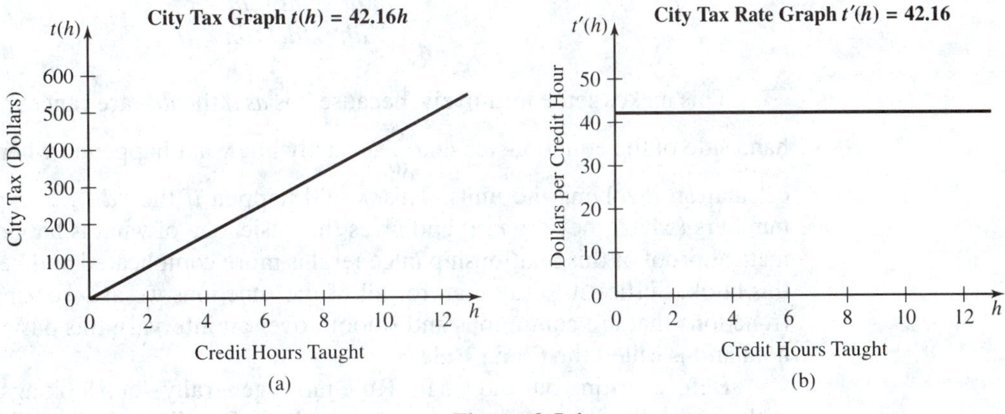

Figure 2.5-1

□

Notice that in the above case, t is a *linear* function of h, so the *approximation* based on the derivative is *exact*, since the *tangent* and the *curve* are the same. Thus, again the answer was evaluated directly by finding $f[g(h)]$ for $h = 3$, so the answer again is \$126.48.

The Basic Concept of the Chain Rule

Let's see how symbolic calculus would give us the above answers for the rate of change of the wage tax with respect to the number of credit hours taught.

Notice that we now have t as a direct function of h: $t = 42.16h$. The rate of change of t with respect to h (the derivative of t with respect to h) would be found in the usual way, so we would have

$$t = 42.16h \Rightarrow \frac{dt}{dh} = 42.16$$

To see how we could have derived this from the original separate functions, notice that

$$i = 850h \Rightarrow \frac{di}{dh} = 850$$

$$t = 0.0496i \Rightarrow \frac{dt}{di} = 0.0496$$

Looking back, we see that the 42.16 came from the calculation of 0.0496(850). In other words,

$$\frac{dt}{dh} = 42.16 = 0.0496(850) = \frac{dt}{di} \cdot \frac{di}{dh}$$

Just looking at the symbols, then, we see that

$$\frac{dt}{dh} = \frac{dt}{di} \cdot \frac{di}{dh}$$

This makes sense intuitively, because it is *as if* the di's are canceling out on the right-hand side of the equation, leaving $\frac{dt}{dh}$, exactly like what happened when we looked at the calculation involving the units. This would happen *if* the "d . . ." expressions were all numbers (which they are *not*) and gives the basic idea of what is happening. The mathematical proof of this relationship in general is more complicated and beyond the scope of this book. Suffice it to say that, for all of the functions we are working with in this text (functions that are continuous and smooth over an interval), this basic relationship does hold and is called the **Chain Rule**.

Before writing out the Chain Rule more generally, let's look again at what we did in the example and introduce some terminology. Recall our basic definitions:

$$t = f(i) = 0.0496i,\ i = g(h) = 850h$$

so

$$t = f[g(h)] = f[850h] = 0.0496(850h)$$

Remember that we called $i = g(h) = 850h$ the *inner function*. Thus we can call $i' = g'(h) = 850$ the **inner derivative**. Similarly, we called $t = f(i) = .0496i$ the *outer function*, and so we can call $t' = f'(i) = 0.0496$ the **outer derivative**.

Let's express the Chain Rule more generally now:

The Chain Rule: If $y = f(u)$ and $u = g(x)$, so $y = f[g(x)]$, then

$$\frac{dy}{dx} = \frac{dy}{du} \cdot \frac{du}{dx}$$

$$= f'(u) \cdot u' \qquad \text{(since } u = g(x),\ u' \text{ is another symbol for } g'(x))$$

$$= f'(g(x)) \cdot g'(x) \qquad \text{(substituting } g(x) \text{ for } u, \text{ and } g'(x) \text{ for } u')$$

In other words,

$$D_x(f[g(x)]) = (f'[g(x)]) \cdot g'(x)$$

A more explicit version of the verbal form on the last line (needed when the outer function and its derivative are more complicated) would say that **the derivative of a composite function equals**

(outer derivative, with the inner function plugged in) · (inner derivative)

Let's see how this works with a simple example.

Sample Problem 2: If $y = 0.2e^{-0.2x}$, find $\frac{dy}{dx}$.

Solution: In this case, the inner function, $g(x)$, is

$$u = g(x) = -0.2x$$

You can usually recognize **the inner function** because it **is set apart from everything else in some way—either in parentheses or** in a situation where it *could* be in parentheses, but doesn't *have* to be, as **in an exponent, radical, numerator, or denominator**. The inner function should also involve the variable, because if it is just a constant, it can be handled much more simply. It is called the *inner* function because, if it is placed within parentheses, the parentheses would be nested *within* the overall expression.

The outer function, $f(u)$, is what you get by replacing the inner function by u in the expression:

$$y = f(u) = 0.2e^{u}$$

You can think of this as "the big picture," or the basic form of the function. It is called "outer" because the inner function is nested within *it*.

Now, if we think of u as the variable in the outer function and take the simple derivative (treat it as we would treat x in a standard simple function), then the outer derivative is given by

$$y = f(u) = 0.2e^{u} \Rightarrow \frac{dy}{du} = f'(u) = 0.2e^{u}$$

In the more detailed verbal version of the Chain Rule, we said we want the outer derivative *with the inner function plugged in*. In this case, that means we want to replace the u with the inner function, since $u = -0.2x$. Thus we get

$$\frac{dy}{du} = f'(u) = 0.2e^{u} \text{ and } u = -0.2x = g(x)$$

$$\Rightarrow \frac{dy}{du} = f'(g(x)) = 0.2e^{(-0.2x)}$$

The inner derivative is much simpler and more straightforward: Just take the simple standard derivative of the inner function, using u instead of y for the name of the output:

$$u = g(x) = -0.2x \Rightarrow \frac{du}{dx} = g'(x) = -0.2 = u'$$

Now, putting the pieces together, we see that

$$\begin{aligned}\frac{dy}{dx} &= \frac{dy}{du} \cdot \frac{du}{dx}\\ &= \text{(outer derivative, with inner function plugged in)} \cdot \text{(inner derivative)}\\ &= (0.2e^{-.2x})(-0.2)\\ &= -0.04e^{-.2x}\end{aligned}$$

Remember that this calculation is simply $(f'([g(x)])(g'(x))$, as we saw earlier. When doing one of these calculations, you don't need to write out *anything* involving u or the intermediate steps, but you *should* go through those steps in your head. For this problem, your thinking might go something like the following:

> I see that the inner function, u, must be $-0.2x$, so the outer function is $0.2e^{u}$. That means the outer derivative is $0.2e^{u}$, and I need to plug back in the $-0.2x$ for u, so I will write '$(0.2e^{-0.2x})$' down for the outer derivative. This needs to be multiplied by the inner derivative, which is just -0.2, so now I put in the '$\cdot(-0.2)$.'

Your only writing would then be

$$\frac{d}{dx}(0.2e^{-.2x}) = (0.2e^{-.2x}) \cdot (-0.2) = -.04e^{-0.2x} \quad \square$$

Let's generalize what we just did:

$$\mathbf{D_x(e^{kx}) = (e^{kx})(k) = ke^{kx}}$$

since here the inner function is $u = kx$, so the outer function is e^{u}, and so the outer derivative is $D_u(e^{u}) = e^{u} = e^{kx}$ and the inner derivative is $D_x(kx) = k$.

The Power Rule

Probably the most commonly used form of the Chain Rule is a version of it for situations when the outer function is of the form au^{n}. Let's look at some sample problems.

Sample Problem 3: When using calculus to *derive* a best-fit least-squares regression model, you have to take the derivative of a function of the form

$$f(x) = (5 - 3x)^2$$

Find $f'(x)$.

Solution: The inner function is the expression set apart from everything else, in this case in parentheses:

$$u = (5 - 3x) = \text{inner function}$$

This means that the outer function is given by

$$u^2 = \text{outer function}$$

and so the outer derivative is

$$2u = 2(5 - 3x) = \text{outer derivative}$$

Since the inner function is $(5 - 3x)$, the inner derivative is given by

$$u' = D_x(5 - 3x) = 0 - 3 = -3 = \text{inner derivative}$$

So now we can put together the overall derivative:

$$\begin{aligned} D_x[(5 - 3x)^2] &= (\text{outer derivative})(\text{inner derivative}) \\ &= [2(5 - 3x)](-3) \\ &= -6(5 - 3x) \\ &= -30 + 18x \end{aligned}$$

The above calculation is a good model for how to think about and find this kind of derivative. Just for the sake of seeing how this relates to function notation, we have $u = g(x) = 5 - 3x$ is the inner function and $y = f(u) = u^2$ is the outer function, so

$$(5 - 3x)^2 = f[g(x)],\ g'(x) = -3, \text{ and } f'(u) = 2u$$

so

$$\begin{aligned} D_x[(5 - 3x)^2] &= D_x(f[g(x)]) \\ &= (f'[g(x)])[g'(x)] && \text{(Chain Rule in function notation form)} \\ &= (f'([5 - 3x])(-3) && \text{(substituting for } g(x) \text{ and } g'(x)) \\ &= [2(5 - 3x)](-3) && \text{(since } f'(u) = 2u) \\ &= -30 + 18x && \text{(simplifying as above)} \end{aligned}$$

as before. □

Sample Problem 4: Find the derivative of $a[g(x)]^n$ in general.

Solution: This is just a straight application of the Chain Rule, where the inner function is $u = g(x)$ and the outer function is au^n. This means that the outer derivative is anu^{n-1} and the inner derivative is $g'(x)$, so

$$\frac{d}{dx}\{a[g(x)]^n\} = \{an[g(x)]^{n-1}\} \cdot [g'(x)] \qquad \text{(outer derivative times inner derivative)}$$

This is often expressed more succinctly as

$$\frac{d}{dx}au^n = anu^{n-1} \cdot u'$$

and is often called the **power rule**:

$$\mathbf{D_x\{a[g(x)]^n\} = \{an[g(x)]^{n-1}\}[g'(x)]}$$

$$\mathbf{D_x(au^n) = (anu^{n-1})(u')} \quad \square$$

Sample Problem 5: In Sample Problem 5 of Section 1.5, we found a formula for T-shirt sales (in shirts) after t days to be

$$N(t) = \frac{6803}{1 + 190.5e^{-1.050t}}$$

At what rate are T-shirt sales changing after 4 days?

Solution: Since the numbers in this example are a bit messy, it is actually easier to find the derivative of a *general* logistic function first, which could be written

$$f(x) = \frac{L}{1 + Ae^{-Bx}}$$

The key to finding this derivative is recognizing that it can be written in the form

$$\frac{L}{1 + Ae^{-Bx}} = L(1 + Ae^{-Bx})^{-1}$$

Thus we can think of this as a composite function where the inner function is

$$u = 1 + Ae^{-Bx}$$

and so the outer function is Lu^{-1}. Thus the outer derivative will be

$$D_u(Lu^{-1}) = (L)(-1)u^{-1-1} = -Lu^{-2}$$

To find the inner derivative, we want

$$D_x(1 + Ae^{-Bx}) = 0 + (A) \cdot [D_x(e^{-Bx})] = (A)[(-B)(e^{-Bx})] = -ABe^{-Bx}$$

since $D_x(e^{kx}) = ke^{kx}$. This then gives us the derivative

$$\frac{d}{dx}\left(\frac{L}{1 + ae^{-Bx}}\right) = \frac{d}{dx}\left[L(1 + Ae^{-Bx})^{-1}\right]$$

$$= \{L(-1)(1 + Ae^{-Bx})^{-1-1}\} \cdot [0 + A(-B) \cdot (e^{-Bx})] \quad \text{(outer derivative times inner derivative)}$$

$$= \{-L(1 + Ae^{-Bx})^{-2}\} \cdot [-ABe^{-Bx}] \quad \text{(simplifying)}$$

$$= \left\{ -L \cdot \frac{1}{(1 + Ae^{-Bx})^2} \right\} \cdot [-ABe^{-Bx}] \qquad \left(\text{since } x^{-n} = \frac{1}{x^n} \right)$$

$$= \frac{LABe^{-Bx}}{(1 + Ae^{-Bx})^2} \qquad \text{(simplifying)}$$

Notice again that the inner derivative itself required *another* application of the Chain Rule, where the inner function was $v = -Bx$ and the outer function was e^v.

In our example, then, the derivative is

$$\frac{6803(190.5)(1.050)e^{-1.050t}}{(1 + 190.5e^{-1.050t})^2} = \frac{1360770e^{-1.050t}}{(1 + 190.5e^{-1.050t})^2}$$

If we now plug in 4 for t, we find that $N'(4) \approx 1372$. This means that sales are increasing at a rate of about 1372 shirts per day after 4 days. In other words, you can expect to sell about 1372 shirts on the fifth day. □

If you ever need to calculate the derivative of a logistic function by hand, you can always refer to the above general formula, although you should be able to derive it on your own.

Derivative of ab^x

Sample Problem 6: We saw in Sample Problem 3 of Section 1.4 that the total accumulated value (in dollars) of \$1000 invested at an effective annual interest rate of 12% after t years is given by

$$A(t) = 1000(1.12)^t \text{ for } t \geq 0$$

What is the instantaneous rate at which your money is growing after t years? After 5 years?

Solution: $A(t)$ has the general form of ab^x (with the variable being t instead of x). After working on Sample Problem 3 of Section 1.4, we mentioned that an exponential function of the form ab^x can be written in the form ce^{kx}. If we do this, we can find the derivative we want, since we already know how to find the derivative of e^x and can use the Chain Rule as we did in Sample Problem 2 to find the derivative of ce^{kx}.

Recall that ln b means $\log_e b$, which means "the exponent you would have to raise e to, in order to get an answer of b." In other words,

$$\text{If } \ln b = \log_e b = y, \text{ then } e^y = b$$

But since $y = \ln b$, if we substitute ln b in for y in the last expression above, we get

$$e^{\ln b} = b$$

This result can also be thought through, since it is saying that "e, raised to the power that you have to raise e to in order to get an answer of b, must give an answer of b!" From a different perspective, we mentioned that the natural logarithm function is the **inverse** of the exponential function (like cubing and taking the cube root are inverses)—just as when you take the cube root of something cubed, you get back to what you started with,

$$\mathbf{e^{\ln x} = \ln e^x = x}$$

Or, to beat this to death: y is the exponent you have to raise e to in order to get b, so when you raise e to that power, you by definition get b.

Now, since $e^{\ln b} = b$, we can rewrite

$$ab^x = a(e^{\ln b})^x = ae^{(\ln b)x}$$

since

$$(e^k)^x = e^{kx} \text{ (to raise a power to a power, multiply the exponents)}$$

We can think of $ae^{(\ln b)x}$ *as a composite function where the inner function is*

$$u = (\ln b)x$$

and so the outer function is ae^u. This means the outer derivative is

$$D_u(ae^u) = ae^u = ae^{(\ln b)x}$$

and the inner derivative is

$$D_x[(\ln b)x] = \ln b$$

Thus the derivative of ab^x is given by

$$\begin{aligned}
\frac{d}{dx}(ab^x) &= \frac{d}{dx}(ae^{(\ln b)x}) && \text{(rewriting } b \text{ as } e^{(\ln b)}\text{)} \\
&= [ae^{(\ln b)x}] \cdot (\ln b) && \text{(outer derivative times inner derivative)} \\
&= \{a[e^{\ln b}]^x\} \cdot (\ln b) && \text{(rewriting, since } (e^k)^x = e^{kx}\text{)} \\
&= \{a[b]^x\} \cdot (\ln b) && \text{(again, since } e^{\ln b} = b\text{)} \\
&= a(\ln b)b^x && \text{(regrouping)}
\end{aligned}$$

Notice that this is very similar to the derivative of ae^x, which is ae^x, except that there is the added $(\ln b)$ factor. Also, notice that if $b = e$, then $\ln e = 1$, so we get back the derivative of ae^x.

Thus, we have a general derivative rule for the **derivative of ab^x**:

$$\frac{d}{dx}(ab^x) = a(\ln b)b^x$$

or

$$\mathbf{D_x(ab^x) = a(\ln b)\, b^x}$$

For our example, $A(t) = 1000(1.12)^t$, *so* $A'(t) = 1000(\ln 1.12)(1.12)^t$.

Plugging in $t = 5$, we get $A'(5) \approx \$199.72$/year, so after 5 years your money is increasing in value at a rate of about \$199.72 per year. In other words, you should be getting approximately \$200 in interest in the fifth year. □

Sample Problem 7:[11] You are deciding what hair dryer you want to buy. You have already collected data about prices of different models and tried out friends' hair dryers to see how long each model takes to dry your hair. After working with the data and fitting some models, you have determined that if you define

x = the cost of a new hair dryer (in dollars)

$u = g(x)$ = the average time (in minutes) to dry your hair using a hair dryer that costs \$$x$

$y = f(u)$ = the dollar equivalent to you of your wasted time spent drying your hair over the life of the hair dryer if it takes u minutes to dry your hair each time

From your data, you determine that

$$u = g(x) = 4 + 4.05(0.941)^x$$

and

$$y = f(u) = 1.76u^2 - 5.06u + 14.3$$

for $10 \le x \le 50$ and $u \ge 4$.

a. Find the rate of change of the cost of your wasted time with respect to x.

b. Now find the rate of change of your *total* cost with respect to x.

c. You are considering buying a hair dryer for \$16. Find the rate of change of the total cost at \$16. Would a more expensive one be better?

d. Find an expression for y (the cost of your wasted time) in terms of x.

e. Find the *total* cost as a function of x, evaluate it at $x = 16$, and use this value and your answer to (c) to *approximate* the total cost of an \$18 hair dryer.

Solution: **a.** The cost of your wasted time is defined to be y, so the rate of change of that with respect to x (the cost of the hair dryer) will be $\frac{dy}{dx}$. But the information we know gives us y in terms of u ($y = f(u)$) and u in terms of x ($u = g(x)$), so to find y in terms of x requires chaining these together into a composite function:

$$y = f(u) = f[g(x)]$$

This means that to find the derivative we want, $\frac{dy}{dx}$, we need to use the Chain Rule:

$$\frac{dy}{dx} = \frac{dy}{du} \cdot \frac{du}{dx}$$

Since $y = 1.76u^2 - 5.06u + 14.3$, we see that

$$\frac{dy}{du} = 3.52u - 5.06$$

[11] Based on a student project.

and since $u = 4 + 4.05(0.941)^x$, we see that

$$\frac{du}{dx} = 0 + 4.05(\ln 0.941)(0.941)^x \approx -0.246(0.941)^x$$

the derivative we want is thus given by

$$\begin{aligned}\frac{dy}{dx} &= \frac{dy}{du} \cdot \frac{du}{dx} \\ &\approx (3.52u - 5.06)[-0.246(0.941)^x] \\ &= (3.52[4 + 4.05(0.941)^x] - 5.06)[-0.246(0.941)^x]\end{aligned}$$

This expression could be simplified somewhat (multiplying out constants, for example), but if you are evaluating it using technology, it is probably faster to just enter it as is.

b. Let's define the total cost function to be

$C(x)$ = your total cost (cost of the hair dryer *and* your wasted time), in dollars, if you buy a hair dryer for x dollars

Since the cost of the hair dryer is simply x, and the cost of your wasted time is y, we can get a simple expression for $C(x)$, although it is not yet in terms of just x:

$$C(x) = x + y$$

The rate of change of this total cost function will then be

$$\begin{aligned}C'(x) &= D_x(x + y) = 1 + \frac{dy}{dx} \\ &\approx 1 + \{3.52[4 + 4.05(0.941)^x] - 5.06\}[-0.246(0.941)^x]\end{aligned}$$

c. To see if a hair dryer that costs more than \$16 is better than one that costs \$16, we can calculate the rate of change of the total cost at $x = 16$. If the rate is *positive*, it will mean that the total cost is *increasing* as the price increases, and so a more expensive one would *not* be better. Let's see what happens:

$$\begin{aligned}C'(16) &\approx 1 + \{3.52[4 + 4.05(0.941)^{16}] - 5.06\}[-0.246(0.941)^{16}] \\ &\approx -0.34\end{aligned}$$

This means that, at a price of \$16, the total cost is *changing* at a rate of −\$0.34 per dollar of selling price for the hair dryer, so the total cost is *decreasing* at a rate of about 34 cents per dollar of selling price. This suggests that a more expensive hair dryer (costing more than \$16) *would* be better.

d. The cost of your wasted time was defined to be y, and we know that

$$y = f(u) = f[g(x)]$$

so this question involves evaluating a composite function at a general input value. Since $u = g(x) = 4 + 4.05(0.941)^x$ and $y = f(u) = 1.76u^2 - 5.06u + 14.3$, we get

$$\begin{aligned}y &= f[g(x)] = f[4 + 4.05(0.941)^x] \\ &= 1.76[4 + 4.05(0.941)^x]^2 - 5.06[4 + 4.05(0.941)^x] + 14.3\end{aligned}$$

e. The total cost is given by

$$C(x) = x + y = x + \{1.76[4 + 4.05(0.941)^x]^2 - 5.06[4 + 4.05(0.941)^x] + 14.3\}$$

so the total cost over its lifetime for a hair dryer priced at \$16 is

$$C(16) = 16 + \{1.76[4 + 4.05(0.941)^{16}]^2 - 5.06[4 + 4.05(0.941)^{16}] + 14.3\}$$
$$\approx 16 + 40.2 = 56.2$$

In other words, although the hair dryer itself only costs \$16, the time wasted by getting a cheap one (because it takes longer to dry your hair) costs you about an additional \$40.20 over the life of the hair dryer, for a total cost of about \$56.20.

Now, we just found that $C(16) \approx \$56.20$, and in part **c** we estimated that

$$C'(16) \approx -\$0.34$$

We can use these values to estimate the total cost of an \$18 hair dryer. Since the derivative indicates that, at a selling price of \$16, the total cost is decreasing at a rate of about 34 cents per additional dollar of selling price, and since \$18 is \$2 more than \$16, the change in your total cost should be approximately

$$\left(\frac{-0.34 \text{ cost dollars}}{\text{selling price dollar}}\right)(2 \text{ selling price dollars}) = -0.68 \text{ cost dollars}$$

In other words, your total cost should *decrease* by about \$0.68. Since $C(16)$ was about \$56.20, this means you would project your total cost for a \$18 hair dryer to be about

$$56.20 - 0.68 \approx 55.5$$

or about \$55.50. Let's see how close this approximation is by calculating $C(18)$:

$$C(18) = 18 + \{1.76[4 + 4.05(0.941)^{18}]^2 - 5.06[4 + 4.05(0.941)^{18}] + 14.3\}$$
$$\approx 18 + 37.7 = 55.7$$

or about \$55.70. So our estimate was only about 20 cents off, out of about \$56, which is less than half a percent! In Chapter 3, we will learn about how to find the optimal solution to a problem like this (here, the price of a hair dryer that will have the smallest possible total cost) using calculus. □

Sample Problem 8: Find $\frac{d}{dv}e^{-\frac{1}{2}v^2}$.

Solution: In this case, the inner function is $u = -\frac{1}{2}v^2 = g(v)$, so the inner derivative is $u' = g'(v) = \frac{d}{dv}\left(-\frac{1}{2}v^2\right) = \left(-\frac{1}{2}\right) \cdot (2) \cdot v^{(2-1)} = -1v^1 = -v$. The outer function is e^u, so the outer derivative is also $e^u = e^{-\frac{1}{2}v^2}$. Thus we get

$$\frac{d}{dv}e^{-\frac{1}{2}v^2} = (\text{outer derivative, with inner function plugged in}) \cdot (\text{inner derivative})$$
$$= \left(e^{-\frac{1}{2}v^2}\right) \cdot (-v) = -ve^{-\frac{1}{2}v^2}$$

We will use this result in the next problem. We can also generalize it, in a form that is really just a special case of the Chain Rule:

$$D_x(e^u) = (e^u)(u') \quad \square$$

Sample Problem 9: The relative frequency of an SAT score of x is given by the formula

$$f(x) = \frac{1}{100\sqrt{2\pi}}e^{-\frac{1}{2}\left(\frac{x-500}{100}\right)^2}$$

Find the derivative of this, calculate $f'(400)$, and interpret the meaning of the *sign* (plus or minus) of your answer.

Solution: This function is very complicated and in fact will require two applications of the Chain Rule to find its derivative. For the first application of the Chain Rule, the inner function will be

$$u = -\frac{1}{2}\left(\frac{x-500}{100}\right)^2$$

so the outer function will be

$$y = \frac{1}{100\sqrt{2\pi}}e^u$$

Remember that when you are deciding what to choose for the inner function (u), the idea is to pick an inner expression set off from the rest that, when replaced by u, you know how to differentiate (find the derivative of). We already know how to take the derivative of the function e^u, and the outer function above is just a constant times that, so we can easily find its derivative.

Thus the needed derivative will be of the form

$$\frac{dy}{dx} = \frac{dy}{du} \cdot \frac{du}{dx}$$

$$= \left(\frac{1}{100\sqrt{2\pi}}e^u\right) \cdot \frac{du}{dx}$$

On the other hand, we do not have a simple rule for the derivative of the inner function above. Notice that this inner function *itself* has parentheses within *it*, so we will apply the Chain Rule *a second time* to find $\frac{du}{dx}$. For this *second* Chain Rule, the inner function (which we will call v this time, to keep the two straight) would be

$$v = \frac{x-500}{100}$$

$$= \frac{x}{100} - \frac{500}{100} \quad \text{(separating terms)}$$

$$= \frac{1}{100}x - 5 \quad \text{(simplifying)}$$

The outer function would be

$$u = -\frac{1}{2}v^2$$

This looks a lot more manageable! The outer derivative of this second Chain Rule is thus

$$\frac{du}{dv} = \frac{d}{dv}\left(-\frac{1}{2}v^2\right) = -v$$

and the inner derivative is

$$\frac{dv}{dx} = \frac{d}{dx}\left(\frac{1}{100}x - 5\right) = \frac{1}{100} - 0 = \frac{1}{100}$$

So we can now use the second Chain Rule to write out $\frac{du}{dx}$:

$$\begin{aligned}
\frac{du}{dx} &= \frac{du}{dv} \cdot \frac{dv}{dx} = (-v)\left(\frac{1}{100}\right) \\
&= \left(-\left(\frac{1}{100}x - 5\right)\right)\left(\frac{1}{100}\right) && \text{(substituting in for } v\text{)} \\
&= \frac{-1}{10{,}000}x + \frac{5}{100} && \text{(multiplying through)} \\
&= -0.0001x + 0.05 && \text{(simplifying)}
\end{aligned}$$

This is now the inner derivative of the original Chain Rule, $\frac{du}{dx}$. Now let's go back to the original Chain Rule:

$$y = \frac{1}{100\sqrt{2\pi}}e^u \quad \text{and} \quad u = -\frac{1}{2}\left(\frac{x - 500}{100}\right)^2$$

so

$$\frac{dy}{dx} = \frac{dy}{du} \cdot \frac{du}{dx} = \left[\frac{1}{100\sqrt{2\pi}}e^u\right] \cdot (-0.0001x + 0.05)$$

$$= \left\{\frac{1}{100\sqrt{2\pi}}e^{\left[-\frac{1}{2}\left(\frac{x-500}{100}\right)^2\right]}\right\} \cdot (-0.0001x + 0.05)$$

Wow! This can be simplified a little more by evaluating the constant in front of the e as a decimal, but it will never be simple—just like taking the SATs!

Another way to approach this derivative, which may become easier as you gain more experience with these operations, would look something like the following:

$$\frac{d}{dx}\left[\frac{1}{100\sqrt{2\pi}}e^{\left[-\frac{1}{2}\left(\frac{x-500}{100}\right)^2\right]}\right]$$

$$= \left[\frac{1}{100\sqrt{2\pi}}e^{\left[-\frac{1}{2}\left(\frac{x-500}{100}\right)^2\right]}\right] \cdot \frac{d}{dx}\left[-\frac{1}{2}\left(\frac{x-500}{100}\right)^2\right]$$

$$= \left\{\frac{1}{100\sqrt{2\pi}}e^{\left[-\frac{1}{2}\left(\frac{x-500}{100}\right)^2\right]}\right\} \cdot \left[\left(-\frac{1}{2}\right)(2)\left(\frac{x-500}{100}\right)^{2-1} \cdot \frac{d}{dx}\left(\frac{x-500}{100}\right)\right]$$

$$= \left\{\frac{1}{100\sqrt{2\pi}}e^{\left[-\frac{1}{2}\left(\frac{x-500}{100}\right)^2\right]}\right\} \cdot \left[(-1)\left(\frac{x-500}{100}\right)^1 \cdot \frac{1}{100}\right]$$

$$= \left\{\frac{1}{100\sqrt{2\pi}}e^{\left[-\frac{1}{2}\left(\frac{x-500}{100}\right)^2\right]}\right\} \cdot [-0.0001x + 0.05]$$

In fact, once the idea is clear, you could go straight to the third line (rather than *writing* the $\frac{d}{dx}$ expressions, just *find* them).

Now, we need to find $f'(400)$. Plugging 400 in for x in the expression above, we find that $f'(400) \approx 0.0000242$. Since the sign is positive, it means that the graph of the relative probability is *increasing* at 400, which suggests that the highest frequency score is *more* than 400. □

Section Summary

Before you proceed to the exercises, be sure that you

- Know how to evaluate composite functions by substituting the inner function for the input variable of the outer function.
- Understand that the domain of a composite function can contain only values from the domain of the inner function whose corresponding output values are in the domain of the outer function.
- Understand that, if a single name is given to the composite function, it should be *different* than the name of the outer function.
- Understand how to recognize when the Chain Rule might be needed to find a derivative (when no basic rules seem to apply), and how to identify the inner function (set apart from the rest of the expression, such as in parentheses, an exponent, a radical, a numerator, or a denominator) and the outer function, so that the outer function is a basic form for which you know the derivative rule.
- Understand the basic idea of the Chain Rule: If $y = f(u)$ and $u = g(x)$, so that $y = f[g(x)]$, then $\frac{dy}{dx} = \frac{dy}{du} \cdot \frac{du}{dx}$ (as if the du's cancel). In other notation, this can be expressed as $D_x\{f[g(x)]\} = \{f'([g(x)]\}[g'(x)]$. In words, this means that the derivative of a composite function is (the outer derivative, with the inner function plugged into it) times (the inner derivative).

- Understand when and how to apply special cases of the Chain Rule, such as the Power Rule ($D_x(au^n) = (anu^{n-1})(u')$), and derivatives where the variable is in the exponent ($D_x(e^{kx}) = ke^{kx}$ and $D_x(ab^x) = a(\ln b)b^x$, or in more general form, $D_x(e^{ku}) = (ke^{ku})(u')$ and $D_x(ab^u) = [a(\ln b)b^u](u')$).

EXERCISES FOR SECTION 2.5

Warm Up

1. If $y = (x - 10)^2 + 15$, then $y' =$ _______.

2. If $y = (3 - 5x)^2$, then $\dfrac{dy}{dx} =$ _______.

3. $\dfrac{d}{dx}\sqrt{25 + x^2} =$ _______.

4. $D_x\sqrt{15 + (x - 5)^2} =$ _______.

5. $D_p\dfrac{2}{p + 3} =$ _______.

6. $\dfrac{d}{dq}\left(\dfrac{5}{q^2 + 4q}\right) =$ _______.

7. $\dfrac{d}{dt}1000e^{0.06t} =$ _______.

8. $D_t 2500e^{-0.08t} =$ _______.

9. If $f(x) = \dfrac{2}{3 + 4e^{-5x}}$, find $f'(x)$ by hand. Verify your solution by finding $f'(1)$ both from the formula you derived by hand and using the numerical derivative operation with your favorite technology.

10. If $G(t) = \dfrac{95}{1 + 4.1e^{-0.42t}}$, find $G'(t)$ by hand. Verify your solution by finding $G'(6)$ both from the formula you derived by hand and using the numerical derivative operation with your favorite technology.

11. $\dfrac{d}{dx}250(0.953)^x$ _______.

12. If $q = 1000(0.9987)^p$, then $\dfrac{dq}{dp} =$ _______.

13. $D_x[\ln(2x^3 + x)] =$ _______.

14. $D_x(5e^{3x^2+2x+5}) =$ _______.

Game Time

15. For the demand data for selling sweatshirts given in Sample Problem 8 of Section 1.3:

Selling Price	\$10	\$15	\$20	\$25
Quantity Sold	40	25	13	5

(a) Fit an exponential function to obtain a model of the quantity sold as a function of the selling price, $q = D(p)$, and *fully define* this model.
(b) Find the derivative of your function at \$20 and interpret your answer in words.
(c) Calculate $D(20)$ and use this value and your answer to part (b) to estimate the demand at \$22.

16. In Sample Problem 4 of Section 1.5, we found that we could model the percentage of U.S. households with cable TV as follows (rounded to 3 significant digits):

Verbal Definition: $P(x)$ = the percentage of U.S. households with cable TV x years after Dec. 31, 1977.

Symbol Definition: $P(x) = \dfrac{65.7}{1 + 3.53e^{-0.262x}}, 0 \le x \le 21.$

Assumptions: Certainty and divisibility. Certainty implies the relationship is *exact*. Divisibility implies that *any* fractions of years or percentage points are possible.

(a) Find the derivative of this function at $x = 10$, and interpret your answer in words.

(b) Calculate $P(10)$, and use this value and your answer to part (a) to estimate the percentage of U.S. households with cable TV at the end of 1989.

17. You earn \$5.50/hr at a work-study job. Your state has a flat 2.2% income tax (no deductions or exemptions).

(a) Define a model to express your income from the job as a function of the number of hours you work.

(b) Define a model to express the state income tax you pay for this job as a function of the income from the job.

(c) Now define a model to express the amount of state income tax you pay for this job as a function of the number of hours you work, and show in symbols how it is related to your answers to (a) and (b). What is the technical name for such a function?

(d) How much state income tax do you pay *per hour worked*? What does this correspond to mathematically? Express it in symbols.

(e) Show how you can use symbols to derive the answer to (d) in a way that explains the basic idea of the Chain Rule. Explain in words the concepts involved.

18. A friend of yours has just invented a new lozenge that has been clinically shown to reduce the frequency and duration of the common cold in people. You are helping him market the lozenge. From two separate series of market surveys of samples of people in your metropolitan area, you obtained the following data:

Advertising (\$10,000s)	% of People Have Heard
0	1
1	3
2	5
3	8
4	12
5	17
6	23
7	30

% of People Have Heard	% of People Have Bought
5	1
10	1.5
15	2.3
20	3.4

(a) Fit and define a quadratic model expressing the percentage of people who have heard of the lozenge as a function of the money spent on advertising (the table gives cumulative values). Round your parameters to 3 significant digits.

(b) Fit and define an exponential model expressing the percentage of people who have *bought* the lozenge as a function of the percentage of people who have *heard* of it. Round your parameters to 3 significant digits.

(c) Now define a model expressing the percentage of people who have bought the lozenge as a function of advertising expenditure, and show in symbols how it is related to your answers to (a) and (b). What is the technical name for such a function?

(d) Now suppose you have started an advertising campaign in a different metropolitan area, about the same size as yours. You have spent \$20,000 so far in this new area. How could you *approximate* the effect on the percentage of people buying the lozenge of another \$10,000 in advertising? What mathematical concept can you use to answer this? Show a symbolic way to represent your answer.

(e) Using the same concept as in (d), estimate the effect of an additional \$5000 (in addition to the current \$20,000) in advertising on the percentage of people buying the lozenge. Is this approximation or the one in (d) likely to be more accurate? Explain.

(f) Using your model, what would be your estimate of the *exact* effect on the percentage of people buying the lozenge for the situations in (d) and (e)? Does this confirm your answer at the end of (e)?

(g) Suppose that from additional research you find that the following model holds quite well:

Verbal Definition: $S(p)$ = the average sales (in dollars) per person in a year if p percent of the people buy the lozenge.

Symbol Definition: $S(p) = 4.2 + 0.23p$, for $0 \le p \le 25$.

Assumptions: Certainty and divisibility. Certainty implies the relationship is *exact*. Divisibility implies *any* fractions of percentage points and sales are possible.

Find a model for average sales per person in a year as a function of advertising expenditure. Find and interpret its derivative.

2.6 THE PRODUCT RULE

In Section 1.4 on quadratic and exponential models, we found models for the demand as a function of the price charged. Some of the models that we fit were linear, some quadratic, and some exponential. We then found models for the revenue as a function of price charged by multiplying the selling price times the demand function (quantity sold as a function of price). Revenue equals price times quantity, and quantity is a function of price: $R(p) = (p)[Q(p)]$. We also found models for the price as a function of the quantity (a different form for a demand function), in which case revenue again equals price times quantity, but this time price is a function of quantity: $R(q) = [P(q)](q)$. In either case, revenue is the product of two functions. We have formulas for finding the derivatives of several different types of functions, even composite functions such as e^{3x}. In this section we will learn how to find the rate of change of functions that involve products of two functions, called the Product Rule.

Here are the kinds of problems that this section will help you solve:

- You help run a local coffeehouse and have had the same musical group perform three years in a row. You have experimented with different admission prices. Find the rate at which your revenue is changing as the price for admission is changed.
- You are selling mugs as part of an environmental project. You hope to achieve two things: communicate the idea of reusing resources to as many people as possible and raise money for the environmental effort. How would your revenue change as you change the price of the mugs?

By the end of this section you should be able to answer questions like these and should

- Recognize when a function is the product of two separate functions.
- Know how to use the Product Rule to find the derivative of a product of two functions.

- Know that the revenue function is the product of the price times the quantity, and that one of these variables is often given in terms of the other, price as a function of quantity, or quantity as a function of price.
- Be able to find the marginal revenue function.
- Understand the idea of the product rule.

Sample Problem 1: Recall Example 4 from Section 1.1. In the last three years you had the same musical group perform at the same coffeehouse. The first year you charged \$7 admission and 100 people came. The next year you decided to charge \$8 admission and sold 80 tickets. Last year you decide to raise admission to \$9 to help cover rising costs, even though you were disappointed by the turnout the last time (70). The best-fit linear model of the number of tickets sold as a function of the price, to 3 significant digits, is

Verbal Definition: $T(p)$ = the number of tickets sold at a price of p dollars per ticket.

Symbol Definition: $T(p) = -15p + 203, 0 \le p \le 13.50.$

Assumptions: Certainty and divisibility. Certainty implies the relationship is *exact* (no sales or discounts, for example). Divisibility implies that *any* fractions of dollars or numbers of tickets are possible.

How is your revenue changing as the price charged changes?

Solution: We must first find the revenue function and then find the derivative of that function:

Verbal Definition: $R(p)$ = the revenue in dollars from the sale of tickets at p dollars per ticket.

Symbol Definition: $R(p) = p(-15p + 203), 0 \le p \le 13.50.$

Assumptions: Certainty and divisibility. The implications are the same as above.

In this case we can easily multiply the price times the quantity and get a simple quadratic model for the revenue: $R(p) = -15p^2 + 203p$. As we learned in Section 2.4, the derivative of this function is $R'(p) = -30p + 203$. This tells us that the rate of change of the revenue depends on the price charged.

It would seem quite reasonable that the derivative of the product of two functions equals the product of the derivatives of the functions. Let's try it and see. The derivative of a variable raised to the first power is 1, so the derivative of p is 1. The derivative of $-15p + 203$ is -15. The product of the two derivatives is -15, which would mean that the revenue is decreasing at the rate of 15 dollars for every dollar change in price, no matter what the price is. This is nothing like what we got before. The "easy way" may be easy, but it is *not* correct. □

Sample Problem 2: We also discussed a quadratic model to the coffeehouse data, since the points did not lie in a straight line:

Verbal Definition: $T(p)$ = the number of tickets sold at p dollars per ticket.

Symbol Definition: $T(p) = 5p^2 - 95p + 520, 0 \le p \le 9.5.$

Assumptions: Certainty and divisibility. Certainty implies the relationship is *exact* (no sales or discounts, for example). Divisibility implies that *any* fractions of dollars or numbers of tickets are possible.

The model for revenue, $R(p)$ equals the revenue in dollars when the price per ticket is p is: $R(p) = p(5p^2 - 95p + 520), 0 \le p \le 9.5$. How is the revenue changing (what is the marginal revenue[12]) when the price is \$3 and when the price is \$8?

Solution: We already found out that we can't just take the derivatives and multiply them. It doesn't work. Instead, we will multiply the two functions and then take the derivative of the product:

$$R(p) = 5p^3 - 95p^2 + 520p, 0 \le p \le 9.5$$

Using the rules we learned in Section 2.4 we find that the derivative is

$$R'(p) = 15p^2 - 190p + 520$$

Then $R'(3) = 15(3)^2 - 190(3) + 520 = 85$. When the price per ticket is \$3, the revenue is increasing at the rate of \$85 per \$1 increase in ticket price. The actual increase in revenue when the price increased from \$3 to \$4 was $R(4) - R(3) = \$40$. This is not a very good estimate, even though it is the correct derivative. Our estimate was too high. Is it any better at \$8? $R'(8) = 15(8)^2 - 190(8) + 520 = -40$. The actual change in revenue when the price went from \$8 to \$9 was $-\$10$. Our estimate was too low. Since we are finding the *instantaneous* rate of change at a price of \$3, it will not be exactly the same as the *average* rate of change of the revenue from \$3 to \$4. A quick look at the graph of the function will show us why (Figure 2.6-1).

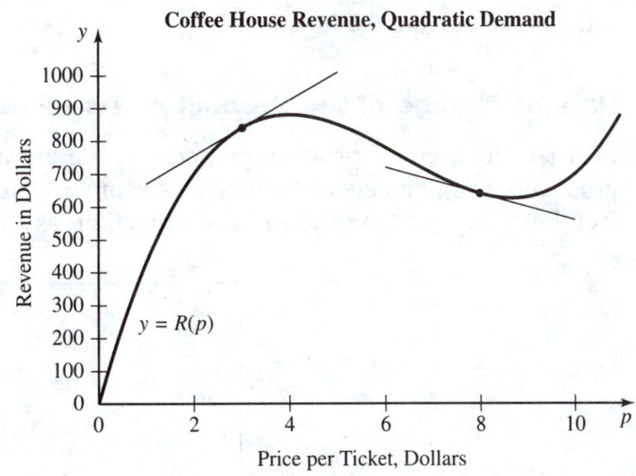

Figure 2.6-1

[12]Recall from our discussion in Section 2.3 that in economics, "marginal revenue" usually assumes that *quantity* is the input variable, rather than price, as we have here. We are using the term "marginal" in a more general way in this text, essentially synonymous with "derivative of." The two will coincide for the standard economic assumption.

The tangent line drawn at 3 is *above* the curve because the curve is *concave down*. The estimate at 4 will be *too high*, since the tangent is higher than the curve at 4. The tangent line drawn at 8 is *below* the curve because the curve is *concave up*. The estimate at 9 will be *too low*, since the tangent is lower than the curve at 9. □

An exponential model could also be fit to the coffeehouse data for a third model:

Verbal Definition: $T(p)$ = the number of tickets sold at p dollars per ticket.
Symbol Definition: $T(p) = 343(0.837)^p$, $p \geq 0$.
Assumptions: Certainty and divisibility. The implications are the same as above.

The function for revenue, where $R(p)$ equals the revenue in dollars when the price per ticket is p, is given by

$$R(p) = p[343(0.837)^p], p \geq 0$$

Ordinarily we rewrite this as

$$R(p) = 343p(0.837)^p, p \geq 0$$

simply moving the p after the initial constant.

Now we have a real problem. Our original idea isn't working: We can't multiply this function out and use the rules we learned in Section 2.4 to find the derivative. We need a way to find the derivative of the product of two functions, in this case $f(p) = 343p$ and $g(p) = 0.837^p$. How do we find the derivative of the product of two functions without actually multiplying the functions?

Rate of Change of the Product of Two Functions

First let's look at the problem graphically. We are all familiar with the concept of the product of two numbers as an area: for example a 9 foot × 12 foot rug covers 108 square feet. Let's think of the product of two functions as an area (Figure 2.6-2).

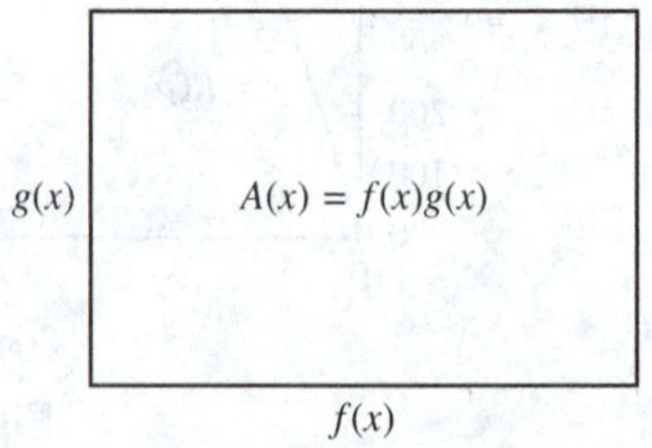

$g(x)$ $A(x) = f(x)g(x)$

$f(x)$

Figure 2.6-2

We want to know how the product of these two functions $f(x)g(x)$ is changing; in other words, how is the area changing? One of the ways we used to indicate the change in a variable was by the Greek letter delta, Δ. The change in the x would be Δx, and the change in y, Δy. We'll use those symbols here because it is easier to distinguish between the changes in the two functions. With this notation, as the input value changes from x to $x + \Delta x$, Δf will indicate the change in $f(x)$ and Δg will indicate the change in $g(x)$ (Figure 2.6-3).

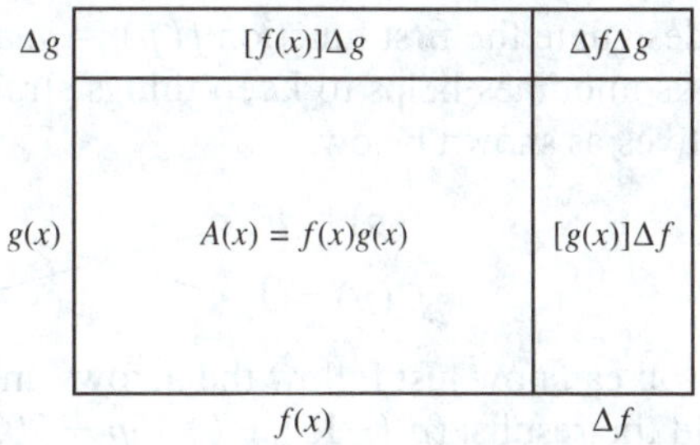

Δg	$[f(x)]\Delta g$	$\Delta f \Delta g$
$g(x)$	$A(x) = f(x)g(x)$	$[g(x)]\Delta f$
	$f(x)$	Δf

Figure 2.6-3

The new area would be equal to $[f(x + \Delta f] \cdot [g(x) + \Delta g]$, the base times the height. The **change** in the area would be the parts that were *added*: the L-shape on the top and right of the rectangle. Intuitively, as Δg and Δf get smaller and smaller, the upper-right area, $\Delta f\, \Delta g$ will get very small and approach 0. We are left with the other two pieces of the L-shape, at the top and on the right, $f(x)\Delta g$ and $g(x)\Delta f$, representing the change in the area. This is the sum of the first function times the change in the second function plus the second function times the change in the first function:

$$\Delta A \approx f(x)\Delta g + g(x)\Delta f$$

Now, since the derivative of a function is approximately equal to the change in the function when the input increases one unit, it can be shown[13] that the above relationship leads to a similar relationship involving derivatives:

$$A'(x) = f(x)g'(x) + g(x)f'(x)$$

The Product Rule

The derivative of the product of two functions is the first function times the derivative of the second plus the second function times the derivative of the first.

Derivative of product = (1st)(Derivative of 2nd) + (2nd)(Derivative of 1st)

In symbols

$$\frac{d}{dx}[f(x)g(x)] = f(x)g'(x) + g(x)f'(x)$$

[13]See the Appendix for a proof.

Let's go back to our first two problems and see if this works.

Example 1: In Sample Problem 1, the revenue function was

$$R(p) = p(-15p + 203),\ 0 \le p \le 13.50$$

We'll designate the first function $f(p) = p$, and the second function $g(p) = -15p + 203$.

It sometimes helps to keep things straight if you write out the functions and their derivatives as shown below:

$$\begin{array}{ll} f(p) = p & g(p) = -15p + 203 \\ f'(p) = 1 & g'(p) = -15 \end{array}$$

You can now just follow the arrows: multiply p times -15 and $-15p + 203$ times 1 and add the results: $(p)(-15) + (-15p + 203)(1) = -30p + 203$. This is exactly what we obtained before. □

Example 2: In Sample Problem 2, $R(p)$ equals the revenue in dollars when the price per ticket is p: $R(p) = p(5p^2 - 95p + 520),\ 0 \le p \le 9.5$. We'll designate the first function $f(x) = p$, and the second function $g(x) = 5p^2 - 95p + 520$:

$$\begin{array}{ll} f(p) = p & g(p) = 5p^2 - 95p + 520 \\ f'(p) = 1 & g'(p) = 10p - 95 \end{array}$$

Cross-multiplying and adding we get $(p)(10p - 95) + (5p^2 - 95p + 520)(1) = 10p^2 - 95p + 5p^2 - 95p + 520 = 15p^2 - 190p + 520$. The derivative of the revenue function, the marginal revenue[14] is $R'(p) = 15p^2 - 190p + 520$. Once again, this is exactly what we got before. □

Sample Problem 3: For the exponential demand model of the coffeehouse, where the revenue is given by

$$R(p) = 343p(0.837)^p,\ p \ge 0$$

find the derivative of the revenue function.

Solution: Since $R(p)$ is a product of 2 functions, we can use the Product Rule. We'll designate the first function $f(p) = 343p$, and the second function $g(p) = 0.837^p$:

$$\begin{array}{ll} f(p) = 343p & g(p) = 0.837^p \\ f'(p) = 343 & g'(p) = (\ln 0.837)(0.837)^p \end{array}$$

[14]Once again, recall from our discussion in Section 2.3 that in economics, "marginal revenue" usually assumes that *quantity* is the input variable, rather than price, as we have here. We are using the term "marginal" in a more general way in this text, essentially synonymous with "derivative of." The two will coincide for the standard economic assumption.

We can simplify $g'(x)$ by finding the natural log of 0.837, which is -0.178. Cross multiplying and adding we get

$$(343p)[-0.178(0.837^p)] + (0.837^p)(343)$$

This can be left as it is or rewritten: The exponential expression and 343 appear in both terms and can be factored out:

$$343(0.837^p)(-0.178p + 1)$$

It is a good idea to note this because it will be of help later on. The derivative of the revenue function is $R'(p) = 343(0.837^p)(-0.178p + 1)$. □

Using this model, the change in revenue when the price per ticket is \$8 equals -35.20. When the price is \$8 per ticket, the revenue is *decreasing* at the rate of approximately \$35.20 per dollar increase in ticket price. If we calculate $R(9) - R(8)$, the model shows a decrease of \$38.80 for a ticket price change from \$8 to \$9. The *estimate* of the decrease is fairly close to the *actual* decrease.

Graph of the Revenue and Marginal Revenue Using the Exponential Model

By this time you should be fairly familiar with graphs of quadratic and cubic functions, such as the revenue functions we found in Sample Problems 1 and 2. The graph of the revenue function for Sample Problem 3 is not one of the general types that we studied. Let's look at the graph of the revenue function: $R(p) = 343p(.837)^p$, $p \geq 0$ (Figure 2.6-4).

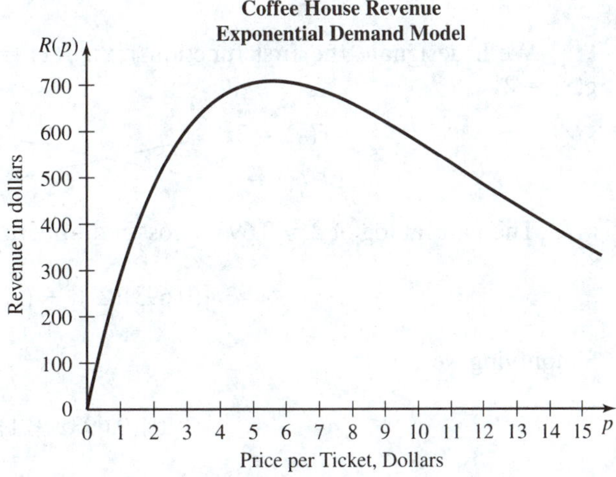

Figure 2.6-4

From looking at the graph of the revenue function we can see that the function is increasing from 1 to about 5, rapidly at first and then more slowly. From 5 on, the function is decreasing. The rate of decrease seems to be slowing a bit toward 15, but it is still decreasing. From this we would expect the derivative function to be positive but decreasing from 1 to 5, and then negative.

The graph of the derivative function is shown in Figure 2.6-5.

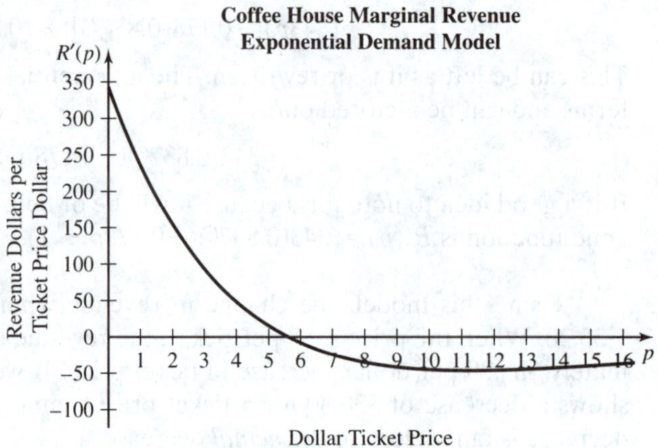

Figure 2.6-5

The numbers in the last problem were sort of "ugly," but they are much more realistic than the "pretty" numbers we get from contrived problems. Don't let the ugly numbers put you off. When you can use technology, they are really no harder to work with than the simple, pretty numbers. They may take a half second more to type in, that's all. We'll run through a simple example of the same type:

$$y = h(x) = 3x(2^x)$$

We'll designate the first function $f(x)$: $f(x) = 3x$, and the second function $g(x)$, $g(x) = 2^x$

$$f(x) = 3x \qquad g(x) = 2^x$$

$$f'(x) = 3 \qquad g'(x) = (\ln 2)(2^x)$$

The natural log of 2 ≈ 0.693. Cross-multiplying and adding we get

$$3x[(0.693)(2^x)] + (2^x)(3)$$

Simplifying we get

$$3(2^x)(0.693x + 1)$$

This does not look nearly so bad, but the process was exactly the same.

Derivative of $ax(b^x)$

★ **Discovery Question:**

If you have a function of the form $y = h(x) = ax\,b^x$, can you find a "shortcut" or formula for finding the derivative? A shortcut or formula for this type of function is very useful because this type of function occurs quite often. For instance, if an exponential function has been used to model demand, the revenue will be a function of this type. Try to answer this before you read on.

Answer: Let's designate the first function $f(x)$, $f(x) = ax$, and the second function $g(x)$, $g(x) = b^x$:

$$\begin{array}{ccc} f(x) = ax & \longleftrightarrow & g(x) = b^x \\ f'(x) = a & \longleftrightarrow & g'(x) = (\ln b)(b^x) \end{array}$$

So

$$y' = (ax)[(\ln b)b^x] + (b^x)(a)$$

Factoring ab^x from both terms gives

$$y' = ab^x[x(\ln b) + 1]$$

Since $\ln b$ is a real number we could rewrite it as

$$y' = ab^x[(\ln b)\,x + 1]$$

Thus, **the derivative of axb^x is $ab^x[(\ln b)\,x + 1]$**. In symbols,

$$\mathbf{D_x(axb^x) = ab^x[(\ln b)\,x + 1]} \qquad \text{★}$$

After going through three very similar examples above, perhaps you got bored with the repetition. Certainly by the third example it was very clear what the pattern was! Notice that the general rule with lots of letters, especially the a and b, is simply a concise and clear way to show the pattern. Once you are comfortable with this type of process, you can look at a general formula and realize that it is just a summary of such a pattern, and you could substitute *any* constant values for the parameters. You may also get to the point where you can derive a formula like this *directly*, without any numerical examples first. That's great! When you get to that point, you are truly thinking mathematically.

Derivative of a Quotient

Sample Problem 5: In Section 1.2 we formulated a model for the average cost of shirts:

Verbal Definition: $A(x)$ = the average cost, in dollars per shirt, if we order x T-shirts.

Symbol Definition: $A(x) = \dfrac{7x + 50}{x}$, for $x \geq 0$.

Assumptions: Certainty and divisibility. Certainty implies the relationship is *exact*. Divisibility implies *any* fractions of shirts and dollars are possible.

How fast was the average cost changing as the number of shirts bought changed?

Solution: This function is expressed as a quotient of two functions, $7x + 50$ divided by x. The actual division can be simplified by splitting up the fraction:

$$\frac{7x + 50}{x} = \frac{7x}{x} + \frac{50}{x} = 7 + \frac{50}{x}$$

We saw in Section 2.4 that we can write the second term as $50x^{-1}$, since in general $x^{-n} = \frac{1}{x^n}$, and then use the basic derivative rules from that section to find the derivative:

$$f(x) = 7 + 50x^{-1}$$

so

$$f'(x) = 0 + (50)(-1)x^{-1-1} = -50x^{-2} \quad \square$$

Sample Problem 6: Your organization has written a cookbook to help raise funds for scholarships. You have contacted a number of small publishing houses that specialize in this type of book. They quoted you prices for publishing your books. They have a volume discount program, and you developed the following model estimating their costs for producing the books:

Verbal Definition: $c(q)$ = the cost (in dollars) to produce q books.
Symbol Definition: $c(q) = 750\ln(0.1q + 50)$, for $q \geq 0$.
Assumptions: Certainty and divisibility. Certainty implies the relationship is *exact.* Divisibility implies *any* fractions of books and dollars are possible.

The average cost to produce the books is $A(q) = \frac{750\ln(0.1q + 50)}{q}$. How is the average cost per book changing when 500 books are produced?

Solution: In this case we cannot divide the numerator by the denominator and simplify the expression. We can, however, rewrite the expression as

$$A(q) = [750\ln(0.1q + 50)](q^{-1})$$

This is now a product, and we can use the product rule to find the rate of change of the average cost per book with respect to the number of books ordered. We will designate the first function $f(q) = 750\ln(0.1q + 50)$ and the second function $g(q) = q^{-1}$:

$$f(q) = 750\ln(0.1q + 50) \qquad g(q) = q^{-1}$$

$$f'(q) = \left[750\left(\frac{1}{0.1q + 50}\right)\right](0.1) = \frac{75}{0.1q + 50} \qquad g'(q) = -1q^{-1-1} = -q^{-2}$$

Notice that to get $f'(x)$, we had to use the Chain Rule, with an inner function of $u = 0.1q + 50$ and an outer function of $750\ln u$, so the outer derivative was $750\left(\frac{1}{u}\right)$ and the inner derivative was 0.1.

By the product rule we have

$$A'(q) = [750\ln(0.1q + 50)](-q^{-2}) + \left(\frac{75}{0.1q + 50}\right)(q^{-1})$$

Simplifying, we get

$$A'(q) = \frac{-750\ln(0.1q + 50)}{q^2} + \frac{75}{q(0.1q + 50)}$$

This derivative of the average cost function is sometimes called the **marginal average cost function** for obvious reasons. If we plan to evaluate this function using technology, then it is probably not worth simplifying it any further, so let's do it:

$$A'(500) = \frac{-750\ln(0.1(500) + 50)}{(500)^2} + \frac{75}{(500)(0.1(500) + 50)} \approx -0.012315\ldots \approx -0.012$$

When 500 books are produced, the average cost per book is decreasing at the rate of approximately 0.012 dollars (about 1.2 cents) per book for each additional book produced. We could say the average cost is decreasing at a rate of (1.2 cents per book) per book. □

Let's double-check this method of finding the derivative of a quotient with Sample Problem 5, where we already got an answer:

$$D_x\left(\frac{7x + 50}{x}\right) = D_x(7x + 50)(x^{-1}) = (7x + 50)(-x^{-2}) + (x^{-1})(7)$$

$$= \frac{-7x - 50}{x^2} + \frac{7}{x} = \frac{-7x - 50}{x^2} + \frac{7x}{x^2} = \frac{-7x - 50 + 7x}{x^2}$$

$$= \frac{-50}{x^2} = -50x^{-2}$$

The same answer, but it was a *lot* easier the first time! **If you're finding the derivative of a function that includes division by an expression involving the variable, see if you can simplify it first by breaking it up or canceling.**

We found the derivative of a quotient by expressing the denominator using negative exponents and making a product. If we do this in general and simplify, we obtain something commonly called the **Quotient Rule**.

$$D_x\left(\frac{f(x)}{g(x)}\right) = D_x([f(x)][g(x)]^{-1}) = [f(x)](-1[g(x)]^{-2}\,g'(x)) + [g(x)]^{-1}[f'(x)]$$

$$= \frac{-f(x)g'(x)}{[g(x)]^2} + \frac{f'(x)}{g(x)} = \frac{-f(x)g'(x)}{[g(x)]^2} + \frac{f'(x)g(x)}{[g(x)]^2}$$

$$= \frac{g(x)f'(x) - f(x)g'(x)}{[g(x)]^2}$$

In words, this can be described as

$$\text{derivative of a quotient} = \frac{\text{(bottom)(derivative of top)} - \text{(top)(derivative of bottom)}}{\text{(bottom)}^2}$$

Check that this works for Sample Problem 5 and Sample Problem 6. The derivative of a quotient can be found using any of the methods we have discussed here.

Section Summary

Before you start the exercises be sure that you

- Recognize a function that consists of the product of two terms involving the variable, so can be written in the form $f(x)g(x)$.
- Know that the revenue function is the product of the price times a demand function of price ($Q(p)$) or a demand function of quantity ($P(q)$) times the quantity: $p \cdot Q(p)$ or $P(q) \cdot q$.
- Know that the derivative of the product of two functions is the first function times the derivative of the second *plus* the second function times the derivative of the first. In other words,

$$(\text{Derivative of product}) = (\text{1st})(\text{Derivative of 2nd}) + (\text{2nd})(\text{Derivative of 1st})$$

- Know the product rule in symbols: $\frac{d}{dx}[f(x)g(x)] = f(x)g'(x) + g(x)f'(x)$.
- Know that the product rule can be used to find the derivative of a quotient, since $\frac{f(x)}{g(x)} = [f(x)] \cdot ([g(x)]^{-1})$, and that a special Quotient Rule also exists.
- Know that a short-cut formula can be derived for the derivative of a function of the form $y = f(x) = ax\, b^x$. The derivative is $ab^x[(\ln b)\, x + 1]$.

EXERCISES FOR SECTION 2.6

Warm Up

For Exercises 1–10, find the derivatives of the following functions, using the product rule. Where possible, check your answers by multiplying the functions and using one of the polynomial rules.

1. $y = f(x) = x(3x + 5)$

2. $g(r) = r(-2r + 5)$

3. $P = (q + 5)(q - 30)$

4. $T(s) = s^2(s^{0.5} + 1)$

5. $R(p) = p(0.35^p)$

6. $T(c) = 0.95^c 5c^2$

7. $f(x) = \dfrac{x}{x + 1}$

8. $T(c) = \dfrac{c - 3}{c}$

9. $A(q) = \dfrac{40 \ln q}{q}$

10. $f(x) = \dfrac{35(0.955^x)}{x}$

Game Time

11. You have been selling sodas at local sporting events. The number of sodas that you sell depends on several different things, such as the weather (you sell more on hot dry days) and the attendance. These are things over which you have no control, but you can change the price that you charge. You have collected data over a period of time and tried to keep the other variables such as weather and popularity of team as constant as possible. You have found that you can model the number of sodas sold as a function of the price charged per soda:

Verbal Definition: $N(p)$ = the number of sodas (thousands) sold in one day at p dollars per soda.

Symbol Definition: $N(p) = 1.0924(0.3178)^p$, $0.5 \le p \le 5$.

Assumptions: Certainty and divisibility. Certainty implies the relationship is *exact.* Divisibility implies *any* fractions of numbers of soda and dollars are possible.

Find the rate at which the number of sodas sold is changing when the price is
(a) \$1.00 per can.
(b) \$3.00 per can.
(c) \$5.00 per can.
(d) What conclusions, if any, could you draw from these answers?

12. Using the model in Exercise 11, find a model for the revenue from the sale of sodas as a function of the price charged. Find the rate of change of the revenue with respect to the price charged and *interpret your answer* when soda is sold at
(a) \$1.00 per can.
(b) \$3.00 per can.
(c) \$5.00 per can.
(d) What conclusions, if any, can you draw from these answers?

13. Using the model in Exercise 11, find a model for the total cost of the sodas sold as a function of the price charged per soda if the sodas cost the vendor \$0.38 each. Find the rate of change of the vendor's cost with respect to the price charged for the sodas when the price charged per soda was
(a) \$1.00.
(b) \$3.00.
(c) \$5.00.
(d) What conclusions, if any, can you draw from these answers?

14. Using the revenue function from Exercise 12 and the cost function from Exercise 13, find a model for the profit from the sale of the sodas as a function of the price charged for the sodas. Find the rate of change of the vendor's profit with respect to the price charged for the sodas when the price charged per soda was
(a) \$1.00.
(b) \$3.00.
(c) \$5.00.
(d) What conclusions, if any, can you draw from these answers?

15. The number of tickets to a local sporting event when the price is p dollars per ticket can be modeled by

Verbal Definition: $N(p)$ = number of tickets sold at p dollars per ticket.
Symbol Definition: $N(p) = 3165.6(0.94919^p)$, $p \geq 0$.
Assumptions: Certainty and divisibility. Certainty implies the relationship is *exact* (no discounts, everyone pays full price). Divisibility implies that *any* fractions of dollars in price and tickets sold are possible.

Find the rate of change of the number of tickets sold when the ticket price is
(a) \$18 per ticket.
(b) \$20 per ticket.
(c) \$22 per ticket.
(d) What conclusions, if any, can you reach from these answers?

16. Find a model for the revenue from the sale of tickets in Exercise 15. How fast is the revenue changing
(a) When the price is \$18 per ticket?
(b) When the price is \$20 per ticket?
(c) When the price is \$22 per ticket?
(d) What conclusions, if any, can you reach from these answers?

17. The manager of the local sports arena charges a flat fee of \$1000 for the use of the arena. A variable charge of \$1.00 per ticket sold is added to cover the cost of cleanup. Using the model for the number of tickets sold in Exercise 15, find a model for the cost of using the arena as a function of the price charged per ticket. Find the rate of change of the cost when the price is
(a) \$18 per ticket.
(b) \$20 per ticket.
(c) \$22 per ticket.
(d) What conclusions, if any, can you reach from these answers?

18. Using the models for revenue and cost in Exercises 16 and 17, find a model for the profit as a function of the price of the tickets. Find the rate of change of the profit when the price is
(a) \$18 per ticket.
(b) \$20 per ticket.
(c) \$22 per ticket.
(d) What conclusions, if any, can you reach from these answers?

19. The cost of producing a small component for computers can be modeled by

Verbal Definition: $C(q)$ = the cost, in dollars, to produce q hundred units.
Symbol Definition: $C(q) = -2347 + 1573\ln(q + 10)$ for $q \geq 0$.
Assumptions: Certainty and divisibility. Certainty implies the relationship is *exact*. Divisibility implies that *any* fractions of units produced and cost dollars are possible.

Find the average cost function per 100 units and determine how fast it is changing:
(a) When 800 units are made.
(b) When 1000 units are made.
(c) When 5000 units are made.

20. The cost of manufacturing a new toy can be modeled by

Verbal Definition: $C(q)$ = the cost, in dollars, to produce q thousand units.
Symbol Definition: $C(q) = -1529 + 9785\ln(q + 4)$ for $q \geq 0$.

Assumptions: Certainty and divisibility. Certainty implies the relationship is *exact*. Divisibility implies that *any* fractions of units produced and cost dollars are possible.

Find the average cost function and determine how fast the average cost is changing

(a) When 8000 units are made.

(b) When 15,000 units are made.

(c) When 50,000 units are made.

CHAPTER 2 SUMMARY

The **average rate of change of $y = f(x)$ over the interval $[x_1, x_2]$** (from x_1 to x_2) is given by

$$\frac{\Delta y}{\Delta x} = \frac{\text{change in output}}{\text{change in input}} = \frac{y_2 - y_1}{x_2 - x_1} = \frac{f(x_2) - f(x_1)}{x_2 - x_1}$$

Graphically, this corresponds to the **slope of the secant line** drawn from the first point through the second point. The **units** for the average rate of change over an interval are (output units) per (input unit).

The **percent rate of change in $f(x)$ over the interval $[x_1, x_2]$** in pure decimal form is

$$\frac{\left(\dfrac{f(x_2) - f(x_1)}{x_2 - x_1}\right)}{f(x_1)}$$

This can be thought of as the average rate of change expressed as a percent of the initial output value, or as the **percent change** in the output over the interval, averaged over the input interval. The **units** for percent rate of change in % form are (%) per (input unit).

The **instantaneous rate of change of a function at a point** is the limiting value of the average rate of change of the function between that point and a second point, as the second points approach, but never reach, the original point from *both* sides of it. This is called the **derivative of the function at the point**. It tells the change in the output value **per additional unit** of the input (from the point); its **units** are (output units) per (input unit). Graphically, the derivative corresponds to the **slope of the tangent line to the graph of the function at the point**. The **tangent line** to the curve at the point is the straight line through the point that most closely resembles the curve *near* the point. The slope of the tangent can be found by finding the *limit* of the slope of the secant lines joining that point and a second point on the curve as the second point approaches, but never reaches, the first point from *both* sides.

The **limit of a function $f(x)$ as x approaches a**, denoted $\lim_{x \to a}$, means the value that $f(x)$ approaches as x gets closer and closer to a from *both* sides, ignoring what happens when $x = a$.

Notations for the derivative of $y = f(x)$ at $x = a$ include

$$f'(a) \quad \text{or} \quad \frac{dy}{dx} \text{ at } x = a \quad \text{or} \quad \frac{df}{dx} \text{ at } x = a$$

The algebraic definition of the derivative at a point is

$$f'(a) = \lim_{x \to a} \frac{f(x) - f(a)}{x - a} = \lim_{h \to 0} \frac{f(a + h) - f(a)}{h}$$

How fast a function is changing at a point is called by several different names:

- **the rate of change of the function at the point**
- **the instantaneous rate of change of the function at the point**
- **the slope of the curve at the point**
- **the slope of the tangent to the curve at the point**
- **the derivative of the function at the point**

Marginal revenue is the **derivative of the revenue function**, which in economics is ***normally*** **the rate of change of revenue with respect to the *quantity* sold**. When the marginal revenue is the rate of change of revenue at a given *quantity*, the marginal revenue at a point is interpreted as the **approximate revenue per *additional* item produced and sold.**

The derivative of a function at a point can be used to approximate the change in the output for a given change in the input:

$$(\text{change in output}) \approx (\text{derivative})(\text{change in input})$$

If the output value of the function at the point is also known, then the *new* output value (after the change in input) can be approximated by

$$\begin{aligned}(\text{new output}) &\approx (\text{original output}) + (\text{change in output}) \\ &= (\text{original output}) + (\text{derivative})(\text{change in input}).\end{aligned}$$

These approximations are based on using the **tangent line as a *linear approximation* of the curve near that point**, and are best when the change in input is small. This process is called **marginal analysis.**

The derivative of the function $y = f(x)$ at x **is a *function* and can be denoted**

$$\frac{dy}{dx} \quad \text{or} \quad f'(x) \quad \text{or} \quad y' \quad \text{or} \quad \frac{d}{dx}[f(x)] \quad \text{or} \quad D_x f(x)$$

The algebraic definition of the derivative *function*, $f'(x)$, is

$$f'(x) = \lim_{h \to 0} \frac{f(x + h) - f(x)}{h} = \lim_{\Delta x \to 0} \frac{\Delta y}{\Delta x} = \lim_{\Delta x \to 0} \frac{f(x + \Delta x) - f(x)}{\Delta x}$$

General rules for finding derivatives:

$f(x)$	$f'(x)$	
c	0	The derivative of a constant is 0.
$c[f(x)]$	$c[f'(x)]$	The derivative of a constant times a function is the constant times the derivative of the function.
$f(x) \pm g(x)$	$f'(x) \pm g'(x)$	The derivative of the sum or difference of two functions is the sum or difference of the derivatives of the functions.
ax^n	$an(x^{n-1})$	The derivative of a constant coefficient times x raised to a constant exponent is the coefficient times the exponent times (x raised to one less than the exponent).
e^x	e^x	The derivative of e raised to the x power is e raised to the x power: e^x is its own derivative.
ab^x	$a(\ln b)(b^x)$	The derivative of a constant times b raised to the x power is the constant times the natural logarithm of b times b raised to the x power.
$\ln x$	$\frac{1}{x}$	The derivative of the natural logarithm of x is the reciprocal of x.
$y = f(u), u = g(x)$ $(y = f(g(x)))$	$\frac{dy}{dx} = \frac{dy}{du} \cdot \frac{du}{dx}$ $= f'(g(x))g'(x)$	The derivative of a composite function is the outer derivative with the inner function plugged in times the inner derivative (the **Chain Rule**).
$f(x)g(x)$	$f(x)g'(x) + g(x)f'(x)$	The derivative of the product of two functions is the first function times the derivative of the second plus the second function times the derivative of the first (the **Product Rule**).
$a[g(x)]^n$	$(an[g(x)]^{n-1})[g'(x)]$	The **Power Rule**.
$e^{kg(x)}$	$[ke^{kg(x)}][g'(x)]$	A special case of the Chain Rule.
$\ln[g(x)]$	$\frac{g'(x)}{g(x)}$	Another special case of the Chain Rule.

CHAPTER 3

Single-Variable Optimization and Analysis

INTRODUCTION

In the first chapter we learned how to model real-life problems by fitting curves to data. In the second chapter we learned how to find the rate of change, or derivative, of these models. In this chapter we will learn how to optimize and analyze the models using the derivative. **Optimization** means finding the maximum or minimum value of a function. In Section 3.1, we will show how to analyze the graph of a smooth function and the graph of its derivative, which we will call its **slope graph**. We will see that either graph gives us information about the other and will discuss the relationships between the two. In Section 3.2 we will define the concepts of **local** and **global maximum** and **minimum** points, which collectively are referred to as **local and global extrema**. We will then show how to use the derivative of a function to find **critical points**, which are *potential* extrema. To determine whether a critical point is an extremum, in most cases we can simply use technology to graph the function and figure it out visually. In Section 3.3, we talk about **concavity** and how it relates to the derivative of the derivative, or the **second derivative**, and how these concepts can be used to test a critical point to see if it is an extremum. We will also see how to recognize and calculate **points of inflection**, which are places where concavity changes, as we discussed in Chapter 1. We call the graph of the second derivative the original function's **concavity graph** and discuss the relationships between the concavity graph, the slope graph, and the original graph of a smooth function. These first three sections describe how to perform optimization *mathematically*. When solving real-world optimization problems, the initial mathematical optimal solution is often only the *starting* *point* rather than the end of the investigation. In Section 3.4, we discuss **post-optimality analysis**, which includes the concepts of **verification**, **validation**, **sensitivity analysis**, and estimation of a **margin of error**, as well as the process of putting all of these together to form a **conclusion** or recommendation for action in the real-world problem. Putting the recommended plan of action into practice is called **implementation** of the analysis. Section 3.5 is an optional supplementary section focusing on some particular applications of these concepts, including the economic concept of **elasticity** and how it relates to maximization of revenue, and the conditions for maximizing profit and for minimizing average cost.

Here are some of the types of problems that the material in this chapter will help you analyze:

- You want to sell mugs as part of an ecology project. How should you price the mugs to maximize the profit?
- How many practice games should you play on the computer so you will be at the peak of your game before you challenge your friend?
- How much should you discharge a battery each cycle in order to maximize its useful lifetime?
- How much should you charge for tickets to a particular show at a coffee house that you run to realize the most profit?
- How much time should you spend studying each week to optimize the balance between academics and the rest of your life?
- A company's profits have been declining. You were hired to "turn the company around." One year after you were hired the profits are still declining. You have to go before the board of directors and justify your policies and your very large salary. How can you possibly do this?

Most of these problems refer to optimizing something. We are always trying to get the most out of something: our energy level from exercising, the profit from an enterprise, satisfaction from time spent, performance from equipment. By the end of this section you should be able to solve these and similar problems and should

- Be able to analyze the graph of a function and roughly sketch its slope and concavity graphs.
- Be able to recognize local extrema on a graph and from a slope graph.
- Know how to determine global maxima and minima using derivatives and graphs.
- Know how to find and interpret points of inflection.
- Understand post-optimality analysis, and how to perform verification, validation, sensitivity analysis, and a rough calculation of a margin of error.
- Know how to synthesize mathematical optimization and post-optimality analysis to come to a conclusion and recommendations for action in a real-world problem.

3.1 ANALYSIS OF GRAPHS AND SLOPE GRAPHS

In Chapter 1 we learned how to choose what type of function to fit to data by examining a scatter plot of the data. The general shape of the pattern and/or its apparent concavity affected the types of functions that we tried to fit. If the scatter plot had no concavity, we fit a linear function. If the scatter plot had one concavity and approached a limiting value or did not change direction, we fit an exponential function or a power function. If the scatter plot had one concavity and changed direction, we fit a quadratic function. If the scatter plot was both concave up and concave down and had one inflection point, we fit a cubic or logistic model. If the scatter plot changed concavity twice, and so had two inflection points, we fit a quartic function. In Chapter 2 we learned how to find the rate of change, or derivative, of a function, which corresponds to the slope of the tangent line to the curve. We also looked at the graphs of some of these derivatives, which we call **slope graphs**. In this section we will learn how to analyze the graph of a smooth function, visually determining on what intervals the function is increasing or decreasing, where the function changes direction to indicate a local maximum or minimum, and where any points of inflection occur. We will use this analysis to *sketch* the slope graph of the function.

Here are the kinds of problems that the material in this section will help you to solve:

- If you saw a graph of a smoothed model of the inflation rate [rate of change of the Consumer Price Index (CPI)] before, during, and after the Depression of the 1930s, how could you visually estimate where the CPI peaked and bottomed out over that period? How could you estimate the point where the CPI was experiencing the sharpest decline?
- In economics, the marginal cost function for a product as a function of the quantity produced is used in a number of kinds of analysis. If you had a graph of the total cost function, how could you draw a rough sketch of the marginal cost function from it?
- Suppose you were given a graph of the total profit for an item your company makes as a function of the quantity produced and sold (using the demand function to determine the selling price). What would the marginal profit function look like? What if you had a way to estimate the marginal profit graph directly and wanted to predict from it roughly the quantity that would maximize your profit?

When you have finished this section you should be able to solve these and similar problems and should be able to

- Look at a graph and determine approximately where it is increasing and where it is decreasing.
- Look at a graph and determine approximately where local and global maximum and minimum points are.
- Recognize approximately where the slope of a curve is 0.
- Look at a graph and determine approximately where inflection points are.
- From looking at the graph of a function, roughly sketch the graph of its slope function.

- Estimate a function's local maxima or minima and points of inflection from its slope graph.

Slope Graphs of Linear Functions

Sample Problem 1: Given the graphs of linear functions shown in Figure 3.1-1, sketch the graphs of the slope functions.

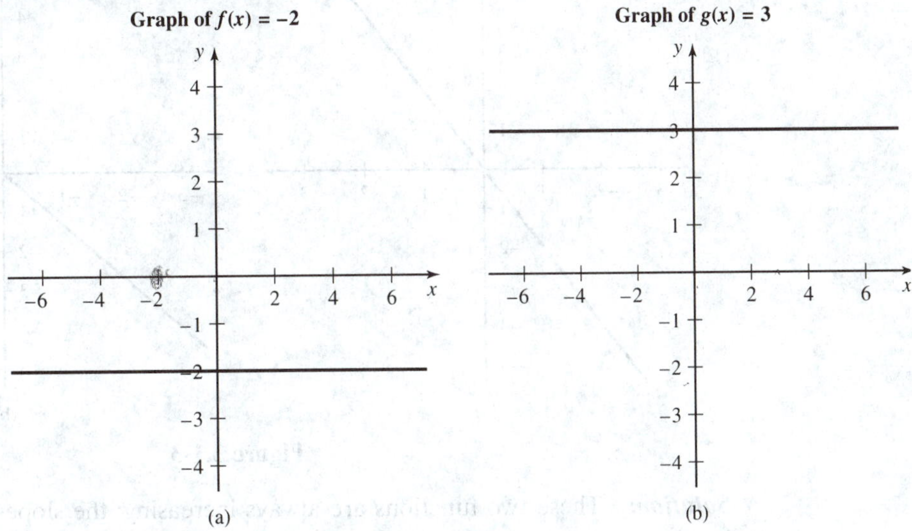

Figure 3.1-1

Solution: As we noted in Section 2.4, these linear graphs are not sloping, that is, the slope is zero. So the slope graph for each of these graphs is as shown in Figure 3.1-2. The functions in Figure 3.1-1(a) and (b) are neither increasing nor decreasing, and have no concavity. **The derivative of a constant function is 0.** □

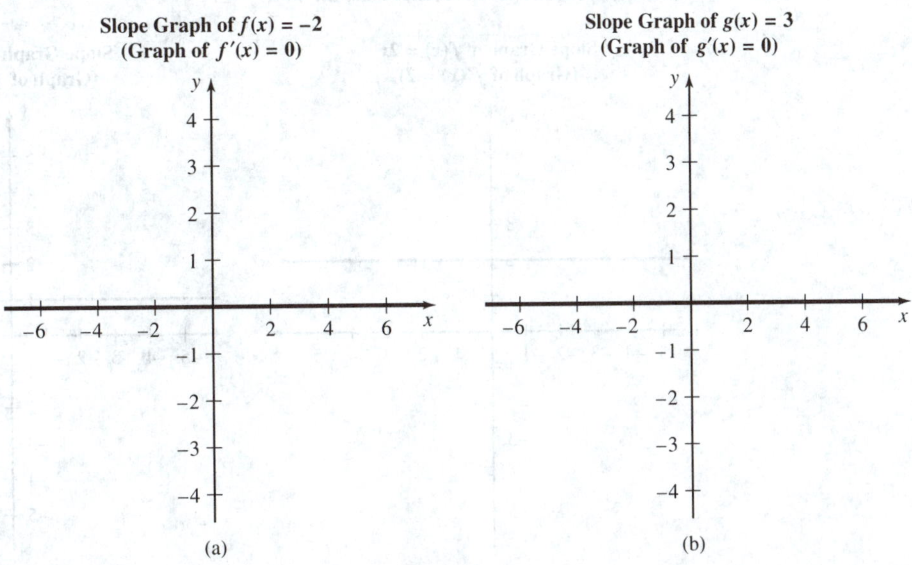

Figure 3.1-2

Sample Problem 2: Given the graphs of the linear functions in Figure 3.1-3, draw the graphs of the slope functions.

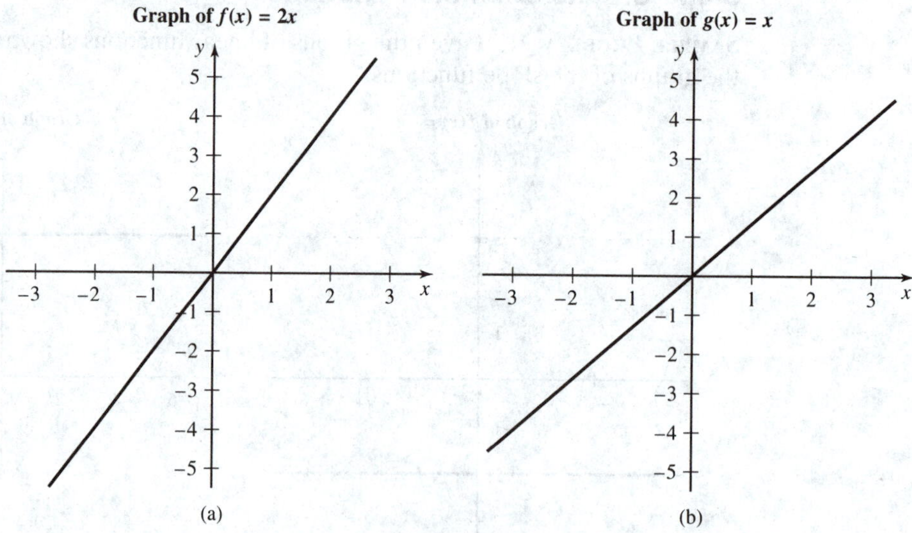

Figure 3.1-3

Solution: These two functions are always increasing; the slopes are positive. The two functions are not increasing at the same rate, as the two lines do not have the same slope. They are both positive, but one is steeper than the other. In the first function, $f(x) = 2x$, the dependent variable, y, is rising 2 units for every 1 unit rise in the independent variable, x. In the second function, $g(x) = x$, the dependent variable is rising 1 unit for every 1 unit increase in the independent variable. The horizontal line showing the first slope graph is at 2 units above the x-axis, and the second is at 1 unit above the x-axis (Figure 3.1-4). In general, **the derivative of ax is a.** □

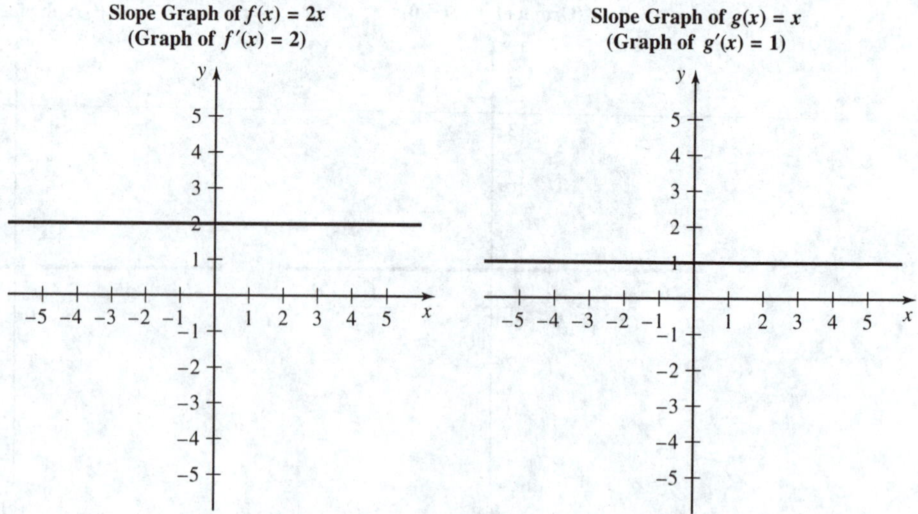

Figure 3.1-4

Sample Problem 3: Given the graphs in Figure 3.1-5, draw the slope graphs.

Graph of $f(x) = -4x$
Graph of $g(x) = -4x + 5$
y
x
(a)
(b)

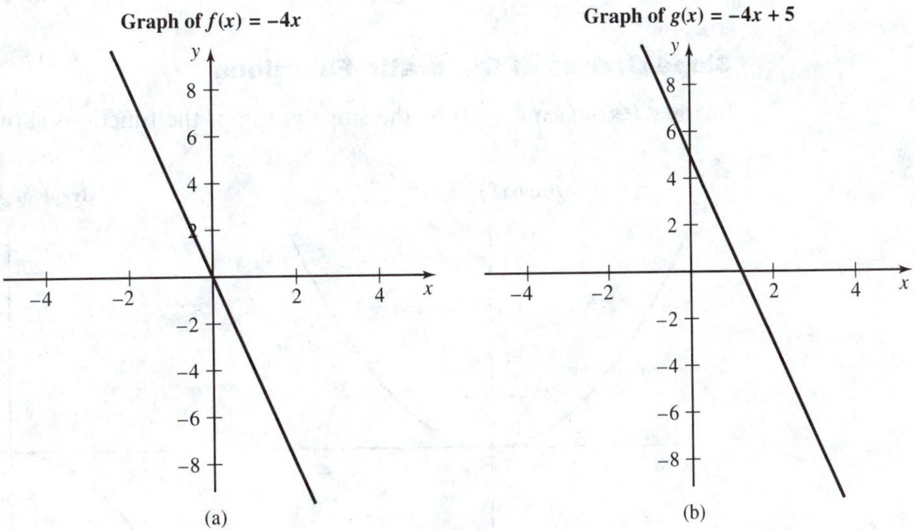

Figure 3.1-5

Solution: In these two graphs the functions are decreasing everywhere; the slopes are negative. The functions are decreasing at the same rate, 4 units down for every 1 unit to the right; the slopes are the same. The fact that the first one goes through the origin (the y-intercept is 0) and the second one does not (the y-intercept is 5) does not affect the slope at all. The slope graph for both functions is shown in Figure 3.1-6. **Adding a constant to a function does not change the derivative of the function.** □

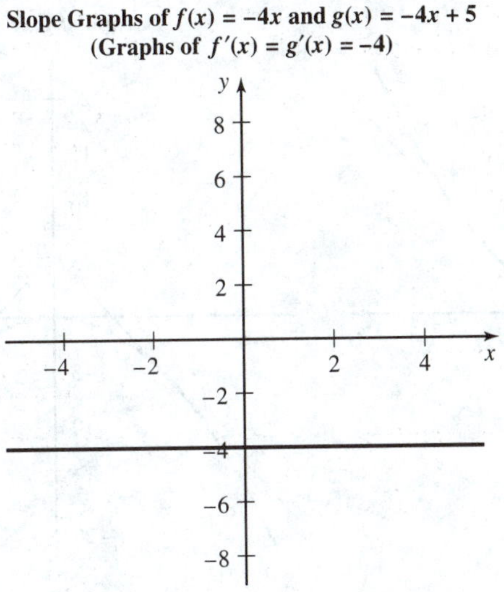

Slope Graphs of $f(x) = -4x$ and $g(x) = -4x + 5$
(Graphs of $f'(x) = g'(x) = -4$)
y
x

Figure 3.1-6

Slope Graphs of Linear *Functions:* Sketching the slope graphs of *linear* functions isn't difficult. The slope graphs **will always be horizontal lines. If the function is increasing, the slope is positive and the slope graph is a horizontal line *above* the x-axis. If the**

function is decreasing, the slope is negative and the slope graph is a horizontal line *below* the x-axis. Linear functions never change direction and have no concavity.

Slope Graphs of Quadratic Functions

Sample Problem 4: Draw the slope graph of the functions shown in Figure 3.1-7.

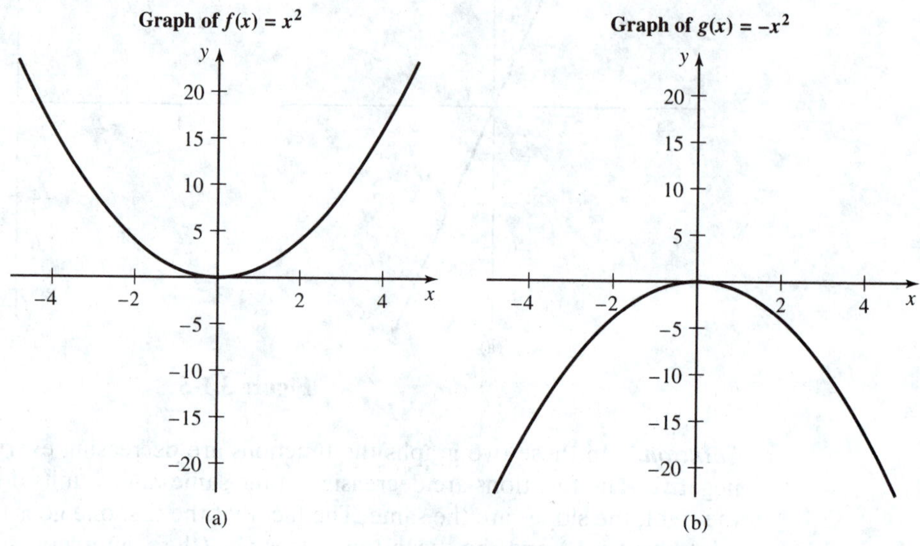

Figure 3.1-7

Solution: Let's draw in some tangent lines to get a feel for the slopes (Figure 3.1-8).

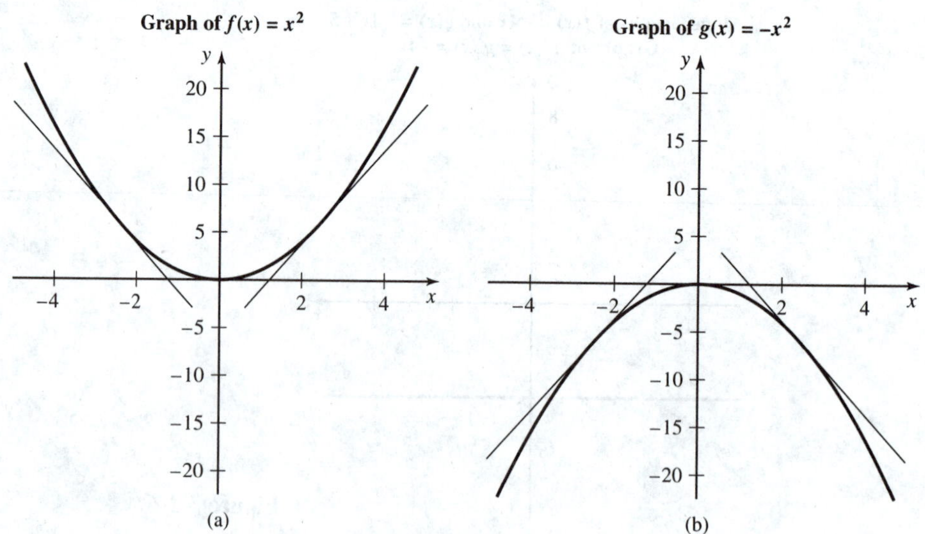

Figure 3.1-8

The first graph is decreasing for all values $x < 0$ and increasing for all values $x > 0$. The slope graph will be negative for all values $x < 0$ and positive for all values $x > 0$. When $x = 0$ the function is neither increasing or decreasing; the slope is 0. Something important has happened at the point where $x = 0$. The function is decreasing rapidly at first and then as $x \to 0$ it is decreasing more slowly. The slope graph will have large negative values and then smaller negative values. For positive values of x the situation is reversed: The function is increasing slowly at first and then more rapidly. The slope graph will have small positive values and then larger positive values. This curve has the general shape of a quadratic function. As we saw in Section 2.4 the derivatives (slope functions) of quadratic functions are straight lines, just like those for linear functions, but they are not horizontal.

The slope graph for the first curve is shown in Figure 3.1-9. It is just as we described: negative for all values $x < 0$ and positive for all values $x > 0$. At $x = 0$ the slope is 0. The point at which a function levels off, the point where the slope is 0, is called a **critical point**. When the function changes from decreasing to increasing, the critical point is a **minimum point**.

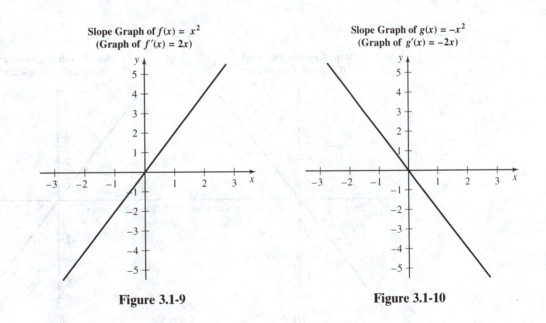

What about the second graph? It is the same as the first graph, only turned upside down. It increases for all values $x < 0$ and decreases for all values $x > 0$. It increases rapidly at first and then the rate of increase slows down. At $x = 0$ the function is neither increasing nor decreasing; the slope is 0. The slope graph of the second function is shown in Figure 3.1-10. When the function changes from increasing to decreasing, the critical point is a **maximum point**.

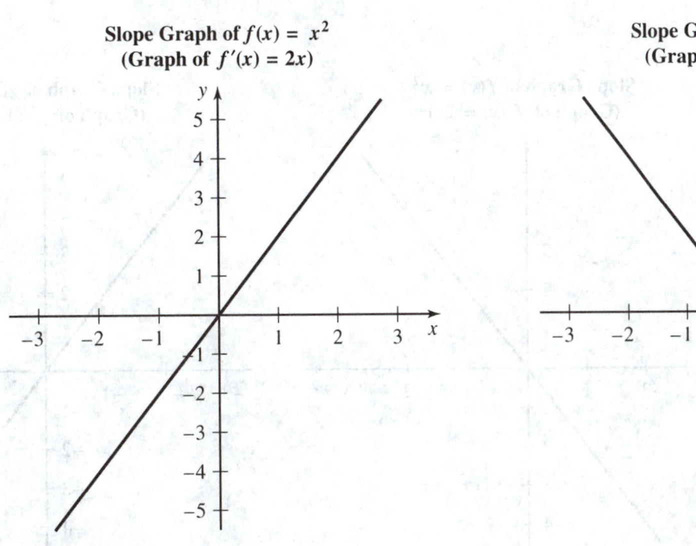

Figure 3.1-9

Figure 3.1-10

Let's look at these two function graphs and their slope graphs more closely (Figure 3.1-11).

Quadratic functions have one **critical point**. At this critical point the slope levels off, equals 0, and there is a maximum or minimum. □

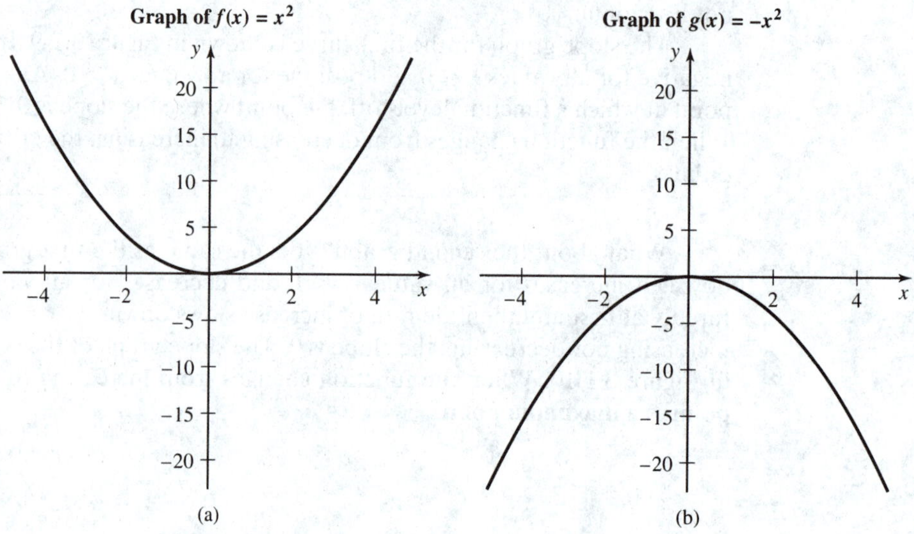

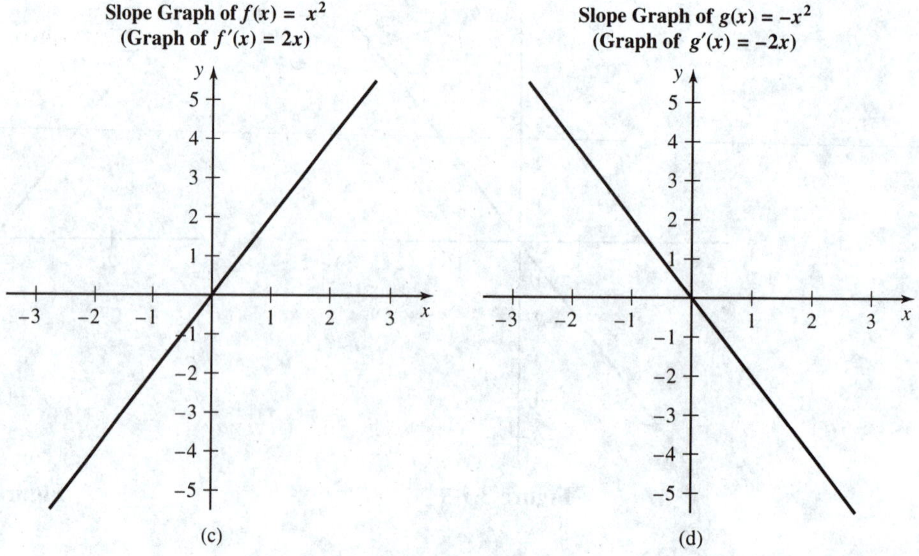

Figure 3.1-11

Sample Problem 5: Draw the slope graphs of the functions shown in Figure 3.1-12.

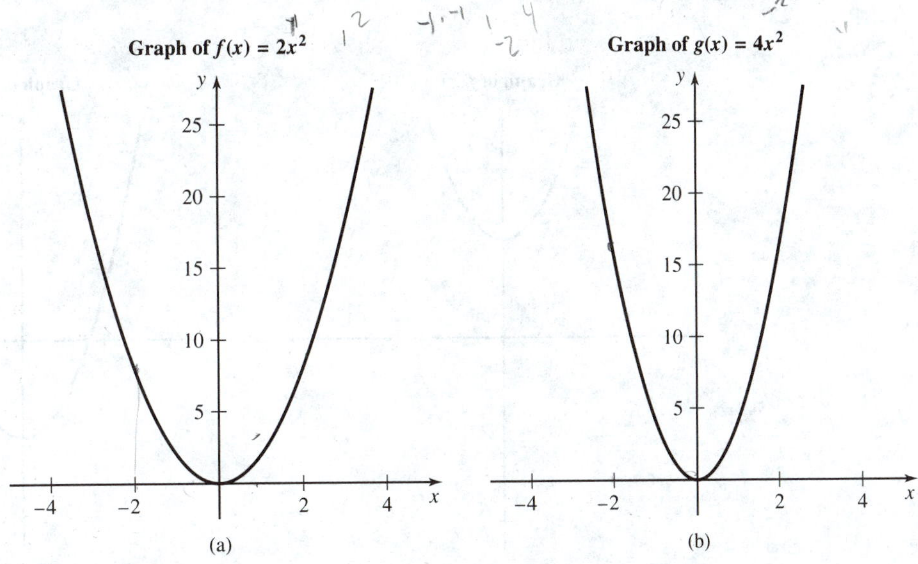

Figure 3.1-12

Solution: Unlike the graphs in Sample Problem 4 both of these graphs are decreasing, rapidly at first and then more slowly for all values $x < 0$. They both level off at $x = 0$ and then start increasing, slowly at first and then more rapidly. The slope graphs will be negative for all values $x < 0$, be 0 at $x = 0$ and be positive for all values $x > 0$. But these two graphs are not the same—the second is much "steeper" than the first. Since they are the graphs of quadratic functions, we know that their slope graphs will be straight lines. The slope graph of the second function will have a "steeper" slope than that of the first. The slope graphs are shown in Figure 3.1-13.

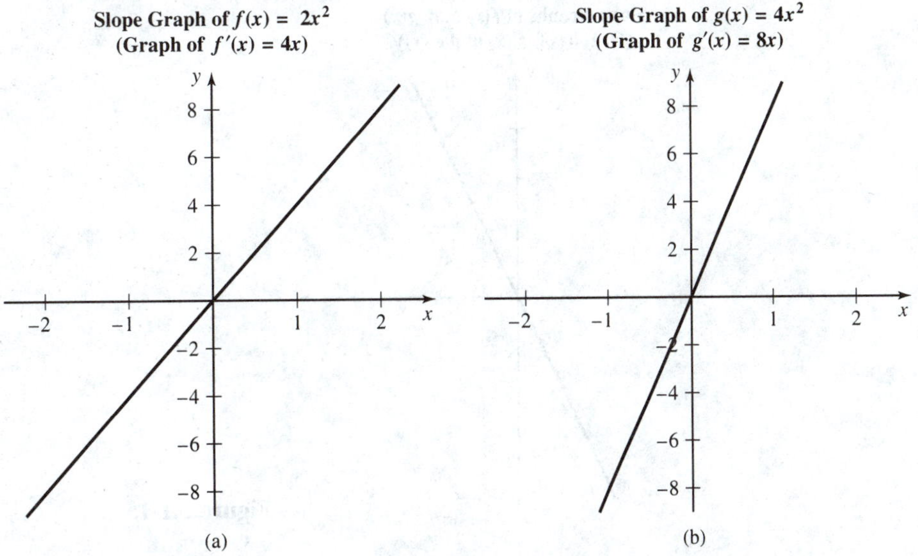

Figure 3.1-13

SAMPLE PROBLEM 6: *Sketch* the slope graphs of the functions shown in Figure 3.1-14.

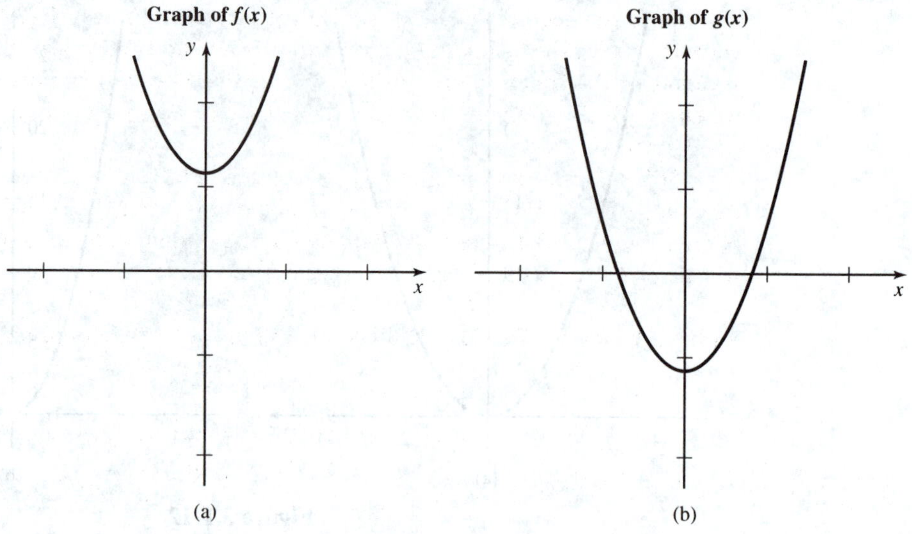

Figure 3.1-14

Solution: These two graphs have no scale! How in the world can you get a slope graph if you have no numbers to go by? Note the change from the word *draw* to the word *sketch*. The slope graphs won't have scales either. They will just indicate over what intervals the functions are increasing and decreasing and approximately at what rate, that is, rapidly (a large value) or slowly (a smaller value). They will also indicate where the function *changes its direction, or has a critical point.*

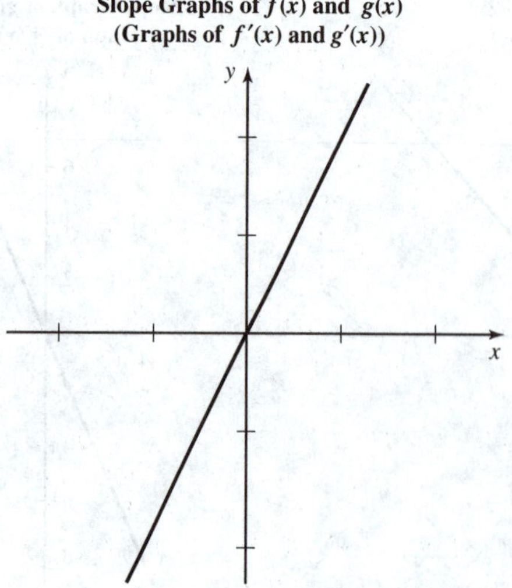

Figure 3.1-15

These two functions look very much alike: both decreasing for all values $x < 0$, equal to 0 at $x = 0$ and increasing for all values $x > 0$. The actual slopes appear to be the same. But what about the fact that the first function is always positive, always above the x-axis and the second function is negative for some values of x? We are not concerned with the values of the original function, only how they are changing. **Whether the y-values of the function are positive or negative does not change the slope graph of the function.** Both of these functions have exactly the same slope graph, shown in Figure 3.1-15.

In a *sketch* of a slope graph, just note where the slope of the original function is positive, negative and equal to 0. These correspond to where the original graph is increasing, decreasing, and flat, respectively. The corresponding output values on the slope graph will be positive, negative, and 0, respectively. □

Sample Problem 7: Sketch the slope graphs of the functions shown in Figure 3.1-16.

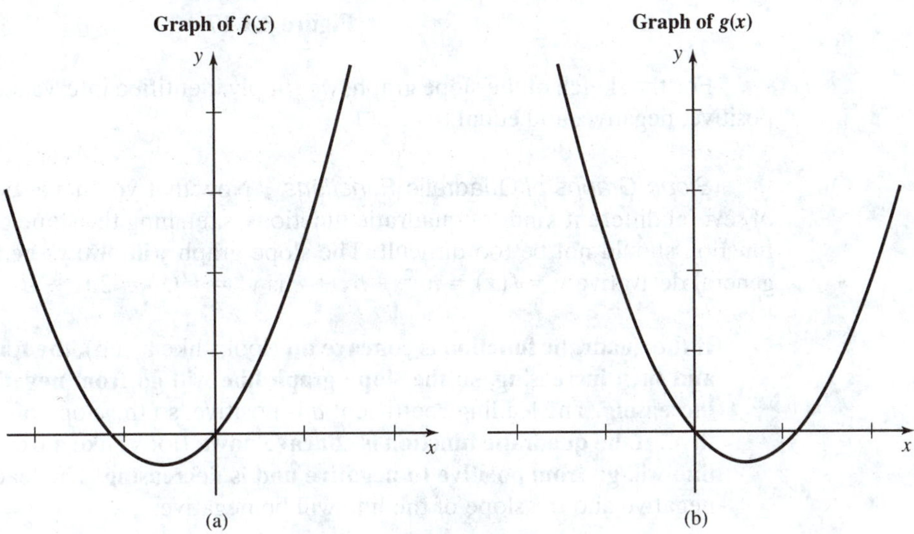

Figure 3.1-16

Solution: These functions have the same shape; they both appear to be quadratic functions. The slopes seem to be decreasing and increasing at approximately the same rate. They decrease, level off, and then start increasing again; their slope graphs will be negative, equal to 0, and then positive. However, the place where the change from decreasing to increasing occurs is *not* the same. On the first graph it is someplace where $x < 0$. In the second graph it is someplace where $x > 0$. We assume that the slope graphs will be straight lines, negative at first and then positive. We do not know how steep to make the slope of the lines because we have no scales (Figure 3.1-17).

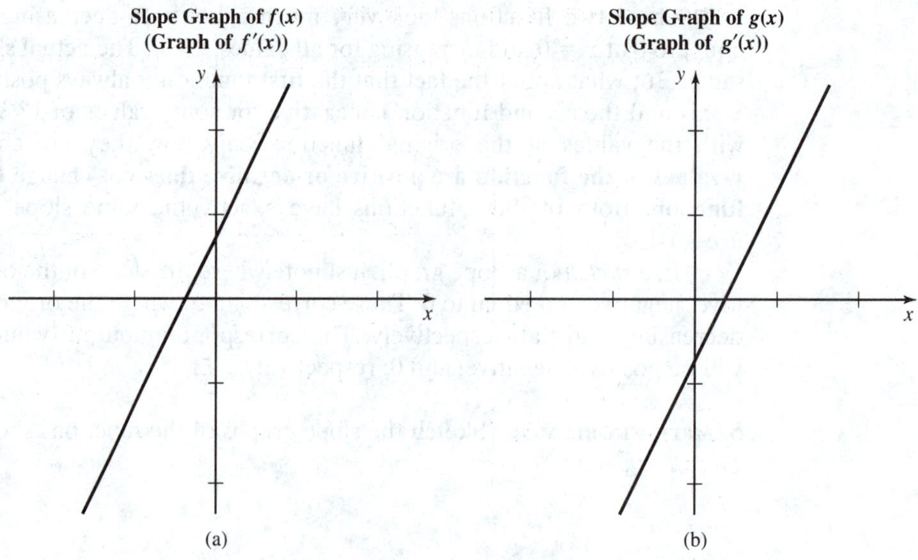

Figure 3.1-17

For the sketch of the slope graph, we simply identified intervals where the slope was positive, negative, and equal to 0. □

Slope Graphs of Quadratic *Functions:* Now that you have seen the slope graphs of several different kinds of quadratic functions, sketching the slope graph of a quadratic function should not be too difficult. The slope graph **will always be a straight line**: The general derivative $y = f(x) = ax^2 + bx + c$ is $y' = f'(x) = 2ax + b$.

> **If the quadratic function is concave up** (looks like a cup)**, the function is decreasing and then increasing, so the slope graph line will go from negative to positive; it is increasing.** The leading coefficient a is positive, so the slope of the line will be positive. **If the quadratic function is concave down** (looks like a frown)**, the slope graph line will go from positive to negative and is decreasing.** The leading coefficient a is negative and the slope of the line will be negative.

In both cases the slope line crosses the x-axis where the quadratic function *changes vertical direction* (from decreasing to increasing or vice versa). **Where the quadratic function *changes vertical direction* there will be a critical point, either a maximum or minimum. The function will have a maximum or a minimum point where the slope function crosses the x-axis. If the function is concave *up*, the critical point will be a *minimum*. If the function is concave *down*, the critical point will be a *maximum*.**

Where a Smooth Function	Its Slope Graph
is increasing	is positive
is decreasing	is negative
has a maximum or minimum (is flat)	equals zero

Slope Graphs of Cubic Functions

Sample Problem 8: Sketch the slope graph of the function shown in Figure 3.1-18.

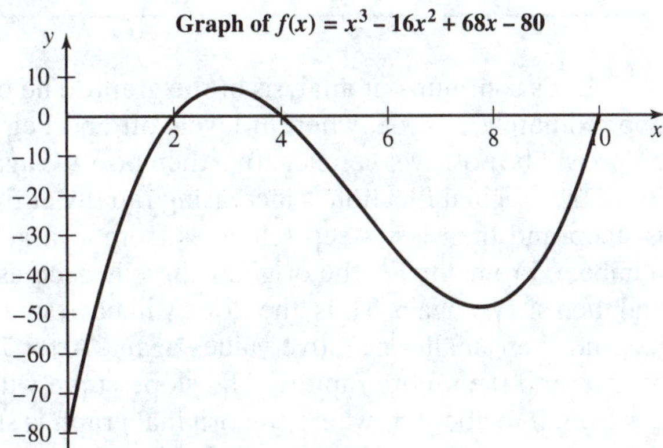

Figure 3.1-18

Solution: Let's analyze the graph. Initially, approximately from 0 to 3 (on the interval $0 \le x < 3$), the function is increasing. At approximately $x = 3$, it levels off and changes direction. A point at which a function changes vertical direction is a critical point. For quadratic functions we said that a point at which the function changed from increasing to decreasing was a maximum point. This is also true for cubic and other functions, but it may not be *the* maximum point. Since these functions change vertical direction more than one time, there may be other points that are higher than this critical point. To indicate this, we call such points ***local* (relative) maximum points**. This point is **the highest point in the immediate vicinity**; there will not be any points higher than this until the function changes direction again. The same is true about critical points where the function changes from decreasing to increasing: These critical points will be ***local* (relative) minimum points**. Look again at Figure 3.1-18. The value of the function at $x = 0$ is approximately -80. The value at the critical point where the function changes from decreasing to increasing, $x = 7.5$, is approximately -50. So $x = 7.5$ is only a local minimum.

The *global* (absolute) minimum of a function over the domain simply means the point where the output value is the smallest over the entire domain, and here it occurs at $x = 0$. **The global *maximum* of a function over the domain is the point where the output value is the *largest* over the entire domain**, and here it seems to occur at around $x = 3$. **The general term for a maximum or minimum point is an *extremum*. The technical *plurals* of these terms are: *maxima, minima,* and *extrema.* In this book, we will *not* call endpoints of an interval local extrema**, so we consider the point at $x = 0$ to be a global minimum, but *not* a local minimum. This convention is arbitrary, but we need it for clarity. To determine global or absolute maximum or minimum points, we must consider the *entire* domain of the function, including the endpoints.

Let's make a table summarizing what we know about the original graph and the slope graph so far:

Approximate Interval of Inputs	Direction of Original Graph	Sign of Slope Graph
0 to 3	Increasing	Positive
3 to 7.5	Decreasing	Negative
7.5 to 10	Increasing	Positive

Let's continue our analysis of the graph. The original graph decreases from $x = 3$ to approximately $x = 7.5$, where it levels off and begins to increase again. Thus, the slope graph will be positive then negative then positive again. On the first positive interval from 0 to 3 the original function is increasing rapidly at first and then more slowly, so the slope is steep and then less steep (changes from a large positive number to smaller positive number). From 3 to 7.5 the original curve is decreasing slowly at first, then more rapidly, and then slowly again. Thus, the slope will have small negative values, larger negative values and then smaller negative values again. After 7.5 the slope starts to increase, slowly at first, and then more rapidly. The slope curve will be decreasing at first, for input values from 0 to about 5 (where the original graph is steepest in the negative direction) and then increasing from about 5 to 10. The changes in the slope seem relatively gradual throughout.

Ignoring the vertical scale on the original graph and sketching in a horizontal axis, we might want to try to sketch the slope graph something like that shown in Figure 3.1-19. We drew in vertical lines at the three landmark points: the local maximum point

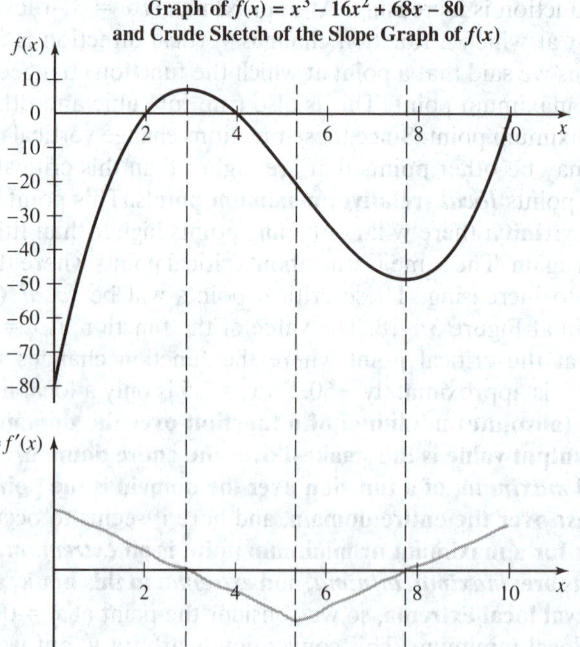

Figure 3.1-19

where the function changed from increasing and started decreasing, the inflection point where the function changed from concave down to concave up, and the local minimum point where the function changed from decreasing to increasing.

Somehow this doesn't look quite right. The slope of the original graph does change someplace around $x = 5$, but it doesn't change that abruptly. Recall from our discussion of cubic models that they are distinguished by a change in concavity. Somewhere around $x = 5$ the curve changes from concave down to concave up. We have already seen that this is also the point where the slope is steepest in the negative direction. This point is called **an inflection point, the point where the slope of the function is locally the steepest or shallowest.** To verify that this is true, look at Figure 3.1-20. The slope of the leftmost tangent line is fairly steep. The slope of the middle tangent line, approximately at the inflection point, is even steeper. The slope of the rightmost tangent line is not so steep. The slopes are changing gradually. We will discuss concavity and inflection points further in Section 3.3.

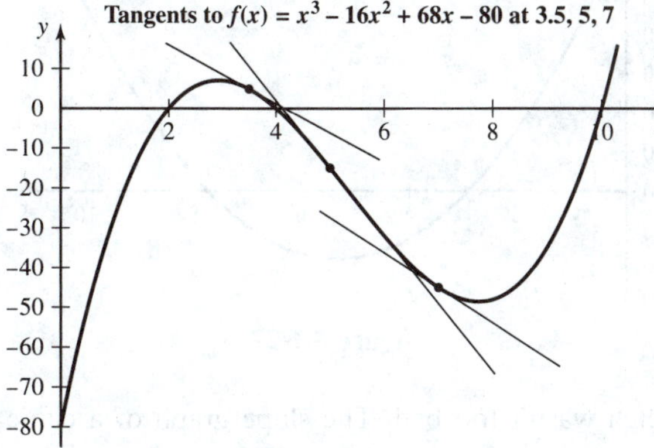

Figure 3.1-20

Since the change in the slope of the tangent line is gradual, we should make the change in the slope graph gradual, that is, round it off. Let's make another attempt at a sketch (Figure 3.1-21). This is a freehand drawing and not very elegant, but you can prob-

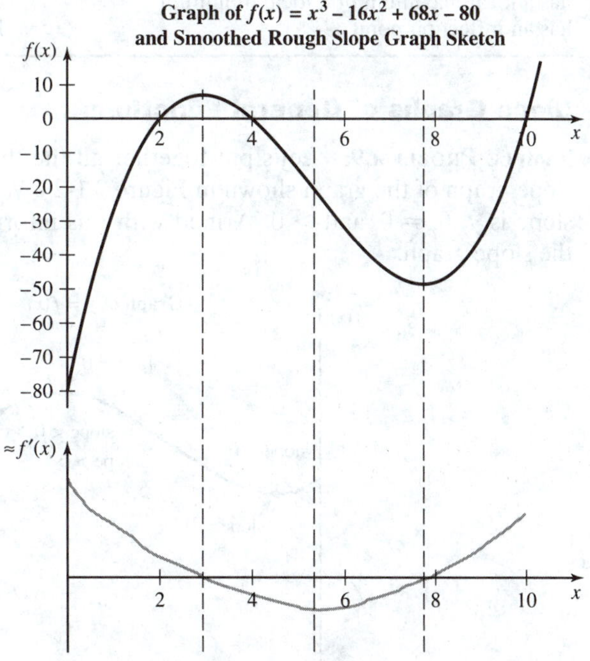

Figure 3.1-21

ably get the idea. After all, it is just a sketch. The slope is positive from 0 to about 3, then becomes negative, has the largest negative value around 5 at the inflection point, and then becomes positive after around 7.5. The actual graph of the slope function is shown in Figure 3.1-22.

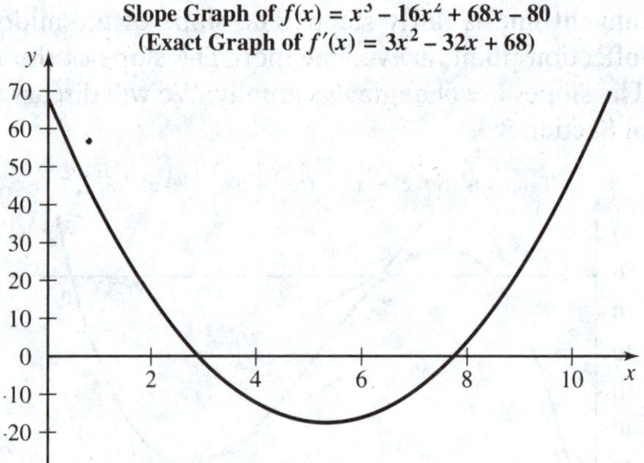

Figure 3.1-22

Our rough sketch wasn't too bad. The slope graph of a cubic function will be a parabola. □

Where the Original Smooth Function	Its Slope Graph
is increasing	is positive
is decreasing	is negative
has a local maximum or a local minimum	crosses the horizontal axis (is 0)
has an inflection point	has a local maximum or a local minimum

Slope Graphs of General Functions

Sample Problem 9: Let's put together all the things we have learned and sketch the slope graph of the graph shown in Figure 3.1-23. We have identified all places where the slope is < 0, $= 0$, and > 0. Armed with this information you should be able to sketch the slope graph.

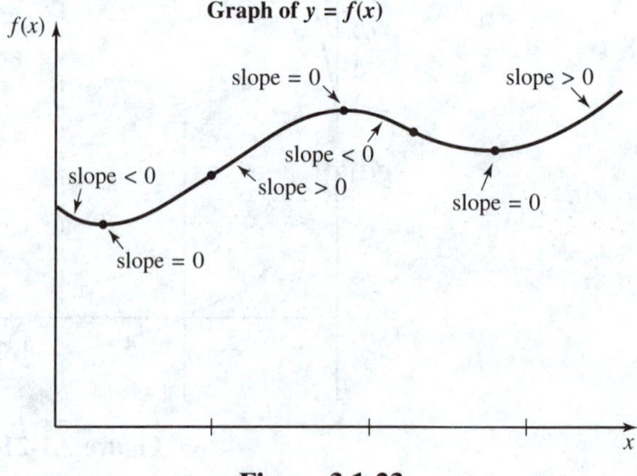

Figure 3.1-23

First, draw in vertical lines where there are landmark points: at critical points where the slope is 0, and where the concavity changes—at points of inflection, where the slope is locally the steepest or shallowest. Sketch in an x-axis for the slope graph. The points where the vertical lines are drawn corresponding to places where the slope is 0 will be the places where the slope graph crosses the x-axis. Next, put a mark on the vertical lines drawn from the inflection points, above the x-axis if the slope of the original curve is positive at this point and below the x-axis if the slope of the original curve is negative at this point. These will be the local maximum and minimum points for the slope graph. Now join these points, trying to make the curve as smooth as possible (Figure 3.1-24). The sketch shown in Figure 3.1-24 is not beautiful, but it really isn't all that bad. The actual slope graph is shown in Figure 3.1-25.

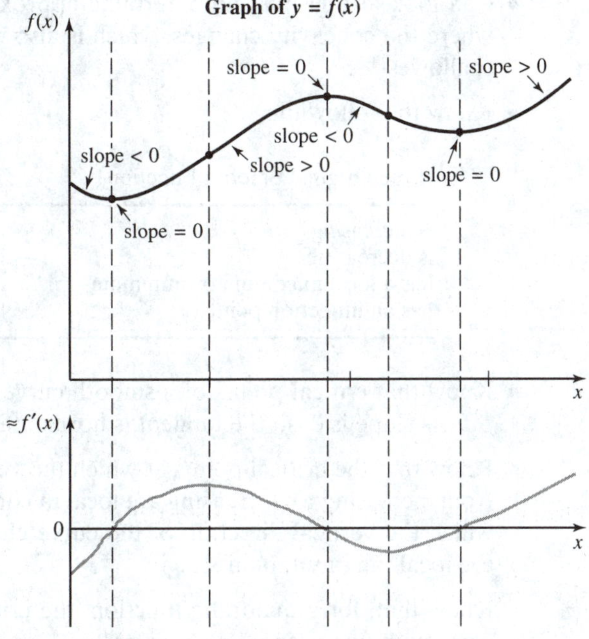

Figure 3.1-24

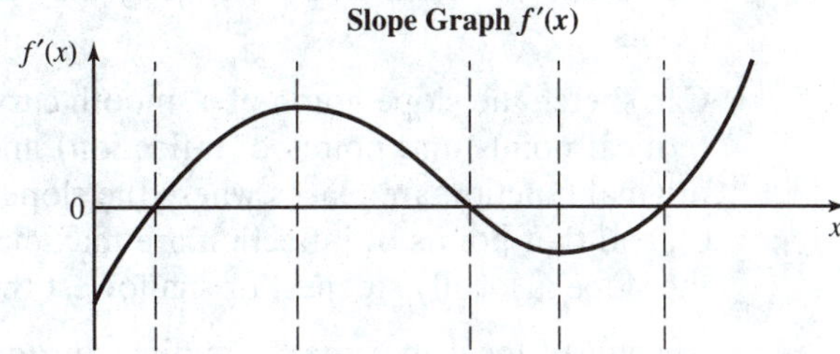

Figure 3.1-25

Our rough sketch wasn't too bad. □

Section Summary

Before you begin the exercises be sure that you

- Know that a function is increasing where its graph is sloping up from left to right.
- Know that a function is decreasing where its graph is sloping down from left to right.
- Know that a *local* (relative) extremum is a point where the output value is highest or lowest *over some interval on either side* of the point, while a *global* (absolute) extremum is a maximum or minimum *over the entire domain.*
- Can look at a graph and determine approximately where local and global maximum and minimum points are.
- Can look at a graph and determine approximately where inflection points are: where the concavity changes, which is also where the slope is locally steepest or shallowest.
- Know the following:

If the Original Smooth Function	Then the Slope Graph
is increasing	is positive
is decreasing	is negative
has a local maximum or minimum	crosses the horizontal axis (is 0)
has an inflection point	has a local maximum or minimum

- Know that critical points of a smooth curve are points at which the curve levels off (its slope is 0, so the tangent is horizontal).
- Know that the critical points at which the vertical direction of the curve changes from increasing to decreasing are local maximum points, and the critical points at which the vertical direction of the curve changes from decreasing to increasing are local minimum points.
- Know that, for a quadratic function, the critical point is *both* a local and global extremum (maximum or minimum).
- Know that, for polynomials of degree three or higher and general curves, the critical points at which the direction of the function changes are local or relative maximum or minimum points. They may also be global maximum or minimum points.
- Can sketch the slope graph of a smooth curve by identifying the landmark points (critical points and points of inflection) and knowing that critical points of the original function are places where the slope graph crosses the x-axis (the slope is 0), and that points of inflection are maxima or minima of the slope graph, since the slope is locally steepest or shallowest there.
- Recognize local maxima or minima of an original curve from its slope graph (where the slope graph goes from negative to positive, from left to right, will be a local minimum, and where the slope graph goes from positive to negative will be a local maximum).

EXERCISES FOR SECTION 3.1

Warm Up

1–8. For each of graphs in Figure 3.1-26 estimate

(a) The intervals on which the function is increasing and/or decreasing.

(b) The approximate local maximum and minimum points, if any.

(c) The approximate global maximum and minimum over the domain shown on the graph.

(d) The approximate inflection points, if any.

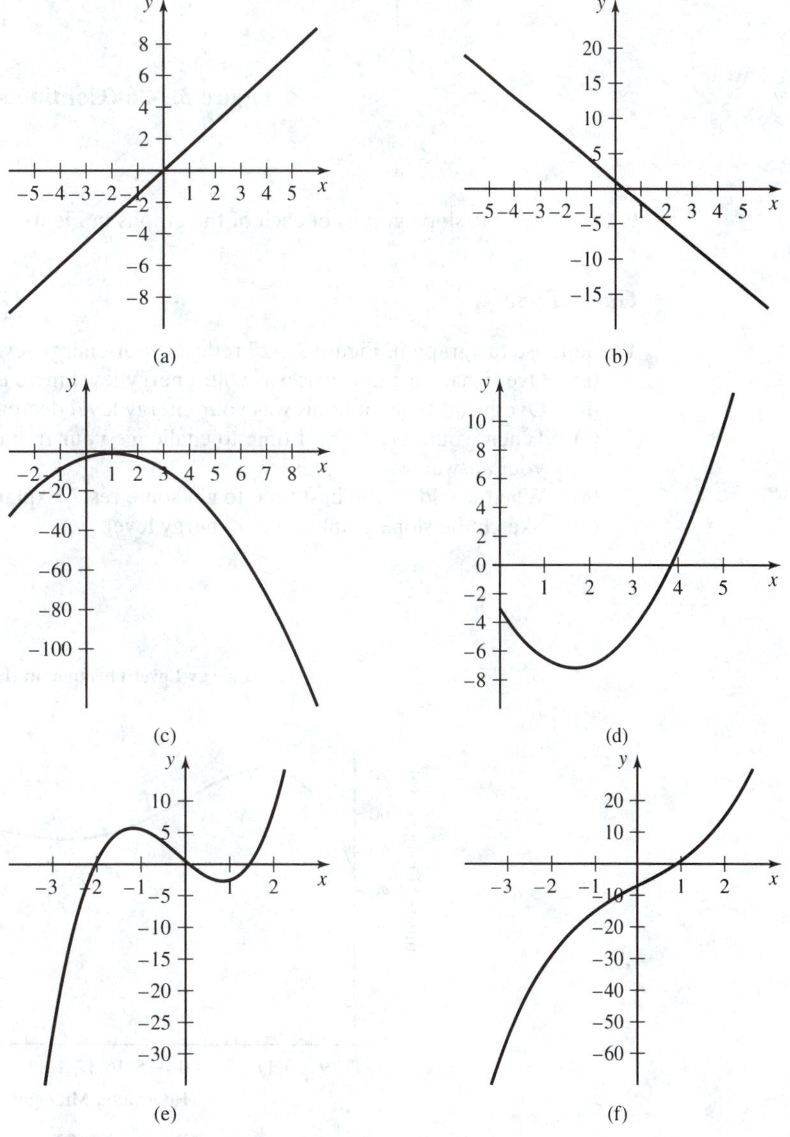

Figure 3.1-26
Continued

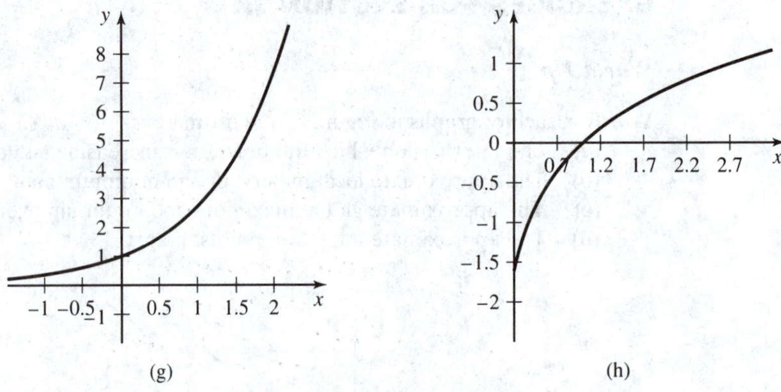

Figure 3.1-26 (Continued)

9–16. Sketch the slope graphs of each of the graphs in Figure 3.1-26.

Game Time

17. Suppose the graph in Figure 3.1-27 reflects your energy level on an average day.

(a) Over what time intervals was your energy level increasing?

(b) Over what time intervals was your energy level decreasing?

(c) When would be the best time to challenge your friend to a game of basketball? Explain your answer.

(d) When would be the best time to get some rest? Explain your answer.

(e) Sketch the slope graph of your energy level.

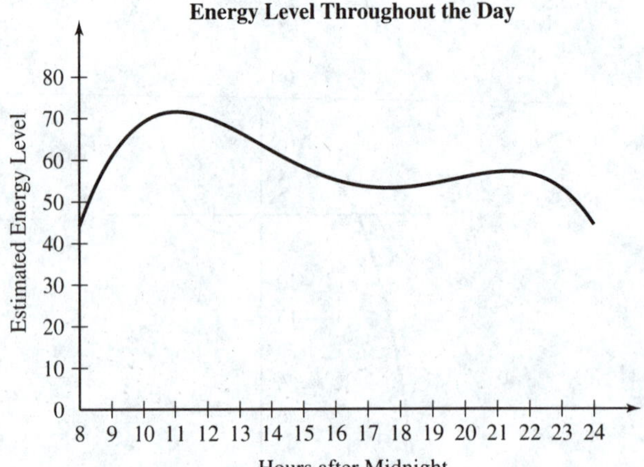

Figure 3.1-27

18. Figure 3.1-28 is a graph of a model of the crime index in the United States.

(a) Over what intervals was the crime index in the United States increasing?
(b) Over what intervals was the crime index in the United States decreasing?
(c) At what time was the crime index in the United States increasing most rapidly?
(d) Can you think of any ways that this information could be of use to someone?
(e) Sketch the slope graph of the crime index in the United States.

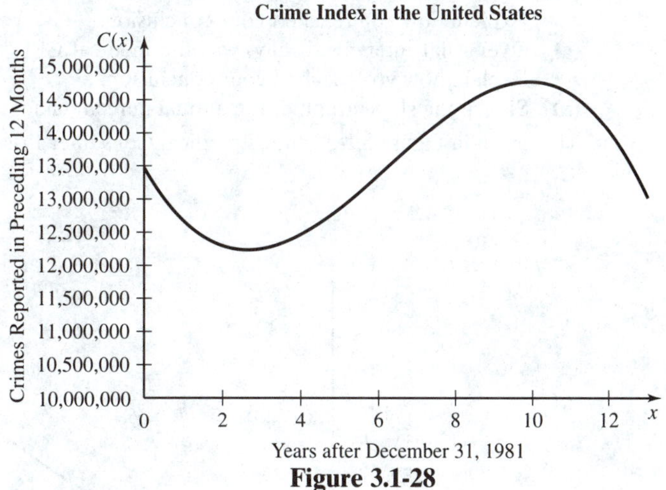

Figure 3.1-28

19. The graph in Figure 3.1-29 shows the annual percent change in the medical Consumer Price Index.

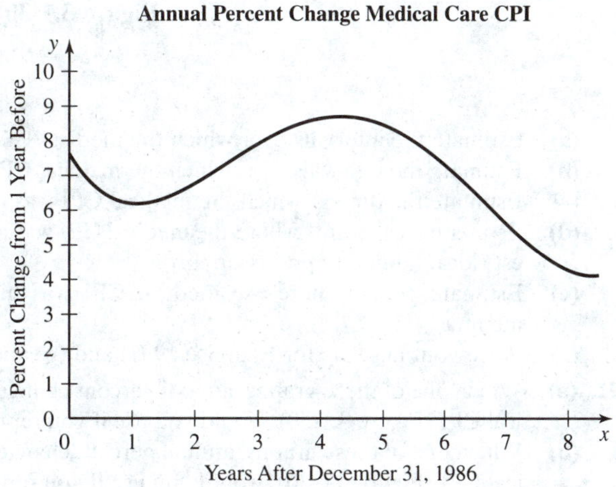

Figure 3.1-29

(a) Estimate the intervals over which the percent change in the medical CPI was increasing.
(b) Estimate the intervals over which the percent change in the medical CPI was decreasing.
(c) Estimate the time at which the percent change in the medical CPI was increasing most rapidly.
(d) Estimate local points where the percent change in the medical CPI was highest (local maximum points) and lowest (local minimum points), if any.
(e) Estimate points where the percent change in the medical CPI was highest and lowest over the domain shown on the graph (global maxima and minima).
(f) Over what intervals, if any, was the medical CPI (not the percent change) increasing? Explain how you reached your conclusion.
(g) Over what intervals, if any, was the medical CPI (not the percent change) decreasing? Explain how you reached your conclusion.
(h) Sketch the slope graph of the annual percent change in medical CPI.

20. The graph in Figure 3.1-30 shows medical Consumer Price Index.

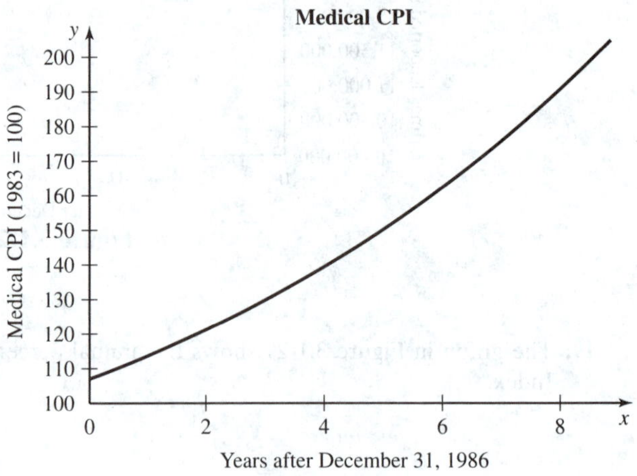

Figure 3.1-30

(a) Estimate the intervals over which the medical CPI was increasing.
(b) Estimate the intervals over which the medical CPI was decreasing.
(c) Estimate the time at which the medical CPI was increasing most rapidly.
(d) Estimate local points where the medical CPI was highest (local maximum points) and lowest (local minimum points), if any.
(e) Estimate points where the medical CPI was highest and lowest (global maxima and minima).
(f) Were your answers for Exercise 19 (f) and (g) above correct?

21. (a) Which one of these graphs, annual percent change in medical CPI (Figure 3.1-29) or medical CPI (Figure 3.1-30), would you use if you represented the medical community? Why?
(b) Which one of these graphs, annual percent change in medical CPI (Figure 3.1-29) or medical CPI (Figure 3.1-30), would you use if you represented a consumers' advocacy group? Why?

22. The graph in Figure 3.1-31 shows the annual percent change in the Consumer Price Index.
    (a) Estimate the intervals over which the percent change in the CPI was increasing.
    (b) Estimate the intervals over which the percent change in the CPI was decreasing.
    (c) Estimate the time at which the percent change in the CPI was increasing most rapidly.
    (d) Estimate local points where the percent change in the CPI was highest (local maximum points) and lowest (local minimum points), if any.
    (e) Estimate points where the percent change in the CPI was highest and lowest (global maxima and minima).
    (f) Over what intervals, if any, was the CPI (not the percent change) increasing? Explain how you reached your conclusion.
    (g) Over what intervals, if any, was the CPI (not the percent change) decreasing? Explain how you reached your conclusion
    (h) How does the annual percentage change in medical CPI (Figure 3.1-29) compare to the annual percentage change in CPI?
    (i) Sketch the slope graph of the annual percent change in CPI.

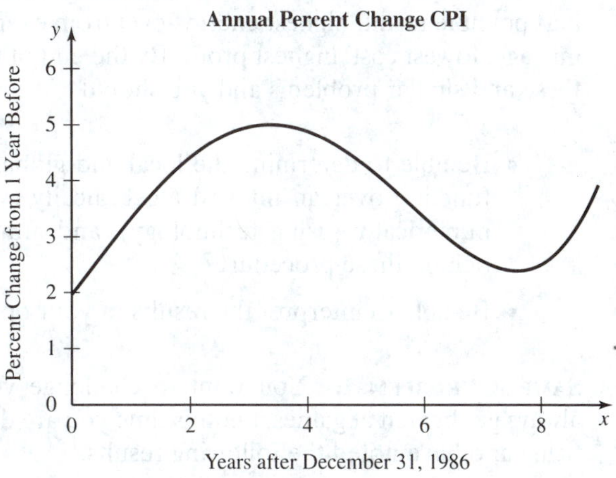

Figure 3.1-31

Source: Statistical Abstract of the United States, 1995, No. 761

3.2 *OPTIMIZATION—ALGEBRAIC DETERMINATION OF MAXIMA AND MINIMA*

In the previous section we learned how to sketch the slope graphs of smooth functions from the graphs of the functions. In order to do this we first had to analyze the graph of the function. We noted the intervals on which the function was increasing and decreasing. A point at which the function changed direction was a local maximum or minimum. We *estimated* the local maximum and minimum points from the graph. You can also estimate local maximum and minimum points from a graph on your calculator or computer.

In this section we will learn how to use calculus and algebra to find these points more precisely.

Here are some types of problems that the material in this chapter will help you to solve:

- How many practice games should you play on the computer so that you will be at the peak of your game when you challenge your friend?
- How much should you charge for tickets to your show at a coffeehouse to realize the most profit?
- How should you divide up your study time between two subjects to maximize your expected average grade?
- You want to sell mugs as part of an ecology project. How should you price the mugs to maximize the profit and how many mugs should you order?

These problems all refer to optimizing something. We are always trying to get the best possible result: highest energy level from exercise, best results in a game, highest gas mileage, lowest cost, highest profit. By the end of this section you should be able to solve these and similar problems and you should

- Be able to determine the local and global maximum and minimum points for a function over an interval algebraically (with hand calculations using calculus), numerically (using technology), and graphically, and understand the concepts behind these procedures.
- Be able to interpret the results of your optimization.

Sample Problem 1: You want to challenge your friend to a computer game. You always get beaten at games, but this time you are determined to win. You have been practicing and have noted the following results:[1]

Minutes	10	20	30	40	50	60
Tiles Left	540	260	80	15	75	380

The object of this game is get rid of all the tiles, so the lower the score the better. You can plainly see that your best score came after you practiced for 40 minutes. Then your scores started to rise again because you were getting tired. Each game takes only about a minute. Your friend is really good at this game, so you want to be sure that you are at your best. Exactly how many minutes should you practice before you challenge?

[1]Data adapted from a student project.

Solution: First let's look at a graph of the data in Figure 3.2-1. This looks as if a quadratic model would be a good fit. Using technology to find the best-fit least-squares quadratic function and rounding to 4 significant digits, we get the following model:

Verbal Definition: $T(x)$ = the number of tiles left after practicing for x minutes, on average.

Symbol Definition: $T(x) = 0.6940x^2 - 52.62x + 1015$, for $0 \le x \le 60$.

Assumptions: Certainty and divisibility. Certainty implies the relationship is *exact.* Divisibility implies that *any* fractions of minutes or tiles are possible.

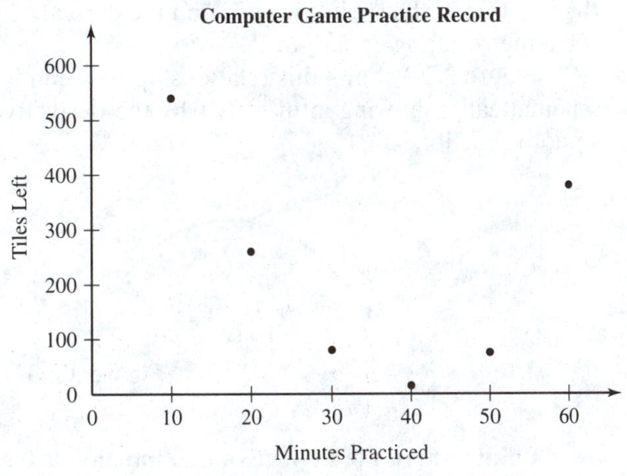

Figure 3.2-1

The graph of the data with the quadratic model is shown in Figure 3.2-2. From the graph it looks like your best practice time is somewhere between 37 and 40 minutes. If you are happy with "somewhere between," that is fine. If you want to be more precise,

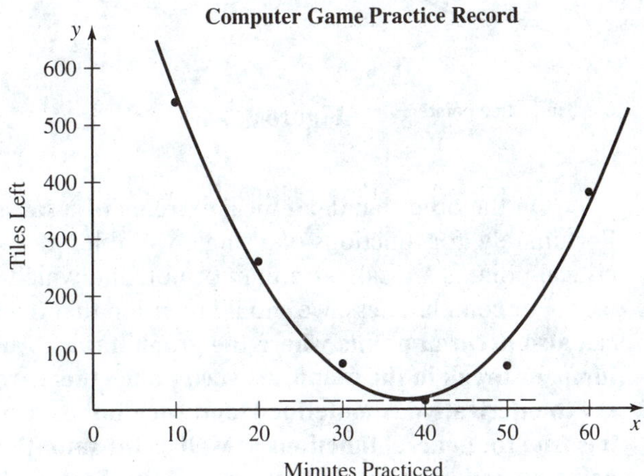

Figure 3.2-2

you can calculate the exact number of minutes of practice at which the number of tiles left will be at a minimum, *based on the model*. Then, since each game lasts about a minute, you can stop just one minute before you reach this time, so that when you play, you will be at your best.

The function that represents the number of tiles left is a quadratic function. In the last section we stated that the minimum point of a quadratic function opening up is the point at which the curve changes direction, from decreasing to increasing. For a quadratic function, this minimum is *both* a global minimum *and* a local minimum. At this point (see Figure 3.1-2) the tangent line is horizontal, so the slope of the function is equal to 0. We can *estimate* this point from the graph or we can *calculate* the point precisely using calculus and algebra. Since the slope of the tangent to a function at a point is the derivative of the function at that point, we can find the derivative of the function, set it equal to 0, and solve the resulting equation.

Figure 3.2-3 shows this relationship between local extrema and horizontal tangents schematically, showing intuitively why the derivative will be 0 at a local extremum of a smooth function.

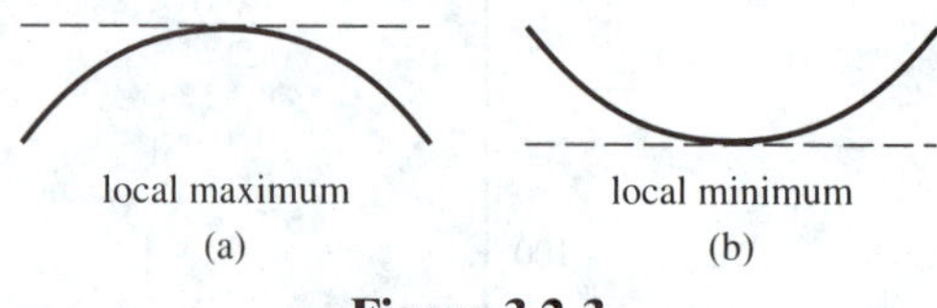

Figure 3.2-3

Points where the derivative of a function is 0 are called *critical points* of the function. These are *candidates* for local extrema. Unfortunately, **not all critical points are local extrema**. For example, an inflection point where a curve levels off instantaneously (as in $f(x) = x^3$ at $x = 0$) will also be a critical point. Figure 3.2-4 shows an example of this.

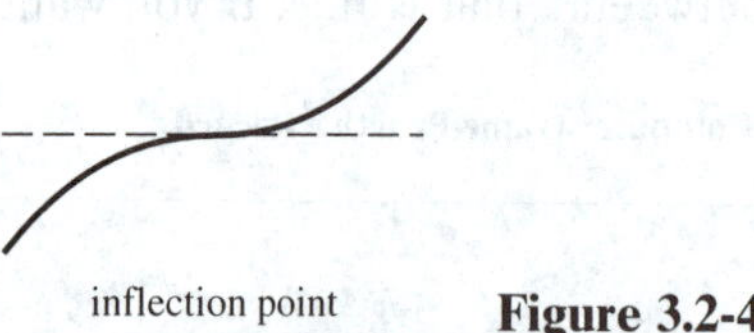

Figure 3.2-4

On the other hand, **all local extrema of a smooth curve will occur at critical points**. Fortunately, for functions of a single variable, we can simply look at the graph to see if a critical point is a local extremum or not, and which kind of extremum it is.

For completeness, we should mention that **if a function is not smooth, local extrema can also occur at points where the graph has a corner or angle, or where there is a hole, jump, or break in the graph. At such points there would be no unique tangent line, so we say the derivative is undefined, and such points are also considered critical points. Thus it is true for general functions as well as for smooth functions that *all* local extrema must occur at critical points.** Much more complicated situations can arise as well, including functions that have no local or global extrema, such as when a domain interval does not include both endpoints. For example $f(x) = x^2$ over all real numbers has no maximum (local or global). It is proved in advanced calculus books that **all smooth functions with domain intervals that include the endpoints *must* have *both* global maxima and minima. This is also true for continuous functions.**

The derivative of the tile function is

$$T'(x) = 1.398x - 52.62$$

From the above discussion, we can find the local minimum by finding out where this derivative is equal to 0. Thus, we set the derivative equal to zero:

$$1.398x - 52.62 = 0$$

Solving this equation we have

$$1.398x = 52.62$$

or

$$x = \frac{15.62}{1.398} \approx 37.64$$

Translation: Your best score will occur when you have been practicing for about 37.64 minutes. As we discussed earlier and can see from the graph, this local minimum is also the global minimum over the domain of the model. **Global extrema can only occur at local extrema or at the endpoints of domain intervals.** If you can graph the entire domain, then it is easy enough to see where the global maximum and minimum points are, at least approximately. If the domain is unbounded, then you need to know the behavior of the graph as the inputs approach positive or negative infinity, which we discussed in Chapter 1.

This value we obtained algebraically is pretty close to the visual estimate we obtained from looking at the graph. Realistically, 37.64 is far too precise for these circumstances, but it would make sense to say that the optimal time seems to be about 38 minutes, perhaps give or take 2 to 3 minutes. (This idea will be discussed in more detail in Section 3.4.) Since each game takes about a minute, you will want to quit practicing after about 37 minutes. Maybe then you will have a good chance of winning the challenge. □

Sample Problem 2: A group that you belong to is trying to raise money by selling sweatshirts, which cost \$9 each, with a fixed order cost of \$25. You did a survey in your dorm to estimate the demand. In Section 1.3 we used the data from the survey to derive a demand function model, given below (rounded off to 3 significant digits):

Verbal Definition: $p = d(q)$ = the price you would have to charge for each sweatshirt, in dollars, in order to sell *exactly* q sweatshirts.

Symbol Definition: $p = d(q) = -0.420q + 26.2$, for $0 \le q \le 45$.

Assumptions: Certainty and divisibility. Certainty implies the relationship is *exact*. Divisibility implies that *any* fractions of sweatshirts and dollars are possible.

Find the optimal number of sweatshirts to order to maximize profit, and the price you should charge for each.

Solution: As we saw in Section 1.3 , the cost function, in dollars, is given by

$$c(q) = 9q + 25$$

and the revenue function, in dollars, is given by

$$r(q) = pq = [d(q)](q) = (-0.420q + 26.2)(q) = -0.420q^2 + 26.2q$$

so the profit function, in dollars, is given by

$$\pi(q) = r(q) - c(q) = (-0.420q^2 + 26.2q) - (9q + 25) = -0.420q^2 + 26.2q - 9q - 25$$
$$= -0.420q^2 + 17.2q - 25$$

Expressed as a model, we have

Verbal Definition: $\pi(q)$ = the profit, in dollars, if you order and sell exactly q sweatshirts and charge $p = d(q)$ dollars per sweatshirt.

Symbol Definition: $\pi(q) = -0.420q^2 + 17.2q - 25$, for $0 \le q \le 45$.

Assumptions: Certainty and divisibility. Certainty implies that the demand function is *exactly* right, and that all sweatshirts are sold at the same price, and so the profit is given *exactly* by the function. Divisibility implies that you can order and sell fractions of sweatshirts, charge any fraction of dollars for the sweatshirts, and get a profit of any fraction of dollars.

(See Section 1.4 to recall how the domain was determined.)

To see the relationship between the cost, revenue, and profit graphically, we have graphed all three on the same graph in Figure 3.2-5. Notice Table 1 after Figure 3.2-5, showing the q values marked on the graph and the values of each of the functions at each q value. For example the narrow vertical lines at $q = 30$ correspond to a revenue of \$408,

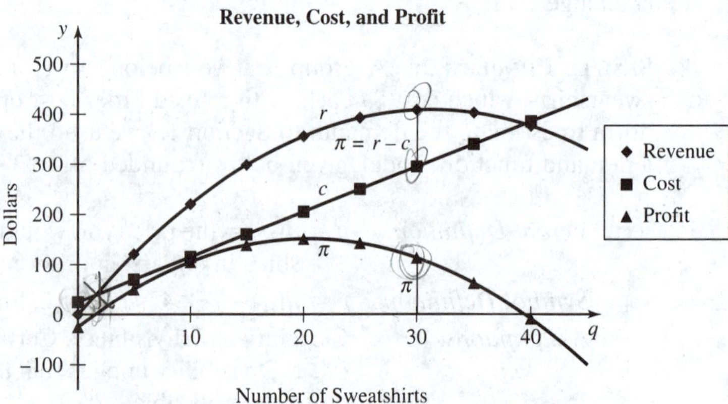

Figure 3.2-5

TABLE 1

Sweatshirts	Revenue	Cost	Profit
0	0	25	−25
5	120.5	70	50.5
10	220	115	105
15	298.5	160	138.5
20	356	205	151
25	392.5	250	142.5
30	408	295	113
35	402.5	340	62.5
40	376	385	−9
45	328.5	430	−101.5

a cost of \$295, and a profit of \$113. Notice that **the height of the profit curve is the same as the vertical distance from the cost curve to the revenue curve**. When $q = 0$, the revenue (\$0) is *below* the cost (\$25), so the profit is negative (−\$25). Recall our discussion of the **break-even point** in Section 1.3: The break-even point is the point where you *first* start to turn a profit, which happens at around $q = 2$ here. **To find the precise break-even point algebraically, simply set the profit function equal to 0 and solve for *q*. This is equivalent to setting the revenue equal to the cost, as we discussed in Section 1.3.** Thus, you are making a profit wherever your revenue curve is *above* your cost curve, in which case the profit is the vertical distance from the cost curve to the revenue curve. This means that our problem of maximizing profit corresponds to finding where this vertical distance from the cost curve to the revenue curve is largest, within the interval of input values for which the revenue curve is above the cost curve.

From our earlier discussion and from looking at Figure 3.2-5, you can see that the profit will be maximized at the point where the tangent line to the profit curve is horizontal, which is where the derivative of the profit (the **marginal profit**) is equal to 0. So let's find the marginal profit function, set it equal to 0, and solve for q:

$$\pi(q) = -0.420q^2 + 17.2q - 25$$

$$\pi'(q) = -0.840q + 17.2$$

$$-0.840q + 17.2 = 0$$

$$-0.840q = -17.2$$

$$q = \frac{-17.2}{-0.840} \approx 20.5$$

So this model is telling us that you can maximize your profit by ordering about 20.5 sweatshirts. At that value, the maximum profit would be

$$\pi(20.5) = -0.420(20.5)^2 + 17.2(20.5) - 25 \approx 151$$

Translation: The maximum profit would be about \$151. The price to charge for the sweatshirts is given by the demand function value at $q = 20.5$:

$$p = d(20.5) = -0.420(20.5) + 26.2 \approx 17.6$$

Translation: The selling price our solution assumes is about \$17.60

At this point, the assumptions of our model become important. The model assumes any fraction of sweatshirts can be ordered and sold, but this does not make sense in real life, and the price is not likely to be *any* value either, since prices tend to run in even dollars, or end in .99 or a multiple of \$0.25. These concerns are related to the concept of **validation**, which is checking a mathematical solution against the real-world problem you want to solve. We will discuss this further in Section 3.4. In this section, we are primarily concerned with mathematical optimization, which basically means just working with the model function directly. □

At this point, let's look at the problem of maximizing profit more generally. Since we know that profit is revenue minus cost, we can write profit as

$$\pi(q) = r(q) - c(q)$$

If we want to maximize this function, we first find the critical points by finding the derivative and setting it equal to 0. We get

$$\pi'(q) = D_q[r(q) - c(q)] = r'(q) - c'(q)$$
$$r'(q) - c'(q) = 0$$

Solving this equation by adding $c'(q)$ to both sides, we get

$$r'(q) = c'(q)$$

Recall that the derivative of the revenue function is called **marginal revenue**, and the derivative of the cost function is called **marginal cost**. What we have derived here is that, for a smooth profit function that is not maximized at an endpoint, **profit is maximized where marginal revenue equals marginal cost.**

Looking back at Figure 3.2-5, this is saying that the profit is maximized at the point where the slope of the revenue function and the slope of the cost function are the same. Notice that, to the left of the optimal profit, the revenue is steeper than the cost, so each additional sweatshirt pulls in *more* revenue than it costs (increasing the profit), and thus you have *undershot* the optimum. Graphically, this means the distance from the cost curve to the revenue curve is still getting *larger* from left to right. To the right of the optimal point, the slope of the revenue curve is shallower than the slope of the cost curve, so each additional sweatshirt brings in *less* than it costs (lowering the profit), suggesting that you have *overshot* the optimum. Graphically, the vertical distance from the cost curve to the revenue curve is now getting *smaller* from left to right. At the optimum itself, there is no mismatch.

Sample Problem 3: In Sample Problem 6 in Section 1.5 we found a model for the temperature over a period of time (rounded to 3 significant figures here) to help a citrus grower decide whether or not to cover her trees when faced with a possible frost:

Verbal Definition: $T(x)$ = the temperature, in degrees Fahrenheit, x hours after 21:00 (9:00 P.M.).

Symbol Definition: $T(x) = 0.0581x^3 - 0.759x^2 + 1.03x + 42.8$, for $0 \le x \le 10$.

Assumptions: Certainty and divisibility. Certainly implies the relationship is *exact.* Divisibility implies that any fractional number of hours or degrees is possible.

The citrus grower is particularly interested in the minimum temperature. Is it likely to drop below freezing before the sun starts to warm things up?

Solution: We now have the tools to answer this question. Let's first graph the function to get an approximate idea of when the minimum temperature will occur. Figure 3.2-6 shows the temperature graph over its domain. We can see from the graph that the minimum seems to occur somewhere around $x = 8$. Let's use the model to figure out the precise time and temperature at the minimum.

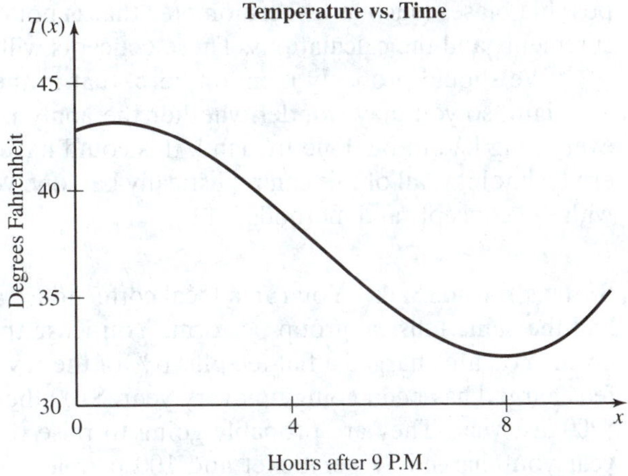

Figure 3.2-6

The minimum is a local minimum, so it will be a critical point where the slope of the tangent to the curve is 0. The derivative of the function gives the slope of the tangent, so we will find the derivative, set it equal to zero, and solve the resulting equation. The derivative is

$$T'(x) = D_x(0.0581x^3 - 0.759x^2 + 1.03x + 42.8) = 0.1743x^2 - 1.518x + 1.03$$

Setting this equal to 0, we get

$$0.1743x^2 - 1.518x + 1.03 = 0$$

Solving this quadratic equation (either using the quadratic formula or using technology[2]), we get the roots

$$x \approx 7.97 \text{ and } x \approx 0.742$$

[2]See your technology supplement for how to use technology to solve an equation involving one variable. The process will usually be called something like "solve" or "Solver," but could also be contained within something called "Optimizer." Some require the equation to be in the form "(expression) = 0" while others allow it to be in a more natural form.

Which is the local minimum, and how low does the temperature go? From the graph, we can see that the local minimum must be the first value, $x \approx 7.97$, at which time the model would predict the temperature to be

$$T(7.97) = 0.0581(7.97)^3 - 0.759(7.97)^2 + 1.03(7.97) + 42.8 \approx 32.2$$

So the lowest temperature will occur a little before 5:00 A.M. (a little less than 8 hours after 9 P.M.) and be just above freezing (32.2 degrees Fahrenheit). So according to the model, the citrus grower would not need to cover her trees. On the other hand, the minimum is *very close* to freezing, so further post-optimality testing would seem highly desirable. This would include validating how well the model tracks reality, checking the assumptions, performing sensitivity analysis (seeing how the results might change if any of the data were questionable and were adjusted), and assessing the margin of error and possible biases (such as a thermometer that is not calibrated perfectly) in the data measurements and the calculations. These concepts will all be discussed in Section 3.4.

We should probably mention here, that in the situation of the citrus grower, time was tight, so you may wonder whether the analysis could really be done in time. When everything had to be done by hand, this could have been a real problem, but with modern technology, all of this can realistically be done very quickly, once someone is familiar with the concepts and methods. □

SAMPLE PROBLEM 4: You run a local coffeehouse and for the past three years you have had the same musical group perform. You lease the place where the performances are given. You are charged a flat fee plus 6% of the revenue from the sale of tickets. The flat fee charged has been going up every year: \$100 the first year, \$150 the second year, and \$200 last year. They are probably going to raise the flat fee to \$250 this year. The first year you charged \$7 per ticket and 100 people came. The second year you raised the price to \$8 per ticket because the fee had gone up by \$50, and 80 people came. Last year you raised the ticket price again this time to \$9 per ticket, but only 70 people came. How much should you charge for tickets to the show at the coffeehouse to realize the most profit? Assume for this problem that the band has donated their performance gratis.

Solution: We discussed the coffeehouse problem in Sample Problem 4 of Section 1.2. The data are given by

Ticket Price (\$)	7	8	9
Paid Attendance	100	80	70

These data points are graphed in Figure 3.2-7.

We want to fit a model to this data, a demand model. We could fit a linear function, but the data seem to have a bit of curvature. We could fit a quadratic model but would have to be careful to restrict our domain, since the function would start to increase at some point, and it is not likely that attendance will go up if the price increases (see the discussion at the end of Section 1.4). An exponential model would seem to make sense, since it would predict that the attendance would gradually approach 0 as the price continued to increase.

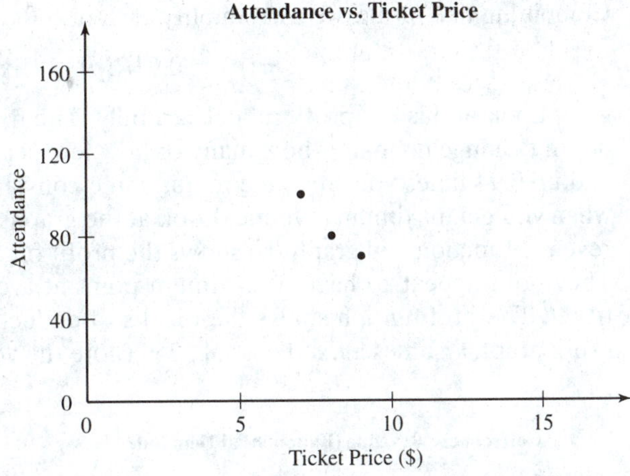

Figure 3.2-7

To fit an exponential function to these data points, most technologies would give the following model[3] (rounded to 3 significant digits):

Verbal Definition: $q = T(p)$ = the number of tickets sold when the ticket price is p dollars.

Symbol Definition: $q = T(p) = 343(0.837)^p$, for $p \geq 0$.

Assumptions: Certainty and divisibility. Certainty implies the relationship is *exact*. Divisibility implies that *any* fractions of dollars and tickets are possible.

From this demand model, we can get a revenue function:

$$R(p) = pq = (p)[T(p)] = (p)[343(0.837)^p] = 343p(0.837)^p, \text{ for } p \geq 0.$$

Your costs consist of the flat charge (fixed cost) plus the percentage of the revenue. The percentage of the revenue does not change; it is 6%. Thus the cost function is given by

$$C(p) = (\text{fixed cost}) + 0.06R(p)$$

Your profit will be equal to the revenue minus cost. In symbols,

$$\pi(p) = R(p) - C(p) = R(p) - [(\text{fixed cost}) + 0.06R(p)]$$
$$= R(p) - (\text{fixed cost}) - 0.06R(p)$$

[3]Most technologies do not do a true least-squares fit of an exponential model but take a logarithm of the data so that the resulting model will be linear (since $\ln(ab^x) = (\ln a) + (\ln b)x$), then transform back using the inverse function (raising e to the power of the linear parameters).

Combining like terms, we can simplify this to

$$\pi(p) = 0.94R(p) - (\text{fixed cost})$$

Look at this last profit model carefully: The fixed cost is just as the name implies, it doesn't change no matter how many or how few tickets are sold. Your profit will actually end up 0.94 times your revenue minus some constant. You will get the maximum profit when you get maximum revenue. Look at the graphs in Figure 3.2-8: Graph (a) shows the revenue function and graph (b) shows the profit function. They have very similar shapes. They both appear to have a maximum point at around 5 or 6, or a price of \$5–\$6 per ticket. The profit graph shows the results of deducting the costs from the revenue. The profit graph assumes a fixed cost of \$250 (note the vertical intercept on the profit graph).

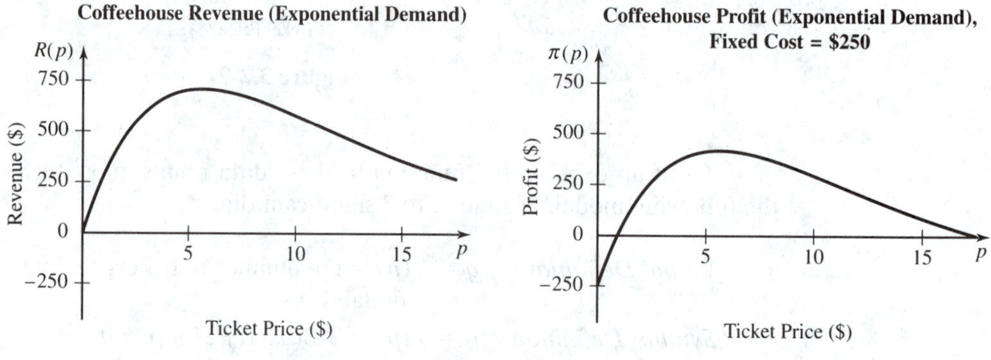

Figure 3.2-8

We will work out the problem using both the revenue and the profit models so that you can see that we get the same results.

For the revenue function, the maximum appears to be around \$5 or \$6 per ticket, as we said earlier. Let's find it algebraically by taking the derivative (recall the Product Rule!):

$$R'(p) = D_p([343p][(0.837)^p]) = (343p)[(\ln 0.837)(0.837)^p] + [(0.837)^p](343)$$

Calculating the natural logarithm of 0.837 (≈ -0.178) and factoring we get

$$R'(p) \approx 343(0.837)^p[-0.178p + 1]$$

Setting this function equal to zero we get

$$[343(0.837)^p][-0.178p + 1] = 0$$

For this equation to be equal to zero, either the first factor, $343(0.837)^p$, must be equal to zero or the second factor, $[-0.178p + 1]$, must be equal to zero, or both factors must be equal to zero. In general, **to solve an equation of the form of a product of factors equaling 0, set each factor equal to zero individually and solve. If $xyz = 0$, then $x = 0$ or $y = 0$ or $z = 0$ (or some combination).** Let's look at the first factor. This is an exponential function of the type $y = ab^x$. We know that such an exponential function is *never* equal to zero (the graph approaches 0 as a limit, but never reaches it). Therefore, since we can see from the graph that there *is* a critical point, the second factor *must* be equal to zero. This is a simple linear function. Let's set it equal to 0 and solve:

$$-0.178p + 1 = 0$$
$$-0.178p = -1$$
$$p = \frac{-1}{-0.178} \approx 5.62$$

Translation: According to this model, the optimal price of a ticket to maximize revenue is about \$5.62. The optimal revenue would then be given by

$$R(5.62) = 343(5.62)(0.837)^{(5.62)} \approx 709$$

Translation: The maximum revenue should be about \$709. To find the number of tickets you would expect to sell, we can plug the optimal p value into the demand function:

$$T(5.62) = 343(0.837)^{(5.62)} \approx 126$$

This means you would expect to sell about 126 tickets.

Now let's analyze the profit function and assume the fixed cost is \$250. If we take the derivative, we get

$$\pi(p) = 0.94R(p) - (\text{fixed cost}) = 0.94R(p) - 250$$

so

$$\pi'(p) = D_p[0.94R(p) - 250] = 0.94R'(p) - 0 = 0.94R'(p)$$

If we set this equal to 0 and divide both sides by 0.94, we get

$$0.94R'(p) = 0$$
$$R'(p) = 0$$

In other words, this is exactly the same condition, $R'(p) = 0$, that we had for maximizing the revenue! This means that our optimal values for p and the number of tickets sold will be exactly the same as before (ticket price of \$5.62, at which you would expect to sell 126 tickets), but the profit will have to be calculated:

$$\pi(5.62) = 0.94R(5.62) - 250 \approx 709 - 250 = 459$$

(since the optimal revenue was about \$709 when $p = 5.62$). In other words your maximum profit should be about \$459. □

How can we understand graphically why the optimal solutions for both the revenue and the profit occurred at the same input value? Recall that the profit function involved multiplying the revenue function by a positive constant (0.94) and adding another constant (the fixed cost, \$250). **Multiplying a function by a positive constant stretches (if the constant is more than 1) or shrinks (if the constant is less than 1) its graph vertically, as if the graph plane were made of rubber and *anchored* along the horizontal axis. Adding a constant to a function shifts its entire graph vertically up (if the constant is positive) or down (if the constant is negative) uniformly.**

So to form your profit function from the revenue function, we first *shrank* the revenue curve slightly (since the constant was 0.94, *less* than 1) in the vertical direction to 94% of its original height everywhere, and then shifted it *down* 250 units (since the constant added was −250, which is *less* than 0). If you think about it, you may realize that **multiplying a function by a positive constant and then adding a constant to the result will not change the location (input value) of any local or global extrema over the same domain.** These operations are like changing units (such as between Fahrenheit and Celsius) or scaling for the vertical axis, so it makes sense that such changes would not affect the optimal input value.

We should also mention that, even for the above problem where revenue and cost were in terms of *price* (not the standard in economics, where they are usually a function of *quantity*), it was still the case that the profit was maximized when the derivative of revenue equaled the derivative of cost, marginal revenue equaled marginal cost, since

$$\begin{aligned}
&\pi(p) = R(p) - C(p), \text{ so}\\
&\pi'(p) = R'(p) - C'(p), \text{ so}\\
&\pi'(p) = 0 \text{ means that}\\
&R'(p) - C'(p) = 0, \text{ and so}\\
&R'(p) = C'(p)
\end{aligned}$$

SAMPLE PROBLEM 5: In Sample Problem 8 in Section 1.2 we formulated a model for the monthly cost of a special cheese:

Verbal Definition: $C(Q)$ = the total cost per month, in dollars per month, if you order Q ounces of cheese every trip.

Symbol Definition: $C(Q) = 16 + \dfrac{160}{Q} + 0.10Q$, for $Q > 0$.

Assumptions: Certainty and divisibility. Certainty implies that no shortages are allowed, that we go to the store as soon as we run out of cheese, that our consumption is continuous and uniform, that there are no discounts or sales and that all of the costs remain exactly the same, that the storage cost is directly proportional to the amount of cheese and the time it is held, and that the cost estimates are exactly right. Divisibility implies that we can buy *any* quantity of cheese and costs can be *any* fractional amount.

How much cheese should we order each trip to minimize the cost?

Solution: Figure 3.2-9 shows a graph of the function. It is difficult to even estimate from this graph exactly how many ounces to order each trip to minimize the cost. Let's find an exact solution for this problem by taking the derivative of the function and finding at what point the derivative is equal to 0.

$$C(Q) = 16 + \frac{160}{Q} + 0.10Q = 16 + 160Q^{-1} + 0.10Q$$

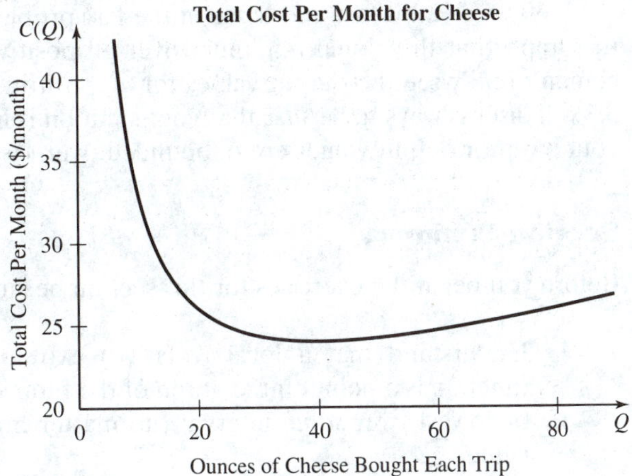

Figure 3.2-9

so

$$C'(Q) = \frac{-160}{Q^2} + 0.1$$

Once again, at the minimum, the tangent will be horizontal, so we set the derivative equal to 0, and we get

$$\frac{-160}{Q^2} = -0.1$$

Multiplying both sides of the equation by $-Q^2$ (keeping in mind that Q cannot be 0), we have

$$160 = 0.1Q^2$$

Dividing both sides of the equation by 0.1 we have

$$1600 = Q^2, \text{ so } Q = \pm 40$$

For this problem, we are interested in the positive solution, so you should order 40 ounces each trip to minimize your monthly cost. We can see from the graph that this is both a local minimum and the global minimum over all positive values of Q. Your minimum monthly cost will thus be

$$C(40) = 16 + \frac{160}{40} + 0.10(40) = 16 + 4 + 4 = 24$$

or \$24 per month.

Another question of interest in this problem is how often you will end up going to the store to buy the cheese. In the original description of the problem, your consumption was given as 32 ounces per month. So if you buy 40 ounces at a time it will last you

$$\frac{40 \text{ ounces}}{32 \text{ ounces/month}} = 1.25 \text{ months}$$

so you will only need to go to the store about every 5 weeks or so. □

We were a bit fast and loose in the last problem. We said that you could see from the graph that the global minimum over all positive Q values was at $Q = 40$. But we couldn't really see the output values for *all* positive values. If you are a skeptic, Section 3.3 will discuss ways to be *sure* that your solution is in fact a global extremum, even when your input or output values are unbounded over the domain.

Section Summary

Before you begin the exercises for this section be sure that you

- Understand that a local (relative) extremum (maximum or minimum) of a function is a point on the graph of the function where the output value is highest or lowest over *some* interval (no matter how small) on either side of its input value.
- Recognize that local maximum and minimum points of a *smooth* function *must* occur where the tangent to the curve is horizontal, and so where the derivative of the function is equal to zero (at critical points).
- Know that to find critical points for a smooth function (points where the tangent to its graph is horizontal), you can set the derivative equal to zero and solve the resulting equation for the input variable (using technology if necessary).
- Know that critical points are *candidates* for local extrema but don't *have* to be local extrema (they could be flat points of inflection, as happens with $f(x) = x^3$ at $x = 0$).
- Know that you can often determine whether a critical point of a smooth function is a local maximum or local minimum (or neither) by looking at its graph, using technology if necessary.
- Recognize that a global (absolute) maximum or minimum point for a function over an interval, *if* one exists, *must* occur at a local extremum *or* at an end point of the interval.
- Know that to find the global maximum or minimum of a smooth function over an interval, you first find the critical points (set the derivative equal to 0 and solve), then look at the graph to determine whether the global extremum, if it exists, occurs at a critical point (local extremum) or at an endpoint of the interval.

EXERCISES FOR SECTION 3.2

Warm Up

Sketch the graphs of the following functions (Exercises 1 and 2) using technology. Then use calculus and algebra ***by hand*** *to find all local and global minimum and maximum values on the given intervals:*

1. $f(x) = x^3 - 12x$ **(a)** $[-3,0]$ **(b)** $[0,3]$ **(c)** $[3,5]$

2. $g(t) = 2t^3 + t^2 - 20t + 4$ **(a)** $[-2,-1]$ **(b)** $[-1,0.5]$ **(c)** $[0.5,3]$

Use calculus and algebra ***by hand*** *to find all local and global maxima and minima of the following functions (Exercises 3–10):*

3. $f(x) = 4x - 1, x \le 7$

4. $g(x) = 2x + 5, x \ge -1$

5. $r(p) = 4p^2 - 2p - 1$

6. $s(t) = 3t - 2t^2$

7. $h(x) = 4x^3 + 6x^2 - 15x + 10$

8. $c(t) = t^3 + 6t^2 - 2t + 6$

9. $f(x) = 0.01x\,e^{-0.1x}$, for $x \ge 0$

10. $f(x) = e^{-0.5(x-3)^2}$

Game Time:

For each of Exercises 11–20, find the derivative by hand to find extrema. If you can solve by hand the equation resulting from setting the derivative equal to 0, do so. If not, you can use technology to solve the equation. If you find your own model, you can round off the parameters in your symbol definition to 3 or 4 significant digits to make the hand calculations more manageable.

11. Exercise 9 in Section 1.5 gave data for unemployment in the United States. One possible model is

Verbal Definition: $U(x)$ = the unemployment, in thousands, in the U.S. civilian population x years after Dec. 31, 1983.

Symbol Definition: $U(x) = -13.72x^3 + 305.7x^2 - 1901x + 10704, 0 \le x \le 11$.

Assumptions: Certainty and divisibility. Certainty implies the relationship is *exact*. Divisibility implies that both of the variables can be *any* fractions.

(a) According to the model, when was unemployment at its highest and lowest levels between 1983 and 1994?

(b) Go back and check the data given for the problem. Do the answers you found in part (a) agree with the actual figures? If not, by approximately how many months are they off?

Another possible model is

Verbal Definition: $U(x)$ = the unemployment (thousands) in the U.S. civilian population x years after Dec. 31, 1983.

Symbol Definition: $U(x) = -3.910x^4 + 72.30x^3 - 286.9x^2 - 586.6x + 10{,}262, x \ge 0$.

Assumptions: Certainty and divisibility. Certainty implies that the relationship is exact. Divisibility implies that both of the variables can be fractions

(c) Use technology to determine, according to the model, when unemployment was at its highest and lowest levels between 1983 and 1994.

(d) Go back and check the data given for the problem. Do the answers you found in part (c) agree with the actual figures? If not, by approximately how many months are they off?

(e) Which model do you think should be used?

12. The following table shows charitable giving in the United States.

Year	Contributions	Year	Contributions
1984	108.83	1990	130.77
1985	109.88	1991	130.94
1986	119.55	1992	128.90
1987	123.13	1993	129.16
1988	127.72	1994	129.88
1989	132.00		

Contributions are in billions of inflation-adjusted dollars.
Source: The World Almanac, 1996, quoting American Assn. of Fund-Raising Counsel, Inc. AAFRC Trust for Philanthropy.

(a) Find a model for these data, after aligning the data.
(b) Using the model, find the local maximum and minimum total contributions for 1984 to 1994 [0,10].
(c) Do the answers you found in (b) agree with the data in the table? If not, how can this be explained?

13. You have determined that you can sell 500 mugs if you charge \$3.50, 350 mugs at \$5, and only 200 mugs at \$6.50. The supplier charges \$2.25 per mug.
(a) Find a demand model for the price as a function of the number of mugs for this situation ($p = d(q)$ = the price you'd have to charge to sell q mugs) and *fully define* it.
(b) Find the revenue and cost functions for this situation; then *fully define* a profit model.
(c) How many mugs should you order and sell to maximize your profit? What is your optimal profit?
(d) How much should you charge for each mug?
(e) On *one* set of axes, graph the revenue, cost, and profit functions, and show how they are related to each other.

14. Consider the mug problem from Exercise 13, but now suppose that the mug distributor offers a price break for every mug *over* 100. The cost of the first 100 is still \$2.25 each, but if you order more than 100, the cost of the *additional* mugs is only \$2 each.
(a) Formulate the cost function for the domain $q \geq 100$.
(b) Find the new profit function over the same domain, $q \geq 100$.
(c) How many mugs should you order and sell to maximize your profit over the domain $q \geq 100$? What is your optimal profit?
(d) How much should you charge for the mugs?
(e) What would be the optimal solution over the full domain, $q \geq 0$? (*Hint:* Use your answer to Exercise 13 to find the optimal solution over the interval [0,100).)

15. A model has been developed for the profit from the sale of sodas as a function of the price charged per soda:

Verbal Definition: $\pi(p)$ = profit in dollars from the sale of sodas at p dollars per soda.
Symbol Definition: $\pi(p) = (p - 0.38)(1.0924)(0.3178)^p$, $0.5 \leq p \leq 5$.
Assumptions: Certainty and divisibility. Certainty implies the relationship is *exact*. Divisibility implies that *any* fractions of sodas or dollars in profit are possible.

(a) Determine the price per soda that will maximize profit.
(b) In reality, do you think this mathematical solution is a good price, or should the price be rounded? Explain your answer.

16. Previously, we developed a model for the profit from the sale of tickets to a local sporting event as a function of the price charged per ticket:

Verbal Definition: $\pi(p)$ = profit in dollars from the sale of tickets at p dollars per ticket.
Symbol Definition: $\pi(p) = 3165.6(p - 1)(0.94919)^p - 1000,\ p \geq 0.$
Assumptions: Certainty and divisibility. Certainty implies the relationship is *exact.* Divisibility implies that *any* fractions of tickets or dollars in profit are possible.

(a) Determine the price per ticket that will maximize profit.
(b) In reality, do you think this mathematical solution is a good price, or should the price be rounded? Explain your answer.

17. In Section 1.0 there were three different models given for Meg's exercise problem. Each of the models was used to identify an optimum time for Meg to exercise each week. Verify the optimum figures given for each model. (In the exercises for Section 1.4 you were asked to verify these models.)

18. In Section 1.5 we fit a model for the crime index in the United States:

Verbal Definition: $C(x)$ = crimes reported in the preceding year in the United States x years after Dec. 31, 1981.
Symbol Definition: $C(x) = -13{,}150x^3 + 247{,}560x^2 - 1{,}030{,}143x + 13{,}471{,}438,\ x \geq 0.$
Assumptions: Certainty and divisibility. Certainty implies the relationship is *exact.* Divisibility implies that *any* fractions of years or reported crimes are possible.

(a) From the model, determine at what time the crime index in the United States was the lowest between Dec. 31, 1981 and Dec. 31, 1993, and what the crime index was then.
(b) From the model, determine at what time the crime index in the United States was the highest between Dec. 31, 1981 and Dec. 31, 1993 and what the crime index was then.
(c) A friend of yours was elected to a two-year term in November 1990. If you were running a reelection campaign for this friend, would you use these data? Why or why not?
(d) A friend of yours was elected to a two-year term in November 1992. If you were running a reelection campaign for this friend, would you use these data? Why or why not?
(e) If your friend's political party had been in power since the elections in November 1988, would you use these data? Why or why not?

19. The following table shows the energy level throughout the day (measured on a scale of 0 for absolutely no energy to 100 for totally full of energy, 1:00 A.M. = 1, 2 A.M. = 2, and so on, 12 midnight = 24):

Time	Energy Level	Time	Energy Level
8	50	17	45
9	55	18	50
10	60	19	55
11	70	20	60
12	80	21	55
13	75	22	50
14	60	23	45
15	55	24	40
16	50		

(a) Fit a model to the energy level throughout the day. Is the model a good fit?
(b) Use the model and technology to determine at what time during the period the energy level was the highest and what that energy level was.
(c) Use the model and technology to determine at what time during the period the energy level was the lowest and what that energy level was.
(d) Use the model and technology to determine at what time during the period the energy level reached a relative (local) high and what that energy level was.
(e) Use the model and technology to determine at what time during the period the energy level reached a relative (local) low and what that energy level was.

20. We fit a model to the annual percentage change in the medical CPI:

Verbal Definition: $I(x)$ = annual percentage change in the medical CPI x years after Dec. 31, 1986.
Symbol Definition: $I(x) = 0.0193x^4 - 0.3547x^3 + 1.9314x^2 - 3.0161x + 7.6229,\ x \geq 0.$
Assumptions: Certainty and divisibility.

(a) Use the model to determine at what time during the period Dec. 31, 1986, to Dec. 31, 1994, the annual percentage change in the medical CPI was the highest and how high it was.
(b) Use the model to determine at what time during the period Dec. 31, 1986, to Dec. 31, 1994, the annual percentage change in the medical CPI was the lowest and how low it was.
(c) Use the model to determine at what time, if any, during the period the annual percentage change in the medical CPI reached a relative high and how high it was.
(d) Use the model to determine at what time, if any, during the period the annual percentage change in the medical CPI reached a relative low and how low it was.
(e) Name at least two ways this information could be used.

3.3 TESTING CRITICAL POINTS, CONCAVITY, AND POINTS OF INFLECTION

In the preceding section we learned how to find local and global maximum and minimum points algebraically. A smooth function has a local maximum or minimum point where the slope changes from positive to negative or negative to positive. At the exact point where this occurs, the tangent line is horizontal, and so its slope (the derivative) is 0. Points where the derivative is zero are called critical points, but not all critical points are local extrema. To find such critical points, we found the derivative of the function and found where it was equal to zero by setting it equal to 0 and solving. This gave us the input value(s) for the critical point(s). Since critical points do not have to be local extrema, we looked at the graph of the function to determine whether critical points were local maxima, local minima, or neither. But sometimes we cannot graph a function over its entire domain, especially if the input or output values are unbounded (go off to infinity). Is there another way to determine if a critical point is a local maximum or local minimum? Besides maximum and minimum points, we have also discussed another important type of point: the inflection point, where a function changes concavity. Is there a way to determine exactly where such points are located? What determines concavity algebraically? In this section we will learn how to identify critical points as maximum or minimum points or neither, how to find the concavity of a graph in different places, and how to identify inflection points. We will see how all of these operations are related to the concept of the derivative of the derivative of a function, called the **second derivative**, which can be thought of as the rate of change of the slope of the graph.

Here are some types of problems that the material in this section will help you solve:

- Timing is really critical. You want to use the crime index in the United States for a political campaign platform. When was it highest? When the lowest? When did the situation *really* start to improve?
- A company's profits have been declining. You were hired to "turn the company around." One year after you were hired the profits are still declining. You have to go before the board of directors and put a "spin" on the data to justify your policies and your very large salary. Can you possibly do this?
- You work for an advertising firm and have been handling the account for a new product. You launched the product with a really great ad campaign. When should you approach the company to start a second-wave ad campaign for the product?

When you have finished this section you should be able to solve these and similar problems and you should

- Understand the idea behind the second derivative of a function and know how to find it.
- Understand what the second derivative corresponds to graphically.
- Know how to use the second derivative to determine the concavity of a graph in different places.
- Know how to use the second derivative to determine if a critical point is a maximum or minimum.
- Know how to use the second derivative to find inflection points precisely.
- Know how to draw the concavity graph (graph of the second derivative) for a function, and how it relates to the original graph and the slope graph.

Concavity and the Second Derivative

Sample Problem 1: In Sample Problem 5 of Section 3.2, we wanted to find the amount of a special kind of cheese to buy each time you go to the cheese store in order to minimize the total cost per month (balancing the fixed cost of going to the cheese store each time versus the costs of storing the cheese). The total cost per month is given by the model:

Verbal Definition: $C(Q)$ = the total cost per month, in dollars per month, if you buy Q ounces of cheese every trip

Symbol Definition: $C(Q) = 16 + \dfrac{160}{Q} + 0.10Q$, for $Q > 0$.

Assumptions: Certainty and divisibility. Certainty implies that no shortages are allowed, that we go to the store as soon as we run out of cheese, that our consumption is continuous and uniform, that there are no discounts or sales and that all of the costs remain exactly the same, that the storage cost is directly proportional to the amount of cheese and the time it is held, and that the cost estimates are exactly right. Divisibility implies that we can buy *any* quantity of cheese and costs can be *any* fractional amount.

In Section 3.2 , we found that the only critical point in the domain was at $Q = 40$, and from the graph this seemed to be the global minimum, but we could not graph the entire function. Figure 3.3-1 shows the graph we used, which seemed to give strong evidence that the critical point is the global minimum. Use calculus and algebra to prove that, in fact, the critical point is the global minimum.

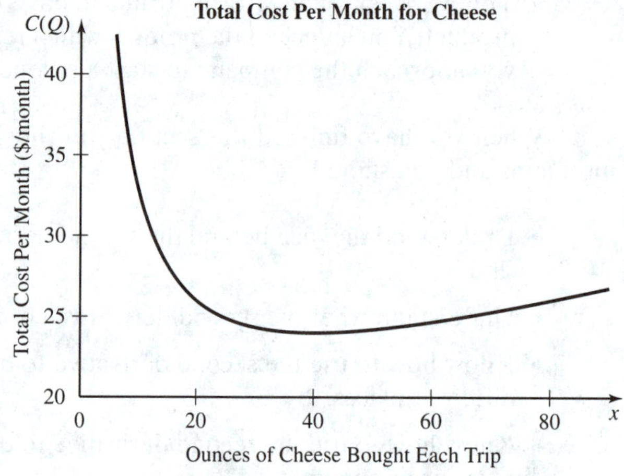

Figure 3.3-1

Solution: It appears from the graph that the function is concave up over its domain. What exactly does concavity *mean*, mathematically speaking? One way to understand this is to draw generic concave up and concave down graphs and draw in tangent lines at a few strategic points, as shown in Figure 3.3-2.

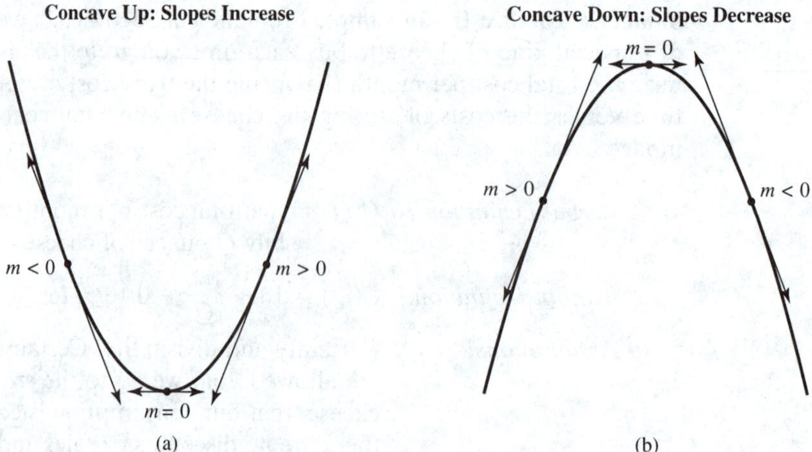

Figure 3.3-2

Figure 3.3-2(a) shows a curve that is concave up over the entire domain shown. Notice that the tangent line on the left has a *negative* slope, the middle one has a slope of *zero*, and the right one has a *positive* slope. Proceeding from left to right, then, the slopes are *increasing*. Notice that if you drew in more tangent lines, this would hold for them as well. For instance, if you drew a tangent to the right of the rightmost tangent in graph (a), it would be steeper than the rightmost one, so its slope would be a larger positive number.

It is true in general that **a curve being concave up over an interval corresponds to its slope *increasing* (from left to right) over that interval. This is the same as saying that the *rate of change* of the slopes is *positive***, since when something (the slope in this case) is increasing, its rate of change is positive. Now we're "speaking calculus!" We know that a *rate of change* corresponds to a derivative, and that *slope* also corresponds to a derivative. This means that "the *rate of change* of the *slope*" of the function can be thought of as "the *derivative* of the *derivative*" of the function. We already know how to take the derivative of a function, and that the result is itself a function. So we can go ahead and take the derivative *of* the derivative. In our example the derivative of the cost function was

$$C'(Q) = D_Q\left(16 + \frac{160}{Q} + 0.1Q\right) = D_Q(16 + 160Q^{-1} + 0.1Q) = -160Q^{-2} + 0.1$$

If we now take the derivative of this, we get

$$D_Q(-160Q^{-2} + 0.1) = 320Q^{-3} = \frac{320}{Q^3}$$

What do we call this? The derivative of the derivative of a function is called the **second derivative** of the function, and the simplest notation for it is to use a *double* prime:

$$C''(Q) = D_Q[C'(Q)] = \frac{320}{Q^3}$$

If we plug in $Q = 40$ to determine the sign of the second derivative at $Q = 40$, we get

$$C''(40) = \frac{320}{40^3} > 0$$

so the rate of change of the slope is positive, as we discussed. In fact, the second derivative will be positive for *any* value in the domain (any positive value of Q), since Q^3 will also be positive, so the graph is concave up over the entire domain. We have shown that **a curve being concave *up* over an interval corresponds to the second derivative being *positive* over that interval**. From a memory point of view, it is easy to remember that *up* and *positive* go together. If you forget which way it goes, just draw a sketch like Figure 3.3-2 to figure out the sign of the rate of change of the slopes from left to right.

It is proved in advanced calculus books that **if a smooth curve is concave up[4] over its entire domain, then there can be only one critical point in that domain, and that critical point *must* be the only local minimum, and *must* be the unique global minimum over the domain**. Intuitively, to have another point be a local or global minimum, the curve

[4]When we say "concave up" here, we technically mean "*strictly* concave up," which simply means that the curve is never flat or linear anywhere, so being *strictly* concave up corresponds to the second derivative being *strictly* positive (not 0).

would have to turn back down, which would require a change in concavity or a corner or cusp or a break in the graph, none of which are possible for smooth functions. This gives us the mathematical proof we wanted that in fact our critical point *is the* (unique) global minimum for our problem. □

You have probably figured out by now from looking at Figure 3.3-2(b), that **a curve being concave *down* over an interval means that the slopes of the tangent lines are *decreasing* (from left to right), so the rate of change of the slope is *negative*, so it corresponds to the second derivative being *negative*.** In Figure 3.3-2(b) notice that, from left to right, the slopes are *positive*, then *zero*, then *negative*, so they are in fact *decreasing* over the domain. Again, from a memory perspective, it makes sense that *down* and *negative* go together.

Analogous to the principle above, **if a smooth curve is concave down[5] over its entire domain, then there can only be at most one critical point in that domain, and that critical point *must* be the only local maximum, and *must* be the unique global maximum over the domain.**

When we want to clearly distinguish between the original derivative of a function, $f'(x)$, and the second derivative, $f''(x)$, we can call the original derivative the ***first* derivative.**

The Second Derivative Test for Local Extrema

Sample Problem 2: In Sample Problem 3 of Section 3.2, we were trying to help a citrus grower decide whether or not to cover her trees in the face of a possible frost. The model that we had found in Section 1.4 was

Verbal Definition: $T(x)$ = the temperature, in degrees Fahrenheit, x hours after 21:00 (9:00 P.M.).

Symbol Definition: $T(x) = 0.0581x^3 - 0.759x^2 + 1.03x + 42.8$, for $0 \leq x \leq 10$.

Assumptions: Certainty and divisibility. Certainly implies the relationship is *exact*. Divisibility implies that *any* fractional number of hours or degrees is possible.

We used calculus and algebra (set the derivative equal to 0 and solved) to find the critical points

$$x \approx 7.97 \text{ and } x \approx 0.742$$

Is there a way to use calculus and algebra to determine whether these critical points are local maxima or local minima *without* graphing the function?

Solution: We know that at the critical points of a smooth function the slope of the tangent is 0, so the curve is instantaneously horizontal. What if we knew that the curve was

[5] When we say "concave down" here, we technically mean "*strictly* concave down," which simply means that the curve is never flat or linear anywhere, so being *strictly* concave down corresponds to the second derivative being *strictly* negative (not 0).

also concave *up* or concave *down* at that point? Recall our sketches of generic local extrema for smooth curves, shown again here in Figure 3.3-3.

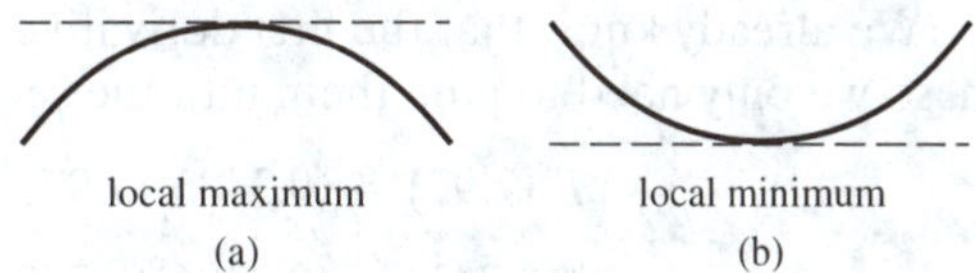

Figure 3.3-3

From these sketches, we can see that **a point where the tangent is horizontal (derivative is 0) and where the curve is concave *down* (second derivative is *negative*) must be a local maximum, and a point where the tangent is horizontal (derivative is 0) and where the curve is concave *up* (second derivative is *positive*) must be a local minimum.**

This suggests a way of testing critical points. ***The Second Derivative Test* for testing critical points of a smooth function $f(x)$ to see if they are local extrema** involves the following concepts:

> **If $f'(x) = 0$ (x is a critical point) and if $f''(x) > 0$ (f is concave *up* at x), then there is a local *minimum* at x.**
>
> **If $f'(x) = 0$ (x is a critical point) and if $f''(x) < 0$ (f is concave *down* at x), then there is a local *maximum* at x.**
>
> **If $f'(x) = 0$ (x is a critical point) and if $f''(x) = 0$ (f is flat at x), then the test is *inconclusive* (you can't be sure if it's an extremum).**

The above concepts lead naturally to a procedure for testing critical points when solving a real-world problem. Once you have set the derivative equal to 0 and solved for x to find critical points, suppose there is a critical point at $x = k$. Then, apply the **Second Derivative Test to test a critical point at $x = k$:**

1. **Find the general second derivative, $f''(x)$,**
2. **Plug the input value of the critical point, k, into the second derivative to find $f''(k)$, and**
3. **Use the criteria:**
   a. **If $f''(k) > 0$, then the critical point is a local minimum.**
   b. **If $f''(k) < 0$, then the critical point is a local maximum.**
   c. **Otherwise, the test is inconclusive.**

We will see examples later which show that in the third case (when $f'(k) = 0$ and when $f''(k)$ is *neither* positive not negative, such as when $f''(k) = 0$), the critical point *could* be a local extremum (such as for $f(x) = x^4$ at 0, which is a local minimum), but it *might not* be (such as for $f(x) = x^3$ at 0, which is a point of inflection but not a local extremum).

Let's apply the Second Derivative Test to our temperature problem. The first derivative is given by

$$T'(x) = D_x(0.0581x^3 - 0.759x^2 + 1.03x + 42.8) = 0.1743x^2 - 1.518x + 1.03$$

so the second derivative is

$$T''(x) = D_x[T'(x)] = D_x(0.1743x^2 - 1.518x + 1.03) = 0.3486x - 1.518$$

We already know that the first derivative is 0 at the critical points given in the question, so we only need to plug them into the second derivative:

$$T''(7.97) = (0.3486)(7.97) - 1.518 \approx 1.26 > 0$$
$$T''(0.742) = (0.3486)(0.742) - 1.518 \approx -1.26 < 0$$

This tells us that the second derivative is *positive* at $x = 7.97$, so the graph is concave *up*, so there must be a local *minimum* there. Similarly, at $x = 0.742$, the second derivative is *negative*, so the curve is concave *down* there, so there must be a local *maximum* there. Let's take another look at the graph, shown in Figure 3.3-4, to confirm this conclusion. The graph does indeed verify our test. There is a local minimum at $x \approx 7.97$ and a local maximum at $x \approx 0.742$. □

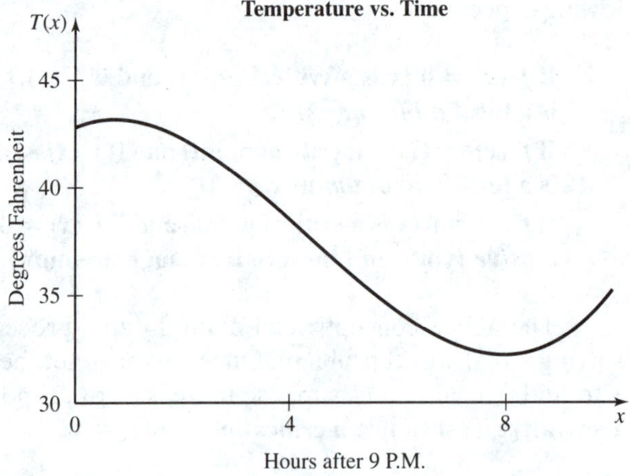

Figure 3.3-4

In some ways, the signs for the second derivative in the Second Derivative Test might seem the opposite of what you would expect. The best way to remember them is to think through the logical sequence of how the *sign* of the second derivative determines the *concavity* (in the direction you *would* expect, or by thinking of Figure 3.3-2); then picture *that* concavity (as in Figure 3.3-3) to determine what kind of extremum you get.

Using Calculus and Algebra to Find Points of Inflection Precisely

Recall that in Section 1.5 we introduced the idea of a point of inflection as a point on a curve where the concavity changes between concave up and concave down. Then in Section 3.1 when we discussed slope graphs, we noticed that points of inflection corresponded to points where the slope of the curve was locally steepest or shallowest. Let's try to integrate these two perspectives by looking at a sample problem.

Sample Problem 3: You are running a reelection campaign. Your candidate campaigned on an anti-crime platform in 1985 and first took office in 1986. Could you use the crime index in the United States for the current reelection campaign to show that your candidate had some degree of success? The model for crimes reported in the United States, the Crime Index, is

Verbal Definition: $C(x)$ = the crime index in the United States x years after Dec. 31, 1980 (in the *preceding* year).

Symbol Definition: $C(x) = -13150x^3 + 247506x^2 - 1030143x + 13471438,$ $0 \le x \le 12.$

Assumptions: Certainty and divisibility. Certainty implies that the relationship is exact. Divisibility implies that fractions of crimes and fractions of years are possible.

When was the crime index the lowest? When was it the highest? When did the situation start to improve?

Solution: We already know how to find the local and global extrema for this problem. First let's look at the graph (Figure 3.3-5).

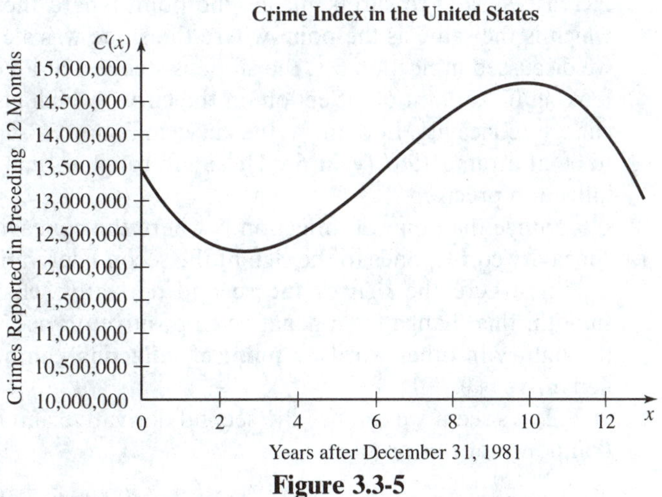

Figure 3.3-5

From the graph we can estimate that crime was lowest around 1984 (year 3), and highest around 1991 (year 10). To determine exactly when the crime index was the highest and when it was the lowest, we must find the first derivative:

$$C'(x) = -39450x^2 + 495012x - 1030143$$

Setting this equal to zero, we get

$$-39450x^2 + 495012x - 1030143 = 0$$

Using technology or the quadratic formula to help us solve this equation we find that the derivative is equal to zero when $x \approx 2.68$ and when $x \approx 9.91$. We can see from the graph which of these critical points gives a local maximum and which gives a local minimum, but let's verify this by taking the second derivative to test our critical points:

$$C''(x) = -78900x + 495012$$

Plugging our first critical point, $x = 2.68$, in the second derivative we have

$$C''(2.68) = (-78900)(2.68) + 495012 = 283560 > 0$$

This is a positive number. Since the second derivative is positive at this point, the curve is concave up and the point at $x \approx 2.68$ must be a local minimum point. Plugging our second critical point, $x = 9.91$, in the second derivative, we get

$$C''(9.91) = (-78900)(9.91) + 495012 = -286887 < 0$$

This is a negative number. The second derivative is negative at this point, so the curve is concave down, and the point at $x \approx 9.91$ must be a local maximum point.

Since the minimum crime index occurred before your candidate took office and the maximum five years later, this does not look too promising as a campaign platform. Maybe you should keep on looking. The question asks when the situation started to improve. This could be interpreted as being the point when the rate of increase in the crime index started to shrink. What does this mean? Well, if something is just starting to get smaller, then it must have just reached its peak. So the point where the rate of increase started to shrink means the point where the rate of increase was at its peak, which is the same as the point where the slope was steepest (in a positive direction). As we discussed in Section 3.1, the slope is steepest at a point of inflection, so we are really looking for a point of inflection on the curve, which also means a point where the curve changes concavity. Looking at the curve in Figure 3.3-5, the only point of inflection seems to be at around 1987 (year 6). These are rough estimates. How can we find the point of inflection precisely?

Since the point of inflection is where the concavity of the curve changes and since concavity corresponds to the sign of the second derivative, the point of inflection must be a point where the *sign* of the second derivative changes. If the second derivative is smooth, this change from negative to positive or vice versa could only occur by passing through 0. In other words, **a point of inflection can occur at a point where the second derivative is 0**.

Let's see if we can use the second derivative and this observation to find the inflection point for this problem:

$$C''(x) = -78900x + 495012$$

Setting this equal to zero we get

$$-78900x + 495012 = 0$$

Solving this equation we get

$$x = \frac{-495012}{-78900} \approx 6.27$$

We can see from Figure 3.3-5 that this is in fact an inflection point. We can also check that the second derivative is positive to the left of 6.27 and negative to the right of it, verifying that this is a point of inflection.

Will *all* points where the second derivative is 0 be points of inflection? Let's look at an example: $y = f(x) = x^4$. The first derivative of this function is $f'(x) = 4x^3$ and the second derivative is $f''(x) = 12x^2$. The second derivative will be equal to 0 when $x = 0$. Can we then assume that the function has an inflection point at $x = 0$? Look at the graph of this function (Figure 3.3-6).

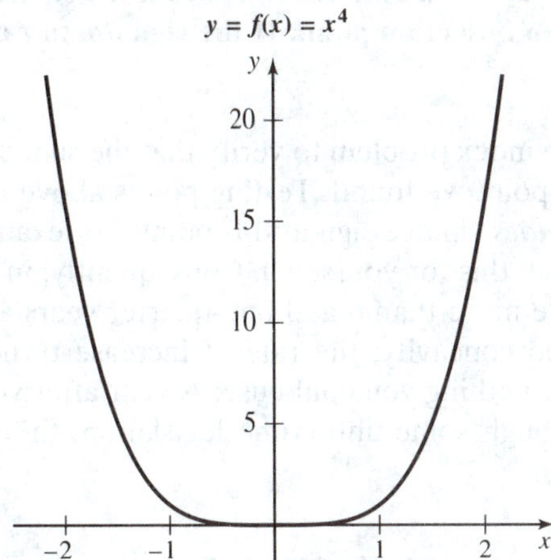

Figure 3.3-6

This function does *not* have an inflection point at $x = 0$. We will test points to the left and right of $x = 0$. For $x = -1$,

$$f''(-1) = 12(-1)^2 = 12$$

For $x = 1$,

$$f''(1) = 12(1)^2 = 12$$

Since the second derivative is $f''(x) = 12x^2$, it will be nonnegative (positive or 0) everywhere. The second derivative is equal to 0 at $x = 0$, but it does *not* change signs at this point. Looking at the graph of the function we can see that it is concave up everywhere (except at $x = 0$). It never *changes* concavity. There is no inflection point. Intuitively, the unusual case here is related to the fact that the graph flattens out so much around $x = 0$.

Notice that this function satisfies the third case of the conditions of the Second Derivative Test at $x = 0$: $f'(x) = 0$ and $f''(x) = 0$, and the critical point there is a local minimum.

The conclusion from this discussion is that, just as a point where the first derivative is 0 is not *necessarily* a local extremum, **a point where the second derivative is 0 is not *necessarily* a point of inflection**. In most cases, you can determine whether or not such a point is a point of inflection by looking at the graph. If necessary, you can check the sign of the second derivative on either side, as we did above: If the sign changes, it is a point of inflection; if the sign doesn't change it is *not* a point of inflection.

To find points of inflection of the smooth function $y = f(x)$:[6]

1. **Find $f'(x)$.**
2. **Find $f''(x)$.**
3. **Set $f''(x) = 0$ and solve for x.**
4. **To determine whether or not each answer is an inflection point, look at the graph or test the second derivative on either side to see if the sign changes. If the sign *changes*, the point *is* an inflection point. If the sign *doesn't* change, it is *not* an inflection point.**

Let's go back to the crime index problem to verify that the sign of the second derivative changes at the inflection point we found. Testing points above and below 6.27, we see that the second derivative *does* change sign at this point. For example, $C''(6)$ is positive and $C''(7)$ is negative (check this for yourself!). Consequently, in approximately the beginning of April 1987 (a little more than 6 and one-quarter years after December 31, 1980), the crime index changed concavity; the rate of increase turned the corner and started to shrink. Now *this* is something you could use: A year after your candidate took office and started pushing through some anti-crime legislation, the *growth rate* of the index started to turn back down. □

Connections between Graphs, Their Slope Graphs, and Their Concavity Graphs

From the earlier discussion, if a function is concave *up* over an interval, then the slope is *increasing*. This means that the slope *graph* should be increasing over that interval as well. If the function is concave *down* over an interval, then the slope, and so the slope *graph*, is *decreasing* over the interval. Let's take a closer *graphical* look at two simple functions and their first and second derivatives. Figure 3.3-7 shows the original functions we will study:

$$f(x) = x^2 \text{ and } g(x) = -x^2$$

The graphs of the *derivatives* of the original functions (the *slope graphs* of the original Figure 3.3-7 graphs) are shown in Figure 3.3-8.

In general, we have several different symbols to denote the second derivative, which correspond to the symbols we used for the first derivative:

First derivative:	y'	$f'(x)$	$\dfrac{dy}{dx}$
Second derivative:	y''	$f''(x)$	$\dfrac{d^2y}{dx^2}$

[6]The only exception to this procedure would be a graph that had a point of inflection where there was a unique *vertical* tangent line (as at the middle of a sideways "S"). At such a point, the slope of the tangent would be undefined, so the derivative would be undefined. The standard technical definition of a smooth function is one where the derivative is continuous over the domain, which would exclude such an exception. In a real-world context, this would imply an *infinite* rate of change, which will usually not make sense. Of the models we have discussed, this could only happen for a power function where the domain included negative numbers, which we did not consider.

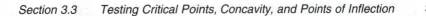

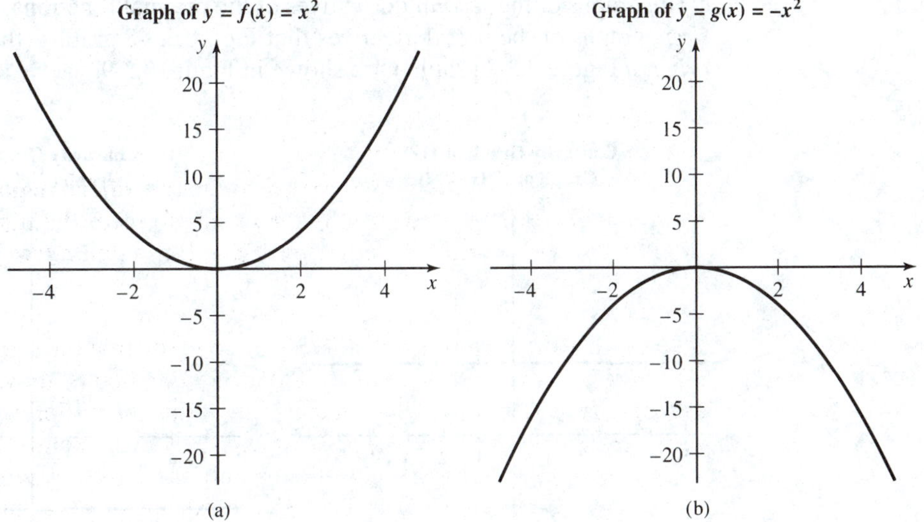

Figure 3.3-7

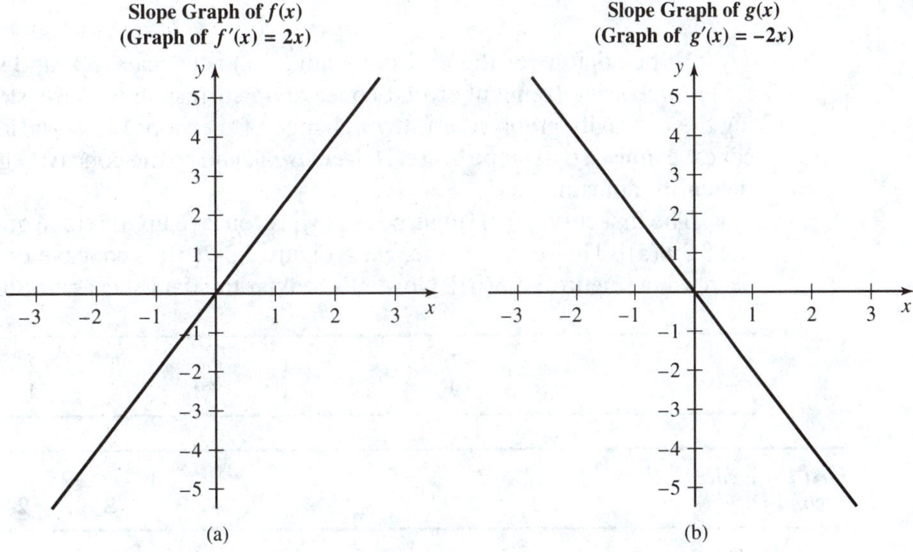

Figure 3.3-8

Remember that **the sign of the second derivative of a function determines the concavity of the original function**. For this reason, **we will call the *graph* of the second derivative the *concavity* graph.**

The graphs of the *second* derivatives of the original functions, which are themselves the *slope* graphs of the *first* derivatives that Figure 3.3-8 graphs (the *concavity* graphs of the *original* Figure 3.3-7 graphs) are shown in Figure 3.3-9.

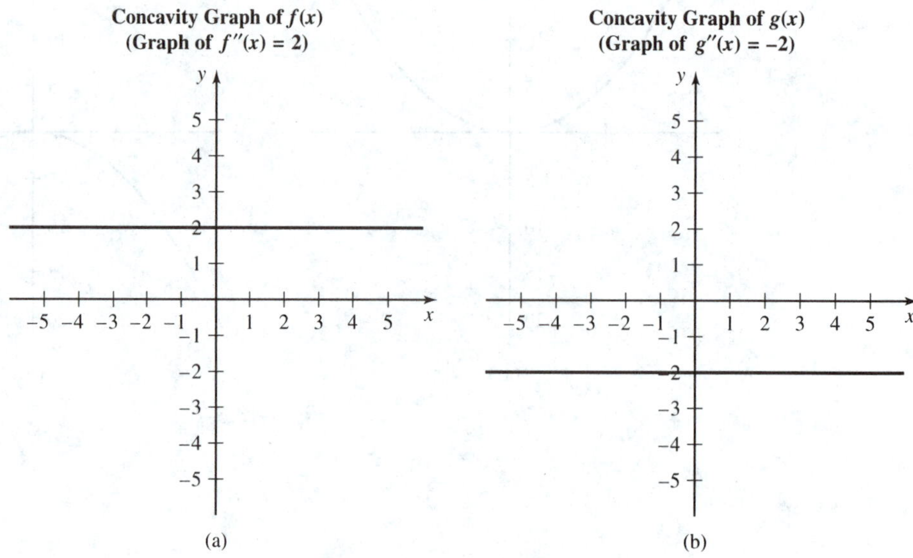

Figure 3.3-9

Notice that, over the entire domain, $f(x)$ is concave *up*, and so the slope graph of $f(x)$ is *increasing* (remember that concave up corresponds to the slope increasing), causing the concavity graph (the rate of change of the slope) to be *positive*. Similarly, $g(x)$ is concave *down*, so its slope graph is *decreasing*, and so the concavity graph is *negative*, over the entire domain.

The first curve, $f(x)$ [Figure 3.3-7(a)] is concave up. Its slope graph is increasing [Figure 3.3-8(a)]. The second curve, $g(x)$ [Figure 3.3-7(b)] is concave down. Its slope graph is decreasing [Figure 3.3-8(b)]. Now let's look at the data sets associated with these graphs.

x	−5	−4	−3	−2	−1	0	1	2	3	4	5
$f(x)$	25	16	9	4	1	0	1	4	9	16	25

First Differences	−9		−7		−5		−3		−1		1		3		5		7		9
Second Differences		2		2		2		2		2		2		2		2		2	

The first differences correspond to slopes (of secant lines), and the second differences correspond to (average) rates of change of these slopes. So the second differences being positive corresponds to the rate of change of the slope being positive (the slopes are increasing), which corresponds to the graph being concave up, as we have already discussed. Note that the first differences are increasing in a linear way, corresponding to the slope graph being linear and increasing. The second differences are positive and constant, just like the concavity graph.

x	-5	-4	-3	-2	-1	0	1	2	3	4	5
$g(x)$	-25	-16	-9	-4	-1	0	-1	-4	-9	-16	-25

First Differences	9		7		5		3		1		−1		−3		−5		−7		−9
Second Differences		−2		−2		−2		−2		−2		−2		−2		−2		−2	

This time the first differences are decreasing in a linear way, so the slopes are decreasing, so the rate of change of the slope is negative. This is reflected in the fact that the second differences are all negative and constant. As we saw before, decreasing slopes correspond to the function being concave down. The slope graph is decreasing and linear, like the first differences. The concavity graph is negative and constant, like the second differences.

In general, then, **positive second differences correspond to a graph that is concave up, and negative second differences correspond to a graph that is concave down.**

Now, let's consider the graphs for the functions

$$h(x) = x^3/3 \quad \text{and} \quad k(x) = -x^3/3$$

shown in Figure 3.3-10.

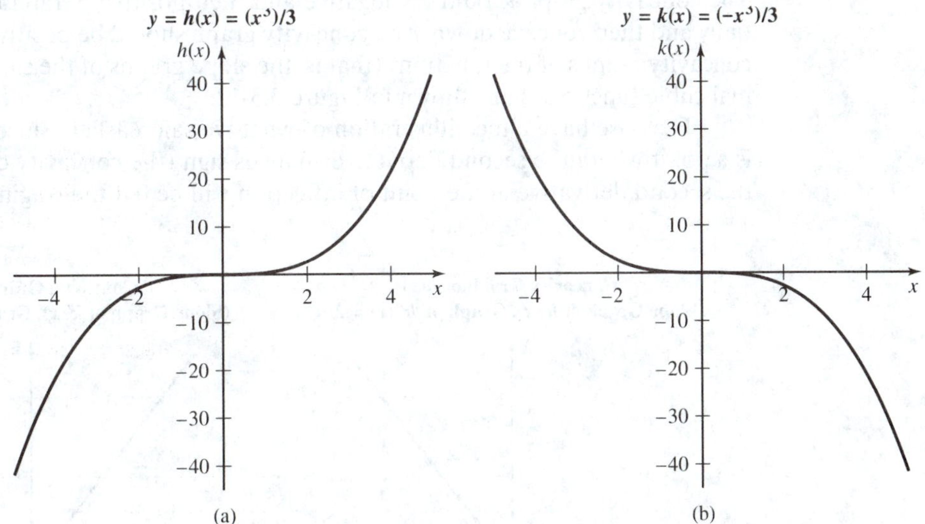

Figure 3.3-10

Let's discuss the characteristics of the functions in Figure 3.3-10. The first function, graph (a), is always increasing from left to right, except that it is instantaneously flat at $x = 0$. The slope graph should be positive everywhere except at $x = 0$, where it should be 0. The second function, (b), is always decreasing from left to right, except at 0. The slope graph should be negative everywhere, except for being 0 at 0. The slope graphs of these functions are shown in Figure 3.3-11. Just as we expected! Let's look at the concavity of the functions in Figure 3.3-10. Graph (a) is concave down initially, and then it is concave up.

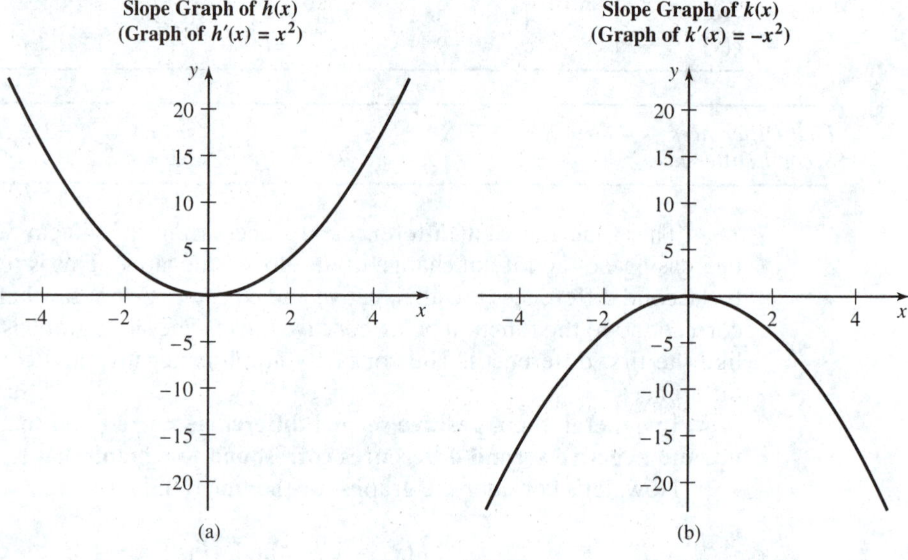

Figure 3.3-11

The concavity graph should be negative and then positive. Graph (b) is concave up initially and then concave down. The concavity graph should be positive, then negative. The concavity graphs of the functions (that is, the slope graphs of the slope graphs of the original cubic functions) are shown in Figure 3.3-12.

Here we have a nice illustration of what we said earlier: Since a point of inflection is a point where the second derivative changes sign (the concavity changes), the value of the second derivative at the point of inflection will be 0 if the original function is smooth

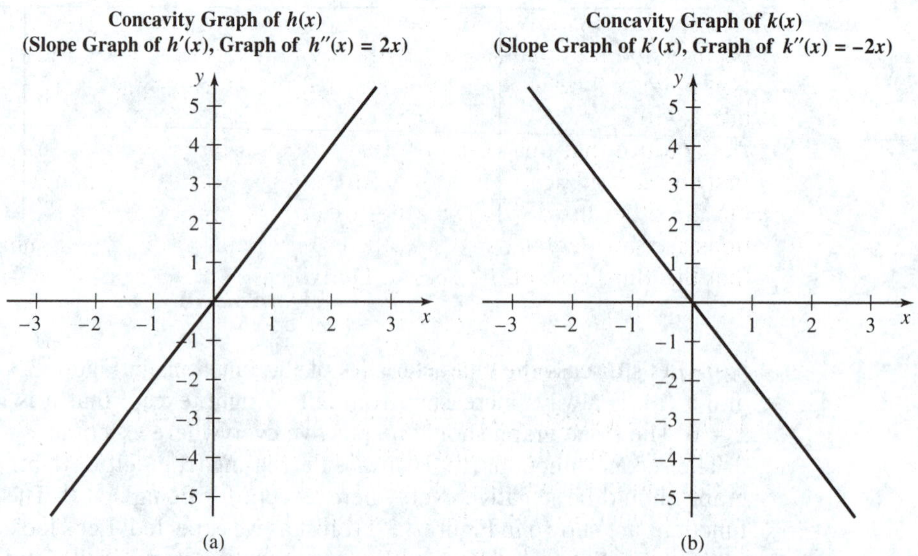

Figure 3.3-12

(the only smooth way to go from positive output values to negative output values is to pass through 0, which will mean the concavity graph crosses the horizontal axis).

The function we looked at in Figure 3.3-10(a) was

$$y = h(x) = \frac{x^3}{3}$$

The first derivative of this function is

$$y' = h'(x) = \frac{3x^2}{3} = x^2$$

The second derivative of this function is

$$y'' = h''(x) = 2x$$

Setting the second derivative equal to zero, we have $2x = 0$. This is only true when $x = 0$. Testing to be sure that the second derivative changes signs at this point we see that

$$h''(-0.01) = -0.02 \quad \text{and} \quad h''(+0.01) = +0.02$$

In fact, for any value $x < 0$, the second derivative will be negative and for any value $x > 0$ the second derivative will be positive. The second derivative crosses the x-axis at $x = 0$. The only inflection point occurs when $x = 0$.

For the second function, $y = k(x) = \frac{-x^3}{3}$, the second derivative will be

$$y'' = k''(x) = -2x$$

The second derivative $= 0$ when $x = 0$. For any value $x < 0$ the second derivative will be positive and for any value $x > 0$ the second derivative will be negative. Again the inflection point will be at $x = 0$. We saw that this is true on the graphs of the functions (Figure 3.3-10).

Notice that this example satisfies the third case of the Second Derivative Test (the first and second derivatives are both 0), and the critical point at 0 is *not* a local extremum. On the other hand we saw earlier that for $f(x) = x^4$ (see Figure 3.3-6), the same conditions are satisfied at $x = 0$, and the critical point *is* a local minimum. This is why we say that this third case of the Second Derivative Test is *inconclusive*.

Inflection Points: Extrema of the Slope Graph

This is also a good place to emphasize the connection between a point of inflection being where the concavity changes and being where the slope is locally shallowest or steepest (being a local maximum or minimum of the slope graph). Since the concavity graph of the original function is the slope graph *of the slope graph* of the original function, we can show that the conditions for a local extremum of the slope graph of the original function correspond to the conditions for an inflection point of the original curve. For example, in

Figure 3.3-11(a), the *slope graph* of the original function is decreasing to the left of 0, flat at 0, and increasing to the right of 0, which means there is a local minimum at 0. This corresponds to *its* slope (the value of the *second* derivative, shown on the concavity graph in Figure 3.3-12(a)) being negative, then 0, then positive, which were the conditions for a point of inflection (the second derivative is 0 and the second derivative changes sign).

Let's check this out on another cubic function and view the three graphs (original, slope graph, and concavity graph) lined up vertically, as shown in Figure 3.3-13. When we draw a vertical line through the inflection point of the original function in graph (a), it goes through the minimum point of the slope graph in graph (b) and through the x-intercept of the concavity graph in graph (c). Graph (b) is the slope graph of the original function, or the graph of the first derivative of the original function. Graph (c) is the concavity graph of the original function, or the graph of the second derivative of the original function, which can also be thought of as the slope graph of the first derivative graph.

A function has an inflection point when the concavity changes. When a function is concave down, its slope graph is decreasing. When a function is concave up, its slope graph is increasing. So at an inflection point, the slope graph of the function will change direction (from increasing to decreasing or vice versa); that is, it will have a maximum or a minimum point. When the slope graph is smooth and has a maximum or a minimum point, *its* slope graph, the second derivative, will be equal to zero.

Original Graph	Slope Graph (1st Derivative)	Concavity Graph (2nd Derivative)
Inflection Point	Local Maximum or Minimum	Crosses the horizontal axis, output = 0

Let's take a look at the function we graphed in Figure 3.3-13:

$$y = f(x) = x^3 - 16x^2 + 68x - 80$$

The first derivative of this function is

$$y' = f'(x) = 3x^2 - 32x + 68$$

The second derivative of this function is

$$y'' = f''(x) = 6x - 32$$

Setting this equal to zero we have

$$6x - 32 = 0$$

Solving for x we have

$$x = 32/6 \approx 5.333$$

For values of $x < 5.332$, the second derivative will be negative (for example, $f''(5) = -2$). For values of $x > 5.334$ the second derivative will be positive (for example, $f''(6) = 4$). Looking at the graph in Figure 3.3-13 we can verify that this is an inflection point.

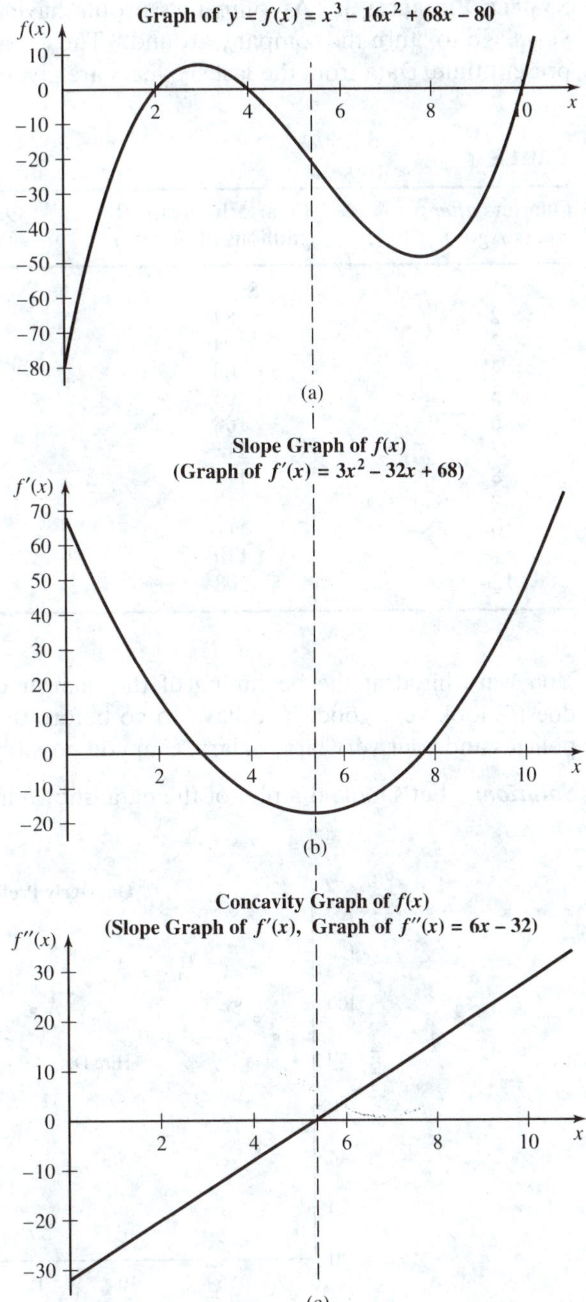

Graph of $y = f(x) = x^3 - 16x^2 + 68x - 80$
$f(x)$
10
0
−10
−20
−30
−40
−50
−60
−70
−80
2
4
6
8
10
x
(a)
Slope Graph of $f(x)$
(Graph of $f'(x) = 3x^2 - 32x + 68$)
$f'(x)$
70
60
50
40
30
20
10
0
−10
−20
2
4
6
8
10
x
(b)
Concavity Graph of $f(x)$
(Slope Graph of $f'(x)$, Graph of $f''(x) = 6x - 32$)
$f''(x)$
30
20
10
0
−10
−20
−30
2
4
6
8
10
x
(c)

Figure 3.3-13

Sample Problem 3: A company's profits have been declining. You were hired two years ago to "turn the company around." The company's profits are still declining at the present time. Data from the last six years are given in Table 1.

TABLE 1

Quarters *after* 6 Years Ago, t	Quarterly Profit, P (millions of dollars)	Quarters *after* 6 Years Ago, t	Quarterly Profit, P (millions of dollars)
1	\$ 78	13	\$105
2	87	14	102
3	94	15	98
4	100	16	94
5	105	17	90
6	108	18	86
7	111	19	81
8	112	20	77
9	112	21	73
10	111	22	69
11	110	23	66
12	108	24	63

You were hired at the beginning of the quarter corresponding to $t = 17$. Frankly, it doesn't look very good. You have to go before the board of directors and justify your policies and your very large salary. Can you possibly do this?

Solution: Let's look at a plot of the data, shown in Figure 3.3-14.

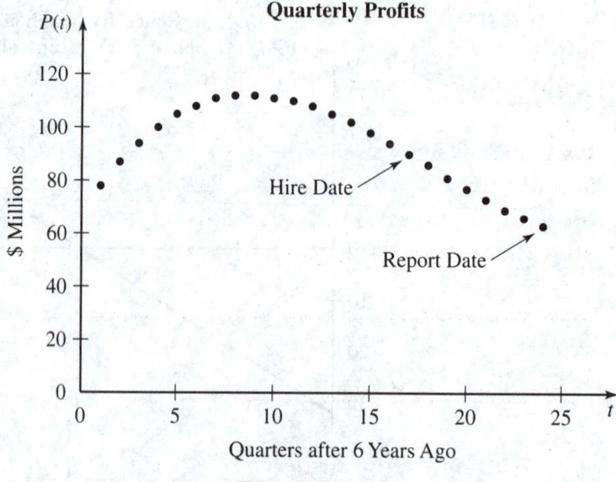

Figure 3.3-14

It does look as if your hire date is very close to being a point of inflection. As in the crime problem (Sample Problem 1), this would mean there was a turning point at that time, since the decline was steepest but started to get better then. Let's analyze the situation mathematically to support our visual analysis.

First let's fit a model to the data (rounded to 3 significant figures here):

Verbal Definition: $P(t)$ = the profit in millions of dollars for the quarter *preceding* the point in time that is t quarters *after* 6 years ago.

Symbol Definition: $P(t) = 0.0158t^3 - 0.865t^2 + 11.5t + 67.2, t \geq 0$.

Assumptions: Certainty and divisibility. Certainty implies that the relationship is exact. Divisibility implies that there can be fractions of dollars and fractions of years.

The first derivative of this function is

$$P'(t) = 0.0474t^2 - 1.730t + 11.5$$

You could use this to show that the maximum profit occurred when $t = 8.74$, or a little more than 2 years (8 quarters) after 6 years ago (a little less than 4 years ago). But this doesn't really help your cause; you weren't in charge then. The second derivative of the function is

$$P''(t) = 0.0948t - 1.730$$

Setting this equal to zero and solving we get

$$t = 1.730/0.0948 \approx 18.2$$

This suggests that there is an inflection point at around $t = 18$ quarters, just a quarter or two after you arrived. The profits were still decreasing, but at that point, the *rate* of decrease started to lessen. The profit curve becomes concave up, after having been concave down. About six months after you were hired, the decline in company profits started to slow down. Your policies are showing results. Perhaps not quite as dramatic as you hoped, but definitely there. This company had some very large problems. Many of them could not be corrected overnight. □

EXAMPLE 1: Suppose you work for an advertising firm and have been handling the account for a new product, No-Refrigeration Microwave Pizza. You launched the product with a really great ad campaign. The sales figures for the new product for the 8 weeks after the start of the ad campaign are shown below:

Weeks	1	2	3	4	5	6	7	8
Sales ($100's)	3	9	28	82	238	652	1553	2876

Normally, after a vigorous new campaign, sales will pick up slowly at first as the ad reaches the public. Then the effect of the ad and word-of-mouth publicity will make sales grow quite rapidly. Then the growth rate will start to slow down. You need a strategy for being able to answer the following questions: Has the slowdown happened yet? When will it happen? Can you catch it early so you can recommend a new campaign?

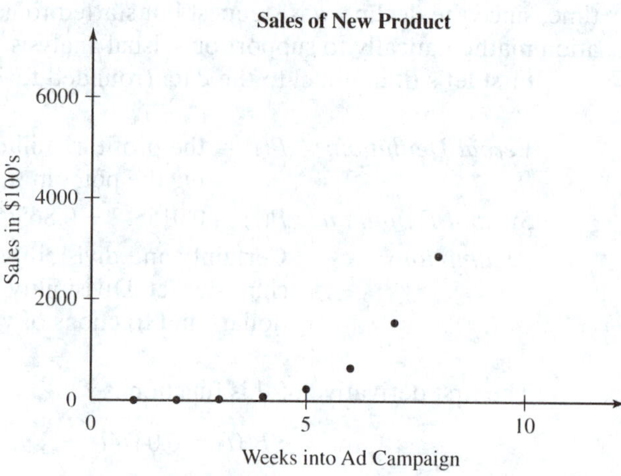

Figure 3.3-15

To answer these questions, let's look at a graph of the data, shown in Figure 3.3-15. It is not easy to tell from the graph whether the slowdown has started yet or not. Let's try looking at the first differences shown in the table below:

First Differences	+6	+19	+54	+156	+414	+901	+1323

The first differences, the average rate of change over each week, are increasing very nicely. You might be tempted to fit either a quadratic or exponential model, but you are enough of a realist to recognize that sales cannot continue to climb at this rate. You know that this is not the general sales pattern for new products. Let's look at the second differences:

Second Differences	+13	+35	+102	+258	+487	+422

So far, so good. The function is still concave up at this point, but it is starting to flatten out. The second differences are still positive, but the last second difference was lower than the week before. Suppose you decide to wait another week and get one more data point: Sales for the ninth week after the initial campaign turn out to be about $401,300, so the output value is 4013. The new graph is shown in Figure 3.3-16.

Is the growth starting to level out? The added first difference for the new data point is +1137, and the added second difference for the new data point is −186. Notice from the plot of the data points that, if the points were connected by a smoothed curve, it looks as if that curve would have switched from concave up to concave down somewhere around the last interval, just before the new data point. It is not a coincidence that the last second difference was negative! Just as the first differences are closely related to the first derivative (average rate of change versus instantaneous rate of change), **the second dif-**

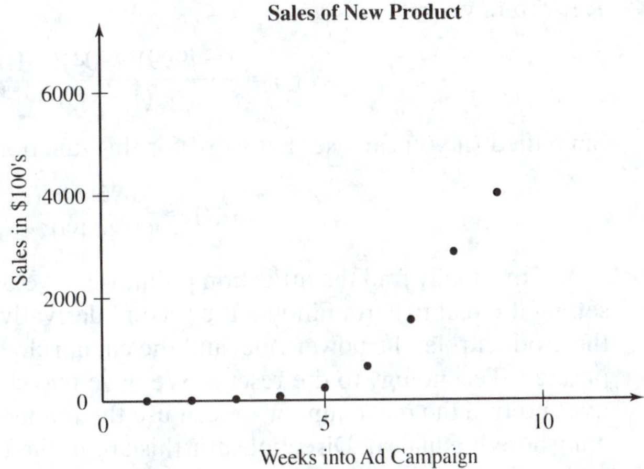

Figure 3.3-16

ferences are closely related to the second derivative, and so their signs give us information about concavity (positive second differences correspond to concave up and negative second differences correspond to concave down), as we discussed earlier. You had gotten a hint of this change in concavity when the second differences began to decrease. Here is the confirmation. We can check this algebraically by fitting a model, finding the second derivative, and setting the second derivative equal to zero to look for potential points of inflection.

A logistic curve would seem to be a good model for this data since sales are likely to slow down and level off with no additional major advertising campaigns. Using technology, a reasonable model (rounded to 4 significant figures) for this data is

Verbal Definition: $S(t)$ = sales in hundreds of dollars for the preceding week t weeks after the start of the ad campaign.

Symbol Definition: $S(t) = \dfrac{4999}{1 + 4904e^{-1.100t}}, t \geq 0.$

Assumptions: Certainty and divisibility. Certainty implies that the relationship is exact. Divisibility implies that there can be fractions of dollars in sales and fractions of weeks.

To find the derivative of this function we first rewrite it as

$$S(t) = 4999(1 + 4904e^{-1.100t})^{-1}, t \geq 0$$

We now have to apply the Chain Rule, (Power Rule) and then the Chain Rule again. The initial inner function is

$$u = 1 + 4904e^{-1.100t}$$

and the initial outside function is $4999u^{-1}$. So the derivative is

$$S'(t) = [4999(-1)(1 + 4904e^{-1.100t})^{-2}][(4904e^{-1.100t})(-1.100)]$$

Rewritten, we get

$$S'(t) = \frac{(-4999)(4904)(-1.100)(e^{-1.100t})}{(1 + 4904e^{-1.100t})^2}$$

Simplified (if you can use that word for this function), this is[7]

$$S'(t) \approx \frac{26966606(e^{-1.100t})}{(1 + 4904e^{-1.100t})^2}$$

To actually find the inflection point, we have been taking the second derivative and setting it equal to zero. Finding the second derivative of this function requires the use of the product rule, the power rule, and the chain rule twice. In other words, it is quite complicated. Technology to the rescue! We have two choices: We can graph the first derivative and find the maximum or we can use the numerical derivative function twice, to find out where it equals 0. Directions for this are in the technology supplement. The solution comes out to approximately

$$t \approx 7.7$$

Translation: The slowdown in growth started about 7.7 weeks after the beginning of the ad campaign. Anywhere around that time would seem to be a good time to launch the second major ad campaign. Start now, if you haven't already! □

Section Summary

Before you start the exercises for this section be sure that you

- Understand that the second derivative is simply the derivative of the derivative of a function, and can be denoted with symbols such as $f''(x)$, y'', and $\frac{d^2y}{dx^2}$.
- Know how to find the second derivative of the basic kinds of functions we have studied, both by hand (except possibly for the logistic!) and using technology (at least to find its value at a particular input value).
- Understand that, since a derivative corresponds to a rate of change and also to a slope, the second derivative (derivative of the derivative) can be thought of as the rate of change of the slope.
- Understand that if a curve is concave up over an interval, then its slope is increasing from left to right, so the rate of change of the slope is positive, so the second derivative is positive over the interval. Similarly, if a curve is concave down, the slopes are decreasing, so the rate of change of the slope is negative, so the second derivative is negative.
- Understand that if a smooth function is concave up over its domain, then a critical point (there can only be at most one) must be the unique local and global min-

[7]In Section 2.5 we found a general form for the derivative of a logistic function: if $f(x) = \frac{L}{1 + Ae^{-Bx}}$, then $f'(x) = \frac{LABe^{-Bx}}{(1 + Ae^{-Bx})^2}$.

imum of the function over the domain. Similarly, a critical point for a concave down function must be the only local and global maximum.

- Understand that, at a particular point, if the first derivative is 0 (so it is a critical point, with a horizontal tangent) and if the second derivative is negative (the curve is concave down there), then the point must be a local maximum. Similarly, if the first derivative is 0 and the second derivative is positive (the curve is concave up there), then the point must be a local minimum. If both derivatives are 0, the point could be a local extremum (like x^4 at 0) or a point of inflection (like x^3 at 0).
- Understand how the above concepts are the basis for the Second Derivative Test to test whether or not critical points are local extrema (plug them into the second derivative; if the result is negative, you have a local maximum; if positive, you have a local minimum; and if neither, you cannot be sure).
- Understand that a point of inflection is a point where the concavity changes, so for a smooth function, it will occur at a point where the second derivative is 0 and changes from negative to positive or vice versa. Thus to find a point of inflection, find all points where the second derivative is 0, and check them to see if the sign of the second derivative changes. If it does, you have a point of inflection. If not, you don't.
- Understand also that a point of inflection is a point where the slope is steepest or shallowest, so it is a local extremum of the first derivative.
- Know how to sketch the concavity graph of a smooth function, based on the principles in the following chart (assuming the second derivative is defined over the domain).

Function	Slope Graph (1st Derivative)	Concavity Graph (2nd Derivative)
Concave up	Increasing	Positive
Concave down	Decreasing	Negative
Maximum point	Crosses horizontal axis (equals 0)	≤ 0
Minimum point	Crosses horizontal axis (equals 0)	≥ 0
Inflection point	Maximum or minimum point	Crosses horizontal axis (equals 0)

EXERCISES FOR SECTION 3.3

Warm Up

Find the second derivative of the following functions in Exercises 1–6:

1. $f(x) = 3x^4 + 2x^2 - x - 8$

2. $R(s) = 6s^3 - 6s^2$

3. $g(x) = -7x^5 + 3x^3 - 9x$

4. $d(t) = \frac{-2}{3}t^4 + 2t^2 - 6$

5. $f(x) = 2xe^{-3x}$

6. $D(p) = 168(0.863)^p$

For the functions in Problems 7 through 12 use the second derivative to identify

(a) *the local maximum and minimum points;*

(b) *the coordinates of the points of inflection; and*

(c) *the intervals over which the function is concave up and concave down.*

7. $f(x) = 3x^5 - 5x^3$

8. $g(t) = t^4 - 5t^2$

9. $r(s) = s^3 - s$

10. $g(x) = 3x^4 - 4x^3 + 6$

11. $f(x) = 0.01xe^{-0.1x}$, for $x \geq 0$

12. $R(p) = 168p(0.863)^p$, for $p \geq 0$

Game Time

13. Exercise 9 in Section 1.5 discussed a model for unemployment in the United States. One possible model is

Verbal Definition: $U(x)$ = the unemployment (thousands) in the U.S. civilian population x years after Dec. 31, 1983.

Symbol Definition: $U(x) = -13.72x^3 + 305.7x^2 - 1901x + 10{,}704$, $x \geq 0$.

Assumptions: Certainty and divisibility. Certainty implies that the relationship is exact. Divisibility implies that the unemployment rate can be a fraction and the number of years can be a fraction.

(a) According to this model, when was unemployment at its highest and lowest levels between 1983 and 1994?

(b) Use the Second Derivative Test to test whether your answers actually were local maximum or minimum points.

(c) Find any points of inflection. Describe what was happening to the unemployment rate in the United States at this time. Explain how this information could be used.

14. A possible model for unemployment in the United States, different from the one given in Exercise 13, is

Verbal Definition: $U(x)$ = the unemployment (thousands) in the U.S. civilian population x years after Dec. 31, 1983.

Symbol Definition: $U(x) = -3.910x^4 + 72.30x^3 - 286.9x^2 - 586.6x + 10{,}262$, $x \geq 0$.

Assumptions: Certainty and divisibility. Certainty implies that the relationship is exact. Divisibility implies that the unemployment rate can be a fraction and the number of years can be a fraction.

(a) According to this model, when was unemployment at its highest and lowest levels between 1983 and 1994?

(b) Use the Second Derivative Test to test whether your answers were actually local maximum or minimum points.

(c) Find any points of inflection. Describe what was happening to the unemployment rate in the United States at this time. Explain how this information could be used.

15. The following table shows charitable giving in the United States.

Year	Contributions	Year	Contributions
1984	\$108.83	1990	\$130.77
1985	109.88	1991	130.94
1986	119.55	1992	128.90
1987	123.13	1993	129.16
1988	127.72	1994	129.88
1989	132.00		

Contributions are in billions of inflation-adjusted dollars.

Source: The World Almanac, 1996, op. cit., quoting the American Association of Fund-Raising Counselors, Inc., Trust for Philanthropy.

(a) If you have not already done so, find a model for these data (and align them).

(b) Using the model, find the local maximum and minimum total contributions for 1984 to 1994 [0,10].

(c) Use the Second Derivative Test to test whether the points are actually maximum or minimum points.

(d) Find any points of inflection. Describe what was happening to the charitable giving in the United States at this time. Explain how this information could be used.

16. Sample Problem 2 of this section referred to the danger of destroying a crop if the temperature fell below freezing. A model for the temperature was

Verbal Definition: $T(x)$ = temperature in degrees Fahrenheit x hours after 9:00 P.M.

Symbol Definition: $T(x) = 0.0581x^3 - 0.759x^2 + 1.03x + 42.8,\ 0 \le x \le 10.$

Assumptions: Certainty and divisibility. Certainty implies that the relationship is exact. Divisibility implies that there can be fractions of degrees and hours.

From the model, find any points of inflection. Describe what was happening to the temperature at this time. Explain how this information could be used.

17. In Sample Problem 4 in Section 1.4 we developed a model for the profit from the sale of tickets to a local sporting event as a function of the price charged per ticket:

Verbal Definition: $\pi(p)$ = profit in dollars from the sale of tickets at p dollars per ticket.

Symbol Definition: $\pi(p) = 3166(p - 1)(0.9492)^p - 1000,\ p \ge 0.$

Assumptions: Certainty and divisibility. Certainty implies that the relationship is exact, no discounts or special prices. Divisibility implies that fractional tickets and dollars are possible.

(a) Find the price per ticket that will maximize the profit. Use the Second Derivative Test to verify that this is a maximum point.

(b) Find any points of inflection. Describe what was happening to the profit at this time. Explain how this information could be used.

18. A model was fit to the energy level throughout the day:

Verbal Definition: $E(t)$ = average energy level on a 0–100 scale t hours after midnight on a typical day.

Symbol Definition: $E(t) = -0.0137t^4 + 0.914t^3 - 22.1t^2 + 228.27t - 777,\ 8 \le t \le 24.$

Assumptions: Certainty and divisibility. Certainty implies the relationship is *exact*. Divisibility implies that *any* fractions of energy level points and hours are possible.

(a) Use the model to determine the local extrema for this function. Use the Second Derivative Test to verify that they are local extrema and to identify what kind (local maximum versus local minimum) they are, and then find the energy level at that point.

(b) Now, using your answers to part (a), find the global extrema for the function over its domain and explain how you know they are the global extrema. Again, find the energy level at each.

(c) Find any points of inflection. Describe what was happening to the energy level at these times. Explain how this information could be useful.

19. We fit a model to the annual percentage change in the medical CPI:

Verbal Definition: $I(x)$ = annual percentage change in the medical CPI x years after Dec. 31, 1986 (over the preceding year).

Symbol Definition: $I(x) = 0.0193x^4 - 0.355x^3 + 1.93x^2 - 3.02x + 7.62, 0 \le x \le 8.$

Assumptions: Certainty and divisibility. Certainty implies that the relationship is exact. Divisibility implies that fractions of percentages and fractions of years are possible.

(a) Use the model to determine the local maximum and minimum points during the period Dec. 31, 1986, to Dec. 31, 1994. Use the Second Derivative Test to verify the kind of local extremum each is and find the value of the annual percent change in the medical CPI for each.

(b) Use the model and your answers to part (a) to determine the global maximum and minimum points during the period Dec. 31, 1986, to Dec. 31, 1994, and find the value of the annual percent change in the medical CPI for each.

(c) At what time during the period that the medical CPI was increasing was it increasing most rapidly?

(d) At what time during the period that the medical CPI was decreasing did the rate of decrease start to slow down?

(e) Name at least two ways that the above information could be used.

20. The following table shows the Consumer Price Index (1982–84 = \$1.00) for the years 1985 through 1994:

Year	1985	1986	1987	1988	1989	1990	1991	1992	1993	1994
Index	107.6	109.6	113.6	118.3	124.0	130.7	136.2	140.3	144.5	148.2

Source: *Statistical Abstract of the United States 1995*, op. cit., U.S. Department of Commerce, No. 760.

(a) Calculate the change and the percent change in the index for the years 1986–1994.

(b) Find a model for the annual percent change in the CPI for the years 1986 through 1994.

(c) Use the model to determine the local extrema during the period Dec. 31, 1986, to Dec. 31, 1994, use the Second Derivative Test to verify the type of extrema they are, and find the output value for each of them.

(d) Use the model to determine the global extrema during the period Dec. 31, 1986, to Dec. 31, 1994, and find the output value for each of them. Explain how you know that your answers are the global extrema.

(e) At what time during the period that the percent change in the CPI was increasing did the rate of increase start to slow down?

(f) At what time during the period that the percent change in the CPI was decreasing was it decreasing most rapidly?

(g) Compare your answers to this exercise with your answers to Exercise 19 for the *medical* CPI. What does this tell you about the medical costs in the United States? How could this information be used?

3.4 POST-OPTIMALITY ANALYSIS

You have now seen how to mathematically analyze functions of one variable that are continuous and smooth: to understand their rates of change and how to optimize them over an interval. We have also talked quite extensively about setting up and formulating models to do such analysis. In our plane analogy, you have learned how to take off, gain altitude, and cruise at 5000 feet. Now it is time to descend and land. In Section 1.0, we discussed most of the concepts of this section and saw how they apply to the example of finding Meg's optimal exercise plan. We have also referred to many of the ideas as we have solved other problems along the way. In this section, we will elaborate on the concepts of **verification**, **validation**, **sensitivity analysis**, and **margin of error** and give further examples to illustrate them.

This section should help you to complete the analysis of problems like the following:

- You want to determine, to a reasonable level of precision, the optimal amount of sleep for you to get on a school night so you can accomplish the most possible the next day.
- You want to figure out how often to go to the local grocery store to get a favorite snack food and how much of it to buy each time.
- You want to know how fast to drive your car on the highway to get the most miles per gallon of gas.
- You are trying to raise money for a campus group by selling T-shirts. How many shirts should you order, and how much should you charge for each?
- You want to determine the point along the 3-point line in basketball where you tend to achieve your highest shooting percentage.

After studying this section, in addition to being able to solve problems like those above, you should also

- Understand the idea of verifying (double-checking) models and solution calculations, including when and how to do it.
- Understand the idea of validating assumptions, models, and solutions (checking them against reality), including when and how to do it.
- Understand the idea of performing sensitivity analysis (seeing the effects of different plausible data values or even alternative models on the analysis and solution), including when and how to do it.
- Understand how to determine a rough margin of error for a solution, using knowledge about the precision and accuracy of the data, validation results for the assumptions and the model, sensitivity analysis, and the results of reasonable alternative models, as well as guidelines for precision and accuracy of calculations and common sense.
- Know how to use *all* of the above to come to meaningful conclusions and make suggestions for solving real-world problems.

Sample Problem 1: You love to play basketball, and want to perfect your three-point shot. You want to determine what angle is best for you to obtain the highest shooting percentage. You first decide to take shots from multiples of 45° (where 0° is a shot from the baseline on the left side of the basket, from the perspective of looking at the basket from the center of the court). The data are as follows (taking 30 shots from each angle):

Shot Angle	Shooting Fraction	Shooting Percentage (%)
0°	10/30	33.33
45°	12/30	40
90°	21/30	70
135°	8/30	26.67
180°	6/30	20

From this, you realize that an angle in the neighborhood of 90° seems best, so you gather more data in that vicinity to try to narrow down the solution more precisely:

Shot Angle	Shooting Fraction	Shooting Percentage (%)
70°	14/30	46.67
80°	20/30	66.67
90°	21/30	70
100°	18/30	60

What is your best angle for shooting? How can you verify your calculations and validate your assumptions and model? How could you perform sensitivity analysis? What conclusions can you come to with respect to precision, accuracy, and a final rough margin of error for your solution?

Solution: According to your definition of the problem, we are interested in what is the *best* shooting angle. The graph of the data is given in Figure 3.4-1. Visually, we can see that the optimal angle seems to be about 80–90 degrees. This in itself is a good ballpark answer, but let's do the mathematical analysis as well.

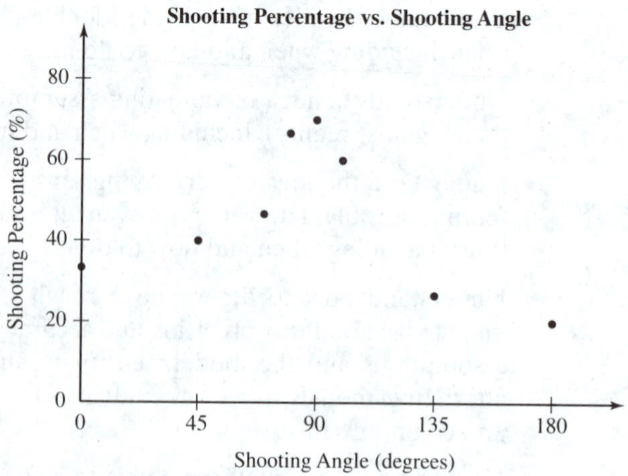

Figure 3.4-1

If we want to fit a model to all of the data, let's first do a crude hand sketch of a smooth curve that might come close (Figure 3.4-2).

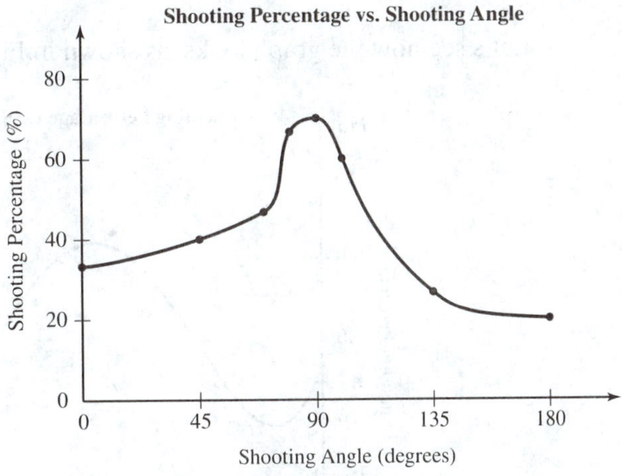

Figure 3.4-2

Now, if we want to fit a model to this data, notice that it has *two* points of inflection. A quadratic model has *no* inflection point. A cubic function can have only *one* point of inflection, so we need to go one step further, to a **quartic** model. The **degree of a polynomial** is the exponent of the highest power of the variable that has a *nonzero* coefficient: A quadratic is of degree two, cubic degree three, and quartic degree four. **In general, for a polynomial function to have n inflection points, it must be of at least degree $(n + 2)$.** Put differently, **a polynomial of degree n can have at most $(n - 2)$ points of inflection.** If we think of a local maximum or minimum (local optimum) as a **turning point** in the graph (where it turns from increasing to decreasing or vice-versa), we can make a similar statement: **In order for a polynomial to have n turning points, it must be of at least degree $(n + 1)$.** Alternatively, **a polynomial of degree n can have at most $(n - 1)$ turning points.**

In our problem, we have two inflection points, so we should fit a model of degree at least 2 + 2, or degree of at least 4. Let's find the best-fit quartic model.

A Review of Model Optimization and Analysis

Using technology, we find that the best-fit quartic model (rounded to 4 significant digits) is given by

Verbal Definition: $P(a)$ = the shooting percentage (successful shots out of total attempts) on average, expressed as a percent, when the shooting angle along the 3-point line is a degrees (measured from the left baseline, as viewed from the center of the basketball court).

Symbol Definition: $P(a) = 0.000001836a^4 - 0.0006586a^3 + 0.06914a^2 - 1.891a + 33.66$, for $0 \leq a \leq 180$.

Assumptions: Certainty and divisibility. Certainty implies that the model function gives the *exact* relationship between the shooting angle and the shooting percentage. Divisibility implies that the angle can be *any* real number of degrees between 0 and 180, and that *any* shooting percent is possible.

Let's see how the graph looks, as shown in Figure 3.4-3. Not too bad!

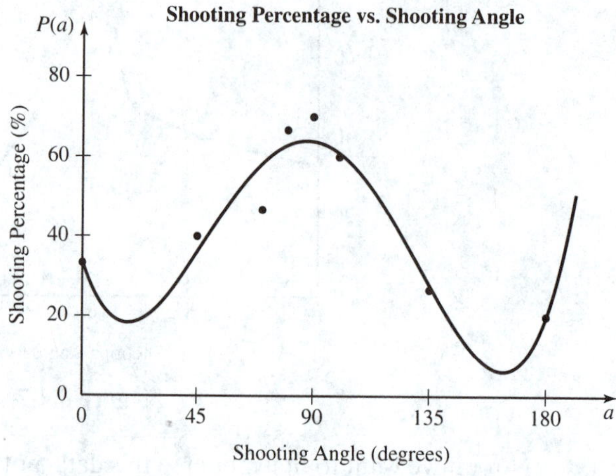

Figure 3.4-3

What does this model predict would be the best shooting angle? Let's take the derivative, set it equal to 0, and solve for a:

$$P'(a) = 0.000007344a^3 - 0.0019758a^2 + 0.13828a - 1.891 = 0$$

How can we solve this for a? An algorithm does exist for solving cubic equations exactly by hand, but since our data and model are only approximations anyway, we might as well use technology to solve this numerically. First, we can graph the derivative to get an idea of where the **roots** of this equation (sometimes called the **zeros** of the first derivative) are (Figure 3.4-4).

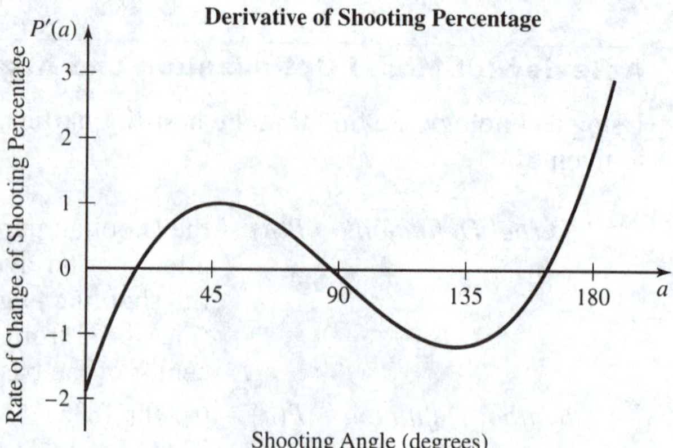

Figure 3.4-4

We can see from the graph that there are three roots: one at around 20, one at around 90, and one at around 160. Notice that they correspond to the local extrema on the original graph, as we know they should from the earlier sections of this chapter. Using technology, with the above ballpark estimates as initial guesses, we get solutions of

$$a \approx 17.99$$
$$a \approx 87.53$$
$$a \approx 163.5$$

We can use the Second Derivative Test to determine which points are local maximum points and which points are local minimum points, but this is not necessary since we have graphs of the original function and the derivative function. The maximum occurs at

$$a \approx 87.53$$

Translation: The optimal shooting percentage should occur at around 87.53 degrees. This seems reasonable, in light of our visual ballpark estimate of 80–90 degrees. It is saying that our model predicts that our best shooting angle along the 3-point arc is 87.53° from the left baseline.

Post-Optimality Analysis

All the above is a review of the methods discussed in previous sections. This carries us through the step we called the **analysis (mathematical solution)** in our 12-step program. In most traditional math classes, our solution above would have been the end of the problem. But **when solving a real-world problem, the mathematical optimization or analysis of your initial model is only the beginning!** We have found *an* optimal solution to a promising model, and now we are ready for **post-optimality analysis**. This means we want to make sure we haven't made any calculation mistakes, that our solution matches the real-world problem in the best possible way, and that we come to meaningful and realistic conclusions and suggestions about how to solve the real problem.

Verification

Let's start with **verification**. Verification simply refers to double-checking your mathematical calculations, to make sure that you have correctly applied the method you have chosen, and have not made any careless or other calculation errors, typos, and so on. There are many possible ways to verify your calculations.

Possible Verification Approaches

- **Do the calculation(s) both by hand (pencil and paper) and using technology.**
- **Do the calculation(s) using two different technologies (for example, on a graphing calculator and on a spreadsheet), or using the same technology in two different ways.**
- **Have another person do the calculation (by any method) independently. For best results, *don't show them your work or tell them what answer you obtained!!***

If your analysis involved several calculations, be sure to verify the *entire* process, either by verifying each calculation separately, or by using a method that verifies the entire solution.

In our example, we had three major calculations: fitting the model, finding the derivative, and finding where the derivative was equal to 0. If you know how to use a second type of technology to find the best-fit curve, do so. At this point, you may know only one way, so you may not be able to verify that. You certainly can look at the data points and the fit curve, and at least visually check that the fit is good. In Chapters 6 and 8 you will learn how to find the best-fit curve by hand and on a spreadsheet.

To verify the derivative, you could go back to the definition of the derivative [involving $(x + h)$ and the limit as h goes to 0], as we did in Section 2.4. If you are working on a student-generated project, your instructor may suggest that you do this to reinforce the definition. For a quartic model, this would be very long. There is another way that we can verify *both* the form of the derivative *and* where it is 0 at the same time: by using technology to find the maximum of our model function. On a graphing calculator, this comes out to

$$a \approx 87.53$$

again (to 4 significant digits), although the values are *not* exactly the same when taken out to 5 significant figures.

If we had wanted to verify the solution of the equation resulting from setting the derivative equal to 0, we could have used a *different* technology than we did the first time (*both* graphing calculators and spreadsheets normally have such a function), or we could have had a friend solve it independently. Another technological approach would be to plug your answer into a numerical derivative function (for example, on a graphing calculator) to verify that the value of the derivative there is 0 (or rounds off to 0).

Validation: Reconciling Models and Solutions with Reality

Having verified your solution does not necessarily mean it is good! It simply means you did not make any calculation errors. For example, if we had fit an exponential function to the basketball data, we could have verified that our model parameters were correct, but the model itself would be terrible, since it can have no turning points. This is why the idea of **validation** is crucial. Now that we have an answer, and some confidence that we did our calculations correctly, we want to know **how closely do our model and solution fit the real problem?**

There are in fact *many* ways to validate your model and solution. It is always best to use as many of these methods as you possibly can to get the best possible picture of how good your results are for your problem.

Possible Ways to Validate a *Model*

- **See how well the model fits the original data**, keeping in mind its **context** and how it is to be **used** (such as for optimization, interpolation, or extrapolation). This can be done visually by simply **graphing the model and data together**, or by using statistical measures of fit, such as R^2, which we will discuss in Chapter 6.

- **Try alternative models**, and **see which fits the original data the best** (again, keeping in mind the *context* and *use* for which the model is needed).
- **Use the model to predict the output value for a given input** (such as the global optimal solution value); **then *try* that input value in the real world**, and **see how the *actual* output compares to the output value *predicted by the model*.** Alternatively, you can ***withhold* some data** from your given data (such as the most recent observation) when formulating your model and then see how well the model predicts the withheld value(s). In either situation, you can also **compare alternative models** to see which best predicts what actually happened.
- **Analyze how well the assumptions of the model hold for the specific real-world situation.**

Possible Ways to Validate a *Solution*

- **Ask yourself if the solution makes good common sense to you.** If not, try to understand what is going on: Is your intuition flawed, or is the model faulty, or is there some other calculation or logical error?
- **Before formulating a model, make a ballpark or intuitive guess** at an optimal solution that makes sense to you and then **compare your mathematical result to this. If there are significant differences, try to understand *why*, and don't rest until you can reconcile the differences!**
- **Try finding a solution using *alternative* models, and see how close the results are to each other.** In general, the more *different* the modeling *approaches*, the better. The more *similar* the alternative *solutions* are to the original solution, the more confidence you have in the original solution.

Model Validation: Checking the Fit of a Model Considering Its Context and Use

The first type of validation mentioned in the list above is **how well does the model fit the data, considering the context of the problem and how the model will be used?** The idea of **how well the model fits the data** is very intuitive and doesn't need much elaboration. Looking at Figure 3.4-3, you can see that the fit of the model to the data is pretty good, but certainly not perfect. But the fit of the model to *all* of the data is not always the only criterion for evaluating the validity of a model. This is where the ideas of the model's *context* and *use* come in, and they are more subtle.

The **context** for a model refers to the real-world situation the model is supposed to represent. The context is important, because it **determines what kind of values make sense for the variables, what shape seems reasonable for the graph, and how much randomness we would expect**. These factors all have an important role in selecting and defining a model.

In our problem, the context (shooting basketballs from the 3-point line) determines that the only **values that make sense** for the domain are 0 to 180 (because of the way we have defined the input variable). Similarly, the range for the output values *must* be between 0 and 100, since they are percents. In other problems, you might not have such clearly defined domains and ranges, but you will often at least know that certain values (such as negative numbers) do not make sense, and perhaps that certain *combinations* of values don't make sense.

Regarding the **shape of the graph**, in our problem it would seem that the shape should be smooth, although it could certainly have plenty of ups and downs. These ups and downs would occur if you have several different "hot spots" with higher shooting percentages than points around them. In general, the context could determine the overall shape and end behavior of a graph in a way that would affect the *type* of model that makes most sense. For example, if you have weekly sales data for a product after an ad campaign that increased at first and then reached a point where it leveled off, the context could determine whether you expect it to stay at the high point (if there is an ongoing maintenance-level advertising effort) or to drop back down (if there is no ongoing marketing effort). In the first case, a logistic model would make sense, while in the second case, a cubic would probably be better.

By the nature of basketball, you would expect a fair amount of **randomness** and inherent uncertainty. If you make 30 attempts from the same angle several different times (even if it's at the same time of day and after the same amount of warming up, etc.), you are likely to get somewhat different numbers of baskets made. The number of baskets made for an average player could vary by plus or minus one or two baskets out of 30, and so the shooting percentages would also vary (plus or minus 1 or 2 out of 30 would correspond to plus or minus 3–7%). This means that we wouldn't expect a model to fit the data *perfectly*. Looking at Figure 3.4-3, most of the vertical differences (errors) seem within this range, although the point at 70 degrees seems as if it might be more. If possible, you might want to try taking more shots at 70 degrees to see if your data value is accurate or if the data point was unusually low because of a random fluctuation. On the other hand, the data from Sample Problem 1 of Section 2.1 from tossing a ball, since it is bound by the laws of physics, would have very little uncertainty and randomness, so the model should fit nearly perfectly (as it did).

There are two major uses for models: optimization and forecasting. Optimization means finding the *best* way to do something, looking for a decision (input variable value) that is a global maximum or minimum of some objective over a specified domain. **Forecasting can be separated into two categories: interpolation, which is forecasting output values for input values that are *within* the interval from the lowest to the highest input value of the data, and extrapolation, which is forecasting output values for input values that are *outside* the interval from the lowest to the highest input value of the data.** In some ways, optimization is *more* than forecasting, since it involves *both* forecasting (saying what the effect of different decisions would be) *and* optimizing based on those forecasts. From that perspective, when optimizing, it can still be helpful to think about whether the forecasts implicit in your optimization involve interpolation or extrapolation. Most people and businesses are more interested in optimization, but since optimization involves forecasting, we need to be able to validate models for *both* uses.

Since **forecasting** is involved in both cases, let's first look at how the *type* of forecasting (interpolation versus extrapolation) can affect how valid a model is for a particular problem. The general principle here is **when you are *extrapolating*, you are going *beyond* the data, so there is more uncertainty, and the *context* of the problem tends to be *relatively* more important than with interpolation**. The example above about the weekly sales data after an ad campaign is an example of extrapolation, and we saw how the context could have a major effect on the choice of a model. In our problem, since our data included shots from 0 and 180 degrees, the forecasting implicit in our problem would all be interpolation. Obviously, the context is still important, but **with *interpolation*, the *fit* of the model to the data tends to be *relatively* more important than with extrapolation.** How perfect the fit should be will still be determined by the randomness inherent in the context of the problem. For our example, again, the fit is pretty good within the randomness we would expect, except perhaps at 70 degrees.

When you are optimizing, in addition to the forecasting considerations above, **it is *most* crucial that the model fit very well *near* the optimal solution; it is less essential that the model fit well at points which are clearly not optimal and which are far away.** What does this mean for our example? We mentioned earlier that the description of the problem initially suggests that we are, in fact, only interested in the *optimal* shooting percentage, not necessarily in the best model for *all possible* shooting percentages. Thus we are really most interested in modeling reality near the optimal angle. Let's look at the data and the model again (Figure 3.4-5).

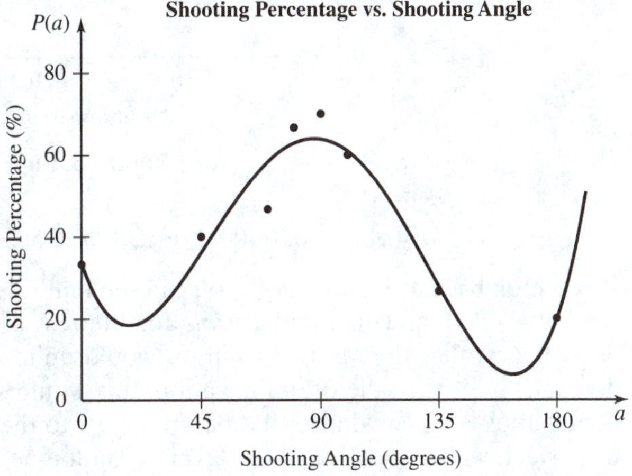

Figure 3.4-5

The model seems pretty good overall, but in some ways seems worst around the points we are most interested in! One strategy would be to fit a curve just to the points nearest the optimum. If we looked only at the second group of data collected ($a = 70, 80,$ 90, and 100), we see that we could fit a parabola to them. We could even fit a parabola just to the top 3 points (80, 90, and 100), but that parabola would then fit *exactly*, since a parabola is completely determined by 3 points. Given our earlier comments about the level of randomness and uncertainty we would expect in this problem, such a perfect fit does not seem justified.

These comments lead us to the next aspect of validation: trying alternative models and seeing which seems best for the context and use of the problem.

Model Validation: Selecting the Best Model for the Context and Use from Alternative Models

Given the somewhat limited toolbox of functions we have studied in this course, **have we modeled the situation in the best possible way?** From our most recent comments, it would seem to be a good idea to fit a parabola to the four newest data points and see if it suits the context and use even better than our original model.

Let's try fitting a parabola to the second group of 4 data points. Figure 3.4-6 shows a parabola fit to these points. The quadratic function is given by

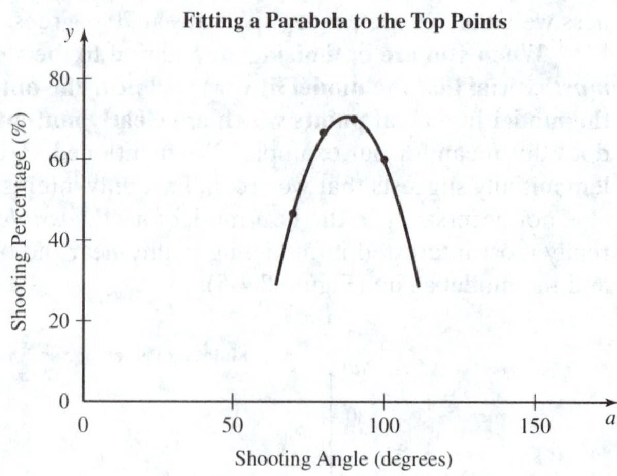

Figure 3.4-6

$$P_2(a) = -0.07500a^2 + 13.18a - 508.5, \text{ for } 70 \leq a \leq 100$$

If you look back at Figure 3.4-5, you can see that this model is *excellent* for the input values between 70 and 100, and *terrible* for the rest of the original domain. However, the data suggest that the optimal solution would quite clearly seem to lie in this restricted domain, so for the sake of optimization, this would seem to be the best model. As mentioned above, we could also fit a parabola just to the top 3 points, but then the fit would be perfect, which does not fit the level of randomness we would expect from the context of this problem.

Based on our earlier comments, it might make sense at this stage of solving this problem to go out and get more data at 70 degrees, which might help decide even more reliably between our original model and the 4-point parabola.

If you look back to Section 1.0 when we solved Meg's optimal exercise problem, we took a similar subset of the original data in the region where the optimal solution quite clearly seemed to lie. Notice that for that problem, we preferred to fit a parabola to 3 data points, whereas in the basketball example we preferred the 4-point parabola. This is because the **context** of the two problems is different. Since the basketball situation has more inherent randomness, a perfect fit is less desirable, but Meg's data seemed very smooth and stable, so a perfect fit made good sense.

Model Validation: Comparing Model Predictions to Actual Independent Data

Often, the most crucial question in validation is: **How well does my model predict reality?** In our example, we would be interested in how well our model predicts your actual shooting percentage at different angles. An excellent way to validate your model and solution for a problem like this is to test the *optimal* solution that you have calculated by gathering more data *after* your analysis. For this problem, this would be easy. You can try shooting from the original model's predicted optimal angle of 87.53 degrees (or as close as you can get to it), and see what your shooting percentage turns out to be. The model would predict

$$P(87.53) \approx 64.0$$

In other words, it predicts that if you shoot from an angle of 87.53 degrees on the 3-point line, your shooting percentage should be about 64.0%. Suppose you go out and shoot 30 more shots, this time from an angle of 87.53 degrees, and you get 21 baskets, for 21/30 = 70%. This is a pretty good validation of your model. Nineteen baskets would have been the closest possible result, but 21 is still pretty close, and within the plus or minus 1–2 baskets we mentioned as a reasonable level of randomness for the context of the problem. In fact, you could have shot 50 or 100 shots to get better precision for your data value, which you would expect to give an actual shooting percentage that was even closer to the model prediction.[8]

Since we have already found the 4-point parabola model, we can use the same actual data to validate that model and see whether it predicts better than the original model. Plugging into the $P_2(a)$ function, we get

$$P_2(87.53) \approx 70.5$$

This gives another strong argument for favoring the 4-point parabola model over the original! The actual value was 70%, and the prediction was 70.5%! We have not yet found the optimal solution for this model, but we might want to try to validate that in the same way.

It should be clear that you could gather more data in other ways to validate your model in other areas of the domain, especially if you were more interested in a forecasting model for the entire domain. For example, the quartic model predicts two local minima. You could gather data at those points. If all you care about is the maximum, this is less important.

For some situations, especially for forecasting problems involving **time series** (values of something over time, like the Dow Jones Industrial Average or quarterly profits of a company), it may not be possible to gather new data immediately. If you are projecting quarterly profits, you might have to wait up to three months to get the next figure. In such a case, one strategy for validation is to **withhold** one or more data points (usually from the *end* of the series, which would be the most recent values) when fitting your model, and then use the withheld values as if they were new data points for the *actual* values to compare with your model *predictions*. This can even be done sequentially using historical data (use the data up through 1980 to predict 1981 and then use the data through 1981 to predict 1982, etc.) to provide actual and predicted values for a particular forecasting method. The differences are called *errors*, similar to what is done in least-squares regression. You can then use a measure of the total error over some number of years to choose between different forecasting methods, by choosing the method with the least total error. We will discuss this more in Chapter 6.

The critical condition for this type of validation is that **the data used to compare the predicted and actual values *should not be used in the creation of the model*.** That is what

---

[8]You might also notice that your shooting percentage improves with practice. This is an example of how measurement can *change* the quantity you are trying to measure!

we mean by saying that the data should be *independent* of the model. **It is a good idea to consider how you plan to validate your model and solution even at the very beginning of defining a problem, deciding how to gather data for it, and formulating a model for it.** For example, if you are going to withhold data for validation, you need to do that *before* formulating your model and performing your analysis.

Model Validation: Evaluating Model Assumptions

Another aspect of validation is **validating your modeling assumptions**. In our basketball example, our modeling assumptions, as usual, were certainty and divisibility. The **divisibility** assumption, that the shooting angle can be *any* real number between 0 and 180 degrees, and that the shooting percentage can similarly be *any* real number between 0 and 100, can both be said to hold almost exactly. For the angle, the assumption is truly exact (*any* angle is theoretically possible). For the shooting percentage, we need to realize that, if we shoot *only* 30 shots, it is not really true. But conceptually, we could shoot as many attempts as we want, and so the shooting percentage could be at least any *fraction* (rational number). This is as close to "any real number" as we normally would ever want in a realistic problem like this. So we can think of divisibility as holding essentially exactly.

Certainty, on the other hand, is another matter. We have already discussed at some length the fact that, in this situation, we would expect a fair amount of random variation in the number of baskets you would get out of 30 at a given angle, even under virtually identical conditions, perhaps as much as plus or minus 1–2 baskets, or 3–7%. The bottom line is that the assumption of certainty does *not* hold *exactly*, by any means, although in this situation, it may not be too bad an assumption.

Notice that our definition of $P(a)$ was the shooting percentage *on average* at a given angle. This definition makes the assumption of certainty much more reasonable, since the shooting percentage out of a *particular* 30 shots would be much more variable than the *average* would be. It also makes the assumption of divisibility even more realistic.

We should also clarify here the difference between *stating* assumptions and *validating* them. When a model is fully defined, its assumptions need to be stated. In choosing a model, we go through many of the same considerations as those we have discussed here under validation to try to ensure from the start that we have the best model possible. But in *defining* the model, we *only* need to *state* the assumptions as fully as possible. There is no need to *comment* on the assumptions at that time, although there is certainly nothing wrong with it. The act of *commenting* on how well the assumptions hold is really a *validation* step.

Solution Validation: Common Sense

Once you have a fair amount of confidence in your *model*, you can move on to validate your *solution*. The best initial validation is common sense: Does your solution, when translated back to the real situation, make sense? Does it agree with your intuition?

In our basketball example, an optimal angle of about 87 degrees certainly sounds reasonable. It is close to a straight-on shot to the basket (90 degrees), which many people feel is best for them, partly because the extra shots that may go in because of bouncing off the backboard are likely to be greatest at around that angle.

If our mathematical analysis had come up with an angle that was negative or more than 180, we would know we had a problem (in fact, it would suggest we did the mathematical analysis wrong, since we should have restricted our optimization to the domain). Similarly, if the optimal shooting percentage came out to be negative or over 100, we would know there was a problem. Even if the optimal shooting percentage was 4% or 98%, we would suspect a problem, since these would not be typical maximum values for most basketball players.

If your solution violates your common sense, try to figure out the problem. Is the basketball player just incredibly bad, or is your intuition wrong, or did you make some mistakes in your math or in recording or entering your data? This is a place to check that you are being consistent with your *units*, since such errors can cause solutions that don't make any sense. Of course, you should also make sure that you have verified all of your calculations and information.

Solution Validation: Comparing to an Independent Ballpark Solution Using the Data

In many problems, you can look at the raw data (or a graph of it) and get a good ballpark estimate of an optimal solution. In our basketball example, just looking at the data suggests that the optimal angle should be somewhere around 90 degrees, perhaps a little less. This gives good validation of our mathematical solution, which was 87.53 degrees. Once again, if there is a discrepancy between your intuitive ballpark estimate of the optimal solution and your mathematical solution, try to understand where the discrepancy comes from. **Don't rest until you have resolved the discrepancy!**

Solution Validation: Solutions to Alternative Models

So far in the course of our discussion of the basketball example, we have formulated one other model, the 4-point parabola fit to the top 4 shooting angles. Let's now calculate the optimal solution for that model. The shooting percentage function we obtained was

$$P_2(a) = -0.07500a^2 + 13.18a - 508.5, \text{ for } 70 \leq a \leq 100$$

In this case, the derivative is

$$P_2'(a) = -0.15a + 13.18$$

Setting this equal to 0, we get

$$-0.15a + 13.18 = 0$$

$$a = 13.18/.15 \approx 87.87$$

Our second model gives an optimal shooting angle of 87.87 degrees, which is amazingly close to the 87.53 obtained from the first model. Clearly, the fit is very good at the top points, although it would be terrible for the rest of the data.

Out of curiosity, if we also try fitting a parabola just to the top 3 points, we get

$$P_3(a) = -0.06665a^2 + 11.66a - 439.9, \text{ for } 80 \leq a \leq 100$$

with an optimal angle (using technology) of

$$a \approx 87.50$$

The fact that all three solutions are so close gives good validation to this optimal angle, although each would give a different prediction as to the optimal shooting percentage. The second model predicts around 70.5%, and the last model predicts around 70.4%. From the validation data given above (supposing that you tried shooting from the optimal 87.53 degrees from the original model and achieved a shooting percentage of 70%), either of the last two models would seem very good, and we can have a high level of confidence that the optimal angle is somewhere in the 87.5–88-degree range.

Sensitivity Analysis

What is left to be done? Well, suppose you didn't have equal confidence in all of your data values. For example, suppose that the sun was in your eyes for the last one or two of the 100-degree shots (it was behind the clouds for all the other shots). You believe that you could have easily gotten 19/30 instead of 18/30 for $a = 100$, which would be 63.33% instead of 60%. How would this affect the solution?

Such a question is exactly the idea behind **sensitivity analysis**. You want to know how different plausible data values might affect your solution and analysis. For the particular change above, we can simply change the data value, fit a new curve, and find its maximum, using the shortest and/or easiest method possible. In this case, using the 4-point quadratic model, the new shooting percentage function is

$$P_4(a) = 0.06668a^2 + 11.87a - 457.0$$

and the optimal solution comes out to be

$$a = 89.00$$

This is a bit different, but not very different, again giving us confidence in our solution.

Another possible sensitivity analysis target in this problem is the data point for 70 degrees. If you look at the graph of all the data, that point seems to stand out as not fitting in with the rest. For sensitivity analysis, we could try just removing that point or give that input value an output value that seemed to fit the pattern better (assuming that the shooting percentage was unusually low just due to randomness). If we removed the data point, we would actually end up with our third model, the 3-point parabola using the original data (or with the changed data point at 100 degrees, if we made *both* sensitivity changes at the same time), so the results are what we have already seen.

In general, for sensitivity analysis, it is best to *first* ask yourself, **"Which data points and values do I have *least* confidence in, and what *different* values could they conceivably have been?"** Try making those changes, both individually and in combinations, to get a feel for the magnitude of change they cause, both in your optimal input value, and in the optimal output value.

At the same time, **just because a point does not fit a nice pattern does not *necessarily* mean it should be removed or adjusted!** There is no guarantee that your margin of error will be, or *should* be, low. It could simply mean that there is a good amount of random fluctuation or there are variables you are not accounting for yet in the situation you are studying. For example, in Meg's exercise example, if she had not gotten a consistent amount of sleep every night, her energy level could have varied a great deal with respect to her amount of exercise.

In the validation part of the discussion, we looked at other possible models. This could also be thought of as being related to sensitivity analysis, because it is seeing how sensitive the results are to the *model* used, as opposed to sensitivity to changes in *data*. It is better categorized as validation but has some similarities to sensitivity analysis.

Precision and Accuracy

When you find a mathematical solution, it can often be given to 10 or more significant digits, and it usually is if you used technology. But how accurate and precise is your answer *really*, with respect to the real world? Let's start by defining these terms, using an example of a bathroom scale used to weigh yourself in the morning.

Suppose when you weigh yourself at a given time, the reading from the scale normally varies as much as 2 pounds (it was on sale; you get what you pay for), so that if you stepped on and off the scale repeatedly, you might get measurements of 154, 155, 156, 155. We could call this measurement a weight of 155, plus or minus 1 pound, so it would have a **measurement error** of ±1 pound. Now suppose that you dashed off immediately to your doctor's office, or somewhere else where you know there is an expensive and carefully calibrated scale, so it should give a very true reading of your weight. With the same amount of clothing and no other changes that would affect your weight, this scale gives a reading of 152.0 pounds and is known to have a measurement error of only ±0.1 pounds.

If you think of a reading as consisting of a central value plus or minus some measurement error, *precision* refers to the *size of the measurement error* (either in absolute or relative terms), and *accuracy* refers to the *difference or bias* between the *central value* and the *true* value of the quantity being measured. A measurement is *more precise* if its measurement error is *smaller*. We say a measurement has a *positive* bias if the measured central value is *higher* than the true value. To be consistent with this convention, **we define the *bias* to be the central value minus the true value.**

In the case of the bathroom scale, we could say that it is adequately precise (±1 pound, which out of about 150 is about $\pm\frac{2}{3}\%$), but not very accurate (has a positive **bias** of $155 - 152 = 3$ pounds, which is about 2%). There is probably not much you can do about the precision, but most scales do have a way to correct a known bias, in the form of a knob or dial to twist that more or less uniformly shifts the central value of the measurement. In fact, if you look at your scale when there is nothing on it, it is likely to read somewhere around 3 pounds, reflecting the bias. After you make the adjustment, it will probably read close to 0, and your weight measurement should now read with a central value of 152, still with an error of ± 1 pound. Thus, after the adjustment, your measure ments should be very accurate (probably within one half of a pound, or about $\frac{1}{3}\%$), but still only adequately precise.

Mathematically, there are two ways we can express precision. One is to explicitly write a margin of error in absolute or relative terms, as we did above for the plus or minus quantities. The second is a bit rougher, but simpler, and that is to use significant digits. For our scale example, it would make sense to express the weight with 3 significant digits. After adjusting for the bias, then, we could express the weight as 152 pounds. The implied precision here would be ±0.5 pounds, since any value in the interval [151.5,152.5], including 151.5, but excluding 152.5, rounded off to 3 significant digits would round off to 152. If we wrote the answer as 152.0, this would imply a precision of ±0.05 pounds, which would be more appropriate for the doctor's scale, but not yours. And if we rounded to 2 significant digits, we would have to call it 150 pounds, with an implied precision of ±5 pounds, which does not reflect the precision of your scale as well as 3 significant digits.

As a general rule, it's best to use the number of significant digits which reflect your precision just right or are slightly more precise. In the scale example, 2 significant digits were not precise enough and 3 significant digits were a little too precise, so 3 significant digits would be the recommended precision.

In most real-world problems, you are lucky if the precision of your data is 3 or 4 significant digits.

Accuracy: Roughly Estimating and Adjusting for Bias

We said above that bias is the central measured value minus the true value of a quantity. But what is the true value? In most cases, we don't have the equivalent of the doctor's scale to assess our bias. To make matters more complicated, in a real-world problem, "the true value" is not always a clearly defined concept. In certain situations, it may be the case that getting better and better information would close in on a "true value" (although even in physics there are limitations to how much we can know), but there could also be inherent random fluctuations. In such a case we can talk about the true *expected value* (or average), but not the true value in itself.

If there are ways to get measurements that are likely to be more accurate (and hopefully more precise as well), then it is definitely worth trying if the required time and/or money are not prohibitive. Even if you can do this for only one measurement, it could give you an idea of what adjustment might be appropriate for all of your data.

The other alternative is to simply make a subjective assessment of possible biases in your data, and use your intuitive judgment to adjust the data for any such biases. In either case, re-solve the problem with the adjusted data. In essence, this is what we were doing in sensitivity analysis, but focusing more on individual data values rather than systematic biases.

In either case, our bottom-line interest is in what effect any biases have on the input and output values of our solution (optimal values or forecast values).

Let's go back to our basketball example. Suppose that all of your data were gathered at the very beginning of the season, before you had played very much. Then you might expect your shooting percentages to be biased on the low side. If you are doing your analysis right after gathering the data, then there may not be a way to assess this bias empirically, so you might have to simply estimate that you expect you would do about 10% better in your shooting percentages.

What would be the new solution if you made this adjustment for the beginning-of-season bias? The effect on the data would be that all of the second values would get increased by 10% (multiplied by 1.1), and the effect on the model will be that it also gets multiplied by 1.1. We saw earlier that multiplication by a positive constant greater than 1 will have the effect of vertically stretching the graph, anchored along the horizontal axis.

Thus, the maximum of the original graph and the stretched graph will occur at the same input value, and the corresponding output value will simply be 1.1 times what it was for the original problem. Using the full quartic model, then, the optimal angle would still be 87.53 degrees, and the corresponding shooting percentage would be approximately

$$1.1(64.0) = 70.4$$

or 70.4%. If the validating data collection at 87.53 degrees also happened at the beginning of the season, then *its* output value would also get adjusted by 10%, giving

$$1.1(70) = 77$$

or 77%, so the validation assessment does not change significantly. The adjustments for the other models would be similar.

On the other hand, if the validation data came from the *middle* of the season, then we would *not* adjust its output value, so it would stay at 70%, and this would argue for the *original* model being better.

If your analysis did not occur until the middle of the season, then you could *replicate* one of your original data points, perhaps with a larger number of attempts (to improve the precision, like the doctor's scale). For example, suppose you shot 50 shots from 100 degrees (in mid-season), and made 34 baskets, for a shooting percentage of

$$\frac{34}{50} = 68\%$$

If we assume the bias is proportional at all angles, then the proportionality factor would be

$$\frac{68}{60} \approx 1.133$$

(an increase of about 13.3%). Our analysis would then be similar to what we did before, but using this factor of 1.133 instead of 1.1. Presumably, the validation data would also then have come from the middle of the season, so would not be adjusted, and this would again suggest that the original quartic model was better.

If you anticipate this kind of bias in advance, and if circumstances allow, you could avoid the problem completely by waiting to gather your data until the early-season warming up/relearning curve has stabilized and come close to some kind of steady state, not likely to change a lot more. In Meg's exercise problem in Section 1.0, she had already been running on an ongoing basis when she started collecting data, so there was no significant training effect. If she had just been starting, the relationship between her exercise time and energy level could have changed significantly. For example, she might likely have been more tired (lower energy level) for longer exercise times early in the data collection period compared to later, based on her changing level of fitness.

Another possible bias in the basketball example would be the amount of warm-up *on the day of collecting data* for each particular data point. It could well be that your shooting percentage starts off on the low side until you get warmed up, and at the end of a session, it may also deteriorate as you get tired or sore. If you collected the data in sets of 10 attempts on 3 different days, and always did the smaller angles first and worked your way up in order, this could be a significant bias. It would mean a low bias for the data at the smaller angles (from not being warmed up), and possibly the larger ones too (due to tiredness). It could be complicated to try to adjust for this, although it could be done. Much better would be to **design your data collection to control for or eliminate biases**. For example, you could randomly pick the order of your angles for each set of 10 attempts, or establish a pattern so that each angle occurs near the beginning, near the middle, and near the end of a session.

For a different example, if you ask people what is the most they would pay for a T-shirt you want to sell, they may not tell you the truth for strategic reasons. Think about it: If they are honest, you might end up charging more than you would if they give you a deceitful lower value. This would probably bias your data on the negative side, since their answers would be *below* their true values.

When gathering data using **surveys** of people, it is important to try to select people in some **randomized** manner, and in such a way that they are **as representative of the population you are concerned about as possible**. For example, if you want information about all students at a college, just selecting one dormitory and talking to people who are in their rooms could be biased in several ways. If the dorm is special in some way (such as mostly athletes, say), it may not be representative of students in general. And by talking only to students in their rooms, you are biased by *not* including students who tend to have active lives (social life or many activities), who could behave quite differently than students who stay in their rooms more. These topics are all covered in more depth in statistics.

Precision: Roughly Estimating Margin of Error for Data

Just as we would like to know the effect of biases on our solution, we also want to know the effect of precision on our solution. We said earlier that the precision of a data value is the measurement error (the amount added or subtracted to the central value to give the full interval of possible values). Since our analysis involves numerous calculations (fitting models, taking derivatives, solving equations, plugging values into functions, etc.), these errors could get magnified (or reduced) in the process of finding a solution. We will call the error that results from these operations the **margin of error** of our solution, and we will often use margin of error as the generic term for the error in a quantity (whether measurement error or error from calculations). **Assessing the margin of error for a solution involves roughly estimating the margin of error in the input and output values of the data, and then applying some basic rules governing margin of error and precision in calculations to obtain rough estimates of the margin of error in the solution.**

There are **three major sources of imprecision in your solution**:

- **Inherent probabilistic randomness or uncertainty.**
- **Measurement errors** (such as due to roundoff or limited resolution in obtaining data).
- **Roundoff errors** in performing calculations.

These sources are listed roughly in order of increasing control *you* have over them. You can't really change inherent probabilistic uncertainty. You may have to use data values collected by someone else, or know for other reasons that they are limited in precision, although you may be able to do something about the situation. However, you have a great deal of control over the precision and rounding done in your calculations.

As we mentioned earlier, margin of error can be handled directly, or more crudely using significant digits. We will discuss both. When dealing with errors, we will normally express them as percents, since these are more comparable when doing calculations involving several different quantities with different units and scales.

Let's go back again to our basketball example. The first values for our data pairs were the angle of the shots. Realistically, even using a good protractor and string, you probably can't measure an angle like 70 degrees more accurately than within 1–2 degrees. Since our optimal value came out close to 90 degrees, let's express this roughly as a percentage. One out of 90 is close to 1 out of 100, so would be about 1%, and similarly 2 out of 90 is about 2%. Thus we can say our margin of error for the first values of our data pairs (input variable) is roughly 1–2%. This is purely a **measurement error**.

What about the second values of our data pairs? Given that our data values were all based on only 30 attempts, the shooting percentage can only be accurate to within an interval that is $1/30 \approx 0.0333 = 3.33\%$ in total width. This suggests a margin of error of one-half that size, or 1.67 percentage points (since $\pm 1.67\%$ *on either side* of an estimate gives an interval with a *total* width of 3.33%). The following table below may help to clarify this point:

Shots out of 30	20		21		22
Percentage	66.67		70		73.33
Midpoint of Range		68.33		71.67	

For example, 20 out of 30 is a percentage of 66.67%, 21 out of 30 would be 70%, and 22 out of 30 works out to 73.33%. You would expect to get 21 out of 30 if your *true* percentage were between 68.33% (halfway between 66.67% and 70%) and 71.67% (halfway between 70% and 73.33%), which is 70% plus or minus 1.67%

Our model predicts an optimal shooting percentage of 64.0%, so the possible error on either side, expressed as a percent, would be

$$1.67/64.0 \approx 2.6\%$$

This suggests that, at least at a rough level, we estimate about 2–3% error in our output values due to limiting our number of attempts (sample size) to 30. This is also a **measurement error**.

If we attempted 100 shots, we could cut our error as calculated above to about one-third of the above (plus or minus about 0.5%, instead of plus or minus 1.67%). Thus, there *is* something we could do to improve the precision of our output values. Even so, it is more than likely that if we did several sets of 100 attempts, we would not get the exact same number of baskets every time. This aspect of uncertainty is related to the idea of **probability and statistics**. It reflects the fact that, sometimes, under the same circumstances (such as rolling a die), different outcomes occur in a manner we think of as **random**. In fact, if we did several trials at each angle, we could get a meaningful estimate of the uncertainty or variability of our shooting percentage. In Chapter 6, in fact, we will make this notion much more precise and study it at some length.

In this example, we said earlier that, out of 30 attempts, we might expect a fluctuation of plus or minus 1 or 2 baskets made. This would correspond to plus or minus 3–7% roughly. We would call this **error due to inherent randomness** in the problem.

Now, we have roughly a 2–3% error due to the relatively small sample size (number of shots attempted), and roughly a 3–7% error due to the nature of basketball. How do we combine them?

The Combined Effect of Different Errors. If the errors are completely independent of each other, (the size of one is unrelated to the size of the other), then the *squares* of the errors would be additive, so the combined error is like finding the hypotenuse of a right triangle given the two legs, which means the errors themselves combine in a way that is *less* than additive, but bigger than either individual value. If the errors are directly related (called **positively correlated** in technical statistical language), so that one tends to be large when the other is, and with the same sign, then the combined error is *more* than when they are independent. For rough approximation purposes in this case, you could assume the errors are roughly additive. When the errors are **negatively correlated** so that the signs of the errors tend to be *opposite* rather than the same, the combined error is *less* than when they are independent. For rough estimation purposes, you could use the *average* of the errors to reflect the combined effect. **As a general rule of thumb, we will assume that the combined effect of different errors is more than the largest error, but less than the sum of the errors, unless we have strong reason to believe the errors are highly correlated.**

In our example, there is no reason to believe that our two errors are correlated, so let's assume they are more or less independent of each other. If we added the two errors (2–3% plus 3–7%) we would get 5–10%. If we averaged them, we would get 2.5–5%, or roughly 3–5%. Since the case of independent errors should lie somewhere in between, let's assume our margin of error for our output values is roughly 4–8%. This fits our rule of thumb, since it is more than the larger error, but less than the sum of the two errors.

We have roughly estimated 1–2% error in our input values and 4–8% error in our output values. If we again use our rule of thumb for the combined overall error in our data, we could use a rough range of 5–9%, since adding would yield 5–10% and the largest error is 4–8%.

Since our optimal angle is on the order of 90 degrees, 5–9% of that would be about plus or minus 5–9 degrees, which would be less than 1 significant digit. This is because 1 significant digit would mean an answer of 90, versus 80 or 100, degrees. Any value between 85 and 95 would be rounded off to 90, so 1 significant digit corresponds to plus or minus 5 degrees. Since our estimated rough error is more than this, it suggests even less precision than 1 significant digit. On the other hand, our input measurement error of 1–2% (1 or 2 degrees out of about 90) is closer to 2 significant digits of precision, which would be like an answer of 88 versus 87 or 89. Values between 87.5 and 88.5 would round off to 88, so 2 significant digits would correspond to an error of plus or minus 0.5 degrees. Thus, the margin of error for our solution would correspond to 1 or 2 significant digits. This would suggest that, in our model(s), if we round off to 3 or 4 significant digits, we should capture the precision reflected in our data. If you look back over this section, that is exactly what we have done.

Propagation of Errors in Calculations

The study of how errors can get magnified in the process of making calculations is a very complicated subject, and we won't get into great detail here. We can give a simple introduction, however and some basic rules of thumb that should be adequate for the kinds of problems you are likely to be dealing with in the real world.

The general principle of precision is that **the precision of the result of a calculation is only as precise as the *least* precise of the values involved.**

When working with **addition and subtraction**, it is **decimal places** (the number of places after the decimal point) that count. For example, when adding 23.308 and 1.9, if we assume both numbers are rounded off, then 23.308 could be any value between 23.3075 and 23.3085 and 1.9 could be any value between 1.85 and 1.95. Let's look at all the possible combinations for the sums:

$$\begin{array}{r}23.3075\\+1.85\\\hline 25.1575\end{array}\qquad\begin{array}{r}23.3075\\+1.95\\\hline 25.2575\end{array}\qquad\begin{array}{r}23.3085\\+1.85\\\hline 25.1585\end{array}\qquad\begin{array}{r}23.3085\\+1.95\\\hline 25.2585\end{array}$$

How could we give an answer that reflected all of these possibilities simply? Probably the best answer would be 25.2, which is exactly what you would get if you added the original numbers and rounded off to the number of decimal places of the least precise value (to the tenths place here, since the 1.9 is only rounded to 1 decimal place, while 23.308 is rounded to 3 decimal places):

$$\begin{array}{l}23.308\\+1.9\\\hline 25.208\text{, which we round to 25.2}\end{array}$$

If you look carefully at the example showing the four combinations, you can get a feel for why this rule works as stated.

Here is another example illustrating this principle:

$$\begin{array}{l}12.9075\\-0.05\\\hline 12.8575\text{, which we round to 12.86}\end{array}$$

In this case, the least precise value was the 0.05, rounded to 2 decimal places, so we also rounded the answer to 2 decimal places.

When working with multiplication and division and other operations, it is *significant figures* (also called *significant digits*, the number of meaningful digits *including and after the first nonzero digit*) that are most important. To see this, let's try an example as we did for addition. Suppose we are multiplying 3.7 times 2.016. If both are rounded off, then 3.7 could be anywhere between 3.65 and 3.75, and 2.016 could be anywhere between 2.0155 and 2.0165. Again, let's try multiplying all four combinations:

$$(3.65)(2.0155) = 7.356575$$

$$(3.65)(2.0165) = 7.360225$$

$$(3.75)(2.0155) = 7.558125$$

$$(3.75)(2.0165) = 7.561875$$

What answer would come closest to representing all of these answers? It looks as if either 7.4 or 7.5 would be close, and between the two, 7.5 is a little better (closer to the middle of all the answers). This suggests that we have no more than 2 significant digits of precision in our answer. Since 3.7 has 2 significant digits and 2.016 has 4 significant digits, we see that again it is the *least* precise value that determines the precision of the answer. If we multiplied the original values and rounded to 2 significant digits, we would get

$$(3.7)(2.016) = 7.4592, \text{ which we would round to } 7.5.$$

If you try a similar exercise **with powers or roots**, you will see that **the precision depends on significant figures**, as it does for multiplication and division.

When calculations involve a mixture of operations, significant figures tend to be most important. If we use our above guidelines for individual calculations, we can see how this works. If the individual terms are guided by significant figures and these are then added or subtracted, the tendency will be that the larger terms will have the smaller number of decimal places of precision, so their precision will dominate, and the answer will tend to have a similar number of significant figures (and decimal places) to these larger terms. For example,

$$\begin{aligned}(2.34)(3.56)^2 + 15.2(3.56) + 0.00378 &= (2.34)(12.6736) + 54.112 + 0.00378\\ &\approx (2.34)(12.7) + 54.1 + 0.00378\\ &= 29.718 + 54.1 + 0.00378\\ &\approx 29.7 + 54.1 + 0.00378\\ &= 93.80378\\ &\approx 93.8\end{aligned}$$

In this example, we have shown the exact result of each calculation (which you can recognize by the equal sign, =) and then rounded the result of those calculations to the appropriate level of precision according to the rules given above (which you can recognize by the approximately equal sign, ≈). For example, (15.2)(3.56) = 54.112, but since the numbers both have 3 significant digits and the operation is multiplication, the answer should be rounded to 3 significant digits, to 54.1. Normally, in this text, we will not write out the steps with the exact results and will often only give the final answer. Notice that the final answer has 3 significant figures, corresponding to the least precise of the original values, and that the calculations are a mixture of addition, multiplication, and powers. This reinforces our comment that in such mixed calculations, it is best to use significant digits as your main guide for the precision of your final answer.

The situation that is most "dangerous" is when mixing subtraction and higher-order operations (multiplication, division, powers, roots). For example, consider the calculation

$$\frac{15.6789 - 15.6743}{54.32}$$

At first glance, this looks as if the numerator will have 6 significant digits, and the denominator 4 significant digits, so the answer should have 4 significant digits. But if all of these numbers are rounded, the numerator simplifies to give us

$$\frac{15.6789 - 15.6743}{54.32} \approx \frac{0.0046}{54.32}$$

This means that the numerator actually has only 2 significant figures, so our answer should have only 2 significant figures. In essence, **the subtraction in the numerator made us lose 4 significant figures of precision!** This situation actually does arise when we are calculating average rates of change (slopes of secant lines), but for the most part we only used such calculations to *understand* the concepts of limit and derivative. For finding extrema and function values, **such complications are quite rare, but it is good to be aware of this possible erosion of precision in calculations if you get some surprising results.**

To finish our example, then, we would have

$$\frac{15.6789 - 15.6743}{54.32} \approx \frac{0.0046}{54.32} \approx 0.000085$$

A good general rule of thumb is to **carry out your hand calculations with at least *one or two* more decimal places or significant figures of precision than you expect to need for your final result.** This will usually mean carrying 3 or 4 significant digits. You can then round your final answer to the level of precision justified by your data and the principles given above. The reason for carrying the extra digit or two of precision is that the **roundoff error during the calculations** tends to erode the precision slightly, so the extra digit is a hedge against this erosion.

When working with technology, carry as many places or digits of accuracy as possible throughout your calculations, and then round only at the very end. In this text, when we have given you all the data and other information you need to be able to do the complete calculations using your technology, we have written numbers by showing the first 4 or 5 digits followed by an ellipsis (3 periods or dots, which you probably think of as "dot-dot-dot"), as we discussed in Section 1.2. The number is *truncated* (chopped off) after the last digit, rather than being rounded, but the number can then be *rounded* to one less digit. For instance,

$$37.56192758367$$

on a calculator could be truncated after 5 digits and written as

$$37.561\ldots$$

and, from that form, could be rounded to 4 significant digits and written as

$$37.56$$

The processes of **verification and validation help to protect you from potential problems that inadequate precision could cause**. Our general approach in this text is to carry 3 or 4 significant figures of precision in our hand calculations, and round answers to 2 or 3 significant figures. For example, we might use 0.004569 (4 significant digits) as a parameter in our calculations but round a final answer to 37.2 (3 significant digits). If a particular problem has more accurate and precise data, we may sometimes go beyond this level, but not as a rule. As an additional caveat, **errors can get magnified much more easily in exponential functions, so additional digits can be carried in such cases to hedge against possible roundoff errors**. Also, in the second half of the text, **major problems can occur when working with matrices**. One of the best ways to try to protect yourself from such problems is to be sure to verify your results. One final comment: **Be aware of which values are *exact* in calculations.** For example, when finding the perimeter of a rectangle with the formula $2l + 2w$, the 2's have only one digit, but the values are *exact*, so technically have an *infinite* number of significant digits. If your width value had 3 significant digits, then the value of $2w$ should also have 3 significant digits, not 1. This is true of the exponents in best-fit polynomial models, for example.

Roughly Estimating Margin of Error for Solutions

To get a rough ballpark estimate of the margin of error in your solution, you can incorporate all of the above comments by applying your estimates of the relative (percent) error in your data to your input and output solution values, and by using the number of significant digits justified by the precision of your data.

In our basketball example, our rough estimate of the overall error of our data was about 1–2% for the input values, 4–8% for the output values, and 5–9% overall, and we talked about using 2 or 3 significant figures of precision.

A second way to get a feel for the margin of error of your solution is to look at the results of trying alternative models and doing sensitivity analysis. Our optimal angles varied from 87.5 degrees to 89.0 degrees, and our optimal shooting percentages ranged from 64.0% to 70.5%. This is quite consistent with our original estimates of 1–2% error for the input values and 4–8% error for the output values

Final Conclusions: Putting It All Together

After going through all of the above stages of analysis, we can try to come up with **final conclusions**. These **include estimates of optimal input and output values, at an appropriate level of precision, with the specification of a rough margin of error for each value to reflect measurement error, randomness, calculation and rounding error, modeling error, and possible biases**. **A good conclusion also specifies the solution in a language that would make sense with respect to the way the decision maker thinks about the problem.**

In our basketball example, we said that the model that seems to make the most sense is our 4-point parabola model. Mathematically, we could say our best guess at the optimal angle is about

88 degrees, plus or minus 1–2 degrees

expecting an optimal shooting percentage of about

70%, plus or minus 3–6 percentage points

(4–8% of 70 is 3–6).

To put this in words that reflect the way a basketball player would probably think about the problem, we could say

> *Shoot from slightly to the left of a straight-on shot from the 3-point line for your highest possible shooting percentage, which should be about 70 percent, give or take 6 percentage points.* □

Now let's perform post-optimality analysis for a problem from beginning to end for a second example, using all of the principles we derived above.

Sample Problem 2: An organization you belong to is trying to raise money by selling sweatshirts, which cost \$9 each, with a fixed order cost of \$25. You did a survey in your dorm to estimate the demand, given in Table 1.

TABLE 1

Selling Price (\$)	10	15	20	25
Quantity Sold	40	25	13	5

In Sample Problem 2 of Section 3.2, we did an initial optimization analysis by modeling the price as a linear function of the quantity and then formulating expressions for the revenue, cost, and profit in terms of the quantity and optimizing the profit function. The solution obtained was to plan to order and sell 20.5 sweatshirts, for a profit of \$151, planning to sell them for \$17.60 each. Perform post-optimality analysis for this problem.

Solution: *Verifying the Models.* Using a different technology, we can verify that the linear demand function is given by

$$d(q) = -0.4198\ldots q + 26.21\ldots$$

If you had a partner working with you, you could have them then verify the derivation of the cost, revenue, and profit functions, and they could confirm that the profit function comes out to be

$$-0.4198\ldots q^2 + 17.21\ldots q - 25$$

Verifying the Solutions. To verify the maximum, you could use technology to find the maximum of the profit function, and obtain an answer of

$$q = 20.498\ldots$$

Substituting this answer into the profit function using technology, you would get

$$\pi(20.498\ldots) = 151.401\ldots$$

verifying the profit. We could also plug our optimal quantity into the demand function to verify the optimal price:

$$p = 17.605\ldots$$

Thus we have verified the results of our initial analysis, and we can use similar methods to verify all of the other analyses performed in the post-optimality analysis.

Validating the Fit of the Models. Next we come to the step of *validation*. First we want to evaluate the *fit* of our model, considering the *context* of the problem and the *use* we want from the model. In this case, we want to optimize the profit, so it is an *optimization* problem, and from our initial analysis, the answer seems to lie between the lowest and highest data values for the quantity, so the implicit forecasting we are doing corresponds to *interpolation*. This suggests that we would like a good fit to the data, especially near the optimal solution. We can evaluate the fit by visually examining the graph of the demand function with the data, shown in Figure 3.4-7.

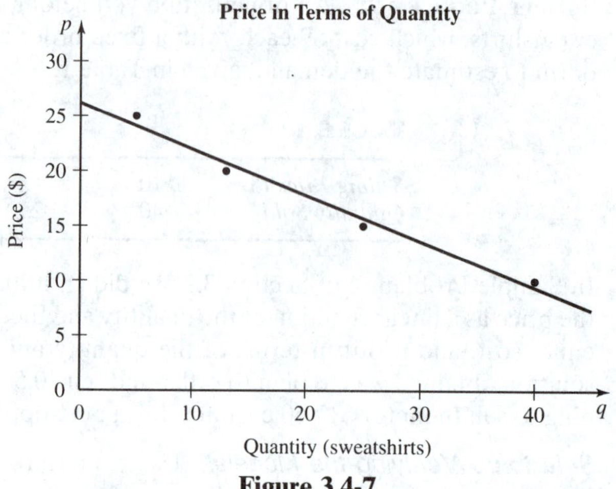

Figure 3.4-7

From looking at the graph, we can see that the fit is pretty good overall, and pretty good around the optimal quantity (20.5), although the data do seem to have a bit of curvature to them, and the model does not.

Validation by Comparing the Fit of Alternative Models. This validation of the fit immediately raises the question of whether we have chosen the best possible model for the problem. One alternative model we could consider would be to make *price* the input and *quantity* the output variable. In many ways, this is more intuitive for most of us (who are not economists). To just evaluate the fit of this alternative model, let's look at its graph, shown in Figure 3.4-8.

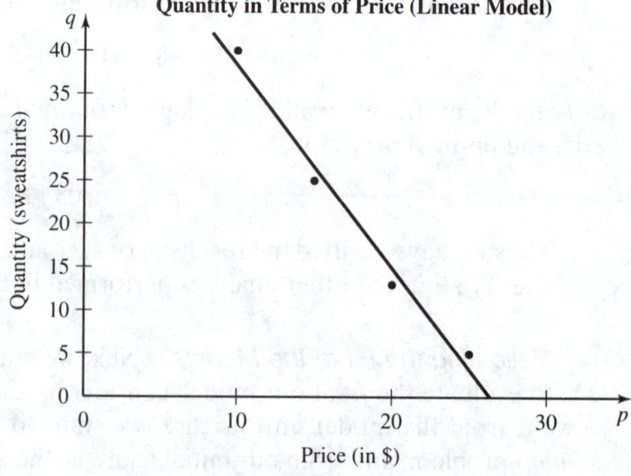

Figure 3.4-8

Since this is very close to a simple flip of the axes (but not exactly, as shown in Section 1.3), the fit is very similar: pretty good overall, and pretty good around the original optimal quantity of 20.5. Again, the data suggest some curvature not reflected by the linear model.

What if we tried to fit a quadratic model to the data in this form? Figure 3.4-9 shows the resulting graph. This looks good! It reflects the curvature nicely and fits the data better than the linear model. This is where the context and use become important, however, because if we were *extrapolating* (if the optimal solution had been a price over $25), the parabola could be a real problem, as mentioned briefly in Section 1.4. This is because the parabola would decrease to a minimum, then start *increasing*! In other words, beyond some point, charging more money per sweatshirt would result in selling *more* sweatshirts. This *could* be possible, if somehow the higher price made people *think* the sweatshirts were worth more, but in this situation that is quite doubtful and goes against our basic intuition and experience of price versus demand. If we use the quadratic model, we will have to be sure to restrict our domain to get meaningful results.

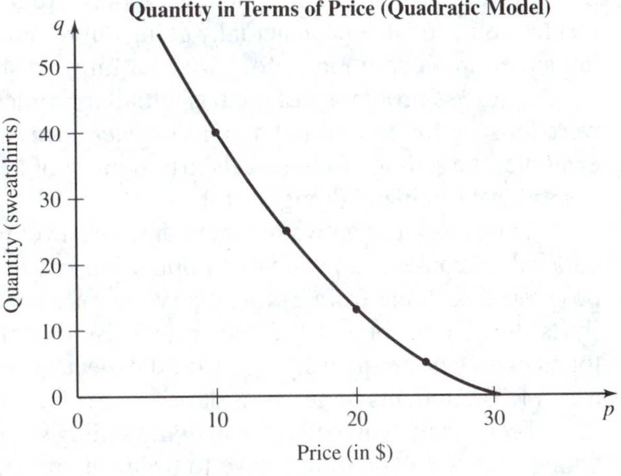

Figure 3.4-9

What other models might make sense? We could go to a cubic model, but the quadratic fits so well, that does not seem justified. Remember KISS! Notice that our data are concave up everywhere and decreasing. The other models we have studied that would seem to make sense here are the exponential and power models. Let's see what their graphs would look like, shown in Figure 3.4-10.

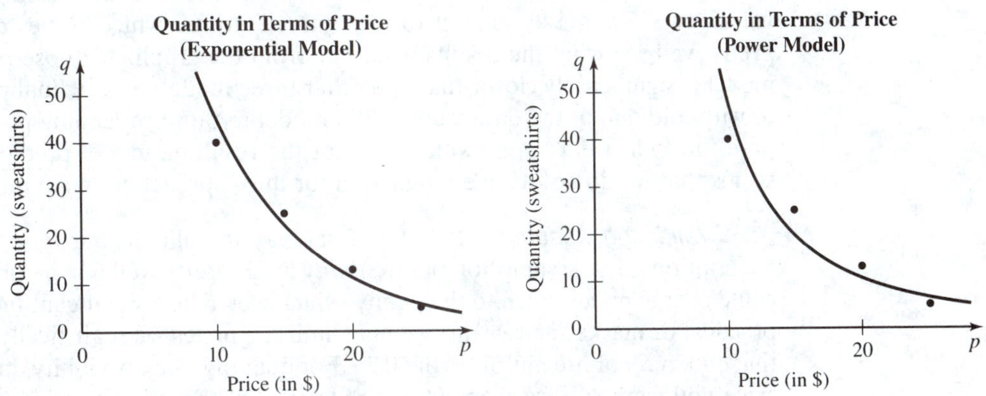

Figure 3.4-10

Neither of these models looks as good as the quadratic model for interpolation (fit to the data). However, once again, *if* our optimal price had been more than \$25, one of these models would probably be better than the quadratic model, because both of them continue decreasing down to a limiting value of 0 (if you charge too much, you won't sell any sweatshirts). In other words, their **shape** and **end behavior** would then fit the **context** of the problem better than the quadratic model, which is relatively more important for *extrapolation*.

From all of these considerations, it would seem that the *quadratic* demand function is the best model for the context and use in *this* problem.

Validation by Using Independent Data to Compare with Model Predictions. The next step in the validation process for our model would be to gather new data and compare the actual results to what the models predict. In this particular problem, actual results would ideally mean actually going out to sell sweatshirts, but then this would mean implementing a solution before we had fully finished our post-optimality analysis. For Meg's exercise problem and the basketball examples, this was less of an issue, because we were looking for an optimal action *in general* for an ongoing activity. For the sweatshirt example, the selling of the sweatshirts is more of a one-time big deal, at least in terms of the way we originally thought of it.

This raises a point we haven't considered yet in this problem: Exactly where did the data values come from, and what population and time period do they refer to? Let's suppose the data came from asking everyone on your floor of your dorm how many sweatshirts they would buy at the four prices listed in the data, and you simply added them together to get the quantity you would expect to sell at each price. Suppose you asked a total of 100 students to get your data.

Technically, you *could* try actually selling sweatshirts to 10 other students (perhaps taking orders, so you didn't have to order them yet, but with a picture of the design and color) at the optimal price and see if the demand is *proportional* to that predicted by the model. For instance, at \$10, 40 of the 100 students indicated they would buy, so you would expect 4 out of the 10 to buy. This prediction using the model then would just be to multiply the output by 0.1 (divide by 10). If you did this by orders, there could be a bias that you might have to take into account: People might be more willing to buy with a real sweatshirt in front of them than to order an abstract one.

In our problem, let's suppose this would be too much trouble. Another validation of this type would be to compute the model predictions for the various models at the price values of \$15 and \$20 (closest to the optimal) to see which comes closest. But we can pretty well see what the results would be from the graph: At those points the quadratic model is significantly closer than the other three models are. A final possibility would be to withhold one of the data values, fit a model to those remaining points, and then compare the fourth data point value to what the resulting model predicted. In this case, it seems that we do not have enough data for this approach to make much sense.

Validating Assumptions. The final way to validate our *model* is to evaluate its assumptions. The assumption of *divisibility* holds pretty well for the price (technically, we could charge prices down to the penny, which would be 4 significant figures), although for practical or marketing reasons we may limit the prices we realistically consider to values that end in .99 or are multiples of \$0.25. For quantity, the divisibility does not hold as well, since you can't sell fractions of a sweatshirt. On the other hand, when we defined the demand function in Section 1.3, we talked about the number of sweatshirts you could *expect* to sell, which could be interpreted as meaning the number *on average* (if you did it for a number of floors of 100 students each). In such a case, fractions do make sense, and so divisibility could hold very well. This kind of interpretation (that the output is an *expected* result, *on average*) often will make the most sense when using continuous models.

The assumption of *certainty* is somewhat harder to assess. There is a good chance that people's actual behavior will not match exactly what they say verbally; in this case, actual sales are probably likely to be *less* than what the survey data suggest. This is actually a *bias*, and you could try to adjust for it by gathering validation data by actually trying to sell sweatshirts at a price, as discussed earlier. If your ultimate goal is to sell to the same 100 students you spoke to in your survey, then the assumptions of certainty could be quite strong (especially if you got some validation data and found they matched the model well). If you planned to sell to a *different* floor, or to a larger population (like your whole dorm), then the level of certainty would be less. Overall, you probably would feel that certainty holds pretty well in this case (assuming you just plan to sell to your floor), but quite far from perfectly.

Intuitive Validation of Solution. Now we are ready to validate our *solution*. Let's start by seeing if our original solution makes sense. A little under \$18 sounds reasonable for a sweatshirt, and from the data, this would seem to go with a demand of between 13 and 25, so everything seems to make sense.

How could we make a ballpark guess? Well, we could estimate the profit at the data values. For example, at \$10, you'd sell 40, for a revenue of \$400, and the cost would be $(9)(40) + 25 = \$385$, so your profit would be \$15. At the other prices, the profit would be

$$(15)(25) - [(9)(25) + 25] = 375 - 250 = 125$$

$$(20)(13) - [(9)(13) + 25] = 260 - 142 = 118$$

$$(25)(5) - [(9)(5) + 25] = 125 - 70 = 55$$

These calculations suggest an optimal price somewhere around \$15, perhaps a little more. This validates our original solution quite nicely.

Validating the Solution Using Alternative Models. Now let's try validating our solution by looking at the solutions for some of the alternative models we have discussed. So far we only have a solution using the linear demand model (with quantity as the independent variable), but our validation suggested that the quadratic model (with price as the independent variable) would be the best, and we could also try the linear demand model with price as the input.

In Section 1.3 we used the data from the survey to derive a linear demand function model with price as the input (in this case, the parameters are actually exact):

Verbal Definition: $D(p)$ = the expected number of sweatshirts sold (on average) if you charge p dollars per sweatshirt.

Symbol Definition: $D(p) = -2.34p + 61.7$ for $9 \le p \le 26.4$.

Assumptions: Certainty and divisibility. Certainty implies the relationship is *exact*. Divisibility implies that *any* fractions of sweatshirts and dollars are possible.

What price should you charge for the sweatshirts to realize the most profit?

In Section 1.4 (Sample Problem 6) we used the demand function to find the revenue function and used the data given to find the cost function. We finally determined the profit function by subtracting the cost function from the revenue function. The revenue function was given by

$$R(p) = pq = (p)[D(p)] = (p)(-2.34p + 61.7) = -2.34p^2 + 61.7p$$

and the cost function was given by

$$\begin{aligned} C(p) &= 9q + 25 = (9)[D(p)] + 25 = (9)(-2.34p + 61.7) + 25 \\ &= -21.06p + 580.3 \end{aligned}$$

so the profit function was

$$\begin{aligned} \Pi(p) &= R(p) - C(p) = (-2.34p^2 + 61.7p) - (-21.06p + 580.3) \\ &= -2.34p^2 + 82.76p - 580.3 \end{aligned}$$

Our profit model was thus:

Verbal Definition: $\Pi(p)$ = the expected profit, in dollars, if you charge p dollars per sweatshirt and order and sell exactly $q = D(p)$ sweatshirts (on average).

Symbol Definition: $\Pi(p) = -2.34p^2 + 82.76p - 580.3$ for $9 \le p \le 26.4$.

Assumptions: Certainty and divisibility. Certainty implies that the demand function is exactly right. Divisibility implies that you can order and sell fractions of sweatshirts.

(See Section 1.4 to recall how the domain was determined.)

The profit function is a quadratic function. The leading coefficient is negative, so the function will be concave down, first increasing, leveling off, and then decreasing. If a function changes from increasing to decreasing, there will be a local maximum at the point where the derivative is 0 (Figure 3.4-11).

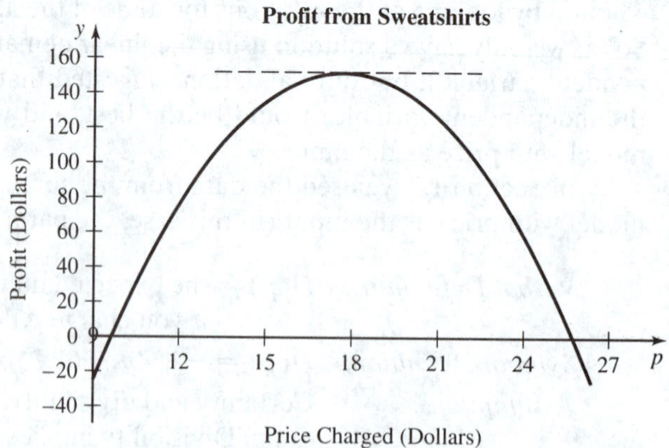

Figure 3.4-11

From the graph we can see that the local maximum is also a global maximum. Thus, it appears that a maximum profit of approximately \$150 can be realized if you charge about \$18 for the sweatshirts. To find the actual optimum price and profit we can find the derivative of the profit function, set it equal to zero, and solve the resulting equation:

$$\Pi'(p) = -4.68p + 82.76$$
$$-4.68p + 82.76 = 0$$
$$4.68p = 82.76$$
$$p = \frac{82.76}{4.68} = 17.683\ldots$$

Our estimate from the graph wasn't too bad. You should charge approximately \$17.68 per sweatshirt to optimize your profit. In reality, you would probably round off the price to \$17.75 or even \$18. If you actually charged \$17.68, the predicted profit would be

$$-2.34(17.68)^2 + 82.76(17.68) - 580.3 \approx 151.5$$

or approximately \$151.50. The quantity of sweatshirts you would expect to sell, on average, in this case would be

$$D(17.68) \approx 20.33$$

These results are very similar to our original solution, which makes sense, since both used a linear model for the demand.

What about the quadratic model? The demand function in this case turns out to be

$$D_2(p) = 0.07p^2 - 4.79p + 80.95, \text{ for } 10 \leq p \leq 25$$

The revenue function is then given by

$$\begin{aligned} R_2(p) = pq = (p)[D_2(p)] &= (p)(0.07p^2 - 4.79p + 80.95) \\ &= 0.07p^3 - 4.79p^2 + 80.95p \end{aligned}$$

The cost function is given by

$$\begin{aligned} C_2(p) = 9q + 25 = 9[D(p)] + 25 &= 9(0.07p^2 - 4.79p + 80.95) + 25 \\ &= 0.63p^2 - 43.11p + 728.55 - 25 = 0.63p^2 - 43.11p + 703.55 \end{aligned}$$

Thus, the profit function is given by

$$\begin{aligned} \Pi_2(p) = R_2(p) - C_2(p) &= (0.07p^3 - 4.79p^2 + 80.95p) - (0.63p^2 - 43.11p + 703.55) \\ &= 0.07p^3 - 5.42p^2 + 124.06p - 703.55 \end{aligned}$$

To find the optimal price, we now set the derivative equal to 0 and solve:

$$\Pi_2'(p) = 0.21p^2 - 10.84p + 124.06$$
$$0.21p^2 - 10.84p + 124.06 = 0$$
$$p \approx 17.13 \text{ or } p \approx 34.49$$

Let's check these critical points graphically, shown in Figure 3.4-12.

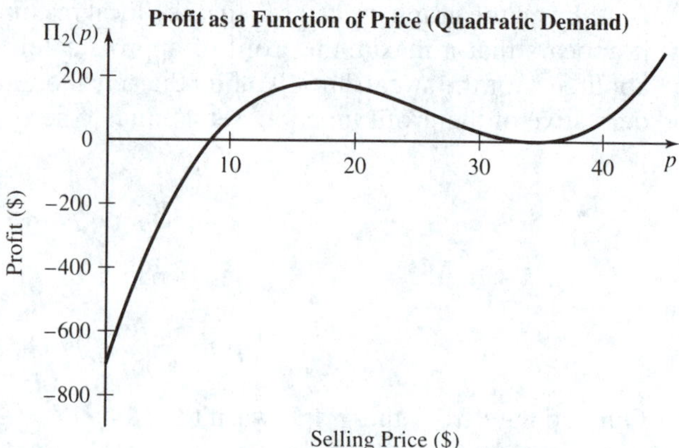

Figure 3.4-12

We see that a price of $17.13 is the global maximum over our domain of interest ($10 to $25). Notice that, because of the unrealistic behavior of the quadratic demand model, the local minimum at $34.49 is where the profit starts to increase again, and it continues to increase to positive infinity after that. Without a domain restriction, the model would predict that unlimited profit is possible!

This time, then, our optimal price comes out to be about $17.13. Plugging back into the profit function, we get an expected profit of $133.03 and a demand of 19.44 sweatshirts.

If we wanted, we could also try the quadratic demand model that would be a function of quantity, but we leave that as an exercise.

These answers help validate each other, since all come up with a selling price between $17 and $18, and predict a demand of about 20 sweatshirts at the optimal price, with a profit in the neighborhood of $140.

Sensitivity Analysis. What about sensitivity analysis? If you felt that one or more of the people you surveyed didn't understand your question or were generally clueless, then you might remove or adjust those data points, and resolve the problem using technology. We leave this for an exercise, since we aren't given any information to do this in the question.

Accuracy and Biases. Next, let's look at accuracy and possible biases. We have already discussed the fact that there is likely to be a bias in our data that makes it seem as if we will sell more than we actually will, because people's verbal estimates of how much they would pay at most are likely to be high. This means that, for a given price, the demand from the data is likely to be high. On the other hand, people who are thinking strategically may quote a lower price than they would really pay at the maximum, hoping it will push you to lower your selling price. Chances are that this strategic bias will probably be less than the words versus deeds bias. How could we adjust for this words versus deeds bias? We already mentioned how we could gather validation data by asking 10 students to place orders for sweatshirts. If we offered them at $15, and got orders for 2 (versus the 2.5 the model would predict), we could assume that people's actual demand would be the same proportion (2.0/2.5 = 80%) at all of the selling prices. However, this adjustment itself is likely to be understated, since ordering abstract sweatshirts is not as enticing as the real thing. In fact, this could be good, because the understatement could just compensate for the strategic bias we mentioned above.

How could we find the new solution after adjusting for this bias? Let's look at this using our best model, the quadratic model in terms of price. Since the demand would be 80% of the original demand, we can simply replace $D_2(p)$ with $0.8D_2(p)$. If you work through the algebra, you will see that this is equivalent to multiplying the profit function by 0.8, followed by the addition of a constant (−5):

$$\Pi_3(p) = p[0.8D(p)] - (9[0.8D(p)] + 25) = (p - 9)[0.8D(p)] - 25$$
$$0.8\Pi_2(p) = 0.8[(p - 9)[D(p)] - 25] = (p - 9)[0.8D(p)] - 20,$$

so

$$\Pi_3(p) = 0.8\Pi_2(p) - 5$$

As we mentioned in Section 3.3 and in the basketball example, multiplication by a positive constant and addition of a constant do not change the locations (input values) for extrema, so our optimal prices would come out the same. For the quadratic model, this would again be a selling price of $17.13. The profit would be given by

$$(0.8)(133.03) - 5 \approx 101.42$$

($101.42) and the demand would be 80% of 19.44, or about 15.55.

Margin of Error: Precision of Data. What about the precision of our input and output variables? Our input values, since they were given exactly to the people responding to your survey, have essentially perfect precision. On the other hand, the output values (quantity) could easily be anywhere between the given values and 80% of the given values, based on our earlier scenario. This would suggest it might make most sense to use 90% of the original data values as a central value, with a margin of error of 10% on either side (and probably less, since neither extreme is very likely). There is also likely to be some level of rounding error, since people were giving answers in whole numbers of sweatshirts, but, if we are considering the expected demand on average, fractions could also be possible. This could conceivably contribute up to 5% margin of error, but this should be quite independent of the other error, and the net effect will probably still be around plus or minus 10–11%. This would suggest no more than 2 significant digits for the solution, since for a price of about $17, 10% would be about $1.70. Rounding to 1 significant digit would correspond to plus or minus $5, and rounding to 2 significant digits would correspond to plus or minus $0.50.

Margin of Error: Alternative Solutions. As another assessment of the margin of error, we can use technology to work out the solutions using the exponential and power models, just to see how far away they are. For the exponential model, the solution comes out to be a price of $16.25, with a demand of 18.96, for a profit of $112.53. For the power model, the optimal price is $16.61, with a demand of 15.83, for a profit of $95.53. These results suggest a margin of error of a dollar or two in the price and of 2 or 3 sweatshirts in demand, and up to $30 in profit.

Overall Margin of Error for the Solution. How can we put the pieces together? The numbers suggest values of about \$17, give or take \$1, for the selling price, ordering about 19 sweatshirts to meet the demand (which could fluctuate by up to 2 or 3 sweatshirts), expecting a profit of about \$125, plus or minus as much as \$30.

Conclusions. What's the bottom-line conclusion? You should aim to order about 19 sweatshirts (maybe 20 if you want a round number) and plan to sell them for about \$17, expecting a profit of about \$125, give or take up to \$30. If the verbal bias is especially strong, then your demand could be as little as 15, so you might need to resort to a clearance sale price, or expand your target market. If you do the latter, your profit should end up on the high end of the range. If you did not gather validation data (such as the 10 students), your margin of error would be even higher; if you validated with a bigger sample or real sales, your margin of error should be lower. □

Section Summary

Before you begin the exercises, be sure that you

- Can verify your model and solution by checking all of the calculations, either by hand or by using a second technology. If this is not possible, get someone else to do the calculations independently. Try to verify your calculations from beginning to end (check for typos in data, and all stages of calculations).
- Know how to validate your model and alternative models by visually evaluating the fit graphically, keeping in mind the context and use of the model in the problem, as well as by evaluating the assumptions of the models.
- Know how to validate your *solution* by checking it with your common sense and your initial ballpark guess at a solution, and by comparing the solutions of reasonable alternative models.
- Know how to do sensitivity analysis by thinking about which data values are most questionable and seeing what effect eliminating or adjusting the problematic data values has on your solution. Remember that not fitting a nice pattern is not enough reason *in itself* to change a data value but can suggest *possible* problematic data.
- Have tried to identify any biases in your data or model and attempted to eliminate them (ideally, by planning your data collection carefully, or possibly by adjusting your data if necessary and seeing the effect on your solution) to maximize the accuracy of your model and solution.
- Know how to estimate the margin of error from measurement error or inherent randomness in your input and output data, and how to roughly estimate the combined effect of different errors (usually more than the largest error, but less than the sum).
- Understand the nature of rounding errors in calculations and how they can propagate, and how to choose an appropriate number of significant digits in model parameters, calculations, and solutions. The number of significant figures needed

for your final solution is determined by your overall margin of error, with an extra hedge of precision. It is a good idea to carry at least one extra significant figure from the beginning in your model parameters and intermediate calculations (by hand) than is needed in your final solution. When doing calculations with technology, use all possible precision until the very end, when you can round to the appropriate number of significant figures.

- Remember that, in general, the precision of an answer is determined by the *least* precise of the components. With addition and subtraction, it is *decimal places* that count; with other operations, it is *significant figures* that count. Subtraction combined with other operations can erode precision, but this is rare. As a general rule, use significant figures as a guide.
- Understand that another way to estimate the margin of error in your solution is to examine the solutions using reasonable alternative models.
- Understand how to put the above considerations together to form final conclusions about a solution and how to implement it in real life. Use your margin of error explorations to express your answer with a reasonable level of precision (both number of significant figures and a margin of error for all relevant variables), and give the solution in words that reflect the way the decision maker thinks about the problem.

If you are working on a student-generated project, answering the following questions can help your analysis and conclusions:

- How accurate and precise are your data values (what biases are likely, can you avoid or estimate them, how many digits make sense; what kind of margin of error is likely)?
- Can you guesstimate an optimal solution?
- What models make sense for your data and context and purpose? Pick one that you think is best, and explain why.
- Use your model to find an optimal solution (or a forecast). How can you verify your answer?
- How can you validate your model and solution? What assumptions did you make? How valid are they?
- Which data values are most questionable? If you plug in values that in your opinion are more reasonable or representative, or are at least equally plausible, what happens to your solution?
- How much inherent randomness does there seem to be in your problem? Why?
- Can you find a solution with 1 or more other models? How do the results compare? Which do you have most confidence in?
- What level of precision and margin of error make sense for your solution and conclusions?

EXERCISES FOR SECTION 3.4

Warm Up

For Exercises 1–8, assume that the given numbers are only as precise as the given digits indicate (are rounded off), and round off each answer to an appropriate level of precision:

1. $1.23 + 0.0367$

2. $63.25 - 0.371$

3. $(6.893)(4.2)$

4. $15 \div 3.14159$

5. $3.2e^{-0.0511}$

6. $2.37 \ln(6.2)$

7. $(2.837)(97.5) + 5.2$

8. $3.7e^{-0.0863} - 2.367$

9. For the calculation $2.016^{3.7}$, assuming both numbers are rounded, figure out the interval of values each could represent, do the calculation for each of the four possible combinations of the extremes of these intervals, and indicate what level of rounding would seem to make most sense. What does this suggest for a rule for precision when working with powers?

10. For the calculation $\sqrt[3.705]{4.2}$, assuming both numbers are rounded, figure out the interval of values each could represent, do the calculation for each of the four possible combinations of the extremes of these intervals, and indicate what level of rounding would seem to make most sense. What does this suggest for a rule for precision when working with roots?

Game Time

11. You are working on a typing project for one of your professors and are being paid a penny for every 5 words typed. You want to see what this corresponds to in dollars per hour as you work, so you keep careful records of your time and use the word count facility on your word processor to count the words. In the first stretch of working, you work from 10:02 to 11:27, and type 4311 words. Your watch keeps track of minutes, but not seconds.

(a) *Without using your calculator or writing anything down*, try to get a ballpark estimate of how much money you earned, and roughly how much you were earning per hour. Explain your thinking. Does your answer seem to make sense from a real-life perspective?

(b) How much money did you earn in that period of time? How should this answer be rounded (if at all)? Explain.

(c) What is your estimate for the number of minutes you worked? Since your watch tracks only minutes, if it read 10:02 at the start and 11:27 at the end, what are the *extreme* cases of how few or how many minutes you *could* have technically been working? What *error* does this represent (in minutes), on either side of your estimate?

(d) Roughly, what percent error does your answer to (c) represent (to the nearest percentage point), on either side of your estimate?

(e) When the time discussed in part (c) is expressed in hours, how much would you have to round your answer to be correct for all possible values between the extremes?

(f) Do the straightforward calculation of your earnings in dollars per hour, taken out to the greatest precision you can get from the calculator display (as if all of your numerical values were exact with unlimited precision).

(g) Now try the calculation again, using the extreme values for the number of minutes worked.

(h) What would be a reasonable way to round your answer in (f) to give a meaningful value?

(i) What rough margin of error would seem to make sense for your answer realistically? What percent does this correspond to, roughly?

12. You have just filled the gas tank in your car at an old gas station and want to calculate how many miles per gallon you achieved since your last fill-up. You remembered to set your trip odometer at your last fill-up, so you know that you traveled 283.25 miles (it's halfway between 283.2 and 283.3). Since the gas pump is old, it only has an analog circular readout of the number of gallons you put in (similar to the odometer), and it looks to be about 8.35 gallons (halfway between 8 3/10 and 8 4/10). You were as careful as you could be to "top off" the tank to the same level, *both* at the *last* fill-up *and* just *now*.

(a) *Without using your calculator or writing anything down*, try to get a ballpark estimate of how many miles per gallon you achieved. Explain your thinking. Does your answer seem to make sense from a real-life perspective?

(b) Use your calculator to do the basic mileage calculation (in miles per gallon), showing all of the digits given by the calculator display.

(c) Based on the principles in this section, how should you round off your answer so that the digits are meaningful? Explain *why* you rounded the way you did.

(d) Since the amount of gas was rounded to 8.35 gallons, it could have been anywhere between 8.345 and 8.355. Try the calculation with each of these values and see to what level of precision the answers agree (in other words, how could you round your answer to be accurate for any value between the extremes?). Does this agree with your answer in part (b)?

(e) Since the number of miles is rounded off to 283.25, what range of values could the exact value lie within (similar to what was done in part (c))? Try the calculation again with these extremes and the original 8.35 gallons (two calculations), then with every possible combination of the extremes of both measurements (4 calculations). What level of rounding will give the same answer for all six of these combinations?

(f) Suppose that you decide that topping off is so imprecise that your estimate of the number of gallons used since your last fill-up is really only accurate to the nearest tenth of a gallon. How would you express the estimate of your miles per gallon in this case?

13. You are selling T-shirts for a school club. Your dorm has 200 people in it, and you pick 20 at random. You show them the design and ask questions about how much they would pay for a T-shirt. The results are given below.

Price ($)	5	10	15	20
Number of Buyers	7	4	2	1

(The number who would buy is out of the sample of 20.) The shirts cost $6 each, with no fixed cost (you can avoid shipping by picking them up).

(a) From the table, make a ballpark guess at the optimal price to charge to maximize profit.

(b) Find a quadratic model for the demand (from the *entire* dorm, not just the sample) as a function of the price charged and use it to express the cost of the shirts as a function of the price charged (assume you will order the predicted demand).

(c) Find a model for your profit as a function of the selling price and find the optimal price and quantity to order.

(d) Given your profit function, state at least three ways you could find the optimal price, to verify your answer, and do so.

(e) What assumptions are you making? Comment on their likely validity.

(f) Now try fitting an exponential demand function and find the optimal price and quantity using *that*.

(g) After asking a few more people, suppose you believe that the demand at $5 would be more like 90 (rather than 70). Use a quadratic demand model to find a new solution. How sensitive is your answer to this change?

(h) How accurate and precise do you think your data are? What level of precision makes sense for your answer?
(i) Considering all of the above, what seems to be a reasonable margin of error for your answer?

14. Problem 1 of Section 3.2 involved finding the best amount of time to warm up with a computer game before playing a match with a friend. The game involves removing tiles, so you want to have as few tiles left at the end of the game as possible. From data you gathered, the relation between warm-up time (in minutes) and the result of the game (in tiles left) is given below:

Minutes	10	20	30	40	50	60
Tiles Left	540	260	80	15	75	380

(a) Find a quadratic model for this relationship and find the optimal warm-up time.
(b) Given your model, state at least four different ways in which you could find the optimal warm-up time to verify your answer; then do so and compare the results.
(c) What would be the best way to validate your model and answer?
(d) Comment on the validity of your assumptions.
(e) Try fitting a cubic model to your data and see what optimal solution it gives. Which model do you think is better? Why?
(f) Suppose the last data point (the 60-minute warm-up) was biased because the phone rang, and you believe a more representative output value would have been 250 tiles left. Find a new quadratic model and solution. How sensitive is the solution to such a change?
(g) How accurate and precise do your data seem to be? What level of precision makes sense for your answer?
(h) Considering all of the above, what seems to be a reasonable margin of error for your answer?

15. For the sweatshirt problem in Sample Problem 2 of this section, find quadratic, exponential, and power models for the demand in terms of *quantity*, derive the revenue, cost, and profit models for each, and find the optimal quantity, profit, and price in each case using technology.

16. For the sweatshirt problem in Sample Problem 2 of this section, suppose after talking to one of the people you had surveyed, you realized they misunderstood what you were asking, and so the first data point should have been a demand of 41 rather than 40 corresponding to a price of \$10. Redo the original analysis (linear demand in terms of quantity) using technology and see how much the solution is affected.

17. For the sweatshirt problem in Sample Problem 2 of this section, assuming that the true demand function is 80% of the one fit from the survey (use the quadratic model, in terms of the price), derive the full adjusted functions for the demand, revenue, cost, and profit and verify the result stated in the text. If you can, fill in the details for the explanation in the text of how the new profit function is 80% of the original one, minus 5.

18. Look back at Meg's exercise problem in Section 1.0 and fill in the details of the post-optimality analysis discussed there, now that you have a more complete understanding of the different steps.

19. For the coffeehouse problem in Sample Problem 4 of Section 3.2, perform post-optimality analysis to come up with a final conclusion about the best price to charge for the concert and the expected attendance and profit, as best you can with the given information.

20. If you are working on a student-generated project, answer the questions at the end of the section summary for your project (do a post-optimality analysis for your problem).

3.5 PERCENT RATE OF CHANGE AT A POINT, ELASTICITY, AVERAGE COST*

We have seen a number of examples of optimization problems by now, including applications to personal life, sports, politics, and business. One of the major application areas for optimization is the field of **economics**. Economics looks at optimization from the perspective of individual businesses and other organizations (**microeconomics**) and from the perspective of an entire region or country or group of countries (**macroeconomics**). The basic mathematics of optimization are exactly the same as we have discussed earlier in this chapter. In this section, we will examine a couple of specific applications of optimization in economics, and some of the new concepts and definitions that are involved. One of these concepts is the idea of **price elasticity of demand**, which can be thought of intuitively as the marginal effect of price on demand, expressed in percentage terms for both. In other words, it is the percentage change in demand per percentage point change in price. You can think of it as the percent change in the quantity that will be sold if the price increases 1%. We will define this for both intervals and at a point, as we did for rates of change, and see how this tells us about the effect of price on the total revenue. We will also look at the problem of **minimizing average cost** (the cost per unit produced), which would correspond to maximizing profit per unit if your revenue per unit (price) is fixed and constant, as long as the demand is sufficient. We will see that the condition for minimizing average cost is simple and intuitive.

Here are some of the kind of problems that the material in this section can help you solve:

- You are putting on a concert, and your total cost is fixed in advance. You have estimated a model for the demand for tickets as a function of the price you charge. How much should you charge to maximize your profit from the concert?
- You have a small business making juggling beanbags. You have several people who help you make them, and have made an estimate of your total cost in a month as a function of the number of beanbags you make. The market is fairly competitive for beanbags, so you need to charge the going price to sell a reasonable number and make a decent profit. How many should you make in a month, if demand is not a realistic constraint?
- A group you belong to has bought a certain number of sweatshirts and will not be buying any more. You have done a small market survey to estimate the demand function for the population you plan to sell them to. What price should you charge to maximize your profit?

By the end of this section, you should be able to solve the above problems and others like them, and you should

- Understand how price elasticity of demand over an interval and at a point are defined, and know how to calculate each.

*This section is optional and may be omitted with no loss of continuity or logical flow in later sections.

- Understand the idea behind the percent rate of change at a point, how to calculate it, and how to interpret it.
- Understand how to interpret the magnitude of the elasticity at a point.
- Understand the relationship between the magnitude of the elasticity at a point and the effect on the revenue of raising the price a small amount.
- Understand why revenue is maximized when elasticity has a magnitude of 1.
- Understand the relationship between the value of the marginal cost compared to the average cost at a point and the effect of making one more unit on the average cost.
- Understand why average cost is minimized when the marginal cost equals the average cost.

EXAMPLE 1: In Sample Problem 2 of Section 3.4, we looked at a problem involving the sale of sweatshirts. As part of our solution, we used the linear demand function given by

Verbal Definition: $D(p)$ = the expected number of sweatshirts sold (on average) if you charge p dollars per sweatshirt.

Symbol Definition: $D(p) = -2.34p + 61.7$ for $9 \leq p \leq 26.4$.

Assumptions: Certainty and divisibility. Certainty implies the relationship is *exact*. Divisibility implies that *any* fractions of sweatshirts and dollars are possible.

Suppose someone in the group had already gone ahead and ordered 35 sweatshirts before you got involved in the operation. Now, the only question left is: How much should you charge for the sweatshirts?

Review of Rates of Change

Let's start by seeing what selling price would get rid of all of the sweatshirts, according to the demand model. This means finding the value of p for which the demand, $D(p)$, is 35. So we can set $D(p)$ equal to 35 and solve:

$$\begin{aligned} D(p) &= 35 \\ -2.34p + 61.7 &= 35 \\ -2.34p &= -26.7 \qquad \text{(subtracting 61.7 from both sides)} \\ p &= -26.7/-2.34 \approx 11.4 \end{aligned}$$

Translation: In order to sell exactly 35 sweatshirts, we would charge about \$11.40 per sweatshirt. If we charge less than this, we could sell more sweatshirts than 35 if we had them, but let's assume here that the fixed cost for the sweatshirts is high enough that the group will not consider ordering more, so would sell only 35.

Of course, if you charge more than \$11.40, you would expect to sell fewer than 35 sweatshirts, as given by the demand function. Let's consider for a moment the possibility of charging \$12 for the sweatshirts (to round up the 11.4 to a value that would mean you don't have to worry about coins for change). At \$12, you would expect to sell

$$D(12) = -2.34(12) + 61.7 \approx 33.6$$

sweatshirts. What if you charged a dollar more and made the selling price \$13? Since the demand function is linear, we know we would expect the demand to change by −2.34. We could say that the **marginal demand** at $p = 12$ is −2.34 sweatshirts per dollar of price increase. Because the demand is linear, this is both the instantaneous rate of change at 12 and the average rate of change over the interval [12,13]. If we had a curved demand function, we could use the derivative at 12 to estimate the effect of a dollar increase (marginal analysis), or we could calculate the demand at 12 and 13 and calculate the average rate of change over that interval. With a linear demand function, this is greatly simplified.

From the above comments, we can figure out that the approximate expected demand if you charge \$13 would be about

$$33.6 - 2.34 \approx 31.3 \text{ sweatshirts.}$$

What would be the **percent rate of change** in the demand over the interval [12,13]? Recall that one way of thinking of this quantity is that it is the regular rate of change, expressed as a percentage of the initial output value. In other words, we know that the expected demand at \$12 is about 33.6 sweatshirts, and that it is decreasing by 2.34 sweatshirts per dollar of price increase. What percentage of the 33.6 sweatshirts does this decrease of 2.34 sweatshirts correspond to? The easiest way to do this calculation is to write it as a fraction first, then express it as a decimal, and then multiply by 100% (move the decimal point 2 places to the right and tack on the % sign):

$$\begin{aligned}\frac{-2.34 \text{ sweatshirts/dollar of price increase}}{33.6 \text{ sweatshirts}} &\approx -0.0696/\text{dollar of price increase}\\ &= -6.96\%/\text{dollar of price increase}\end{aligned}$$

Translation: Between a selling price of \$12 and a selling price of \$13, the expected demand (number of sweatshirts you would sell) is decreasing at a rate of 6.96% per dollar of price increase. If you look at the units in the above calculations, the sweatshirt units cancel.

What about the instantaneous rate of change? We could do the same thing (**divide the rate of change by the initial output value, which in this case is the *only* output value, the output value at the given point**) and call it the **instantaneous percent rate of change at a point.**

In our example, we have already mentioned that the instantaneous rate of change is −2.34 sweatshirts per dollar increase in price, which we can verify by taking the derivative and evaluating it at \$12:

$$\begin{aligned}D(p) &= -2.34p + 61.7\\ D'(p) &= D_p(-2.34p + 61.7) = -2.34 && \text{(taking the derivative)}\\ D'(12) &= -2.34 && \text{(the derivative is constant)}\end{aligned}$$

We already know that the demand at 12 is about 33.6, so to express the instantaneous rate of change as a percent of this, we would do the exact same calculation as before and obtain an instantaneous percent rate of change at 12 of −6.96% per dollar of price increase.

Price Elasticity of Demand over an Interval

Recall that we originally defined the percent rate of change over an interval as the average rate of change over the interval divided by the initial output value. Let's do some mathematical manipulation to see another way to think of this:

$$\text{Percent rate of change over } [x_1, x_2] = \frac{\text{average rate of change over } [x_1, x_2]}{y_1}$$

$$= \frac{\frac{\Delta y}{\Delta x}}{y_1} \qquad \left(\text{average rate of change over } [x_1, x_2] \text{ is } \frac{\Delta y}{\Delta x}\right)$$

$$= \frac{\Delta y}{y_1 \Delta x} \qquad (\text{multiplying top and bottom by } \Delta x)$$

$$= \frac{\frac{\Delta y}{y_1}}{\Delta x} \qquad (\text{dividing top and bottom by } y_1)$$

$$= \frac{\text{percent change in } y \text{ over } [x_1, x_2]}{\Delta x} \qquad \left(\text{percent change in } y \text{ over } [x_1, x_2] \text{ is } \frac{\Delta y}{y_1}\right)$$

In other words, the percent rate of change over an interval is also the percent change in the output value over the interval, averaged out per unit of the input variable over the interval. For example, if the output went down by 16% over an input interval with a width of 2 units, the percent rate of change over that interval would be

$$\frac{16\%}{2 \text{ input units}} = 8\% \text{ per input unit}$$

What if we followed this logic a step further? In moving from the *average* rate of change over an interval (change in output over change in input) to the *percent* rate of change over an interval, we have expressed the change in the *output* as a percent of its initial value instead of in its original units. This made the answer independent of the output units. What if we made the same substitution in the average rate of change formula for the change in the *input* as well, so our answer was independent of *all* units? We would then have

$$\frac{\text{percent change in } y \text{ over } [x_1, x_2]}{\text{percent change in } x \text{ over } [x_1, x_2]} = \frac{\frac{\Delta y}{y_1}}{\frac{\Delta x}{x_1}} \qquad \left(\%\text{ change in } y \text{ is } \frac{\Delta y}{y_1}; \%\text{ change in } x \text{ is } \frac{\Delta x}{x_1}\right)$$

$$= \frac{x_1}{y_1} \cdot \frac{\Delta y}{\Delta x} \qquad (\text{multiplying top and bottom by } x_1 y_1)$$

In our example, if we define the output variable to be $q = D(p)$, this expression becomes

$$\frac{\text{percent change in } q \text{ over } [p_1, p_2]}{\text{percent change in } p \text{ over } [p_1, p_2]} = \frac{\dfrac{\Delta q}{q_1}}{\dfrac{\Delta p}{p_1}} = \frac{p_1}{q_1} \cdot \frac{\Delta q}{\Delta p}$$

This is exactly the definition of the **price elasticity of demand** (which we will sometimes shorten to just **elasticity** for convenience) over the interval $[p_1, p_2]$, often denoted with the Greek letter η (written "*eta*" and pronounced "AY-tah"):

$$\eta = \text{price elasticity of demand over the interval } [p_1, p_2]$$

$$= \frac{\text{percent change in } q \text{ over } [p_1, p_2]}{\text{percent change in } p \text{ over } [p_1, p_2]} = \frac{\dfrac{\Delta q}{q_1}}{\dfrac{\Delta p}{p_1}} = \frac{p_1}{q_1} \cdot \frac{\Delta q}{\Delta p} = \frac{p_1}{q_1} \cdot \frac{(q_2 - q_1)}{(p_2 - p_1)}$$

A good way to think of the price elasticity of demand is as the **percent change in demand per percentage point increase in price**.

Let's make this more tangible by applying it to our example. Let us consider again a reference starting point of a selling price of \$12. We know that the demand at \$12 is 33.6 sweatshirts, and at \$13 it is 31.3 sweatshirts. So

$$p_1 = 12, q_1 = 33.6, p_2 = 13, q_2 = 31.3$$

$$\Delta p = p_2 - p_1 = 13 - 12 = 1, \text{ and } \Delta q = q_2 - q_1 = 31.3 - 33.6 = -2.3$$

(Notice that we have lost a digit of precision here because of the subtraction, as mentioned in Section 3.4. Δq should be -2.34, but by rounding our earlier values to 3 significant digits and subtracting, our answer is now only accurate to 2 significant digits. We could put back the full value of -2.34, but let's continue to see how to handle the precision when you have rounded your answers, which should only be when you are working by hand.)

Now we can find the price elasticity of demand over the price interval [12,13]:

$$\eta = \frac{p_1}{q_1} \cdot \frac{\Delta q}{\Delta p} = \frac{12}{33.6} \cdot \frac{-2.3}{1} \approx -0.82$$

Translation: If you raise the price from \$12 to \$13, the demand will be *decreasing* at a rate of about 0.82% of the initial demand per percentage point increase in price.

Let's look at this calculation from another perspective. The percent change in price is given by

$$\frac{13 - 12}{12} = \frac{1}{12} = 0.08333\ldots \approx 8.33\%$$

The percent change in demand is given by

$$\frac{31.3 - 33.6}{33.6} = \frac{-2.3}{33.6} \approx -0.068 = -6.8\%$$

So the percent change in demand per percentage point increase in price is simply the ratio of these values

$$\eta = \frac{\%\text{ change in demand}}{\%\text{ change in price}} = \frac{-6.8}{8.33} \approx -0.82$$

Thus, over this interval from \$12 to \$13, for each 1% increase in price, the demand *decreases* by about 0.82%.

So far we have only been discussing changes over an *interval* of prices, as we did in Section 2.1 with *average* rates of change. What if we wanted to make analogous definitions *at a point*, corresponding to *instantaneous* rates of change?

(Instantaneous) Percent Rate of Change at a Point

Recall that the average and instantaneous rates of change can be defined as

$$\text{Average rate of change over an interval} = \frac{\Delta y}{\Delta x}$$

$$\text{Instantaneous rate of change at a point} = \lim_{\Delta x \to 0} \frac{\Delta y}{\Delta x} = \frac{dy}{dx}$$

In other words, the rate at a point is the limit of the rate over an input interval, as the interval shrinks down to that point. Let's apply the same logic to the idea of the percent rate of change:

$$\text{Percent rate of change over an interval} = \frac{\frac{\Delta y}{\Delta x}}{y_1}$$

$$\text{Percent rate of change at a point} = \lim_{\Delta x \to 0} \frac{\frac{\Delta y}{\Delta x}}{y_1} = \frac{\lim_{\Delta x \to 0} \frac{\Delta y}{\Delta x}}{y_1} = \frac{\frac{dy}{dx}}{y_1} = \frac{f'(x)}{f(x)}$$

In the last line we have made the standard assumption that $y = f(x)$ and have used x for the point at which we are finding the rate of change (in the interval we would have called it x_1). Thus $y_1 = f(x_1) = f(x) = y$, and $\frac{dy}{dx} = f'(x)$. We can move the limit into the numerator since, as the interval shrinks, the value of y_1 is not affected, so it acts like a constant. In other words **the percent rate of change at a point is just the rate of change at that point expressed as a percent of the output value there, which is equal to the rate of change at that point divided by the output value there. To find it, we simply divide the derivative at the point by the value of the function there.**

Let's calculate the percent rate of change for the demand function at \$12. We know that the percent rate of change at 12 is simply the derivative at 12 divided by the function value at 12, so we get

$$\text{Percent rate of change of } D(p) \text{ at } 12 = \frac{D'(12)}{D(12)} = \frac{-2.34}{33.6} \approx -0.0696 = -6.96\%$$

Translation: At a price of \$12, the demand is decreasing at the rate of about 6.96% per dollar increase in price. If this value sounds familiar, it is the same numerical value as the percent rate of change over the interval [12,13]. These are identical because the demand function is linear, so the curve and the tangent are the same, as we pointed out earlier.

(Instantaneous) Price Elasticity of Demand at a Point

Now, we have calculated everything except the elasticity at a point. If we follow the analogy of taking the limit of the interval definition as the input interval shrinks to the point we are studying, then we would get

$$\text{price elasticity of demand at a point} = \lim_{\Delta p \to 0}\left(\frac{p_1}{q_1} \cdot \frac{\Delta q}{\Delta p}\right)$$

$$= \frac{p_1}{q_1} \cdot \lim_{\Delta p \to 0}\left(\frac{\Delta q}{\Delta p}\right) \qquad (p_1 \text{ and } q_1 \text{ act as constants})$$

$$= \frac{p_1}{q_1} \cdot \frac{dq}{dp} \qquad \text{(definition of derivative)}$$

$$\eta = \frac{p}{q} \cdot D'(p) \qquad (p = p_1, q = q_1)$$

Again, we have called the price and quantity at the desired point p and q, respectively, and recognized that, since $q = D(p)$, it follows that $\frac{dq}{dp} = D'(p)$.

Let's apply this calculation to our example at a price of \$12:

$$\eta = \frac{12}{33.6} \cdot D'(12) \approx (0.357)(-2.34) \approx -0.835$$

Translation: At the point where the selling price is \$12, the demand is *decreasing* at an instantaneous rate of about 0.835 percentage points per percentage point increase in the price.

Recall that before we obtained −0.82. In fact, the two answers should be identical in this case, again because the demand function is linear. Why did we get different values? Rounding errors! With a higher level of precision, the equality would be easier to see.

Now, why are we spending so much time on this idea of elasticity? Elasticity involves the relationship between price and quantity. What else involves just those two quantities (other than the demand function)? See if you can think of an answer.

One answer is the revenue function, since revenue is price times quantity. Let's calculate the revenue when the price is \$12:

$$R = pq = (12)(33.6) \approx 403$$

Translation: When we charge \$12, we expect to sell 33.6 sweatshirts and expect a revenue of about \$403 (on average).

Maximizing Revenue and Unit Elasticity

What would happen if we raised the price? Intuitively, since our elasticity value was about −0.835, for every percentage point increase in the price, the demand will go down by *less* than a percentage point. In other words, **the amount that the demand *decreases* is**

***less, proportionately*, than the amount the price increases. This would suggest that the *benefit* from raising the price *outweighs the cost* (in lost demand), suggesting that our revenue should *increase*.** If you think about it, it would seem that **this is true as long as the *magnitude* (absolute value) of the elasticity is smaller than 1**.

Let's digress for a moment to talk about the **sign of the elasticity value**. Of course, in any realistic situation, the price and quantity values should be positive. So the expression that determines the sign of the elasticity is the sign of the derivative of the demand function. As we have discussed on several occasions, the normal expectation of a demand function is that the higher price you charge, the fewer items you will sell. In other words, **we normally expect demand functions to be *decreasing* functions (whether expressed in terms of the price or the quantity). Therefore, the sign of the derivative of the demand function, and so the sign of the elasticity, should be *negative*.**

Since the elasticity is normally assumed to be negative, some economists only use its absolute value or magnitude when talking about it. **We will try to be careful to talk about the *magnitude* of the elasticity when we want to refer to its absolute value, to avoid any ambiguity.**

Before the digression, we observed that, if the magnitude of the elasticity were less than 1, then raising the price seemed as if it should *increase* the revenue. Let's see if we can use calculus to explore this hypothesis.

If we express revenue in terms of price, we get

$$R = pq = pD(p)$$

We are interested in how the price affects the revenue, so we could find the rate of change of the revenue in terms of the price. This would simply be the derivative of $R(p)$, which (using the Product Rule) would be

$$R(p) = pD(p)$$
$$R'(p) = D_p[pD(p)] = (p)[D'(p)] + [D(p)](1) = pD'(p) + D(p)$$

In the original question for this problem, your group is interested in maximizing their profit. The problem stated that the sweatshirts had been ordered already, and that no more would be ordered, so the cost is essentially fixed and constant, sometimes referred to as a **sunk cost**. This means that the profit is just the revenue minus a constant, and so the maximum for the profit will occur at the same price as the maximum for the revenue. In other words, we can find the optimal price by maximizing the revenue. Since we already found the derivative of the revenue, now we just need to set that equal to 0 and solve for p, so the profit will be maximized when

$$R'(p) = 0$$
$$pD'(p) + D(p) = 0$$

Now, recall that the formula for elasticity is

$$\eta = \frac{p}{q} \cdot D'(p) = \frac{p}{D(p)} \cdot D'(p) = \frac{pD'(p)}{D(p)}$$

Notice that the first term in our equation above (when we set marginal revenue equal to 0) also involves the expression $pD'(p)$, but in the formula for elasticity, it is divided by $D(p)$. This means that if we divide both sides of our equation by $D(p)$, we should get an expression involving the elasticity, η. Let's try it:

$$pD'(p) + D(p) = 0 \qquad \text{(to maximize revenue, } R'(p) = 0\text{)}$$

$$\frac{pD'(p) + D(p)}{D(p)} = 0 \qquad \text{(dividing both sides by } D(p)\text{)}$$

$$\frac{pD'(p)}{D(p)} + \frac{D(p)}{D(p)} = 0 \qquad \text{(splitting up the fraction)}$$

$$\frac{p}{D(p)} D'(p) + 1 = 0 \qquad \text{(simplifying each term)}$$

Now, notice that the first term in the last equation is now the formula for the elasticity value, as we had contrived, so the condition to find critical points of the revenue function is equivalent to

$$\eta + 1 = 0$$
$$\eta = -1$$

In other words, **the condition for the revenue being maximized is that the magnitude of the elasticity is 1**. We say that such a point has **unit elasticity**.

Elasticity and the Direction of Change in Revenue (Elastic and Inelastic Demand)

What about our hypothesis that if the magnitude of the elasticity is less than 1, then increasing the price will increase the revenue? Let's look again at the expression we got for the derivative of the revenue:

$$R'(p) = pD'(p) + D(p)$$

If we assume that the demand will always be a positive quantity, the *sign* of this marginal revenue function will not be affected by dividing by $D(p)$, so we will see that

$$\text{the } sign \text{ of } R'(p) = \text{the sign of } \frac{R'(p)}{D(p)}$$
$$= \text{the sign of } \frac{pD'(p) + D(p)}{D(p)}$$
$$= \text{the sign of } \eta + 1$$

The sign of $\eta + 1$ changes at $\eta = -1$, which is the value at which the revenue is maximized. In our example, we calculated the value of the elasticity at 12 to be $\eta = -0.835$. This means that

$$\eta + 1 = -0.835 + 1 = 0.165 > 0$$

so the sign of $\eta + 1$ is *positive*, which means that the sign of $R'(p)$ is also positive, which means that the revenue is *increasing* at 12, which means that to maximize your revenue, you should charge *more* than \$12. If you think about it, this will be true as long as η is between -1 and 0, or

$$-1 < \eta < 0$$

since this will mean that $\eta + 1$ will be between 0 and 1 (so will be positive). This condition is equivalent to saying that the *magnitude* of η is less than 1.

In words, we say in this situation that the **demand is inelastic**. By this, we mean that the demand is *not very responsive* to a price increase (is smaller proportionately). For a visual image to help understand the term *inelastic*, you could think of price and demand (quantity) as connected by an *elastic* (rubber band) or spring on a table top. When you raise the price, it is like pulling the rubber band a bit, which will pull the demand along in response. When the demand is inelastic, it's like saying the rubber band is not very springy (as if it is very old and dried out), so the response of the changing demand is *less* proportionately than the original change (the friction of the table top is stronger than the pull of the rubber band). To make this analogy work, ignore the *direction* of the changes and simply focus on their *magnitude*. Or think of it as a flexible seesaw in a thick fluid like molasses, so that when one quantity goes up, the other goes down. Elastic demand is like the fluid being thin and the bar being very springy, so the response is *more* than the initial shove. Inelastic demand is like the fluid being thick and the bar being *not* very springy, so the response is *less* than the initial shove.

Now we know that our optimal price should be more than \$12. Suppose we made a guess that \$15 might be the price at which the revenue (and so the profit here) is maximized. Let's evaluate the demand, marginal demand, and elasticity to help see if we are right.

$$D(15) = -2.34(15) + 61.7 = 26.6 \text{ and } D'(15) = -2.34$$

so

$$\eta = \frac{p}{D(p)} D'(p) = \frac{15}{26.6}(-2.34) \approx -1.32$$

Translation: Demand is decreasing at a rate of 1.32 percentage points per percentage point of increase in the price. In other words, **demand is decreasing *faster* proportionately than the price is**, at the reference price of \$15. This suggests that, **if we consider raising the price**, this time **the *cost* of raising the price (lower demand) *outweighs the benefit* (higher revenue per sweatshirt), so raising the price would actually *lower* the revenue**. You may have heard similar discussions related to taxes: Sometimes raising taxes can actually *lower* revenues if the **demand** (for cigarettes, mass transit, or whatever) **is elastic**.

Numerically, this will happen **when the elasticity is less than −1 (when the magnitude of the elasticity is greater than 1)**. Using our rubber band analogy, this means that the rubber band is so springy that an increase in the price leads to a *larger* demand response proportionately, which you can think of as a sort of slingshot reaction.

We should mention at this point that our comments about maximizing revenue so far really only involved finding *critical points*, which technically could be local maxima or local minima or neither (such as horizontal points of inflection). In our example, the demand function was linear and the revenue function was quadratic and concave down, so we know that the only critical point will be the local and global maximum. Normally, since the demand function is typically decreasing and smooth, the revenue function will be concave down, but this is not absolutely guaranteed. For any particular problem, you should convince yourself that you have found the optimal value that you want using the concepts in this chapter, as when we checked the sign of $R'(p)$. In economics, this is called **checking second order conditions**, since one way to test a critical point is with the *Second* Derivative Test. The condition for finding a critical point would similarly be called a **first-order condition**, since it involves setting the *first* derivative equal to 0.

We already know how we could maximize the revenue function directly. For this problem, how could we find the optimal price *directly using elasticity*? We know that the first-order condition for a critical point is that the elasticity should be -1, so let's write out this condition in general (in terms of p) and solve it.

$$\eta = -1$$

$$\frac{pD'(p)}{D(p)} = -1 \qquad \text{(plugging in the general formula for } \eta\text{)}$$

$$\frac{(p)(-2.34)}{-2.34p + 61.7} = -1 \qquad \text{(plugging in the functions for } D(p) \text{ and } D'(p)\text{)}$$

$$\frac{-2.34p}{-2.34p + 61.7} = -1 \qquad \text{(simplifying the numerator)}$$

$$-2.34p = (-1)(-2.34p + 61.7) \qquad \text{(multiplying both sides by the denominator } D(p) > 0\text{)}$$

$$-2.34p = 2.34p - 61.7 \qquad \text{(multiplying out the right-hand side)}$$

$$-4.68p = -6.17 \qquad \text{(subtracting } 2.34p \text{ from both sides and simplifying)}$$

$$p = \frac{-61.7}{-4.68} \approx 13.2 \qquad \text{(dividing both sides by } -4.68\text{)}$$

Translation: Revenue (and profit, in this example) should be maximized when the price of the sweatshirts is approximately \$13.20.

Remember that we can only come to this conclusion because we have already convinced ourselves that this unique critical point is in fact the local and global maximum for this problem.

Notice that when we multiplied both sides of the equation by the denominator, the denominator was just $D(p)$, the demand function, which we have been assuming is *positive* for all values of p in the domain ($p \geq 11.4$). This means that we don't have to worry about multiplying both sides by 0, which could add spurious solutions (solutions that seem to solve the problem but would create a 0 in a denominator in the original equation). As a simple example if you had to solve the equation

$$\frac{x}{x} = 0$$

and you blindly multiplied both sides by x, you would get the equation $x = 0$, which would appear to be a solution to the original equation. But if you plug this solution in to the original equation you get

$$\frac{0}{0} \stackrel{?}{=} 0$$

As we discussed earlier, $\frac{0}{0}$ is undefined (is not *uniquely* defined, since *any* number times 0 equals 0), so $x = 0$ is *not* a solution to the original equation (it has no solution).

The lesson from this is that **when you multiply both sides of an equation by a variable expression, if the expression can be zero, then you need to *check* (verify) all of your final answers to make sure that they satisfy the *original* equation. In particular, make**

sure that no answer would lead to a 0 in a denominator of the original equation. If the expression *cannot* be 0 (at least over your domain of interest), then you need not worry about this.

How could we *verify* this answer of an optimal selling price of \$13.20? As we said above, we could optimize the revenue function directly. Let's do so, in abbreviated form.

$$\begin{aligned}
R &= pq && \text{(revenue is selling price times quantity sold)}\\
R(p) &= pD(p) && (q = D(p)\text{; we get revenue as a function of } p)\\
&= (p)(-2.34p + 61.7) && (\text{since } D(p) = -2.34p + 61.7)\\
&= -2.34p^2 + 61.7p && \text{(multiplying out)}\\
R'(p) &= -4.68p + 61.7 = 0 && \text{(take the derivative and set it equal to 0)}\\
-4.68p &= -61.7 && \text{(subtract 61.7 from both sides)}\\
p &= \frac{-61.7}{-4.68} \approx 13.2 && \text{(divide both sides by } -4.68\text{; round to 3 digits)}
\end{aligned}$$

As we would hope, we got the same answer. In fact, working with the revenue function is actually *easier*! The advantage of working with **elasticity**, on the other hand, is that **it is independent of the units used for the input and the output variables**, so is, in a sense, a more *universal* value associated with the relationship between price and demand. It also has the advantage of telling us about *both* the relationship between price and demand *and* the *direction* of the effect of raising price (at least a small amount) on revenue, all in one value.

For our sweatshirt example, how can we find the prices at which the demand is inelastic, and the prices at which it is elastic?

Solving Inequalities

We know that at \$12, the demand was inelastic, since $\eta = -0.835 > -1$. Remember that "greater than" means "to the right of, along a standard horizontal axis." So values of η that are between -1 and 0 are *greater* than -1, even though their *magnitude* is *less* than the *magnitude* of -1. This can get confusing. Let's look at a number line to help make this clearer (Figure 3.5-1).

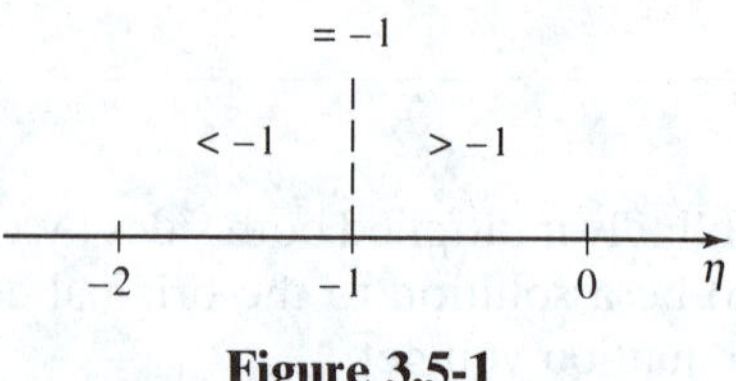

Figure 3.5-1

How can we determine for what values of p the demand is inelastic? Instead of setting η *equal* to -1, now we want to determine when η is *greater than* -1. Thus we want to set up the *inequality* $\eta > -1$ in terms of p and solve for p. In fact, the steps are virtually identical to what we did before to solve the equation. The main difference in working with inequalities is that when you multiply or divide both sides by a number, the *sign* of the number has an effect on the result.

If you multiply (or divide, since division can be thought of as multiplication by the reciprocal) both sides of an inequality by a *positive* number, the direction of the inequality (for example > versus <) stays the same. This is like saying that one distance is bigger than another whether you measure it in feet or inches (changing units corresponds to multiplying by a positive constant).

On the other hand, **if you multiply both sides of an inequality by a negative number (< 0), the direction of the inequality is *reversed* (for example, $<$ switches to $>$). This is because when you change the signs of two numbers, their order (which is greater) gets switched.** For example, 5 is *greater* than 2, but when we switch the signs, -5 is *less* than -2. Similarly, -5 is *less* than 2, but 5 is *greater* than -2. Try some more examples to convince yourself that this is true in general. Since multiplying by a negative number is like multiplying by its absolute value (a positive number, which keeps the direction of the inequality the same) and then switching the sign (which reverses the inequality), the net effect of multiplying by any negative number is to reverse the direction of the inequality.

Thus, when working with inequalities, if you multiply both sides by a number (except 0), you simply apply the rules above. If you multiply by an *expression* (involving one or more variables), you need to know the sign of the expression to know which direction you need for the resulting inequality. If you know the sign will always be the same, you can apply the appropriate rule. If the sign can be different, the simplest approach is to split your solution into parts corresponding to the intervals over which the sign stays the same and solve separate inequalities for each piece. As with equations, if the expression can be 0, you could get spurious solutions, so you need to check your answers to make sure you can never get a 0 in a denominator.

Let's apply these concepts to finding where the demand in the sweatshirt example is inelastic.

$$\eta > -1 \qquad \text{(values where demand is inelastic)}$$

$$\frac{pD'(p)}{D(p)} > -1 \qquad \text{(plugging in the general formula for } \eta\text{)}$$

$$\frac{(p)(-2.34)}{-2.34p + 61.7} > -1 \qquad \text{(plugging in the functions for } D(p) \text{ and } D'(p)\text{)}$$

$$\frac{-2.34p}{-2.34p + 61.7} > -1 \qquad \text{(simplifying the numerator)}$$

$$-2.34p > (-1)(-2.34p + 61.7) \qquad \text{(multiplying both sides by the denominator } D(p) > 0\text{)}$$

$$-2.34p > 2.34p - 61.7 \qquad \text{(multiplying out the right-hand side)}$$

$$-4.68p > -61.7 \qquad \text{(subtracting } 2.34p \text{ from both sides and simplifying)}$$

$$p < \frac{-61.7}{-4.68} \approx 13.2 \qquad \text{(dividing both sides by } -4.68\text{)}$$

Notice that when we multiplied both sides by the denominator, we knew that it was positive (by assumption, since it is the demand), so the direction of the inequality stayed the same ($>$). On the other hand, when we divided both sides by -4.68 $\left(\text{multiplied both sides by } -\frac{1}{4.68}\right)$, the direction of the inequality was reversed (from $>$ to $<$).

Thus we can see that for prices *below* \$13.20, the demand is inelastic.

Now if we want to find the prices where the demand is *elastic*, we would apply the same approach, but the initial inequality would be in the *other* direction ($\eta < -1$). This means that the results would be exactly the same, except that *all* of the inequalities would go in the other direction. In other words, the solution would be $p > 13.2$, so the demand is elastic for all prices *over* \$13.20. If you feel unsure about solving inequalities, you can get some practice in the exercises for this section.

Let's look at a graph that helps show the relationships of some of the quantities we have discussed here (Figure 3.5-2). The graph shows how the price where the demand has unit elasticity (\$13.20 in the example) corresponds to the point where revenue is maximized. To the left of that, the revenue is increasing and the demand is inelastic, and to the right of it, the revenue is decreasing and the demand is elastic.

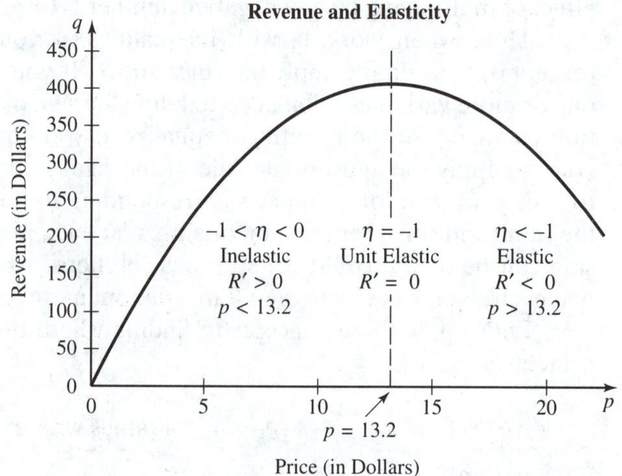

Figure 3.5-2

So what should your group do? We know that the optimal price is \$13.20. The demand we would expect at this price is

$$D(13.2) = -2.34(13.2) + 61.7 \approx 30.8$$

and the revenue should be

$$R(13.2) = (13.2)(30.8) \approx 407$$

If we assume the cost function is the same as before (\$9 per sweatshirt plus a fixed order cost of \$25), then the sunk cost from the 35 sweatshirts already bought was

$$C(35) = (9)(35) + 25 = 315 + 25 = 340$$

Thus, the optimal expected profit should be about $407 - 340 = 67$, or about \$67.

Let's make a few validation comments here. We are assuming divisibility for the price, and that holds quite well, depending on what you are willing to charge. If you don't want to worry about coin change, you might decide to charge \$13 (since the revenue graph is a parabola, which is symmetric, rounding the answer will give you the better whole number solution on either side), for which the expected demand would be about 31.2 sweatshirts, and the expected revenue still rounds to \$407.

Whichever price you choose, your model predicts that you are not likely to sell all of your 35 sweatshirts but should come pretty close. You could decide to give the extras to the best salespeople as rewards, or could make a little money by selling them at a discounted price. We calculated before that to sell all 35 sweatshirts, your price would have to be no more than \$11.40. This suggests that if you make the discount price \$11 (or possibly \$10, if you want to get rid of them more quickly), you could probably make another \$30 or so, for a total of about \$97.

Recall from Section 3.4 that, if the analysis had been done *before* the sweatshirts were ordered, a profit of about \$151 could have been obtained, so the hasty ordering had a significant cost. The sunk cost forced you to lower your price from over \$17 down to about \$13. **Mathematical models can help your bottom line!** □

Sample Problem 1: In Sample Problem 4 of Section 3.2, we used historical data to derive a demand model for attendance at a concert as a function of the ticket price charged, given by

Verbal Definition: $q = T(p) =$ the number of tickets sold when the ticket price is p dollars.

Symbol Definition: $q = T(p) = 343(0.837)^p$, for $p \geq 0$.

Assumptions: Certainty and divisibility. Certainty implies the relationship is *exact*. Divisibility implies that *any* fractions of dollars and tickets are possible.

We also formulated a total cost function (in dollars) in terms of the ticket price p (also in dollars) of the form $C(p) = 0.06R(p) + 250$, where $R(p)$ is the revenue function in dollars.

(a) Find the price elasticity of demand for the ticket price interval [4,5], and interpret the result in words.
(b) Find the price elasticity of demand at a ticket price of \$4, and interpret the result in words.
(c) Find the percent rate of change of the demand function at $p = 4$, and interpret the result in words.
(d) Find a general formula for the elasticity at a point, p.
(e) For what p values is the demand inelastic? Elastic? Where is it unit elastic?
(f) What ticket price would maximize the revenue, based on this demand function?
(g) What ticket price would maximize your profit, based on this demand function? Explain your reasoning.

Solution: (Note that the demand function here is called $T(p)$, rather than $D(p)$).

(a) To find the elasticity over an interval, we need to find the demand at the endpoints, so let's make a small table:

p	$q = T(p)$
4	168
5	141

Thus $\Delta p = 5 - 4 = 1$, and $\Delta q = 141 - 168 = -27$, so the elasticity over [4,5] is given by

$$\eta = \frac{\frac{\Delta q}{q_1}}{\frac{\Delta p}{p_1}} = \frac{p_1}{q_1} \cdot \frac{\Delta q}{\Delta p} = \frac{4}{168} \cdot \frac{-27}{1} \approx -0.643$$

Translation: Over the price interval [4,5], for each percentage point of increase in the price, the demand (number of tickets you'll sell) goes down about 0.643 percentage points.

(b) The formula for elasticity at a point is given by

$$\eta = \frac{p}{q} \cdot \frac{dq}{dp} = \frac{p}{T(p)} \cdot T'(p)$$

This means we also need to find the marginal demand function, $T'(p)$:

$$T'(p) = D_p[343(0.837)^p] = (\ln 0.837)[343(0.837)^p]$$

so at $p = 4$,

$$T'(4) = (\ln 0.837)[343(0.837)^4] \approx -30.0$$

Thus we find that, at a price of 4,

$$\eta = \frac{4}{T(4)} \cdot T'(4) = \frac{4}{168} \cdot (-30.0) \approx -0.714$$

Translation: At a price of \$4, the demand is decreasing at an instantaneous rate of 0.714 percentage points per *percentage point* of increase in the price.

(c) The percent rate of change at a point is the instantaneous rate of change (derivative) at that point, expressed as a percent of the output value there, so the formula is given by

$$\text{Percent rate of change at } x = \frac{\text{rate of change at } x}{\text{output value at } x} = \frac{\frac{dy}{dx}}{y} = \frac{f'(x)}{f(x)}$$

For our problem, then, the percent rate of change at 4 is given by

$$\%\text{ rate of change at } 4 = \frac{T'(4)}{T(4)} \approx \frac{-30.0}{168} \approx -0.179 = -17.9\%$$

Translation: At a ticket price of \$4, the demand is decreasing at an instantaneous rate of about 17.9% per *dollar* increase in the price.

(d) A general formula for the elasticity at p is

$$\eta = \frac{p}{T(p)} \cdot T'(p) = \frac{p}{343(0.837)^p}[(\ln 0.837)343(0.837)^p] = p(\ln 0.837) \approx -0.178p$$

(e) Inelastic demand means that demand changes *less* proportionately than a change in price, so corresponds to the magnitude of the elasticity being *less* than 1, which means the elasticity itself is mathematically *greater* than −1. So our condition for inelastic demand is

$$\eta > -1$$

$$-0.178p > -1 \quad \text{(replacing } \eta \text{ with the formula we found for it)}$$

$$p < \frac{-1}{-0.178} \approx 5.62 \quad \text{(dividing both sides by } -0.178 \text{ and so reversing the inequality)}$$

Translation: As long as the price is *less* than \$5.62, demand is inelastic, which means that if you increase the price, you will increase your revenue. Recall that this is because the demand decrease is proportionately *less* than the price increase, so the benefit outweighs the cost.

To find the places where demand is elastic, the direction of all the statements above just gets reversed, as will the solution to the inequality, so the demand will be elastic for all prices *more* than \$5.62, and increasing the price from any value in that interval will *decrease* total revenue, since the loss in demand overpowers the increase in price.

Unit elasticity is simply the point where elasticity *equals* −1, and if you replace the inequality signs with equal signs above, you will see that the answer obtained is \$5.62.

(f) Revenue should be maximized when the elasticity is −1, so the revenue should be maximized when the ticket price is \$5.62. Notice that this agrees with the answer we got in Section 3.2 so is in fact a **verification** of that calculation.

(g) As we mentioned when working with this problem in Section 3.2, since the cost is a constant plus a positive multiple of the revenue function, the profit function also turns out to be a positive multiple of the revenue plus a (negative) constant. We showed there that this implies that maximizing the revenue will also maximize the profit, since they have the same basic shape (the profit is just shrunk vertically a little bit by a factor of 0.94 and then shifted down vertically by 250 dollars). Thus the profit should also be maximized at a ticket price of \$5.62 using this demand function. □

Marginal Cost and Minimizing Average Cost

EXAMPLE 2: You have a small business making and selling juggling beanbags, as well as other items. You have several people who help you make the beanbags and have made an estimate of your total cost each month as a function of the number of beanbags made. From your historical records, you have estimated your total costs for various months during which you produced different numbers of beanbags, given in Table 1.

TABLE 1

Number of Beanbags	Total Cost
57	75
61	80
69	86
76	91
103	115
124	136
137	151

The market is fairly competitive for handmade beanbags, so, in order to sell a reasonable number and make a decent profit, you need to charge the going price, which is \$1.50 per beanbag. Suppose you have not figured all of your own time spent on making beanbags and managing your helpers into the total cost. If you maximize total profit as your objective, you are likely to be working 24 hours a day. You decide that it makes the most sense for you to maximize the profit per beanbag rather than maximizing your total profit to allow you some time to have a decent quality of life. Assuming that demand is not a constraint at this stage in the business (you could sell whatever number you could realistically make), how many beanbags should you make in a month to maximize your profit per beanbag?

Since you know you are going to charge \$1.50 for the beanbags, this is your revenue per beanbag, and it is constant. Since your profit per beanbag will be your revenue per beanbag minus your cost per beanbag, *maximizing your profit* per beanbag will be equivalent to *minimizing your cost* per beanbag. So the essence of this problem is to minimize the cost per beanbag, which is also called the **average cost**, as we discussed in Section 2.6.

To find an average in general, you divide the total by the number of items. When averaging a set of values, this means the sum of the values divided by the number of values. In the case of average cost, we divide the total cost by the number of beanbags to find the cost per beanbag. Let's find a model for the total cost first and then formulate a model for the average cost from it.

To choose a model for the total cost, let's first graph the data points (Figure 3.5-3). This function looks fairly close to being linear, but at the same time it does seem to have some curvature. At the left, it appears somewhat concave down, then to the right it appears to be slightly concave up. This makes some sense, since probably for the smaller numbers of beanbags, you can make them all yourself, and you get more efficient as you make more, so the marginal cost decreases (sometimes called an **economy of scale**). But since you have other things to do for the business, once you need more beanbags, you start going to your helpers. As the number of beanbags continues to increase, you need to use more people, and each additional person is more expensive than the one before. Therefore, as the number of beanbags increases in this higher interval of values, the marginal cost also increases.

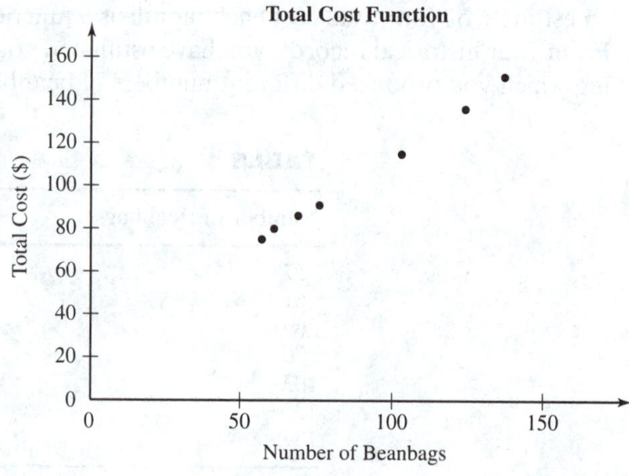

Figure 3.5-3

What **use** do we want to make of this model? Eventually we want to minimize the average cost, so this is an **optimization** problem. However, we do not know yet in what vicinity the optimum will occur, although we can make a guess that it is likely to be somewhere around where the concavity changes, which seems to be somewhere between 75 and 100. Since we do not know where the optimum will be, our use is more similar to **forecasting** than to optimization. We expect in advance that the optimum should probably fall between the smallest and largest data input values, which would mean we are **interpolating**, but we are not even absolutely sure of that. Thus, in general, we'd like the model to fit the data as closely as possible over the entire data domain, at least for now.

We know from the **context** of this problem that we expect the graph to be concave down for the smaller input values and concave up for the larger input values, so it should change concavity once. This would suggest either a cubic or logistic function. In this case, the logistic does not fit the context, since the concavity of our data does not match that of the logistic graph. In addition, we would not expect the total cost to level off at *either* end, let alone at both. The total cost would only level off for large numbers of beanbags if the marginal cost actually approached 0. This would mean that additional beanbags were essentially cost free to you beyond some point. But since the *materials* are not free, this is not possible, even if you can keep cutting labor costs. Thus, the logistic function would not be a good choice. Let's try fitting a cubic.

Using technology, the least-squares best-fit cubic model comes out to be

Verbal Definition: $C(q)$ = the expected total cost, in dollars, if you make q beanbags in a month.

Symbol Definition: $C(q) = 0.00003191\ldots q^3 - 0.006861\ldots q^2 + 1.321\ldots q + 16.78\ldots$ for $50 \le q \le 150$.

Assumptions: Certainty and divisibility. Certainty implies the relationship is *exact*. Divisibility implies that *any* fractions of beanbags and dollars of total cost are possible.

Figure 3.5-4 shows this model with the data. The fit seems very good. It is a little hard to see the concavity, since it is subtle, but it does appear to be going the way we want. If it were important, we could, of course, check this using the second derivative.

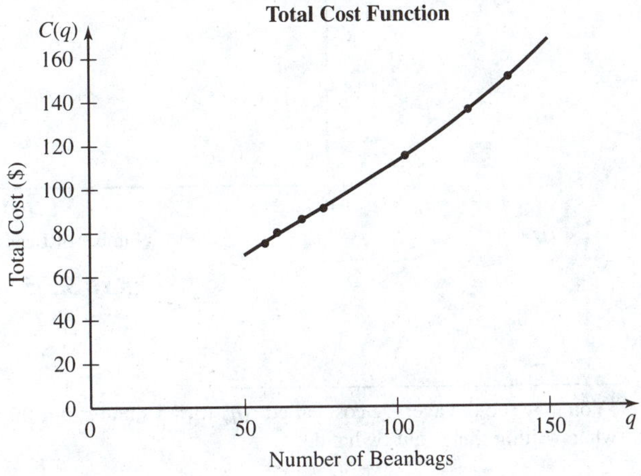

Figure 3.5-4

Assuming that this is a reasonable model for the total cost, let's now formulate a model for the average cost. We have already said that the average cost is the total cost, which is $C(q)$, divided by the number of beanbags, which is q. If we call the average cost function $\overline{C}(q)$, the function will be

$$\overline{C}(q) = \frac{C(q)}{q} = \frac{0.00003191\ldots q^3 - 0.006861\ldots q^2 + 1.321\ldots q + 16.78\ldots}{q}$$

$$= 0.00003191\ldots q^2 - 0.006861\ldots q + 1.321\ldots + \frac{16.78\ldots}{q}$$

A bar over a variable is often used to denote its average.[1] Thus we can define a model for the average cost:

Verbal Definition: $\overline{C}(q)$ = the expected average cost, in dollars per beanbag, if you make q beanbags in a month.

Symbol Definition: $\overline{C}(q) = 0.00003191\ldots q^2 - 0.006861\ldots q + 1.321\ldots + \frac{16.78\ldots}{q}$ for $50 \le q \le 150$.

Assumptions: Certainty and divisibility. Certainty implies the relationship is *exact.* Divisibility implies that *any* fractions of beanbags and dollars per beanbag of average cost are possible.

Let's see what this average cost function looks like graphically (Figure 3.5-5). It looks as if the minimum average cost occurs somewhere around 125 beanbags. Let's try to calculate a more precise answer.

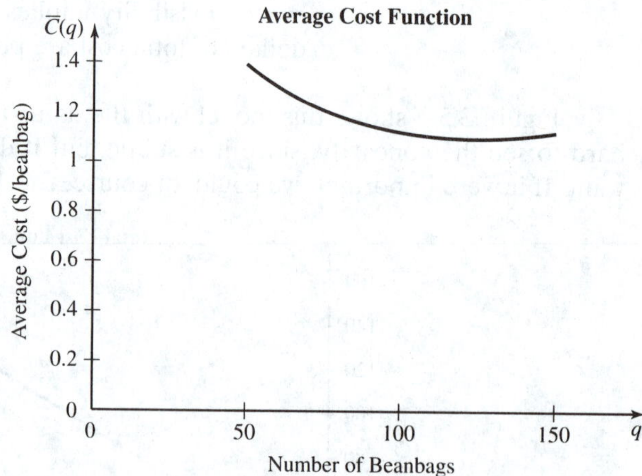

Figure 3.5-5

---

[1] You may see the average cost called $c(q)$ with a small c, but this is harder to distinguish from $C(q)$, especially when writing things out by hand.

If we want to minimize the average cost by hand, we first take the derivative and set it equal to 0:

$$\overline{C}(q) = 0.0000319q^2 - 0.00686q + 1.32 + 16.8q^{-1}$$

$$\overline{C}'(q) = 0.0000638q - 0.00686 - 16.8q^{-2} = 0.0000638q - 0.00686 - \frac{16.8}{q^2} = 0$$

Now we want to solve this for q. We could multiply both sides of the equation by q^2, but then we would end up with a cubic equation, which would be extremely difficult to solve by hand. At this point, let's use technology. In fact, we might as well just use technology to minimize the average cost function directly, although we could **verify** our answer by solving the above equation using technology. If we use the full precision of our technology and minimize the average cost function, we get an answer of

$$q \approx 124$$

which is very close to our visual estimate of 125.

We saw before that the condition for maximizing revenue could be expressed in terms of the elasticity. Can we make any similar kind of general condition for minimizing the average cost? We know that

$$\overline{C}(q) = \frac{C(q)}{q} = C(q) \cdot \frac{1}{q} = [C(q)](q^{-1}) \qquad \left(\frac{a}{b} = a \cdot \frac{1}{b}; \frac{1}{x} = x^{-1}\right)$$

$$\overline{C}'(q) = [C(q)](-q^{-2}) + (q^{-1})[C'(q)] = 0 \qquad \text{(Product Rule; set derivative = 0)}$$

$$-\frac{C(q)}{q^2} + \frac{C'(q)}{q} = 0 \qquad \left(\text{since } x^{-n} = \frac{1}{x^n}\right)$$

Now, remember that the average cost is $\overline{C}(q) = \dfrac{C(q)}{q}$. If we multiplied both sides of the above equation by q, we would get

$$(q)\left[-\frac{C(q)}{q^2} + \frac{C'(q)}{q}\right] = 0$$

$$-\frac{C(q)}{q} + C'(q) = 0 \qquad \text{(multiplying out and simplifying)}$$

$$-\overline{C}(q) + C'(q) = 0 \qquad \left(\text{since } \overline{C}(q) = \frac{C(q)}{q}, -\frac{C(q)}{q} = -\overline{C}(q)\right)$$

$$C'(q) = \overline{C}(q) \qquad \text{(adding } \overline{C}(q) \text{ to both sides)}$$

In other words, the average cost will be minimized when the *marginal* cost is equal to the *average* cost. Fortunately, there is also a very intuitive way to understand and remember this condition.

Consider the sequence of numbers: 9,7,5,7,9,11. Let's suppose these were the costs to make the first, second, third, and so on, items for a business. Now let's make a table to calculate the accumulating total cost, average cost, and marginal cost one item at a time

(Table 2). Since the marginal cost is approximately the cost of the next item, we will use that value for simplicity to see what is happening.

TABLE 2

i	Cost of Item i	Total Cost of the First i Items	Average Cost of the First i Items	Cost of the Next Item (≈ Marginal Cost)
1	9	9	9	7
2	7	16	8	5
3	5	21	7	7
4	7	28	7	9
5	9	37	7.4	11
6	11	48	8	

Notice what happens here. The average cost for the first item is 9, and the cost of the next item is 7, which is going to bring down the average cost, to 8. The cost of the *next* item (the marginal cost after 2 items) is 5, which is again less than the average up to that point (for the first 2 items), and so it brings down the average, this time to 7. Now, the next item costs exactly the same as the average up to that point (the marginal cost equals the average cost), so it will keep the average the same, at 7. At that point (after the first 4 items), the marginal cost is now 9, which is *more* than the average cost, and so it brings the average *up*, to 7.4. And for the last iteration, the marginal cost (11) is again more than the average up to that point (7.4), so it raises the average once more, this time to 8.

The point here is that, **as long as the marginal cost is *less* than the average cost at a point, the average cost continues to get smaller. Once the marginal cost is *more* than the average, the average starts to go up. And at the point where the average achieves its minimum, the marginal cost was *equal* to the average cost**, since this kept the average the same (so the graph of the average cost would be horizontal there, corresponding to the derivative being 0).

In our beanbag example, the marginal cost (taking the derivative of the total cost function, where the total cost function is rounded to 3 significant figures) is given by

$$C'(q) = D_q(0.0000319q^3 - 0.00686q^2 + 1.32q + 16.78)$$
$$= 0.0000957q^2 - 0.01372q + 1.32$$

If we set this equal to the average cost function and solve using technology, we get

$$C'(q) = \overline{C}(q)$$

$$0.0000957q^2 - 0.01372q + 1.32 = 0.0000319q^2 - 0.00686q + 1.32 + \frac{16.8}{q} \quad \text{(substituting)}$$

$$0.0000638q^2 - 0.00686q - \frac{16.8}{q} = 0 \quad \text{(combining like terms)}$$

Notice that this equation is exactly the same as the equation we obtained earlier, multiplied by q. Solving this equation using technology, we get a solution of

$$q \approx 125$$

(slightly different than the 124 obtained earlier, due to rounding error). If we used more precision or used a numerical derivative with a small margin of error, the answers would agree to a higher level of precision. Still, the calculation validates the theoretical result.

If we graph the marginal and average cost functions together, we get the graph shown in Figure 3.5-6.

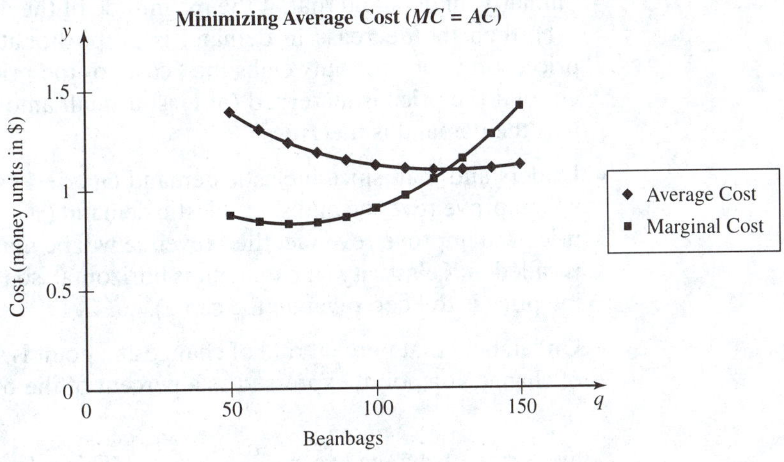

Figure 3.5-6

As in our small numerical example, notice that the marginal cost starts out below the average cost, pulling it down at first, but then at around 125 it becomes larger than the average cost, so the average cost turns back up at that point. **The minimum average cost occurs at exactly the point where the marginal cost and the average cost are equal (where they cross).** □

Section Summary

Before you try the exercises, be sure that you

- Understand that the price elasticity of demand over an interval is the percentage points of change in demand per percentage point increase in price, so is given by

$$\frac{\text{\% change in demand}}{\text{\% change in price}} = \frac{\frac{\Delta q}{q}}{\frac{\Delta p}{p}} = \frac{p}{q} \cdot \frac{\Delta q}{\Delta p}$$

- Understand that the price elasticity of demand at a *point* is analogous to the elasticity over an *interval* and has the same interpretation but is an *instantaneous* rate of change and uses the *derivative* at the point instead of the *slope of the secant* over an interval: $\eta = \frac{p}{q} \cdot \frac{dq}{dp} = \frac{p}{D(p)} \cdot D'(p)$, where $D(p)$ is the demand function.
- Understand that, by its nature, the elasticity is normally negative, since demand generally *decreases* as price increases.

- Understand that if the magnitude of the elasticity is greater than 1 (so $\eta < -1$), then the decrease in demand is proportionately *more* than the increase in price, so the "cost" outweighs the benefit of the price increase, and revenue will go *down* if the price is increased (at least a small amount). In this situation, we say that the demand is **elastic**.
- Similarly, understand that, if the magnitude of the elasticity is less than 1 (so $\eta > -1$), then the decrease in demand is proportionately *less* than the increase in price, so the benefit outweighs the "cost" of the price increase, and revenue will go *up* if the price is increased (at least a small amount). In this situation, we say that the demand is **inelastic**.
- Understand that, since inelastic demand ($\eta > -1$) suggests that *increasing* price will improve revenue and since elastic demand ($\eta < -1$) suggests that *decreasing* price will improve revenue, then revenue will be maximized when $\eta = -1$, which is called **unit elasticity** (the tangent is horizontal, so the marginal effect of increasing price is to keep revenue the same).
- Understand that percent rate of change at a point is simply the instantaneous rate of change at a point, expressed as a percent of the output value there, so is given by % rate of change at $x = \dfrac{\frac{dy}{dx}}{y} = \dfrac{f'(x)}{f(x)}$. This calculation will give the answer in pure decimal form, which can be converted to percent form by multiplying by 100 (moving the decimal point 2 places to the right) and appending the % sign. The result will then tell you the instantaneous rate of change of the output (in percentage points) per unit increase in the input value.
- Understand that, when the marginal cost is less than the average cost, the next item made will be cheaper than the average cost of the units up to that point, so making more units will *lower* the average cost. Similarly, if the marginal cost is *more* than the average cost, making more will *raise* the average cost (so it is better to make *fewer*). The average cost will be minimized at a point where the marginal cost *equals* the average cost (again, this means where the tangent is horizontal; the marginal effect of making more is to keep the average cost the same).

EXERCISES FOR SECTION 3.5

Warm Up

For Exercises 1–4 find the percent rate of change at the given point.

1. $y = x^2$ at $x = 5$

2. $y = 3x^2 + 5x - 2$ at $x = 3$

3. $y = 258(0.130)^x$ at $x = 2$

4. $q = 578.4(0.8932)^p$ at $p = 10$

For Exercises 5–8, find the price elasticity of the demand equations at the indicated values of p and determine whether demand is elastic, inelastic, or unit elastic.

5. $q = D(p) = -2.5p + 100; p = 7$

6. $q = D(p) = -0.357p + 1478; p = 100$

7. $q = D(p) = \frac{2}{p}; p = 25$

8. $q = D(p) = \frac{5}{p^2 + 4p}; p = 10$

9. Suppose $q = D(p) = 347.0(0.8793)^p$.

(a) Find the price elasticity of demand for the price interval [15,20].

(b) Find the price elasticity of demand at $p = 15$. How does this answer compare to your answer for part (a)? Why should they differ the way they do?

(c) What kind of elasticity does the demand at $p = 15$ represent?

(d) Based on your answer to part (c), what would be the effect on the revenue (in direction only) of raising the price a little bit from 15? Explain why in words.

(e) Based on your answers to (c) and (d), should the price at which revenue is maximized be less than or greater than 15? Explain your reasoning.

(f) Use *elasticity* to determine where revenue is maximized, and where demand is elastic and inelastic over all positive prices.

10. Suppose $q = D(p) = 1569(0.9004)^p$.

(a) Find the price elasticity of demand for the price interval [5,8].

(b) Find the price elasticity of demand at $p = 5$. How does this answer compare to your answer for part (a)? Why should they differ the way they do?

(c) What kind of elasticity does the demand at $p = 5$ represent?

(d) Based on your answer to part (c), what would be the effect on the revenue (in direction only) of raising the price a little bit from 5? Explain why in words.

(e) Based on your answers to (c) and (d), should the price at which revenue is maximized be less than or greater than 5? Explain your reasoning.

(f) Use *elasticity* to determine where revenue is maximized, and where demand is elastic and inelastic over all positive prices.

Game Time

11. In Section 2.6 we found the following demand function for tickets to a show at a coffeehouse:

Verbal Definition: $T(p)$ = the number of tickets sold at p dollars per ticket.
Symbol Definition: $T(p) = 5p^2 - 95p + 520$ for $0 \leq p \leq 9.50$.
Assumptions: Certainty and divisibility. Certainty implies that the relationship is *exact*, divisibility implies that the number of tickets sold and the price per ticket can take *any* value in the domain.

(a) Find the price elasticity of demand for the price interval [7,9]. Translate your answer into words.

(b) Find the price elasticity of demand at $p = 7$. Write out what your answer means in everyday language.

(c) What kind of elasticity does the demand at $p = 7$ represent?

(d) Based on your answer to part (c), what would be the effect on the revenue (in direction only) of raising the price a little bit from 7? Explain why in words.

(e) Based on your answers to (c) and (d), should the price at which revenue is maximized be less than or greater than 7? Explain your reasoning.

(f) Use *elasticity* to determine where revenue is maximized, and where demand is elastic and inelastic over all positive prices.

(g) Under what circumstances would your answer to part (f) also be the price where *profit* would be maximized?

12. In Section 2.6 we found an exponential model for the demand for tickets for a show at a coffeehouse:

Verbal Definition: $T(p)$ = the number of tickets sold at p dollars per ticket.
Symbol Definition: $T(p) = 343.3(0.8367)^p$ for $p \geq 0$.
Assumptions: Certainty and divisibility. Certainty implies that the relationship is *exact*, divisibility implies that the number of tickets sold and the price per ticket can take *any* value in the domain.

(a) Find the price elasticity of demand for the price interval [7,9]. Translate your answer into words.
(b) Find the price elasticity of demand at $p = 7$. Write out what your answer means in everyday language.
(c) What kind of elasticity does the demand at $p = 7$ represent?
(d) Based on your answer to part (c), what would be the effect on the revenue (in direction only) of raising the price a little bit from 7? Explain why in words.
(e) Based on your answers to (c) and (d), should the price at which revenue is maximized be less than or greater than 7? Explain your reasoning.
(f) Use *elasticity* to determine where revenue is maximized, and where demand is elastic and inelastic over all positive prices.
(g) Under what circumstances would your answer to part (f) also be the price where *profit* would be maximized?
(h) How do your answers compare to the answers for Exercise 11 (which are in the back of the text, if you didn't do them yourself)?

13. In the Exercises of Section 2.6 we found the following model for the demand for sodas:

Verbal Definition: $N(p)$ = the number of sodas (thousands) sold at p dollars per soda.
Symbol Definition: $N(p) = 1.092(0.3178)^p$ for $p \geq 0$.
Assumptions: Certainty and divisibility. Certainty implies that the relationship is *exact*, divisibility implies that the number of sodas and the price per soda can take *any* value in the domain.

(a) Find the price elasticity of demand for the price interval [0.60,0.70]. Translate your answer into words.
(b) Find the price elasticity of demand at $p = 0.60$. Write out what your answer means in everyday language.
(c) What kind of elasticity does the demand at $p = 0.60$ represent?
(d) Based on your answer to part (c), what would be the effect on the revenue (in direction only) of raising the price a little bit from 0.60? Explain why in words.
(e) Based on your answers to (c) and (d), should the price at which revenue is maximized be less than or greater than 0.60? Explain your reasoning.
(f) Use *elasticity* to determine where revenue is maximized, and where demand is elastic and inelastic over all positive prices.

14. The number of tickets to a local sporting event sold when the price is p dollars per ticket can be modeled by

Verbal Definition: $T(p)$ = the number of tickets sold at p dollars per ticket.
Symbol Definition: $T(p) = 3165(0.9492)^p$ for $p \geq 0$.
Assumptions: Certainty and divisibility. Certainty implies that the relationship is *exact;* divisibility implies that the number of tickets sold and the price per ticket can take *any* value in the domain.

(a) Find the price elasticity of demand for the price interval [10,15]. Translate your answer into words.

(b) Find the price elasticity of demand at $p = 10$. Write out what your answer means in everyday language.
(c) What kind of elasticity does the demand at $p = 10$ represent?
(d) Based on your answer to part (c), what would be the effect on the revenue (in direction only) of raising the price a little bit from \$10? Explain why in words.
(e) Based on your answers to (c) and (d), should the price at which revenue is maximized be less than or greater than \$10? Explain your reasoning.
(f) Use *elasticity* to determine where revenue is maximized, and where demand is elastic and inelastic over all positive prices.

15. The demand for fudge at a candy store can be modeled by

Verbal Definition: $N(p)$ = the expected (average) number of pounds of fudge sold in a day at p dollars per piece.
Symbol Definition: $N(p) = -18.2p + 208$ for $0 \le p \le 11$.
Assumptions: Certainty and divisibility. Certainty implies that the relationship is *exact*; divisibility implies that the number of pounds sold and the price per pound can take *any* value in the domain.

(a) Find the price elasticity of demand for the price interval [7,9]. Translate your answer into words.
(b) Find the percent rate of change of the demand with respect to price at a price of \$7 per pound. Translate your answer into words.
(c) Find the price elasticity of demand at $p = 7$. Write out what your answer means in everyday language.
(d) What kind of elasticity does the demand at $p = 7$ represent?
(e) Based on your answer to part (d), what would be the effect on the revenue (in direction only) of raising the price a little bit from \$7? Explain why in words.
(f) Based on your answers to (d) and (e), should the price at which revenue is maximized be less than or greater than \$7? Explain your reasoning.
(g) Use *elasticity* to determine where revenue is maximized, and where demand is elastic and inelastic over all positive prices.

16. The number of pounds of fudge sold when the price per pound is p dollars can be modeled by

Verbal Definition: $N(p)$ = the expected (average) number of pounds of fudge sold at p dollars per pound.
Symbol Definition: $N(p) = 210.8(0.9028)^p$ for $p \ge 0$.
Assumptions: Certainty and divisibility. Certainty implies that the relationship is *exact*; divisibility implies that the number of pounds sold and the price per pound can take *any* value in the domain.

(a) Find the price elasticity of demand for the price interval [7,9]. Translate your answer into words.
(b) Find the percent rate of change of the demand with respect to price at a price of \$7 per pound. Translate your answer into words.
(c) Find the price elasticity of demand at $p = 7$. Write out what your answer means in everyday language.
(d) What kind of elasticity does the demand at $p = 7$ represent?
(e) Based on your answer to part (d), what would be the effect on the revenue (in direction only) of raising the price a little bit from \$7? Explain why in words.
(f) Based on your answers to (d) and (e), should the price at which revenue is maximized be less than or greater than \$7? Explain your reasoning.

(g) Use *elasticity* to determine where revenue is maximized, and where demand is elastic and inelastic over all positive prices.

(h) How do your answers compare to the answers for Exercise 15 (which are in the back of the text, if you didn't do them yourself)? Describe a strategy for dealing with any differences in the answers to help come to a final conclusion.

17. The number of T-shirts sold when the price per shirt is p dollars can be modeled by

Verbal Definition: $N(p)$ = the number of shirts sold at p dollars per shirt.

Symbol Definition: $N(p) = -1.38p + 26.7$ for $0 \le p \le 19$.

Assumptions: Certainty and divisibility. Certainty implies that the relationship is *exact*; divisibility implies that the number of shirts sold and the price per shirt can take *any* value in the domain.

(a) Find the price elasticity of demand for the price interval [9,11]. Translate your answer into words.

(b) Find the percent rate of change of the demand with respect to price at a price of \$9 per shirt. Translate your answer into words.

(c) Find the price elasticity of demand at $p = 9$. Write out what your answer means in everyday language.

(d) What kind of elasticity does the demand at $p = 9$ represent?

(e) Based on your answer to part (d), what would be the effect on the revenue (in direction only) of raising the price a little bit from \$9? Explain why in words.

(f) Based on your answers to (d) and (e), should the price at which revenue is maximized be less than or greater than \$9? Explain your reasoning.

(g) Use *elasticity* to determine where revenue is maximized, and where demand is elastic and inelastic over all positive prices.

18. The number of T-shirts sold when the price per shirt is p dollars can be modeled by

Verbal Definition: $N(p)$ = the number of shirts sold at p dollars per shirt.

Symbol Definition: $N(p) = 62.1(0.859)^p$ for $p \ge 0$.

Assumptions: Certainty and divisibility. Certainty implies that the relationship is *exact*; divisibility implies that the number of shirts sold and the price per shirt can take *any* value in the domain.

(a) Find the price elasticity of demand for the price interval [9,11]. Translate your answer into words.

(b) Find the percent rate of change of the demand with respect to price at a price of \$9 per shirt. Translate your answer into words.

(c) Find the price elasticity of demand at $p = 9$. Write out what your answer means in everyday language.

(d) What kind of elasticity does the demand at $p = 9$ represent?

(e) Based on your answer to part (d), what would be the effect on the revenue (in direction only) of raising the price a little bit from \$9? Explain why in words.

(f) Based on your answers to (d) and (e), should the price at which revenue is maximized be less than or greater than \$9? Explain your reasoning.

(g) Use *elasticity* to determine where revenue is maximized, and where demand is elastic and inelastic over all positive prices.

(h) How do your answers compare to the answers for Exercise 17 (which are in the back of the text, if you didn't do them yourself)? Describe a strategy for dealing with any differences in the answers to help come to a final conclusion.

CHAPTER 3 SUMMARY

Function (Original Graph)	First Derivative (Slope Graph)	Second Derivative (Concavity Graph)
$y = f(x)$	$y', f'(x), \frac{dy}{dx}$	$y'', f''(x), \frac{d^2y}{dx^2}$
Increasing	Positive	(Not determined)
Decreasing	Negative	(Not determined)
Concave up	Increasing	Positive
Concave down	Decreasing	Negative
Local maximum	$= 0$, sign goes from $+$ to $-$ (left to right)	Negative
Local minimum	$= 0$, sign goes from $-$ to $+$ (left to right)	Positive
Inflection point	Local extremum	$= 0$ and sign changes

To Optimize a Smooth Function of One Variable, $f(x)$, over an Interval $[a,b]$

1. Look at the graph, if possible, to visually determine approximate local and global extrema.
2. To find precise local extrema, take the derivative, set it equal to 0, and solve for x to get critical points.
3. Test critical points to see what kind of local extrema they might be, visually from the graph or using the Second Derivative Test.
4. Check the critical points and endpoints (a and b) to determine the global extrema.

(These steps can be done by hand or using technology.)

The Second Derivative Test to See if a Critical Point at $x = k$ Is a Local Extremum of a Smooth Function $f(x)$

1. Take the derivative of the first derivative to find the general second derivative ($f''(x)$), and plug in k for x.
2. If $f'(k) = 0$ and $f''(k) > 0$, then the tangent is horizontal and the curve is concave up, so there is a local minimum at $x = k$.
3. If $f'(k) = 0$ and $f''(k) < 0$, then the tangent is horizontal and the curve is concave down, so there is a local maximum at $x = k$.
4. If $f'(k) = 0$ and $f''(k) = 0$, then the test is inconclusive.

To Find the Points of Inflection for a Smooth Function $f(x)$

1. Find the general second derivative, $f''(x)$, set it equal to 0, and solve for x.
2. If the second derivative changes sign on either side of that value, then it is a point of inflection. If not, it is not (like for $f(x) = x^4$ at 0).

Post-Optimality Analysis

Verification means double-checking your model and solution calculations, such as by hand and using technology, or using two technologies, as independently as possible. If you get different answers, keep re-solving until you are sure what the correct answer is.

Validation means seeing how closely your model and solution match to the real-life situation. One method is visually looking at the fit between the model and the data from a graph. Another is to gather new data (such as trying your optimal value) or test withheld data and see how the actual outcome compares to what the model predicts. A third method of validation is to examine and comment on the assumptions you have made and their implications in the context of your problem. Remember that the best model depends on the use for which the model is needed (interpolation/extrapolation, optimization/forecasting) to help determine the relative importance of pure fit versus the shape matching the context. Solving alternative reasonable models can help validate your solution.

Sensitivity Analysis involves determining which data values (if any) seem questionable and seeing the effect on the solution of possible different data values you think might be more representative.

Accuracy has to do with lack of biases. In any problem, think of what biases are likely in your data, and how you might prevent or adjust for them.

Precision involves the number of digits in your answer. Try to assess the appropriate level of precision for your data and your best guess at a **margin of error**. When doing calculations by hand, try to carry 1 or 2 extra significant figures through intermediate calculations; then round your final answer. When doing calculations with technology, carry as much precision as possible through intermediate calculations; then round off appropriately for your final answer. The general principle is that the precision of your answer is determined by the *least* precise value in the calculation. For adding and subtracting, decimal places count. For other operations (and in general), significant figures are best. Use the same principle for a rough estimate of the margin of error in your solution. In general, when combining errors, the overall error is likely to be a little more than the worst error (between the worst and the sum of the errors). Solving alternative reasonable models and sensitivity analysis can also help assess the overall margin of error for your solution.

Conclusions should be based on all of the above considerations and should include a statement of your best estimate of a numerical solution and margin of error at an appropriate level of precision that matches the way the decision maker thinks of the problem.

Percent rate of change at a point is the instantaneous rate of change at a point expressed as a percent of the output value there:

$$\%\text{ rate of change of } f(x) \text{ at } k = \left[\frac{f'(k)}{f(k)} \cdot 100\right]\% \text{ per input unit}$$

Price elasticity of demand over an interval $[p_1, p_2]$ is the average percent change in demand per percentage point increase in price over the interval:

$$\frac{\frac{\Delta q}{q}}{\frac{\Delta p}{p}} = \frac{p}{q} \cdot \frac{\Delta q}{\Delta p}$$

Price elasticity of demand at a point p is the instantaneous percent change in demand per percentage point increase in the price:

$$\eta = \frac{p}{q} \cdot \frac{dq}{dp} = \frac{p}{D(p)} \cdot D'(p)$$

If demand is elastic ($\eta < -1$, so $|\eta| > 1$), then if you increase price, demand decreases even *more* proportionately, so **revenue decreases. If demand is inelastic ($-1 < \eta < 0$, so $|\eta| < 1$), then if you increase price**, demand decreases *less* proportionately, so **revenue increases.** Thus **revenue is maximized when demand is unit elastic ($\eta = -1$).**

If marginal cost is *less* than average cost, then making one more unit will bring *down* average cost, and if marginal cost is *greater* than average cost, then making one more unit will *increase* average cost, so **average cost is minimized when marginal cost *equals* average cost.**

CHAPTER 4

Continuous Probability and Integration

INTRODUCTION

Life is full of uncertainties: the performance of the stock market; the weather; playing the lottery; gathering data; the holdout price of someone with whom we are negotiating; whether someone will accept (or make) a marriage proposal; the distribution of things in the world. Much of what we do involves situations where we are not certain of the outcome. In many of these situations,[1] it is possible to describe or characterize this uncertainty to help us make decisions. The fields of probability, statistics, and (stochastic) operations research (also called management science, or decision science) exist largely to help us to deal with these situations.

In this chapter, we will focus on situations where the uncertain quantity can take on **any** numerical value in an entire interval, called a **continuous** random variable. Of the examples of uncertainty listed above, the one that is most clearly **not** continuous (called **discrete**) is the marriage proposal, since the response within a given period of time will be a yes, or it will not be a yes. However, one could think of the **time** until the person responds as a continuous random variable. The lottery example is discrete by nature (there are only a finite number of possible outcomes that can be randomly drawn) but could also be analyzed approximately using a continuous random variable, since there are so many possible outcomes. Note that our use of continuous (equivalent to **divisible**) and

[1]Some people would in fact claim that this is possible in *all* such decision situations.

discrete here are consistent with our definitions in the Preface and when we discussed the divisibility assumption.

We will see in this chapter that the calculation of probabilities in this situation is related to the problem of finding the **area under a curve** graphically. In calculus, this operation is done by a technique called **integration**. In some ways, integration can be thought of as a continuous version of the process of **summation** of discrete quantities. It involves **accumulating** a total when the values being accumulated can vary in a continuous way. An example of this would be if you have figures for daily sales revenues, and you want to project the total sales revenues over a period of time into the future. The answer involves finding an **antiderivative** of the daily sales model (a function whose derivative equals the model for daily sales). These concepts underlie virtually all of the statistics you will be studying and using throughout your career, as well as many concepts in economics, finance, and operations management. We will show that there is a natural and elegant inverse-like relationship between derivatives and integrals (areas under curves) that is summarized in the **Fundamental Theorem of Calculus**. We will also show how calculus can be used to find statistical measures like the **mode** and **median** of a continuous probability distribution and to calculate useful measures in economics such as **Consumer Surplus** and **Producer Surplus** that help describe how a competitive free market benefits society. In addition, we will show how to calculate areas even when the interval of interest goes to infinity, called an **improper integral**.

Here is a selection of examples to give you a feel for the kind of problems you should be able to analyze after studying this chapter:

- Suppose you work for a start-up firm with a radical new concept for cold-weather gloves. You don't have enough capital to make all possible glove sizes, but your design is flexible enough to fit a **range** of glove sizes. You need to decide which range of glove sizes to produce.
- You want to estimate what fraction of people in a certain population scored within a certain range of SAT scores to get a feel for your chances of receiving a particular scholarship.
- Your car odometer is broken. You have been traveling for several hours, and you have a pretty good idea of what your speed has been over the time of the trip (in fact, it has been increasing as you have gotten increasingly bored and eager to get to your destination). Now you want to estimate the distance you have traveled so far.
- You are considering buying an oil well. An engineer has given you an estimate of the rate of production of oil over time from this well. To help you decide how much you'd be willing to bid for the well, you want to estimate the total monetary value of the well.
- You run a business, and you know a major competitor is about to come out with a new product. In order to plan some financial strategies in advance, you want to estimate the most likely release time for the new product, the median (50th percentile) time before its release, and the probability of release in a given interval.
- As a sideline from your main job, you produce a novel item for performers: a cupholder that attaches to a microphone stand. You don't have a lot of free time to make them, so the number you are willing to make in a month depends on how

much money you get for them. You also have a feel for how many you could sell in a month for different selling prices. If you charged the selling price where the number you want to make matches the number people would want to buy, how much *extra* revenue will you get over and above the *minimum* you would have sold each cupholder for, and how much money will your customers *save* compared to the *maximum* they would have paid (on average) for each?

In addition to being able to solve problems like those above, by the end of this chapter, you should also:

- Understand the idea behind a continuous random variable, including what its probability density function means and how it can be used to calculate probabilities, modes (most likely values), and medians (50th percentile values).
- Understand that finding the change in an original function given its nonnegative rate function corresponds to finding the area under the rate curve and above the horizontal axis.
- Understand that an area can be approximated using rectangles, trapezoids, or parabolas drawn on individual subintervals of equal widths to match the curve as closely as possible.
- Understand that an area can be calculated with great accuracy and precision by taking the limit of the sums from any of the subinterval area approximations, as long as the curve is continuous.
- Understand that the exact value of the limit of the sums of a subinterval method is called a definite integral, and corresponds to the area above the horizontal axis minus the area below it between the curve and the axis over the interval of interest.
- Understand that an original function can be recovered from its rate function by finding a general antiderivative with an arbitrary constant, then using a known value of the original function to find the specific value of the constant.
- Understand that the net change in an original function over an interval can be obtained from its rate function by finding the original function as above (or by using *any* antiderivative of the rate function, since the constant eventually cancels out), and subtracting its value at the upper endpoint minus its value at the lower endpoint.
- Understand that *any* exact net area between a curve and the horizontal axis over an interval can also be found by finding *any* antiderivative and evaluating it at the upper endpoint minus the lower endpoint.
- Understand how derivatives and integrals are very nearly inverse operations on functions.
- Understand how to find areas where one or both endpoints are infinite by using variable limits of integration and taking limits.
- Understand how to find the social benefit to consumers and producers of a product that sells at the equilibrium price in a competitive free market.

4.1 CONTINUOUS PROBABILITY DISTRIBUTIONS

When there is uncertainty about a continuous quantity, such as snow accumulation or future profits, we can describe that uncertainty numerically using a function called a **probability density function**. The most familiar example of such a function is the normal (Gaussian) distribution's "bell-shaped curve." For example, you can think of the distribution of grades on a test in a large class, or the related idea of "grading on a curve." The intuitive idea of such a curve is that the height (y-value) of the function expresses the **relative likelihood** (frequency, probability, chance, odds) of obtaining the corresponding x-value (or values in a small neighborhood of it) compared to all the other x-values. Thus, the peak of the bell-shaped curve corresponds to the x-value that can be thought of intuitively as the most likely or most common (called the **mode**). In this section, we will study these probability density functions. We will also discuss how to find the mode and how to calculate, for certain simple examples of density functions, the probability that the uncertain quantity will lie within a given interval. This will serve as a lead-in to doing similar calculations for more complicated cases in later sections.

Here are some sample problems to give you a feel for what is involved:

- You have estimated the probability distribution for the time before a competitor will be releasing a new product. What is the most likely value for the release time?
- You have conducted a survey of students to see what is the most they would pay for a certain T-shirt. Estimate the probability that a student picked at random from the same population would pay at most a price within a specified price range (interval).
- Based on past performance and trends, you have estimated the probability distribution for your profit this year. To assess your current situation and make plans for possible borrowing during the year, you want to find your most likely profit, and the probability that your profit will fall within a given interval of values.

After studying this section, in addition to being able to solve problems like those above, you should also

- Understand what is meant by a continuous random variable.
- Understand the idea of a continuous probability density function, and how it describes the relative likelihood of the possible values of a continuous random variable.
- Understand that the mode is the most likely value of a random variable and can be found by finding the maximum point on the probability density function.
- Be able to calculate the probability that the value of a random variable lies within a given interval for uniform (constant) and histogram (bar chart) distributions.
- Understand that the probability that the value of a general continuous random variable lies within a given interval corresponds to the area under its probability density function (and above the horizontal axis) over that interval.

The unknown value of an uncertain but well-defined quantity, such as the time until the competitor's new product is released and the profit value in the examples above, **is called a continuous random variable and is usually designated with an uppercase *X*.** We will only consider random variables where the possible values are *real numbers*, although more general definitions are possible. As discussed above, the nature of a continuous random variable can be described by a function called its **probability density function.** This function is sometimes called a density function, or a probability distribution function, and is sometimes written in abbreviated form as a **pdf**. The idea behind the name is that the functional value, $y = f(x)$ gives the relative **density** (intensity, level) of the probability at that x-value. **The height of the probability density function gives the *relative* likelihoods, or comparative chances, that the random situation will result in the different *x*-values** (considered as possible actual values of the random variable X, or values in the **domain** of its probability density function).

Uniform Probability Density Functions

The simplest example of a density function occurs when all of the x-values (values of the random variable) are equally likely over an entire finite interval. For example, when computers generate "random" numbers,[2] usually between 0.0000 and 0.9999, they assume such a **uniform distribution** over the interval $0 \leq x < 1$, which we can write as $[0, 1)$.[3] **Since the relative probability of all values is the same in a uniform distribution, the density function is just a horizontal line over the entire domain.** In this example, this would mean a horizontal line above the interval between 0 and 1, with a value of 0 everywhere else.

Before proceeding further, let us make the comment that, with continuous random variables, we can be a little sloppy about whether or not to include the endpoints of the interval in the domain of the density function. This is the case because **the actual probability of any one exact real number (or even a finite number of them) is really zero**, and so including it or not will not affect any probability calculations. We will go into this further when we have more concepts to help explain the idea.

It is the convention in probability (as in weather forecasts) to state probability values from 0% (impossible, or at least essentially so) to 100% (certain, or at least essentially so); that is, from 0 to 1.

If we want to simulate the flip of a coin, we could say that if the random number generated is between 0.0000 and 0.4999, then we will consider that "Heads" occurred, and if it is between 0.5000 and 0.9999, we will consider that "Tails" occurred. We define our random variable on $[0, 1)$ as X, where

$$X = \text{the random number generated between 0.0000 and 0.9999}$$

Thus, we want the probability of X being in the interval $[0, 0.5)$ to be 1/2, and the probability of X being in the interval $[0.5, 1)$ to likewise be 1/2. Note that the probability of each interval is the same as the length of the interval in this case.

---

[2] Actually, they are better described as "**pseudorandom**," since they are calculated according to a deterministic formula, but they come out in a way that *appears* random and satisfies statistical tests of randomness. These can be used to **simulate** coin flips and other random phenomena.

[3] In this standard **interval notation**, a square bracket indicates that the endpoint is *included* in the interval (like 0 in the example) and a round parenthesis indicates the endpoint is *not* included (like 1 in the example).

Now suppose that we used a random variable X_2 that has a uniform distribution on the interval [0, 10) instead of [0, 1) to simulate our coin toss. In this case, we would consider any value in the interval [0, 5) to correspond to Heads and any value in [5, 10) to correspond to Tails, each of which should again have a probability of 50%, or 1/2. Recall that it is the convention in probability to state probability values from 0% to 100%, that is, from 0 to 1. Thus for X we can no longer say the probability in each case is *equal* to the length of the interval, but we **can** say that the **probabilities are *proportional* to the lengths of the intervals for a uniform distribution** (the intervals are the same length as each other, and the probabilities are also the same as each other).

Since the density function **for a uniform distribution** is a horizontal line, the **probability of an interval can also be thought of as being proportional to the area under the density function and above the x-axis between the endpoints of that interval**. This is because these areas will all be rectangles with the same height (as long as the intervals are all within the domain), and so the areas will simply be proportional to the lengths of the intervals. We can then choose the height of the density function in such a way that these areas will be exactly the probabilities of the corresponding intervals. In fact, since these areas will be proportional to the probabilities of the corresponding x-value intervals, **we can define the height of the density function in such a way that the probability of an interval is *equal* to the area under the density function and above the horizontal axis over that interval**. To decide what that height should be, we can simply pick an interval for which we know what we want the probability to be (such as the domain of all the values that can actually occur, which should have a probability of 1), and figure what height would make its area come out right.

For our original uniform distribution over [0, 1), we know that the area under the horizontal line and above the x-axis from 0 to 1 should be 100%, or 1. Since the area is a rectangle, and its base is $1 - 0 = 1$, we would need a height of 1 also to make the area come out to be 1:

$$\begin{aligned}(\text{base})(\text{height}) &= 1 \\ (1 - 0)(h) &= 1 \\ h &= 1\end{aligned}$$

Thus, the density function will have the graphical form shown in Figure 4.1-1. Note that the **total area under the density function (and above the x-axis) over its domain (where the relative probability is not 0) is 1**. This is related to the fact that a sure thing (like a forecast for rain being given during a downpour) corresponds to a probability of 100%, or 1.

For the random variable X_2 that we defined earlier over the interval [0, 10), the probability of the entire domain should again be 100%, or 1. We can see that we would need a height of 1/10, or 0.1, to make this happen. Algebraically, if we call the height h, we could say that we want

$$\begin{aligned}(\text{base})(\text{height}) &= 1 \\ (10 - 0)(h) &= 1 \\ 10h &= 1 \\ h &= \frac{1}{10}\end{aligned}$$

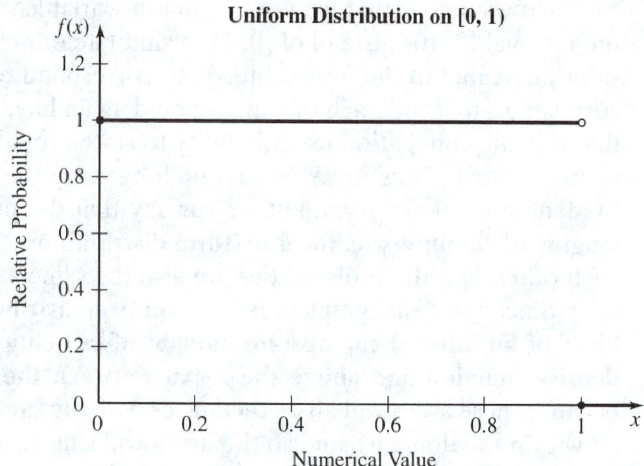

Figure 4.1-1

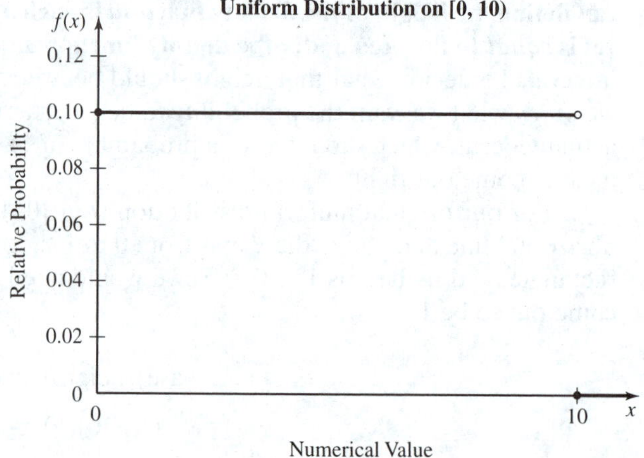

Figure 4.1-2

Thus the graph of the density function for X_2 is shown in Figure 4.1-2. and its algebraic form is given by

$$f(x) = \begin{cases} \dfrac{1}{10} & \text{for } 0 \leq x < 10 \\ 0 & \text{otherwise (for } x < 0 \text{ or } x \geq 10) \end{cases}$$

Suppose we wanted to find the probability that X falls between 3 and 7. Since [3, 7] lies within [0, 10), the desired probability is simply the area of the rectangle under $f(x)$ and above the x-axis between $x = 3$ and $x = 7$. Graphically, we have Figure 4.1-3.

We can see that the base of this rectangle is the length of the interval [3, 7], which is equal to $7 - 3 = 4$. The height of the rectangle is the height of $f(x)$, which is 1/10, or 0.1. Thus the area of the rectangle is

$$(4)(0.1) = 0.4$$

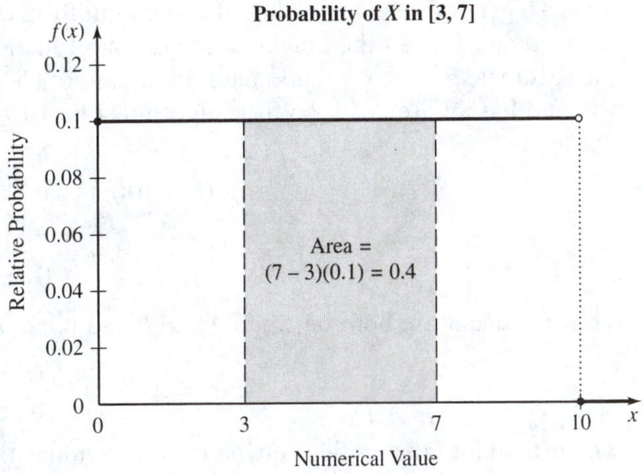

Figure 4.1-3

and so the probability of X being between 3 and 7 is 0.4 (or 40%). Notationally, we write

$$P(3 \leq X \leq 7) = 0.4$$

Translation: The probability that X lies between 3 and 7 is 0.4. X could be the distribution of, say, the number of days before your kid brother's loose tooth will fall out, which you figure could be any time in the next ten days, all about equally likely. In the loose tooth scenario, the statement would mean there was a 40% chance that the tooth would fall out at a point between 3 days and 7 days from now.

In general, then, the probability of this same X falling in any specified interval $[c, d]$ is the area under $f(x)$ and above the x-axis between $x = c$ and $x = d$. If c and d both lie between 0 and 10, then the region of interest is a rectangle, with a base of $(d - c)$ (the length of the interval) and height of 0.1 (Figure 4.1-4), and so

$$P(c \leq X \leq d) = 0.1(d - c)$$

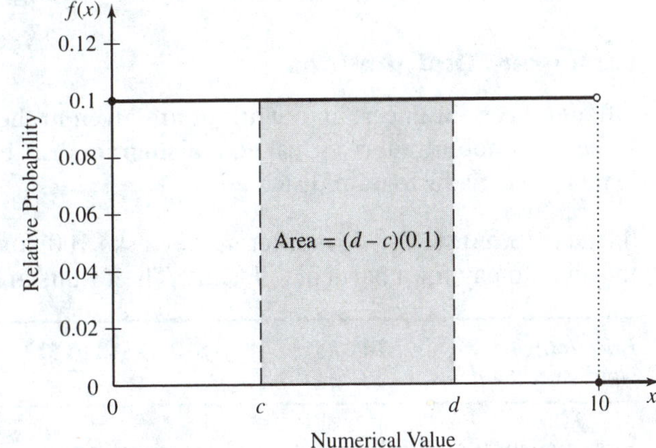

Figure 4.1-4

To generalize even further, if X has a uniform distribution on any interval with endpoints a and b, then the height of the density function will be $1/(b - a)$. This is because the rectangle for the whole domain has a base of $(b - a)$, the length of the domain interval, and has an area of 1, so the height must be $1/(b - a)$. Derivation: If h is the height, we need

$$(b - a)(h) = 1$$

$$h = \frac{1}{(b - a)}$$

Thus if c and d are both between a and b and if $c \le d$, then

$$P(c \le X \le d) = \frac{(d - c)}{(b - a)}$$

The probability that X lies between c and d equals the length of that interval divided by the length of the domain interval.

Sample Problem 1: Suppose that you know there is going to be a fire drill sometime today between 8:30 A.M. and 4:30 P.M., and that you think the exact time is equally likely to be any time in that interval. What is the probability that the fire drill will occur during your big math test, between 11:30 A.M. and 12:45 P.M.?

Solution: Let us define X = the number of hours after 8:30 A.M. that the fire drill occurs (so the domain is [0, 8]). Then the density function for X is given by

$$f(x) = \begin{cases} \dfrac{1}{(8 - 0)} = \dfrac{1}{8} = 0.125 & \text{for } 0 \le x \le 8 \\ 0 & \text{otherwise} \end{cases}$$

The time 11:30 A.M. corresponds to $x = 3$ and 12:45 P.M. corresponds to $x = 4.25$. So the desired probability is

$$P(3 \le X \le 4.25) = \frac{(4.25 - 3)}{(8 - 0)} = \frac{1.25}{8} = 0.15625$$

Translation: There is a little less than a 1 in 6 chance that the fire drill will occur during your big math test. Of course, Murphy's law would give a different answer! □

Histogram Distributions

Often data about uncertain quantities are given in the form of a table of frequencies for values in different intervals, called a **histogram** distribution in statistics. Let's do a problem involving a histogram distribution:

Sample Problem 2: A market survey asks 100 students at random what is the most they would pay for a particular T-shirt. The results are as follows:

Price Range	\$6 to \$8	\$8 to \$10	\$10 to \$12	\$12 to \$14	\$14 to \$16	\$16 to \$20
Number of Students	11	32	27	19	9	2

First, we should clarify how we are interpreting the overlapping values at the endpoints of the subintervals. Let's assume that each interval includes the left endpoint and not the right, as we have in the earlier examples (so the first interval is [6, 8)).

What is the probability that the most a student picked at random would pay for the T-shirt is between \$10 and \$16? Explain in words what your answer corresponds to graphically. What price ceiling response is most likely from the random student?

Solution: The standard assumption in a case like this is that, within each interval, all values are equally likely. This is like assuming that the distribution within each interval is uniform. Thus, the density function for a histogram looks like a bar chart (although technically the vertical lines are not part of the graph of the density function), with horizontal line segments of different heights reflecting the different relative probabilities. From looking at the frequency data above, it is intuitive that the probability a randomly selected person would pay a maximum of \$10–\$12 is 3 times the probability that they would pay \$14–\$16, since the ratio of the frequencies is 27/9 = 3. It is also intuitive that the probability that the maximum the randomly chosen person would pay (their **reservation price**) is between \$8 and \$10 is 32/100 = 0.32. Thus, if we define X to be the reservation price of a person selected at random from the population of interest, we are saying that

$$P(8 \le X < 10) = 0.32$$

We want this probability to correspond to the area under the density function, as before. If h is the height of the density function on [8, 10), we want

$$\begin{aligned}(10 - 8)(h) &= 0.32\\ 2h &= 0.32\\ h &= 0.16\end{aligned}$$

Thus we would have

$$f(x) = 0.16 \text{ for } 8 \le x < 10$$

Applying the same procedure to the interval [10, 12), we would get

$$\begin{aligned}(12 - 10)(h) &= \frac{27}{100} = 0.27\\ 2h &= 0.27\\ h &= \frac{0.27}{2} = 0.135\end{aligned}$$

Thus we get

$$f(x) = 0.135 \text{ for } 10 \le x < 12$$

Similarly, for [14, 16), we get

$$f(x) = \frac{0.09}{2} = 0.045 \text{ for } 14 \le x < 16$$

We pointed out above that [10, 12) is 3 times as likely as [14, 16); note that the density function is three times as much. However, now let us look at the interval [16, 20). Here we get

$$\begin{aligned}(20 - 16)(h) &= 0.02\\ 4h &= 0.02\\ h &= \frac{0.02}{4} = 0.005\end{aligned}$$

From the frequencies, note that the ratio of the frequency of [8, 10) to the frequency of [16, 20) is 32/2, or 16. But the ratio of the density function values for those intervals is 0.16/0.005 = 32 (32 times as much, not 16). In a sense, **the density function is really a probability *rate per unit of the random variable* within each interval**. Since the frequency of 2 is spread over an interval that is twice as large as the others, the rate per unit is half what it would have been if the interval had been the same width as the others. Thus, if the interval widths of a histogram are all the same, then the height of the density function for each interval is proportional to the frequency. However, if the intervals are not all the same width and if the frequencies are nonzero, this will *not* be the case.

From the above discussion, we see that the density function for X is given by

$$f(x) = \begin{cases} 0.055 & \text{for } 6 \le x < 8 \\ 0.16 & \text{for } 8 \le x < 10 \\ 0.135 & \text{for } 10 \le x < 12 \\ 0.095 & \text{for } 12 \le x < 14 \\ 0.045 & \text{for } 14 \le x < 16 \\ 0.005 & \text{for } 16 \le x < 20 \\ 0 & \text{otherwise} \end{cases}$$

Graphically, the density function looks like Figure 4.1-5.

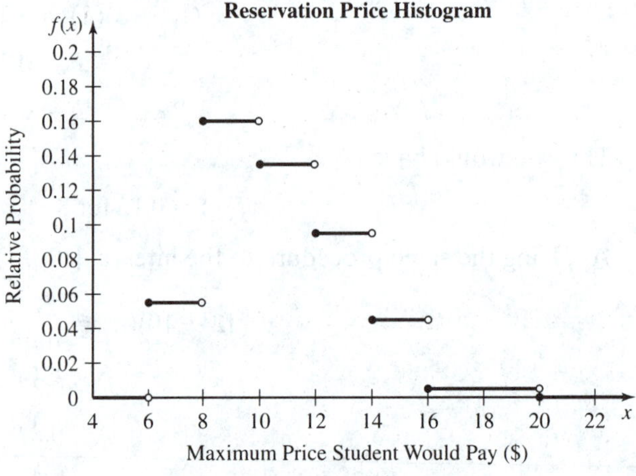

Figure 4.1-5

Using the same intuition as above, we can see that the probability that X falls between 10 and 16 is 55/100 = 0.55, since a total of 55 out of the sample of 100 fell within that interval. Graphically, we can see that this corresponds to the area under the density function and above the x-axis between $x = 10$ and $x = 16$:

$$\begin{aligned}(12 - 10)(0.135) + (14 - 12)(0.095) + (16 - 14)(0.045) &= 2(0.135) + 2(0.095) + 2(0.045) \\ &= 0.27 + 0.19 + 0.09 \\ &= 0.55\end{aligned}$$

Since the probability density function intuitively gives the relative probability or likelihood of the different x-values of a random variable, the **most likely value** (called the **mode**) corresponds to the **global maximum of the density function** over its domain.

From looking at the density function, we can say that the mode of this distribution is any value between \$8 and \$10. If we wanted to give a unique answer, we could use the midpoint of this interval, or \$9. But, in general, we are not even guaranteed of being able to define a unique mode. For example, if a density function had the shape of an "M" such as shown in Figure 4.1-6, (this could be the distribution of the heights of a mixed-gender group of people, measured in inches *over* 5′4″), then it would have two modes, and taking the midpoint or average of these would not make sense. Given these considerations, it makes the most sense to allow for multiple modes, and to call all global (and sometimes even local) maximum points modes. Thus, if we assume that the earlier *histogram* density function is exactly right, it makes sense to say that *all* values between \$8 and \$10 are modes.

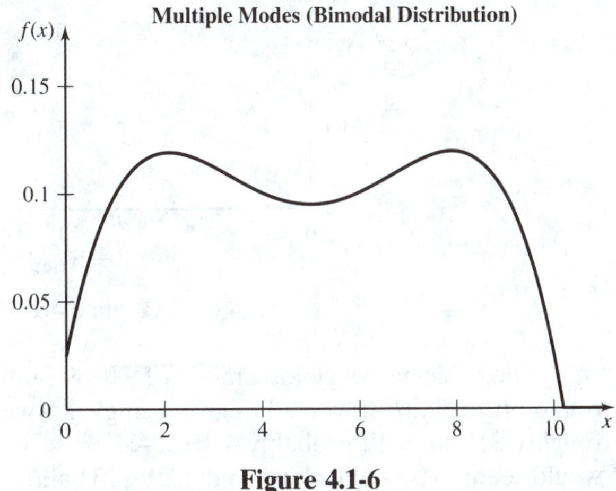

Figure 4.1-6

Smooth Continuous Distributions

On the other hand, considering the real problem, it is very unlikely that the histogram density function is exactly right. More realistically, it is probably an approximation of a smooth continuous density function. We could try to fit a curve to the given information. We could use the midpoint of each interval and its density function value as data points, fit a curve to them, then make an adjustment if necessary to force the area under the curve over the domain to be 1 by dividing the entire function by a constant. That curve then would be likely to have a single mode, called a **unimodal** distribution. The "M"-shaped distribution in Figure 4.1-6 is called **bimodal** because it has two modes, and the histogram density could be called **multimodal**. In such a case, the value of \$9 mentioned above might be the best simple approximation of this single mode based on the histogram of the continuous model. For a more exact estimate, we could perform the actual curve fitting, and find the location of the global maximum.

Let's try this latter operation. Using the midpoints of the intervals from our histogram density function, our data points are

Price (x)	7	9	11	13	15	18
Probability Density (y)	0.055	0.16	0.135	0.095	0.045	0.005

A plot of these data points is shown in Figure 4.1-7. These data points look mostly (although not *completely*) concave down, and *both* increase *and* decrease. This would seem to suggest a quadratic model, but the graph also makes it clear that the function is not **symmetric** around its peak like a parabola—it is in fact **skewed** to one side. This suggests that we will need at least a cubic model to capture the shape of the data. The graph in Figure 4.1-8 shows the fit of the cubic model.

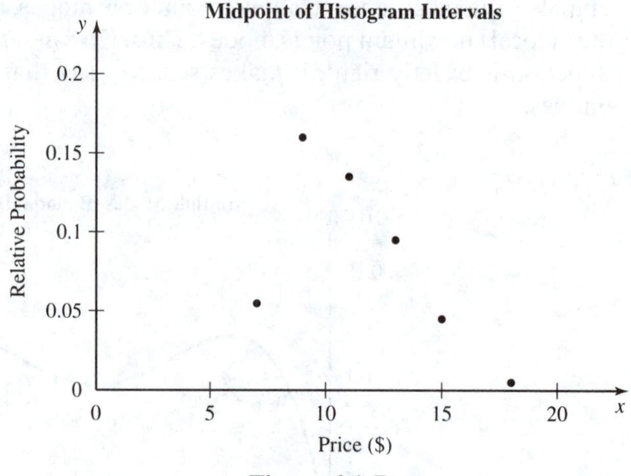

Figure 4.1-7

The cubic model yields a mode of 9.9689 (global and local maximum), which we can round off to \$9.97. If we had to give a single answer, it might make the most sense to say roughly \$10.00, with probably a \$0.50–\$1 or so margin of error, since visually the peak would seem to be somewhere in the \$9–\$10 ballpark. It looks as if a polynomial of higher dimension (such as a quartic) might fit better and give a better estimate, with a smaller margin of error. □

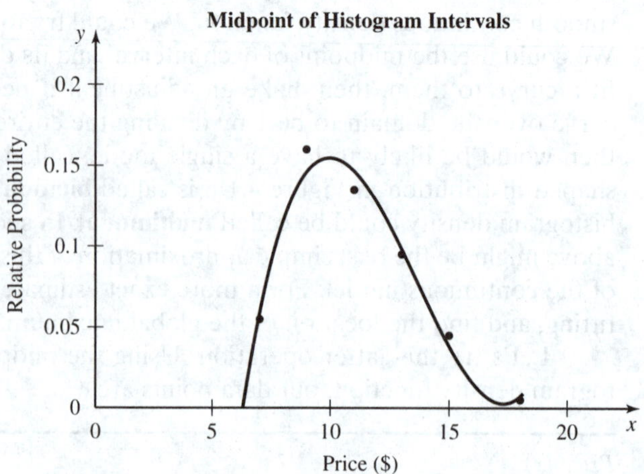

Figure 4.1-8

Many times teachers are asked if they are going to "grade on the curve." This refers to the practice of assigning grades to approximate what is called a normal distribution: A certain percent of the students will get an "A," another percent a "B," and so on. In fact, if the class were very, very large, this would probably come very close to holding true, just as the distribution of heights of male or female college students would very nearly follow what is called a normal distribution if the sample were large enough. The curve that represents this probability density function is sometimes called a "bell curve," because it has the general shape of a bell (Figure 4.1-9). The simplest version of the bell-shaped curve, often called the **standard normal distribution**, has the following density function:

$$f(x) = \frac{1}{\sqrt{2\pi}} e^{-x^2/2}$$

(You can imagine what the more complicated versions look like!) To give this a plausible context, let's consider the following problem.

Sample Problem 3: Based on past performance and trends, you have estimated the probability distribution for the change in the profit of your business from last year to this year. To assess your current situation and make plans for possible borrowing during the year, you want to find your most likely change in your profit. You define

X = the *change* in the profit of your business from last year (fiscal year) to this year, in millions of dollars.

$$f(x) = \frac{1}{\sqrt{2\pi}} e^{-x^2/2}$$

Solution: Since no domain is specified, we assume the domain is all real numbers. This is an example of ***unconstrained*** **global optimization** (maximization in this case) of a single-variable function. The graph of the density function is shown in Figure 4.1-9.

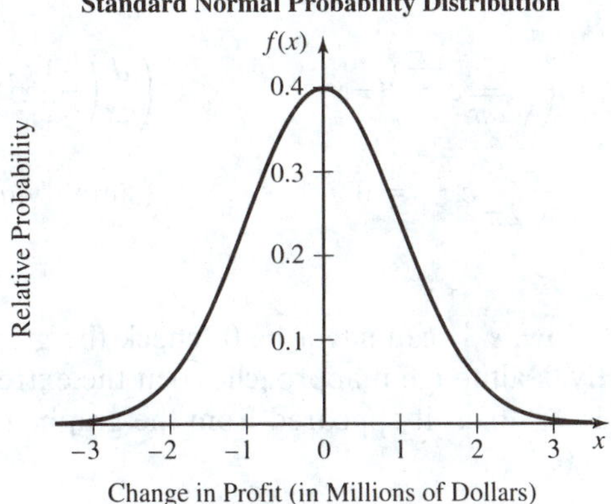

Figure 4.1-9

To interpret this graph, let us do some rough visual estimates. When $x = 0$, the graph suggests that $f(x)$ is approximately 0.40, and when $x = 1.5$, $f(x)$ is approximately 0.13. This means that it is roughly three (0.40/0.13) times as likely that our change in profit will be 0 compared to 1.5. In other words, it is about three times as likely that our profit will stay about the same as last year compared to increasing by about \$1.5 million. To be more precise, we have seen that the probability of X falling in an interval is the area under $f(x)$ over that interval. If you picture very skinny (but equal in width) intervals near 0 and 1.5, the area around 0 would be about 3 times the area around 1.5 (Figure 4.1-10).

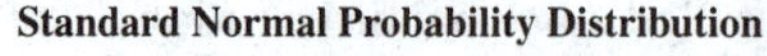

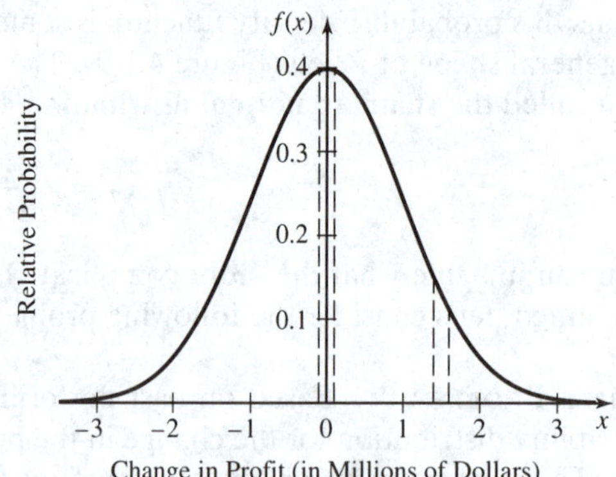

Figure 4.1-10

From looking at the graph, it looks as if the global maximum occurs at 0. Let's check this algebraically. As we learned in Chapter 3, to find a local maximum of a smooth function (and we know the global maximum must also be a local maximum or an endpoint), we can find the derivative, set it equal to 0, and solve for x. Here we get

$$f'(x) = \frac{d}{dx}\left(\frac{1}{\sqrt{2\pi}}e^{-x^2/2}\right)$$

$$= \left(\frac{1}{\sqrt{2\pi}}e^{\frac{-x^2}{2}}\right) \cdot \frac{d}{dx}\left(-\frac{1}{2}x^2\right) \qquad \text{(Chain Rule } \left(u = -\tfrac{1}{2}x^2\right)\text{: Outer derivative times inner derivative)}$$

$$= \left(\frac{1}{\sqrt{2\pi}}e^{\frac{-x^2}{2}}\right)(-x) \qquad \left(\frac{d}{dx}\left(-\frac{1}{2}x^2\right) = \left(-\frac{1}{2}\right)(2)x^{2-1} = -x\right)$$

$$= \frac{-x}{\sqrt{2\pi}}e^{\frac{-x^2}{2}} = 0 \qquad \text{(simplifying and setting derivative = 0)}$$

Now, $e^{\frac{-x^2}{2}}$ can never be 0 (check the graph to verify that the output value is never exactly 0, although it approaches 0 at the extremes). Thus the only solution to this equation is $x = 0$, as it appeared from the graph. The mode of the standard normal distribution is 0.

Translation: It is most likely that there will be no change in the profit of your business from last fiscal year to this year. □

Sample Problem 4: By subjectively estimating the relative likelihood of different possibilities and fitting a curve, you estimate that the number of months (X) before a competitor will come out with a new product can be modeled with the following **exponential probability density function**:

$$f(x) = 0.2e^{-0.2x}, \text{ for } x \geq 0 \text{ (0 otherwise)}$$

Find the mode (most likely value) of the distribution of the release time of the new product.

Solution: First we note that the domain over which $f(x)$ is positive is $x \geq 0$, so this is an example of **constrained global optimization** (maximization here), or optimization of a single-variable function over an interval (in this case for $x \geq 0$, which we can denote as the interval $[0, \infty)$). The graph of the density function for X is shown in Figure 4.1-11.

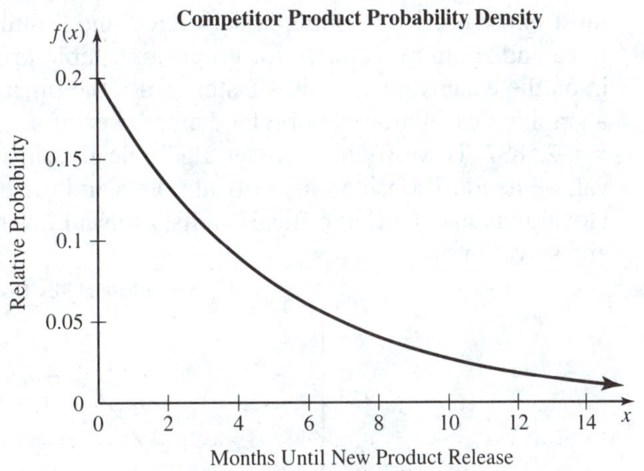

Figure 4.1-11

By inspection, we see that the global maximum occurs at $x = 0$. To check this algebraically, we can again find the first derivative:

$$f'(x) = (0.2e^{-0.2x})(-0.2) = -0.04e^{-0.2x}$$

Note that in this case, the derivative is always negative for any value of x in the domain (which is $x \geq 0$). This means, first of all, that it will never equal 0 for any real number, so there are no critical values (no potential local optima). As x tends toward infinity, $f'(x)$ does *approach* 0. However, as we can see from the graph, this corresponds to what could be thought of as a "local minimum *at infinity*," because the graph heads down toward 0 as x increases, in the limit. But we are looking for the global maximum, not the minimum. Since the derivative is always negative, this means that the function is always **strictly decreasing** over the entire domain. And since the domain is a single interval, this means that the global maximum must occur at the left endpoint, which is $x = 0$. Put another way, since there are no critical points (places where the derivative is 0 or is undefined), the global maximum must occur at one of the endpoints of the domain. So we can find the value of the function at the endpoints and take the larger of the two. Here $f(0)$ is 0.2, while "at infinity," $f(x)$ is 0; that is, $f(x)$ approaches 0 as x approaches infinity. So the global maximum value of the density function is 0.2, and occurs at $x = 0$. Thus we have confirmed that the mode is $x = 0$.

Translation: The most likely time for your competitor to come out with the new product is right now. □

Sample Problem 5: Using curve-fitting techniques, you estimate that the probability density function for your profit for this year (X, in millions of dollars) has the form

$$f(x) = 0.0021699x^4 - 0.022319x^3 + 0.038670x^2 + 0.097920x + 0.048556$$

for $-1 \leq x \leq 6$. What is the mode of this probability distribution?

Solution: This is another constrained optimization problem. Let's start by looking at the graph of the density function over its domain. Using technology, you can graph the density function for x values between -1 and 6. A graph is shown in Figure 4.1-12. We see that the global maximum seems to be somewhere between $x = 2$ and $x = 3$ (so the most likely profit, or mode, is between 2 and 3 million dollars). If your technology has trace and zoom procedures for graphs or a table procedure, you could use them to zero in on the exact value, but it is faster to use the function maximization procedure:[4] Using a graphing calculator, we obtained an answer of $x = 2.588687467$, which we round off to $x = 2.5887$. To verify this answer algebraically, since this is maximization over an interval, we need to consider any critical points and the endpoints of the interval for possible global maxima. To find critical points, as usual, we take the derivative, set it equal to 0, and solve for x:

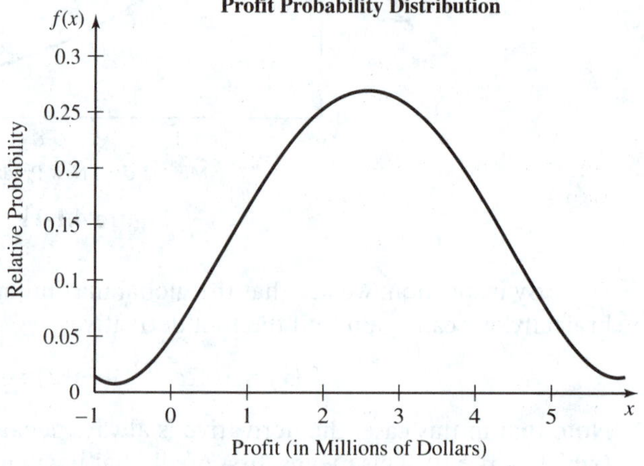

Figure 4.1-12

$$f'(x) = D_x(0.0021699x^4 - 0.022319x^3 + 0.038670x^2 + 0.097920x + 0.048556)$$
$$= 0.0086796x^3 - 0.066957x^2 + 0.077340x + 0.097920$$
$$0.0086796x^3 - 0.066957x^2 + 0.077340x + 0.097920 = 0$$

Using technology to solve this equation with an initial guess of 2.5,[5] we get an answer of

$$x \approx 2.5887$$

Note that, looking at the graph, and realizing that the derivative is a cubic function, there could be up to two other critical values where the derivative is 0. In fact, if you zoom out, you can see that these will be local minima occurring near the endpoints of the domain, at the values

[4]See your technology supplement for details about how to use these procedures.

[5]Consult your technology supplement to see how to solve an equation in this form.

$$x \approx -0.74265 \text{ and } x \approx 5.8683$$

Thus the guess used within the equation-solving procedure is very important. If we used values near -1 or 6, we would have gotten very different answers than what we really wanted. This is why it is important to look at the graph after obtaining a solution to make sure the solution you get is the one you want. In this case we can see that the mode is 2.5887. In other words, the most likely profit for this year is about \$2.59 million. Notice that we have obtained this result by two different methods, so we have **verified** (double-checked) our calculations, as well as seen that the answer makes sense from looking at the graph. □

Probability and Area Under the Density Function

The main purpose of the discussion in this section is to lead into the next sections of the chapter. Given our discussion about histogram distributions, what are the implications for how to deal with calculating the probability over an interval for smooth continuous density functions? The intuitive answer is that, as with the other density functions we have discussed, the probability over an interval can be found by finding the area under the density function and above the x-axis.

> **The probability of a continuous random variable over an interval is the area under the density function and above the x-axis between the endpoints of the interval.**

Sample Problem 6: Using curve-fitting techniques, you estimate that the probability density function for your profit for this year (in millions of dollars) has the form

$$f(x) = 0.0021699x^4 - 0.022319x^3 + 0.038670x^2 + 0.097920x + 0.048556$$

(the same density function as Sample Problem 5) for $-1 \leq x \leq 6$. Explain in words what finding the probability of your profit being between \$1 million and \$4 million corresponds to graphically.

Solution: The probability corresponds graphically to the area under the density function $f(x)$ and above the x-axis between $x = 1$ and $x = 4$ (Figure 4.1-13). □

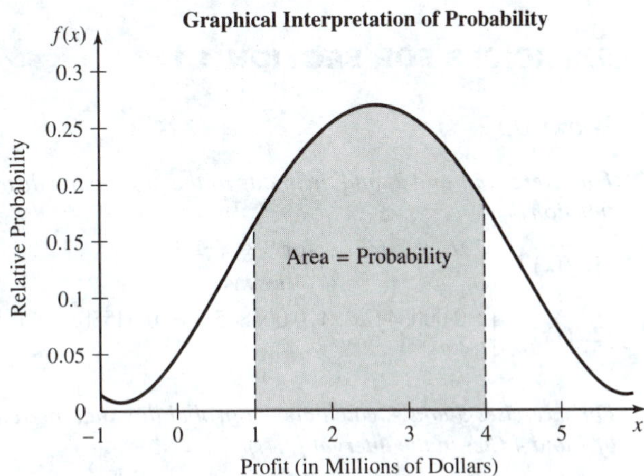

Figure 4.1-13

This then raises the question: How do we find the area under a general curve? The second half of single-variable calculus, involving what is called an **integral**, is designed to answer exactly that question.

Section Summary

Before trying the exercises, be sure that you

- Understand that a **continuous random variable** X is a variable representing a quantity whose value is clearly defined but uncertain, where the value it can take on can be any real number in a given interval (its domain).
- Understand that the **probability density function** for a continuous random variable is a function defined for the domain of the random variable whose value (height of the curve above the x-axis) expresses the *relative* likelihood of that value (or a tiny interval around it).
- Understand that the **mode** is the *most likely* value of a random variable and is a maximum point of the probability density function.
- Understand that the probability that the value of a random variable X with a **uniform distribution on [a, b]** will fall inside a particular interval that is within the domain is given by the width of the interval over the width of the domain:

$$P(c \leq X \leq d) = \frac{d - c}{b - a}$$

- Are able to generalize the above concept to calculate the probability that the value of a random variable lies within a given interval for **histogram** (bar chart) distributions.
- Understand that the **probability** that the value of a general continuous random variable lies within a given interval corresponds to the **area under its probability density function** (and above the horizontal axis) over that interval.

EXERCISES FOR SECTION 4.1

Warm Up

For Exercises 1 and 2, find the mode of the probability distribution with the given probability density function.

1. $f(x) = \begin{cases} 20x^3 - 20x^4, & \text{for } 0 \leq x \leq 1 \\ 0, & \text{otherwise} \end{cases}$

2. $f(x) = \begin{cases} -0.0003422x^4 + 0.006845x^3 - 0.04558x^2 + 0.1136x + 0.02448, & \text{for } 0 \leq x \leq 10 \\ 0, & \text{otherwise} \end{cases}$

For Exercises 3 and 4, calculate the probability that the random variable X falls between the values of 2 and 4 (lies in the interval [2,4]).

3. $f(x) = \begin{cases} 0.2, & \text{for } 0 \leq x \leq 5 \\ 0, & \text{otherwise} \end{cases}$

4. $f(x) = \begin{cases} 0.2, & \text{for } 1 \leq x \leq 3 \\ 0.3, & \text{for } 3 < x \leq 5 \\ 0, & \text{otherwise} \end{cases}$

Game Time

5. Suppose you run a construction contracting business, and you are bidding for a big project. You have one major competitor. Based on past experience, you believe that your competitor's bid for the project could be anywhere between \$8 million and \$10 million, and that all values in between are equally likely.
   (a) Define the random variable for this problem.
   (b) Define the probability density function for this problem, and sketch its graph.
   (c) What is the probability that your competitor's bid will be between \$8.9 million and \$9.3 million?
   (d) To what does your calculation in (c) correspond graphically?
6. To help decide what you want to plant in your garden and when, you want to estimate when the last frost will occur. From historical records, you determine that in your area it could be anywhere between March 28 and April 19, but that any date within that interval is about equally likely to be the last frost.
   (a) Define the random variable for this problem.
   (b) Define the probability density function for this problem, and sketch its graph.
   (c) What is the probability that the last frost will be between March 29 and April 5?
   (d) What assumptions are you making in your calculations above? Comment on them.
   (e) To what does your calculation in (c) correspond graphically?
7. You work for a company that is developing a drug to help treat people with AIDS. The research is moving along well, but there is a lot of uncertainty in this kind of project. You estimate that the distribution of time (in years) from the present before the drug can go into production, X, is given by the probability density function

$$f(x) = \begin{cases} 0.25xe^{-0.5x}, & \text{for } x \geq 0 \\ 0, & \text{otherwise} \end{cases}$$

   (a) Graph the density function. Sketch the curve on your paper.
   (b) Find $f(1)$ and $f(7)$, and explain in words how to interpret their relative values.
   (c) Using the graph, visually estimate the value of the mode (to the nearest year or so).
   (d) Use calculus to find the mode, by hand if possible. Explain what it means in everyday terms.
   (e) Verify your answer to (d) in *three* different ways using technology if possible, and explain each of them briefly.
   (f) If you wanted to estimate the probability that the drug will go into production in the next 3 years, what would this calculation correspond to graphically?
8. You are in the middle of a lawsuit. You have had some experience with similar suits, and estimate that the eventual court settlement will result in a net benefit to you of X hundred thousand dollars, with a probability density function given by

$$f(x) = \begin{cases} -0.0208x^2 + 0.0417x + 0.208, & \text{for } -2 \leq x \leq 4 \\ 0, & \text{otherwise} \end{cases}$$

   (a) Graph the density function. Sketch the curve on your paper.
   (b) Find $f(0)$ and $f(4)$, and explain in words how to interpret their relative values.
   (c) Using the graph, visually estimate the value of the mode (to the nearest \$100,000 or so).
   (d) Use calculus to find the mode by hand. Explain what it means in everyday terms.

(e) Verify your answer to (d) in *three* different ways using technology if possible, and explain each of them briefly. Can you think of a *fourth* way as well?

(f) If you wanted to calculate the probability that you would end up with a net loss from the lawsuit, what would the calculation correspond to graphically?

9. You have a scholarship that requires you to maintain an average of at least a B of some kind (2.5 GPA or more). In your math class, your professor's past history has resulted in the following grade distribution:

Grade (GPA)	F ([0, 0.5))	D ([0.5, 1.5))	C ([1.5, 2.5))	B ([2.5, 3.5))	A ([3.5, 4])
Number	7	23	49	51	32

(a) What is the probability of a student picked at random getting an A in this class? What would be the probability **per unit** in this interval (so the area under a line at that height between 3.5 and 4 would give the correct probability)?

(b) Define a histogram probability density function for this situation. Be sure to define your random variable.

(c) Graph your probability density function.

(d) What is the probability that your grade in this class will be at least at the minimum level needed overall for your scholarship? What assumption are you making in doing this calculation?

(e) What does your answer in (d) correspond to graphically?

(f) Suppose you are considering applying for a scholarship that requires at least a GPA of 3.0. What would be your probability of getting at least that GPA in this math class? What assumptions did you make in getting your answer?

(g) Since only letter grades are given at your school, with pluses and minuses corresponding to 1/3 of a GPA point, what actual scores on the 0 to 4.0 scale would be rounded off to a B (3.0) or better? What is your probability of getting such a grade, from this perspective? Is this answer more appropriate than your answer to (f)? Explain.

(h) What could you use as data points to fit a continuous smooth curve as a probability density function for this problem? See if you can find a reasonable model.

10. You are considering starting a small business selling bagels in your dorm on campus. You buttonhole 40 random students in your dorm and ask what is the most they would pay for a bagel and cream cheese, and the results are as follows:

Maximum Price ($)	[0.40, 0.65)	[0.65, 0.90)	[0.90, 1.15)	[1.15, 1.40)	[1.40, 1.65)
Number	9	7	6	4	3

(the other 11 indicated a value below 40 cents).

(a) What is the probability of a student picked at random in your dorm being willing to pay between 40 and 65 cents maximum for a bagel? What would be the probability **per unit** in this interval (so the area under a line at that height between 0.40 and 0.65 would give the correct probability)?

(b) Define a histogram probability density function for this situation. Be sure to define your random variable.

(c) Graph your probability density function.

(d) You figure that, for the business to be worth the effort, you need to be able to sell them for at least 65 cents, and you need at least half of the people in the dorm to be willing to spend that much. Are your conditions satisfied? What percentage of the people in the dorm are willing to spend at least 65 cents for a bagel? How could you think of this value in terms of probability?

(e) What does your answer in (d) correspond to graphically?

(f) You would like to be able to sell the bagels for at least 75 cents. At 75 cents, what is your estimate of the percentage of people who would be willing to buy? What assumption did you make in getting your answer?

(g) What could you use as data points to fit a continuous smooth curve as a probability density function for this problem? See if you can find a reasonable model.

4.2 APPROXIMATING AREAS UNDER CURVES (SUBINTERVAL METHODS)

In the last section, we saw that with continuous random variables, calculating a probability corresponds to finding the area under the probability distribution curve and above the horizontal axis between two interval endpoints. In this section, we will see another situation calling for finding the area under a curve: finding the net change in a function given its rate of change function over an interval. Examples of this include finding distance traveled from a velocity function and finding the total profit over a time interval given a weekly profit function. How do we find this kind of area under a general continuous, smooth curve? We can start by trying to find ways to **approximate** this area. The basic idea developed in *this* section is to do essentially the opposite of what we discussed in Section 4.1. There we tried to infer a smooth density function from a histogram density function. Now we will show ways of constructing a histogram to approximate a smooth density function. More generally, we will develop ways to construct rectangles, or other shapes, the areas of which we can add together to approximate the area under a curve.

Here are the kind of problems that the material in this section will help you solve:

- You have estimated the probability density function for the number of months before a competitor will release a new product. Now you want to use this density function to approximate the probability that the new product release will be in the next six months.
- Based on past performance and trends, you have estimated the probability distribution for your profit this year. How can you approximate the probability that your profit will fall within a given range?
- Given a weekly revenue function, approximate the total sales revenue over a particular interval of time.

By the end of this section, in addition to being able to solve the above kinds of problems, you should

- Understand how the *change* in a function over an interval given its nonnegative *rate of change* function (derivative) corresponds to the area under the rate function over that interval.
- Understand how to divide up an interval $[a, b]$ into n equal subintervals and be able to calculate the width of each subinterval (Δx) and the endpoints of the subintervals $x_0, x_1, x_2, \ldots, x_n$.

- Be able to draw in various geometric shapes, such as rectangles and trapezoids, to approximate the area under a curve and above the horizontal axis within a subinterval.
- Understand that the vertical height of a drawn-in rectangle corresponds to the function value at the point where the height was drawn (if the height was drawn in at x, then the height is $f(x)$).
- Be able to express an approximation of the area under a curve using the sum of areas of rectangles or other shapes for the different subintervals.
- Understand how to write and interpret expressions involving summation (sigma) notation.
- Understand how to approximate the area under a curve using the methods known as Right Rectangles, the Trapezoid Rule, and Simpson's Rule, and the concepts behind them.

Net Change and Area under a Constant Derivative Function

In the last section, we saw that the area under a curve and above the horizontal axis over an interval can correspond to a probability calculation. Now consider the following problem.

Sample Problem 1: Your car odometer is broken. You're driving along the highway on a trip. You set your cruise control to a speed of 62 miles per hour (mph) as soon as you got on the highway, and your speed has stayed at 62 mph for the last 2 hours. How far have you traveled on the highway, and what does this calculation correspond to graphically?

Solution: You may recall from previous math courses that

$$\text{Distance} = (\text{Rate})(\text{Time}), \text{ or } d = rt$$

If you examine the units, this makes sense, since (miles/hr)(hrs) = miles. It's as if the hour units cancel out, analogous to $(3/2)(2) = 3$. In this example, the rate is 62 miles/hr, and the time traveled is 2 hours, so the distance traveled is (62 miles/hr)(2 hrs) = 124 miles. You probably already knew how to do this calculation, since the context is a fairly familiar one.

How can we interpret this calculation graphically? Let's define

t = time (in hours) after you got on the highway

$v(t)$ = your velocity (speed, in mph) t hours after you got on the highway

In everyday usage, the term **velocity** is identical to speed, except that its *sign* indicates direction (such as forward versus backward), while speed is always positive. In fact, speed can be thought of as the *magnitude*, or absolute value, of velocity. In this example, since we are always traveling in the same direction, the two functions are identical. The graph of this velocity function is shown in Figure 4.2-1.

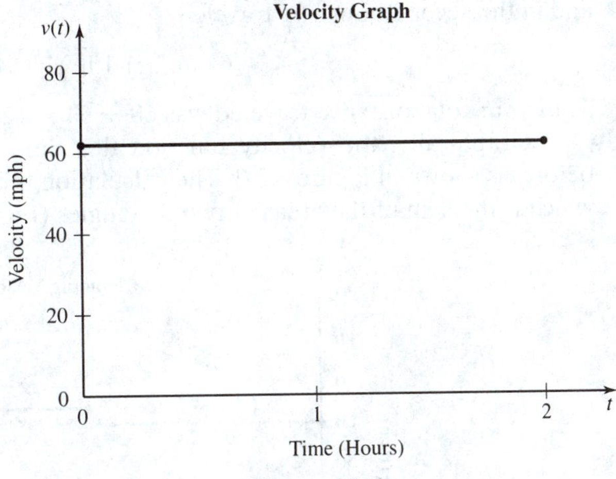

Figure 4.2-1

What does the calculation (62 miles/hr)(2 hrs) correspond to on this graph? It is the area under the velocity graph and above the horizontal axis between 0 and 2 (for t in the interval $[0, 2]$). Graphically, we show this in Figure 4.2-2. □

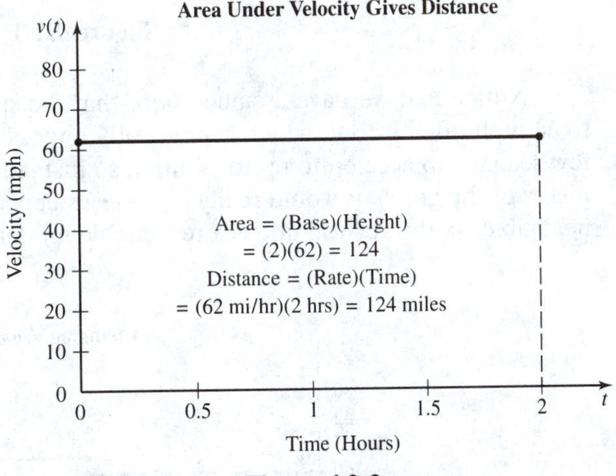

Figure 4.2-2

Sample Problem 2: In the same car trip situation, suppose you had traveled at 60 mph for the first hour, then raised it to 64 mph for the second hour. This time, how much distance have you traveled, and what does it correspond to graphically?

Solution: Since your speed is constant in each hour, the calculation for each hour is the same idea as above. So in the first hour you traveled

$$(60 \text{ mi/hr})(1 \text{ hr}) = 60 \text{ miles}$$

and in the second hour you traveled

$$(64 \text{ mi/hr})(1 \text{ hr}) = 64 \text{ miles}$$

Thus your total distance traveled was $60 + 64 = 124$ miles.

Graphically, the velocity function this time (with the same verbal definition as before) is shown in Figure 4.2-3. The calculation this time was $(60)(1-0) + (64)(2-1)$, which is the sum of the areas of two rectangles (Figure 4.2-4).

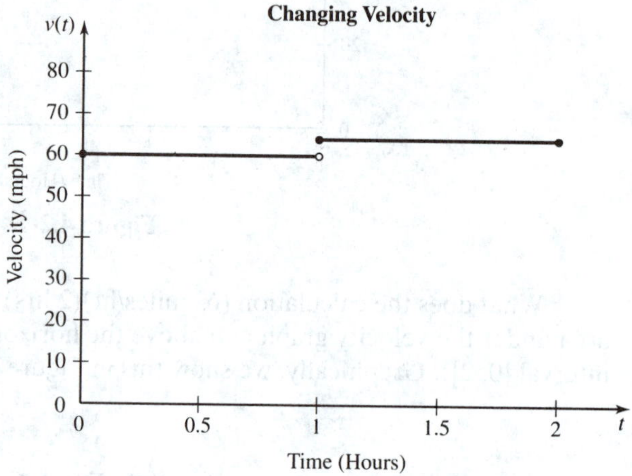

Figure 4.2-3

Notice that we have assumed here that the speed jumped *instantaneously* from 60 to 64 mph after 1 hour, which is physically impossible. In reality, it would have taken a few seconds to accelerate up to 64 mph, so instead of a vertical line connecting the two pieces of the graph, it would really be a *very* steep angled line. The difference, though, is negligible, so the vertical line is a reasonable model of what happened. □

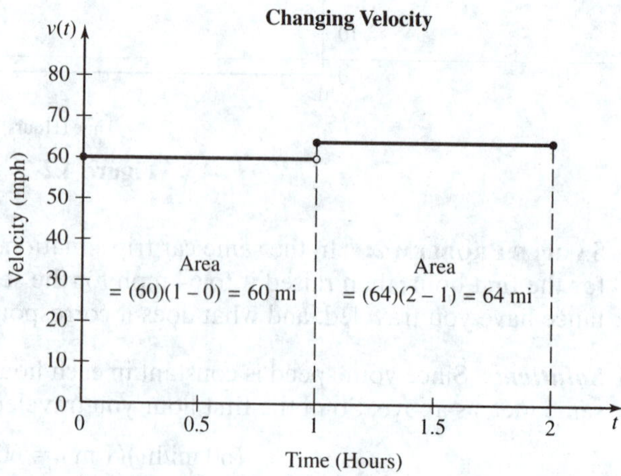

Figure 4.2-4

The Area Under Any Nonnegative Derivative Function as the Change in the Original Function

The same calculations and graphical interpretation would have applied if $v(t)$ had been a yearly revenue function, in units such as thousands of dollars per year, and t had been in years. Thus if revenues were coming in at a constant rate of \$60,000 per year for the first year and then jumped to a constant rate of \$64,000 per year for the second year, then the total revenue for those two years would be \$124,000. The same idea applies to **any** rate of change (derivative) function.

> **The area under a nonnegative derivative function over an interval will give the change in the value of the *original* function (the function the derivative comes from) over the specified interval.**

This interpretation also applies to the probability case. You may remember that when we derived the histogram probability density function for the T-shirt demand example in Section 4.1 (Sample Problem 2), we said that the value of the density function could be thought of as the probability **rate** (the amount of probability *per unit*), and so the area under the density function is the probability (the change in the accumulated probability) over that interval.

Now suppose that $v(t)$ corresponds to a smooth curve that is not a line. How can we do the same kind of calculation as in the last example to approximate the distance traveled or the revenue over an interval? Take a moment to see if you can think of an approach, before you continue to read further.

It may have occurred to you that one approach would be to define a step function like the one above to **approximate** $v(t)$ and then use the simple rectangular area calculations to approximate the area under the curve. That is exactly the basic concept developed in the rest of this section.

Suppose that you are trying to estimate how to model the random variable X, the number of months before a competitor will come out with a new product. Your advisors estimate that there is about a 20% chance that the new product will come out within the next month, and then the probability decreases for larger values of X. This corresponds to an exponential function like those we studied in Section 1.4. The following exponential probability density function models this situation well:

$$f(x) = 0.2e^{-0.2x}, \text{ for } x \geq 0$$

To begin the process of approximating the probability that the delay will be less than 6 months, divide the interval [0, 6] into 6 equal subintervals. We will use these as the breakpoints of a step function.

We are asked to find the probability that the number of months before a competitor will come out with a new product, the random variable X, is less than six months,

$$P(0 \leq X \leq 6)$$

Figure 4.2-5 is a graph of the density function for this problem. On it, we have indicated the area desired in the problem: the area under the density function and above the x-axis, between $x = 0$ and $x = 6$.

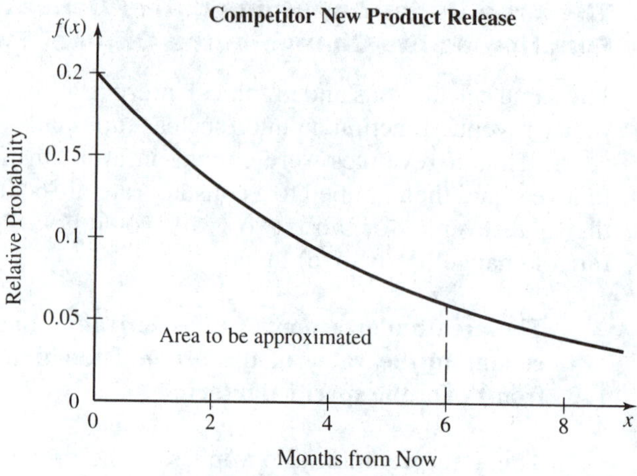

Figure 4.2-5

How can we approximate this area? Using the idea mentioned above, we can construct a histogram distribution that comes reasonably close to the exact density. Then, since the area under a histogram density is just the sum of areas of rectangles, we can easily calculate it. This number will give us an approximation to the area under the exact density. The accuracy of the approximation will roughly depend on how well the histogram approximates the exact density.

Definition of Subintervals

Unless we have good reason to do otherwise, for simplicity we usually use rectangles (histogram subintervals) that are all the **same width**. (This was not the case in Sample Problem 2 in Section 4.1, where one subinterval was bigger than the others.) The width of each rectangle will then depend only on the width of the full interval over which we want the area and the number of subintervals. In our example, [0, 6] is the full interval, with a width of $6 - 0 = 6$. The number of rectangles (subintervals) we chose was also 6. In general, if we call the full interval $[a, b]$, the width is $(b - a)$. The number of rectangles (or subintervals of the full interval), is usually denoted n.

In the problem above, we discussed using $n = 6$; in other words, we want to create a histogram of 6 subintervals with equal widths, which we will use to form 6 rectangles.

★ **DISCOVERY QUESTION 1:**

To divide up the interval [0, 6] into $n = 6$ equal subintervals, how wide should each of the subintervals (rectangles) be? What would the actual endpoints of the subintervals be? (Try to answer the question on your own before reading on. If you need it, there is a hint given before the solution.)

Hint: How wide is the full interval for this question? How many rectangles (subintervals) do we want? Therefore, how wide should each subinterval be? (Once again, see if you can answer the question on your own before reading on.)

Answer: The full interval [0, 6] has a length of 6 − 0 = 6. (Recall that in general, the length of an interval is simply the right/larger endpoint minus the left/smaller endpoint.) We want 6 subintervals. So each subinterval must be 6/6 = 1 unit wide (the total width divided by the number of equal subintervals, since it will be broken into that many equal pieces). Since they start at $x = 0$ and end at $x = 6$, the endpoints must be 0,1,2,3,4,5, and 6. In other words, the 1st subinterval is [0, 1], the 2nd is [1, 2], and so on, and the 6th (last) is [5, 6]. In general we call the leftmost endpoint (that is, the left endpoint of the full interval, a, which is also the left endpoint of the 1st subinterval) x_0, and the other endpoints x_1, x_2, and so on, through x_n. In other words, x_n is the rightmost endpoint (the right endpoint of the full interval, b, *and* the right endpoint of the nth subinterval). Thus,

the 1st subinterval is $[x_0, x_1]$,
the 2nd subinterval is $[x_1, x_2]$, and
the last (nth) subinterval is $[x_{n-1}, x_n]$.

In our example, then, $x_0 = 0, x_1 = 1, \ldots$, and $x_6 = 6$. **Notice that the number of each subinterval (1st, 2nd, etc.) is the same as the subscript of its *right* endpoint.** This will come in handy later.

Notationally, let's call the width of each subinterval (rectangle) Δx. In our example, $\Delta x = 1$. In general, if we call the full interval $[a, b]$, notice first that x_0 will always correspond to a and x_n will always correspond to b. Since the width of the full interval is $(b - a)$, and we are dividing this up into n equal pieces, the width of each subinterval is

$$\Delta x = \frac{b - a}{n}$$

Furthermore, notice that each successive endpoint after x_0 $(x_1, x_2, \ldots, x_n)$ is obtained by simply adding Δx. Thus,

$$x_1 = a + \Delta x$$
$$x_2 = x_1 + \Delta x = (a + \Delta x) + \Delta x = a + 2(\Delta x) = a + 2\Delta x,$$
$$x_3 = x_2 + \Delta x = (a + 2\Delta x) + \Delta x = a + 3\Delta x$$
$$\ldots$$
$$x_n = a + n(\Delta x) = a + n\Delta x$$

This is shown in Figure 4.2-6.

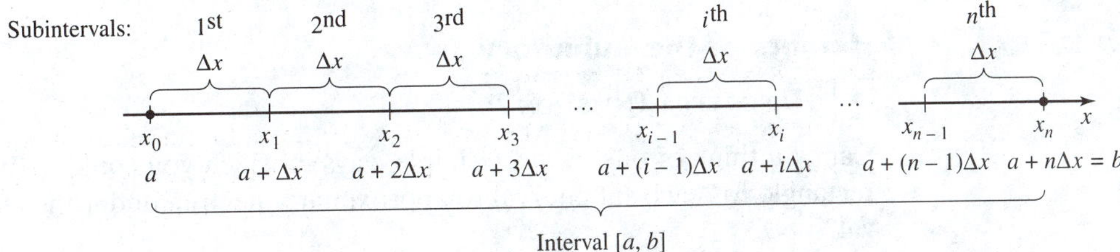

Figure 4.2-6

In other words, if x_i denotes the *right* endpoint of the ith subinterval, then

$$x_i = a + i(\Delta x) = a + i\Delta x$$

The right endpoint of the ith subinterval is $i\ \Delta x$'s to the right of the left endpoint of the full interval (which is a, or x_0). If you are not familiar with the idea of the "ith subinterval," it is just a way of talking about a generic subinterval. When $i = 1$, the ith subinterval refers to the 1st subinterval, when $i = 2$, it refers to the 2nd subinterval, and so on. We use this notation to describe patterns that are the same in all problems. To illustrate a use of this notation, the ith subinterval would be $[x_{i-1}, x_i]$. For example, the 1st subinterval ($i = 1$) is

$$[x_{1-1}, x_1] = [x_0, x_1]$$

Convince yourself that this works for $i = 2$, $i = n$, and all of the other possible values of i in between as well. Simply pick a value of i, substitute that value everywhere for i, simplify, and see that you get a result you already know to be true. ★

Having set the stage with a realistic example concerning a competitor's new product release, let's explore rectangle approximation with a simpler function first to help make it easier to understand the concept. You can think of this more mathematical example in the context of its application, but we will not focus on that aspect right now.

EXAMPLE 1: Suppose you are running a business that is just taking off. Six months ago (at the beginning of the current fiscal year) you just started to turn a profit. Suppose that if we define

x = number of months since the beginning of this fiscal year

and

$f(x)$ = monthly profit (the instantaneous rate of change of the accumulated profit) in \$10,000's per month

our data from the last six months are modeled quite well by the function $f(x) = x^2$. Use rectangles with $n = 6$ to approximate the total profit from the past six months.

Since the full interval ([0, 6]) and the number of subintervals ($n = 6$) is the same as in Question 1, all of our work there applies to this example as well, so the endpoints of the subintervals will again be 0, 1, 2, 3, 4, 5, and 6.

As developed above, we now have a way to construct subintervals that will serve as the widths (bases) of the rectangles corresponding to the histogram distribution. The question that remains is: How should we construct the heights of the rectangles?

Heights of the Subintervals

★ **DISCOVERY QUESTION 2:**

Can you think of one or more simple ways in which you could define the height of each rectangle for each subinterval to approximate the area under the curve in that subinterval?

(Try to answer this question on your own before reading on. After you have tried, a hint is given before the solution if you need it.)

Hint: Since we want to approximate $f(x)$, the simplest way to choose a height for the rectangle is to use the value of $f(x)$ (the y value of the given function) for some x-value in the subinterval. Graphically, this is the height (above the x-axis) of the original

curve above a particular x-value. Once we pick an x-value, we can find the height graphically by looking vertically above that x-value to find the corresponding y-value on the curve that is directly over it. We can then draw a horizontal line at that height, spanning the entire subinterval. For a given subinterval, what x-values could be used that would be convenient? What choice of x-value would yield a rectangle that might be likely to give a reasonable approximation of the area under the curve for that subinterval?
(Again, try to answer the question yourself before reading on.)

Answer: There is no *one* correct answer to this question. Since we have already found the left and right endpoints of each of the subintervals for this problem, they would be the easiest to use. These two different choices give us two different methods, which we will call the Left Rectangle method and the Right Rectangle method. Let's examine the Right Rectangle method. ★

The Right Rectangle Method

We have already sketched out the idea of this method. The idea is that each rectangle has a width (base) of 1 unit, with the length (height) given by the height of the curve (value of the function) at the right endpoint of each subinterval. The first subinterval is [0, 1], so the right endpoint of the 1st subinterval (x_1) is 1. The height of the first rectangle then is the value of $f(x) = x^2$ when $x = 1$: $f(1) = (1)^2 = 1$. Figure 4.2-7 shows that the height at $x = 1$ corresponds to starting at 1 on the x-axis, then moving up vertically until you hit the parabola at $y = f(1) = 1$.

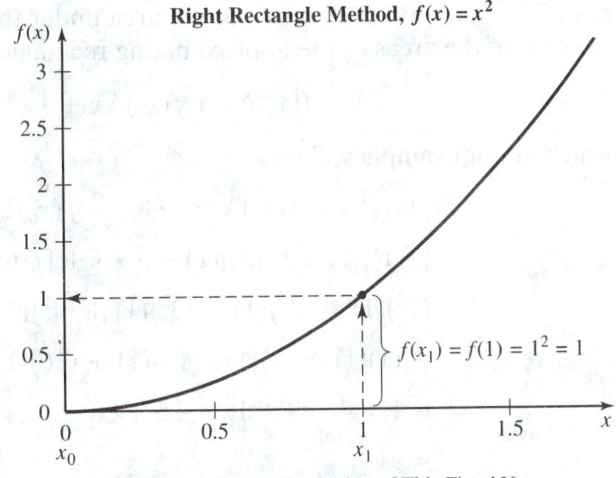

Figure 4.2-7

You then draw a horizontal line at that height of 1 over the entire subinterval [0, 1], which in the case of right endpoints means you draw the horizontal line a length of $\Delta x = 1$ units to the left. Notationally, we could say the height of this first rectangle is $f(x_1)$, and the width (base) is Δx. So

$$\text{the area of this 1st rectangle is (height)(base)} = (1)(1) = 1 = f(x_1)\Delta x$$

We do the same thing for the 2nd subinterval, [1, 2]. This time the right endpoint (x_2) is 2, so the length/height is $f(2) = 2^2 = 4$ (the width/base is again $\Delta x = 1$, as it is for all of the subintervals). The area of the 2nd rectangle is then $(4)(1) = 4 = f(x_2)\Delta x$ (Figure 4.2-8). In a similar way, the area of the 3rd rectangle is $(3^2)(1) = 9 = f(x_3)\Delta x$.

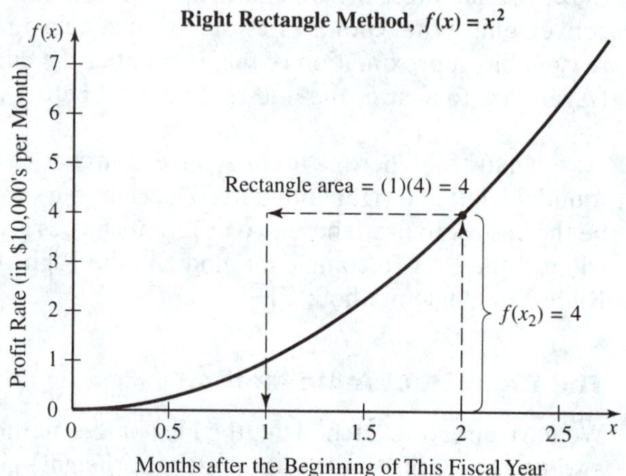

Figure 4.2-8

Notice the pattern that is developing here. The area of the 1st rectangle is $f(x_1)\Delta x$, the area of the 2nd rectangle is $f(x_2)\Delta x$, and the area of the 3rd rectangle is $f(x_3)\Delta x$. For the ith rectangle, the pattern suggests that the area is $f(x_i)\Delta x$, and this will be true for $i = 1, 2, \ldots, n$. So the estimate of the area under the curve over the full interval [0, 6] is the sum of the areas of the approximating rectangles:

$$f(x_1)\Delta x + f(x_2)\Delta x + \cdots + f(x_n)\Delta x$$

which in our example will be

$$f(x_1)\Delta x + f(x_2)\Delta x + \cdots + f(x_6)\Delta x$$
$$[f(1)](1) + [f(2)](1) + \cdots + [f(6)](1)$$
$$(1^2)(1) + (2^2)(1) + (3^2)(1) + (4^2)(1) + (5^2)(1) + (6^2)(1)$$
$$= (1)(1) + (4)(1) + (9)(1) + (16)(1) + (25)(1) + (36)(1)$$
$$= 1 + 4 + 9 + 16 + 25 + 36$$
$$= 91$$

Thus our approximation of the area under $y = x^2$ and above the x-axis between $x = 0$ and $x = 6$ using the Right Rectangle method with $n = 6$ is 91 (Figure 4.2-9). □

Notice in the symbolic form of the sum of the areas of the rectangles above, each term being added (the area of each rectangle) looked exactly the same except for the subscript of the x. Generically, we pointed out that the ith term would be written $f(x_i)\Delta x$. Mathematicians have a notation for this kind of summation, in which just one variable (like i, usually standing for a whole number, or integer) changes from term to term. The notation used is called **sigma notation** (sigma is the Greek letter equivalent to s in

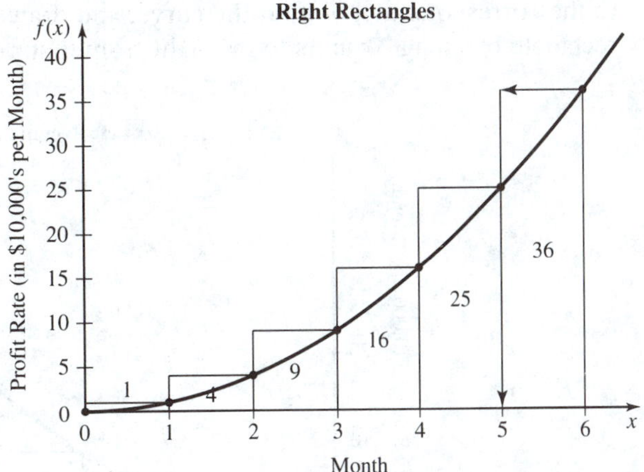

Figure 4.2-9

English, and is the first letter of the Greek word for "sum"). For our expression above we can write

$$\sum_{i=1}^{n} f(x_i)\Delta x = f(x_1)\Delta x + f(x_2)\Delta x + \cdots + f(x_n)\Delta x$$

The "Σ" is the uppercase letter sigma. When reading the notation, you can say "sigma, $i = 1$ to n, of f of (x sub i) times delta x" or "the summation, i going from 1 to n, of ..." The idea of the notation is that the variable to the left of the equal sign under the sigma is the "index variable," and its value starts at the number to the right of the equal sign under the sigma and goes up by increments of 1, until you reach the final value given above the sigma sign. For each term, you simply plug in the appropriate value for the index variable. Then you add up all of the terms.

Here are some further examples:

$$\sum_{i=1}^{n} x_i f(x_i) = x_1 f(x_1) + x_2 f(x_2) + \cdots + x_n f(x_n)$$

$$\begin{aligned}\sum_{i=2}^{4}(i^3 + 4i) &= (2^3 + 4(2)) + (3^3 + 4(3)) + (4^3 + 4(4)) \\ &= (8 + 8) + (27 + 12) + (64 + 16) \\ &= 16 + 39 + 80 \\ &= 135\end{aligned}$$

Left Rectangle and Midpoint Methods

We could have decided to use the left endpoint of each interval instead of the right endpoint. This is exactly the same idea as the Right Rectangle method, except that the height of the rectangle for each subinterval is determined by the **left** endpoint of the subinterval instead of the right endpoint. Graphically, we start at the left endpoint, trace straight up to the corresponding point on the curve, and draw the horizontal line for the top of the rectangle by going Δx units to the right from that point (Figure 4.2-10).

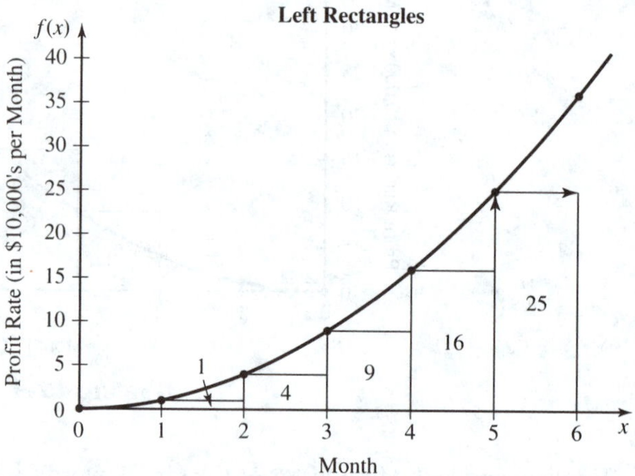

Figure 4.2-10

If we had used the left endpoint of each rectangle to determine the height, called the Left Rectangles method,[6] our approximation of the area under the curve over the full interval would be 55, whereas for the Right Rectangle method it was 91. This is quite a difference! Look at Figure 4.2-10: Since the function is increasing, the left endpoint is always the lowest point on the subinterval, so the area of the rectangle is always at *most* as big as the area under the curve over the subinterval. This means that 55 is a **lower bound** on the exact area over the full interval [0, 6]. In a similar way, 91 is an **upper bound**.

This also suggests that both the Right Rectangle and Left Rectangle methods are not very accurate. What x-value in each subinterval would give us a better approximation? The most obvious answer would be to use the **midpoint** of each subinterval, which would give a rectangle height somewhere in between the heights from the Right Rectangle and Left Rectangle methods. We will call this method (what a surprise!) the Midpoint Rectangle method.

For this method, we graphically start at the midpoint of each subinterval, trace vertically above it until we hit the curve, then draw the top of the rectangle at that height, extending $(1/2)\Delta x$ horizontally to the right and also the same distance to the left from that point, as shown in Figure 4.2-11. If we did these calculations,[7] we would get an approximation of 71.5, which is in between the lower bound of 55 from the Left Rectangle method and the upper bound of 91 from the Right Rectangle method, although not exactly in between.

[6]The calculation formula for the Left Rectangles method is given by $\sum_{i=0}^{n-1} f(x_i)\Delta x = \sum_{i=1}^{n} f(x_{i-1})\Delta x$.

[7]The calculation formula for the Midpoint Rectangles method is $\sum_{i=1}^{n} f\left(\frac{x_{i-1} + x_i}{2}\right)\Delta x = \sum_{i=1}^{n} f(\bar{x}_i)\Delta x$, where $\bar{x}_i = \frac{x_{i-1} + x_i}{2}$ is the midpoint of the ith subinterval.

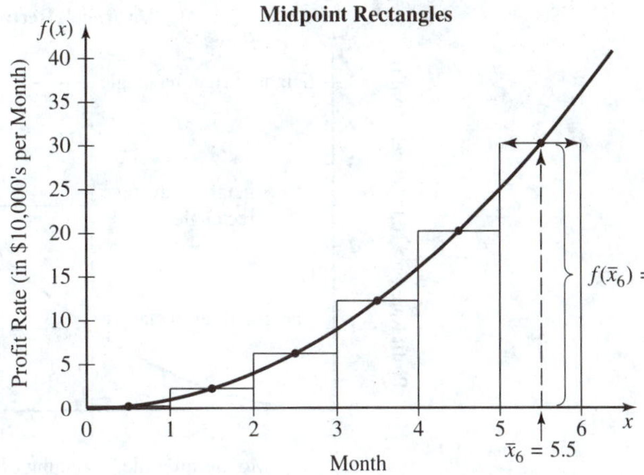

Figure 4.2-11

Trapezoid Method

Our last comment raises an interesting question: What method *would* result in an approximation that was exactly halfway in between the lower and upper bounds we got for the area under the curve (for both the full interval, and for each subinterval)?

The simple answer would be to take the **average** of the lower bound (left endpoint in this example) and upper bound (right endpoint) areas. At the level of the full interval [0, 6], we could simply take the average of our results from the Right Rectangle and the Left Rectangle methods. For this example, that would be $(55 + 91)/2 = 146/2 = 73$. Note that this is close to the approximation from the midpoint method (71.5), but not the same.

How could we apply this same idea of averaging the left and right rectangles at the subinterval level? Let's look at the 2nd subinterval, [1, 2]. Recall that our function is

$$y = f(x) = x^2$$

The area of the left rectangle is $(1^2)(1) = 1$, and the area of the right rectangle is $(2^2)(1) = 4$. The average of these areas is $(1 + 4)/2 = 5/2 = 2.5$. Graphically, this corresponds to the area of a rectangle with a height of 2.5 (and the same width, 1). This is exactly halfway between the heights of the left and right rectangles, as in the Figure 4.2-12.

As mentioned above, notice that the average rectangle is close to, but not the same as, the midpoint rectangle, which would go through the data point (1.5,2.25), since when $x = 1.5, f(x) = f(1.5) = (1.5)^2 = 2.25$.

The average of the areas of the left and right rectangles in this example is given by

$$\frac{f(x_1)\Delta x + f(x_2)\Delta x}{2} = (\Delta x)\left[\frac{f(x_1) + f(x_2)}{2}\right] = \frac{\Delta x}{2}[f(x_1) + f(x_2)]$$

$$= \frac{(1)}{2}[f(1) + f(2)] = \frac{1}{2}[1^2 + 2^2] = \frac{1}{2}[1 + 4] = \frac{1}{2}[5] = \frac{5}{2} = 2.5$$

which agrees with our earlier calculation.

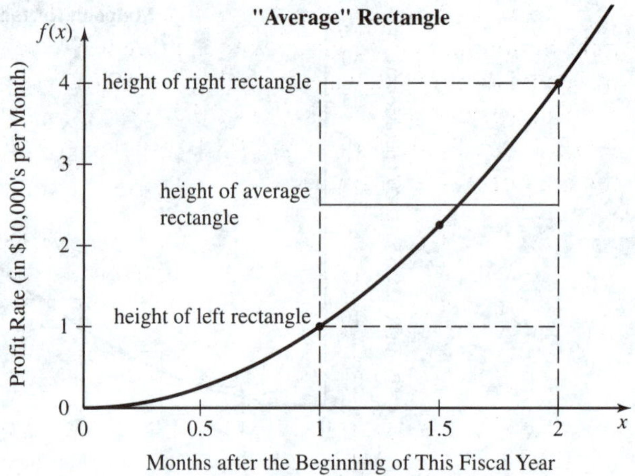

Figure 4.2-12

To generalize for the ith subinterval, the left rectangle area is given by $f(x_{i-1})\Delta x$, and the right rectangle area is given by $f(x_i)\Delta x$. Thus, the average of the two areas is given by

$$\frac{f(x_{i-1})\Delta x + f(x_1)\Delta x}{2} = (\Delta x)\left[\frac{f(x_{i-1}) + f(x_i)}{2}\right] = \frac{\Delta x}{2}[f(x_{i-1}) + f(x_i)]$$

Notice that in the algebra above, all we did was factor out the Δx and then combine it with the division by 2, since both will always be the same (as long as we use rectangles of equal widths).

Let's take a moment now to look at the first factorization of the expression above

$$(\Delta x)\left[\frac{f(x_{i-1}) + f(x_i)}{2}\right]$$

If you remember your high school geometry (which may be a big if), this expression may look vaguely familiar, as in "the average of the bases times the height (altitude)," or $\left(\frac{b_1 + b_2}{2}\right)h$. This is the formula for the area of a trapezoid, as in Figure 4.2-13.

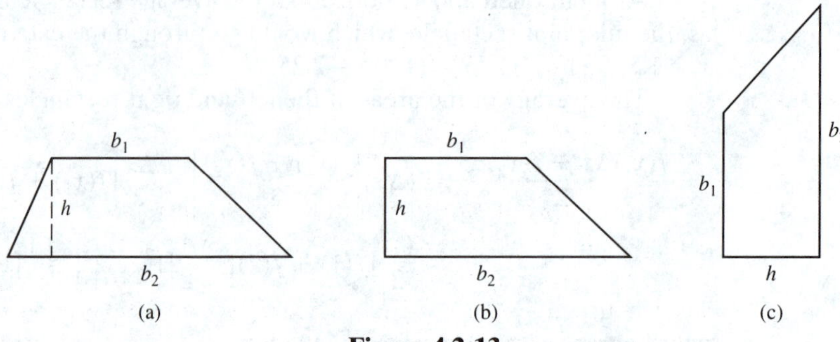

Figure 4.2-13

How does this connect with our subinterval calculations? If you connect the left and right endpoints on the curve with a straight line segment (Figure 4.2-14), the result is a trapezoid (sort of sideways from the way we usually picture a trapezoid). The two "bases" of this trapezoid are the heights of the curve at the left and right endpoints, which for the ith subintervals are $f(x_{i-1})$ and $f(x_i)$, respectively, and the "altitude" (height) is Δx. Thus the area of this trapezoid is

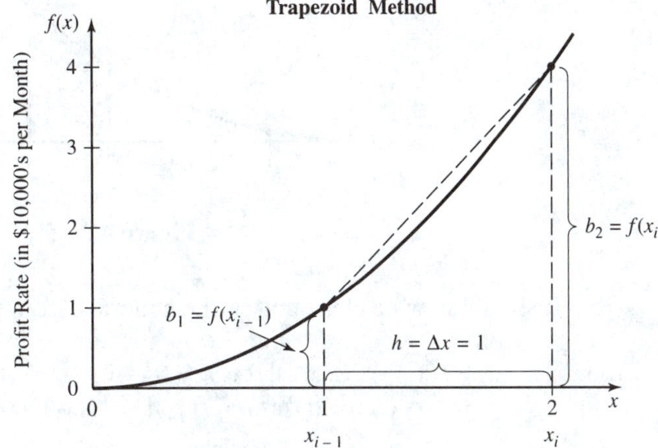

Figure 4.2-14

$$\left(\frac{b_1 + b_2}{2}\right)h = \left[\frac{f(x_{i-1}) + f(x_i)}{2}\right]\Delta x = \frac{\Delta x}{2}[f(x_{i-1}) + f(x_i)]$$

the same expression we obtained before.

Thus, this method is often called the **Trapezoid Rule**, or the Trapezoid method.

We have already shown that the average area (area of the trapezoid) for subinterval 2 ([1, 2]) is $(1/2)[f(1) + f(2)]$. For subinterval 1 ([0, 1]), then, the average area (area of the trapezoid) will be $(1/2)[f(0) + f(1)]$. The pattern here is simple and continues through subinterval 6 ([5, 6]), for which the area of the trapezoid is $(1/2)[f(5) + f(6)]$. Thus the sum of the areas of the trapezoids is

$$\begin{aligned}(1/2)[f(0) + f(1)] + (1/2)[f(1) + f(2)] + \cdots + (1/2)[f(5) + f(6)]& \\ = (1/2)\{[f(0) + f(1)] + [f(1)+ f(2)] + [f(2) + f(3)] + [f(3) + f(4)]& \\ + [f(4) + f(5)] + [f(5)+ f(6)]\}& \\ = (1/2)\{f(0) + 2f(1) + 2f(2) + 2f(3) + 2f(4) + 2f(5) + f(6)\}&\end{aligned}$$

Notice that all of the subinterval endpoints except the first and last (0 and 6) occur as heights in two different trapezoids, so have coefficients of 2, while the first and last occur in only one trapezoid each, so only have coefficients of 1. Thus the pattern of the coefficients is 1 2 2 2 2 ... 2 2 1. To help picture what is going on, Figure 4.2-15 is a graph with the trapezoids drawn in:

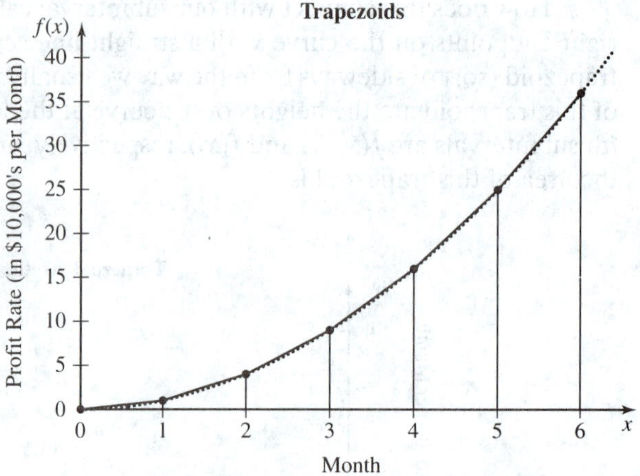

Figure 4.2-15

It should now be clear that in the general case we would get

$$\begin{aligned}(\Delta x/2)[f(x_0) + f(x_1)] + (\Delta x/2)[f(x_1) + f(x_2)] + \cdots + (\Delta x/2)[f(x_{n-1}) + f(x_n)]\\ = (\Delta x/2)\{[f(x_0) + f(x_1)] + [f(x_1) + f(x_2)] + \cdots + [f(x_{n-1}) + f(x_n)]\}\\ = (\Delta x/2)\{f(x_0) + 2f(x_1) + 2f(x_2) + \cdots + 2f(x_{n-1}) + f(x_n)\}\end{aligned}$$

Looked at another way, the Trapezoid method fits a straight line model to the two endpoints of each subinterval as an approximation of the exact curve over that subinterval. Then it finds the area under that line over the subinterval (the area of the trapezoid) to approximate the area under the curve. In our example for the subinterval [1, 2] (Figure 4.2-16), see how the curve between the points at 1 and 2 is approximated by a line drawn connecting them:

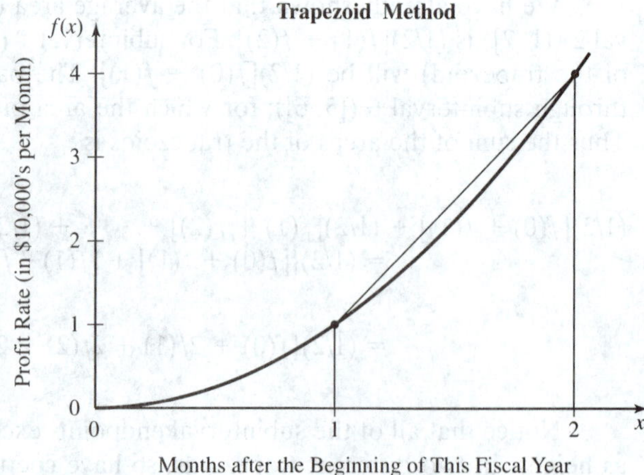

Figure 4.2-16

For Example 1, since the curve is concave up, the figure for the Trapezoid Rule (Figure 4.2-16) shows that even the trapezoid approximation is an upper bound (that is, the straight-line approximation always lies *above* the actual curve, so our answer of 73 is *still* an upper bound). This makes the midpoint approximation look promising, since it is close to, but a little below, that upper bound. Can we do any better?

The answer, of course, is "Yes!" (or we probably would not have asked). Picking up on the idea of the Trapezoid method corresponding to fitting a linear model to the endpoints of each subinterval, a parabola (quadratic model) should give us an even better fit to a curve than a line. The method that does this is called **Simpson's Rule**, or the parabola method. But to fit a parabola perfectly, we need three points rather than two. This is because a line needs only two points, since it has only two parameters, m and b, but a parabola has three parameters, a, b, and c, so requires three points. To get three endpoints, though, we will have to use **two** subintervals at a time. Thus Simpson's Rule requires that n be an even number. We will not derive the formula for Simpson's Rule, but you can use technology to approximate the area under a curve using Simpson's Rule.[8]

The graph for Simpson's Rule in our example would look exactly like the original curve, with vertical lines from the endpoints of the subintervals up to the curve. For our example, the approximation using Simpson's Rule is 72.

In fact, since our original function $f(x) = x^2$ is a parabola, Simpson's Rule should give us the exact area! This squares with our earlier discussion of accuracy, which said that the Trapezoid Rule estimate of 73 is an upper bound, and that the Midpoint Rule estimate of 71.50 was likely to be close to the exact answer.

Having approximated the area under $f(x) = x^2$ between 0 and 6 for $n = 6$, let's consider how we would answer the same question for $n = 12$. The full interval is still $[0, 6]$, so its width is still $6 - 0 = 6$, but this time we want 12 equal subintervals (rectangles), so the width of each should be the full width divided by the number of equal subintervals, or $6/12 = 1/2 = 0.5$. Thus the endpoints of the subintervals will be $x_0 = 0$, $x_1 = 0.5$, $x_2 = 1, \ldots, x_{12} = 6$.

In general, if our full interval is $[a, b]$, then its width is $(b - a)$, as we said earlier. If we want to break this up into n equal subintervals, then the width of each subinterval will again be the total width divided by the number of equal subintervals, so we get

$$\Delta x = \frac{(b - a)}{n}$$

Let's now summarize the two different approximation methods we have discussed in detail:

To approximate the area under the smooth curve $y = f(x)$ and above the x-axis over the interval $[a, b]$ (between $x = a$ and $x = b$) using n equal subintervals, define

$$\boldsymbol{\Delta x = \frac{(b - a)}{n}}$$

---

[8]The calculation formula for Simpson's Rule is given by $\frac{\Delta x}{3}\,[f(x_0) + 4f(x_1) + 2f(x_2) + 4f(x_3) + 2f(x_4) + \cdots + 2f(x_{n-2}) + 4f(x_{n-1}) + f(f(x_n)]$, where the pattern of the coefficients is 1 4 2 4 2 4 2 4 2 . . . 2 4 2 4 1 , and where n must be an even number. See Appendix A for a derivation.

Right Rectangle method: The approximate area is given by

$$\sum_{i=1}^{n} f(x_i)\Delta x = \Delta x\{f(x_1) + f(x_2) + \cdots + f(x_{n-1}) + f(x_n)\}$$

Trapezoid Rule: the approximate area is given by

$$\frac{\Delta x}{2}\{f(x_0) + 2f(x_1) + 2f(x_2) + \cdots + 2f(x_{n-1}) + f(x_n)\}$$

Sample Problem 3: Approximate the area under the curve $f(x) = x^2$ over the interval [3, 5] using $n = 4$, by the Right Rectangle and Trapezoid methods discussed in this section.

Solution: In this case $a = 3$ and $b = 5$, so

$$\Delta x = \frac{(b-a)}{n} = \frac{(5-3)}{4} = \frac{2}{4} = \frac{1}{2} = 0.5$$

This means the endpoints are $x_0 = 3$, $x_1 = 3.5$, $x_2 = 4$, $x_3 = 4.5$, and $x_4 = 5$. Just to reinforce a point made earlier, we said that in general $x_i = a + i(\Delta x)$. For example, here

$$x_3 = 3 + 3(0.5) = 3 + 1.5 = 4.5$$

Notice that this works even for $i = 0$.

Right Rectangle Method. Our approximation will be given by

$$\begin{aligned} &f(3.5)(0.5) + f(4)(0.5) + f(4.5)(0.5) + f(5)(0.5) \\ &= (0.5)\{f(3.5) + f(4) + f(4.5) + f(5)\} \\ &= 0.5\{3.5^2 + 4^2 + 4.5^2 + 5^2\} \\ &= 0.5\{12.25 + 16 + 20.25 + 25\} \\ &= 0.5\{73.50\} = 36.75 \end{aligned}$$

Trapezoid Rule. Our approximation is

$$\begin{aligned} &\frac{0.5}{2}\{f(3) + 2f(3.5) + 2f(4) + 2f(4.5) + f(5)\} \\ &= (0.25)\{3^2 + 2(3.5^2) + 2(4^2) + 2(4.5^2) + 5^2\} \\ &= (0.25)\{9 + 2(12.25) + 2(16) + 2(20.25) + 25\} \\ &= (0.25)\{9 + 24.50 + 32 + 40.50 + 25\} \\ &= (0.25)\{131\} = 32.75 \quad \square \end{aligned}$$

Estimating Area From Data

Sample Problem 4: A friend has to deliver a sailboat across a large lake. The wind has been consistently out of the north, so you can pretty much sail straight across the lake. However, wind speed has been variable, and so your boat speed over the water has been variable, too. You have been trying to keep a record of your speed, so have recorded your speed at different times.

Boat Speed mph	4.3	5.2	4.5	3.8	4.1	4.3
Time	09:00	09:15	09:30	9:45	10:00	10:15

Estimate how far you have traveled.

Solution: You want to estimate the distance you have traveled. You have values for the rate of change of your distance with respect to time (your instantaneous velocity) and the times. From what we said earlier, we could try to fit a curve to the data points to get a graph of the instantaneous velocity and then try to estimate the area under that curve. But if we selected $n = 5$ and used the Trapezoid Rule, then all we would need would be the subinterval endpoints and the height of the curve at each of them. But that is exactly what we are given in the data! **The Trapezoid Rule is ideally suited for area approximations from discrete data.** Let's identify each subinterval. Note that the subintervals are all the same width, so we can use the formula, but remember that it *only* applies when the subintervals are all the same size.[9]

$$\begin{aligned}\text{Area} &\approx \frac{\Delta x}{2}[f(x_0) + 2f(x_1) + 2f(x_2) + 2f(x_3) + 2f(x_4) + f(x_5)]\\ &= \frac{0.25}{2}[4.3 + 2(5.2) + 2(4.5) + 2(3.8) + 2(4.1) + 4.3]\\ &= 0.125(4.3 + 10.4 + 9.0 + 7.6 + 8.2 + 4.3)\\ &= 0.125(43.8)\\ &= 5.475\end{aligned}$$

Adding the estimates for the areas of the subintervals we have 5.475. You have traveled approximately $5\frac{1}{2}$ miles across the lake since you left at 9:00 A.M. □

Notice that we used the data values as the values of the "function" here. If we had fit a curve, the data values could be somewhat different from the values of the function.

Negative Areas

So far, we have only worked with functions $f(x)$ that are nonegative (always above the x-axis), like probability density functions. But the above formulas also make sense if $f(x)$ takes on negative values as well.

★ **DISCOVERY QUESTION 3:**

If $y = f(x) = x^2 - 6$, find the approximation given by the Right Rectangle method for the interval [0, 6] with $n = 6$, and interpret it graphically (both what the approximation means, and what exact value it is approximating).
(See if you can answer before reading ahead.)

Hints: Calculating the approximation is just plugging into a formula, so the real challenge is the graphical interpretation. Think of how we constructed the rectangles for this method originally. How could you generalize what we did? What would be the analogous interpretation in terms of exact areas?
(Try again to answer yourself before reading the solution.)

---

[9] If the subintervals had been different widths, we could still use the principle behind the Trapezoid Rule, but we would find the average height of each subinterval, multiply that by the width of the subinterval to get an estimate of the area of that subinterval, and then add these results. Fortunately, we can simply use the formula here.

Answer: If you think about the original Right Rectangle problem, we drew our rectangles by starting at the right endpoint of each subinterval, drawing a vertical line to the curve from the horizontal axis, then drawing a horizontal line segment over the subinterval at that level, and connecting the other end back to the horizontal axis with another vertical line. If we apply that concept to the first subinterval in this example, we get Figure 4.2-17:

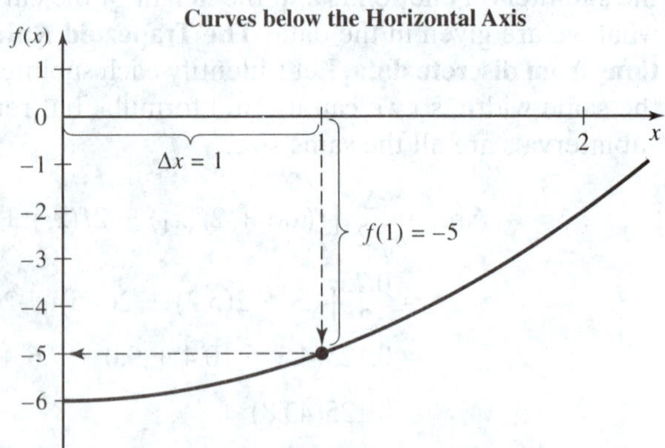

Figure 4.2-17

The effect of the calculation, then, is to treat the rectangles that go **below** the x-axis as **negative** areas. In other words, that calculation finds **the sum of the areas of the rectangles above the *x*-axis minus the sum of the areas of the rectangles below the *x*-axis.** In a similar way,

> **the exact area being approximated is the area between the curve and the horizontal axis *above* the horizontal axis *minus* the area between the curve and the horizontal axis *below* the horizontal axis, between the endpoints of the full interval.**

In this example the endpoints are $x = 0$ and $x = 6$, or between $x = a$ and $x = b$ in general.

In the original Right Rectangle method, we used the formula

$$f(x_1)\Delta x + f(x_2)\Delta x + \cdots + f(x_n)\Delta x$$

For this example, we once again have $\Delta x = (6 - 0)/6 = 1$, so $x_0 = 0$, $x_1 = 1$, . . . , and $x_6 = 6$, so we get

$$f(1)(1) + f(2)(1) + \cdots + f(6)(1)$$
$$= [(1)^2 - 6](1) + [(2)^2 - 6](1) + [(3)^2 - 6](1) + [(4)^2 - 6](1) + [(5)^2 - 6](1) + [(6)^2 - 6](1)$$
$$= [1 - 6](1) + [4 - 6](1) + [9 - 6](1) + [16 - 6](1) + [25 - 6](1) + [36 - 6](1)$$
$$= [-5](1) + [-2](1) + [3](1) + [10](1) + [19](1) + [30](1)$$
$$= -5 + (-2) + 3 + 10 + 19 + 30 = 62 - 7 = 55$$

Graphically, the calculation corresponds to Figure 4.2-18. The 62 is the sum of the areas of the rectangles above the x-axis (the last 4), and the 7 is the sum of the areas of the rectangles below the x-axis (the first 2). The exact value (the value being approximated) is illustrated in the Figure 4.2-19.

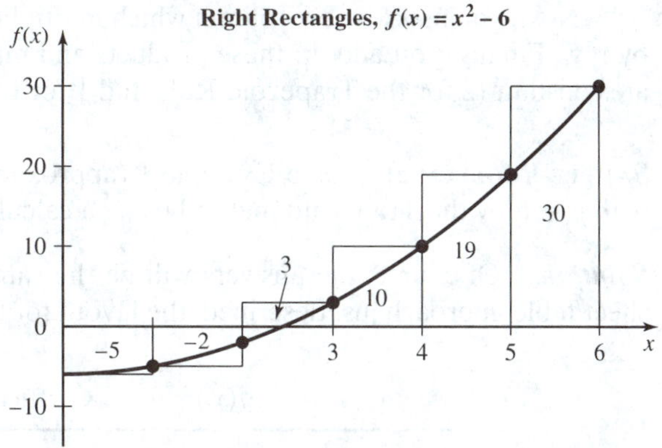

Figure 4.2-18

If we call the area below the x-axis A_1 and the area above the x-axis A_2, then the value being approximated is $(A_2 - A_1)$. ★

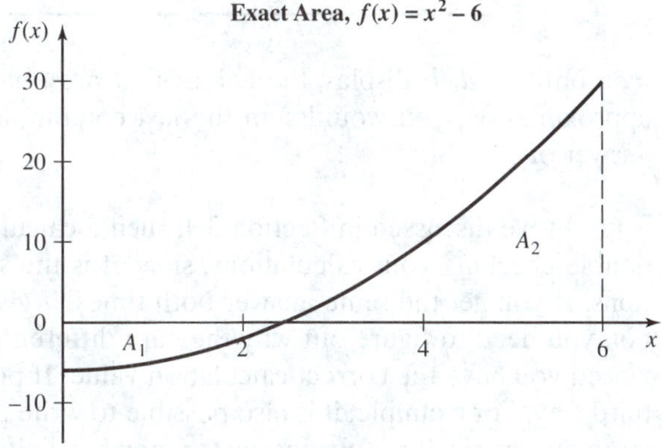

Figure 4.2-19

Clearly, the calculations for these methods are extensive and laborious. That's one of the best uses of technology: to save us time and frustration! (It also makes fewer careless mistakes.) Let's see how we can do these calculations more quickly and easily. If data are entered correctly, technology will normally give you a correct answer. Of course, if you enter the data wrong, the results could be meaningless ("garbage in, garbage out").

On either a calculator or a spreadsheet, you can make a column of the endpoints of your subintervals (if you know the technologies well, you can find a shortcut for this, but you could also just enter them individually). It is then easy to apply the function to these endpoints and make a second column of these y values. You can then make a column of the coefficients on each term (1 2 2 . . . 2 1) for the Trapezoid Rule. For the Right Rectangle method you put a 0 on the term (subinterval endpoint) that should not be included, which is the x_0 term for right rectangles. The coefficient pattern will be

0 1 1 1 . . . 1 1. This assumes you have made a row for *every* subinterval endpoint, from x_0 to x_n. Of course, you could also just leave off the row for the unneeded endpoint for these rectangle methods.

Next, you make a new column which multiplies the y-values by the coefficients, row by row. Finally, you add up these products and multiply by Δx divided by the appropriate constant (2 for the Trapezoid Rule and 1 for the Right Rectangle Method).

Sample Problem 5: Solve Example 1 (approximate the area under $f(x) = x^2$ on [0, 6] with $n = 6$ by the Trapezoid method using a calculator or spreadsheet.

Solution: Of course, the answers will be the same as before. To illustrate the spreadsheet/table approach just described, the layout for the Trapezoid Method would look like

i	x_i	$f(x_i)$	Coefficient	(Coefficient) $[f(x_i)]$
0	0	0	1	0
1	1	1	2	2
2	2	4	2	8
3	3	9	2	18
4	4	16	2	32
5	5	25	2	50
6	6	36	1	36

You only *need* to display the values that are *not* in bold in the table. To get the final approximation, you would sum the last column, and multiply by $\Delta x/2 = 1/2$ to get the answer of 73. □

As we discussed in Section 3.4, such a calculation is an example of **verification**, or double-checking your calculations, since it is the second time doing these same calculations. If you get the same answer both times, it gives you confidence in your answers. If not, you need to figure out why they are different, and keep working until you are convinced you have the correct calculation value. If possible, this could then be verified in a third way. For example, it is also possible to write programs to do these calculations even more automatically. Your instructor may have given you programs that you can run for this purpose.

Sample Problem 6: If you have such programs, use them to again approximate the area under $f(x) = x^2$ on [0, 6] for $n = 6$ by *all* the methods discussed in this section to verify your earlier calculations in a different way.

Solution: Once again, the numerical answers are all the same, so we will not repeat them here. Programs may vary. See your technology supplement for how to use those available to you. □

Sample Problem 7: Suppose that the number of months before a competitor will come out with a new product (the random variable X) can be modeled with the probability density function $f(x) = 0.2e^{-0.2x}$, for $x \geq 0$. Approximate the probability that the competitor will release the new product within the next 6 months, using all of the methods discussed in this section, with $n = 6$.

Solution: The quickest and easiest way to get your answers here is to use the programs, with a lower limit of 0 and an upper limit of 6. The results are

Left Rectangles: 0.771
Right Rectangles: 0.631
Midpoint Rectangles: 0.698
Trapezoid Rule: 0.701
Simpson's Rule: 0.699 □

We mentioned earlier in this section that the probability density function can be thought of as a probability **rate** per unit. One way to think of this is to **think of the height of the density function at x (the value of $f(x)$) as approximately the probability of an interval with a width of 1 unit that includes x, such as $[x, x + 1]$**. This is similar to the interpretation of marginal cost we discussed in Chapters 2 and 3. Or, put differently, think of a left rectangle of width 1 unit drawn in for the interval $[x, x + 1]$, which gives an approximation of the probability of obtaining a value between x and $x + 1$.

Section Summary

Before trying the exercises, be sure that you

- Understand how the *change* in a function over an interval given its nonnegative *rate of change* function corresponds to the area under the rate function (derivative) curve and above the horizontal axis over that interval.
- Understand how to divide up an interval $[a, b]$ into n equal subintervals and be able to calculate the width of each subinterval ($\Delta x = (b - a)/n$) and the endpoints of the subintervals $x_0, x_1, x_2, \ldots, x_n$ are given by $x_i = a + i\Delta x$).
- Are able to draw in various geometric shapes, such as rectangles and trapezoids, to approximate the area under a curve and above the horizontal axis within a subinterval, corresponding to the Right Rectangle method, the Trapezoid method, the Midpoint Rectangle method, and the Left Rectangle method.
- Understand that the vertical height of a drawn-in rectangle corresponds to the function's value at the point where the height was drawn (if the height was drawn in at x, then the height is $f(x)$).
- Are able to express an approximation of the area under a curve using the sum of areas of rectangles for the different subintervals for the Right Rectangle method, both to calculate the answer numerically for a given function and a given value of n and to derive the general formula yourself and understand what it corresponds to graphically:

$$\text{Right Rectangles: } S_n = \sum_{i=1}^{n} f(x_i)\Delta x = \Delta x[f(x_1) + f(x_2 + \cdots + f(x_n)]$$

- Given the formulas for the Right Rectangle and Trapezoid methods and Simpson's Rule, are able to set up calculator lists or a spreadsheet to approximate the area under the curve for a given value of n.

$$\text{Trapezoids: } S_n = \frac{\Delta x}{2}[f(x_0) + 2f(x_1) + 2f(x_2) + \cdots + 2f(x_{n-1}) + f(x_n)]$$

$$\text{Simpson's Rule: } S_n = \frac{\Delta x}{3}[f(x_0) + 4f(x_1) + 2f(x_2) + \cdots + 4f(x_{n-1}) + f(x_n)]$$

- Can follow the logic of the derivation of the formula for the Trapezoid method approximation.
- Understand how to write and interpret expressions involving summation (sigma) notation.
- Understand how to approximate the area under a curve using at least the methods known as Right Rectangles and the Trapezoid Rule by hand and using technology, and the concepts behind them.

EXERCISES FOR SECTION 4.2

Warm Up

For Exercises 1–4, an interval ([a, b]) is specified, along with a value of n (the number of equal subintervals into which you are to subdivide the interval). For each exercise, first find the value of Δx, then find all of the endpoints of the subintervals, identifying them symbolically (using the variable with subscripts). If you can, also write out a general expression for the ith endpoint, x_i. If a variable other than x is specified, adjust your answers appropriately.

1. $[0, 8]$; $n = 4$

2. $[0, 4]$; $n = 8$

3. $a = 2$, $b = 20$; $n = 6$; independent variable is t.

4. $a = 4$, $b = 6$; $n = 8$; independent variable is q.

For Exercises 5–8, evaluate the following summations as fully as you can.

5. $\sum_{i=1}^{5}(i^2 + 4i)$

6. $\sum_{i=3}^{6}[(i-1)^2(3)]$

7. $\sum_{i=1}^{4} f(x_i)\Delta x$

8. $\sum_{i=1}^{6} f(t_{i-1})\Delta t$

*For Exercises 9–12, estimate **by hand** the area under the given function over the given interval using the given value of n. Do this by hand for both the Right Rectangle and Trapezoid methods discussed in this section and sketch a graph for each method, showing what the calculations correspond to.*

9. $f(x) = 2x$; $a = 2$, $b = 20$; $n = 6$

10. $f(x) = x^2/2$; $a = 0$, $b = 8$; $n = 4$

11. $R(q) = 8q - q^2$; $a = 0$, $b = 8$; $n = 4$

12. $P(t) = 10 - 2t$; $a = 4$, $b = 6$; $n = 8$

For Exercises 13–16, use a spreadsheet or use the lists (L_1, L_2, etc.) on your graphing calculator as a spreadsheet to estimate the area under the given function over the given interval using the given value of n using the indicated method. Write out the spreadsheet values that you obtain, and explain what you did. The exercises are the same as 9–12, so these can serve to ***verify*** *your answers from before.*

13. $f(x) = 2x$; $a = 2$, $b = 20$; $n = 6$. Right Rectangles.

14. $f(x) = x^2/2$; $a = 0$, $b = 8$; $n = 4$. Right Rectangles.

15. $R(q) = 8q - q^2$; $a = 0$, $b = 8$; $n = 4$. Trapezoids.

16. $P(t) = 10 - 2t$; $a = 4$, $b = 6$; $n = 8$. Trapezoids.

Now, if you were given a calculator program or spreadsheet template by your instructor to automatically estimate areas, use it to estimate the areas specified in Exercises 17–20 using both Simpson's Rule and the indicated method. Once again, these are the same as in the earlier two sets, so this is another way to verify your results.

17. $f(x) = 2x$; $a = 2$, $b = 20$; $n = 6$. Right Rectangles.

18. $f(x) = x^2/2$; $a = 0$, $b = 8$; $n = 4$. Right Rectangles.

19. $R(q) = 8q - q^2$; $a = 0$, $b = 8$; $n = 4$. Trapezoids.

20. $P(t) = 10 - 2t$; $a = 4$, $b = 6$; $n = 8$. Trapezoids.

21. For the function $f(x) = 2x$ over the interval $[-2, 4]$ with $n = 6$, find the value given by the Right Rectangle method. Draw in the rectangles and explain what your answer corresponds to graphically (pay particular attention to *signs*).

22. For the function $P(t) = 12 - 2t$ over the interval $[4, 10]$ with $n = 6$, find the value given by the Right Rectangle method. Draw in the rectangles and explain what your answer corresponds to graphically (pay particular attention to *signs*).

Game Time

23. You have been biking in a very flat area of Ohio, with virtually no traffic. You have a speedometer on your bike. As an exercise, for the last 2 hours, you worked to maintain a steady speed of 16 miles per hour (mph).

(a) How far did you travel? Show how units can help to understand this calculation.

(b) Draw in your velocity (speed) curve on a graph, and show what your calculation in (a) corresponds to on that graph.

24. You are stuffing envelopes for a mailing by a campus group you belong to. You estimate that you can stuff about 6 envelopes in a minute. Suppose you work for half an hour at this rate.

(a) How many envelopes would you expect to stuff? Show how units can help to understand this calculation.

(b) On a graph, sketch the graph of your rate (in envelopes per minute) versus time (in minutes). To what does your calculation in (a) correspond?

25. Back to biking in Ohio (as in Exercise 23). Suppose that you traveled at 14 mph for an hour and a half, then pushed it to 18 mph for the last half hour.

(a) How far did you travel? Explain how you did your calculation.

(b) Show what your calculation in (a) corresponds to graphically.

26. Back to envelope stuffing (as in Exercise 24). Suppose that you worked for an hour at a rate of 6 envelopes per minute, and then figured out a better way to do the operation that enables you to do 8 envelopes per minute, which you did for another half an hour before taking a break.

(a) How many envelopes did you stuff in that hour and a half? Explain your calculation.

(b) Show what your calculation in (a) corresponds to graphically.

27. You work for a company that is developing a drug to help treat people with AIDS. The research is moving along well, but there is a lot of uncertainty in this kind of project. You estimate that the distribution of time (in years) from the present before the drug can go into production, X, is given by the probability density function

$$f(x) = \begin{cases} 0.25xe^{-0.5x}, & \text{for } x \geq 0 \\ 0, & \text{otherwise} \end{cases}$$

(a) Use Midpoint Rectangles and the Trapezoid Rule with $n = 6$ to estimate the probability that the drug will go into production in the next 3 years. Sketch a graph to show what each method corresponds to.

(b) Which answer do you think is more accurate? Why?

28. You are in the middle of a lawsuit. You have had some experience with similar suits and estimate that the eventual court settlement will result in a net benefit to you of X hundred thousand dollars, with a probability density function given by

$$f(x) = \begin{cases} -0.0208x^2 + 0.0417x + 0.208, & \text{for } -2 \leq x \leq 4 \\ 0, & \text{otherwise} \end{cases}$$

(a) Use Right Rectangles and the Trapezoid Rule with $n = 6$ to estimate the probability that the settlement will result in a net loss. Sketch a graph to show what each method corresponds to.

(b) Which answer do you think is more accurate? Why?

29. Back to biking in Ohio (as in Exercises 23 and 25). Suppose your velocity (in miles per hour) after t hours of biking is given by

$$v(t) = 14 + 2t, \text{ for } 0 \leq t \leq 2$$

(a) Sketch a graph of your velocity function and describe in words what the function says about your trip. What assumptions are implied by this function?

(b) What would finding your total distance traveled in the 2 hours correspond to graphically?

(c) What method(s) of approximation should give you the exact correct answer?

(d) Pick one of the methods you specified in (c) and an appropriate value of n to find the distance traveled.

(e) Use the Trapezoid method with $n = 4$ to approximate the answer and explain what each of the calculations in the formula corresponds to in the context of this problem. Use sigma notation and the expanded form of the summation in your explanation.

30. Back to envelope stuffing (as in Exercises 24 and 26). Suppose your rate (in envelopes per minute) after t minutes is given by

$$r(t) = 6 + \frac{t^2}{4050}, \text{ for } 0 \leq t \leq 90$$

(a) What is $r(0)$? What does this mean?

(b) What is $r(90)$? What does this mean?

(c) Graph your rate function. What story does the function tell you about what happened?

(d) What would the number of envelopes you stuffed in an hour and a half correspond to graphically?

(e) What approximation method would give you an exact answer?

(f) Use the Right Rectangle method with $n = 6$ to approximate the number of envelopes stuffed in an hour and a half. Sketch what it is doing on a graph and explain what the individual terms of the summation correspond to. Also, write out the sigma notation for the approximation.

31. In Exercise 9, you estimated the area under $f(x) = 2x$ between $x = 2$ and $x = 20$.
(a) Geometrically, what shape is this region? Find its exact area using geometry.
(b) How accurate were the various methods of approximation? Why?
(c) Using Right Rectangles with $n = 3$, what is the approximation for the area?
(d) What is the approximation using Right Rectangles with $n = 6$?
(e) Which value of n (3 or 6) gives a more accurate approximation? Draw a graph to explain why.
(f) What could you say about the accuracy of the approximation you should get by using Right Rectangles with $n = 12$?

4.3 FINDING EXACT AREAS USING LIMITS OF SUMS

We have seen that probability calculations in statistics, and calculating the net change in a function (like profit) over an interval, can be thought of graphically as areas under curves. In the last section we saw how to approximate these areas using rectangles and other shapes. We broke down the original interval into equal subintervals, approximated the area under the curve for each subinterval, and added the results. The key word here is **approximate**. Can we get **more accurate and precise** approximations?

The basic idea of this section is that more subintervals will, in general, yield a better approximation. Do the results of a given approximation method get closer and closer to some number (a **limit**)? If it does, do *all* of the approximation methods approach the *same* limit? If so, then to find the exact area, we can look for the **limit** of the results of any approximation method as the number of subintervals (n) gets bigger and bigger (as n approaches infinity). Since the approximation methods all involve sums, we can call this idea the calculation of areas using **limits of sums**. In this section we will be using calculator programs and spreadsheet templates to calculate these limits of sums.

Here are some problems to illustrate why you might need to know the concepts and skills in this section:

- You have good weekly sales data and want to find a good estimate for total sales revenue in the next quarter.
- You know that a competitor business is about to release a new product that could have a major impact on your business. You want to obtain a good estimate of the probability that the release of the competitor's new product will be within the next six months.

After studying this section, in addition to being able to analyze problems like those above, you should also

- Understand and be able to explain why more subintervals in general give a better approximation of an area.
- Understand how to (and be *able* to) use technology to estimate the limiting value of a particular approximation method by finding the approximation using that method for larger and larger values of n (the number of subintervals) until the answers converge to the desired level of precision.

- Understand why, for a curve that is continuous and smooth, *any* of the approximation methods should result in the *same* limit.

Sample Problem 1: We return to Example 1 of Section 4.2, finding the area under the curve defined by $y = f(x) = x^2$ over the interval [0, 6], which corresponded to the total profit (in \$10,000's) over six months. Approximate the exact area using technology (calculator program or spreadsheet template) and all five methods discussed in Section 4.2, using $n = 6$ and $n = 12$. Which estimate is more accurate? In general, what kind of change in the value of n gives you more accurate answers with each method?

Solution: Using technology, we obtain the following results:

(Using $f(x) = x^2$ on [0, 6])	Approximate Area with $n = 6$	Approximate Area with $n = 12$
Left Rectangles	55	63.25
Right Rectangles	91	81.25
Midpoint Rectangles	71.5	71.875
Trapezoids	73	72.25
Simpson's Rule	72	72

For this particular function, as we have discussed before, we know the exact area should be obtained by Simpson's Rule, since it fits parabolas to the curve, which is itself a parabola. Thus we know that the exact area is 72. Of all the other methods, the Midpoint Rule gives the next most accurate answer, then the Trapezoid Rule, then the Left Rectangle method, and finally the Right Rectangle method. This order holds pretty well in general, although the left and right rectangle methods are essentially equivalent, just depending on the function as to which is better. For example, if we had used $f(x) = 36 - x^2$, the order of the two would have been reversed.

Notice that, for any of the individual methods, the approximation using $n = 12$ is better (closer to the exact answer, 72) than the one with $n = 6$. It is easy to see why by looking at Figure 4.3-1. This illustration is for the Right Rectangle method. Notice how, with 12 subintervals, the errors (the parts of the rectangles sticking up above the curve) are much smaller. There are twice as many rectangles, and the total error is about cut in half compared to the case with 6 subintervals.

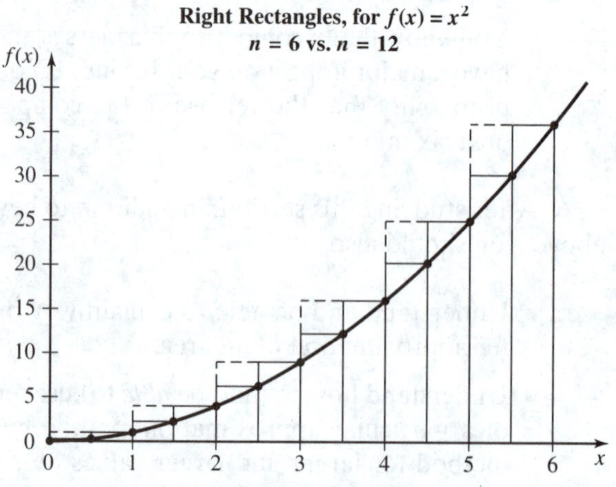

Figure 4.3-1

Intuitively, it makes sense that using more subintervals makes it possible to get a better approximation of the exact area. This is related to the fact that computer screens with better resolution (more dots or pixels per square inch) can draw curves more accurately than old computers (or graphing calculators), which look rather jagged for curves.

Thus, **for any method, as long as our function is continuous (has no jumps, gaps, or holes) and smooth (has no sharp angles), we should be able to get as close as we want to the exact area by using more and more subintervals (larger and larger values of *n*).**[10] □

Sample Problem 2: Find the exact area under $f(x) = x^2$ on [0, 6] to 3 significant figures using the idea of finding the limit of sums with Right Rectangles and the Trapezoid Rule.

Solution: We have already seen the approximations using Right Rectangles with $n = 6$ and $n = 12$. Use technology (your calculator program or spreadsheet template) to find the approximations for $n = 24$, 48, and so on, until your answers seem to level off and your answer seems accurate to 3 significant figures. You should get the following:

TABLE 1 Right Rectangles

n	Approximate Area	Rounded Approximation (3 significant figures)
6	91	91
12	81.25	81.3
...	...	...
384	72.28149414	72.3
768	72.14068604	72.1
1536	72.07032776	72.1
3072	72.03516006	72.0
6114	72.01757908	72.0
12,288	72.00878930	72.0

TABLE 2 Trapezoid Rule

n	Approximate Area	Rounded Approximation (3 significant figures)
6	73	73
12	72.25	72.3
24	72.0625	72.1
48	72.015625	72.0
96	72.00390625	72.0

[10] In advanced calculus books, it is proved that, as long as the function being integrated is continuous over $[a, b]$, then all of the methods will converge to the same limit. One way to show this is to define two methods we could call Upper Rectangles (draw the height for each subinterval at the maximum value over that subinterval) and Lower Rectangles (same idea, using the minimum point for the height). Since the function is continuous, it can be proved that there will be a unique global maximum and minimum output value over each subinterval. The Upper Rectangle method will always be an upper bound on the exact area (and on the other methods we have discussed), and the Lower Rectangle method will always be a lower bound. Thus, if both converge to the same limit, then so will all of the methods. It can then be shown that as long as the function is continuous, the Upper and Lower Rectangle methods will converge to the same limit. Intuitively, the continuity makes the approximation get more and more accurate in general as the number of subintervals increases. Formulas exist to calculate upper limits on the error for each method as a function of *n*.

TABLE 3 Simpson's Rule

n	Approximate Area	Rounded Approximation (3 significant figures)
6	72	72
12	72	72
24	72	72

We can see from Tables 1–3 that all the methods result in the same limit, although some took more subintervals to show the limit clearly. The "Approximate Area" column gives the precise result (to 10 significant figures, anyway) of the approximation formula for each rule, which we then round off to 3 significant figures in the last column. Clearly, the value levels off at 72.0, which we know to be the exact answer in this particular example. □

Sample Problem 3: You estimate that the number of months before a competitor will come out with a new product (X) can be modeled with the following exponential probability density function: $f(x) = 0.2e^{-0.2x}$, for $x \geq 0$. Using limits of sums, find the probability (to 4 significant figures of accuracy) that the release will be in less than 6 months.

Solution: We know that the most accurate of the approximation methods is Simpson's Rule, since it fits parabolas to pairs of subintervals, which fit curves better than horizontal lines or general (slanted) lines. Thus, we will use our calculator program or spreadsheet template for Simpson's Rule, starting with $n = 6$, then $n = 12$, 24, and so on, until our answer levels off to give us the answer to 4 significant figures of accuracy. Recall that significant figures do not count initial 0's in a decimal fraction (for example, 0.003 has only **1** significant figure of accuracy, not 3).

TABLE 4

n	Approximate Area	Rounded Approximation (4 significant figures)
6	0.6988119702	0.6988
12	0.6988061759	0.6988
24	0.6988058123	0.6988

Notice that the answer levels off immediately. This is because Simpson's Rule is such an accurate approximation. In fact, the value for $n = 12$ is accurate to 6 significant figures!

Let's examine what happens when we use the Trapezoid Rule (Table 5).

TABLE 5

n	Approximate Area	Rounded Approximation (4 significant figures)
6	0.7011335893	0.7011
12	0.6993880292	0.6994
24	0.6989513666	0.6990
48	0.6988421838	0.6988
96	0.6988148871	0.6988

Let's examine what happens if we use the Right Rectangle method. In this case we get the results shown in Table 6.

TABLE 6

n	Approximate Area	Rounded Approximation (4 significant figures)
6	0.6312530105	0.6313
12	0.6644477398	0.6644
24	0.6814812219	0.6815
48	0.6901071115	0.6901
96	0.6944473509	0.6944

Even with 96 subintervals, the Right Rectangle method is not nearly as accurate as Simpson's Rule with 6 subintervals or the Trapezoid Rule with 48 subintervals! We can see that it will get there eventually, but this is not a very effective use of our time! □

The Limit of Sums and the Definite Integral

The one area where the Right Rectangle method has an advantage is notationally. It gives us the simplest way to write out the idea of the limit of a sum. We have already seen that the Right Rectangle method calculation can be written as

$$\sum_{i=1}^{n} f(x_i)\Delta x$$

Now, we are saying that the exact area can be found by taking the **limit** of this quantity as n gets larger and larger, similar to the limit idea we used to find the slope of a tangent line as a limit of the slopes of secant lines. This time, since n is getting larger and larger, we say that we are taking the limit **as *n* approaches infinity**. Notationally, we write

$$\lim_{n\to\infty}\left(\sum_{i=1}^{n} f(x_i)\Delta x\right)$$

This notation helps explain the more concise standard notation for **the area under a curve $y = f(x)$ and above the x-axis minus the area below the x-axis and above the curve between $x = a$ and $x = b$,** called a **definite integral**. The notation is

$$\int_a^b f(x)dx$$

You can see that the dx is analogous to the Δx, the $f(x)$ is analogous to the $f(x_i)$, and the integral sign $\int$ is analogous to the sigma summation sign. You may find it helpful to think of the definite integral symbols as the result of applying the limit operation to the discrete summation symbols. It is sometimes very useful to think of the dx as the width of a very skinny generic rectangle (infinitesimally skinny, in theory) in the interval $[a, b]$, the $f(x)$ as the height of that rectangle, and the integral sign as the (infinite) summation of the areas of such rectangles. (Although we can think of the dx this way, it is actually the symbol that identifies the variable of integration.) In this way, **integration**, the process of evaluating an integral, is the continuous analogy to discrete summation. This idea may take you a little while to get used to.

We said that this $\int_a^b f(x)dx$ is called the definite integral and is the continuous analogy to discrete summation. Like a summation, the definite integral is a number, not a function. Will see in Section 4.5 how to evaluate a definite integral, and that finding the number represented by a definite integral can be useful and important.

As we discussed in Section 4.2, if the function dips below the x-axis, the definite integral calculation looks at the area **between** the *curve and the x-axis*, and **calculates the area above the x-axis minus the area below the x-axis**. This makes good sense, since the y, or $f(x)$, values above the x-axis are positive and the ones below the x-axis are negative.

Remember that we have pointed out that $\Delta x = (b - a)/n$. In our example, $a = 0$ and $b = 6$, so we have $\Delta x = (6 - 0)/n = 6/n$. As n gets bigger and bigger, then, $6/n = \Delta x$ gets smaller and smaller, approaching 0 in the limit. For example,

n	$6/n$
10	0.6
100	0.06
1000	0.006
10,000	0.0006

Clearly, as n approaches infinity, $\Delta x = 6/n$ approaches 0. So sometimes you will see the expression for the exact area, going back to the simplest Right Rectangle method form, written as

$$\lim_{\Delta x \to 0} \left(\sum_{i=1}^{n} f(x_i)\Delta x \right)$$

To further simplify, we could define

S_n = the result of the sum calculation for any of the methods using a particular value of n

or, for the Right Rectangle method,

$$S_n = \sum_{i=1}^{n} f(x_i)\Delta x$$

We could then define the exact area to be

$$\lim_{n \to \infty} (S_n)$$

Before going ahead to try the exercises, try to make sure you can

- Understand and explain why more subintervals, in general, give a better approximation of an area.
- Understand how to (and be *able* to) use a calculator program or spreadsheet template (for example, if one has been given to you by your instructor) to estimate the limiting value of a particular approximation method by finding the approximation using that method for larger and larger values of n (the number of subintervals) until the answers converge to the desired level of precision.

- Understand why, for a curve that is continuous and smooth, *any* of the approximation methods should result in the *same* limit.

EXERCISES FOR SECTION 4.3

Warm Up

For each of Exercises 1–6, use the idea of finding the limit of a sum using a calculator program or spreadsheet template (for example, if one has been given to you by your instructor), for $n = 4, 8, \ldots$, and showing the calculations to find the definite integral of the given function $f(x)$ over the given interval $[a, b]$, using the given method of approximation:

1. $f(x) = 4$, for $1 \le x \le 5$; Trapezoids

2. $f(x) = 3x + 2$, for $0 \le x \le 8$; Right Rectangles (to 2 significant figures)

3. $f(x) = \dfrac{1}{\sqrt{2\pi}}e^{-x^2/2}$, for $-3 \le x \le 3$; Simpson's Rule (to 4 significant figures)

4. $f(x) = \dfrac{5}{x}$, for $1 \le x \le 3$; Trapezoids (to 3 significant figures)

5. $f(x) = 3x - 5$, for $0 \le x \le 2$; Right Rectangles (to 2 significant figures)

6. $f(x) = x^4 - 1$, for $0 \le x \le 1$; Simpson's Rule (to 4 significant figures)

Game Time

7. As in Exercise 29 of Section 4.2, suppose you are biking in Ohio, and your velocity (in mph) after t hours of biking is given by

$$v(t) = 14 + 2t, \text{ for } 0 \le t \le 2$$

(a) Which method(s) of approximation should give an exact answer?
(b) Show on a graph what the calculations using the Right Rectangle method with $n = 2$ and $n = 4$ correspond to, and use the graph to explain which is better.
(c) Use the calculator program or spreadsheet template to find the Right Rectangle approximation for $n = 4,8,16,32,\ldots$ Put the answers into a table as was done in this section, and make a separate column to round off your answers to 3 significant figures. Go as far as you need to, in order to convince yourself of the correct value of the limit, rounded off to 3 significant figures. This is what we mean by finding the **limit of sums**. Write a mathematical expression using the "lim" notation to express your answer.
(d) What is the interpretation of your answer in (c)?
(e) Try doing what you did in (c), but this time use Trapezoids.

8. As in Exercise 30 of Section 4.2, suppose you are stuffing envelopes and your rate (in envelopes per minute) is given by the function

$$r(t) = 6 + \frac{t^2}{4050}, \text{ for } 0 \le t \le 90$$

(a) Which method of approximation should give an exact answer?
(b) Show on a graph what the calculations using the Right Rectangle method with $n = 2$ and $n = 4$ correspond to, and use the graph to explain which is better.
(c) Use the calculator program or spreadsheet template given to you by your instructor to find the Right Rectangle approximation for $n = 4,8,16,32,\ldots$ Put the answers into a table as was done in this section and make a separate column to round off your answers to 3 sig-

nificant figures. Go as far as you need to in order to convince yourself of the correct value of the limit, rounded off to 3 significant figures. This is what we mean by finding the **limit of sums**. Write a mathematical expression using the "lim" notation to express your answer.

(d) What is the interpretation of your answer in (c)?

(e) Try doing what you did in (c), but this time use Trapezoids.

(f) Find the limit of sums using Simpson's Rule as well. Which method seems to get you the answer fastest?

9. You work for a company that is developing a drug to help treat people with AIDS. The research is moving along well, but there is a lot of uncertainty in this kind of project. You estimate that the distribution of time (in years) from the present before the drug can go into production, X, is given by the probability density function

$$f(x) = \begin{cases} 0.25xe^{-0.5x}, & \text{for } x \geq 0 \\ 0, & \text{otherwise} \end{cases}$$

(a) Use Right Rectangles, the Trapezoid Rule, and Simpson's Rule to estimate the probability that the drug will go into production in the next 3 years, using limits of sums with $n = 4,8,16, \ldots$ Find the answer to 3 significant figures.

(b) Which method seems to converge to the limit fastest? How would you put them in order by speed of convergence, with the fastest first?

10. You are in the middle of a lawsuit. You have had some experience with similar suits, and estimate that the eventual court settlement will result in a net benefit to you of X hundred thousand dollars, with a probability density function given by

$$f(x) = \begin{cases} -0.0208x^2 + 0.0417x + 0.208, & \text{for } -2 \leq x \leq 4 \\ 0, & \text{otherwise} \end{cases}$$

(a) Use Right Rectangles, the Trapezoid Rule, and Simpson's Rule to estimate the probability that the lawsuit will result in a net loss, using limits of sums with $n = 4,8,16, \ldots$ Find the answer to 3 significant figures.

(b) Which method seems to converge to the limit fastest? How would you put them in order by speed of convergence, with the fastest first?

4.4 RECOVERING FUNCTIONS FROM THEIR DERIVATIVES

We know that functions such as velocity and marginal profit are rates of change, or derivatives. Velocity is the derivative of the original distance or location function, and marginal profit is the derivative of the original accumulated profit function. What if we had information about the derivative and wanted to find the original function? In such a case, the original function is usually called an **antiderivative** of the derivative function, since this is the opposite direction from taking a derivative. In this section, we will explore this process of finding the original function from its derivative, called **antidifferentiation**. We will also see what the antiderivative of a probability density function means, since we indicated in Section 4.1 that the pdf can also be thought of as a derivative function.

The concepts and techniques in this section can help you solve the following kinds of problems:

- You have been driving on the highway for several hours and have a pretty good idea of what your speed has been the whole time (it has not been constant, but has changed in a steady way). However, you forgot to set your trip odometer to know how many miles you have traveled. How can you figure out this distance traveled just knowing your speed?

- You have good records for your weekly sales revenues so far this year (and for past years) and want to be able to find the accumulated revenues so far this year, or project this figure any number of weeks into the future.
- A competitor business is about to come out with a new product that will affect your business significantly. To assess your risk and plan ahead, you want to be able to estimate the probability that the new product will be released within two, three, four, or any specified number of months from now.

After studying this section, you should be able to solve problems like those above, and should

- Be able to find the most general antiderivative of a given function.
- Know what is meant by an **indefinite integral** and how it differs from a definite integral.
- Know the general antiderivatives for kx^n, ae^{kx}, ab^{kx}, and $\frac{k}{x}$, and how to find antiderivatives for functions involving sums and/or differences of such terms.
- Know when to add $+C$ when finding an antiderivative, and when not to add it.
- Be able to find a specific antiderivative (including the value of C) of a given function (the derivative), given enough information about the original (antiderivative) function.

General Form of the Antiderivative

Sample Problem 1: You have been driving on the highway for 4 hours straight and want to estimate how far you have traveled, because you forgot to set the trip odometer or notice the regular odometer reading when you left. You estimate that your speed has been gradually increasing: You started out at about 56 mph and your speed has inched up steadily, about 2 mph each hour. Find a model for your velocity, which we will use later to estimate the distance traveled in the 4 hours.

Solution: Let's first formulate the velocity function. Recall that in Section 4.2 we stated that in everyday usage, the term velocity is closely related to speed but indicates the *direction* of motion as well (such as by its sign). As we discussed earlier, the first step in formulation is always to define your variables and functions clearly, so there is no ambiguity about what they mean. Before doing this, it is always helpful to think about what units to use for the problem. For this example, the units are clear: miles for distance and hours for time, and so miles per hour for velocity. So let us define

t = the number of hours since you started driving on the highway (4 hours before the time frame of the statement of the problem)

and

$v(t)$ = your velocity (speed) in miles per hour at time t (t hours after you started driving on the highway)

Since velocity is the rate of change of distance with respect to time, it is the derivative of the distance traveled as a function of time, as long as we do not reverse direction at any point. Thus we also want to define

$d(t)$ = the distance (in miles) that you have traveled, after t hours of driving on the highway

From these definitions, we see that $v(t) = d'(t)$. Our goal is to first formulate $v(t)$, then figure what $d(t)$ would have to be in order that its derivative is $v(t)$ and its values match its definition.

For this formulation, we will assume that the estimates given of your speed in the problem statement are exact, which is the *certainty assumption*, as well as the usual *divisibility assumption* that the variables can all take on any real number value within the domain; any fraction or decimal is possible. In this case, divisibility clearly holds exactly by the nature of the problem, but the certainty assumption only holds approximately unless your car is *extremely* high tech!

Thus, with these assumptions, your velocity at time $t = 0$ (when starting to drive on the highway 4 hours ago) was 56 mph, so $v(0) = 56$. We are then told that your speed increases 2 mph each hour. So after 1 hour (at $t = 1$), your velocity should be 58 mph ($v(1) = 58$); after 2 hours ($t = 2$), 60 mph ($v(2) = 60$); 62 mph at $t = 3$ ($v(3) = 62$); and 64 mph at $t = 4$ ($v(4) = 64$). Notice that this pattern represents a constant rate of change of your velocity, supported by the wording "your speed has been gradually increasing" and "your speed has inched up steadily". This means a constant slope (constant first differences for equally spaced inputs) for the velocity function, which implies that the velocity function is linear. As we saw in Section 1.3, this means it has the form $y = mx + b$, or in our example $v = mt + b$, where m is the slope and b is the vertical intercept (the velocity or v intercept in this case).

The v intercept (b) is simply the value of the function when $t = 0$, which we have already seen to be 56, so $b = 56$. And the slope (m) is the rate of change of v with respect to t (in general, change in y over change in x; or here, change in v over change in t). In this case, for each additional hour traveled, the speed *increases* by 2 mph, so the slope is $2/1 = 2$ mph per hour. Incidentally, this quantity is your **acceleration** (the rate of change of your velocity). Thus $m = 2$, and we can now substitute back our values $m = 2$ and $b = 56$ into $v(t) = mt + b$ to get

$$v(t) = 2t + 56$$

To *validate* this symbol definition, let's plug in a value of t and see if we get the desired result. For example, we said earlier that $v(3)$ should be 62. If we plug into the model, we get

$$v(3) = 2(3) + 56 = 6 + 56 = 62$$

so it does agree, and the symbol definition looks good.

To fully answer the first part of the problem (find the velocity function), we need to specify that the domain of this function is $0 \le t \le 4$. The domain could possibly extend beyond 4, but we have no information about this, so we will not assume it. Thus our model for the velocity is

Verbal Definition: $v(t)$ = the velocity of your car (in mph) t hours after you started to drive on the highway.

Symbol Definition: $v(t) = 2t + 56$, for $0 \le t \le 4$.
Assumptions: Certainty and divisibility. Certainty implies that the relationship is exact. Divisibility implies that that the time and velocity can have any fractional value.

This is the model for the velocity that was asked for. □

Example 1: Use the velocity model just formulated to find a formula for the distance, $d(t)$, you have traveled (in miles) as a function of the number of hours since you started driving on the highway (4 hours ago), t, for the problem described in Sample Problem 1.

We are to find your distance function, $d(t)$. As stated earlier (for example, in Chapter 2), we know that velocity is the derivative of the distance function (the rate of change of the distance or position with respect to time), since you are always moving in the same direction, so

$$d'(t) = v(t)$$

so

$$d'(t) = 2t + 56$$

Thus, we are looking for a function whose *derivative* is $2t + 56$. In mathematical terminology, we call this **an antiderivative** of $2t + 56$. From Chapter 2, we know that the derivative of a sum is the sum of the individual derivatives, so we need to find antiderivatives for $2t$ and for 56. Let's start with the easier one. When do we get a constant (a number, like 56) for a derivative? The answer: When the original expression was just a coefficient times the variable (a linear term); for example, the derivative of $3x$ is 3. In this case, the variable is t and the answer is 56, so the expression could have been $56t$, since the derivative of $56t$ is 56.

Graphically, finding an antiderivative is like *starting* with a *slope graph* and trying to find a curve that would go with it, a curve whose slope graph matches the graph you started with. In the case of the 56, the slope graph would be a horizontal constant function, and the antiderivative would be a straight line with a slope of 56, as illustrated in Figure 4.4-1. Notice that we only know the *slope* for the antiderivative. The y-intercept could be *any* value. We have chosen the simplest case, where the y-intercept is 0, so the line goes through the origin.

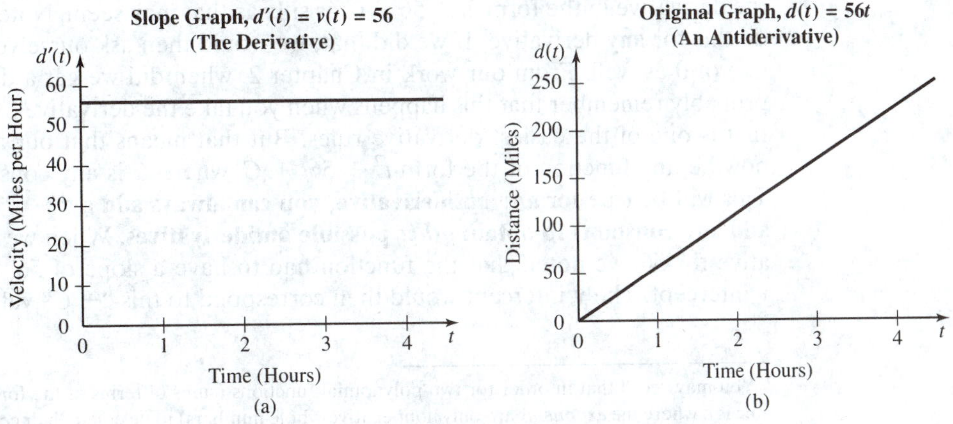

Figure 4.4-1

Now for the $2t$. When do we get a derivative that is a coefficient times the variable to the first power? Intuitively, you probably already know that this happens when you take the derivative of a term involving the square of the variable. Recall that $D_x ax^n = anx^{n-1}$. Or, in this example, $D_t at^n = ant^{n-1}$. In this case, the exponent of the variable in the derivative is 1 ($2t = 2t^1$), which means that $(n - 1) = 1$, so n must be 2. Thus our antiderivative must be of the form at^2. The derivative of at^2 would be $2at$. Since our derivative is $2t$ we want to find a such that

$$2at = 2t$$

so

$2a = 2$ (since we want $2at$ to equal $2t$ for *all* values of t, including 0)[11]
$a = 1$ (dividing both sides by 2)

Thus *an* antiderivative of $2t$ is t^2. Graphically, this is shown in Figure 4.4-2.

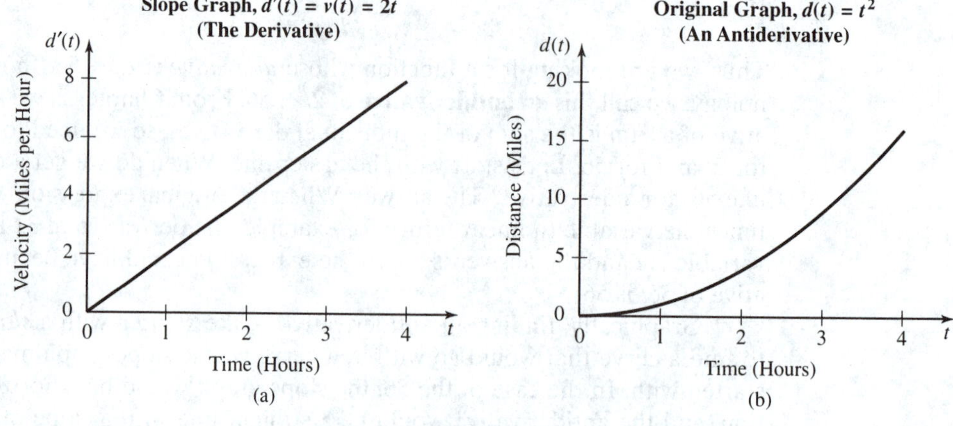

Figure 4.4-2

Putting the pieces together, an antiderivative of $2t + 56$ is $t^2 + 56t$. Is this the only possible antiderivative of $2t + 56$? To answer that, let's note that we could have written the derivative in the form $2t + 56 + 0$, as silly as that may seem. Note that you could also do this for **any** derivative. If we did that, we would then ask ourselves for an antiderivative of 0 as well. From our work in Chapter 2, when did we get a derivative of 0? You probably remember that this happens when you take the derivative of any constant, since that is one of the easiest derivative rules. But that means that our antiderivative could now be any function of the form $t^2 + 56t + C$, where C is any constant (real number). This will be true **for *any* antiderivative; you can always add a "+ C" to the end (that is, add any constant) to obtain *other* possible antiderivatives.** When we found an antiderivative for 56, we noted that the function had to have a slope of 56, but could have *any* y-intercept. The y-intercept would then correspond to this "+ C" value.

[11]You may recall that in order for two polynomial functions (sums of terms of the form ax^n, or at^n if the variable is t, where the *exponents* are only nonnegative whole numbers) to be equal, their corresponding coefficients must be equal, so for $2at$ to equal $2t$, $2a$ must equal 2.

This is, in fact, the most general form of the antiderivative. Put another way, the **only** functions[12] that have a derivative of $2t + 56$ are functions of the form $d(t) = t^2 + 56t + C$. We can verify this to make sure they all work by observing that $D_t(t^2 + 56t + C) = 2t + 56 + 0 = 2t + 56$.

Graphically, the $+ C$ corresponds to shifting the graph vertically by C units, either up or down, depending on the sign of C. But such a shift does not change the *shape* of the graph, and the tangent lines of the new graph have the same slope as the old graph at the same input value; the tangent lines also get shifted vertically by $+ C$. For instance, Figure 4.4-3 shows the graphs of $y = 2t$, $y = 2t + 3$, and $y = 2t - 4$, which are all antiderivatives of 2, since they are all lines with slopes of 2.

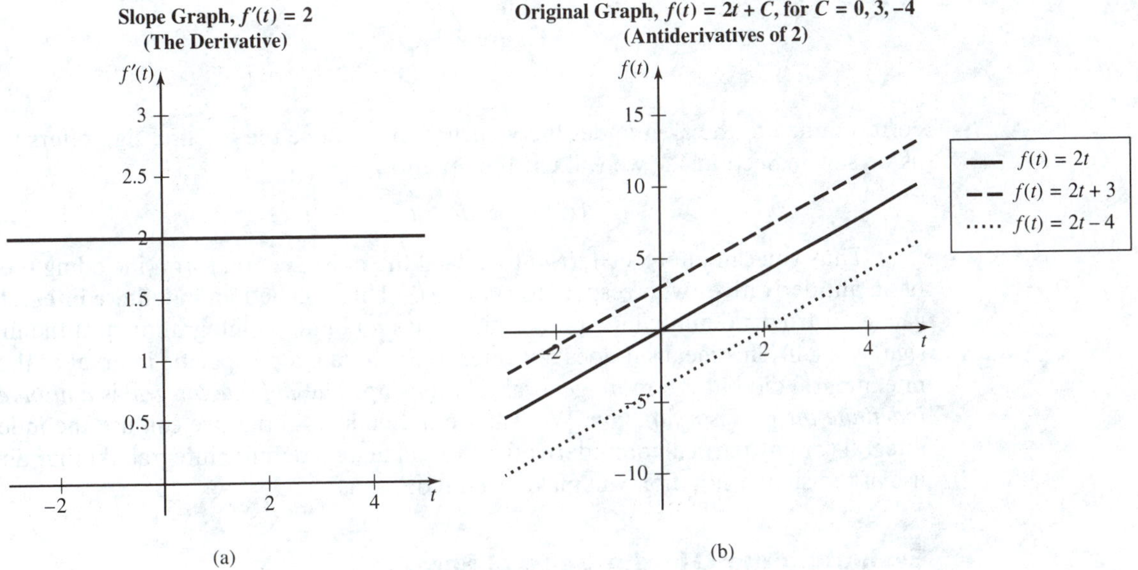

Figure 4.4-3

If you are given the slope graph and asked to find the original graph (the antiderivative), you don't know *which* one to choose. *Anything* of the form $f(t) = 2t + C$ will work.

Let's now look graphically at an antiderivative that is a nonlinear curve. Figure 4.4-4 shows the curves $y = t^2$ and $y = t^2 + 5$, both of which are antiderivatives of $2t$, and draws in the tangents at $t = 2$ (*both* with slopes of 4, consistent with the slope graph) to show that they are parallel. Notice how the different constants in both sets of examples just shift the graphs vertically, so the slope at any given input does not change.

From our earlier work, we can say that **the most general form for an antiderivative** of $2t + 56$ is $t^2 + 56t + C$. Recall that in Section 4.2, we saw that the distance calculation

[12]Mathematically, if another function $f(t)$ had the same derivative as $d(t)$, then the *derivative* of $f(t) - d(t)$ would be $f'(t) - d'(t) = 0$ (since the derivatives are the same). But since the only function that has a derivative of 0 is a constant function, $f(t) - d(t) = C$, so $f(t)$ differs from $d(t)$ only by a constant. Thus any two antiderivatives differ only by a constant, and all of them can be written as any one of them $+C$ (such as $d(t) + C$).

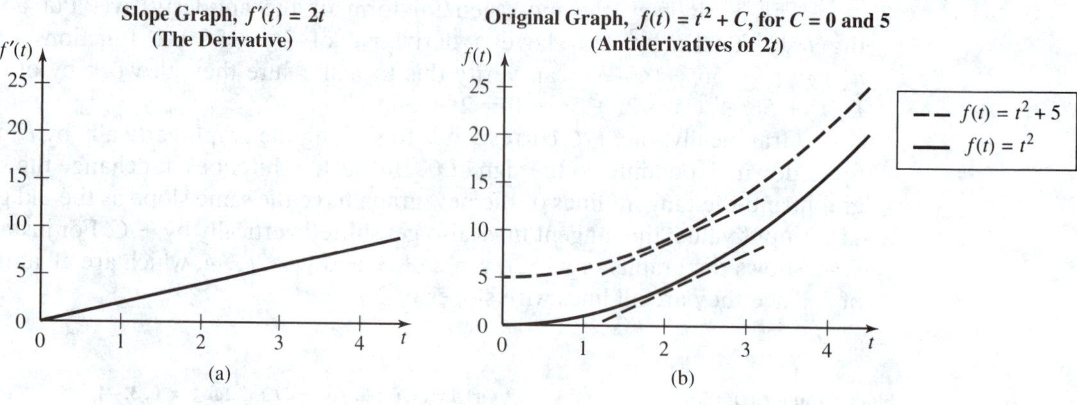

Figure 4.4-4

corresponded to the area under the velocity curve. Since the integral sign refers to area, as we saw in Section 4.3, we will use the notation

$$\int(2t + 56)dt = t^2 + 56t + C$$

Thus you can interpret $\int f(t)dt$ to mean "the most general form (including the $+\ C$) of an antiderivative (with respect to t) of $f(t)$." This is called an **indefinite integral**. You can tell it from a definite integral because it has *no* limits of integration, just the integral sign by itself. This means it does not refer to the area over a specific interval as the definite integral did but is a more general relationship. The *definite integral* is *a number*, the *indefinite integral* is a *function*. We will see in Section 4.5 that we can use the indefinite integral (or any particular antiderivative) to calculate a definite integral. At that time the use of the similar notation will make even more sense. □

Evaluating the *C* in the General Form of the Antiderivative

Now, where are we in our example? We have determined that our distance function must be of the form

$$d(t) = t^2 + 56t + C$$

But we know enough about $d(t)$ to figure out what the constant C must be. By definition, $d(t)$ is the distance we have traveled (in miles) after t hours on the highway.

★ Discovery Question 1:

For what specific value of t do you know the value of $d(t)$? What is the corresponding value of $d(t)$?

(See if you can answer this question on your own without looking below. If you have trouble, try reading the hint that follows below first.)

Hint: What does t stand for? What does $d(t)$ mean? At what point in time do you know how far you have traveled (even if this seems obvious or trivial)?

(Now try again to see if you can answer Question 1 on your own, before reading on.)

Answer: Since $d(t)$ is the distance traveled t hours after getting on the highway, at time $t = 0$ (in other words, at the very beginning of the trip), you will have traveled no distance on that highway, or 0 miles. You know the value of $d(t)$ when $t = 0$, and the value of $d(t)$ there is 0, which is summarized by the statement

$$d(0) = 0 \quad \boxed{\bigstar}$$

Now, since we know that

$$d(t) = t^2 + 56t + C \quad \text{and} \quad d(0) = 0$$

we can plug in 0 for t in the formula for $d(t)$, and set this equal to 0 (the value of $d(0)$) to find out what value C would have to equal (that is, we solve for C):

$$\begin{aligned} d(t) &= t^2 + 56t + C \\ d(0) &= (0)^2 + 56(0) + C \\ &= 0 + 0 + C \\ &= C \end{aligned}$$

Setting this equal to 0 (since we know that $d(0) = C$ *and* that $d(0) = 0$), we get

$$C = 0$$

which is already solved for C, so that is our solution.

We now have the complete form for $d(t)$:

$$d(t) = t^2 + 56t + 0 = t^2 + 56t, \; 0 \leq t \leq 4$$

This, with the original definitions of t and $d(t)$ (and the usual assumptions of divisibility and certainty) is the complete answer to the second question of the example. □

In general, **to find the exact value of C in an antiderivative, you need information about the original function.** The simplest form of this information is knowledge of the value of the function at a specific value of the variable (like $d(0) = 0$ in our example). Given this information, as we did for the example, you can

1. Plug in the specific value of the variable into the general formula for the antiderivative.
2. Set the resulting equal to the known or given **value** of the function at that specific value.
3. Solve for C.
4. Plug this value for C back into the general form of the antiderivative to get your specific antiderivative for the problem.

Sample Problem 2: Back to the highway problem with the velocity function given by $v(t) = 2t + 56$. Suppose that when you started driving on the highway 4 hours ago, you were exactly at mile marker 70 on the highway and that you are driving in the direction in which the numbers on the mile markers are *increasing*. Define a function that expresses your location on the highway (in miles, as measured by the mile markers) at time t, as defined above.

Solution: Let's define

t = number of hours since you began driving on the highway (4 hours ago), and

$L(t)$ = your location on the highway, as measured by the mile markers, at time t

We will make the same assumption as before that the velocity estimates are exact. We will also assume that changing lanes has no effect on location/distance calculations. For example, if you swerved wildly from lane to lane, you would travel more distance than the mile markers would measure; we are assuming this does *not* occur.

With these assumptions, your distance traveled is just the change in your location since time 0. As a result, your velocity can be thought of as the rate of change in your location, as well as the rate of change of your distance traveled. So, similar to before, we get that

$$v(t) = L'(t) = 2t + 56$$

From our earlier discussion, we get that

$$L(t) = \int (2t + 56)dt = t^2 + 56t + C_2$$

We use C_2 here to distinguish from the C used in the original problem. Sometimes we will get a bit sloppy and use the same C for different constants instead of using different subscripts. For now, we will try to be a little more careful, to help make the concepts clearer.

Again, we want to determine the value of C_2, so we need information about $L(t)$, such as its value at a specific time. Again, it is at $t = 0$ that we know the value, because we are told that when we started driving on the highway, we were at mile marker 70. This translates into the equation

$$L(0) = 70$$

As before, to find C_2, we will plug the specific value of t for which we know the value of the function, 0, in for t in the general form above, so we get

$$\begin{aligned} L(0) &= (0)^2 + 56(0) + C_2 \\ &= 0 + 0 + C_2 \\ &= C_2 \end{aligned}$$

So we know that $L(0) = C_2$. We also know that $L(0) = 70$ so we set these two expressions equal:

$$C_2 = 70$$

and solve for C_2 (which is already done for us), so we know that the solution is $C_2 = 70$. With the definitions of t and $L(t)$ above, our function is completely defined by

$$L(t) = t^2 + 56t + 70, 0 \leq t \leq 4. \quad \square$$

You should start to get a feel here for the nature of the "+ C" in the antiderivative. If we think of the distance function $d(t)$ as a location function which starts at 0 when $t = 0$, all of the possible antiderivatives are all of the possible location functions where the location at the start ($t = 0$) is arbitrary (could be 0, or 70, or any real number). This initial location has no effect on the velocity; it is totally arbitrary and relative.

Another comment: We mentioned earlier that the distance function $d(t)$ could be thought of as the change in location from time 0 to time t. Note here that $L(0) = 70$ and $L(t) = t^2 + 56t + 70$, so the *change* in location from time 0 to time t would be

$$L(t) - L(0) = (t^2 + 56t + 70) - (70) = t^2 + 56t$$

which is exactly what we got for $d(t)$, so everything is consistent.

Sample Problem 3: Let's consider a *different*, but related, highway example. Suppose your velocity function was again given by $v(t) = 2t + 56$. But this trip, you didn't notice the mile marker at the beginning of the trip; you only noticed it at the end (when $t = 4$). The value at that time was 368 (mile 368 of the highway), and the mile numbers were increasing in the direction you were traveling. Find the function for the mile marker value at time t (t hours after starting to drive on the highway).

Solution: Let's define this mile marker function to be

$M(t)$ = the mile marker value t hours after starting to drive on the highway ($0 \le t \le 4$)

We know that

$$\begin{aligned} M(t) &= \int (2t + 56)dt \\ &= t^2 + 56t + C \end{aligned}$$

as before. This time, however, we know that $M(4) = 368$. From the antiderivative, we can find another expression for $M(4)$:

$$\begin{aligned} M(4) &= (4)^2 + 56(4) + C && \text{(plugging in 4 for } t\text{)} \\ &= 16 + 224 + C && \text{(simplifying)} \\ &= 240 + C && \text{(simplifying)} \end{aligned}$$

So we see that $M(4) = 240 + C$ and we also know that $M(4) = 368$. Thus we can set these equal to each other to get

$$\begin{aligned} 240 + C &= 368 \\ C &= 368 - 240 && \text{(subtracting 240 from both sides)} \\ C &= 128 && \text{(simplifying)} \end{aligned}$$

In other words, our mile marker function is given by

$$\begin{aligned} M(t) &= t^2 + 56t + C \\ &= t^2 + 56t + 128 \qquad \text{for } 0 \le t \le 4 \end{aligned}$$

This implies that the mile marker at the beginning of this trip was

$$\begin{aligned} M(0) &= (0)2 + 56(0) + 128 \\ &= 0 + 0 + 128 \\ &= 128 \end{aligned}$$

In other words, we started this trip at mile marker 128. □

The General Antiderivative of a Function of the Form kx^n

Sample Problem 4: After fitting a curve to weekly revenue figures for a small business over the last six months, we obtain the following model:

Verbal Definition: $w(t)$ = total revenue (in thousands of dollars) for the week starting t weeks since the beginning of the time period (six months ago).

Symbol Definition: $w(t) = -0.0010t^3 + 0.060t^2 - 0.010t + 3.0, 0 \leq t \leq 26.$

Assumptions: Certainty and divisibility. Certainty implies that the relationship is exact. Divisibility implies that the profit and time can have any fractional values.

Approximate the function of the accumulated total revenue (also in thousands of dollars) from the beginning of the six months to time t (t weeks later than 6 months ago).

Solution: The units of $w(t)$ can be thought of as being in thousands of dollars per week. The word "per" signals that this can be thought of as a rate of change, or derivative. In effect, $w(t)$ is approximately the derivative of the revenue function with respect to time. Specifically, let's define

$R(t)$ = the accumulated total revenue (in thousands of dollars) from the beginning of the six months to time t

Then, as we discussed in Chapters 2 and 3 in the sections on marginal analysis,[13] $R'(t)$ can be thought of as being the **approximate** revenue between time t and time $(t + 1)$. But the span of time between t and $(t + 1)$ is exactly the week starting at time t, so in fact $R'(t)$ is approximately $w(t)$. In what follows, we will use "=" signs, with the understanding that this equality is only approximate in this way.

Thus, we can say that

$$R'(t) = -0.0010t^3 + 0.060t^2 - 0.010t + 3.0$$

Once again, our goal is to find the most general form of an antiderivative of the right-hand side of this equation, to recover the original function ($R(t)$) from its derivative.

---

[13]Earlier we talked about **marginal revenue**, $R'(x)$, which is often defined as the derivative of the revenue function with respect to the *number of units produced*, while here we are looking at the derivative of revenue as a function of *time*, $R'(t)$. Thus, *before*, the derivative $R'(x)$ was the approximate revenue if we make *one more unit* than the given level of production. *Here*, the derivative $R'(t)$ is the approximate revenue realized from *one more week* after the given moment in time.

From our discussion before, we can see that an antiderivative of 3 is $3t$. What about $-.01t$? We know an antiderivative will be of the form at^2, since the derivative of at^2 is $2at$. This means that

$$2at = -0.01t \quad \text{(setting the expressions we want to be the same equal), so}$$

$$2a = -0.01 \quad \text{(for polynomials to be equal, their coefficients must be equal)}$$

$$a = \frac{-0.01}{2} = -0.005 \quad \text{(dividing both sides by 2 and simplifying).}$$

Thus, an antiderivative of $-0.01t$ is $-0.005t^2$. In general, we can see that an antiderivative of kt will always be $\frac{k}{2}t^2$, since the 2 in the denominator will always cancel the exponent 2 when multiplying them to take the derivative.

We are starting now to develop some general rules of antiderivatives:

$$\int k\,dt = kt + C \qquad (\text{or } \int k\,dx = kx + C)$$

$$\int kt\,dt = \frac{k}{2}t^2 + C \qquad \left(\text{or } \int kx\,dx = \frac{k}{2}x^2 + C\right)$$

(Notice that in the first case, the "dt" and the "dx" of the indefinite integral identify what the variable in each problem is.)

Now, we want to find an antiderivative for $0.06t^2$. Remembering the rule for the derivative of a power of the variable: The exponent in the derivative is always 1 less than the exponent in the original function. Thus the exponent in the original function is always one **more** than the exponent in the derivative. In other words, **an antiderivative of a power of a variable will always involve the power of the variable with an exponent that is *1 more* than in the given original function (the derivative)**. So, an antiderivative of $0.06t^2$ will involve a term involving t^3, let us call it at^3. The derivative of at^3 is $3at^2$. If

$$3at^2 = 0.06t^2$$

then

$$3a = 0.06 \quad \text{(equating coefficients)}$$

$$a = \frac{0.06}{3} = 0.02 \quad \text{(dividing both sides by 3)}$$

This means that an antiderivative of $0.06t^2$ is $0.02t^3$. Remember, you can always check an antiderivative by simply taking the derivative of the antiderivative and make sure the answer you get is the same as the derivative function you started with. In this case,

$$D_t\,(0.02t^3) = 3(0.02)t^{(3-1)} = 0.06t^2$$

which is what we wanted, so it checks.

★ Discovery Question 2:

What is the general form for an antiderivative of kx^2? Of kx^n?

(As before, try to answer the question before reading on. If you'd like, after you've tried, you could look at the hint that follows.)

Hint: What kind of function will have a derivative of the form kx^2? How could you express it in general? What would its derivative be? Set this derivative equal to kx^2, and solve for the unknown coefficient. Now see if you can guess the result for an antiderivative of kx^3, by looking at the pattern for kx and kx^2. Finally, try to generalize the pattern for kx^n. See if you can justify this algebraically.

(Now, again see if you can answer the question before you read on.)

Answer: An antiderivative of kx^2 would have to be of the form ax^3, whose derivative is $3ax^2$. So we set

$$3ax^2 = kx^2$$

This will only hold true for all values of x if

$$3a = k$$

which means

$$a = \frac{k}{3}$$

Thus, an antiderivative of kx^2 is $\frac{k}{3}x^3$, or $\frac{kx^3}{3}$. Therefore,

$$\int kx^2 dx = \frac{k}{3}x^3 + C = \frac{kx^3}{3} + C$$

Let's look now at the pattern that is developing:

$$\int kx\,dx = \int kx^1\,dx = \frac{k}{2}x^2 + C = \frac{kx^2}{2} + C$$

$$\int kx^2\,dx = \frac{k}{3}x^3 + C = \frac{kx^3}{3} + C$$

See if you can guess the rule for $\int kx^3 dx$. As above, you can see that an antiderivative will involve x^4, and, following the pattern, should be

$$\int kx^3\,dx = \frac{k}{4}x^4 + C = \frac{kx^4}{4} + C$$

This is easily checked, as usual, by taking the derivative of the antiderivative. Do this to convince yourself it is correct.

So, what is the pattern, and how can we generalize it for finding an antiderivative of kx^n? The general pattern is that the exponent in the antiderivative is 1 more than the exponent in the derivative you start with. The derivative we start with is kx^n, which has an exponent of n. Thus, an antiderivative will have an exponent of 1 more than n, or $(n + 1)$. The other part of the pattern is that the coefficient k gets divided by the *antiderivative's* exponent, so that they cancel out when taking the derivative, to leave just k in the derivative. Thus, for the general case, k will be divided by $(n + 1)$, and the answer will be

$$\int kx^n dx = \frac{k}{n+1}x^{n+1} + C = \frac{kx^{n+1}}{n+1} + C \quad \boxed{\star}$$

Once again, we can take the derivative of the antiderivative to check that this is correct:

$$D_x\left(\frac{k}{n+1}x^{n+1} + C\right) = \left[\left(\frac{k}{n+1}\right)(n+1)\right]x^{(n+1)-1} + 0 = kx^n$$

Returning to Sample Problem 4, we need to find an antiderivative of $-0.0010t^3$. Following the rule we have just developed, we see that an antiderivative of $-0.0010t^3$ is

$$\int -0.0010t^3 dt = \frac{-0.0010t^{3+1}}{3+1} + C = -0.00025t^4 + C$$

Putting the parts together we have

$$R(t) = -0.00025t^4 + 0.02t^3 - 0.005t^2 + 3t + C$$

We know that $R(0) = 0$ because there will be no accumulated revenue at the start of the period. Thus we know that

$$R(0) = -0.00025(0)^4 + 0.02(0)^3 - 0.005(0)^2 + 3(0) + C \text{ and } R(0) = 0$$

so

$$-0.00025(0)^4 + 0.02(0)^3 - 0.005(0)^2 + 3(0) + C = 0$$

$$0 + C = 0$$

$$C = 0$$

Therefore

$$R(t) = -0.00025t^4 + 0.02t^3 - 0.005t^2 + 3t, \text{ for } 0 \le t \le 26. \quad \square$$

The General Antiderivative of Functions of the Form kx^{-1}

What if we try to apply this rule to kx^{-1}? We would get

$$\int kx^{-1}dx = \frac{k}{-1+1}x^{-1+1} + C$$

$$= \frac{k}{0}x^0 + C$$

$$= \frac{k}{0} + C$$

But $k/0$ is undefined, whether k is 0 or any other number.

It turns out that the above rule for $\int kx^n dx$ holds for not only all nonnegative whole numbers (integers), but for **all real numbers except $n = -1$**. This is because when $n = -1$, kx^n means kx^{-1}, which is

$$kx^{-1} = k\left(\frac{1}{x}\right) = \frac{k}{x}$$

As we saw in Chapter 2, the function whose derivative is $1/x$ is $\ln x$, not a power of x, so antiderivatives of terms involving $1/x$ must involve $\ln x$. In fact, since $kx^{-1} = k(1/x)$, the general form for any antiderivative will be $k(\ln x)$, because of the rule that the derivative of a constant times a function is the constant times the derivative of the function. Thus we have the general rule

$$\int kx^{-1}dx = \int \frac{k}{x}\,dx = k(\ln x) + C$$

To check this,

$$D_x\,[k(\ln x) + C] = k\left(\frac{1}{x}\right) + 0 = \frac{k}{x} = kx^{-1}.$$

The General Antiderivative of Functions of the Form e^{kx}

The other major basic derivative rule we studied in Chapter 2 was

$$D_x\,(e^{kx}) = (e^{kx})(k) = ke^{kx}$$

For example, if we wanted to find an antiderivative for e^{2x}, we know that the antiderivative will also involve e^{2x}; let's look at ae^{2x} as a candidate. Taking the derivative of ae^{2x}, we get

$$D_x(ae^{2x}) = a(e^{2x})(2) = 2ae^{2x}$$

Since we want the answer to be $e^{2x} = 1e^{2x}$, this means that we want $2a$ to be 1, so

$$2a = 1 \Rightarrow a = \frac{1}{2}$$

Thus the simple antiderivative of e^{2x} is $\frac{1}{2}e^{2x} = \frac{e^{2x}}{2}$

Notice that the "2" in this example takes exactly the role of the "k" in the general form. Since the pattern of the solution would be exactly the same, we can see that the general antiderivative rule is

$$\int e^{kx}dx = \frac{1}{k}e^{kx} + C = \frac{e^{kx}}{k} + C$$

Similar to what happens for finding an antiderivative of ax^n, the k in the denominator cancels the k in the derivative to make the derivative come out right. Again, as a check,

$$D_x\left[\left(\frac{1}{k}\right)e^{kx} + C\right] = \left(\frac{1}{k}\right)ke^{kx} + 0 = e^{kx}$$

If e^{kx} had a constant in front of it, like ae^{kx}, then this would carry through into the antiderivative (because it does with the derivative), giving us the more general rule:

$$\int ae^{kx}dx = a\left(\int e^{kx}dx\right) = a \cdot \frac{1}{k}e^{kx} = \frac{a}{k}e^{kx} + C = \frac{ae^{kx}}{k} + C$$

The General Antiderivative of Functions of the Form b^{kx}

One additional related derivative rule was

$$D_x\,(b^{kx}) = (\ln b)(b^{kx})(k) = k(\ln b)b^{kx}$$

From this, using the same logic as above (we need to cancel out the $k(\ln b)$ in the derivative, so we put it in the denominator of the antiderivative), we get the antiderivative rule

$$\int b^{kx}dx = \frac{1}{k\ln b}b^{kx} + C = \frac{b^{kx}}{k\ln b} + C$$

Check:

$$D_x\left(\frac{1}{k\ln b}b^{kx} + C\right) = \frac{1}{k\ln b}(k\ln b)b^{kx} + 0 = b^{kx}$$

Similar to the e^{kx} example, if we put in a constant before this function, it carries through into the antiderivative, to give us the general antiderivative rule:

$$\int ab^{kx}dx = \frac{a}{k\ln b}b^{kx} + C = \frac{ab^{kx}}{k\ln b} + C$$

As a final comment for finding antiderivatives, the rules for derivatives when multiplying a function by a constant or when adding/subtracting two functions translate directly in reverse, as follows:

$$\int kf(x)dx = k\int f(x)dx$$

(In words, this says that the integral of a constant times a function is the constant times the integral of the function.) And,

$$\int[f(x) \pm g(x)]dx = \int f(x)dx \pm \int g(x)dx$$

(In words, this says that the integral of a sum or difference of functions is the corresponding sum or difference of the integrals of the functions individually.)

The Chain Rule and Product Rule also have analogous antiderivative rules, but we will not focus on them in this course, since we can use technology to solve problems that would require them when solving by hand. The finding of antiderivatives is not as simple as the finding of derivatives; in fact, sometimes no formula exists for an antiderivative (such as for the normal distribution probability density function). We leave further discussion of these topics for more advanced courses in mathematics.

To practice with all the antiderivative rules we have discussed, let's try a purely mathematical problem.

Sample Problem 5: Find the following indefinite integral:

$$\int\left[3x^4 + \frac{4}{x} - 5e^{-4x} + 7(2^{3x}) - \frac{2}{\sqrt{x}}\right] dx$$

Solution: Applying the above rules, we can first mentally separate the expression in brackets into five separate terms and then find the simple antiderivative for each. We can add a single $+C$ at the end (which could be thought of as the sum of five separate constants, one for each term). The only term that is not a perfectly direct application of a formula is the last term, which can be rewritten as

$$\frac{2}{\sqrt{x}} = \frac{2}{x^{1/2}} = 2x^{-1/2}$$

$$\left[\text{Recall that } x^{m/n} = \sqrt[n]{x^m} \text{ and } x^{-n} = \frac{1}{x^n}\right]$$

Thus,

$$\int\left[3x^4 + \frac{4}{x} - 5e^{-4x} + 7(2^{3x}) - \frac{2}{\sqrt{x}}\right] dx$$

$$= \int 3x^4\,dx + \int\frac{4}{x}\,dx - \int 5e^{-4x}\,dx + \int 7(2^{3x})dx - \int 2x^{-1/2}dx \quad \text{(splitting up the integral)}$$

$$= \frac{3}{(4+1)}x^{(4+1)} + C_1 + 4\ln x + C_2 - \frac{5}{-4}e^{-4x} + C_3 + \frac{7(2^{3x})}{3\ln 2} + C_4 - \frac{2}{\left(\frac{-1}{2}+1\right)}x^{\left(\frac{-1}{2}+1\right)} + C_5$$

$$= \frac{3}{5}x^5 + 4\ln x + \frac{5}{4}e^{-4x} + \frac{7(2^{3x})}{3\ln 2} - \frac{2}{\left(\frac{1}{2}\right)}x^{\left(\frac{1}{2}\right)} + C \quad \text{(simplifying and combining } C\text{s)}$$

$$= \frac{3}{5}x^5 + 4\ln x + \frac{5}{4}e^{-4x} + \frac{7(2^{3x})}{3\ln 2} - 4\sqrt{x} + C \quad \text{(simplifying)} \quad \square$$

Once you get familiar with the process, you could skip the first two steps (the second and third lines above) to save time and space. This means **you can omit the step of writing in a separate *C* for each integral**.

Sample Problem 6: Suppose a competitor is about to come out with a new product that you expect to have a significant impact on your business. You have estimated the probability density function for X, the number of months from now before the new product is released, to be

$$f(x) = 0.2e^{-0.2x}, \text{ for } x \geq 0$$

Find the antiderivative of $f(x)$ and interpret it.

Solution: The general form of any antiderivative of $f(x)$ is given by

$$\begin{aligned}
&\int 0.2e^{-0.2x}dx \\
&= \frac{0.2}{-0.2}e^{-0.2x} + C \qquad \left(\text{applying the rule } \int ae^{kx}dx = \frac{a}{k}e^{kx} + C\right) \\
&= -e^{-0.2x} + C \qquad \text{(simplifying)}
\end{aligned}$$

Let's call the antiderivative $F(x)$, so we know so far that $F(x) = -e^{-0.2x} + C$. Recall that one antiderivative of the velocity function $v(t)$, which we called $L(t)$, gave our location at time t from some arbitrary starting point. $L(t)$ could also be thought of as the starting location at time 0 plus the accumulated distance traveled between time 0 and time t. Analogously, one antiderivative of the density function gives the probability associated with a starting value of 0, plus the accumulated probability of values between 0 and x. In other words, since the probability associated with 0 is 0, it gives the probability that the new product is released within the next x months. To find the value of C, we then need to know the value of the antiderivative function somewhere. In this case, since we know the probability of it happening in 0 months is 0, we can say that $F(0) = 0$. Following the procedure we used earlier, then, we plug in 0 for x in the antiderivative to get

$$F(0) = -e^{-0.2(0)} + C = -e^0 + C = -1 + C$$

Now, since we also know that $F(0) = 0$, we can set the two expressions for $F(0)$ equal to get

$$-1 + C = 0 \Rightarrow C = 0 + 1 = 1$$

Thus the antiderivative is $F(x) = -e^{-0.2x} + C = -e^{-0.2x} + 1 = 1 - e^{-0.2x}$, for $x \geq 0$. This is the probability that the competitor's new product will be released within x months. □

A Word of Caution: In many problems the known value occurs where the input variable is 0, and often the value of the function is 0 there. But this is not always the case, and be aware that plugging in 0 does not always give an answer of 0, as with e^x.

Section Summary

Before trying the exercises, make sure that you

- Know that $F(x)$ is an **antiderivative of** $f(x)$ if $F'(x) = f(x)$.
- Know that the **indefinite integral** of a function, denoted $\int f(x)dx$, means the most general form for any antiderivative of $f(x)$ and so always ends with $+ C$. If $F(x)$ is *any* antiderivative of $f(x)$, then $\int f(x)dx = F(x) + C$.
- Know that an indefinite integral is a *function* and a definite integral with numerical limits of integration is a *number*.
- Know the basic antiderivative rules:

$$\int kx^n dx = \frac{k}{n+1}x^{n+1} + C = \frac{kx^{n+1}}{n+1} + C, \text{ but only for } n \neq -1$$

$$\int kx^{-1}dx = \int \frac{k}{x}dx = k(\ln x) + C$$

$$\int ae^{kx}dx = \frac{a}{k}e^{kx} + C = \frac{ae^{kx}}{k} + C$$

$$\int ab^{kx}dx = \frac{a}{k \ln b}b^{kx} + C = \frac{ab^{kx}}{k \ln b} + C$$

$$\int k\, f(x)dx = k\int f(x)dx$$

$$\int [f(x) \pm g(x)]dx = \int f(x)dx \pm \int g(x)dx$$

and how to find antiderivatives for functions involving combinations of these rules.

- Are able to find a specific antiderivative of a given function (including the value of C), given the value of the antiderivative function at a point, by plugging the known input value into the general form of the antiderivative, setting that equal to the known output value, and solving for C.

EXERCISES FOR SECTION 4.4

Warm Up

For Exercises 1–10, find the indefinite integrals.

1. $\int (x^3 + 2x)dx$

2. $\int (-2x^2 + 5)dx$

3. $\int \left(\frac{3}{x^2} + 2x^{-1} + \frac{4}{\sqrt{x}}\right)dx$

4. $\int \left(\sqrt{x^3} - \frac{4}{3x}\right)dx$

5. $\int 500e^{-0.08t}dt$

6. $\int 3000e^{0.06t}dt$

7. $\int [15(0.9)^q - 10]dq$

8. $\int [20(0.8)^q - (1.2)^q]dq$

9. $\int \left[3e^{-0.05t} + 4t^2 - \frac{5}{2t}\right]dt$

10. $\int \left[6(0.75)^q - \frac{2}{\sqrt[3]{q^4}} + 5\right]dq$

For Exercises 11 and 12, the given graph in Figure 4.4-5 is a slope graph. Roughly sketch three possibilities for the original (antiderivative) graph.

11.

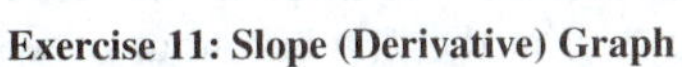

Exercise 11: Slope (Derivative) Graph

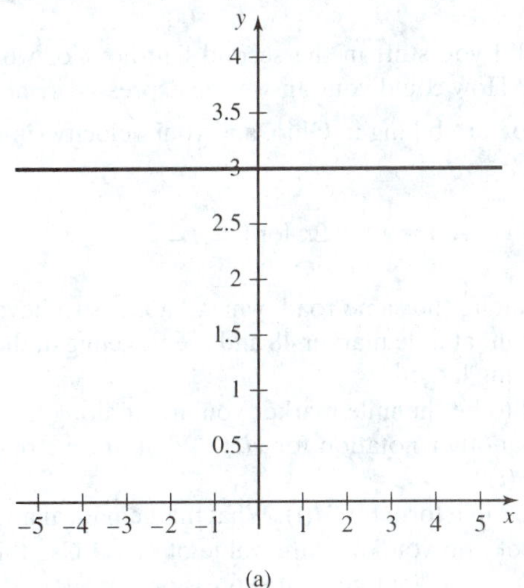

(a)

12.

Exercise 12: Slope (Derivative) Graph

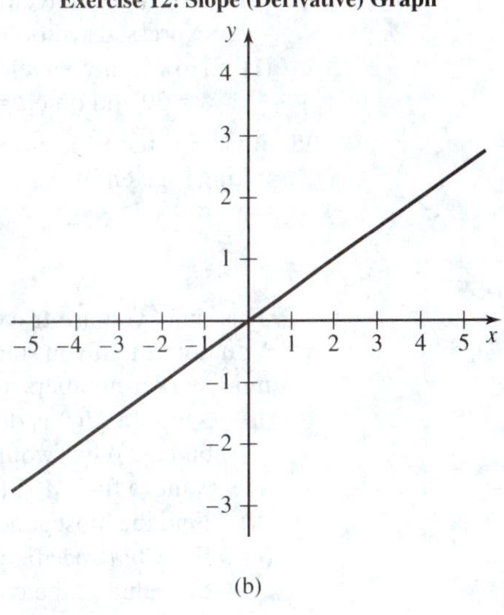

(b)

Figure 4.4-5

Game Time

13. As in Exercise 29 of Section 4.2 and Exercise 7 of Section 4.3, suppose you are biking in Ohio, and your velocity (in mph) after t hours of biking is given by

$$v(t) = 14 + 2t, \text{ for } 0 \le t \le 2$$

(a) Suppose $D(t)$ is defined to be the distance (in miles) that you have traveled on the bike after t hours of biking. What would be another notation for $v(t)$? What are we really looking for if we want to find $D(t)$ from $v(t)$?

(b) Find the most general possible form for $D(t)$. What mathematical notation expresses this?

(c) For what specific value of t do you know the value of $D(t)$? Use this known fact to find the value of the constant (C) in your general expression for $D(t)$.

(d) Now find the value of $D(2)$. What does this mean in the context of the problem?

(e) How many miles did you travel in the first hour of the trip? How could this be expressed symbolically?

(f) How many miles did you travel in the second hour of the trip (between time $t = 1$ and time $t = 2$)? How could your answer be expressed symbolically?

14. As in Exercise 30 of Section 4.2 and Exercise 8 of Section 4.3, suppose you are stuffing envelopes and your rate (in envelopes per minute) is given by the function

$$r(t) = 6 + \frac{t^2}{4050}, \text{ for } 0 \le t \le 90$$

(a) Suppose $E(t)$ is defined to be the total number of envelopes you have stuffed after t minutes of work. What would be another notation for $r(t)$, at least approximately? Mathematically, what are we really looking for if we want to find $E(t)$ from $r(t)$?

(b) Find the most general possible form for $E(t)$. What mathematical notation expresses this?

(c) For what specific value of t do you know the value of $E(t)$? Use this known fact to find the value of the constant (C) in your general expression for $E(t)$.

(d) Now find the value of $E(60)$. What does this mean in the context of the problem?
(e) How many envelopes did you stuff in the first half-hour of working? How could this be expressed symbolically?
(f) How many envelopes did you stuff in the second half-hour of working (between time $t = 30$ and time $t = 60$)? How could your answer be expressed symbolically?

15. As in Exercise 13, suppose you are biking in Ohio, and your velocity (in mph) after t hours of biking is given by

$$v(t) = 14 + 2t, \text{ for } 0 \leq t \leq 2$$

This time, you are traveling along the same road, which happens to have mile markers on it. You notice that you started right at mile marker 48 and are traveling in the direction where the mile marker numbers are getting larger.
(a) Suppose $M(t)$ is defined to be the mile marker you are at along the road after t hours of biking. What would be another notation for $v(t)$? What are we really looking for if we want to find $M(t)$ from $v(t)$?
(b) Find the most general possible form for $M(t)$. What mathematical notation expresses this?
(c) For what specific value of t do you know the value of $M(t)$? Use this known fact to find the value of the constant (C) in your general expression for $M(t)$.
(d) Now find the value of $M(2)$. What does this mean in the context of the problem?
(e) What mile marker are you at after an hour of biking? What mathematical equation would express this?
(f) How many miles did you travel in the first hour of the trip? How could this be expressed symbolically?
(g) How many miles did you travel in the second hour of the trip (between time $t = 1$ and time $t = 2$)? How could your answer be expressed symbolically?
(h) How does your answer to (g) compare to your answer in Exercise 13, part (f)? Why?
(i) How would your velocity function have been different if the mile marker numbers had been getting *smaller* in the direction you were traveling?

16. Suppose your friend Chris is helping you stuff envelopes and Chris' rate (in envelopes per minute) is given by the function

$$r(t) = 6 + \frac{t^2}{4050}, \text{ for } 0 \leq t \leq 90$$

However, this work session is actually Chris' *second* session of envelope stuffing, so t is the number of minutes after Chris started the *second* session. At the first session, Chris was able to stuff 436 envelopes (this is *all* you know about the first session).
(a) Suppose $T(t)$ is defined to be the *total* number of envelopes Chris has stuffed (including *both* the first *and* second sessions) after t minutes of work in the second session. What would be another notation for $r(t)$, at least approximately? Mathematically, what are we really looking for if we want to find $T(t)$ from $r(t)$?
(b) Find the most general possible form for $T(t)$. What mathematical notation expresses this?
(c) For what specific value of t do you know the value of $T(t)$? Use this known fact to find the value of the constant (C) in your general expression for $T(t)$.
(d) Now find the value of $T(60)$. What does this mean in the context of the problem?
(e) What was the *total* number of envelopes (from *both* sessions) Chris had stuffed after 30 minutes of stuffing in the *second* session? Express this fact symbolically.
(f) How many envelopes did Chris stuff in the *first half-hour* of working in the *second* session? How could this be expressed symbolically?
(g) How many envelopes did Chris stuff in the *second* half-hour of working in the second ses-

sion (between time $t = 30$ and time $t = 60$)? How could your answer be expressed symbolically?

(h) How does your answer to (g) compare to your answer in Exercise 14, part (f)? Why?

17. You are waiting for a bus. From past experience, you have figured that the amount of time you have to wait in minutes, X, has the following distribution:

$$f(x) = \begin{cases} 0.10e^{-0.1x}, & \text{for } x \geq 0 \\ 0, & \text{otherwise} \end{cases}$$

(a) Find $\int f(x)dx$. What specific antiderivative $F(x)$ could you define in words that would make sense for this problem?

(b) For what specific value of x do you know the value of $F(x)$? Use this value to find the value of C in your antiderivative (to get the precise form for $F(x)$).

(c) What is the probability that you will have to wait 4 minutes or less for the bus?

(d) You are taking the bus to catch a train. You know that if you have to wait 8 minutes or more, you will miss the train. What mathematical expression will correspond to the probability that you will catch the train? What is the numerical value of this probability?

(e) What is the probability that you will have to wait between 4 minutes and 8 minutes for the bus? Express this quantity symbolically.

18. You are in the middle of a lawsuit. You have had some experience with similar suits and estimate that the eventual court settlement will result in a net benefit to you of X hundred thousand dollars, with a probability density function given by

$$f(x) = \begin{cases} -0.0208x^2 + 0.0417x + 0.208, & \text{for } -2 \leq x \leq 4 \\ 0, & \text{otherwise} \end{cases}$$

(a) Find $\int f(x)dx$. What specific antiderivative $F(x)$ could you define in words that would make sense for this problem?

(b) For what specific value of x do you know the value of $F(x)$? Use this value to find the value of C in your antiderivative (to get the precise form for $F(x)$).

(c) What is the probability that you will receive a net benefit from the lawsuit of \$300,000 or less?

(d) What is the probability that the lawsuit will result in a net loss for you? What mathematical expression will correspond to this probability?

(e) What is the probability that your net benefit from the lawsuit will be between \$0 and \$300,000? Express this quantity symbolically.

(f) What *other* specific value of $F(x)$ do you know? Does your specific antiderivative found in (b) give this answer as it should (to 3 significant figures)?

(g) Try redoing part (b), ignoring everything you did after that, using the *other* known value of $F(x)$ (your first answer to part (f)) to get the value of C and the exact form for $F(x)$ from scratch. Do you get the same answer? (This is *verification*!)

4.5 THE FUNDAMENTAL THEOREM OF CALCULUS

We have already seen how we can recover an original function from its derivative by the process of finding an antiderivative and using a known value of the original function to find the constant C. This is done by substituting the known value of the independent variable (such as x or t) into the most general antiderivative, setting this equal to the known value of the original function at that point, and solving for C. With the highway driving problem we have even done the operation we will focus on in this section: finding the net change in the original function over an interval. We did this in purest form when considering the antiderivative of the velocity function to be the location function. The answer turned out to be the value of the antiderivative (the location) at the *end* of the time period (right endpoint of the interval) minus the value of the antiderivative at the *beginning* of the time period (left endpoint of the interval). This process represents one form of what is called the fundamental theorem of calculus. We will see in this section that the same procedure can be used to find *any* area or definite integral, as long as we can find an antiderivative. We will also see that derivatives and integrals have a special relationship to each other that is almost a perfect inverse relationship.

After studying this section, you should be able to solve problems like the following:

- You have been driving on the highway for several hours and have a pretty good idea of what your velocity has been the whole time (it has not been constant, but has changed in a steady way). However, you forgot to set your trip odometer (and didn't notice your regular odometer) to know how many miles you have traveled. How can you figure out the distance traveled just knowing your velocity?
- Given the rate of change with respect to time of a revenue function (such as a curve fit to weekly sales over the last year), approximate the sales revenue over a particular interval of time (such as the first quarter of this year).
- Based on past performance and trends, you have estimated the probability distribution for your profit this year. How can you approximate the probability that your profit will fall within a given interval of values?
- A competitor business is about to come out with a new product that will affect your business significantly. To assess your risk and plan ahead, you want to be able to estimate the probability that the new product will be released within a specified time window.

In addition to being able to solve problems like those above, after studying this section you should also

- Know that a definite integral of a function can be evaluated by finding any antiderivative of the function, plugging in the endpoints (limits of integration), and calculating the value at the upper limit minus the value at the lower limit.
- Understand why you can use *any* antiderivative for this calculation (any value of C, including 0, or a general C).
- Understand why the above calculation works when the function *is* a derivative of a function with a practical real-world meaning.
- Understand why the above calculation works when the function is *not* a derivative of a function with a practical real-world meaning.
- Understand that the derivative of the indefinite integral of a function is that function itself, so in that sense, derivatives and integrals are like inverses.
- Know what is meant by the Fundamental Theorem of Calculus, and why it is important.

Net Change of a Function over an Interval

Sample Problem 1: You have been driving on the highway for 4 hours straight and want to estimate how far you have traveled, because you forgot to set the trip odometer (and didn't notice the regular odometer) when you left. You estimate that your speed has been gradually increasing: you started out at about 56 mph and your speed has inched up steadily, about 2 mph each hour. How far have you traveled?

Solution: In case you forgot, we already solved this problem in Section 4.4! Recall that we formulated the velocity function $v(t) = 2t + 56$ and said that this is the derivative of the location function $L(t)$, so that $L'(t) = v(t) = 2t + 56$. We then used the idea of finding the antiderivative, or **antidifferentiation**, to find or recover $L(t)$ from its derivative. Finally, since $L(0)$ is the location at the beginning of the highway segment of the trip and $L(4)$ is the location after 4 hours, your distance traveled, the net change in your location on the interval from time 0 to time 4, is $L(4) - L(0)$. The net change in location is the same as the distance in this example because we were always moving in the same direction (we never backed up).

Recall that when we found the general form of an antiderivative of $2t + 56$ (general form of $L(t)$), we obtained

$$L(t) = t^2 + 56t + C$$

In the example, we saw that C corresponds to the value of $L(t)$ when $t = 0$, since plugging in $t = 0$ yields $L(0) = C$. This means that C can be thought of as the initial location. In the mile marker form of the problem, we said this was 70, so $L(t) = t^2 + 56t + 70$. Then

$$\begin{aligned} L(4) - L(0) &= [(4)^2 + 56(4) + 70] - [(0)^2 + 56(0) + 70] \\ &= [16 + 224 + 70] - [0 + 0 + 70] \\ &= 16 + 224 + 70 - 70 \\ &= 16 + 224 = 240 \quad \square \end{aligned}$$

Notice that in this calculation the value of 70 canceled out between the two expressions in brackets. What if we had used the general form of $L(t)$? Then

$$\begin{aligned} L(4) - L(0) &= [(4)^2 + 56(4) + C] - [(0)^2 + 56(0) + C] \\ &= [16 + 224 + C] - [0 + 0 + C] \\ &= 16 + 224 + C - C \\ &= 16 + 224 = 240 \end{aligned}$$

The important point here is that when finding the net change of the original function (antiderivative) over an interval using its derivative, you *do not need to find the value of the constant C*, which is always part of the general form of an antiderivative, since it will get canceled out in the calculation.

Thus, if we are given the derivative $f'(x)$ and are looking for the net change in the original function $f(x)$ on the interval $[a, b]$ (between $x = a$ and $x = b$), we can simply find the original function $f(x)$ by finding the antiderivative of $f'(x)$ and the value of C as we discussed in Section 4.4, then calculate the change in the original function over the interval, $f(b) - f(a)$. **Alternatively, given the derivative $f'(x)$, to find the net change in the original function $f(x)$ on the interval $[a, b]$ (between $x = a$ and $x = b$), we can find *any* antiderivative of $f'(x)$ (such as using a general C, or setting $C = 0$), and take *its* value at b minus its value at a. In general, the latter (alternative) method is quicker and easier.**

At this point, we should mention another notational convention in calculus. When doing a calculation like $L(4) - L(0)$ (that is, evaluating a function at the right endpoint of an interval and subtracting the value at the left endpoint), it is common to use the notation

$$[L(t)]_0^4 = L(4) - L(0)$$

In general, we would write

$$[f(x)]_a^b = f(b) - f(a)$$

and would read the left-hand side as "eff of ex, evaluated from a to b."

SAMPLE PROBLEM 2: You are deciding whether or not to buy the rights to an oil well. You estimate that the *weekly* profit (in $1000's) will gradually decline over time as follows:

Time	Now	In 1 yr.	In 2 yrs.	In 3 yrs.	In 4 yrs.
Weekly Profit ($1000's)	60	45	33	25	19

What is your estimate of the total accumulated dollar value (not adjusted for inflation or the time value of money) of the profit from this oil well over the next 20 years?

Solution: Let's define

t = years from the present time and

$P(t)$ = total accumulated dollar value of the profit from this oil well from time 0 (the present time) to time t (t years from now)

$P'(t)$ would represent the instantaneous rate of change of the profit with respect to time, in units of thousands of dollars per year. If we are trying to calculate the total accumulated dollar value over a period of years, we will want to multiply rate times time as we did for distance so that the time units cancel. Since our input variable is defined in units of years, this means that we would like our rate to be in thousands of dollars per year. Thus the table is really giving us values that are directly related to $P'(t)$ but are given in units of thousands of dollars per **week** instead of thousands of dollars per **year**. Since $P'(t)$ is the **instantaneous** rate of change at time t, we can get a good approximation of this by simply converting the values in the table from units of thousands of dollars per week to units of thousands of dollars per year.

This conversion would be to multiply the values in the table by 52, since there are (approximately) 52 weeks in a year, so the rate per year will be 52 times the rate per week. For example, right now the estimated weekly profit is 60 thousand dollars per week. Thus we want to express 60 thousand dollars per *week* in thousands of dollars per *year*. **As with any conversion, we can simply multiply by fractions that are equal to 1 in a way that cancels the units we want to get rid of and keeps the units we want:**

$$\frac{60 \text{ thousand dollars}}{1 \text{ week}} \cdot \frac{52 \text{ weeks}}{1 \text{ year}} = (60)(52) \text{ thousand dollars/year}$$

$$= 3120 \text{ thousand dollars per year}$$

In this case we wanted to cancel out the weeks and replace them with years, so we put weeks in the numerator of the conversion fraction. Since there are 52 weeks in 1 year, the value of the fraction is 1, so we are not changing the original quantity.

Our table then becomes

Time t	0	1	2	3	4
Profit per Year ($1000's), $P'(t)$	3120	2340	1716	1300	988

The data plot is shown in Figure 4.5-1. This curve is clearly concave up over the entire interval, so our first candidates for a model would be quadratic and exponential functions. The second differences are 156, 208, 104, and the % change values are −25, −26.67, −24.24, −24. Clearly, the % change values are much more stable, which suggests that an exponential model should fit much better. Furthermore, we are going to be **extrapolating** far beyond the limits of our data points (estimates) in this problem, so we need a model whose behavior over the interval [0, 20] (the next 20 years) will fit our common sense and best intuition/guess about what will happen in the long term. A quadratic model would decline for a while, then turn back up at some point. Unfortunately, this is not what we would expect from an oil well. Thus the exponential model would seem best in every way.

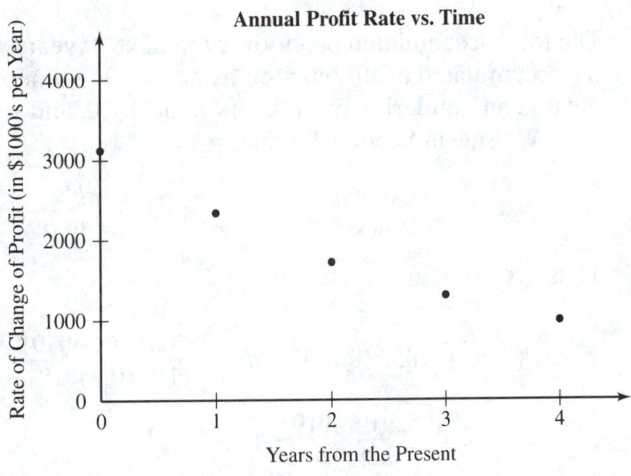

Figure 4.5-1

A best-fit exponential model is

Verbal Definition: $P'(t)$ = the instantaneous rate of change of profit (in thousands of dollars per year) t years from the present time.

Symbol Definition: $P'(t) = 3105.5(0.74919)^t$ for $t \geq 0$.
Assumptions: Certainty and Divisibility. Certainty implies that the relationship is exact. Divisibility implies that the years and dollars can have any fractional value.

As a quick check that this model makes sense, notice that $P'(0)$ would be 3105.5, compared to the data value of 3120, so the fit seems reasonable. The plot of the model with the data is shown in Figure 4.5-2. Clearly, the fit is excellent!

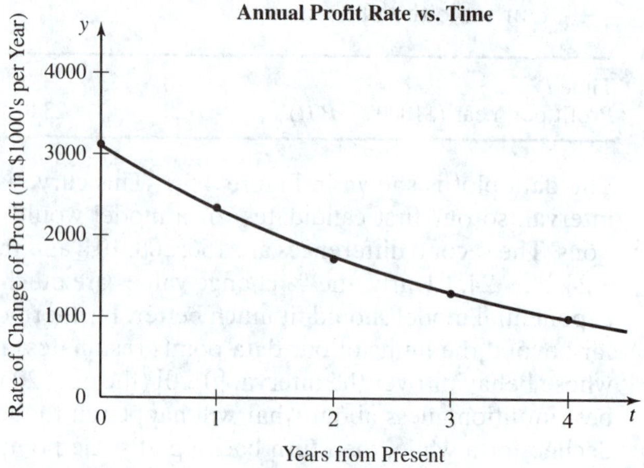

Figure 4.5-2

The total accumulated profit over the next 20 years will be the same as the net change in the accumulated profit function from time 0 to time 20. Thus, for our answer, we can simply find an antiderivative, take its value at 20, and subtract its value at 0.

We saw in Section 4.4 that

$$\int ab^{kx} = \frac{ab^{kx}}{k \ln b} + C$$

Thus, we see that

$$P(t) = \int 3105.5(0.74919)^t dt = \frac{3105.5(0.74919)^t}{(1)\ln(0.74919)} + C \qquad \text{(here, } k = 1\text{)}$$

$$\approx \frac{3105.5(0.74919)^t}{-0.28876} + C$$

$$\approx (-10{,}755)(0.74919^t) + C$$

As we observed before, to find the net change in $P(t)$ from 0 to 20, we do not need to find the constant C, since in calculating $[P(20) - P(0)]$ the Cs will cancel out. Thus the total accumulated profit over the 20 years will be

$$
\begin{aligned}
[P(t)]_0^{20} &= P(20) - P(0) = [(-10{,}755)(0.74919^{(20)}) + C] - [(-10{,}755)(0.74919^{(0)}) + C] \\
&\approx [(-10{,}755)(0.0031034) + C] - [(-10{,}755)(1) + C] \\
&\approx [-33.377 + C] - [-10{,}755 + C] \\
&= -33.377 + C + 10{,}755 - C = 10{,}722 + 0 = 10{,}722
\end{aligned}
$$

Translation: The accumulated total dollar value of the profit from the oil well over the 20 years will be approximately $10,722,000.

Notice that when we substituted 0 into the antiderivative, the answer was *not* 0. When integrating polynomials, you will get 0's, but don't fall into the trap of *assuming* the answer will be 0. Always do the substitution to see what the value should be.

Just to check, let's see what would have happened if we had found the value of the constant C. To find it, we need a value of t where we know the value of $P(t)$. As has been true earlier (and is often, but *not always* the case) we know the accumulated profit when $t = 0$, and the value will be 0 (nothing has accumulated yet at the very beginning of the time period being studied). Thus $P(0) = 0$. So, once again, we plug in 0 for t in the general antiderivative:

$$P(0) = (-10{,}755)(0.74919^0) + C = (-10{,}755)(1) + C = -10{,}755 + C$$

We know that $P(0)$ should also equal 0, so we set the above expression equal to 0 and solve for C:

$$(-10{,}755) + C = 0 \Rightarrow C = 10{,}755$$

Thus our actual accumulated profit function is

$$P(t) = (-10{,}755)(0.74919^t) + 10{,}755$$

The net change in accumulated profit over the 20 years will then be

$$
\begin{aligned}
[P(t)]_0^{20} &= P(20) - P(0) = [(-10{,}755)(0.74919^{(20)}) + 10{,}755] - [(-10{,}755)(0.74919^{(0)}) + 10{,}755] \\
&\approx [(-10{,}755)(0.0031034) + 10{,}755] - [(-10{,}755)(1) + 10{,}755] \\
&\approx [-33.377 + 10{,}755] - [-10{,}755 + 10{,}755] \\
&= -33.377 + 10{,}755 + 10{,}755 - 10{,}755 = 10{,}722 + 0 = 10{,}722
\end{aligned}
$$

This is the same answer we got before. As discussed earlier, the value of C (10,755) just cancels out in the subtraction so was not needed for calculating the net change in accumulated profit over the interval. It *would* be needed, however, if you wanted to directly estimate the value of $P(t)$ at a particular point in time. For example, the accumulated profit through time 10 (after 10 years) would be

$$P(10) = (-10{,}755)(0.74919^{(10)}) + 10{,}755 = 10{,}156$$

This means the accumulated dollar value of the profit from the oil well over the first 10 years is estimated to be $10,156,000. Notice that this means not much profit would come in during the last 10 years (only $10,722,000 − $10,156,000 = $566,000)! As a double-check (verification) of this calculation, we could also calculate

$$[P(t)]_{10}^{20} = P(20) - P(10) = [(-10{,}755)(0.74919^{(20)}) + C] - [(-10{,}755)(0.74919^{(10)}) + C]$$
$$\approx [(-10{,}755)(0.0031034) + C] - [(-10{,}755)(0.055708) + C]$$
$$\approx [-33.377 + C] - [-599.14 + C]$$
$$= -33.377 + C + 599.14 - C = 565.763 + 0 \approx 566$$

This verifies our earlier calculation. □

Interpretation of the Net Change in the Antiderivative of a Probability Density Function

Sample Problem 3: You estimate that the number of months before a competitor will come out with a new product (X) can be modeled with the following exponential probability density function:

$$f(x) = 0.2e^{-0.2x}, \text{ for } x \geq 0$$

Find the probability that the new product will come out in 3–6 months.

Solution: The random variable here has been implicitly defined as

X = the number of months from the present before the competitor releases the new product

The given probability density function is a special case of a general category of random variables which are said to have an **exponential distribution**. The general form for these random variables is given by

$$f(x) = ke^{-kx}, \text{ for } x \geq 0$$

We mentioned in Sections 4.1 and 4.4 that in general a probability density function $f(x)$ can be thought of as the probability *rate* per unit of x, which means it is a rate of change, or derivative.

★ **Discovery Question 1:**

In the context of the other problems we have discussed in this section, what would an **antiderivative** of the density function in this problem represent?

Hint: The antiderivatives in the first two examples were interpreted as accumulated distance and accumulated profit.

Answer: An antiderivative of the probability density function can be thought of as an accumulated or **cumulative probability** that the product will be released up to some point in time. ★

In the example, we are looking for the probability that the competitor will release the new product in 3–6 months, which means we are looking for the probability that X is between 3 and 6. Analogous to the earlier problems, this is $F(6) - F(3)$, the accumulated probability up to 6 months minus the accumulated probability up to 3 months. Technically, this would give us the probability for (3, 6] (not including 3), since the probability at 3 is being subtracted as well, but as we discussed earlier, the probability at a particular point (like 3) for a continuous random variable is 0, so the calculated probability will be correct.

Now, to find $F(x)$, we simply take the antiderivative of $f(x)$:

$$\int 0.2e^{-0.2x}dx = \frac{0.2e^{-0.2x}}{-0.2} + C = -e^{-0.2x} + C$$

Remember that to calculate the probability we want, $F(6) - F(3)$, we do not need to *calculate* the value of the constant C, so we can do the calculation immediately:

$$\begin{aligned} P\{3 \le X \le 6\} &= [F(x)]_3^6 = [-e^{-0.2(6)} + C] - [-e^{-0.2(3)} + C] \\ &= [-e^{-1.2} + C] - [-e^{-0.6} + C] = -e^{-1.2} + C + e^{-0.6} - C \\ &= -e^{-1.2} + e^{-0.6} + 0 = -0.30119 + 0.54881 = 0.24762 \end{aligned}$$

Translation: There is an approximately 25% chance of the competitor's new product coming out in 3–6 months. □

In general, what would an antiderivative of *any* probability density function represent? An antiderivative of any probability density function can be thought of as an accumulated or cumulative probability. The only question then, as in the car distance problem (Sample Problem 1) is, what is your starting point for accumulating?

In Sample Problem 3, we did not specify *where* we *started* accumulating the probability. The standard practice, however, is to use a starting point that would make sense for *any* problem involving that density function. This starting point is the leftmost possible x-value, which can always be considered to be negative infinity. For this problem, we could also think of 0 as the starting point, since negative values are not possible; however, this depends on the problem, so is not general. Thus we can define the antiderivative to be

$F(x)$ = the cumulative probability at x, or the probability that the random variable is less than or equal to x, or the probability of being in the interval $(-\infty, x]$, which is $P[X \le x]$.

The Fundamental Theorem of Calculus

Remember that in Section 4.2, we showed that the net change in the original (antiderivative) function corresponded to the area between the derivative and the x-axis (above – below). In Section 4.3 we introduced the definite integral, which corresponds to this area.

You may have thought about the fact that we have discussed two completely different ways of calculating a probability like the one in Sample Problem 3: limits of sums (definite integral) and taking the value of an antiderivative at the right endpoint minus the value at the left endpoint. The connection between these two calculations is what is known as the **Fundamental Theorem of (Integral) Calculus**. It says that

If $F(x)$ is *any* antiderivative of $f(x)$ (so $F'(x) = f(x)$), then

$$\int_a^b f(x)dx = [F(x)]_a^b = F(b) - F(a)$$

Since the definite integral corresponds to the area above the x-axis minus the area below the x-axis between $f(x)$ and the x-axis over the interval $[a, b]$, this result is saying that we can calculate *any* such area simply by finding *any* antiderivative of $f(x)$ (such as the simplest possible one, with $C = 0$), and subtracting: the value at b minus the value at a.

So far, we have only shown that this relationship is true when $f(x)$ is a derivative function. But, faced with *any* definite integral $\int_a^b f(x)dx$, we could always think to ourselves:

> *Suppose* $f(x)$ were a velocity function. Then I *know* I could use the above Fundamental Theorem of Calculus to find the net change in the original location function (net distance traveled), which would numerically be the value of the definite integral. In this case (the generic definite integral), I also want that same numerical value for the definite integral, so the same procedure will give me the correct answer.

This means that the Fundamental Theorem of Calculus can be used to find *any* definite integral, because we can always think of the function being integrated *as if* it were a derivative of another function, even if we have no idea what the meaning of that antiderivative function would be.

Thus, **the Fundamental Theorem of Calculus always holds, even when the antiderivative used has no clear interpretation, such as a purely mathematical calculation of the area under a curve given by $y = f(x)$.**

Integration and Differentiation as Inverses

In some calculus books, you will see another part for the Fundamental Theorem of Calculus. We will explore later the most common notation used for this, but at this point we can easily give the gist of this other part, via the following problem:

SAMPLE PROBLEM 4: What is $D_x(\int f(x)dx)$?

Solution: If $F(x)$ is any antiderivative of $f(x)$, this means that $F'(x) = f(x)$. Furthermore, we could express the indefinite integral as

$$\int f(x)dx = F(x) + C$$

since any two antiderivatives differ only by a constant. Thus, taking the derivative of both sides, we get:

$$\begin{aligned} D_x(\int f(x)dx) &= D_x[F(x) + C] \\ &= F'(x) + 0 \\ &= f(x) \end{aligned}$$

In words: taking the derivative of the indefinite integral of a function results in that function itself. In this way, these two operations on a function are like *inverses* of each other: One undoes the other, like adding 2 and subtracting 2, or like cubing and finding the cube root. □

Let's now consider the problem in the other direction.

Sample Problem 5: In Sample Problem 4 we took the derivative of an integral of a function, $D_x(\int f(x)dx)$, and found that it was the function $f(x)$. What happens when we take the integral of the derivative of a function?

Solution: We know that

$$D_x[f(x)] = f'(x)$$

so taking the integral of both sides we have

$$\int D_x[f(x)]dx = \int f'(x)dx$$

$f'(x)$ is the derivative of $f(x)$, so $f(x)$ is *an* antiderivative of $f'(x)$.

Clearly, $f(x)$ is an antiderivative of $f'(x)$, but the indefinite integral means the most *general* form for *any* antiderivative, so we need to add the $+ C$:

$$\int f'(x)dx = f(x) + C$$

Taking the derivative of the indefinite integral results in the original function. So does taking the indefinite integral of the derivative of a function, but this time the $+ C$ is tacked on, so it is not a perfect inverse relationship. □

Net Change in a Function: The Limit of Sums and the Antiderivative

Finding that the net change in the original function corresponds to the area under its derivative over the same interval makes sense from a different perspective as well. Let's consider units. For example, in the car travel problem, the units of the derivative are miles/hour, and the units of the independent variable are hours. If you think of the rectangle area calculations (base time height), the units will be

$$\frac{\text{miles}}{\text{hour}} \cdot (\text{hours})$$

or just miles, which are the units we want for the net change in location (distance traveled). Recall from high school the old formula "distance equals rate times time."

Because of the fact that the net change calculation corresponds to a definite integral, which can always be interpreted as an area, we have another way to verify our calculations in the car travel problem. Since the velocity graph is a straight line, the area under it is the area of a trapezoid, and we have already mentioned that

$$\text{the area of a trapezoid} = \left[\frac{\text{base 1} + \text{base 2}}{2}\right](\text{altitude})$$

Again, the trapezoid is sideways from the way we are used to, so the "bases" (the parallel sides) are really the vertical sides, which are the heights of the function at the endpoints, $v(0)$ and $v(4)$. The "altitude" (the height of the trapezoid, which is the perpendicular distance between the parallel sides) is the width of the interval, 4–0. The graph is shown in Figure 4.5-3.

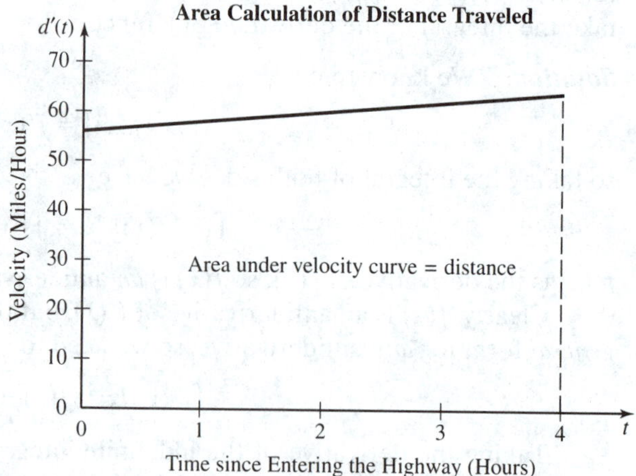

Figure 4.5-3

and the area calculation is

$$\left[\frac{b_1 + b_2}{2}\right](h) = \left[\frac{v(0) + v(4)}{2}\right](4 - 0)$$
$$= \left[\frac{56 + 64}{2}\right](4) = \left[\frac{120}{2}\right](4)$$
$$= [60](4) = 240$$

This confirms our earlier calculations that the total distance traveled is 240 miles.

Having solved problems both ways, you can appreciate the fact that the solution using an antiderivative is much faster and easier than the limit of sums! So, whenever possible, it makes sense to do this. Also, for more advanced problems where the function under which you want the area has symbolic parameters (letters) in it (such as when finding the area under $f(x) = ax^2 + bx + c$ in general; the a, b, and c values are symbolic parameters), the only way to do it is with antiderivatives. However, in most practical examples, the limit of sums (numerical) approach can usually be done for you automatically using technology. All you need to do is input the function and the endpoints of the interval. As a result, both methods are important to know.

Recall that, like $F(x)$, the indefinite integral also represents an antiderivative of a function (in fact, the most general form for any antiderivative). Using this notation, the Fundamental Theorem of Calculus then says

$$\int_a^b f(x)dx = [\int f(x)dx]_a^b$$

which has a certain natural and logical simplicity and shows why the two notations fit together nicely.

Derivative Functions that are Negative Over an Interval: Net Change versus Total Change

Sample Problem 6: Suppose you are playing basketball and that you have the ball at center court and start toward the basket with a defender covering you. You move straight toward the basket with velocity

$$v(t) = 10 - 2t \text{ feet per second, for } 0 \le t \le 8$$

t seconds after starting to move. Explain in words and graphically the nature of your movement. After 8 seconds, how far are you from where you started (at $t = 0$)?

Solution: At the very beginning ($t = 0$), you are moving forward (toward the basket) at

$$v(0) = 10 - 2(0) = 10 \text{ feet per second}$$

A second later ($t = 1$), your velocity is $v(1) = 10 - 2(1) = 8$ ft/sec. Your velocity continues to slow until $t = 5$, when it becomes $v(5) = 0$, so you are momentarily motionless. Then at $t = 6$, $v(6) = -2$ ft/sec, which means you are now moving *backward* along the same line (for example, backing up or turning around), retracing where you have already been at a speed of 2 ft/sec. Finally, after the 8 seconds, you are moving at $v(8) = -6$ ft/sec (even faster backward). The graph of $v(t)$ is shown in Figure 4.5-4.

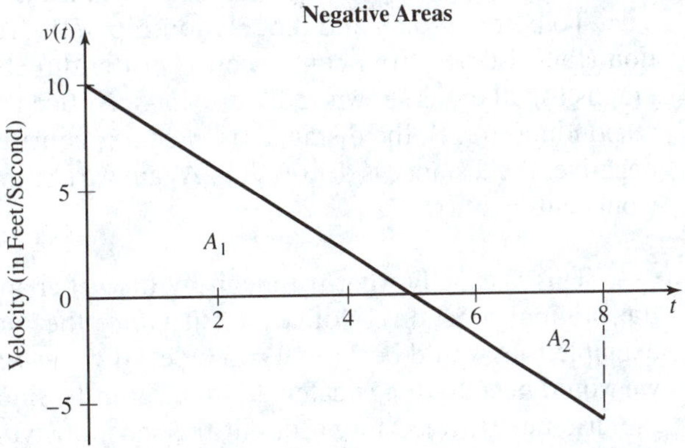

Figure 4.5-4

From our earlier discussions, we know that $\int_0^5 v(t)dt$ corresponds to the distance moved forward, before starting backward. Numerically, this works out to be

$$\begin{aligned}\int_0^5 v(t)dt &= \int_0^5 (10 - 2t)dt \\ &= \left[10t - \frac{2t^2}{2}\right]_0^5 \\ &= [10t - t^2]_0^5 \\ &= [10(5) - (5)^2] - [10(0) - (0)^2] \\ &= [50 - 25] - [0 - 0] \\ &= 25\end{aligned}$$

In other words, you moved 25 feet forward before starting backward. If we calculate $\int_5^8 v(t)dt$, it should be related to the distance you moved backward. Let's do the calculation:

$$\int_5^8 v(t)dt = \int_5^8 (10 - 2t)dt = \left[10t - \frac{2t^2}{2}\right]_5^8$$

$$= [10t - t^2]_5^8 = [10(8) - (8)^2] - [10(5) - (5)^2]$$

$$= [80 - 64] - [50 - 25] = 16 - 25 = -9$$

This means you moved backward 9 feet (the negative sign is what tells you that the direction was backward, as was true for the velocity).

Thus, you moved forward 25 feet, then retraced backward 9 feet, so after the 8 seconds, you end up $(25 - 9) = 16$ feet from where you started. This is the same result that we get if we take the definite integral over the entire interval:

$$\int_0^8 v(t)dt = \int_5^8 (10 - 2t)dt = \left[10t - \frac{2t^2}{2}\right]_0^8 = [10t - t^2]_0^8$$

$$= [10(8) - (8)^2] - [10(0) - (0)^2] = [80 - 64] - [0 - 0] = 16$$

So the net change in your location or position was 16 feet.

To better understand the relationship of the velocity function and the location function (the distance from center court, in the direction of the basket), let's look at both graphs together, as shown in Figure 4.5-5. Notice that where the velocity is positive (corresponding to A_1), the distance from center court is increasing, and when the velocity is negative, the distance is decreasing. Again we see that, after the 8 seconds, you are 16 feet from center court. □

This is exactly what we mean by the *net* change in the location (the net change in the original function). Notice that it is *not* the *total* distance traveled, because in this example that would be $25 + 9 = 34$ feet. If we *wanted* to find the total distance traveled, we would need to do the calculation differently, since the first integral we calculated gave the distance traveled forward, but the second gave the *negative* of the distance traveled backward. Thus, to find the total area, we need to change the sign of the second integral (subtract it), $\int_0^5 v(t)dt - \int_5^8 v(t)dt$.

In general, **to find the *total change* of the original function (or the *total area*, treating all areas as positive and adding them)**, you need to determine where the function being integrated is positive and where it is negative. Normally this will change where that function crosses the horizontal axis, which is where it is 0, so you can find the breakpoints for your limits of integration by finding the zeroes of the function. You then add up all the positive integrals and subtract the negative ones (or **add the absolute values of the individual pieces**).

Sample Problem 7: From records of the last few years, you have estimated that the instantaneous rate of change of your cumulative profit is given by the model:

Verbal Definition: $p(t) = P'(t)$ = the instantaneous rate of change of your cumulative profit (in millions of dollars per year) t years after the beginning of 1990.

Symbol Definition: $p(t) = -0.194t^3 + 1.95t^2 - 2.71t - 2$, for $0 \le t \le 8$.

Assumptions: Certainty and divisibility. Certainty implies that the relationship is exact. Divisibility implies that the profit and time can take on any fractional values.

(a) Explain in words what has been happening to your profit since the beginning of 1990.

(b) What would the definite integral of $P'(t)$ from 0 to 6 tell you?

(c) What is your projection of your total accumulated profit for the next 2 years (from the beginning of 1996 to the beginning of 1998)?

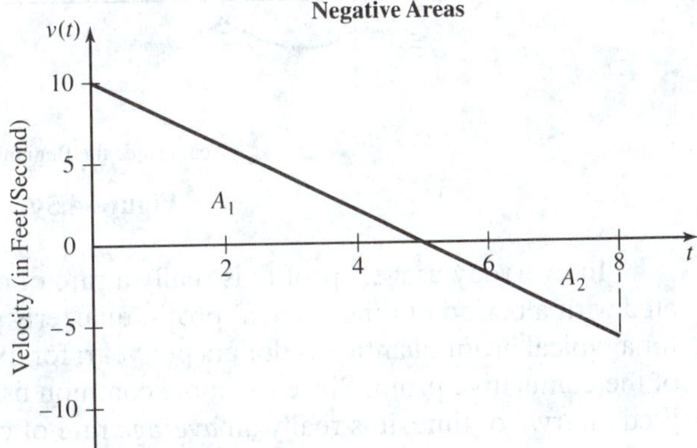

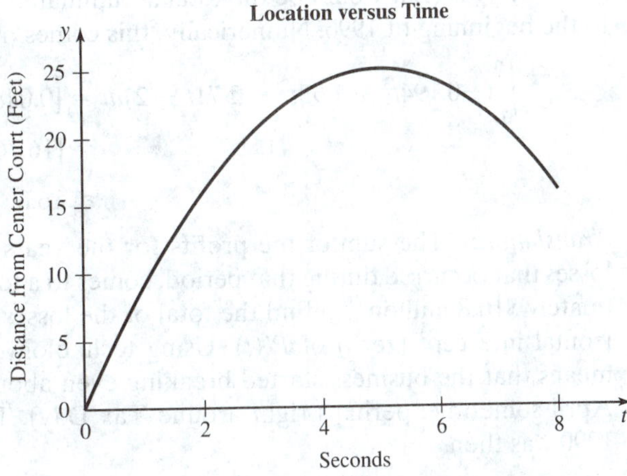

Figure 4.5-5

Solution: **(a)** The graph of the derivative of the cumulative profit function is shown in Figure 4.5-6. This says that at the beginning of 1990, the business was losing money (the rate of change of the cumulative profit was negative), and the loss rate got worse until around the beginning of 1991. Then the loss rate started to decrease until the business started to break even (moved into the black to start making a profit) sometime during

1992. The rate of increase in the profit continued to increase, until it peaked around the beginning of 1996.

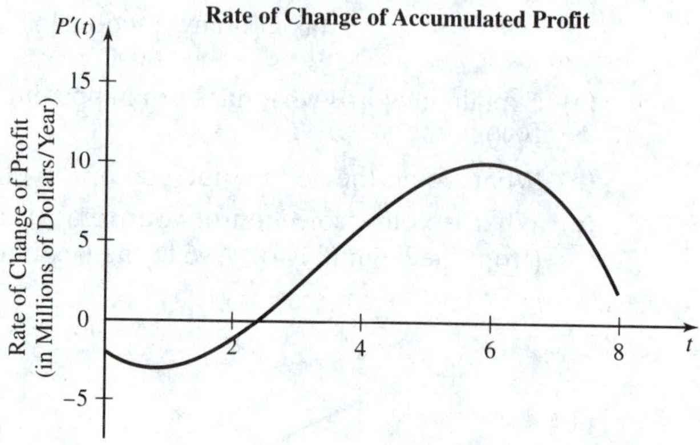

Figure 4.5-6

In everyday usage, "profit" is really a rate of change, because it is normally associated with a period of time (annual profit, quarterly profit, etc.). We could state the units for a typical profit quantity as dollars per year, for example, so it is really a rate of change of the cumulative profit. Since the most common usage just totals the profit over a specified interval of time, it is really an *average* rate of change over that interval. A negative profit means that our cumulative profit is going down, so corresponds to a loss (when costs exceeds revenues).

(b) $\int_0^6 P'(t)dt$ would be the total accumulated net profit from the beginning of 1990 to the beginning of 1996. Numerically, this comes out to

$$\int_0^6 (-0.194t^3 + 1.95t^2 - 2.71t - 2)dt = [0.0485t^4 + 0.65t^3 - 1.355t^2 - 2t]_0^6$$
$$= [16.764 - 0]$$
$$\approx 16.8$$

Translation: The sum of the profits for the years from 1990 through 1995, minus the losses that occurred during that period, comes to a total accumulated net profit of approximately \$16.8 million. To find the total of the losses, we would first have to find the horizontal intercept (zero) of $P'(t)$. Using technology, this comes to about $t = 2.39$. This means that the business started breaking even about one-third of the way into 1992 (in April sometime, perhaps right around Tax Day). The total loss since the beginning of 1990 was then

$$\int_0^{2.39} P'(t)dt \approx -5.23$$

Translation: You sustained a total loss of \$5.23 million from the beginning of 1990 through mid-April 1992. This then means that the total accumulated profit from the break-even point through the beginning of 1996 must have been about

$$16.8 + 5.2 = 22.0 = \int_{2.39}^6 P'(t)dt$$

(c) To estimate profits for the next two years, we could use technology to find

$$\int_6^8 P'(t)dt \approx 14.7$$

Translation: The *total* profit for 1996 and 1997 we would project to be about $14.7 million. If we wanted to get the profit for each year individually, we could calculate the corresponding two integrals:

$$\int_6^7 P'(t)dt \approx 9.34 \text{ and } \int_7^8 P'(t)dt \approx 5.32$$

These sum to 14.66, which rounds off to 14.7 (million). □

Section Summary

Before trying the exercises, you should:

- Know that a definite integral of a function can be evaluated by finding *any* antiderivative of the function, plugging in the endpoints (limits of integration), and calculating the value at the upper limit minus the value at the lower limit. In symbols, if $F'(x) = f(x)$ (if $F(x)$ is *any* antiderivative of $f(x)$), then

$$\int_a^b f(x)dx = [F(x)]_a^b = F(b) - F(a)$$

- Understand why you can use *any* antiderivative for this calculation (any specific value of C, including 0, or a general C), since the Cs cancel out.
- Understand why the above calculation works when the function $f(x)$ *is* a derivative of a real-world function (the definite integral corresponds to the area, which is also the net change in the original, or antiderivative, function).
- Understand why the above calculation works when the function is *not* a derivative of some known function (the definite integral always corresponds to an area, and *if* the function being integrated *were* a velocity function, then the result would hold, so the result holds whether it *is* or not, since it *could* be).
- Understand that the derivative of the indefinite integral of a function is that function itself, $D_x(\int f(x)dx) = f(x)$, so in that sense, differentiation and integration act like inverse operations. However, $\int f'(x)dx = f(x) + C$, so the inverse relationship is not quite perfect.
- Know that the above relationships are different forms of what is called the Fundamental Theorem of Calculus, which is important because it provides a possible quick and exact method for evaluating areas and general definite integrals, and because it communicates the inverse relationship between derivatives and integrals.
- Know that when a smooth function crosses below the horizontal axis, the definite integral corresponds to the area *above* the axis *minus* the area *below* the axis, considering only areas between the axis and the curve within the interval defined

by the limits of integration. This makes intuitive sense, since all of the subinterval formulas involve output values of the function, so will be negative where the curve is below the axis.

EXERCISES FOR SECTION 4.5

Warm Up

For Exercises 1–10, evaluate the following definite integrals using the Fundamental Theorem of Calculus:

1. $\int_0^3 (x^3 + 2x)dx$

2. $\int_2^4 (-2x^2 + 5)dx$

3. $\int_1^2 \left(\frac{3}{x^2} + 2x^{-1} + \frac{4}{\sqrt{x}}\right)dx$

4. $\int_2^3 \left(\sqrt{x^3} - \frac{4}{3x}\right)dx$

5. $\int_0^{10} 500e^{-0.08t}dt$

6. $\int_0^5 3000e^{0.06t}dt$

7. $\int_0^{3.5} [15(0.9)^q - 10]dq$

8. $\int_0^8 [20(0.8)^q - (1.2)^q]dq$

9. $\int_1^8 \left[3e^{-0.05t} + 4t^2 - \frac{5}{2t}\right]dt$

10. $\int_1^4 \left[6(0.75)^q - \frac{2}{\sqrt[3]{q^4}} + 5\right] dq$

Game Time

11. As in previous exercises, suppose you are biking in Ohio, and your velocity (in mph) after t hours of biking is given by

$$v(t) = 14 + 2t, \text{ for } 0 \le t \le 2$$

(a) How many miles did you travel in the first hour of the trip? How could this be expressed symbolically using integral notation?

(b) How many miles did you travel in the second hour of the trip (between time $t = 1$ and time $t = 2$)? How could your answer be expressed symbolically using integral notation?

(c) Use the Fundamental Theorem of Calculus directly to find the number of miles you traveled in the first 2 hours of the trip. How does this relate to your answers in (a) and (b)?

12. As in previous exercises, suppose you are stuffing envelopes and your rate (in envelopes per minute) is given by the function

$$r(t) = 6 + \frac{t^2}{4050}, \text{ for } 0 \le t \le 90$$

(a) How many envelopes did you stuff in the first half-hour of working? How could this be expressed symbolically using integral notation?

(b) How many envelopes did you stuff in the second half-hour of working (between time $t = 30$ and time $t = 60$)? How could your answer be expressed symbolically using integral notation?

(c) Use the Fundamental Theorem of Calculus directly to find the total number of envelopes stuffed in the first full hour of working. How does this relate to your answers in (a) and (b)?

13. You are waiting for a bus. From past experience, you have figured that the amount of time you have to wait in minutes, X, has the following distribution:

$$f(x) = \begin{cases} 0.1e^{-0.1x}, & \text{for } x \ge 0 \\ 0, & \text{otherwise} \end{cases}$$

(a) What is the probability that you will have to wait 4 minutes or less for the bus?

(b) You are taking the bus to catch a train. You know that if you have to wait 8 minutes or more, you will miss the train. What mathematical expression will correspond to the probability that you will catch the train? What is the numerical value of this probability?

(c) What is the probability that you will have to wait between 4 minutes and 8 minutes for the bus? Express this quantity symbolically using integral notation.

(d) Show how to obtain your answer to (c) directly using the Fundamental Theorem of Calculus and verify it by numerical integration using technology.

14. You are in the middle of a lawsuit. You have had some experience with similar suits, and estimate that the eventual court settlement will result in a net benefit to you of X hundred thousand dollars, with a probability density function given by

$$f(x) = \begin{cases} -0.0208x^2 + 0.0417x + 0.208, & \text{for } -2 \le x \le 4 \\ 0, & \text{otherwise} \end{cases}$$

(a) What is the probability that you will receive a net benefit from the lawsuit of \$300,000 or less?

(b) What is the probability that the lawsuit will result in a net loss for you? What mathematical expression will correspond to this probability?

(c) What is the probability that your net benefit from the lawsuit will be between \$0 and \$300,000? Express this quantity symbolically using integral notation.

(d) Show how to obtain your answer to (c) directly using the Fundamental Theorem of Calculus and verify it by numerical integration using technology.

4.6 VARIABLE LIMITS OF INTEGRATION, MEDIANS, AND IMPROPER INTEGRALS

In the last few sections, we saw how to compute a probability by taking the definite integral of a probability density function (pdf). We have also discussed an interpretation of one particular antiderivative of the density function, $F(x) = P\{X \le x\}$, the probability that the value of the random variable is x or less, called the **cumulative distribution function (cdf)** of the random variable. In this section we will examine how this can be defined using a definite integral with a variable (x) for one of the limits of integration. We will then show how this integral relates to the second version of the Fundamental Theorem of Calculus and how it can be used to find the **median** (or 50th percentile/middle value) of a random variable. We will also talk about how to find the area under a curve, such as the exponential probability density function, when one or both of the limits of integration are infinite ($-\infty$ or $+\infty$), a situation called an **improper integral**.

After studying this section, you should be able to solve problems like the following:

- A competitor is working on a new product. You estimate the probability density function for when it will be released. To assess the risks to your business, you want to find the probability that the new product will be released within the next x months, and you want to know the median (50th percentile) release time.
- You have invented a new kind of glove that will fit several different glove sizes but have to pick what range of sizes to target (it can span four glove sizes), since you only have the money for a single production run initially.
- A competitor is coming out with a new product that will affect your business. You have estimated the probability distribution of the number of months before the new product is released and want to estimate the probability that this time will be 6 months or more from now.

In addition to being able to solve problems like those above, after studying this section, you should also

- Understand why, when working with an integral that involves a variable in at least one of the limits of integration, the variable of integration (whatever comes after the "d" in the integral, such as the x in dx) must be *different* than the variable in the limit of integration.
- Understand why the choice of a letter name for the variable of integration is arbitrary.
- Be able to set up an integral to find the cumulative distribution function for a random variable and be able to evaluate it.
- Understand what a median is, and how to set up and solve an integral equation to find it.
- Be able to express and explain an alternative version of the Fundamental Theorem of Calculus using an integral with a variable limit of integration.
- Know how to evaluate a definite integral in which one or both limits of integration is $-\infty$ or $+\infty$ (an improper integral).

Variable Limits of Integration

Sample Problem 1: A competitor is working on a new product. You estimate the probability density function for when it will be released to be

$$f(x) = \begin{cases} 0.2e^{-0.2x} & \text{for } x \geq 0 \\ 0 & \text{otherwise} \end{cases}$$

where x is the number of months from now that the new product is released. Find the probability that the new product will be released within the next T months.

Solution: We know how to do this already, from the last few sections. What we want is the probability that X is in the interval $[0, T]$, which is simply the area under the density function over that interval, given by the definite integral

$$\int_0^T 0.2e^{-0.2x}dx = \left[\frac{0.2}{-0.2}e^{-0.2x}\right]_0^T = [-e^{-0.2(T)}] - [-e^{-.2(0)}] = -e^{-0.2T} + 1 = 1 - e^{-0.2T}$$

Translation: The probability that the new product will be released within the next T months is $1 - e^{-0.2T}$. □

Let's denote the expression we just obtained, the probability that the new product will be released in the interval $[0, T]$ as $F(T)$. Thus, we have

$F(T) =$ the probability the new product will be released in $[0, T]$ months

$F(T) = 1 - e^{-0.2T}$, for $T \geq 0$.

But, when defining a function, we can use any letter we want for the variable, so we could just as easily replace the T with x and define

$F(x) =$ the probability the new product will be released in $[0, x]$ months

$F(x) = 1 - e^{-0.2x}$, for $x \geq 0$,

Now, suppose we had wanted to define our variable, the months within which the product will be released, to be x from the beginning. Then our integral would have been

$$\int_0^x f(x)dx$$

Something is funny here! We can't do this. **We are using the variable x in two *different* ways: both as the variable of integration (indicated by the "dx") and as the upper limit of integration. This is not allowed!** In general, a variable should have a unique and clear meaning in any expression (should be "well defined"). What can we do about it? Well, if our primary question is to find the probability of the release time occurring within x months, then we need to use the x for the upper limit of integration. This suggests that we should use a different letter for the variable of integration. In fact, just as we could use any variable in defining a function, the variable of integration is arbitrary, so we can use any letter we want. In this situation, you will usually see t used, but it could be any letter. If we use t, we get

$$F(x) = \int_0^x 0.2e^{-0.2t}dt = \left[\frac{0.2}{-0.2}e^{-0.2t}\right]_0^x = [-e^{-0.2(x)}] - [-e^{-0.2(0)}] = -e^{-0.2x} + 1 = 1 - e^{-0.2x}$$

Notice that the calculations are exactly the same as before, only the names have been changed to protect the innocent . . . (and avoid errors).

The Fundamental Theorem of Calculus, Part II

In Section 4.5 we mentioned that functions such as the one in Sample Problem 1 can be thought of as the probability that the random variable X, the number of months from now that the new product is released, is less than or equal to x. In symbols, $F(x) = P\{X \leq x\}$. For this example, this probability corresponded to the probability of X being in the interval $[0, x]$, but that is only because this random variable could never be negative. If the probability density function's domain is the interval $[a, b]$, then a more general definition of $F(x)$ is given by

$$F(x) = P\{X \leq x\} = P\{X \text{ is in } [a, x]\} = \int_a^x f(t)dt$$

In fact, for a general function $f(x)$, if $F(x)$ is *any* antiderivative of $f(x)$, then

$$\int_a^x f(t)dt = [F(t)]_a^x = F(x) - F(a)$$

so $\int_a^x f(t)dt$ is an antiderivative of $f(x)$ for any value of a (as long as $f(t)$ is defined over the entire interval $[a, x]$ and is continuous). In fact, different values of a could each yield a different antiderivative, so $\int_a^x f(t)dt$ could, in fact, be equivalent to the indefinite integral, if all constant values are possible. Notice that the "$-F(a)$" term above acts like our "$+C$" in the indefinite integral (as if $C = -F(a)$). Thus, if we take the derivative, we get

$$D_x\left[\int_a^x f(t)dt\right] = D_x\{[F(t)]_a^x\} = D_x\{F(x) - F(a)\} = F'(x) - D_x\{F(a)\} = F'(x) - 0 = F'(x) = f(x)$$

In brief, we get **an alternative form of the second version of the Fundamental Theorem of Calculus:**

If $f(t)$ is continuous (has no holes, breaks, or jumps) on the interval $[a,x]$, then

$$D_x\left[\int_a^x f(t)dt\right] = f(x)$$

Improper Integrals

Sample Problem 2: You estimate that the number of months before a competitor will come out with a new product (X) can be modeled with the following exponential probability density function: $f(x) = 0.2e^{-0.2x}$, for $x \geq 0$. Find the probability that the new product will come out in 6 months or more.

Solution: The new product coming out in 6 months or more corresponds to the value of X being greater than or equal to 6, or in the interval $[6, \infty)$. If we blindly applied the techniques and concepts from earlier in the chapter, we would say that the desired probability corresponds to the integral given by

$$\int_6^{\infty} 0.2e^{-0.2x}dx$$

Such an integral involving $\pm\infty$ as a limit of integration is called an **improper integral.** How can we evaluate it? Let's start with something related but easier.

If the question asked for the probability of the new product coming out in 6–12 months, we already know how to do it. We would simply find the area under the density function between 6 and 12 as follows:

$$\int_6^{12} 0.2e^{-0.2x}dx = \left[\frac{0.2}{-0.2}e^{-0.2x}\right]_6^{12} = -e^{-0.2(12)} - [-e^{-0.2(6)}]$$
$$= -e^{-2.4} + e^{-1.2} \approx 0.21$$

If the interval were [6,18], we would get

$$\int_6^{18} 0.2e^{-0.2x}dx = \left[\frac{0.2}{-0.2}e^{-0.2x}\right]_6^{18} = -e^{-0.2(18)} - [-e^{-0.2(6)}]$$
$$= -e^{-3.6} + e^{-1.2} \approx 0.27$$

If we continued this process, we could obtain the following table:

b	$P\{6 \le X \le b\} = P\{X \text{ in } [6,b]\}$
12	0.210
18	0.274
24	0.293
48	0.301
96	0.301

As we saw earlier with finding instantaneous rates of change and limits of sums to find areas, we can again use the idea of a *limit* to find the answer to this problem. In this case, the limit comes out to a probability of 0.301.

Translation: There is about a 30% chance that the competitor's new product will be released in 6 months or more. □

In general, we evaluate improper integrals by taking a limit. Let's try to express what we just did symbolically. To find $\int_6^{\infty} f(x)dx$, we first found $\int_6^b f(x)dx$ and then took the limit as b approached ∞. Thus in general, we could write

$$\int_a^{\infty} f(x)dx = \lim_{b\to\infty}\left[\int_a^b f(x)dx\right]$$

In practice, first evaluate the integral in the brackets; then take the limit. In Sample Problem 2, this could have been done as follows:

$$\int_6^b 0.2e^{-0.2x}dx = \left[\frac{0.2}{-0.2}e^{-0.2x}\right]_6^b = [-e^{-0.2(b)}] - [-e^{-0.2(6)}] = e^{-1.2} - e^{-0.2b}$$

Now, we take the limit as $b \to \infty$

$$\lim_{b\to\infty}[e^{-1.2} - e^{-0.2b}] = \lim_{b\to\infty}[e^{-1.2}] - \lim_{b\to\infty}[e^{-0.2b}] = e^{-1.2} - \lim_{b\to\infty}\left[\frac{1}{e^{0.2b}}\right] = e^{-1.2} - 0 \approx 0.301$$

It can be proven that, when working with limits and continuous functions

- **You can split up addition or subtraction into separate limits (if each exists);**
- **The limit of a constant is that constant; and**
- **You can factor a constant coefficient out in front of a limit.**

These also make good common sense! In the above calculation, we also used the fact that

$$x^{-n} = \frac{1}{x^n}$$

When evaluating $\lim_{b\to\infty}\left[\frac{1}{e^{0.2b}}\right]$, we simply noted that as b increases, $e^{0.2b}$ also gets bigger and heads toward infinity, so that 1 over it (its reciprocal) is a smaller and smaller fraction, which approaches 0.

There is a temptation in this situation to do the following:

$$\int_6^{\infty} 0.2e^{-0.2x}dx = \left[\frac{0.2}{-0.2}e^{-0.2x}\right]_6^{\infty} = [-e^{-0.2(\infty)}] - [-e^{-0.2(6)}]$$
$$= e^{-1.2} - e^{-0.2(\infty)} = e^{-1.2} - 0 = e^{-1.2} \approx 0.301$$

This is not really wrong, as long as the terms with the ∞ in them are understood to be limits in the more complete form given above. Be sure you can write out and understand the first way before you use the second notation.

By analogy, you can probably see now what to do if the lower limit is $-\infty$:

$$\int_{-\infty}^{b} f(x)dx = \lim_{a\to-\infty}\left[\int_a^b f(x)dx\right]$$

Sometimes you will even encounter problems where both limits of integration involve ∞. This case is best handled by splitting it up into two separate improper integrals and evaluating them individually:

$$\int_{-\infty}^{\infty} f(x)dx = \int_{-\infty}^{c} f(x)dx + \int_{c}^{\infty} f(x)dx$$

It is usually best to use $c = 0$, or some other value that is easy to calculate.

SAMPLE PROBLEM 3: The standard normal density function (the prototype for the "bell curve") is given by

$$f(x) = \frac{1}{\sqrt{2\pi}}e^{-x^2/2}$$

(the domain is all real numbers). Verify that the area under the standard normal density is equal to 1.

Solution: Since the domain is all real numbers, or $(-\infty, \infty)$, the area is

$$\int_{-\infty}^{\infty} f(x)dx = \int_{-\infty}^{0} f(x)dx + \int_{0}^{\infty} f(x)dx \qquad \text{(splitting the integral)}$$

$$= \lim_{a\to-\infty}\left[\int_{a}^{0} f(x)dx\right] + \lim_{b\to\infty}\left[\int_{0}^{b} f(x)dx\right] \qquad \text{(using limits to find improper integrals)}$$

$$= \lim_{a\to-\infty}\left[\int_{a}^{0} \frac{1}{\sqrt{2\pi}}e^{-x^2/2}dx\right] + \lim_{b\to\infty}\left[\int_{0}^{b} \frac{1}{\sqrt{2\pi}}e^{-x^2/2}dx\right] \qquad \text{(substituting in the density)}$$

Unfortunately, as we mentioned before, there *is* no expression for an antiderivative of the normal density function, so we can only solve this problem by numerical integration. For example, for the second integral, you can put an expression for a subinterval approximation of the integral with the variable upper limit into a spreadsheet or table to see what happens to the integral as the upper limit gets bigger and bigger. This expression is actually an approximate antiderivative function for the density (numerical, not symbolic), giving the probability (area) on the interval $[0, b]$. Having put this in as a function, you can then set up a table to read off values of the area for different values of b. You can set up a similar expression for the first integral, but using the interval $[a, 0]$ instead of $[0, b]$. Table 1 gives the results.

TABLE 1

a	area on $[a, 0]$	b	area on $[0, b]$
−1	0.341	1	0.341
−2	0.477	2	0.477
−3	0.499	3	0.499
−4	0.500	4	0.500
−5	0.500	5	0.500

From this table, we can see that each of the two integrals approaches 0.5 in the limit, so the entire improper integral value is $0.5 + 0.5 = 1$. □

The Median Value

The **median**, or 50th percentile of a probability density function, is the value where at least half of the probability lies at that value or above, and at least half of the probability lies at that value or below. In a sample of three test scores, such as {72,85,98}, the median would be 85 ("the middle score"), because two-thirds of the scores (at least half) are ≥ 85 (the 85 and the 98), and two-thirds of the scores (at least half) are ≤ 85 (the 72 and the 85).

When working with a continuous probability density function, the median is simply the value that splits the area under the density function into two equal pieces, each with probability (area) of 0.5. If we call the random variable X, and the median m, then we are saying that $P\{X \leq m\}$ should be 0.5. If $f(x)$ is positive only on $[a, b]$, then

$$P\{X \leq m\} = \int_{-\infty}^{m} f(x)dx = \int_{a}^{m} f(x)dx = 0.5$$

Sample Problem 4: A competitor is working on a new product. You estimate the probability density function for when it will be released to be

$$f(x) = \begin{cases} 0.2e^{-0.2x} & \text{for } x \geq 0 \\ 0 & \text{otherwise} \end{cases}$$

where x is the number of months from now that the new product is released. What is the 50th percentile (median) time for the new product release?

Solution: Graphically, this is shown in Figure 4.6-1. We could say that this is an **integral equation** defining the median, m. Recall, however, that $P\{X \leq m\} = F(m)$, the value of the cumulative distribution function (cdf) at m. In this problem, we already know the functional form for the cdf is $F(x) = 1 - e^{-0.2x}$, so $F(m) = 1 - e^{-0.2m}$. Finding the median is thus equivalent to solving the equation

$$F(m) = 0.5$$

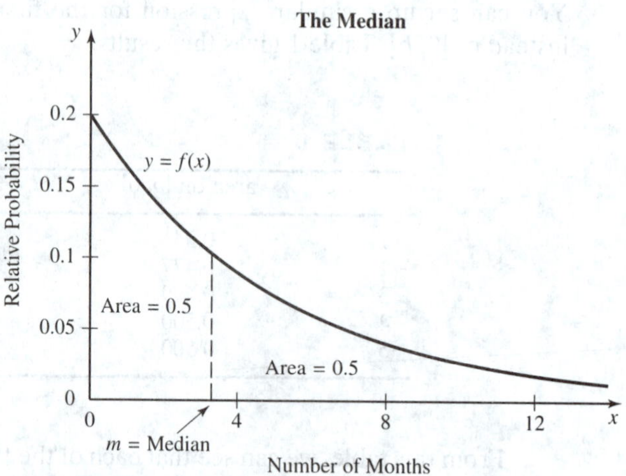

Figure 4.6-1

so in this problem we want to solve

$$1 - e^{-0.2m} = 0.5 \quad \text{(since } F(m) = 1 - e^{-0.2m} \text{ and we want to find where } F(m) = 0.5\text{)}$$
$$0.5 - e^{-0.2m} = 0 \quad \text{(subtracting 0.5 from both sides)}$$

If you are not comfortable working with logarithms, you can solve this using technology. To do it exactly by hand, you get

$0.5 - e^{-0.2m} = 0$

$e^{-0.2m} = 0.5$ (adding $e^{-0.2m}$ to both sides and switching sides)

$\ln(e^{-0.2m}) = \ln(0.5)$ (taking the natural log of both sides)

$-0.2m = \ln(0.5)$ ($\ln(e^x) = x$, since the exponent of e to get e^x is x)

$m = \dfrac{\ln(0.5)}{-0.2} \approx 3.47$ (dividing both sides by (-0.2) and evaluating)

Translation: The median time for the release of the new product is about three and a half months. It is just as likely to be released before three and a half months as after three and a half months. □

Sample Problem 5: Find the median for the standard normal distribution

$$f(x) = \frac{1}{\sqrt{2\pi}} e^{-x^2/2}, \text{ for all real numbers } x$$

Solution: We know that the median satisfies the integral equation

$$P\{X \le m\} = F(m) = \int_{-\infty}^{m} f(x)dx = 0.5$$

For the normal distribution, we can give a quick, smart answer: Since the bell-shaped curve ($f(x)$) is symmetric about $x = 0$ (the vertical axis), the median must be at $x = 0$, since the area under the curve on either side of $x = 0$ must be the same (0.5). We can see this graphically in Figure 4.6-2.

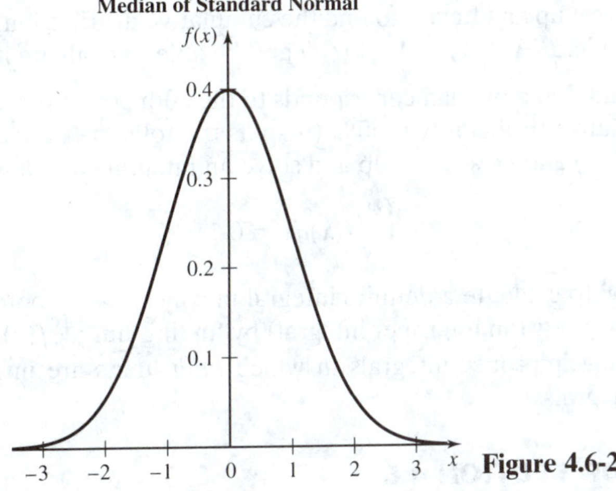

Figure 4.6-2

What if we didn't notice or think about the symmetry? There is no formula antiderivative for $f(x)$, so we can't use the *same* approach as we did for Sample Problem 2, But we can solve the integral equation using technology. If you enter the limits for a numerical integration so that the lower limit is -5 and the upper limit is variable, this is equivalent to finding the definite integral from -5 to x. Virtually all of the area under the standard normal density curve lies between -4 and 4 (see Table 1), so starting at -5 should be very close to what we would get if we could put in $-\infty$. We want to know when this integral will equal 0.5, so we are solving the equation

$$\int_{-5}^{x} \frac{1}{\sqrt{2\pi}} e^{-t^2/2} dt = 0.5$$

Using technology, we get, to 5 significant figures, the correct answer of 0 (it will give an answer such as "7.185282347E-7," meaning $7.185282347(10^{-7}) = 0.0000007185282347$).

Notice that when we were calculating the two improper integrals for Sample Problem 3, one of them corresponded to

$$\int_{-\infty}^{0} f(x)dx = 0.5$$

This verifies our result that the median is at 0, since the integral equation defining the median is satisfied by $m = 0$. □

Before trying the exercises, you should

- Understand why, when working with an integral that involves a variable in at least one of the limits of integration, the variable *of* integration (whatever comes after the "*d*" in the integral) must be *different* than the variable in the *limit* of integration.
- Understand why the variable of integration is arbitrary (like the variable in a function).
- Be able to set up an integral to find the cumulative distribution function for a random variable, $P(X \leq x) = \int_{-\infty}^{x} f(t)dt$ and be able to evaluate it.
- Understand that a median corresponds to the 50th percentile, or the point where the cumulative probability is 50% (it splits the total area under the pdf into two equal pieces) and how to set up and solve an integral equation to find it:

$$\int_{-\infty}^{m} f(x)dx = 0.5$$

- Know how to evaluate a definite integral in which one or both limits of integration is $-\infty$ or $+\infty$ (an improper integral) by finding $\lim_{a\to-\infty} \int_a^c f(x)dx$ or $\lim_{b\to\infty} \int_c^b f(x)dx$ and splitting improper integrals in which *both* limits are infinite into two improper integrals.

EXERCISES FOR SECTION 4.6

Warm Up

Find the answers to Exercises 1 and 2 by first applying the Fundamental Theorem of Calculus to the definite integral and then taking the derivative of the result.

1. $D_x\left[\int_1^x 2t^3 dt\right]$

2. $D_x\left[\int_{-2}^x 2t^3 dt\right]$

For Exercises 3 and 4 apply the version of the Fundamental Theorem of Calculus discussed in this section.

3. $D_x\left[\int_0^x \frac{1}{\sqrt{2\pi}}e^{-t^2/2}\,dt\right]$

4. $D_x\left[\int_1^x 0.25te^{-0.5t}dt\right]$

For Exercises 5 and 6, evaluate the improper integrals using the Fundamental Theorem of Calculus and the idea of a limit.

5. $\int_0^\infty 0.2e^{-0.2x}\,dx$

6. $\int_1^\infty \frac{3}{x^2}dx$

For Exercises 7 and 8, evaluate the improper integrals using technology (for example, use numerical integration on your graphing calculator or spreadsheet).

7. $\int_2^\infty \frac{1}{\sqrt{2\pi}}e^{-x^2/2}\,dx$

8. $\int_0^\infty (5000 + 100t)e^{-0.5t}dt$

Game Time

9. You are in the middle of a lawsuit. You have had some experience with similar suits, and estimate that the eventual court settlement will result in a net benefit to you of X hundred thousand dollars, with a probability density function given by

$$f(x) = \begin{cases} -0.0208x^2 + 0.0417x + 0.0208, & \text{for } -2 \le x \le 4 \\ 0, & \text{otherwise} \end{cases}$$

(a) Write an integral equation with a variable upper limit whose solution would give you the median result of the court settlement.

(b) Evaluate the integral in (a) to get an equation *not* involving an integral whose solution would give you the median.

(c) Solve the equation in (b) to find the median.

(d) Now use technology to solve the integral equation in (a) *without* evaluating the integral first. Do you get the same answer? What step in the 12–Step Program for Plugaholics does this represent?

(e) Can you find the 25th and 75th percentile values (the **quartiles**)?

10. You are waiting for a bus. From past experience, you have figured that the amount of time you have to wait in minutes, X, has the following distribution:

$$f(x) = \begin{cases} 0.1e^{-0.1x}, & \text{for } x \ge 0 \\ 0, & \text{otherwise} \end{cases}$$

(a) Write an integral equation with a variable upper limit whose solution would give you the median waiting time for the bus.

(b) Evaluate the integral in (a) to get an equation *not* involving an integral whose solution would give you the median.

(c) Solve the equation in (b) to find the median. Can you verify your solution in a *different* way (for example, one with technology and the other without)? If so, *do* so!

(d) Now use technology to solve the integral equation in (a) *without* evaluating the integral first. Do you get the same answer?

(e) Can you find the 25th and 75th percentile values (the **quartiles**)?

11. You work for a company that is developing a drug to help treat people with AIDS. The research is moving along well, but there is a lot of uncertainty in this kind of project. You estimate that the distribution of time (in years) from the present before the drug can go into production, X, is given by the probability density function

$$f(x) = \begin{cases} 0.25xe^{-0.5x}, & \text{for } x \geq 0 \\ 0, & \text{otherwise} \end{cases}$$

(a) Write an integral equation with a variable upper limit whose solution would give you the median time before the drug can go into production.

(b) Use technology to solve the integral equation in (a) *without* evaluating the integral first. What is the median time before the drug goes into production?

(c) Can you find the 25th and 75th percentile values (the **quartiles**)?

12. Suppose you run a construction contracting business, and you are bidding for a big project. You have one major competitor. Based on past experience, you believe that your competitor's bid for the project could be anywhere between \$8 million and \$10 million, and that all values in between are equally likely.

(a) Define the random variable for this problem.

(b) Define the probability density function for this problem and sketch its graph.

(c) What is your competitor's median bid? Explain how you got your answer.

13. To help decide what you want to plant in your garden and when, you want to estimate when the last frost will occur. From historical records, you determine that in your area it could be anywhere between March 28 and April 19, but that any date within that interval is about equally likely to be the last frost.

(a) Define the random variable for this problem.

(b) Define the probability density function for this problem and sketch its graph.

(c) What is the median date for the last frost? Explain how you got your answer.

14. Suppose you are considering starting a small business selling bagels in your dorm on campus. You buttonhole 40 random students in your dorm and ask what is the most they would pay for a bagel and cream cheese, and the results are as follows:

Maximum Price (\$)	[0.40, 0.65)	[0.65, 090)	[0.90, 1.15)	[1.15, 1.40)	[1.40, 1.65)
Number	9	7	6	4	3

(The other 11 indicated a value below 40 cents.)

(a) What is the probability of a student picked at random in your dorm being willing to pay between 40 and 65 cents maximum for a bagel? What would be the probability **per unit** in this interval (so the area under a line at that height between 0.40 and 0.65 would give the correct probability)?

(b) Define a histogram probability density function for this situation. Be sure to define your random variable.

(c) Graph your probability density function.

(d) What is the median maximum price a random student would pay, based on your sample? Explain how you got your answer.

15. You work for a company that is developing a drug to help treat people with AIDS. The research is moving along well, but there is a lot of uncertainty in this kind of project. You estimate that the distribution of time (in years) from the present before the drug can go into production, X, is given by the probability density function

$$f(x) = \begin{cases} 0.25xe^{-0.5x}, & \text{for } x \geq 0 \\ 0, & \text{otherwise} \end{cases}$$

(a) Find the area under the density function over $[0, \infty)$. Explain why you get the answer you do, and what it means.
(b) What would the integral over *all* the real numbers be? Explain.
(c) What is the probability that the drug will go into production in 3 years or more? Explain how you got your answer.

16. You are waiting for a bus. From past experience, you have figured that the amount of time you have to wait in minutes, X, has the following distribution:

$$f(x) = \begin{cases} 0.1e^{-0.1x}, & \text{for } x \geq 0 \\ 0, & \text{otherwise} \end{cases}$$

(a) What is the probability that you will have to wait 4 minutes or more for the bus? Find your answer using the Fundamental Theorem of Calculus and verify it using technology (show your calculations in obtaining the limit).
(b) What is the probability you will have to wait 8 minutes or more?
(c) What is the probability you will have to wait *between* 4 minutes and 8 minutes?

4.7 CONSUMER AND PRODUCER SURPLUS*

In the first six sections of this chapter, you have learned about finding the area between a curve and the horizontal axis (above the axis minus below it) over an interval. In some applications, we need **to find the area between two different curves**. An example would be finding the economic benefits, to both consumers and producers, of a competitive free market. These benefits are called the **Consumer Surplus** and **Producer Surplus**, respectively. We will discuss how to find the area between two curves for this example and in general in this section.

After studying this section, you should be able to solve problems like the following:

- You are a craftsperson who makes a unique sculpted candle. Because of other craft items you make, the number of these candles you are willing to make in a given week depends on how much you can sell them for. From your experience, you have information about how many of these candles you could sell in a week at different selling prices. If you choose your price to balance supply and demand, how much will you benefit (compared to your worst case minimum selling price, called your **reservation price**), and how much will your customers benefit (compare to their worst case maximum price, also called a reservation price)?
- You have data about weekly revenues and expenses from which you can fit curves for the revenue rate and cost rate functions (both are derivatives with respect to time). How can you estimate your profit over an interval?

*This section is optional and may be skipped without interrupting the flow or continuity.

After studying this section, in addition to being able to solve the above kinds of problems, you should:

- Be able to explain in real-world terms what demand and supply functions mean, and how you might try to estimate them.
- Understand how to find the equilibrium price, given demand and supply functions.
- Given supply and demand functions, be able to calculate the Consumer Surplus and Producer Surplus and be able to explain what they mean in words and what they correspond to graphically and algebraically.
- Know how to find the area between two curves over an interval.

Let's start with a simple problem that is easy to understand.

Sample Problem 1: You have just invented a contraption that attaches to a microphone stand to hold a glass or cup, so that a performer can have a drink easily available during a performance. You don't have a lot of time to make these (you have a full-time job making other things). You estimate the prices at which you would be willing to make different numbers of these cupholders over the next month. For example, you estimate that for a price between \$18 and \$22, you would be willing to make up to 3 cupholders, so you use \$20, the midpoint, as the price associated with 3. You decide upon the values given in Table 1.

TABLE 1

Number of Cupholders (x)	1	2	3	4	5	6
Price (in \$) at Which You'll Make x	13	16	20	25	31	38

You then estimate the demand for cupholders in the next month, based on recent experience. The same kind of reasoning applies here: If you figure that at a price between \$21 and \$29, you would expect to sell 4 cupholders, then you assign the value of \$25 to 4. Your estimates are given in Table 2.

TABLE 2

Number Sold (x)	1	2	3	4	5	6
Price (in \$) to Charge to Sell x	52	42	33	25	18	12

Find where supply and demand match.

Solution: If you decided to charge \$40 for the cupholders, you'd be willing to make about 6 but would only be able to sell about 2, so there would be a major mismatch between the quantity you'd like to make and the quantity you could sell. If you charged about \$12, the mismatch would go the other way: You could sell about 6, but would only want to make 1. On the other hand, if you charge \$25, the number you'd like to make would *match* the number you could sell. This is called the **equilibrium price**. In a large market, with numerous competitive producers, market forces tend to push the market price toward this equilibrium price. Graphically, this occurs where the two graphs, called the **supply function** and the **demand function**, respectively, intersect. For our simple example, we can show this graphically (Figure 4.7-1). □

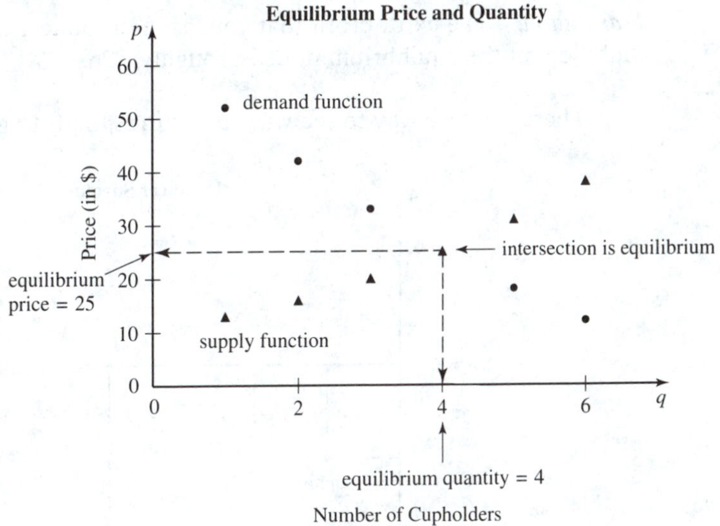

Figure 4.7-1

Producer Surplus

Sample Problem 2: Now that you have determined the equilibrium price, estimate the total monetary benefits to yourself (Producer Surplus) if you charge the equilibrium price, compared to the worst-case prices listed in the tables above for each cupholder. Explain the idea behind your calculations.

Solution: If you make 4 cupholders and charge \$25 for each, you do better than your worst case for the first few cupholders. For example, you were willing to make the first cupholder for as little as \$13. Now that you will be selling it for \$25, that's

$$\$25 - \$13 = \$12$$

more in revenue (a *benefit*, or *surplus*, of \$12) compared to your worst case. For the second cupholder, you would have made it for as little as \$16, for an additional

$$\$25 - \$16 = \$9$$

in revenue (and therefore in profit as well[14]) compared to the minimum you would have sold it for. For the third cupholder, your additional revenue/profit would be \$25 − \$20 = \$5, and for the fourth, it would be \$25 − \$25 = \$0. Thus your total benefit (increased profit) would be

$$\$12 + \$9 + \$5 + \$0 = \$26$$

[14]We have not specificed the cost function. The cost of making q^* units is fixed, whatever form the cost function takes. Thus, an increase in revenue yields an equal increase in profit.

This is called the **Producer Surplus**, since it **is the total amount of money that you, as the producer of the good, get in extra profit by selling each cupholder at the equilibrium price, compared to the worst case (lowest) prices you would have settled for.**

Translation: The extra profit that you, as the producer of the good, get by selling each cupholder at the equilibrium price is, roughly, \$26. □

There is a nice way to see what this corresponds to graphically (Figure 4.7-2).

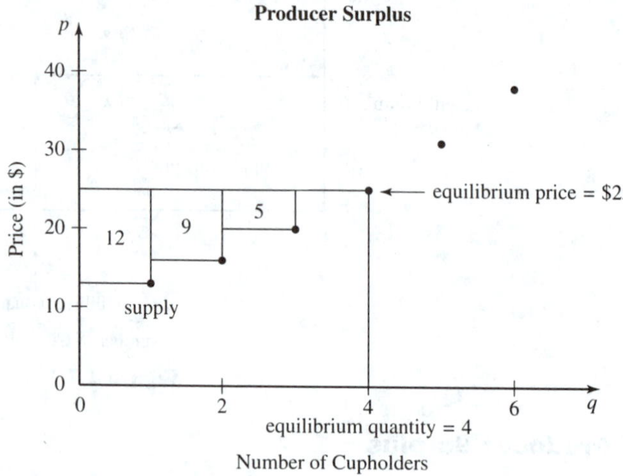

Figure 4.7-2

You can see that the Producer Surplus corresponds to the sum of the areas of rectangles drawn between the selling price (the equilibrium price in this case) and the points of the **supply** function. Typically, the quantities involved will be much larger, and so we can get a good estimate by fitting a curve to the supply function and then finding the area between the horizontal line of the selling price and the supply curve.

If you think about it, this concept of the Producer Surplus is relevant even when a price other than the equilibrium price is charged. However, the analysis gets a bit more complicated, because you need to think carefully about how many units would be made and sold. If you charge less than the equilibrium price, then demand is more than supply (customers want more than you are willing to make), so the quantity made and sold would be determined by the supply function (the quantity value that corresponds to the selling price on it). If you charge more than the equilibrium price, then the supply is more than the demand (you want to make more than you could sell), so the quantity made and sold should be determined by the demand function (again, the quantity that corresponds to the selling price on it). In any of the three cases, you find the Producer Surplus by finding the area below the selling price and above the supply function, from 0 to the quantity made and sold (on the horizontal axis).

Consumer Surplus

Sample Problem 3: In our problem involving the cupholders, you as the producer are not the only one to benefit from the equilibrium price. Those customers who were willing to pay *more* than the equilibrium price will also benefit because they will pay *less* than *their* worst case, the *maximum they* would have been willing to pay. Calculate the **Consumer Surplus**, the total monetary benefit to the *customers* if you charge the equilibrium price.

Solution: In the table of the demand for the cupholders, the fact that you could sell one cupholder at \$52 can be thought of as meaning that there is one customer out there this month willing to pay up to \$52 for the cupholder. At a selling price of \$25, this customer will *save* \$52 − \$25 = \$27 compared to their worst case, the most they would have been willing to pay. This is called the *surplus* (benefit) for that consumer. Notice that the surplus for the producer is *added revenue or profit*, while the surplus for consumers is in *saved money* (as from a coupon). The fact that you could sell 2 cupholders at a selling price of \$42 means there is a second customer out there who is willing to pay up to \$42 for one (this could technically be a second person, or how much the *first* customer would pay at most for a *second* cupholder). In this case, the savings for that second cupholder is \$42 − \$25 = \$17. Analogously, the savings (customer surplus) for the third cupholder is \$33 − \$25 = \$8. The fourth cupholder would have sold for at most \$25, so the surplus is \$25 − \$25 = \$0. Thus the total **Consumer Surplus** for the problem is \$27 + \$17 + \$8 + \$0 = \$52.

Translation: The total monetary benefit that the consumers receive when you sell the cupholders at the equilibrium price is \$52. □

Graphically, this is illustrated in Figure 4.7-3.

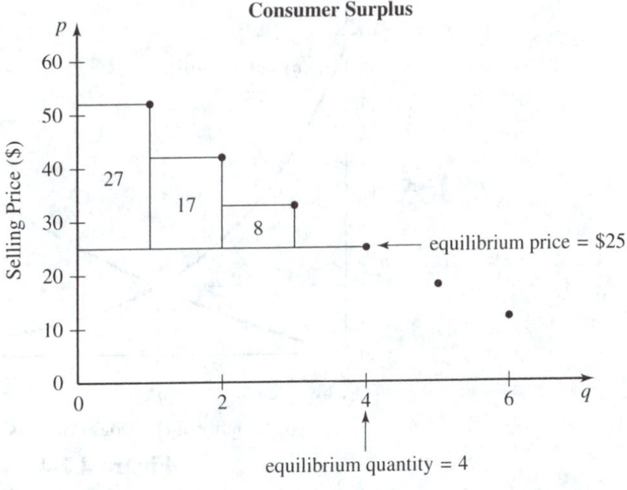

Figure 4.7-3

Analogous to the result for the Producer Surplus, the Consumer Surplus corresponds to the area below the **demand** function and above the selling price (equilibrium price here), between 0 and the quantity made and sold (equilibrium quantity here) on the horizontal axis. Again, we could use fit curves to estimate this quantity, and we could use a selling price different from the equilibrium price, as long as we are careful to reason out what the quantities made and sold should be. □

Calculating Equilibrium Price and Quantity

Sample Problem 4: You manufacture kosher, all-beef hot dogs. Your marketing research department has collected data on supply and demand for these hot dogs in the United States, and fit curves to those data points to obtain estimates of the demand and supply curves for the upcoming year. Let q represent quantity (in units of millions of pounds of hot dogs), and let p represent the price of 1 pound of hot dogs. Suppose that market research has determined that the supply curve is approximately given by

$$p = S(q) = 0.013q^2 + 0.23q, \qquad 0 \le q \le 20$$

and the demand curve is found to be modeled quite well by

$$p = D(q) = 9.7(0.80)^q, \qquad 0 \le q \le 20$$

Find the equilibrium quantity and price.

Solution: As a general comment, the supply curve for an entire industry is a very fuzzy notion. It makes sense that it exists in theory, but since it depends on information that individual firms are likely to feel should be confidential for reasons of competitiveness, it is difficult to obtain a good estimate of it. As usual, we will make our assumption of certainty and divisibility for the variables, as well as the assumption that the functions are continuous. The assumption of certainty will hold less strongly for the supply function than for the demand function. The graphs of the supply and demand are shown in Figure 4.7-4.

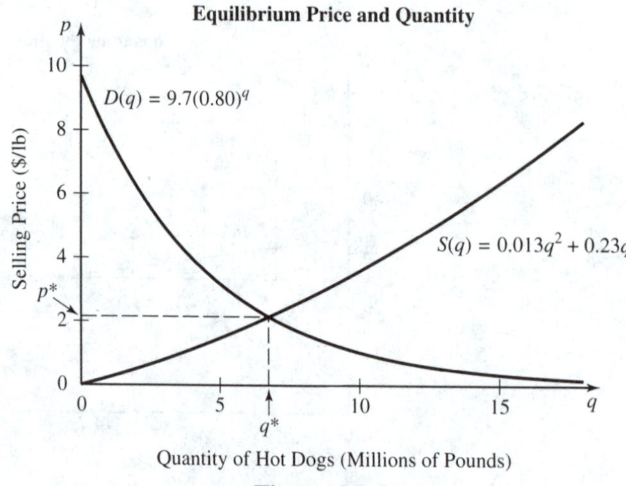

Figure 4.7-4

We need to find the **equilibrium price and quantity, where the supply and demand functions intersect**. A **crude ballpark estimate, just from looking at the graph**, would be a price of about 2.25 ($2.25/lb), and a quantity of about 7 (7 million pounds of hot dogs). How can we find more precise values? We want to know where the two curves intersect. We could use the **tracing** or **table** capabilities of a graphing calculator or other technology, but these are slower and less accurate than solving by hand or using the **equation-solving function** on a graphing calculator or spreadsheet.

The intersection we want is the value of q for which the supply and demand functions are equal:

$$S(q) = 0.013q^2 + 0.23q \text{ and } D(q) = 9.7(0.80)^q$$

so we want

$$0.013q^2 + 0.23q = 9.7(0.80)^q$$

Notice that this equation could also be thought of as simply ***setting the two functions equal to each other***. This is true in general **when looking for the intersection of two functions**.

To solve an equation like this, you can express the equation in a form that has 0 on one side. For our problem, we can do this by subtracting the Right-Hand Side (RHS) of the equation from both sides, to get

$$0.013q^2 + 0.23q - 9.7(0.80)^q = 0$$

Unfortunately, this equation is probably not of a form you have seen before or know how to solve by hand. If the demand function had been linear, we would then have ended up with a quadratic equation, which you could have solved by hand using the quadratic formula, or factoring, or completing the square. But this equation is very difficult, so we will use technology to help solve it. Usually, technology will ask for an initial guess to guide the search for a solution, since there could be more than one solution. Usually, the better your guess, the faster and better quality the answer you get. In this case, we could use our visual graph-based ballpark estimate of 7. Doing so, we get an answer of

$$q^* \approx 6.76$$

Translation: The equilibrium quantity is 6.76 million pounds of hot dogs.

It is common practice to use a star (asterisk) superscript (*) to indicate the equilibrium price and quantity. To check (verify) this calculation, we can substitute $q = 6.76$ in for q into each function and see if we get prices that are the same:

$$S(6.76) = 0.013(6.76)^2 + 0.23(6.76) \approx 2.15$$

$$D(6.76) = 9.7(0.8)^{(6.76)} \approx 2.15$$

Thus, our calculations are verified, and this work also gives us our estimate of the equilibrium price, since that is the common output value for both functions when the input is the equilibrium quantity.

Translation: Our estimate of the equilibrium price (p^*) is \$2.15 per pound. □

Notice that our ballpark guesses of $q = 7$ and $p = 2.25$ were not too bad and help to validate these answers.

Consumer Surplus and Producer Surplus from Models

Sample Problem 5: For Sample Problem 4, find the Consumer and Producer Surpluses at the equilibrium price and quantity and interpret the meaning of each in words. Be sure to verify your answers.

Solution: From the earlier discussion, **the Consumer Surplus is the money saved by consumers due to buying a product at its selling price rather than the maximum they would have been willing to spend. It corresponds to the area below the demand function (the maximum they would pay) and above the selling price, between 0 and *the quantity actually made and sold* on the horizontal axis. If the selling price is p^*, then the quantity made and sold should be q^* (that is, at the equilibrium price, use the equilibrium quantity).**

For our problem, the selling price is the equilibrium price, $p^* = 2.15$, and the quantity made and sold is the equilibrium quantity, $q^* = 6.76$. Graphically, the Consumer Surplus (CS) is represented in Figure 4.7-5. We are looking for the area below $D(q)$ and above p^*, between 0 and q^* on the horizontal axis. Geometrically, this is equal to

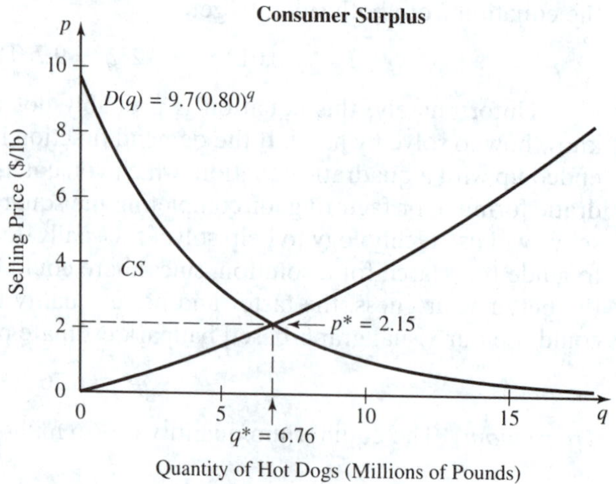

Figure 4.7-5

CS = [the area under $D(q)$ and above the q axis] − [the area under p^* and above the q axis] between 0 and q^*

Roughly, this is like a right trapezoid minus a rectangle, giving a right triangle (Figure 4.7-6). From earlier in this chapter, we know that the areas in the two pairs of brackets in the expression for CS above both correspond to definite integrals. Therefore, we can rewrite the equation (statement) for the CS as

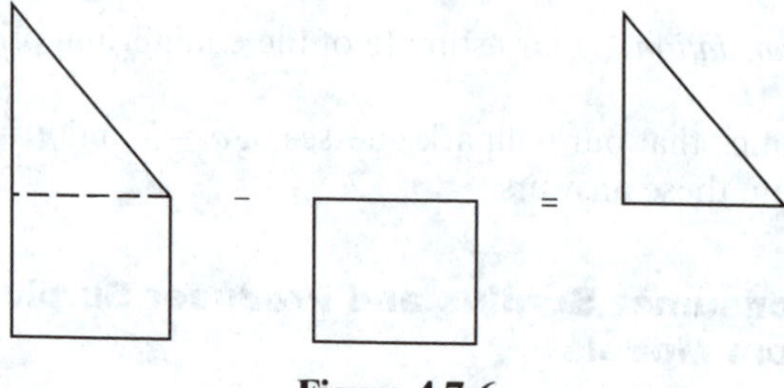

Figure 4.7-6

$$
\begin{aligned}
CS &= (\text{area under demand curve}) - (\text{area under } p^*) \\
&= \int_0^{q^*} D(q)dq - \int_0^{q^*} p^* dq \\
&= \int_0^{6.76} [9.7(0.8)^q]dq - \int_0^{6.76} [2.15]dq \\
&= \left[\frac{9.7}{\ln(0.8)}(0.8)^q\right]_0^{6.76} - [2.15q]_0^{6.76} \\
&\approx \left[\frac{9.7}{-0.223}(0.8)^{(6.76)}\right] - \left[\frac{9.7}{-0.223}(0.8)^{(0)}\right] - \{[2.15(6.76)] - [2.15(0)]\} \\
&\approx [-43.5(0.221)] - [-43.5(1)] - \{14.5 - 0\} \\
&\approx -9.61 + 43.5 - 14.5 \\
&\approx 19.4
\end{aligned}
$$

Translation: Consumers of these hot dogs save a total of \$19.4 million this year if the selling price is set at the equilibrium price and the equilibrium quantity is produced.

To understand how to get the units, think of a rectangle, like that formed by p^* and q^*. Its area is

(base)(height) = (6.76 million pounds)(\$2.15/lb) = \$14.5 million (this year). □

In the integral evaluation, we used the Fundamental Theorem of Calculus (antiderivative method) to get the answer. We could also have used a numerical integration procedure with technology. In general, the Fundamental Theorem is *exact* until you round off decimals in your answers, while numerical integration is always an approximation, although it can be carried to as many places or digits of accuracy as the technology allows (usually 8–16 significant figures). For practical problems, either method is fine. Sometimes one is easier and faster than the other—usually numerical integration. Sometimes you do not know how to find an antiderivative or there is no expression for it, which happens with the normal density function, so your only real choice is numerical integration. But if you are working with symbols to get a general solution, the Fundamental Theorem is the way to go. There are even sophisticated symbolic integration programs that can do this for you (like Maple™ or Mathematica™). These are very useful for more advanced mathematics, science, and engineering but are not crucial for most business and social science applications.

Area Between Curves over an Interval

Just as we did with indefinite integrals, we can use the fact that the integral of a sum or difference equals the sum or difference, respectively, of the integrals of the individual terms *in reverse*; that is, the sum or difference of integrals equals the integral of the sums or differences: $\int f(x)dx - \int g(x)dx = \int [f(x) - g(x)]dx$. Therefore we can say that

$$CS = \int_0^{q^*} D(q)dq - \int_0^{q^*} p^*dq$$

$$= \int_0^{q^*} [D(q) - p^*]dq$$

This leads us to a general principle:

If one function ($f(x)$) is always above a second function ($g(x)$) over an interval ($[a, b]$), then the area between them will be the definite integral of the upper function minus the lower function over that interval. In other words, if $f(x) \geq g(x)$ over the entire interval $[a, b]$, then the area between the curves over that interval is given by $\int_a^b [f(x) - g(x)]dx$.

This is shown graphically in Figure 4.7-7. Notice that, if we think of the x-axis as the function $f(x) = 0$, this gives a different way of understanding why the definite integral of a function $g(x)$ that is below the horizontal axis is negative, since the area between the axis and the curve will be

$$\int_a^b [0 - g(x)]dx = -\int_a^b g(x)dx$$

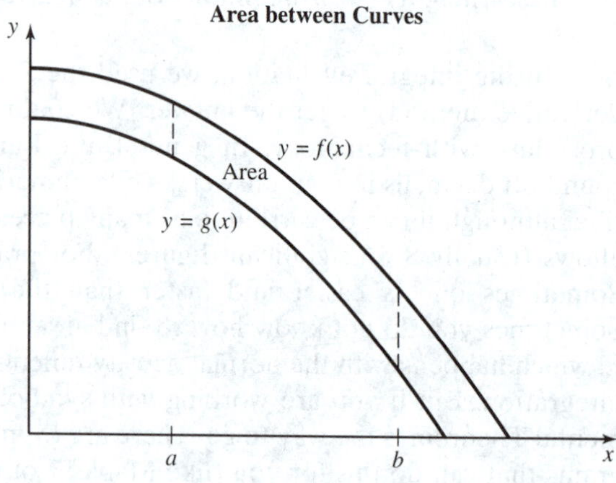

Figure 4.7-7

Since the answer (an area) must be positive, the integral value must be negative. This is because the axis ($f(x)$) is above the curve ($g(x)$) in such a situation.

Since p^* is a constant, the second integral in the first expression for CS above is just the area of the rectangle formed by p^*, q^*, and the axes, which is simply p^*q^*, so in fact we get

$$CS = \int_0^{q^*} [D(q) - p^*]dq = \left[\int_0^{q^*} D(q)dq\right] - (p^*q^*)$$

You can see this from the numerical calculation earlier, where the $(2.15)(6.76) = 14.5$ was subtracted. The first part (the integral) here corresponds to the total amount of money consumers would be willing to pay at most, and the p^*q^* is how much they actually pay, so the difference is their savings or surplus.

Again from our earlier discussion, **the Producer Surplus is the additional revenue or profit received by producers due to a given selling price, over and above the minimum price they would have been willing to sell each unit for, based on the quantity that would be made and sold at that selling price. This corresponds to the area below the selling price and above the supply function, from 0 to *the quantity actually made and sold* on the horizontal axis. If the selling price is p^*, then the quantity made and sold should be q^* (that is, at the equilibrium price, use the equilibrium quantity).**

Graphically, we get Figure 4.7-8.

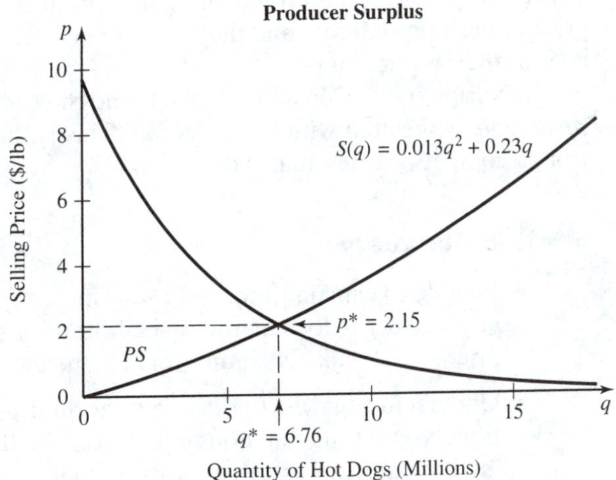

Figure 4.7-8

Using the result we derived earlier about finding the area between two curves, the Producer Surplus (PS) is the area between p^* and $S(q)$, from 0 to q^*. Since p^* is always above $S(q)$ on this interval, the answer is given by

$$PS = \int_0^{q^*} [p^* - S(q)]dq = \int_0^{6.76} [2.15 - (0.013q^2 + 0.23q)]dq$$

$$= \int_0^{6.76} \left[2.15 - 0.013q^2 - 0.23q)\right]dq$$

$$= \left[2.15q - \frac{0.013}{3}q^3 - \frac{0.23}{2}q^2\right]_0^{6.76}$$

$$= \left[2.15(6.76) - \frac{0.013}{3}(6.76)^3 - \frac{0.23}{2}(6.76)^2\right] - \left[2.15(0) - \frac{0.013}{3}(0)^3 - \frac{0.23}{2}(0)^2\right]$$

$$\approx 7.94$$

Translation: The hot dog producers collectively made \$7.94 million in additional revenues this year by selling at the equilibrium price of \$2.15/lb, compared to the minimum price per pound they would have sold the hot dogs for. This assumes they produce the equilibrium quantity of 6.76 million pounds of hot dogs.

We used the Fundamental Theorem of Calculus to get the answer, although we could have used a numerical approximation as well.

Notice that the Producer Surplus can also be thought of geometrically, as the area of the p^*q^* rectangle minus the lower "curved triangle" within it, to give the upper curved triangle within it, or

$$PS = p^*q^* - \int_0^{q^*} S(q)dq$$

This corresponds to the total revenue the producers will actually get at the given selling price minus the total revenue they *would have gotten* if they had sold each pound of hot dogs at their minimum price.

The sum of the Consumer Surplus and the Producer Surplus is sometimes called the *social gain* associated with the situation. This is the total direct social economic benefit from a competitive free market.

Section Summary

- Be able to explain in real-world terms what a demand function (the highest price you could get for each unit made) and a supply function (the cheapest selling price you'd make each unit for) mean, and how you might try to estimate them.
- Understand that to find the equilibrium price, given demand and supply functions, you set the functions equal to each other and solve for the quantity (assuming both are expressed in terms of quantity).
- Understand that Consumer Surplus is the money *saved* by consumers by paying the selling price, rather than the *most* they would have been willing to pay, summed over all consumers and units, and so it is the integral of the demand function minus the selling price. At the equilibrium price and quantity, the result can be expressed in the form

$$CS = \int_0^{q^*} [D(q) - p^*]dq.$$

- Understand that Producer Surplus is the *extra revenue* gained by producers by being able to sell a good at the actual selling price instead of for the *minimum* selling price they would have settled for, summed over all producers and units, and so it is the integral of the selling price minus the supply function. At the equilibrium price and quantity, the result can be expressed in the form

$$PS = \int_0^{q^*} [p^* - S(q)]dq$$

- Given supply and demand functions, be able to calculate the Consumer Surplus and Producer Surplus and be able to explain what they mean in words and what they correspond to graphically and algebraically.

- Know how to find the area between two curves over an interval: If $f(x) \geq g(x)$ over $[a, b]$, then the area between them is given by $\int_a^b [f(x) - g(x)]dx$.

EXERCISES FOR SECTION 4.7

Warm Up

For Exercises 1 through 6, find the equilibrium price and quantity, and the Consumer and Producer Surplus for the given supply and demand functions.

1. $D(q) = 224 - 8q$, $S(q) = 36 + 12q$

2. $D(q) = 75 - 20q$, $S(q) = 40 + 8q$

3. $D(q) = 433 - q^2$, $S(q) = 145 + q^2$

4. $D(q) = 500 - 3q - 10q^2$, $S(q) = 195.5 + 14q^2$

5. $D(q) = 20(0.9)^q$; $S(q) = 2q^{0.5}$

6. $D(q) = 15 - q$, $S(q) = 2q$

Game Time

7. You make embroidered shirts for sale at a gift shop on a small island. You estimate the prices you would have to receive to make it worth your time to make and sell particular numbers of shirts in any given week as follows:

Number of Shirts	1	2	3	4	5	6
Price ($)	14	16	18	20	22	24

You estimate the prices you would have to charge to sell particular numbers of shirts in any given week as follows:

Number of Shirts	1	2	3	4	5	6
Price ($)	32	27	23	20	18	17

(a) At what price would the interests of you and your customers match? What is the economic term for this?

(b) How many shirts would be made and sold at the price you specified in (a)? What is the economic term for this?

(c) At the equilibrium price and quantity, how much *more* money (total) would you make each week, compared to selling *each* shirt at the *minimum* price you would have been willing to make it for (as given in the table)? What is the economic term for this?

(d) At the equilibrium price and quantity, how much money would your customers collectively *save* in a week (total), compared to the *maximum* prices they would have paid for shirts (as given by the table)? What is the economic term for this?

(e) Find a model for the supply function. Be sure to specify it completely (all three parts).

(f) Find a model for the demand function and specify it completely (including the domain).

(g) Using your models, what do you get for your equilibrium price and quantity? Is this the same as before?

(h) Using your models, calculate the Consumer Surplus and the Producer Surplus. How do they compare to before? Why?

(i) Suppose the gift shop wanted to charge \$18 for the shirts. What would happen, and how would it affect the Consumer and Producer Surplus (based on the *tables*)?

8. Suppose that demand and supply values for unicycles in a year can be estimated as follows:

Number of Unicycles (u)	100	200	300	400	500
Price (\$) Where Demand is u	200	160	130	94	72
Price (\$) Where Supply is u	50	60	72	85	100

(a) Find models for the supply and demand functions, and justify your choices.
(b) Find the equilibrium price and quantity using your models.
(c) Find the Consumer Surplus and Producer Surplus at the equilibrium price and quantity using your models, and explain in words what each means.
(d) What would be the *actual* collective *revenue* of the producers at the equilibrium price and quantity?
(e) If unicycle makers colluded to fix the price at \$130, what would happen? What would their actual collective revenue be? Would they *want* to do it? How can a society deal with this situation?
(f) What would be a way to get a different estimate of your answer to (e)? What do you get? What margin of error does this suggest?

9. Students enlisted the help of a friend in the vending machine industry and got the following information:[15]

Number Sodas Sold per Day	221	209	193	177	141	101
Price (\$)	0.50	0.60	0.70	0.80	0.90	1.00

Number Sodas Supplied per Day	97	122	146	175	195	214
Price (\$)	0.50	0.60	0.70	0.80	0.90	1.00

(a) Find models for the supply and demand functions, and justify your choices.
(b) Find the equilibrium price and quantity using your models.
(c) Find the Consumer Surplus and Producer Surplus at the equilibrium price and quantity using your models, and explain in words what each means.
(d) What would be the *actual* collective *revenue* of the producers at the equilibrium price and quantity?
(e) If the suppliers colluded to fix the price at \$0.85, what would happen? What would their actual collective revenue be? Would they *want* to do it? How can a society deal with this situation?
(f) What would be a way to get a different estimate of your answer to (e)? What do you get? What margin of error does this suggest?

CHAPTER 4 SUMMARY

Continuous Random Variables

A **continuous random variable**, usually denoted X, is a variable whose uncertain outcome can take on any real number in a specified interval. The *relative* likelihoods of the different possible outcome values are specified by a nonnegative **probability density function**

[15] Data adapted from a student project.

(pdf), $f(x)$, although any one particular outcome has a probability of 0. The probability that the value of X will lie in an interval $[a, b]$ corresponds to the area under the pdf and above the x-axis between $x = a$ and $x = b$. The total area under $f(x)$ is 1.

Area under a Curve

Given a nonnegative rate of change (derivative) function, the **change in the original function over an interval** corresponds to the area under the rate graph and above the horizontal axis over the interval. This is most obvious when the rate is constant but also holds for general curves.

The **area under a nonnegative curve**, such as $f(x)$, and above the horizontal axis over an interval $[a, b]$ is called a **definite integral** and is symbolized $\int_a^b f(x)dx$. The area can be approximated by dividing the interval into n equal subintervals, each of width $\Delta x = \dfrac{b-a}{n}$, with endpoints $a{=}x_0, x_1, x_2, \ldots, x_n{=}b$, where $x_i = a + i\Delta x$. The approximation formulas for the three major methods used in the chapter are

Right Rectangles: (draws rectangles with heights determined by the *right* endpoints):

$$S_n = \sum_{i=1}^{n} f(x_i)\Delta x = \Delta x[f(x_1) + f(x_2) + \cdots + f(x_n)]$$

Trapezoids: (draws trapezoids by joining adjacent subinterval endpoints):

$$S_n = \frac{\Delta x}{2}[f(x_0) + 2f(x_1) + 2f(x_2) + \cdots + 2f(x_{n-2}) + f(x_n)]$$

Simpson's Rule: (fits parabolas to endpoints of *pairs* of subintervals, so n must be *even*):

$$S_n = \frac{\Delta x}{3}[f(x_0) + 4f(x_1) + 2f(x_2) + 4f(x_3) + \cdots + 2f(x_{n-2}) + 4f(x_{n-1}) + f(x_n)]$$

Two other methods include Left Rectangles (heights determined by the *left* endpoints of the subintervals) and Midpoint Rectangles (heights determined by the height of the curve at the *midpoint* of each subinterval). The three major methods are listed above in order of increasing accuracy. Left Rectangles are similar to Right Rectangles, and Midpoint Rectangles estimate slightly better than Trapezoids. Trapezoids and Simpson's Rule are nicely set up for making estimates directly from discrete data with equal subinterval widths.

For any of the subinterval (numerical integration) approximations, if the curve for $f(x)$ is continuous, then the exact area (definite integral) can be found by taking the **limit of the sums**: $\lim_{n\to\infty} S_n$. This can be done using technology to approximate the limit by constructing a table with increasingly larger values of n until the estimates converge to the desired level of precision (number of significant digits).

Indefinite Integrals, Antiderivatives

Given a rate (derivative) function, the original function from which it came is an **antiderivative** of the rate function (a function whose derivative equals the rate function).

Once you find one antiderivative of a function, you can add any constant (+ C) to it to obtain the most general antiderivative. The most general form for antiderivatives of a given function $f(x)$ is called an **indefinite integral** (different from a definite integral because it has no limits of integration), denoted $\int f(x)dx$. Some common rules for finding antiderivatives include

$$\int ax^n dx = \frac{a}{n+1}x^{n+1} + C = \frac{ax^{n+1}}{n+1} + C$$

$$\int ae^{kx}dx = \frac{a}{k}e^{kx} + C = \frac{ae^{kx}}{k} + C$$

$$\int \frac{k}{x}dx = k \ln x + C$$

$$\int ab^x dx = \frac{a}{\ln b}b^x + C = \frac{ab^x}{\ln b} + C$$

$$\int kf(x)dx = k\int f(x)dx$$

$$\int [af(x) \pm bg(x)]dx = a\int f(x)dx \pm b\int g(x)dx$$

Once you have found the *general* form of any antiderivative of a given rate function, if you know the value of the specific original (antiderivative) function at a particular input value, you can plug in that input value into your formula for the general antiderivative, set the result equal to the known output value at that point, and solve for C. The original function can usually be thought of as a **cumulative** function. Once you know the specific original function, you can then find how much it has changed over a specified interval by plugging in the endpoints and subtracting the value at the upper endpoint minus the value at the lower endpoint. If *all* that you need is the *change* in the original function, you don't even need to find the value of C, since it cancels out in the calculation, so you can use *any* form of the antiderivative.

If a continuous rate function is negative in places, then the definite integral and the above procedures will find the *net* change in the original function, which corresponds to the area *above* the horizontal axis *minus* the area *below* the horizontal axis between the curve and the horizontal axis. This makes sense since all of the subinterval methods involve values of the function, $f(x_i)$.

The Fundamental Theorem of Calculus

Since the function being integrated in *any* general definite integral *could* be a rate or velocity function, in which case the above procedures would apply and would find the desired area calculation, we see that the result is completely general, and is called the **Fundamental Theorem of (Integral) Calculus**: If $F(x)$ is *any* antiderivative of a smooth function $f(x)$ (so that $F'(x) = f(x)$), then

$$\int_a^b f(x)dx = \Big[F(x)\Big]_a^b = F(b) - F(a)$$

The result corresponds to the area above the x-axis minus the area below the x-axis between the function's graph and the x-axis, and between the vertical lines $x = a$ and $x = b$. We also know that

$$D_x[\int f(x)dx] = D_x[F(x) + C] = F'(x) = f(x)$$
$$\int D_x[f(x)]dx = \int f'(x)dx = f(x) + C$$
$$D_x\left[\int_a^x f(t)dt\right] = D_x[F(x) - F(a)] = F'(x) = f(x)$$

These show that derivatives and integrals are very nearly, but not perfectly, inverse operations on functions. The last shows a **variable limit of integration** (the x-limit).

Mode, pdf, cdf

A **mode** (most likely value) of a continuous random variable is a maximum of its pdf. The standard convention is to define a pdf so that the area under it over its domain is one. A pdf can be thought of as a probability *rate* function (probability per unit), so an antiderivative of a pdf can be thought of as a cumulative probability. An antiderivative of the pdf would thus be a cumulative probability function, and the standard form chosen is $F(x) = P\{X \leq x\}$, called the **cumulative distribution function** (cdf).

Improper Integrals

When one or both limits of integration involve infinity, we have an **improper integral**. If both limits are infinite, we break the integral into two pieces. For the single limit case, we replace the infinite limit with a variable, and take a limit

$$\int_a^{\infty} f(x)dx = \lim_{b\to\infty}\int_a^b f(x)dx$$
$$\int_{-\infty}^{b} f(x)dx = \lim_{a\to-\infty}\int_a^b f(x)dx$$
$$\int_{-\infty}^{\infty} f(x)dx = \int_{-\infty}^{c} f(x)dx + \int_c^{\infty} f(x)dx$$

The limit can be estimated using technology, or found exactly by hand.

Median

The cdf of a continuous random variable can be found using $F(x) = \int_{-\infty}^{x} f(t)dt$. If the pdf is only positive on $[a, b]$, this simplifies to $\int_a^x f(t)dt$ for any x in $[a, b]$. The **median**, or 50th percentile, is the middle value of a random variable in the sense that it divides the area under the pdf in half, so that if m is the median, it must satisfy the equation $\int_{-\infty}^{m} f(x)dx = 0.5$. To find the median, set up this equation and solve for m, either by hand or using technology.

Area between Two Curves

If $f(x)$ is always above $g(x)$ on the interval $[a, b]$, then **the area between the two curves** is given by $\int_a^b [f(x) - g(x)]dx$.

Consumer and Producer Surplus

A **demand function** $D(q)$ is the selling price that would have to be charged to sell exactly q units of a good and is usually a decreasing function. A **supply function** $S(q)$ is the selling price at which producers would want to make exactly q units and is usually an increasing function. If the supply and demand functions cross, the intersection corresponds to the **equilibrium price and quantity**, which are the price and quantity that match for *both* producers and consumers. To find the equilibrium price and quantity, set the supply and demand functions equal to each other and solve. In a competitive market, economic forces tend to push the price and quantity toward this equilibrium. At equilibrium, the **Producer Surplus** is the extra revenue gained by producers collectively by selling at that price compared to the *minimum* selling price at which they would have been willing to sell each unit (given by the supply function). Graphically, this is the area between the supply function and the equilibrium price. Similarly, at equilibrium, the **Consumer Surplus** is the collective *savings* by consumers by being able to buy the good at the market price rather than the *maximum* they would have been willing to pay for each unit (given by the demand function). Graphically, this is the area between the equilibrium price and the demand function. To calculate them,

$$PS = \int_0^{q^*} [p^* - S(q)]dq = p^*q^* - \int_0^{q^*} S(q)dq$$

$$CS = \int_0^{q^*} [D(q) - p^*]dq = \int_0^{q^*} D(q)dq - p^*q^*$$

Similar expressions can be derived in nonequilibrium situations.

APPENDIX A

Mathematical Details

EXPONENTS AND LOGARITHMS

$$x^{-n} = \frac{1}{x^n} \quad (x \neq 0)$$

$$x^{m/n} = \sqrt[n]{x^m} = (\sqrt[n]{x})^m, \text{ for } x > 0$$

$$(x^m)^n = x^{mn}$$

$$x^m \cdot x^n = x^{m+n}$$

$$\frac{x^m}{x^n} = x^{m-n}$$

$\log_b x$ means the exponent you would have to raise b to in order to get an answer of x. For example, $\log_2 8 = 3$, since $2^3 = 8$. Thus,

$$y = \log_b x$$

is equivalent to saying that $b^y = x$.

$\ln x$ means $\log_e x$, where

$$e = \lim_{n \to \infty} \left(1 + \frac{1}{n}\right)^n \approx 2.71828\ldots$$

is an irrational number, similar to π (see Section 5.3 of Volume 2 for more details about e). Thus,

$$y = \ln x$$

is equivalent to $x = e^y$, which means that $\ln x$ and e^x are inverse functions, and

$$e^{\ln x} = \ln(e^x) = x.$$

More generally,

$$b^{\log_b x} = \log_b(b^x) = x$$
$$\log_b(xy) = \log_b x + \log_b y$$
$$\log_b\left(\frac{x}{y}\right) = \log_b x - \log_b y$$
$$\log_b(x^n) = n \cdot \log_b x$$
$$\log_b(\sqrt[n]{x}) = \frac{1}{n}\log_b x$$

(when $b = e$, you get equivalent rules for natural logarithms, using "ln" in place of "$\log_b$").

TO CONVERT BETWEEN $y = ab^x$ FORM AND $y = ce^{kx}$ FORM

We can write

$$ce^{kx} = c(e^k)^x$$

so to convert from ce^{kx} form to ab^x form, let $a = c$ and $b = e^k$. To convert in the other direction, we want to find the value of k for which

$$e^k = b$$

Taking the natural logarithm of both sides, we get that

$$\ln(e^k) = \ln b$$

so

$$k = \ln b \qquad (\text{since } \ln(e^k) = k)$$

Thus to convert from ab^x form to ce^{kx} form, let $c = a$ and $k = \ln b$. We will study this relationship more in Section 5.3 of Volume 2.

PROOF THAT $D_x(\ln x) = 1/x$

Let us define $y = f(x) = \ln x$. Then we know that

$$e^y = x, \text{ or } e^{f(x)} = x$$

Taking D_x (the derivative with respect to x) of both sides, we get that

$$
\begin{aligned}
D_x[e^{f(x)}] &= D_x(x) && \text{(taking } D_x \text{ of both sides)}\\
[e^{f(x)}][f'(x)] &= 1 && \text{(chain rule, with } u = f(x))\\
f'(x) &= \frac{1}{e^{f(x)}} && \text{(dividing both sides by } e^{f(x)})\\
&= \frac{1}{e^{\ln x}} && \text{(since } f(x) = \ln x)\\
&= \frac{1}{x} && \text{(since } e^{\ln x} = x)
\end{aligned}
$$

so $f'(x) = D_x(\ln x) = \frac{1}{x}$, as desired.

BINOMIAL THEOREM

A generalization of multiplying out binomials:

$$
\begin{aligned}
&(x + h)^1 = x + h\\
&(x + h)^2 = x^2 + 2xh + h^2\\
&(x + h)^3 = x^3 + 3x^2h + 3xh^2 + h^3\\
&\ldots\\
&(x + h)^n = x^n + nx^{n-1}h + \text{(terms that involve } h \text{ to an exponent of 2 or more)}
\end{aligned}
$$

PROOF THAT FOR ANY POSITIVE INTEGER, n, $D_x(x^n) = nx^{n-1}$

If we define $f(x) = x^n$, then;

$$f'(x) = \lim_{h\to 0}\frac{f(x+h) - f(x)}{h} = \lim_{h\to 0}\frac{(x+h)^n - x^n}{h}$$

Expanding $(x + h)^n$ by the binomial theorem we get

$$f'(x) = \lim_{h\to 0}\frac{[x^n + nx^{n-1}h + \text{(terms that involve } h \text{ to an exponent of 2 or more)}] - x^n}{h}$$

Combining the x^n terms we have

$$f'(x) = \lim_{h\to 0}\frac{nx^{n-1}h + \text{(terms that involve } h \text{ to an exponent of 2 or more)}}{h}$$

Factoring an h out of each term in the numerator we have

$$f'(x) = \lim_{h\to 0}\frac{h[nx^{n-1} + \text{(terms that involve } h \text{ to an exponent of 1 or more)}]}{h}$$

Since h approaches 0 but does not equal zero we can divide the numerator and denominator by h giving

$$f'(x) = \lim_{h\to 0} [nx^{n-1} + (\text{terms that involve } h \text{ to an exponent of 1 or more})]$$

Every term except the first term has an h in it, so every term except the first term will go to zero as h approaches zero and we have

$$f'(x) = nx^{n-1}$$

For the proof that this rule also holds for any n (including any fraction or irrational number), see an advanced calculus text.

PROOF OF THE PRODUCT RULE

We want to prove that, if $f(x)$ and $g(x)$ are smooth functions, then

$$\frac{d}{dx}[f(x)g(x)] = f(x)g'(x) + g(x)f'(x).$$

If we define $y = k(x) = f(x)g(x)$, then

$$\frac{d}{dx}[f(x)g(x)] = k'(x) = \lim_{h\to 0}\frac{k(x+h) - k(x)}{h} = \lim_{h\to 0}\frac{f(x+h)g(x+h) - f(x)g(x)}{h}$$

If we add and subtract the same expression from the numerator, it will not change the numerator. If we subtract and add $[f(x+h)g(x)]$ in the numerator, we get

$$\frac{d}{dx}[f(x)g(x)] = \lim_{h\to 0}\frac{f(x+h)g(x+h) - f(x+h)g(x) + f(x+h)g(x) - f(x)g(x)}{h}$$

Factoring $f(x+h)$ from the first two terms and $g(x)$ from the second two terms we get

$$\frac{d}{dx}[f(x)g(x)] = \lim_{h\to 0}\frac{f(x+h)[g(x+h) - g(x)] + g(x)[f(x+h) - f(x)]}{h}$$

Separating the terms[1], we have

$$\frac{d}{dx}[f(x)g(x)] = \lim_{h\to 0} f(x+h) \cdot \lim_{h\to 0}\frac{g(x+h) - g(x)}{h} + \lim_{h\to 0} g(x) \cdot \lim_{h\to 0}\frac{f(x+h) - f(x)}{h}$$

Look at the first limit. This limit is simply $f(x)$, since we are given that $f(x)$ is smooth (and so continuous). The second limit is the definition of the derivative of $g(x)$, or $g'(x)$.

---

[1]This separation or switching of the order of taking a limit and performing another operation can be done freely with addition, subtraction, multiplication, division, and powers and roots of positive quantities, as long as the individual limits come out to be real numbers.

The third limit is simply $g(x)$, since it does not involve h at all. The fourth limit is the definition of the derivative of $f(x)$ or $f'(x)$. Substituting these for the limits we have

$$\frac{d}{dx}[f(x)g(x)] = f(x)g'(x) + g(x)f'(x)$$

DERIVATION OF SIMPSON'S RULE

We will show that, given the points $(-h, y_1)$, $(0, y_2)$, and (h, y_3), the integral of the parabola fit to these three points on the interval $[-h,h]$ is given by

$$\frac{h}{3}[y_1 + 4y_2 + y_3]$$

Given the general form of the desired parabola, $y = ax^2 + bx + c$, the area we want would be given by

$$\begin{aligned}
\int_{-h}^{h}(ax^2 + bx + c)dx &= \left[\frac{ax^3}{3} + \frac{bx^2}{2} + cx\right]_{-h}^{h} \\
&= \left[\frac{a(h)^3}{3} + \frac{b(h)^2}{2} + c(h)\right] - \left[\frac{a(-h)^3}{3} + \frac{b(-h)^2}{2} + c(-h)\right] \\
&= \frac{ah^3}{3} + \frac{bh^2}{2} + ch - \left(-\frac{ah^3}{3} + \frac{bh^2}{2} - ch\right) \\
&= \frac{ah^3}{3} + \frac{bh^2}{2} + ch + \frac{ah^3}{3} - \frac{bh^2}{2} + ch \\
&= \frac{2ah^3}{3} + 2ch \\
&= \frac{h}{3}(2ah^2 + 6c)
\end{aligned}$$

Again using the general form of the parabola, we can plug in the values of the given points to obtain

$$\begin{aligned}
y_1 &= a(-h)^2 + b(-h) + c = ah^2 - bh + c \\
y_2 &= a(0)^2 + b(0) + c = c \\
y_3 &= a(h)^2 + b(h) + c = ah^2 + bh + c
\end{aligned}$$

Notice that if we add $y_1 + 4\,y_2 + y_3$ by doing the same operations to the right-hand sides above, we get

$$\begin{aligned}
y_1 + 4\,y_2 + y_3 &= (ah^2 - bh + c) + 4(c) + (ah^2 + bh + c) \\
&= 2ah^2 + 6c
\end{aligned}$$

which is the expression we got in the parentheses after factoring out the $h/3$ above. Thus the area under the parabola is given by

$$\frac{h}{3}[y_1 + 4\,y_2 + y_3]$$

This area would be the same whether the interval was centered at 0 or anywhere else, as long as the width of each piece was h, so this result applies to every pair of subintervals, and we get the final formula

$$S_n = \frac{\Delta x}{3}\left\{\begin{array}{l}[f(x_0) + 4f(x_1) + f(x_2)] + [f(x_2) + 4f(x_3) + f(x_4)] + \cdots \\ + [f(x_{n-4}) + 4f(x_{n-3}) + f(x_{n-2})] + [f(x_{n-2}) + 4f(x_{n-1}) + f(x_n)]\end{array}\right\}$$

$$= \frac{\Delta x}{3}[f(x_0) + 4f(x_1) + 2f(x_2) + 4f(x_3) + \cdots + 2f(x_{n-2}) + 4f(x_{n-1}) + f(x_n)]$$

QUADRATIC FORMULA

If $ax^2 + bx + c = 0$, then $x = \dfrac{-b \pm \sqrt{b^2 - 4ac}}{2a}$.

Derivation:

$$ax^2 + bx + c = 0$$

$$x^2 + \frac{b}{a}x + \frac{c}{a} = 0$$

$$x^2 + \frac{b}{a}x = -\frac{c}{a}$$

$$x^2 + \frac{b}{a}x + \frac{b^2}{4a^2} = -\frac{c}{a} + \frac{b^2}{4a^2}$$

$$\left(x + \frac{b}{2a}\right)^2 = \frac{b^2}{4a^2} - \frac{4ac}{4a^2}$$

$$x + \frac{b}{2a} = \pm\sqrt{\frac{b^2 - 4ac}{4a^2}}$$

$$x = -\frac{b}{2a} \pm \frac{\sqrt{b^2 - 4ac}}{2a}$$

$$= \frac{-b \pm \sqrt{b^2 - 4ac}}{2a}$$

APPENDIX B

Student-Generated Projects

BACKGROUND

We strongly recommend that semester-long projects on a topic chosen by each student, possibly working with one or two other students, be a component of this course. The purpose is to give you experience in the *entire* process of problem solving (applying the 12-Step Program for Plugaholics presented in Section 1.0) from beginning to end working on a problem that *you* care about and can use help with. The idea is to pose a question or problem that is interesting, useful, real, personally meaningful and important, appropriately challenging but manageable, and relevant to the topics in the course. For the material in Volume I, this normally means some kind of optimization problem that can be reduced to *one* decision variable (independent variable). This often means trying to find the best way to do something, or the best value for some choice you have to make in order to maximize or minimize some criterion that is important to you. One helpful guideline for what is a good project topic is to see if you can give an initial ballpark guess of the optimal input and output values, or can roughly sketch a graph of the relationship you expect between the input and the output. In general, the topics that usually work best are those where you have an intuitive sense that there's an optimum level *somewhere*, and that a lower or higher input value would be worse on either side of that as-yet unknown point. Mathematically, we'd say that projects work out best when you have an intuitive feel that there is a global extremum which is also a local extremum within your domain of realistic feasibility. Meg's exercise problem in Section 1.0 is a prototype student-generated project followed from start to finish that you should read to get a feel for what this process is all about.

TOPIC IDEAS

Here are some examples of topic ideas that have worked well for other students in the past:

- Find the optimal level of exercise, sleep, studying, time on the phone, and so forth, for a given period of time (day, week, month, etc.). The output (dependent variable) will usually be a subjective scale. It works best to use a scale of 0–100, with verbal definitions to correspond to different numerical values, to encourage as much precision as possible. Try to define the output to be as specific to the problem of interest as possible. For instance, it could simply be a subjective judgment evaluating the *balance* of time spent on the selected activity being studied, compared to everything else, in the time period in question (for example, make 100 be Perfect, 75 be Good, 50 be OK, 25 be Poor, and 0 be the Worst).
- Find the optimal time to wait after a class (or the optimal time of day) for doing the homework from that class. The output variable can be the percentage of questions correct, or perhaps time spent working on the homework, holding other factors constant.
- Find the optimal angle or distance for a particular sports action (shooting a basketball or hockey puck, kicking a soccer ball, etc.) to maximize your scoring percentage. It is best to get observations for at least six values (angles, distances) of the independent (decision) variable, and to try to get 30 or more observations at each value (perhaps broken up, such as 3 groups of 10, to get a measure of variability). Mix the inputs randomly as you gather the data.
- Find the optimal warm-up time before a sports or musical activity. The output can be a subjective evaluation of the quality of performance, or a quantitative measure.
- Find the highway driving speed to maximize car mileage (miles per gallon) on the car you normally drive. This is best with electronic fuel gauges. Otherwise, you need to top off the tank just before and after a trip (the longer the better, for accuracy and precision) to get a good estimate of gas used (the amount added after the trip). You should also get before and after odometer readings for the mileage.
- Find the optimal duration of each practice session (such as, for musical activities) given a fixed total time spent in a given period of time (day, week, month, etc.). For example, if you can spend an hour a day, should you do a 60-minute session (once a day), 30-minute sessions (twice a day), 20-minute sessions (3 times a day), and so forth. The output variable can be the total time to learn comparable pieces (e.g., time before you can play it through without a significant error).
- Find the optimal amount of time to spend studying for each of two tests, given a fixed amount of available time to study total for the two together. This involves first evaluating estimates of the expected grades on each test for different amounts of studying. You then choose the time to study for one of the subjects as the decision variable (so the balance of the time will be spent on the other). For different values of this decision variable, you can then estimate the grades on

each test based on the time allotted, and average the two to get the output score. This is a somewhat difficult topic, so it is better if you feel you have a good understanding of the material.

- Find the optimal amount of time to wait between certain repeated activities (such as stocking up on soda, snacks, or other staples, or doing laundry). This can be a very difficult topic, so is probably best only if you are really looking for a challenge. You can subjectively assess the monetary "cost" of storage each cycle as a function of the length of the cycle (the simplest form is a cost per unit per time unit), including factors like foregone interest (money tied up, not in the bank), electricity, and inconvenience (how much you would have paid over a cycle to *not* have the items stored). You can also assess the "cost" of the activity at the relevant end of the cycle (in direct real money, time, gas, depreciation, etc.), both fixed and variable. You can then divide the total cost of each cycle by its length to get a cost per unit time, which is what you want to minimize, as a function of the cycle time.

If you flip through the text, you will also see numerous sample problems, examples, and exercises that are based on real student projects to give you additional ideas.

WRITING A PROPOSAL, PLANNING FOR DATA COLLECTION

Once you find a topic idea that seems to make sense, you should write up your concept of the project, which is the culmination of Stage A, Clarifying the Problem. First, explain why the problem is interesting, useful, real, and personally meaningful and important for you. Then explain how you are bounding and defining the problem, including definitions of the first and second variables you will be using to gather your data, and the model you expect to formulate as far as you can at this stage (perhaps a verbal definition, a guess at the type of function you expect to fit, a rough idea of the domain, and the assumptions you will be making). You can even draw a rough sketch of what you expect the graph of the function to look like and make a ballpark guess at the optimal input and output values. It is critical to define the scale and units for any variables you are using, especially subjective values. It usually works best to make your subjective scale range from 0 to 100, with verbal descriptions for every multiple of 10, as Meg did in Section 1.0. This gives a natural middle value (50) and allows you to be as precise as you can. Smaller scales are likely to just increase the margin of error in your results.

Part of the difficulty of reducing a real problem to a single decision variable is that there are usually many other factors involved. You need to think about how you can keep these as constant as possible, or keep a record of major factors that might have a strong influence on your results.

This is a good time to start planning for gathering your data. Again, think about how you can try to control for factors other than your decision variable that might influence your output values. As you gather the data, you can keep a journal where you record any unusual circumstances related to these and other factors. Once you have a plan for getting data, think about what biases are likely. Will there be something tending to push all your output values either too high or too low? See the discussion about bias in

Section 3.4 for examples of this. After thinking about these possible biases, see if you can then adjust your data collection plan to eliminate or reduce any of them. **If you are gathering data that involves multiple repetitions of something** (such as a sports shooting percentage problem), **try to get the observations in several groups**, such as 3 or 4 groups of 10 shots each. This will help estimate a margin of error later on. Also, if you have control over your input, **try to mix somewhat randomly the sequence of input values you try**, so that training and fatigue effects don't bias your results. If you are using a random sample (such as a market survey), think about how you can best try to ensure that your sample is truly random and representative of your target population (which you should define clearly). See Section 3.4 for discussion of these topics.

You may not know much about what mathematical techniques you will use to find your model, or to find your optimal solution once you have a model. If you do, you can put these details into your proposal, but it is not crucial.

Your proposal is also a good time to think ahead to how you will verify and validate your model and solutions, perform sensitivity analysis, and estimate a margin of error for your results. Verification is usually best done by doing calculations both by hand and using technology, but at the proposal stage you will not know most of the techniques needed for this, so it is not critical to talk about it. Validation, the reality check that your model and solution make good common sense in the real-world problem, is something that you can anticipate to some extent. Again, see the Meg example in Section 1.0 and the discussion of validation in Section 3.4 to get a feel for what validation is about. The main type of validation to think about at this early stage is whether and how you can test your solution directly. Sometimes this is possible, and sometimes it isn't. If it is, say how you expect to do this; if not, say how you might be able to validate in some other way (such as withholding the last data point and using it to test your model, if that makes sense). For sensitivity analysis, the most important preparation is to plan to keep a running journal to record any unusual circumstances that arise as you gather your data. With respect to margin of error, you could simply give a rough estimate of what you think the errors on your data input and output values are likely to be, and what the final rough margin of error for your results might be based on that alone. You could express your ballpark guess as a tentative conclusion to give an idea of what kind of answer you expect. Again, all of these points are discussed in detail in Section 3.4.

GATHERING DATA

Once you have written your proposal, and possibly gotten feedback on it, you should have a good idea of what you need to do to gather your data. Remember to keep a running journal in which you record any unusual circumstances that may influence each data point. **It is always better to gather too much data than too little!** You can always ignore data, but you usually can't get it after the fact. We mentioned above thinking about all of the other factors that might influence the output value you are studying. It couldn't hurt to record as many of them as you can, even if you are trying to hold them constant. These could possibly be used for a project for Volume II, when we study multivariable models. Remember the comments above about getting data in several groups for repeated activities and about randomly mixing up the input values you try in sequence to prevent training and fatigue biases.

After you have gathered your first couple of data points, check in with your instructor if possible to make sure you seem to be on the right track. It's easy to make adjustments at such an early stage; if you find out later that your data were not appropriate, it may be too late to do anything about it.

THE PROJECT REPORT DRAFT

Once you have an adequate set of data and have covered the appropriate topics in the text (for most projects, that would be through Section 3.4), you can do a full analysis. At this stage, making it beautiful isn't the goal; just focus on the content. If it will make your life easier later to use a word processor now, great, but writing out graphs and tables and equations should be whatever is fastest and easiest for you, all things considered. To get the most out of the project experience, point out the concepts from the course that you are using as you write up your report, especially in the analysis (mathematical) sections.

For the draft and the final report, a suggested structure is given below. Keep in mind that the process of problem solving can have many loops. You may start off with a model that you think is best, but in the process of validation, decide that something else is better. Ideally, for your report, you want to focus your detailed analysis on the model that you think is best (even if it isn't the one you initially thought would be). This could mean a major rewrite of your draft at some point to reflect your new insight. If you made such a change, mention that somewhere, either in the analysis or validation section.

1. **Introduction.** Describe the problem *context*, including what goal you wanted to achieve (usually to find the optimal numerical value of something). Be clear about why the problem is *real*, *interesting*, *important*, and *useful* to *you and your life*, and how it connects to the math topics in the course.
2. **Data.** *Clearly define* the variables you measured in your data, including units. If any data involved subjective scales, write out the scale and *explain* how you actually got your numbers. Explain your *plan* for gathering the data, and what you actually did. **Give all of the raw data values** somewhere (if they're huge, just attach them at the end), as well as summary information. Where possible, use graphs as well as tables to display information. Comment on whether your data points represent a function or not, and how you know. Explain anything you did to avoid or minimize biases in your data. If you used a random sample (such as a market survey), explain how you tried to make sure it was truly *random* and *representative* of your target population. Sections 1.0–1.2 and 3.4 are most relevant to this section of your report.
3. **Analysis.** Talk about how you formulated your model. If you formulated it (or parts of it) from a verbal description, explain the logic of how you derived your final function and show your steps. If you had to fit a curve to data, explain what function categories you considered and why (from the data and the context), why you made the final decision you did (how you balanced fit and shape considerations based on context and use, such as interpolation vs. extrapolation), and *how* you fit the curve. *Fully define your model* (verbal definition, including units, symbol definition, including the domain, and assumptions, including implications), as discussed in Section 1.2. If possible, show the mathematical steps by hand, just rounding parameters to 3 or 4 significant digits. For instance, use the derivative

rules to find the derivative of the function being optimized, set it equal to 0, and solve to find critical points, and check on the graph to find the global optimum, either at a critical point or an endpoint of the domain. As you go through the steps, explain what the meaning of the derivative is, why you set it equal to 0, and so on, to reinforce the concepts of the course. Sections 1.2 through 3.2 are most relevant to this section of your report.

4. **Verification.** The idea of verification is simply to double-check your calculations and make sure you didn't make any typos or other errors. To verify your derivative, try using the $(x + h)$ definition of the derivative involving a limit. You can discuss what this definition corresponds to graphically as well (see Section 2.2). To verify your optimum, you can use technology to maximize or minimize your function over its domain. Be careful, since most technologies require an initial guess, and the value you specify can influence the answer you get (it could be a local extremum but not the global one). To verify that your solution is a local extremum, you can explain and apply the Second Derivative Test and discuss concavity (see Section 3.3). **If any of these verifications gives different results**, work until you find the cause and have full confidence in your results. **Remember to *show* the results of every calculation; don't just say they agree!** Section 3.4 goes over the idea of verification in more detail.

5. **Validation.** This is your reality check, for both your model and your solution. To validate your model, first simply comment on its fit and shape in light of the context of the problem and your use of the model. Also, comment on the validity of the assumptions of certainty and divisibility (and any other assumptions you may have added) and their implications for the real-life context of the problem. To validate your solution, if possible, gather new data *after* your analysis to see how well the output predicted by the model for the new input compares to the actual new output from the data. If you can't do this, you could withhold your last data point from the analysis and then use that last point for validation. For further validation, try other model types that you considered originally and solve them quickly using technology to compare their fit and shape and solutions with your main model. You may realize that something else seems better, which would suggest a major rewrite of the report using that new main model as the centerpiece. Section 3.4 gives more details on validation.

6. **Sensitivity Analysis.** The point of sensitivity analysis is to examine the reliability of your data, and the effects of plausible changes in the data on your solution. Based on your data journal, look back to see if any of your data points corresponded to unusual circumstances that might make them unrepresentative of normal reality. Whether it is because you were sick, or got no sleep, or whatever, identify those points. Decide whether you should just throw out those data points, or modify them in some way to adjust for the unusual circumstances. You can look at the graph of your data to look for points that might suggest something unusual (out of line in some way), but these should really not be changed unless you have a good reason to do so. After making such changes, try re-solving the problem quickly (probably using technology) to see how much your solution is affected. You can try changes individually and/or in groups to get a feel for the sensitivity of your solution to such changes. Section 3.4 gives more details about sensitivity analysis.

7. **Biases and Precision: Margin of Error.** Even after trying to prevent biases, you may believe your data and results still have a bias. At this stage you could try to estimate that by whatever means you can think of, even if it is only subjective, and then adjust your solution accordingly. You can also subjectively try to estimate the margin of error in your input and output data values individually. Besides combining the errors from the data with errors related to calculations and rounding, you can also use solutions from sensitivity analysis and trying alternative models to get a sense of a reasonable margin of error. The final evaluation is very subjective, but very important to communicate. Section 3.4 discusses how to do each component of this in detail, as well as how to then try to estimate an overall margin of error for your solution.
8. **Conclusions.** This stage involves putting all of the pieces together. First and foremost, take your solution and margin of error estimates and, considering the context, **state the optimal value(s) of the decision variable, and the corresponding optimal output value, to an appropriate level of precision, indicating the rough margin of error, in words that match the way that the problem was posed**. Say how you can make practical use of this in real life, and what you have learned from the process. You could also suggest ways you could analyze the problem further (more complicated models, additional data, etc.) or other problems or questions that your analysis has raised that you might like to explore. Section 3.4 is most relevant for this section.

THE FINAL PROJECT REPORT

The final project report should have the same format as the project report draft. Now is the time to **make it beautiful, fun, professional, sexy, high-tech, creative, and impressive!** That means trying to get computer-generated graphs and tables and equations if possible, thinking up a clever title, making an attractive cover, making sure your spelling and grammar and word usage are correct, making your writing style as clear and enjoyable to read as possible, giving full citations for any books or data mentioned from sources other than yourself, and in general doing everything possible to create a quality product. It is a good idea to have at least one other person proofread it for you and give you comments and suggestions; try to leave yourself time for this before the final report is due.

As you write the report, imagine that you are writing it to someone who is just about to take the same course. This means that they would have the *prerequisite* mathematical background but would not know about the concepts of the *course itself*, which is why you are *explaining* the concepts as you use them.

For many projects at this stage, a large part of the value of the project to you personally is the information from the data itself. Often, from looking at a graph of the data, you can get a pretty good feel for the optimal input and output values and even the margin of error. Our main point in recommending these projects is that you experience the full process of using mathematical models to solve real problems that you care about. It is convenient to be able to have a good intuition from the beginning of the problem based on the graph of your data, to put everything in context. In Volume II, covering problems involving two or more decision variables, it is much harder (and often impossible) to graph the data, so the rest of the process becomes even more important. Once you have done this once, it should be that much easier for you to use these skills to solve real problems in your life, whether in other courses, your career, or your personal life.

APPENDIX C

Answers to Selected Exercises

Note: Unless otherwise indicated in the text, the answers are given for most of the **odd**-numbered exercises. For model specifications, our **conventions** are that the assumptions are certainty and divisibility, so they are not always specified explicitly. When working with realistic problems, there is not always exactly one right answer. This may be unfamiliar and a little scary to you in a math class! Part of the purpose of this course is to help you develop independent and critical thinking. When this situation exists in an exercise, we have indicated it by saying "**Answers may vary.**" Sometimes we give one possible solution; sometimes we don't. Either way, you could (and sometimes probably should) give a different answer from ours. The important thing is to **explain your reasoning**; *that* is what makes an answer to this kind of question good or bad.

CHAPTER 1

Section 1.1

1. (a) No, y is not a function of x because there is not exactly one y value for each x value.
(b) Yes, x is a function of y because there is exactly one x value for each y value.

3. (a) Yes, the grade is a function of the name because there is exactly one grade for each name.
(b) No, the name is not a function of the grade because there is not exactly one name for each grade.

5. (a) The graph is a function because it passes the vertical line test.
(b) The graph is not a function because it does not pass the vertical line test.
(c) The graph is a function because it passes the vertical line test.

7. (a) $f(3) = 24$
(b) $f(4.5) = 50.25$
(c) $f(8) = 164$
(d) $f(12)$ is not defined; 12 is not in the domain of the function.

9. (a) $f(0) = 12$
(b) $f(4.5)$ is not defined; 4.5 is not in the domain of the function.
(c) $f(1) = 16$
(d) $f(12)$ is not defined; 12 is not in the domain of the function.

11. (a) $f(2) = 5$
(b) $f(8) = 19$
(c) $f(0) = 5$
(d) $f(22)$ is not defined

13. (a) Yes. The number of points is a function of the game because there is exactly one point value for each game.
(b) No. The game number is not a function of the points scored because the value of 18 points could come from Game 2 or 4.

15.

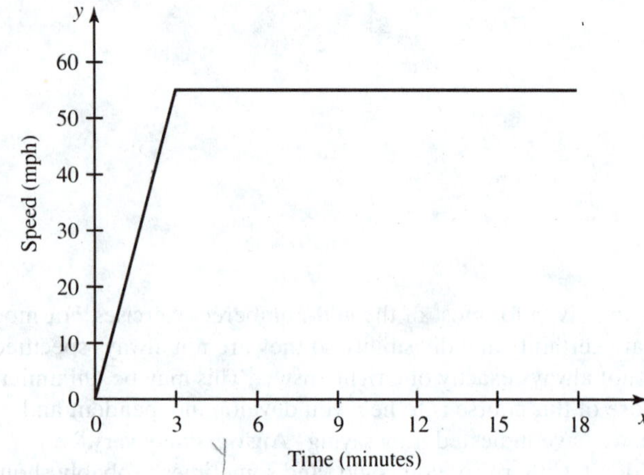

$f(t)$ = the speed (in mph) t minutes after leaving the toll booth.

No, the inverse is not a function because when the speed is 55 mph, the time coul be any value between 3 and 18 minutes after lunch.

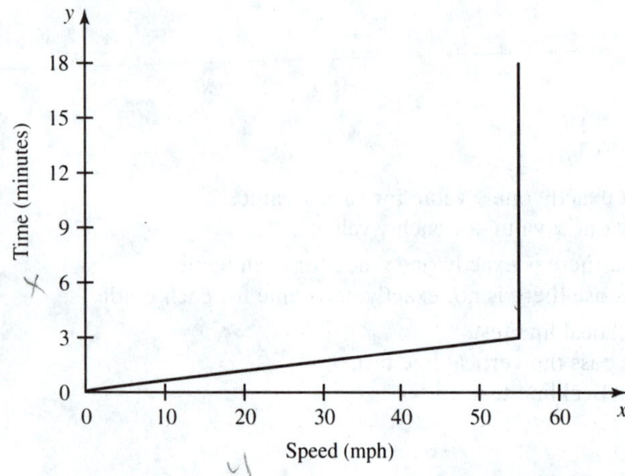

17. 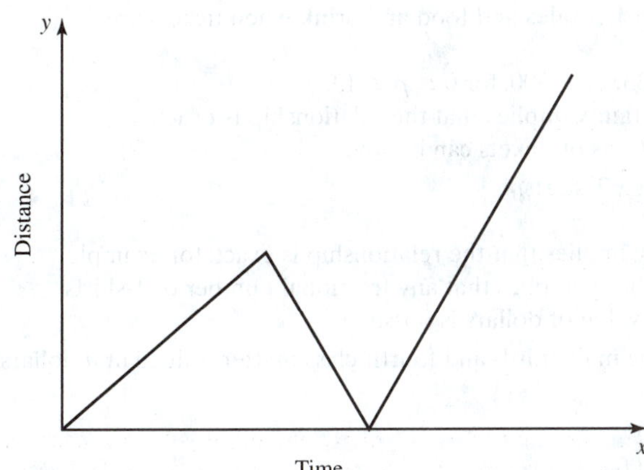

$f(t)$ = the distance (in yards) you are from your room t minutes after you first left your room [other units could be used as well].

19. Yes, the total cost of the oranges will be a function of the number of oranges that you buy.

Verbal Definition: $C(x)$ = total cost (in \$) if you buy x oranges.

Math Expression: $C(x) = 0.35x, x \geq 0$

21. No, a United States Social Security number is not a function of legal name because two people with the same legal name would correspond to two different Social Security numbers.

Yes, your name is a function of your Social Security number since each Social Security number is paired with exactly one legal name.

$f(x)$ = the legal name of the person with a U.S. Social Security number of x.

23. Yes, the marginal tax rate (expressed as a pure decimal) is a function of the taxable income because for a given income (rounded off to the nearest dollar) there is exactly one marginal tax rate.

$f(i)$ = the marginal tax rate that applies to an annual income of \$$i$.

No, income is not a function of marginal tax rate because, for example, a tax rate of 15% is paired with any income between \$0 and \$23,350 (this is also true for other tax rates).

Section 1.2

1. (a) $y = 13.46583x + 0.00376$
(b) $y = 13.466x + 0.0037562$
(c) $y = 13.465\ldots x + 0.0037561\ldots$

3. (a) $y = -0.00239x^2 - 4985.72859x + 8236573.93907$
(b) $y = -0.0023858x^2 - 4985.7x + 8.2366 \times 10^6$
(c) $y = -0.0023857\ldots x^2 - 4985.7\ldots x + 8.2365\ldots \times 10^6$
[Note: On a calculator, the last term may appear as 8.2365E6.]

5. (a) *Verbal Definition*: $r(p)$ = the revenue, in dollars, from food and drink when tickets are sold at p dollars per ticket.
Symbol Definition: $r(p) = 2(-15p + 203.33\ldots)$, for $0 \leq p \leq 13.50$.
Assumptions: Certainty and divisibility. Certainty implies that the relationship is exact. Divisibility implies that fractions of tickets can be sold.

(b) $R(p) = (p + 2)(-15p + 203.33\ldots)$, for $0 \leq p \leq 13.50$.

(c) *Verbal Definition*: $P(p)$ = the total profit from ticket sales and food and drink when tickets are sold at p dollars per ticket.
Symbol Definition: $P(p) = (p + 2)(-15p + 203.33\ldots) - 300$, for $0 \le p \le 13.50$.
Assumptions: Certainty and divisibility. Certainty implies that the relationship is exact. Divisibility assumes that fractions of tickets can be sold.

7. *Verbal Definition*: $C(x)$ = the cost, in dollars, to order x T-shirts.
Symbol Definition: $C(x) = 7.42x + 50$, for $x \ge 0$.
Assumptions: Certainty and divisibility. Certainty implies that the relationship is exact, for example, there are no bulk discounts. Divisibility implies that any fractional number of T-shirts can be ordered and any fractional value of dollars is possible.

9. *Verbal Definition*: $C(d)$ = the cost, in dollars, to insure mail, third- and fourth-class matter, valued at d dollars.

Symbol Definition: $$C(d) = \begin{cases} 0.75, & \text{for } 0 \le d \le 50 \\ 1.65, & \text{for } 50 < d \le 100 \\ 2.50, & \text{for } 100 < d \le 200 \\ 3.40, & \text{for } 200 < d \le 300 \\ 4.30, & \text{for } 300 < d \le 400 \\ 5.20, & \text{for } 400 < d \le 500 \\ 6.10, & \text{for } 500 < d \le 600 \end{cases}$$

[Note: You can also use break points of 50.005, 100.005, etc.]
Assumptions: Certainty and divisibility. Certainty holds exactly. Divisibility implies that any fractional value of dollars is possible.

11. (a) *Verbal Definition*: $C(n)$ = the cost, in dollars, to order n caps.
Symbol Definition: $C(n) = 3.5n + 20$, for $n \ge 0$.
Assumptions: Certainty and divisibility. Certainty implies that the relationship is exact. Divisibility implies that fractions of dollars and caps are possible.

(b) *Verbal Definition*: $\overline{C}(n)$ = the average cost, in dollars per cap, to order n caps.
Symbol Definition: $\overline{C}(n) = 3.5 + \dfrac{20}{n}$, for $n > 0$.
Assumptions: Certainty and divisibility. Certainty implies that the relationship is exact. Divisibility implies that fractions of dollars and caps are possible.

13. *Verbal Definition*: $C(x)$ = the cost, in dollars, to order x cookbooks.
Symbol Definition: $$C(x) = \begin{cases} 100 + 3.95x, & \text{for } 0 \le x \le 100 \\ 119.75 + 3.7525x, & \text{for } x > 100 \end{cases}.$$
Assumptions: Certainty and divisibility. Certainty implies that the relationship is exact, for example, there are no further discounts. Divisibility implies that any fractional number of cookbooks can be ordered and any fractional value of dollars is possible.

15. *Verbal Definition*: $\overline{C}(q)$ = the average cost per day, in dollars, when you buy q liters every trip.
[Note: If you define the variable to be the number of 2-liter bottles bought each time, your variable terms will be different.]
Symbol Definition: $\overline{C}(q) = 0.25 + \dfrac{1}{q} + 0.02q$, for $q > 0$.
Assumptions: Certainty and divisibility. Certainty implies that the relationship is exact. Divisibility implies that any fractional number of liters can be bought and any fractional value of dollars is possible.

17. (a) Yes, the price can be thought of as a function of the number of people buying tickets.
Verbal Definition: $p(n)$ = the price, in dollars, that needs to be charged in order to sell exactly n tickets.
Symbol Definition: $p(n) = -0.06666\ldots n + 13.55\ldots$, for $0 \le n \le 203.33\ldots$
Assumptions: Certainty and divisibility. Certainty implies that the relationship is exact. Divisibility implies that any fractional number of tickets can be sold and any fractional number of dollars is possible.
[Note: If you fit a model directly to the data, your answer will be different.]

(b) *Verbal Definition*: $R(n)$ = the revenue, in dollars, when the price is set so that exactly n tickets are sold.
Symbol Definition: $R(n) = -0.06666\ldots n^2 + 13.55\ldots n$, for $0 \le n \le 203.33\ldots$
Assumptions: Certainty and divisibility. Certainty implies that the relationship is exact. Divisibility implies that any fractional number of tickets can be sold and any fractional number of dollars is possible.

19. Answers may vary.

21. Answers may vary.

Section 1.3

1. Graphs (a), (b), and possibly (c) could reasonably be modeled by a linear function.

3. This data set could reasonably modeled by a linear function.

5. This data set could possibly be reasonably modeled by a linear function.

7. This data set could reasonably modeled by a linear function.

9. This data set could possibly be reasonably modeled by a linear function.

11. Answers may vary. One possible answer is $y = 1.75x + 2$.

13. Answers may vary. One possible answer is $y = -0.6x + 56$.

15. (a) Your pace should be about 9.15 minutes per mile to finish in 4 hours.

(b) $D(t) = \dfrac{t}{9.15}$ ($D(t)$ is the number of miles run in t minutes), for $0 \le t \le 240$.

(c)

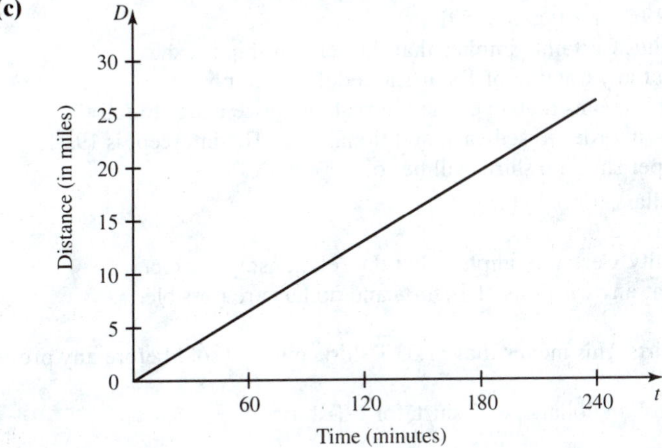

17. (a) *Verbal Definition*: $f(x)$ = the life expectancy at birth, in years, x years after the middle [chosen arbitrarily] of 1982.
Symbol Definition: $f(x) = 0.1013\ldots x + 74.45\ldots$, for $x \ge 0$.
Assumptions: Certainty and divisibility. Certainty implies that the relationship is exact. Divisibility implies that any fractional year is possible.

(b) The intercept is about 74.5 years of life expectancy. This means that at time 0 (mid-1982) life expectancy was about 74 and a half years. The slope is 0.101 years of life expectancy per year. This means that over this period, life expectancy was increasing by about one-tenth of a year every calendar year.

(c) Answers may vary. The model seems decent for interpolation.
(d) mid-1994: $f(12) \approx 75.7$
mid-1995: $f(13) \approx 75.8$
mid-1996: $f(14) \approx 75.9$
mid-2000: $f(18) \approx 76.3$
(e) Answers may vary. For example, since the outputs in the data seem to be leveling off the last few years, these predictions may be a bit high.

19. (a) *Verbal Definition*: $f(x)$ = the number of tickets sold if the price charged is x dollars.
Symbol Definition: $f(x) = -15x + 203$, for $0 \leq x \leq 13.5$.
Assumptions: Certainty and divisibility. Certainty implies that the relationship is exact. Divisibility implies that any fractional number of tickets and dollars are possible.
(b) The intercept is 203. This means that if the concert was free, about 203 people would come. The slope is -15 tickets per dollar of ticket price. This means that you lose about 15 people for each additional dollar charged over this range.
(c) Answers may vary. For example, raising the price was not a good idea, since revenues have gone down ($700 to $640 to $630).
(d) Answer may vary. For example, overexposure of the band could decrease attendance.

21. (a) *Verbal Definition*: $f(x)$ is the number of T-shirts sold if the price is x dollars.
Symbol Definition: $f(x) = -25.34x + 492.8$, for $0 \leq x \leq 19.45$.
Assumptions: Certainty and divisibility. Certainty implies that the relationship is exact. Divisibility implies that any fractional numbers of T-shirts and dollars are possible.
(b) The intercept is 492.8 shirts. This means that if the shirts were given away for free, they could "sell" about 493 shirts. The slope is -25.34 shirts per dollar of selling price. This means that, over this interval, you lose sales of about 25 shirts for each additional dollar of selling price.
(c) Answers may vary. For example, the model seems all right, but a bit too simple. The data seems to have a bit of curvature. Also, the lower bound of the domain is not certain. If the cost of the T-shirts were known, this could be the lower bound for p.
(d) They would be unable to sell any T-shirts when charging approximately $19.45.
(e) *Verbal Definition*: $p(q)$ = the price, in dollars, to charge to sell exactly q T-shirts.
Symbol Definiton: $p(q) = -0.03834q + 19.21$, for $0 \leq q \leq 501$.
Assumptions: Certainty and divisibility. Certainty implies that the relationship is exact. Divisibility implies that any fraction of T-shirts and dollars are possible.
(f) The slope is -0.03834 dollars per shirt. This means that, over the interval, the price needs to fall approximately 0.038 dollars, about 4 cents, in order to sell each additional shirt. The intercept is 19.21. This means that at approximately $19.21 per shirt, no shirts will be sold.

23. (a) *Verbal Definition*: $C(q)$ = the cost, in dollars, for q T-shirts.
Symbol Definition: $C(q) = 5.55q + 45$, for $q \geq 0$.
Assumptions: Certainty and divisibility. Certainty implies that the relationship is exact. Divisibility implies that any fractions of T-shirts and dollars are possible.
(b) $R(q) = 7q$ for $q \geq 0$
(c) The breakeven point is at least 31.03 T-shirts. This means that 31.03 T-shirts must be sold before any profit is realized.
(d) *Verbal Definition*: $\overline{C}(q)$ = the average cost, in dollars per T-shirt, for q T-shirts.
Symbol Definition: $\overline{C}(q) = 5.55 + \dfrac{45}{q}$, for $q \geq 0$.
Assumptions: Certainty and divisibility. Certainty implies that the relationship is exact. Divisibility implies that any fractions of T-shirts and dollars is possible.

25. (a) *Verbal Definition*: $f(t)$ is the total personal income (in billions of dollars) in the 12-month period that starts t years after the beginning of 1980.
Symbol Definition: $f(t) = 250.46\ldots t + 2103.5\ldots$, for $t \geq 0$.
Assumptions: Certainty and divisibility. Certainty implies that the relationship is exact. Divisibility implies that any fractional values for dollars and years are possible.

(b) Answers may vary. The linear model appears to be a fairly good fit, and the first differences are reasonably constant.

(c) 1995: About \$5681 billion
1996: About \$6111 billion
2000: About \$7113 billion

(d) About \$250.5 billion per year

27. (a) $S(t) = 6 + 1.00t$, for $t \geq 0$, where $S(t)$ is the cost of a small pizza with t toppings.
$M(t) = 8 + 1.00t$, for $t \geq 0$, where $M(t)$ is the cost of a medium pizza with t toppings.
$L(t) = 10 + 1.00t$, for $t \geq 0$, where $L(t)$ is the cost af a large pizza with t toppings.

(b) You can take advantage of the discount by ordering 5 large pizzas (at \$65) and save \$1.00 compared to 6 medium pizzas. But this would be a half a large pizza more than you need, so you could also get 4 large pizzas and 1 small pizza at \$61 or get 3 large pizzas and 2 medium pizzas at \$61.

(c) Answers may vary. There are several ways that this can be done. To give everyone a fair share (if you got 5 large pizzas), divide each pizza into 6 slices and each person gets 5 pieces. If you got 4 large and 1 small, cut the larges into 4 pieces each and the small into 2 pieces, and everyone gets 3 pieces. If you got 3 large and 2 medium, cut the larges into 4 pieces each, and give each person 3 pieces from a large or one whole medium pizza.

29. (a)

Percent Change

Year Interval	International Paper	S&P 500
1990–91	0.36 = 36.0%	0.310 = 31.0%
1991–92	−0.0368 = −3.68%	0.0763 = 7.63%
1992–93	0.0458 = 4.58%	0.0993 = 9.93%
1993–94	0.139 = 13.9%	0.0129 = 1.29%
1994–95	0.0256 = 2.56%	0.376 = 37.6%

(b) $f(x) = 0.313x + 0.0520$ (using the pure decimal form).

(c) $\beta \approx 0.313$

(d) The beta value gives a measure of how sensitive the stock is to changes in the market. In this case, β is less than 1, which means that the specific stock (International Paper) tends to change *less* in percentage terms than the fluctuations of the broad market (the S&P 500).

(e) Answers may vary. This would mean the stock is *countercyclical* to the market as a whole, such as perhaps for a collections agency (they get more work when everyone else is hurting).

(f) Answers may vary. Perhaps they did better than the market during the late 1980s.

Section 1.4

1–9. Answers may vary.

1. (a) exponential or quadratic
(b) power or quadratic
(c) power, exponential, or quadratic
(d) quadratic

3. linear

5. quadratic or power, or possibly linear

7. quadratic, power, or possibly even linear

9. linear

11. (a) quadratic or exponential
(b) Answers may vary. A power model with this concavity should go throught the origin.

13. (a) quadratic or exponential
(b) Answers may vary. A power model with this concavity should go through the origin.

15. *Verbal Definition*: $y(x)$ = energy level on average, on a scale of 0 to 100, when Meg ran x minutes per week.
Symbol Definition: $y(x) = -0.0007859\ldots x^2 + 0.8090\ldots x - 123.1\ldots$, for $240 \le x \le 730$.
Assumptions: Certainty and divisibility. Certainty implies that the relationship is exact. Divisibility implies that any fractional values of minutes and energy levels are possible.

17. *Verbal Definition*: $y(x)$ = energy level on average, on a scale of 0 to 100, when Meg ran x minutes per week.
Symbol Definition: $y(x) = -0.001394\ldots x^2 + 1.456\ldots x - 285.6\ldots$, for $420 \le x \le 605$.
Assumptions: Certainty and divisibility. Certainty implies that the relationship is exact. Divisibility implies that any fractional values of minutes and energy levels are possible.

19. Answers may vary. Possible answers are:

(a) *Verbal Definition*: $D(t)$ = the total number of deaths each year from AIDS in the United States, in the 12 months preceding t years after December 31, 1988.
Symbol Definition: $D(t) = -403.119\ldots t^2 + 6264.07\ldots t + 21{,}155.9\ldots$, for $0 \le t \le 6$.
Assumptions: Certainty and divisibility. Certainty implies that the relationship is exact. Divisibility implies that any fractional number of deaths and fractional years are possible.

(b) $D(8) \approx 45{,}469$. The model predicts that there will be about 45,469 deaths from AIDS in the United States in the year ending December 31, 1996.

(c) Answers may vary. The number of deaths is still increasing, so more money is needed to fund research and AIDS hostels.

21. Answers may vary. Possible answers are:

(a) *Verbal Definition*: $F(x)$ = the number of bank failures (closed or assisted) in the 12 months following x years after the beginning of 1978.
Symbol Definition: $F(x) = 3.064\ldots x^2 - 6.549\ldots x + 9.736\ldots$, for $0 \le x \le 9$.
Assumptions: Certainty and divisibility. Certainty implies that the relationship is exact. Divisibility implies that any fractional numbers of bank failures and years are possible.

(b) The plot is concave up, always increasing. A quadratic model or exponential model could fit. The quadratic model fit better for the most recent years.

(c) $F(10) \approx 250.7$. The model predicts that there would be approximately 251 bank failures (closed or assisted) in 1988.

(d) $F(11) \approx 308.5$. The model predicts that there would be approximately 309 bank failures (closed or assisted) in 1989.

23. *Verbal Definition*: $q(p)$ = the quantity of sweatshirts sold when the price is p dollars per sweatshirt.
Symbol Definition: $q(p) = 0.07p^2 - 4.79p + 80.95$, for $10 \le p \le 30.45$.
Assumptions: Certainty and divisibility. Certainty implies thar the relationship is exact. Divisibility implies that any fractional quantities of sweatshirts and dollars are possible.

The lower bound of the domain was chosen because the sweatshirts cost \$9 dollars each and no prices lower than \$10 dollars were considered. The upper bound of \$30.45 was chosen because the function becomes negative at this point, so it makes no sense to consider values over \$30.45.

Verbal Definition: $R(p)$ = the revenue, in dollars, when the price is p dollars per sweatshirt.
Symbol Definition: $R(p) = 0.07p^3 - 4.79p^2 + 80.95p$, for $10 \le p \le 30.45$.
Assumptions: Certainty and divisibility. Certainty implies that the relationship is exact. Divisibility implies that any fractional quantities of sweatshirts and dollars are possible.

Verbal Definition: $C(p)$ = the cost, in dollars, to order p sweatshirts.
Symbol Definition: $C(p) = 0.63p^2 - 43.11p + 753.55$, for $10 \le p \le 30.45$.
Assumptions: Certainty and divisibility. Certainty implies that the relationsip is exact, that there are no discounts. Divisibility implies that any fractional quantities of sweatshirts and dollars are possible.

Verbal Definition: $\pi(p)$ = the profit, in dollars, when the price is p dollars per sweatshirt.
Symbol Definition: $\pi(p) = 0.07p^3 - 5.42p^2 + 124.06p - 753.55$, for $10 \le p \le 30.45$.
Assumptions: Certainty and divisibility. Certainty implies that the relationship is exact. Divisibility implies that any fractional quantities of sweatshirts and dollars are possible.

25. Answers may vary. One possible model is:

(a) *Verbal Definition*: $P(w)$ = productivity, in average number of lawns maintained, with w full-time workers.
Symbol Definition: $P(w) = 0.4285\ldots w^2 + 7.771\ldots w + 7.8$, for $1 \le w \le 9$.
Assumptions: Certainty and divisibility. Certainty implies that the relationship is exact. Divisibility implies that any fractional values of lawns and workers are possible.

(b) $P(8) \approx 42.543$. On average, you could expect to maintain approximately 43 lawns if you had 8 full time workers.

27. Answers may vary. One possible model is:

(a) *Verbal Definition*: $P(m)$ = the percentage of rival products that come out within m months after yours.
Symbol Definition: $P(m) = 20.012\ldots(0.81874\ldots)^m$ or $20.012\ldots e^{-0.19998\ldots m}$, for $m \ge 0$.
Assumptions: Certainty and divisibility. Certainty implies that the relationship is exact. Divisibility implies that any fractional values of percentages and months are possible.

(b) $P(7) \approx 4.9357$. Approximately 4.94 percent of rival products will come out within 7 months of yours.

Section 1.5

1. Answers may vary.

(a) cubic or logistic
(b) logistic or cubic
(c) quartic, cubic, or linear; possibly logistic
(d) cubic

3. Answers may vary: cubic or quartic

5. Answers may vary: cubic or transformed logistic, or possibly linear

7. **(a)** Answers may vary. One possible model is:
Verbal Definition: $f(x)$ = the per capita public debt of the United States, in dollars, x years after mid-1975.
Symbol Definition: $f(x) = 2461.3\ldots(1.1152\ldots)^x$ or $2461.3\ldots e^{0.10903\ldots x}$, for $0 \le x \le 19$.
Assumptions: Certainty and divisibility. Certainty implies that the relationship is exact. Divisibility implies that any fractional values of dollars and years are possible.

(b) Answers may vary.
(c) Answers may vary.
mid-1995: $f(20) \approx 21{,}796$ (\$21,796)
mid-1996: $f(21) \approx 24{,}307$ (\$24,307)
(d) Answers may vary. The predictions do seem to be very high, but public debt has been growing sharply.
(e) Answers may vary as data becomes available.

9. **(a)** Answers may vary. Two possible models are:
Verbal Definition: $f(x)$ = the number of unemployed, in thousands, in the United States civilian population x years after December 31, 1983.
Symbol Definition: quartic: $f_Q(x) = -3.9097\ldots x^4 + 72.297\ldots x^3 - 286.94\ldots x^2 - 586.57\ldots x + 10{,}262.0\ldots$, for $0 \le x \le 11$.
cubic: $f_C(x) = -13.715\ldots x^3 + 305.65\ldots x^2 - 1901.3\ldots x + 10{,}704.3\ldots$, for $0 \le x \le 11$.
Assumptions: For each symbol definition given, certainty and divisibility are assumed. Certainty implies that the relationship is exact. Divisibility implies fractional values of number of unemployed and years are possible.

(b) Answers may vary. Quartic seems to be a better method for interpolation, while the cubic seems better for extrapolation.
(c) 1995: $f_Q(12) \approx 5761$ (5761 thousand)
1996: $f_Q(13) \approx 1315$ (1315 thousand)
1995: $f_C(12) \approx 8201$ (8201 thousand)
1996: $f_C(13) \approx 7508$ (7508 thousand)
(d) Answers may vary.

11. (a) Answers may vary. Possible models are:

Verbal Definition: $f_C(x)$ = the annual average of the foreign exchange rate, Canadian dollars per United States dollars, for the 12 months following x years after the beginning of 1970.

Symbol Definition: $f_C(x) = (3.711\times10^{-5})x^4 - 0.001791x^3 + 0.02631x^2 - 0.09868x + 1.017$, for $0 \le x \le 24$.

Assumptions: Certainty and divisibility. Certainty implies that the relationship is exact. Divisibility implies that any fractional values of dollars are possible.

Verbal Definition: $f_G(x)$ = the annual average of the foreign exchange rate, German marks per United States dollars, for the 12 months following x years after the beginning of 1970.

Symbol Definition: $f_G(x) = (9.129\times10^{-5})x^4 - 0.005040x^3 + 0.09127x^2 - 0.6362x + 3.694$, for $0\le x \le 24$.

Assumptions: Certainty and divisibility. Certainty implies that the relationship is exact. Divisibility implies that any fractonal values of dollars and marks are possible.

Verbal Definition: $f_J(x)$ = annual average of the foreign exchange rate, Japanese yen per United States dollars, for the 12 months following x years after the beginning of 1970.

Symbol Definition: $f_J(x) = -10.421x + 350.89$, for $0 \le x \le 24$.

Assumptions: Certainty and divisibility. Certainty implies that the relationship is exact. Divisibility implies that any fractional values of dollars and yen are possible.

(b) The rate improved in Canada, except 1985–1991, and Germany between 1980 and 1985.

(c) Answers may vary. A possible answer is:

Verbal Definition: Same as in part (a).

Symbol Definition: $f_C(x) = 0.0151x^2 - 0.613x + 7.35$, for $18 \le x \le 24$.

$f_G(x) = 0.00863x^2 - 0.395x + 6.3$, for $18 \le x \le 24$.

$f_J(x) = -2.74x^2 + 111x - 971$, for $18 \le x \le 24$.

Assumptions: Same as in part (a).

These models fit much better.

13. (a) Answers may vary. Possible models are:

Verbal Definition: $k(x)$ = the average kilowatt hours electric use pattern per day, x months after January 1 (0.5 = January average, 1.5 = February average, and so on).

Symbol Definition: $f(x) = 0.1093x^4 - 2.782x^3 + 22.30x^2 - 57.61x + 61.98$, for $0 \le x \le 12$.

Assumptions: Certainty and divisibility. Certainty implies that the relationship is exact. Divisibility implies that any fractional values of kilowatt hours and months are possible.

[Note: This model does not fit particularly well, but is probably the best model using only one formula.]

Verbal Definition: $c(x)$ = the average hundred cubic feet gas use pattern per day, x months after January 1 (0.5 = January average, 1.5 = February average, and so on).

Symbol Definition: $c(x) = 0.2436x^2 - 3.231x + 11.23$, for $0 \le x \le 12$.

Assumptions: Certainty and divisibility. Certainty implies that the relationship is exact. Divisibility implies that any fractional values of cubic feet and months are possible.

(b) The appliances were probably a gas heater and an electric air conditioner

(c) Answers may vary. Possible answers: changes in basic weather patterns, exceptionally hot summers, or exceptionally cold winters.

15. (a) Answer may vary. Possible models are:

Verbal Definition: $f(x)$ = the percent of the population x years of age in 1970.

Symbol Definition: $f(x) = (-1.994 \times 10^{-6})x^4 + (3.430 \times 10^{-4})x^3 - 0.01843x^2 + 0.2426x + 8.563$, for $0 \le x$.

Assumptions: Certainty and divisibility. Certainty implies that the relationship is exact. Divisibility implies that any fractional years and percentages are possible.

Verbal Definition: $f(x)$ = the percent of the population x years of age in 1980.

Symbol Definition: $f(x) = (-2.650 \times 10^{-6})x^4 + (4.969 \times 10^{-4})x^3 - 0.03080x^2 + 0.6315x + 5.075$, for $0 \le x$.

Assumptions: Certainty and divisibility. Certainty implies that the relationship is exact. Divisibility implies that any fractional years and percentages are possible.

Verbal Definition: $f(x)$ = the percent of the population x years of age in 1990.

Symbol Definition: $f(x) = (-1.149\times10^{-6})x^4 + (2.189\times10^{-4})x^3 + 0.01453x^2 + 0.3242x + 5.883$, for $0 \le x$.

Assumptions: Certainty and divisibility. Certainty implies that the relationship is exact. Divisibility implies that any fractional years and percentages are possible.

(b) Answers may vary. Possible answer: 30 to 40

(c) Answers may vary. While the models are decreasing for ages over 60, the percentages of people in these age groups are increasing, indicating that there will be more people using social security and Medicare as the years go on.

CHAPTER 2

Section 2.1

1. (a) 0.9
(b) 1
(c) 1.0333...
(d) 1.04

3.

	average rate of change	percent change	percent rate of change
(a)	-13	-16.7%	$-16.7\%, \dfrac{-16.7\%}{1}$ or $\dfrac{-13}{78}$
(b)	6	10%	$10\%, \dfrac{10\%}{1}$ or $\dfrac{6}{60}$
(c)	0	0%	$0\%, \dfrac{0\%}{2}$ or $\dfrac{0}{60}$

5. Answers may vary. Possible average rates of change are:

(a) $\dfrac{14 - 10}{12 - 10} = 2$

(b) $\dfrac{27 - 10}{14 - 10} = 4.25$

(c) $\dfrac{110 - 14}{20 - 12} = 12$

(d) $\dfrac{90 - 34}{19 - 15} = 14$

7. Answers may vary. Possible answers are:

	average rate of change	percent change	percent rate of change
(a)	$\dfrac{10 - 35}{10 - 5} = -5$	$\dfrac{10 - 35}{35} = -0.7142... \approx -71.4\%$	$\dfrac{-0.7142}{10 - 5} = -0.1428... \approx -14.3\%$
(b)	$\dfrac{35 - 35}{15 - 5} = 0$	$\dfrac{35 - 35}{35} = 0... = 0\%$	$\dfrac{0}{15 - 5} = 0... = 0\%$
(c)	$\dfrac{35 - 60}{5 - 3} = -12.5$	$\dfrac{35 - 60}{60} = -0.4166... \approx -41.7\%$	$\dfrac{-0.4166}{5 - 3} = -0.2083... \approx -20.8\%$

9. (a) 6.5
(b) -1
(c) 15.0
(d) -6.925

11. (a) $\approx$59.1% per unit
(b) -100% per unit
(c) -150% per unit
(d) $\approx$270.2% per unit

13. (a) The average rate of change from 0 to 0.5 years is 5.0 square feet per year.
The average rate of change from 0.5 to 1.0 year is 9.0 square feet per year.

(b) The weed coverage increased at a rate of 500% per year from 0 to 0.5 years.
The weed coverage increased at a rate of about 257% per year from 0.5 to 1.0 years.

15. (a) average rate of change: ≈ –569.7 thousand per year
percent change: ≈ −20.75%
percent rate of change: ≈ –6.916% per year

(b) average rate of change: 952 thousand per year
percent change: 43.75%
percent rate of change: ≈14.58% per year

(c) average rate of change: ≈191.2 thousand per year
percent change: ≈13.92%
percent rate of change: ≈2.321% per year

(d) No, because it shows only an increase rather than an initial decrease followed by an increase.

17. (a) The average rate of change in the number of tiles left from 5 to 10 minutes is about –3 tiles per minute.
The percent rate of change in the tiles left from 5 to 10 minutes is about –5% per minute.

(b) The average rate of change in the number of tiles left from 50 to 60 minutes is 2.7 tiles per minute.
The percent rate of change in the number of tiles left from 50 to 60 minutes is 27% per minute.

(c) The average rate of change in the number of tiles left from 30 to 40 minutes is 0 tiles per minute.
The percent rate of change in the number of tiles left from 30 to 40 minutes is 0% per minute.

19. Answers may vary. Possible answers are:

(a) The average rate of change from the end of 1986 to the end of 1987, [0, 1], is about −1.25 percentage points per year.
The average rate of change from the end of 1988 to the end of 1990, [2, 4], is about 1.0 percentage points per year.

(b) The annual percent change in the medical CPI from the end of 1986 to the end of 1987 was decreasing by about 17% per year. The annual percent change in the medical CPI from the end of 1988 to the end of 1990 was increasing by about 14% per year.

(c) The average rate of change from the end of 1986 to the end of 1987, [0, 1], is –1.4201 percentage points per year. The average rate of change from the end of 1988 to the end of 1990, [2, 4], is 0.9567 percentage points per year. The annual percent change in the medical CPI from the end of 1986 to the end of 1987 was decreasing by about 18.63% per year. The annual percent change in the medical CPI from the end of 1988 to the end of 1990 was increasing by about 14.10% per year. Answers differ since the values from the graph are estimates and do not match the values from the model.

21. Answers may vary. Articles in the business section are good sources for examples.

Section 2.2

1. (a) The slope of the tangent line at (2, 8) is approximately 12.

(b)

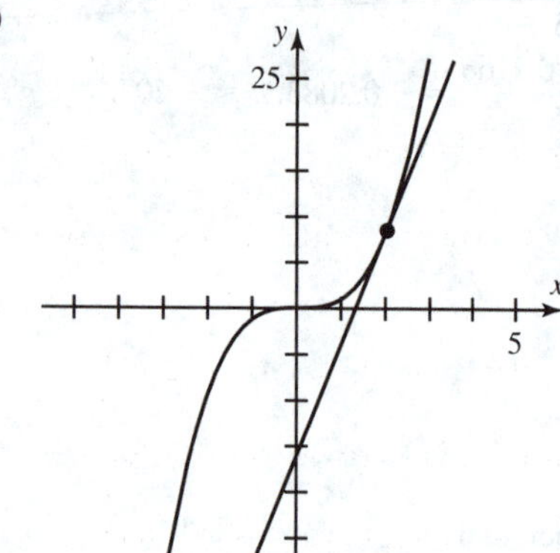

3. The slope of the tangent line at $x = 3$ is approximately 6.

5. The slope of the tangent line at $r = 2.5$ is approximately 8.75.

7. The slope of the tangent line at $x = 5$ is approximately –1.49.

9. **(a)** The average rate of change over [0, 5] is 15.
(b) The slope of the tangent line to the curve at $r = 0$ is approximately 5.
(c) The slope of the tangent line to the curve at $r = 5$ is approximately 50.
(d) The slope of the tangent line to the curve at $r = 2.5$ is approximately 8.75.
(e) The average rate of change over the interval was closer to the slope when the point was in the middle of the interval.

11. Answers may vary. Possible answers are:
(a) The annual percent change in medical CPI was decreasing at the rate of approximately 2.5 percentage points per year on December 31, 1986.
(b) The annual percent change in medical CPI was increasing at the rate of approximately 1.1 percentage points per year on December 31, 1988.
(c) The annual percent change in medical CPI was increasing at the rate of approximately 1.1 percentage points per year on December 31, 1989.

13. Answers may vary. Possible answers are:
(a) On December 31, 1977, the percentage of households with cable TV was increasing at approximately 3.5 percent per year.
(b) On December 31, 1982, the percentage of households with cable TV was increasing at approximately 4.3 percent per year.
(c) On December 31, 1987, the percentage of households with cable TV was increasing at approximately 3 percent per year.

15. **(a)**

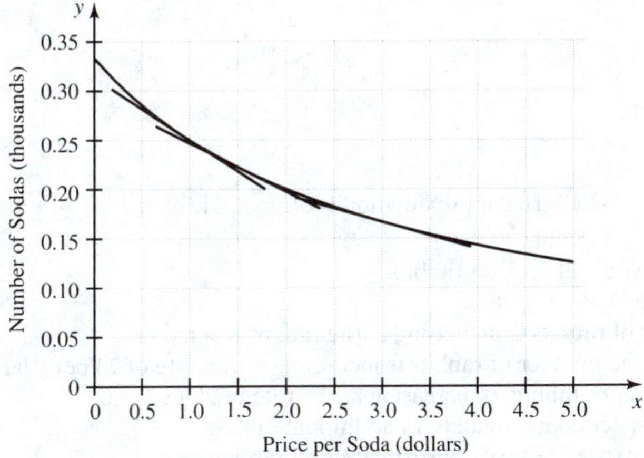

(b) Answers may vary.
The quantity of sodas sold is decreasing at the approximate rate of 70 sodas per dollar of selling price when charging \$1.00 per soda.
The quantity of sodas sold is decreasing at the approximate rate of 50 sodas per dollar of selling price when charging \$1.50 per soda.
The quantity of sodas sold is decreasing at the approximate rate of 20 sodas per dollar of selling price when charging \$3.00 per soda.

17. Answers may vary. Possible answers are:
(a) The slope of the tangent line when the price of soda is \$1.00 is approximately –0.125 (decreasing by 125 sodas per dollar of selling price).
The slope of the tangent line when the price of soda is \$1.50 is approximately –0.099 (decreasing by 99 sodas per dollar of selling price).
The slope of the tangent line when the price of soda is \$3.00 is approximately –0.056 (decreasing by 56 sodas per dollar of selling price).

(b) The average rate of change on the interval [1, 3] is −0.08333... (decreasing by 83 sodas per dollar of selling price).
(c) Answers may vary.

19. Answers may vary. Possible answers are:
(a) The accumulated sales of T-shirts was increasing by approximately 700 shirts per week, three weeks after the sale began.
(b) The accumulated sales of T-shirts was increasing by approximately 1750 shirts per week, five weeks after the sale began.

21. (a) The accumulated sales of T-shirts was increasing by approximately 696 shirts per week, three weeks after the sale began.
(b) The accumulated sales of T-shirts is increasing by approximately 1786 shirts per week, five weeks after the sale began.
(c) The average rate of change over the interval [1, 3] is about 322 (increasing at a rate of 322 T-shirts per week).
(d) Answers may vary.

Section 2.3

1. (a) $f(1) = 5; f(2.5) = 28.25; f(3) = 41$
(b) $f'(1) = 8; f(1.1) \approx 5.8; f(1.1) = 5.85$; The approximation is off by 0.05.
(c) 23
(d) 28

3. (a) $g'(3) = -10.99$
$g'(0.2) = -12.13$
$g'(0) = -12.22$
(b) $g(3) = 312.21$
$g(0.2) = 344.56$
$g(0) = 347$
(c) −12.01
(d) −11.71
(e) $g'(3) = -10.99; g(3.1) \approx 311.11; g(3.1) = 311.12$; The approximation is off by 0.01.

5. (a) After one year of breeding, you will have 5 rabbits.
(b) After two and a half years of breeding, you will have 28 rabbits.
(c) After three years of breeding, you will have 41 rabbits.
(d) After one year of breeding, the number of rabbits is increasing at the rate of 8 per year.
(e) After two and a half years of breeding, the number of rabbits is increasing at the rate of 23 per year.
(f) After three years of breeding, the number of rabbits is increasing at the rate of 28 per year.
(g) Between $t = 3$ and $t = 3.5$ you would expect approximately 14 additional rabbits.
After three and a half years, you would expect to have approximately 55 rabbits.
$R(3.5) = 56.25$. The approximation is off by 1.25 rabbits.

7. (a) The profit from selling 25 thousand items is $3,475,800.00.
(b) When selling 25 thousand items, the profit is increasing at the rate of $2358.00 per thousand items.
(c) The profit from selling 37 thousand items is $3,528,900.00
(d) When selling 37 thousand items, the profit is decreasing at the rate of $145.00 per thousand items.
(e) The profit is not changing when selling 32.5 thousand items.
(f) $P(25.1) \approx 34{,}760.358$.
$P(26) \approx 34{,}781.58$.
$P(28) \approx 34{,}828.74$.
The estimate for 25.1 should be the most accurate because the increment is the smallest.

9. (a) The annual percent change in medical CPI on June 30, 1989 was 7.3656.
(b) The annual percent change in medical CPI on June 30, 1992 was 8.1067.

(c) The annual percent change in medical CPI was increasing at the rate of 0.3503 percentage points per year on December 31, 1990.
(d) The annual percent change in medical CPI was decreasing at the rate of 1.6651 percentage points per year on June 30, 1992.
(e) The annual percent change in medical CPI is stationary 1.07355 years after December 31, 1986 (January 1988).

11. (a) The profit is $4475.00 when charging $5.50 for a mug.
(b) The profit is decreasing by $50.00 per dollar increase in selling price when charging $5.50 per mug.
(c) The profit is increasing by $50.00 per dollar increase in price when charging $5.00 per mug.
(d) The profit is increasing by $150.00 per dollar increase in price when charging $4.50 per mug.
(e) The profit is $4175.00 when charging $7.00 per mug.

13. (a) The revenue is increasing at the rate of $1.94 per ticket sold when 80 tickets are sold.
(b) The revenue is increasing at the rate of $1.29 per ticket sold when 70 tickets are sold.
(c) The revenue is increasing at the rate of $6.30 per ticket sold when 100 tickets are sold.

15. (a) The profit is increasing at the rate of $4.50 per mug when 100 mugs are sold.
(b) The profit is increasing at the rate of $1.50 per mug when 250 mugs are sold.
(c) The profit is decreasing at the rate of $3.50 per mug when 500 mugs are sold.

Section 2.4

1. (a) 0
(b) 0
(c) 1
(d) 4.5
(e) $-4x$
(f) $-32t + 12$

3. (a) $\frac{dy}{dx} = 0$
(b) $\frac{dt}{dx} = 0$
(c) $f'(x) = 3$
(d) $r'(q) = 0.8$

5. (a) $c'(x) = 0.045x^{-0.5} - 0.05x^{-0.9}$
(b) $R'(x) = -0.66p^{1.2} + 40.5p^{-0.1}$
(c) $f'(x) = e^x$
(d) $A' = f'(r) = -0.5e^r$

7. (a) $m'(r) = \frac{3}{r}$
(b) $w' = -\frac{2}{y}$
(c) $\frac{dy}{dx} = 7.5x^{1.5} - 2e^x$
(d) $g'(t) = 0.1158t + 0.0075$

9. Let $k(x) = f(x) - g(x)$

$$k'(x) = \lim_{h\to 0} \frac{[f(x+h) - g(x+h)] - [f(x) - g(x)]}{h}$$

$$= \lim_{h\to 0} \frac{f(x+h) - f(x) \; - \; [g(x+h) \; - \; g(x)]}{h}$$

$$= \lim_{h\to 0} \frac{f(x+h) - f(x)}{h} - \lim_{h\to 0} \frac{g(x+h) - g(x)}{h}$$

$$= f'(x) - g'(x)$$

11. (a) The number of sodas sold was decreasing at the rate of 2000 sodas per \$1 dollar increase when the price was \$1 per soda.
(b) The number of sodas sold was decreasing at the rate of 500 sodas per \$1 dollar increase when the price was \$2 per soda.
(c) The number of sodas sold was decreasing at the rate of about 222 sodas per \$1 dollar increase when the price was \$3 per soda.
(d) The demand is decreasing at a decreasing rate as the price increases.

13. (a) The number of tickets decreases by 15 per dollar increase in ticket price when the price per ticket is \$8.00.
(b) The number of tickets decreases by 5 per dollar increase in ticket price when the price per ticket is \$9.00.
(c) The number of tickets remains the same per dollar increase in ticket price when the price per ticket is \$9.50.
(d) The number of tickets increases by 5 per dollar increase in ticket price when the price per ticket is \$10.00. This answer does not seem reasonable because the model indicates that as the ticket price increases, so does the number of tickets sold, but this is unlikely in reality.

Section 2.5

1. $2(x-10)$

3. $\dfrac{x}{\sqrt{25 + x^2}}$

5. $-\dfrac{2}{(p+3)^2}$

7. $60e^{0.06t}$

9. $f'(x) = \dfrac{40e^{-5x}}{(3 + 4e^{-5x})^2}$
$f'(1) \approx 0.0294$

11. $-12.035...(0.953)^x$

13. $\dfrac{6x^2 + 1}{2x^3 + x}$

15. (a) *Verbal Definition*: $D(p)$ = the number of sweatshirts sold at p dollars per shirt.
Symbol Definition: $D(p) = 178.18...(0.8712...)^p$ or $178.18...e^{-0.1378...p}$ for $10 \le p \le 25$.
Assumptions: Certainty and divisibility. Certainty implies that the relationship is exact. Divisibility implies that any fractional numbers of shirts or dollars are possible.
(b) $D'(20) = -1.559....$ When the selling price is \$20 per shirt, the number of shirts sold is decreasing at the rate of about 1.6 shirts per \$1 increase in the selling price.
(c) $D(20) = 11.312....$ When the selling price is \$22 per shirt, the demand will be approximately 8 sweatshirts.

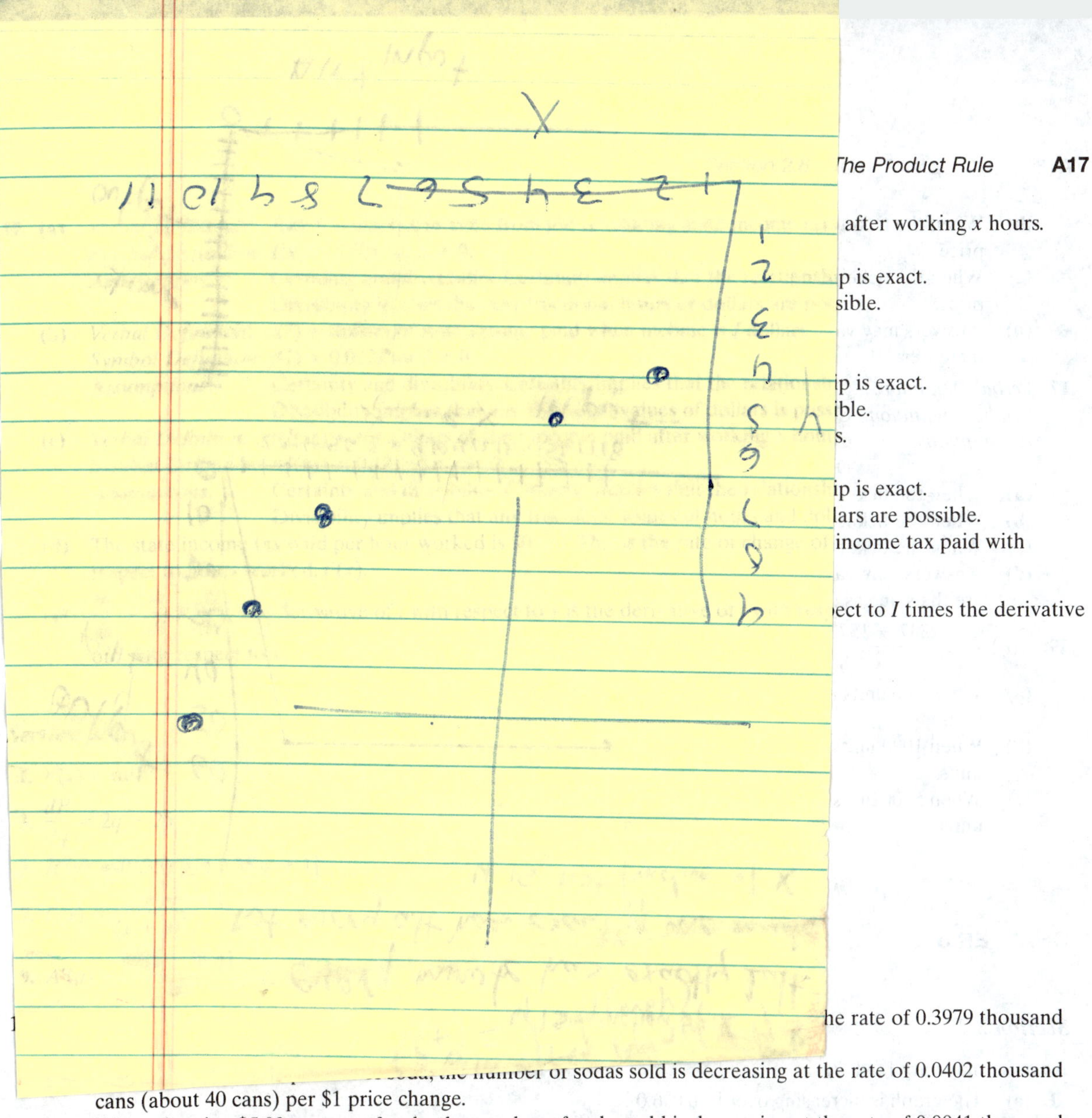

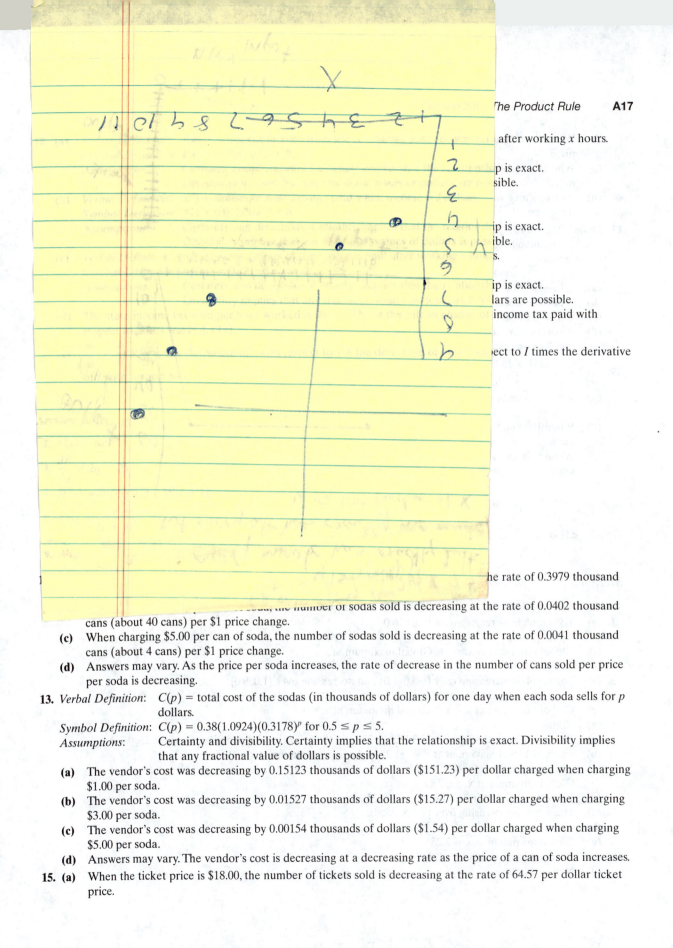

after working x hours.

p is exact.
sible.

ip is exact.
ible.
s.

ip is exact.
lars are possible.
income tax paid with

ect to I times the derivative

he rate of 0.3979 thousand

soda, the number of sodas sold is decreasing at the rate of 0.0402 thousand cans (about 40 cans) per $1 price change.

(c) When charging $5.00 per can of soda, the number of sodas sold is decreasing at the rate of 0.0041 thousand cans (about 4 cans) per $1 price change.

(d) Answers may vary. As the price per soda increases, the rate of decrease in the number of cans sold per price per soda is decreasing.

13. *Verbal Definition*: $C(p)$ = total cost of the sodas (in thousands of dollars) for one day when each soda sells for p dollars.

Symbol Definition: $C(p) = 0.38(1.0924)(0.3178)^p$ for $0.5 \leq p \leq 5$.

Assumptions: Certainty and divisibility. Certainty implies that the relationship is exact. Divisibility implies that any fractional value of dollars is possible.

(a) The vendor's cost was decreasing by 0.15123 thousands of dollars ($151.23) per dollar charged when charging $1.00 per soda.

(b) The vendor's cost was decreasing by 0.01527 thousands of dollars ($15.27) per dollar charged when charging $3.00 per soda.

(c) The vendor's cost was decreasing by 0.00154 thousands of dollars ($1.54) per dollar charged when charging $5.00 per soda.

(d) Answers may vary. The vendor's cost is decreasing at a decreasing rate as the price of a can of soda increases.

15. (a) When the ticket price is $18.00, the number of tickets sold is decreasing at the rate of 64.57 per dollar ticket price.

(b) When the ticket price is \$20.00, the number of tickets sold is decreasing at the rate of 58.18 per dollar ticket price.
(c) When the ticket price is \$22.00, the number of tickets sold is decreasing at the rate of 52.41 per dollar ticket price.
(d) Answers may vary. The number of tickets sold is decreasing at a decreasing rate as the price of a ticket increases.

17. *Verbal Definition*: $C(p)$ = the cost of using arena when charging p dollars per ticket.
Symbol Definition: $C(p) = 1000 + 3165.6(0.94919^p)$ for $p \geq 0$.
Assumptions: Certainty and divisibility. Certainty implies that the relationship is exact. Divisibility implies that any fractions of dollars are possible.
(a) When the ticket price is \$18.00, the cost is decreasing at the rate of \$64.57 per dollar increase in ticket price.
(b) When the ticket price is \$20.00, the cost is decreasing at the rate of \$58.18 per dollar increase in ticket price.
(c) When the ticket price is \$22.00, the cost is decreasing at the rate of \$52.41 per dollar increase in ticket price.
(d) Answers may vary. As the price of the tickets increases, the cost is decreasing at a decreasing rate. The rates are the same as in Exercise 15, since the original functions differ by a constant.

19. $\overline{C}(q) = \dfrac{-2347 + 1573\ln(q + 10)}{q}$ for $q \geq 0$.
(a) When 800 units are made, the average cost is decreasing at the rate of approximately \$23.44 per hundred units.
(b) When 1000 units are made, the average cost is decreasing at the rate of approximately \$15.79 per hundred units.
(c) When 5000 units are made, the average cost is decreasing at the rate of approximately \$1.11 per hundred units.

CHAPTER 3

Section 3.1

Answers may vary for 1–7.

1. (a) The graph is increasing over $[-6.0, 6.0]$.
(b) None
(c) Global minimum at $x = -6$. Global maximum at $x = 6$.
(d) None

3. (a) The graph is increasing over $[-3.0, 1.0)$ and decreasing over $(1.0, 9.0]$.
(b) There is a local maximum at $x = 1.0$.
(c) Global maximum at $x = 1.0$. Global minimum at $x = 9$.
(d) None

5. (a) The graph is increasing over $[-3.25, -1.25)$ and $(0.75, 2.25]$ and decreasing over $(-1.25, 0.75)$.
(b) There is a local maximum at $x = -1.25$ and a local minimum at $x = 0.75$.
(c) Global maximum at $x = 2.25$.
Global minimum at $x = -3.25$.
(d) There is a point of inflection at $x = -0.25$

7. (a) The graph is increasing over $[-1.5, 2.25]$.
(b) None
(c) Global maximum at $x = 2.25$.
Global minimum at $x = -1.5$.
(d) None

9.

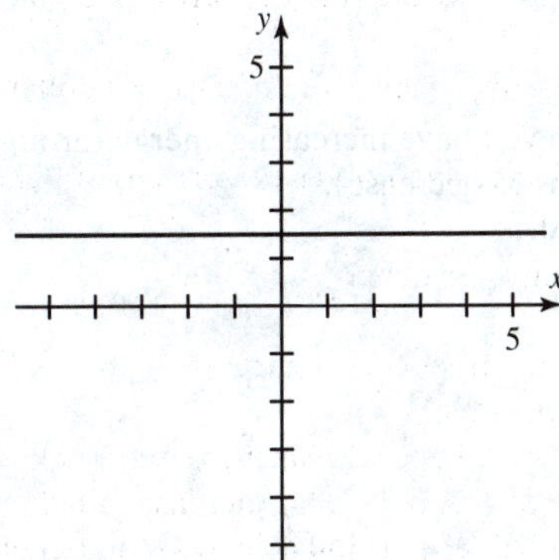

11.

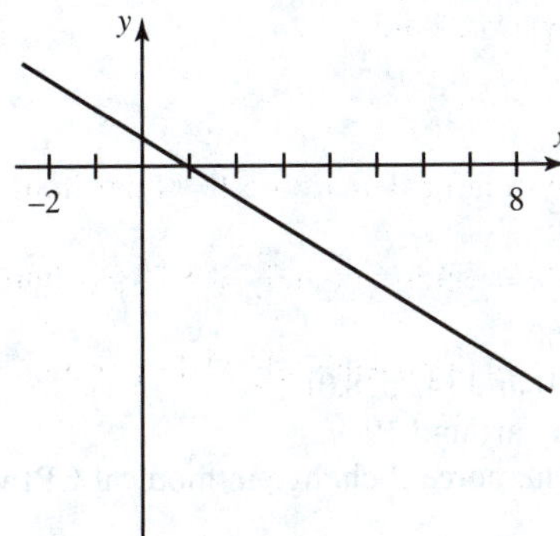

13.

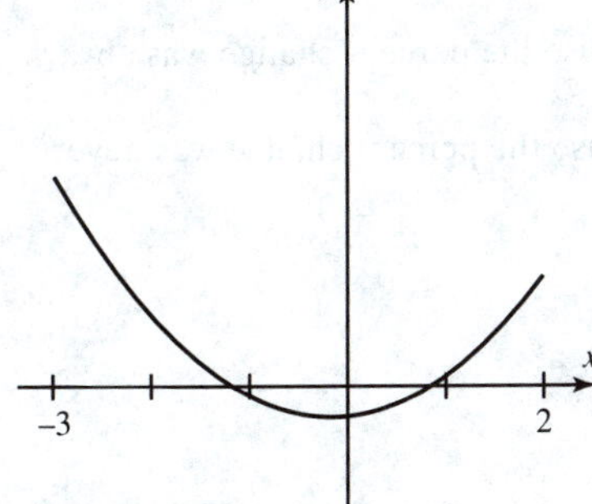

15.

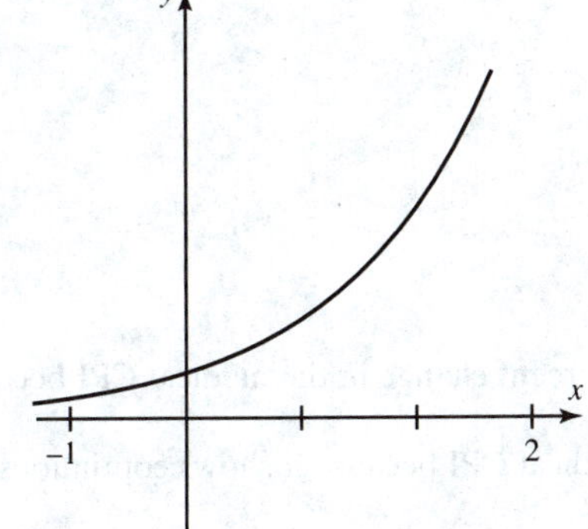

17. Answers may vary.

(a) The energy level is increasing over [8:00, 11:00) and (6:00, 9:00).

(b) The energy level is decreasing over (11:00, 6:00) and (9:00, 12:00].

(c) The best time to challenge your friend would be around 10:00 a.m. so you have increasing energy throughout the game (unless it goes past 11:00 a.m. when your energy level begins to decrease).

(d) Answers may vary. Around 4:00 to 5:00 P.M. or 8:00 to about 11:00 A.M.

(e)

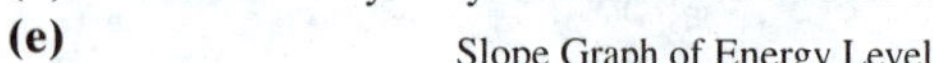

Slope Graph of Energy Level

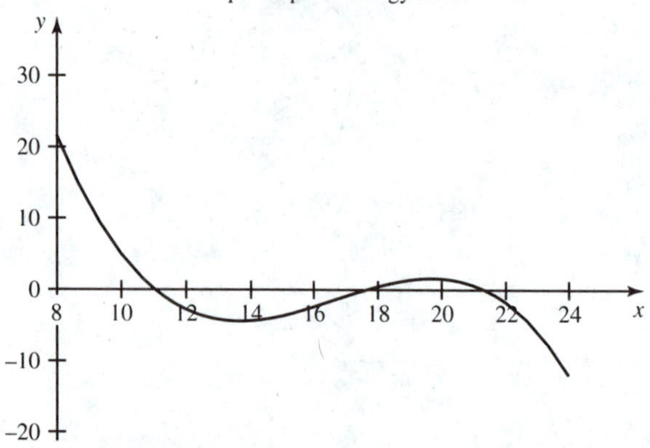

19. (a) The percent change in medical CPI was locally increasing over (1, 4.25)

(b) The percent change in medical CPI was locally decreasing over [0, 1.0) and (4.25, 8.0].

(c) The percent change in medical CPI was locally increasing most rapidly around 1989.

(d) The percent change in medical CPI was locally highest during 1991. The percent change in medical CPI was locally lowest around the end of 1987.

(e) The percent change in medical CPI was globally highest between 1990 and 1991. The percent change in medical CPI was globally lowest around 1994.

(f) The medical CPI was always increasing between 1986 and 1994 because the percent change was always positive.

(g) The medical CPI was never decreasing between 1986 and 1994 because the percent change was never negative.

(h)

Slope Graph of Annual Percent Change in Medical CPI

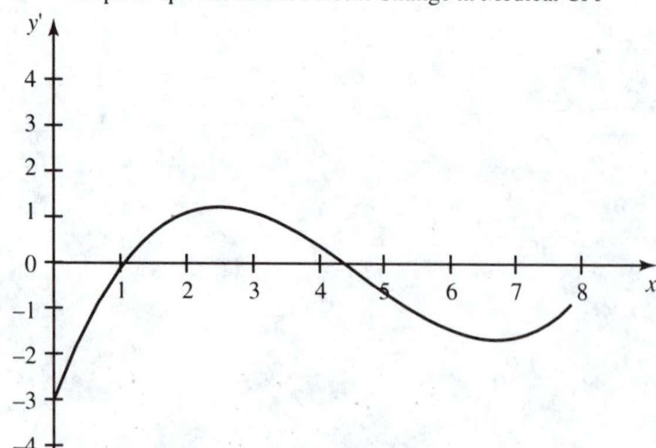

21. Answers may vary. Sample answers are:

(a) A representative of the medical community might use the annual percent change in the medical CPI because it shows a decreasing trend.

(b) A representative of a consumers' advocacy group might use the medical CPI because it shows continuously rising prices.

Section 3.2

1.

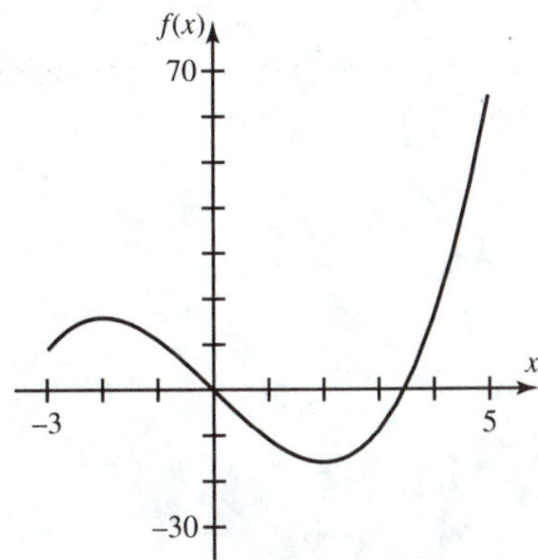

(a) The global minimum value occurs at $x = 0.0$ and the local and global maximum value occurs at around $x = -2.0$.
(b) The local and global minimum value occurs at around $x = 2.0$ and the global maximum occurs at $x = -0.0$.
(c) The global minimum value occurs at $x = 3.0$ and the global maximum value occurs at $x = 5.0$.

3. Global maximum at $x = 7$.

5. There is a local and global minimum at $x = 0.25$.

7. There is a local minimum at $x = 0.725$ and a local maximum at $x = -1.725$.

9. There is a global minimum at $x = 0$. There is a local and global maximum at $x = 10$.

11. (a) Unemployment was highest on December 31, 1983 and lowest around May 6, 1988.
(b) Answers may vary. Among the given data, unemployment was highest on December 31, 1983 and lowest on December 31, 1989. The minimum is off by about 18 months.
(c) Unemployment was highest on December 31, 1983 and lowest around early December 1989.
(d) Answers may vary. Among the given data, unemployment was highest on December 31, 1983 and lowest on December 31, 1989. The minimum is off by less than 1 month.
(e) The second model's figures match the actual maximum and minimum unemployment values more accurately.

13. Answers may vary depending on the model chosen. Sample answers are:
(a) *Verbal Definition*: $p = d(q) =$ the price, in dollars, you would have to charge to sell q mugs.
Symbol Definition: $p = d(q) = -0.01q + 8.5$ for $0 \le q \le 850$.
Assumptions: Certainty and divisibility. Certainty implies that the relationship is exact. Divisibility implies that any fractional number of mugs and parts of dollars are possible.
(b) $R(q) = q(-0.01q + 8.5)$. $C(q) = 2.25q$
Verbal Definition: $P(q) =$ the profit, in dollars, when q mugs are sold at a price of $d(q)$.
Symbol Definition: $P(q) = -0.01q^2 + 6.25q$ for $0 \le q \le 850$.
Assumptions: Certainty and divisibility. Certainty implies that the relationship is exact. Divisiblity implies that any fractional number of mugs and parts of dollars are possible.
(c) You should order and sell 312.5 mugs to maximize your profit. $P(312.5) = 976.5624$ The maximum profit would be approximately \$976.56. If you order and sell 312 mugs, the profit would be \$976.56.
(d) $p = d(312) = 5.38$. You should charge approximately \$5.38 per mug. You might want to round this price to \$5.25 or \$5.50, which would affect the profit.

(e)

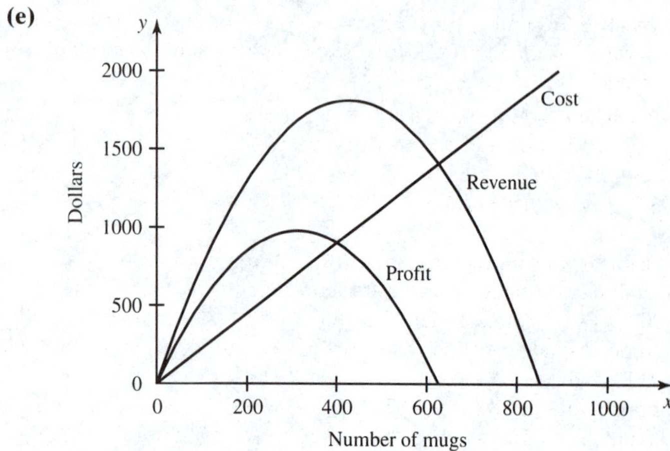

The profit curve is the revenue curve minus the cost curve. (For example, when revenue = cost, the profit is zero.)

15. (a) The profit will be maximized when selling the sodas for \$1.2523 each.
(b) Answers may vary. You would certainly want to round the price to \$1.25.

17. (Answers should agree with the values obtained in Section 1.0)

19. (a) *Verbal Definition:* $f(x)$ = energy level x hours after 8:00 A.M.
Symbol Definition: $f(x) = -0.0137x^4 + 0.4753x^3 - 5.461x^2 + 21.53x + 44.11$, for $0 \leq x \leq 16$.
Assumptions: Certainty and divisibility. Certainty implies that the relationship is exact. Divisibility implies that any fractional values of energy level and hours are possible.
The model is not a good fit. It misses the local extrema and only reflects the general trend.
(b) According to the model, the highest energy level was 71.27 at approximately 11:00 A.M.
(c) According to the model, the lowest energy level was 38.25 at midnight.
(d) According to the model, the energy level reached a relative high of 53.47 at approximately 9:00 P.M.
(e) According to the model, the energy level reached a relative low of 51.41 at approximately 6:00 P.M.

Section 3.3

1. $f''(x) = 36x^2 + 4$

3. $g''(x) = -140x^3 + 18x$

5. $f''(x) = 6e^{-3x}(3x - 2)$

7. (a) There is a local maximum at $x = -1$ and a local minimum at $x = 1$.
(b) There are points of inflection at $\left(-\sqrt{\frac{1}{2}}, -1.24\right)$, $(0, 0)$, and $\left(\sqrt{\frac{1}{2}}, -1.24\right)$.
(c) The function is concave up over $\left[-\sqrt{\frac{1}{2}}, 0\right]$ and $\left[\sqrt{\frac{1}{2}}, \infty\right)$ and concave down over $\left(-\infty, -\sqrt{\frac{1}{2}}\right]$ and $\left[0, \sqrt{\frac{1}{2}}\right]$.

9. (a) There is a local maximum at $s = -\sqrt{\frac{1}{3}}$ and a local minimum at $s = \sqrt{\frac{1}{3}}$.
(b) There is a point of inflection at (0,0).
(c) The function is concave up over $[0, \infty)$ and concave down over $(-\infty, 0]$.

11. (a) There is a local maximum at $x = 10$.
(c) There is a point of inflection at (20, 0.02706…).
(b) The function is concave down over $[0, 20]$ and concave up over $[20, \infty)$.

13. (a) Unemployment was highest on December 31, 1983, and lowest in mid-1988.
(b) $U''(4.431) \approx 246.7$. Since the value is positive, the unemployment figure of mid-1988 is a minimum.

(c) There is a point of inflection 7.43 years after December 31, 1983 (mid-1991). The increase in unemployment is starting to slow down. Answers may vary. Example answer: A politician could claim that his policies, put in place before this date, are taking effect.

15. Answers may vary depending on the model chosen. Sample answers:

(a) *Verbal Definition*: $G(x)$ = charitable giving in the United States, in billions of inflation-adjusted dollars, x years after Dec. 31, 1984.

Symbol Definition: $G(x) = 0.0230798\ldots x^4 - 0.444543\ldots x^3 + 2.15786\ldots x^2 + 2.00129\ldots x + 108.179\ldots$, for $0 \le x \le 10$.

Assumptions: Certainty and divisibility. Certainty implies that the relationship is exact. Divisibility implies that any fractional values of dollars and years are possible.

(b) The local maximum occurs at $x = 5.900\ldots$ (approximately late October 1989) with a contribution of \$131.769… billion. The local minimum occurs at $x = 8.955\ldots$ (approximately mid-November 1992) with a contribution of approximately \$128.330… billions.

(c) $G''(5.9) = -1.78$, so $x = 5.9$ is a local maximum. $G''(8.995) = 2.732$ so $x = 8.995$ is a local minimum.

(d) There are inflection points at $x = 2.05$ and $x = 7.52$. In early 1987, the increase in charitable giving in the United States started to slow down. In mid-1992, the decrease in charitable giving began to slow down. Answers may vary as to how this information could be used, for example, firms that help publicize charitable drives could use this to indicate the success of different strategies.

17. (a) To maximize the profit charge \$20.18… per ticket.

(b) There is an inflection point at $p = 39.36\ldots$. At this point, the decline in profit is starting to slow down. Answers may vary, for example, increasing prices will not have a smaller adverse effect on profit.

19. (a) The local maximum percent change in medical CPI was 8.693, occuring 4.324 years after December 31, 1986. $I''(4.324) < 0$. Since this value is negative, the annual percent change in medical CPI at this time is a maximum. $I''(1.078) > 0$, therefore it is a local minimum. The local minimum percent change in the medical CPI was 6.189 occuring 1.078 years after December 31, 1986.

(b) The global minimum annual percent change in medical CPI was 4.273, occuring 8 years after December 31, 1986. The global maximum occurs at the local maximum, $x \approx 4.324$.

(c) The annual percent change in medical CPI was increasing most rapidly 2.482 years after December 31, 1986.

(d) The annual percent change in medical CPI was decreasing most rapidly 6.715 years after December 31, 1986.

(e) Answers may vary, for example, the continuing decrease of the percentage change in the medical CPI could be used as an argument for increasing the use of HMOs or as political strategy.

Section 3.4

1. 1.27

3. 29

5. 3.0

7. 282

9. $2.0155^{3.65} = 12.912\ldots$
$2.0165^{3.65} = 12.935\ldots$
$2.0155^{3.75} = 13.849\ldots$
$2.0165^{3.75} = 13.875\ldots$
Round to 13. Rounding to two significant figures seems to make the most sense, so precision for powers depends on significant figures

11. (a) $\frac{4300}{5}$ is about 8–9.00 dollars; the time is about 1.5 hours, so about \$6/hr. This seems reasonable.

(b) \$8.62 (to nearest penny)

(c) 85 minutes ± 1 minute (10:02–11:28, for 86 min, or 10:03–11:27, for 84 min)

(d) About 1%

(e) About 1.4 (1.4, 1.41667, and 1.43333)

(f) 6.086117647
(g) 6.015348837 to 6.158571429
(h) About \$6.10 per hour
(i) ±\$0.09 per hour; about 1 or 2%

13. (a) At \$10, profit is \$160; at \$15, it is \$180; at \$20, it is \$140; \$15 looks best, ordering about 20 shirts.
(b) *Verbal Definition*: $q = D(p) =$ the number of T-shirts bought by the dorm at the price of p dollars per shirt.
Symbol Definition: $D(p) = 0.2p^2 - 9p + 110$, for $5 \le p \le 20$.
Assumptions: Certainty and divisibility. Certainty implies that the relationship is exact. Divisibility implies that any fractions of dollars and T-shirts are possible.
(c) *Verbal Definition*: $P(p) =$ the profit, in dollars, at the selling price of p dollars per shirt.
Symbol Definition: $P(p) = 0.2p^3 - 10.2p^2 + 164p - 660$, for $5 \le p \le 20$.
Assumptions: Certainty and divisibility. Certainty implies that the relationship is exact. Divisibility implies that any fraction of dollars is possible.
Maximized when $p = \$13.04$, $D(p) = 26.6$ shirts.
(d) Take the derivative by hand; set it equal to zero and solve by hand. Graph the function and use technology to find the maximum. Take the derivative using technology and find when the derivative is zero.
(e) This assumes that the sample is representative (the *dorm* is proportional), that people's deeds match their words.
(f) $P(p) = 140(p - 6)(0.8775\ldots)^p$, for $5 \le p \le 20$.
Maximized when $p = \$13.66$, $D(p) = 23.52$
(g) The new solution is $p = \$11.51$, $D(p) = 33$.
(h) Precise probably only to 1 or 2 significant figures. May be biased on the high side.
(i) A dollar or two in price, and 5–10 shirts for the order quantity.

15. *Verbal Definition*: $d(q) =$ the price you would have to charge to sell exactly q sweatshirts
Symbol Definition: $d(q) = 0.005403\ldots q^2 - 0.6660\ldots q + 28.05\ldots$, for $5 \le q \le 40$.
$d(q) = 28.353\ldots(0.97438\ldots)^q$
$d(q) = 53.102\ldots q^{-0.42082\ldots}$
Assumptions: Certainty and divisibility. Certainty implies that the relationship is exact. Divisibility implies that the number of sweatshirts and the dollars can have any fractional value. These assumptions hold for all the following models in this exercise.
Verbal Definition: $R(q) =$ the revenue, in dollars, when q sweatshirts are sold.
Symbol Definition: $R(q) = 0.005403\ldots q^3 - 0.6660\ldots q^2 + 28.05\ldots q$, for $5 \le q \le 40$.
$R(q) = 28.353\ldots q(0.97438\ldots)^q$.
$R(q) = 53.102\ldots q^{0.57918\ldots}$.
Verbal Definition: $c(q) =$ the cost, in dollars, to order q sweatshirts.
Symbol Definition: $c(q) = 9q + 25$, for $q \ge 0$.
Symbol Definition: $P(q) =$ the profit, in dollars when q sweatshirts are sold.
Symbol Definition: $P(q) = 0.005403\ldots q^3 - 0.6660\ldots q^2 + 19.05\ldots q - 25$, for $5 \le q \le 40$
$P(q) = 28.353\ldots q(0.97438\ldots)^q - 9q - 25$
$P(q) = 53.102\ldots q^{0.57918\ldots} - 9q - 25$
Quadratic model: $x = 18.44$ for \$133.73 profit, $p = \$17.61$.
Power model: $x = 18.54$ for \$96.26 profit, $p = \$15.54$.
Exponential model: $x = 18.68$ for \$133.07 profit, $p = \$17.46$.

17. Let's call the adjusted quadratic demand function $D_{2b}(p) = 0.8D_2(p) = 0.056p^2 - 3.83p + 64.8$. For all of the functions, the domain will stay $10 \le p \le 25$. The new revenue will be $R_{2b}(p) = 0.056p^3 - 3.83p^2 + 64.8p$. The new cost is $C_{2b} = 0.504p^2 - 34.5p + 608$. So the new profit is $\Pi_{2b}(p) = 0.056p^3 - 4.33p^2 + 99.3p - 608$. To 3 significant digits, the maximum comes to $p = 17.2$. With full precision, the result is again 17.13.

19. A linear demand model yields a price of \$6.78, for a profit of \$389, expecting about 102 people. A quadratic demand model yields a price of \$4.00, for a profit of \$580, expecting about 220 people. An exponential demand model yields a price of \$5.61, for a profit of \$408, expecting about 126 people. All of these prices represent extrapolations (are out of the range of the data inputs). Quadratic gives the best fit within the data input range but is much more optimistic about the benefit of lower prices. Linear is probably too pessimistic. Of the three, exponential is a good middle ground. A price in the \$5 to \$6 range seems most justified. Which price you choose could be affected by the capacity of the space, other revenue (from food sales, etc.) and your intuition about how many more people would come at lower prices. Whatever you do, it looks as if there's a good chance of a decent profit, probably in the range of \$400.

Section 3.5

1. 40%

3. −204%

5. Price elasticity = −0.2121..., inelastic

7. Price elasticity = −1, unit elastic

9. (a) Price elasticity over [15, 20] ≈ −1.42
(b) Price elasticity at 15 = −1.93. They differ because one is over an interval and the other is at a point.
(c) Elastic
(d) Revenue will decrease. The decrease in demand is proportionately more than the increase in price, so the cost outweighs the benefit of the price increase.
(e) Price should be less because revenue is decreasing.
(f) Price of approximately \$7.77 (elastic above, inelastic below).

11. (a) Price elasticity over [7,9] = −1.05. This means that the demand will decrease by approximately 1.05% per percentage point increase in price.
(b) Price elasticity at 7 = −1.75. This means that the demand will decrease by approximately 1.75% per percentage point increase in price.
(c) The demand is elastic.
(d) Revenue will decrease. The decrease in demand is proportionately more than the increase in price, so the cost outweighs the benefit of the price increase.
(e) Price should be less because revenue is decreasing.
(f) Revenue is maximized when the price is \$4 (elastic above, inelastic below).
(g) Profit would be maximized if the cost function were constant, or proportional to revenue.

13. (a) Price elasticity over [0.60, 0.70] is −0.6498. This means that the demand will decrease by approximately 0.65% per percentage point increase in price.
(b) Price elasticity at 0.60 is −0.6879.
(c) Inelastic
(d) The revenue would increase.
(e) The price should be higher because revenue is still increasing.
(f) Revenue is maximized at a price of \$0.87 (elastic above, inelastic below).

15. (a) Price elasticity over [7,9] is −1.5806. This means that the demand will decrease by approximately 1.58% per percentage point increase in price.
(b) −22.6%; The demand for fudge will decrease by about 23% per dollar increase in the price per pound from \$7 per pound.
(c) The price elasticity at 7 is −1.5806. This means that the demand will decrease by approximately 1.58% per percentage point increase in price.
(d) Elastic
(e) Revenue will decrease. The decrease in demand is proportionately more than the increase in price, so the cost outweighs the benefit of the price increase.
(f) Price should be less because revenue is decreasing.
(g) Revenue is maximized when the price is \$5.71 (elastic above, inelastic below).

17. (a) Price elasticity over [9,11] is $-0.8697\ldots$. This means that the demand will decrease by approximately 0.87% per percentage point increase in price.

(b) -9.66%; The demand for shirts will dcrease by about 10% per dollar increase in the price per shirt from \$9 per shirt.

(c) The price elasticity at 9 is $-0.8697\ldots$. This means that the demand will decrease by approximately 0.87% per percentage point increase in price.

(d) Inelastic

(e) Revenue will increase. The decrease in demand is proportionately less than the increase in price, so the benefit outweighs the cost of the price increase.

(f) Price should be more because revenue is increasing.

(g) Revenue is maximized when the price is \$9.67 (elastic above, inelastic below).

CHAPTER 4

Section 4.1

1. 0.75

3. 0.4 (40%)

5. (a) Let X = The competitor's bid (in millions of dollars).

(b) $f(x) = \begin{cases} 0.5, & \text{for } 8 \leq x \leq 10 \\ 0, & \text{otherwise} \end{cases}$

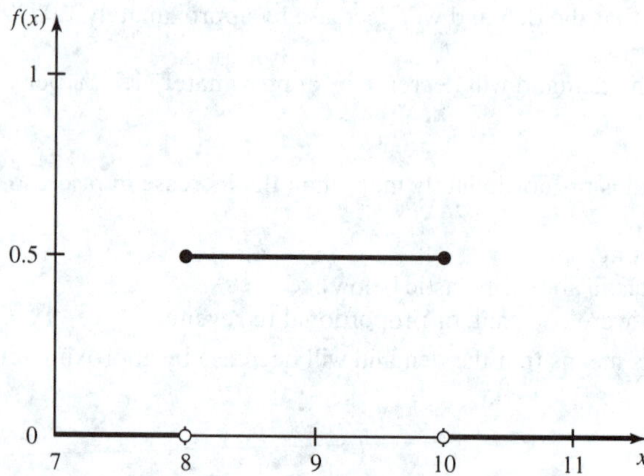

(c) 0.2 (20%)

(d) It is the area under the probability density function (pdf) curve ($f(x)$) and above the x = axis between $x = 8.9$ and $x = 9.3$.

7. (a)

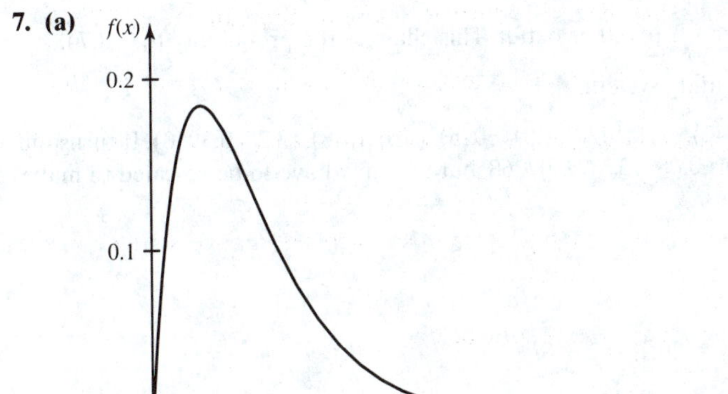

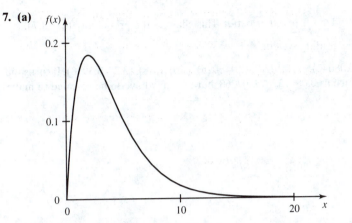

(b) $f(1) \approx 0.15$ and $f(7) \approx 0.05$. Thus, the time until production is about three times as likely to be in a neighborhood around 1 year compared to a neighborhood (of the same size) around 7 years.

(c) The mode appears to occur at about 2 years.

(d) The mode is indeed exactly 2 years, so values in a neighborhood of 2 years are more likely than in a neighborhood (of the same size) anywhere else.

(e) Answers may vary, but all methods give an answer close to 2.

(f) The probability would be the area under $f(x)$ and above the x-axis between $x = 0$ and $x = 3$.

9. (a) The probability of an A is about 0.1975. This works out to a probability of 0.3950 per unit (GPA point).

(b) Let X = the GPA of a student picked at random in the professor's class.

$$f(x) = \begin{cases} 0.3950, & \text{for } 3.5 \le x \le 4 \\ 0.3148, & \text{for } 2.5 \le x < 3.5 \\ 0.3025, & \text{for } 1.5 \le x < 2.5 \\ 0.1420, & \text{for } 0.5 \le x < 1.5 \\ 0.08642, & \text{for } 0 \le x < 0.5 \\ 0, & \text{otherwise} \end{cases}$$

(c)

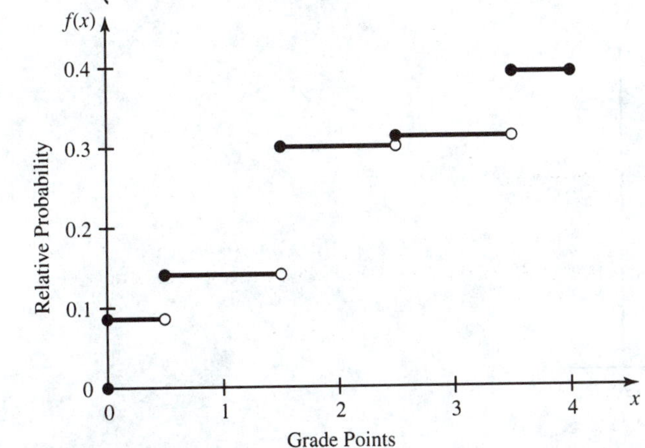

(d) The probability is 0.5123. This assumes you are a *random* student; there is no adjustment for your ability or past history.

(e) The probability is the area under $f(x)$ and above the x-axis between $x = 2.5$ and $x = 4$.

(f) The probability is 0.3549, assuming again that you are a random student, and that grades are equally distributed within each category.

(g) Grades of $2\frac{5}{6} \approx 2.8333$ or more will be rounded off to 3.0 or better. This changes the probability to 0.4074. This answer is more appropriate, given the grading system.

(h) Answers may vary. You could use (0.25, 0.08642), (1, 0.1420), (2, 0.3025), (3, 0.3148), (3.75, 0.3950). It turns out a linear function is a reasonable model, such as $0.08773x + 0.07268$, but this may have to be rescaled to make the probabilities work out right.

Section 4.2

1. $\Delta x = 2; x_0 = 0, x_1 = 2, x_2 = 4, x_3 = 6, x_4 = 8; x_i = 2i$
3. $\Delta t = 3; t_0 = 2, t_1 = 5, t_2 = 8, t_3 = 11, t_4 = 14, t_5 = 17, t_6 = 20; t_i = 2 + 3i.$
5. 115
7. $f(x_1)\Delta x + f(x_2)\Delta x + f(x_3)\Delta x + f(x_4)\Delta x$
9. Right Rectangles: 450

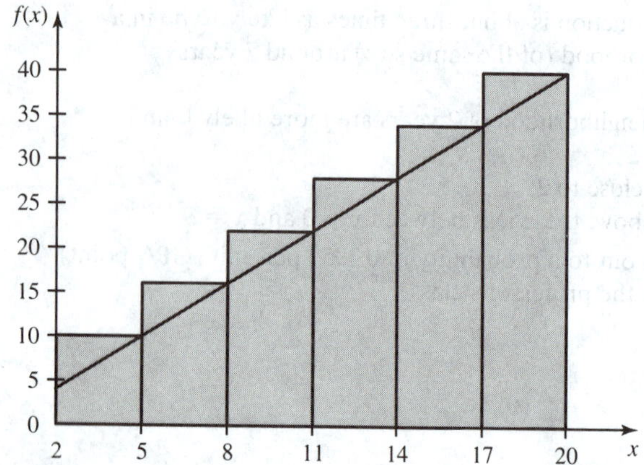

Trapezoids: 396

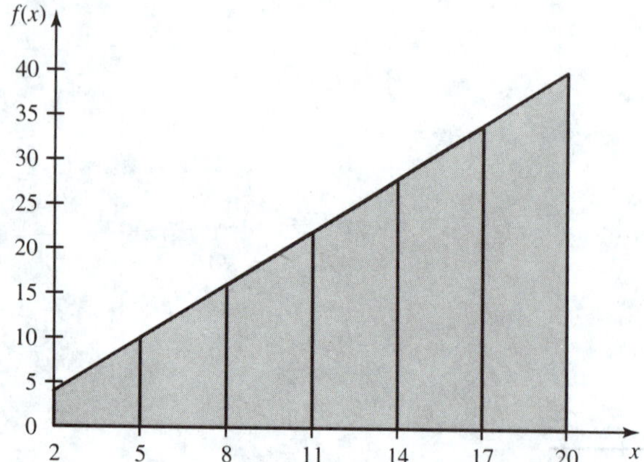

11. Right Rectangles: 80

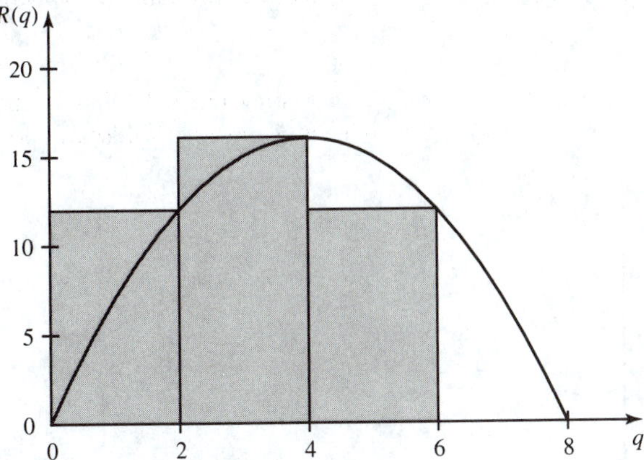

Trapezoids: 80

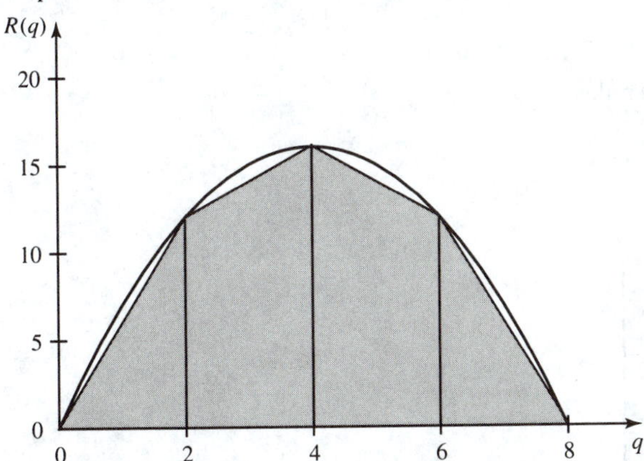

13. 450

15. 80

17. Right Rectangles: 450; Simpson's Rule: 396

19. Trapezoids: 80; Simpson's Rule: 85.3333

21. 18; This is (the area of the rectangles that are *above* the x-axis) – (the area of the rectangles that are *below* the x-axis).

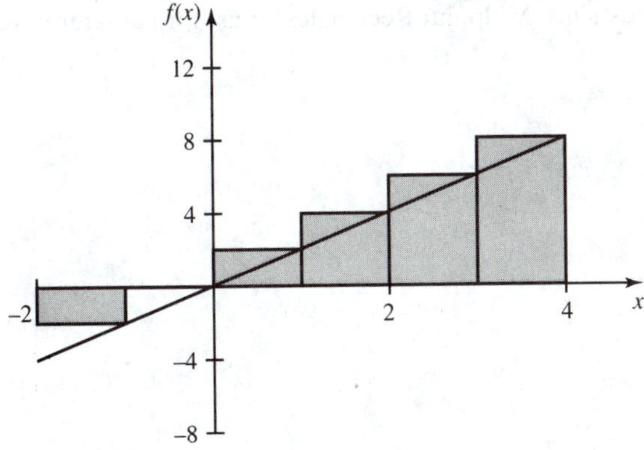

23. **(a)** $\dfrac{16 \text{ miles}}{\text{hour}} \cdot 2 \text{ hours} = 32 \text{ miles}$

(b)

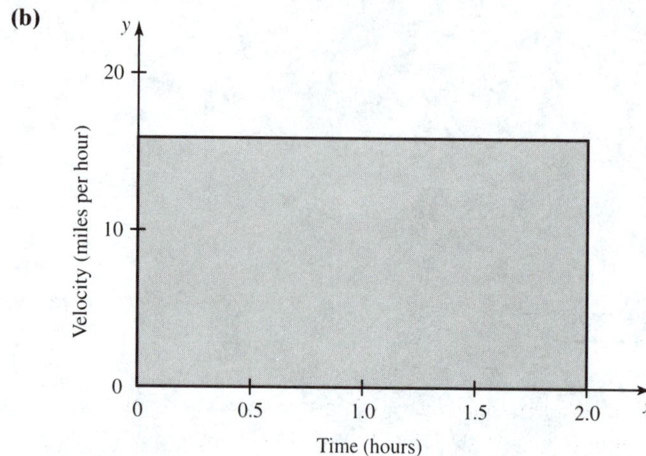

25. **(a)** $\dfrac{14 \text{ miles}}{\text{hour}} \cdot (1.5 \text{ hours}) + \dfrac{18 \text{ miles}}{\text{hour}} \cdot (0.5 \text{ hour}) = 30 \text{ miles}$

(b)

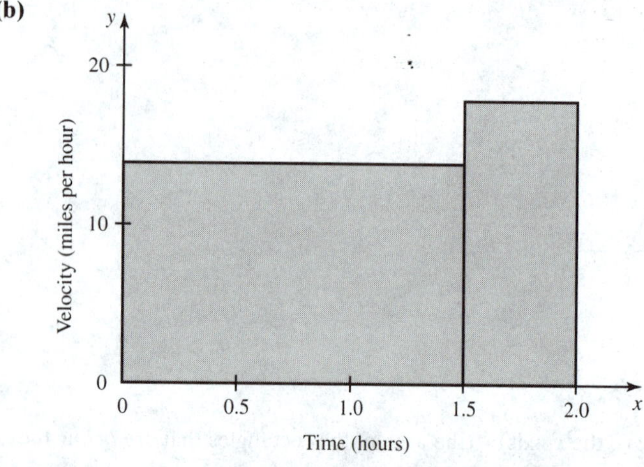

27. **(a)** Midpoint Rectangles: 0.44506; Trapezoids: 0.43640

(b) Midpoint Rectangles are slightly better than Trapezoids, since Midpoint Rectangles adjust for curvature over a subinterval, unlike Trapezoids.

29. (a)

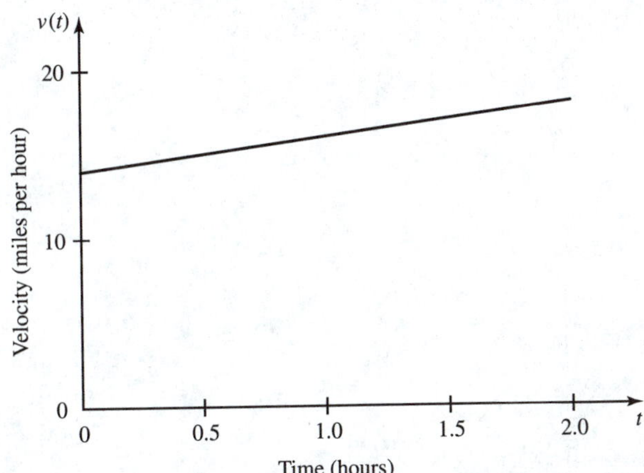

The graph indicates that you started at 14 mph, and then gradually and uniformly raised your speed to 18 mph after 2 hours. It assumes that you did not have to slow down or stop for traffic, lights, intersections, animals, pedestrians, and so on.

(b) The distance would correspond to the area under the velocity curve and above the horizontal axis between $t = 0$ and $t = 2$.

(c) Midpoint Rectangles, Trapezoids, and Simpson's Rule should give the exact correct answer.

(d) Using Trapezoids with $n = 2$: the distance = 32 miles.

(e) $S_4 = \sum_{i=1}^{4} \frac{v(t_{i-1}) + v(t_i)}{2}\Delta t = \frac{v(t_0) + v(t_1)}{2}\Delta t + \frac{v(t_1) + v(t_2)}{2}\Delta t + \frac{v(t_2) + v(t_3)}{2}\Delta t + \frac{v(t_3) + v(t_4)}{2}\Delta t$

$= \left[\frac{14 + 15}{2}\right](0.5) + \left[\frac{15 + 16}{2}\right](0.5) + \left[\frac{16 + 17}{2}\right](0.5) + \left[\frac{17 + 18}{2}\right](0.5) = 32\text{miles}$

Since the change in speed is uniform, $\frac{v(t_{i-1}) + v(t_i)}{2}$ is the *average* speed in the ith subinterval, so multiplying it by Δt gives the distance traveled in that interval.

31. (a) The region is a trapezoid. Its area is

$$A = \frac{b_1 + b_2}{2} \cdot h = \frac{f(2) + f(20)}{2} \cdot (20 - 2) = 396$$

(b) Left and Right Rectangles were both always off in the same direction so had big errors. Midpoint Rectangles, Trapezoids, and Simpson's Rule were all exact because the function was linear.

(c) 504

(d) 450

(e) The result for $n = 6$ is closer (for $n = 3$ the errors stick out more, including and exceeding those for $n = 6$).

(f) The result for $n = 12$ should be even better than using $n = 6$.

Section 4.3

1.

n	S_n
4	16
8	16
16	16

$$\int_1^5 4dx = \lim_{n\to\infty} S_n = 16$$

3.

n	S_n
4	0.92144
8	0.99696
16	0.99728
32	0.99730
64	0.99730

$$\int_{-3}^{3} \frac{1}{\sqrt{2\pi}} e^{-x^2/2} = \lim_{n\to\infty} S_n = 0.9973$$

5.

n	S_n
4	−2.5
8	−3.25
16	−3.63
32	−3.81
64	−3.91
128	−3.95
256	−3.98
512	−3.99

$$\int_0^2 (3x - 5)dx = \lim_{n\to\infty} S_n = -4.0$$

7. (a) Midpoint Rectangles, Trapezoids, or Simpson's Rule should give the exact area.

(b) The version with two rectangles has more area hanging out than that with four rectangles.

(c)

n	S_n
4	33.0
8	32.5
16	32.3
32	32.1
64	32.1
128	32.0
256	32.0

$$\int_0^2 (14 + 2t)dt = \lim_{n\to\infty} S_n = 32.0 = \lim_{n\to\infty} \sum_{i=1}^{n} v(t_i)\Delta t$$

(d) It means that the distance traveled in 2 hours is 32 miles.

(e)

n	S_n
4	32.0
8	32.0

9. (a) Right Rectangles

n	S_n
4	0.492
8	0.470
16	0.457
32	0.450
64	0.446
128	0.444
256	0.443
512	0.443
1024	0.442
2048	0.442

Simpson's Rule

n	S_n
4	0.442
8	0.442

Trapezoids

n	S_n
4	0.429
8	0.439
16	0.441
32	0.442
64	0.442

(b) Simpson's Rule converges the fastest (it had the answer immediately for $n = 4$). Next, in order, come Trapezoids and Right Rectangles.

Section 4.4

1. $\frac{1}{4}x^4 + x^2 + C$

3. $-\frac{3}{x} + 2 \ln x + 8\sqrt{x} + C, x > 0$

5. $-6250e^{-0.08t} + C$

7. $\dfrac{15(0.9)^q}{\ln(0.9)} - 10q + C$

9. $-60e^{-0.05t} + \dfrac{4}{3}t^3 - \dfrac{5}{2}\ln t + C, t > 0$

11. The slope is always 3, so any straight line with that slope will work. In other words, just change the *y*-intercept.

13. **(a)** $v(t) = D'(t)$. We are finding an *antiderivative* of $v(t)$.

(b) $D(t) = 14t + t^2 + C$. It corresponds to $\int v(t)dt$.

(c) $D(0) = 0$, which means that $C = 0$, so $D(t) = 14t + t^2$.

(d) $D(2) = 32$. This means that you have traveled 32 miles in the first 2 hours.

(e) You traveled 15 miles in the first hour. This is $D(1)$ or $\int_0^1 v(t)dt$.

(f) You traveled 17 miles in the second hour. This is $D(2) - D(1)$ or $\int_1^2 v(t)dt$.

15. **(a)** $v(t) = M'(t)$. We are finding an *antiderivative* of $v(t)$.

(b) $M(t) = 14t + t^2 + C$. It corresponds to $\int v(t)dt$.

(c) $M(0) = 48$, which means that $C = 48$, so $M(t) = 14t + t^2 + 48$

(d) $M(2) = 80$. This means that you are at mile marker 80 after 2 hours.

(e) After 1 hour, you are at mile marker 63. This is $M(1) = 63$.

(f) You traveled 15 miles in the first hour. This is $M(1) - M(0)$.

(g) In the second hour, you traveled 17 miles. This is $M(2) - M(1)$.

(h) It is the same. The two functions use different mile marker starting points (the $D(t)$ function starts at 0 and the $M(t)$ function starts at 48), but the distance is the same.

(i) $v(t)$ would have been *negative* (so $M(1) = 48 - 15$, etc.).

17. **(a)** $\int f(x)dx = -e^{-0.1x} + C$. $F(x)$ could be defined to be $P(X \le x)$, the cumulative probability of waiting up to x minutes for the bus.

(b) $F(0) = 0$ (the probability of the wait being 0 minutes or less is 0). This implies that $C = 1$, so

$$f(x) = \begin{cases} 1 - e^{-0.1x}, & \text{for } x \ge 0 \\ 0, & \text{otherwise} \end{cases}$$

(c) Approximately 0.3297

(d) $F(8) \approx 0.5507$

(e) $F(8) - F(4) \approx 0.2210$

Section 4.5

1. 29.25

3. $-6.5 + 2\ln 2 + 8\sqrt{2} \approx 6.200$

5. $6250(1 - e^{-0.8})8 \approx 3442$

7. 8.908

9. 693.0

11. **(a)** $\int_0^1 v(t)dt = 15$ miles

(b) $\int_1^2 v(t)dt = 17$ miles

(c) $\int_0^2 v(t)dt = 32$ miles. This is the sum of the answers to parts (a) and (b).

13. (a) 0.3297

(b) $\int_0^8 0.1e^{-0.1x}dx \approx 0.5507$

(c) $\int_4^8 0.1e^{-0.1x}dx \approx 0.2210$

(d) $\int_4^8 0.1e^{-0.1x}dx = \left[\frac{0.1e^{-0.1x}}{-0.1}\right]_4^8 = [-e^{-0.1x}]_4^8 = -e^{-0.8} + e^{-0.4} \approx 0.2210$

Section 4.6

1. $D_x\left[\int_1^x 2t^3 dt\right] = D_x\left(\left[\frac{2t^4}{4}\right]_1^x\right) = D_x\left(\frac{2x^4}{4} - \frac{2}{4}\right) = 2x^3$

3. $\frac{1}{\sqrt{2\pi}}e^{-x^2/2}$

5. 1

7. 0.02275

9. (a) $\int_{-2}^{m}(-0.0208x^2 + 0.0417x + 0.208)dx = 0.5$

(b) $0.006933m^3 + 0.02085m^2 + 0.208m - 0.2229 = 0$

(c) Approximately 1.004

(d) Yes; verification

(e) To find the 25th percentile (first quartile), set the integral in part (a) equal to 0.25 instead of 0.5. The answer is about −0.1321. For the 75th percentile, use the same idea with 0.75. The answer is about 2.141.

11. (a) $\int_0^m 0.25xe^{-0.5x}dx = 0.5$

(b) Approximately 3.357 years

(c) The 25th percentile is about 1.923 years, and the 75th percentile is about 5.385 years.

13. (a) Let X = the number of days after 12:01 A.M. March 28 when the last frost (temperature below 32°F) occurs.

(b) $f(x) = \begin{cases} \frac{1}{23}, & \text{for } 0 \le x \le 23 \\ 0, & \text{otherwise} \end{cases}$

(c) The median date is April 8th (at noon), since the median is at $x = 11.5$. This is because the probability density function is horizontal, so half the area occurs at the midpoint of the interval [0, 23].

15. (a) The area is 1, as it should be for a probability density function.

(b) It would also be 1, since the integral from $-\infty$ to 0 would be 0.

(c) Using technology, $\int_3^\infty 0.25xe^{-0.5}\,dx \approx 0.5578$. The probability is about 0.56.

Section 4.7

1. $p^* = 148.8$, $q^* = 9.4$, $CS = \$353.44$, $PS = \$530.16$

3. $p^* = 289$, $q^* = 12$, $CS = \$1152$, $PS = \$1152$

5. $q^* = 10.365$; $p^* = 6.5223$; $CS = \$58.56$; $PS = \$23.12$

7. (a) The interests match at \$20. This is *the equlibrium price.*

(b) At \$20, you'd make and sell 4 shirts. This is *the equilibrium quantity*.

(c) You'd earn \$12 more than your minumum acceptable. This is *the Product Surplus*.

(d) Your customers would save an average of \$22. This is *the Consumer Surplus*.

(e) *Verbal Definition*: $S(q)$ = the price, in dollars, that you would have to charge in order to supply an average of q shirts per week.
Symbol Definition: $S(q) = 2q + 12$, for $0 \le q \le 6$.
Assumptions: Certainty and divisibility. Certainty implies that the realtionship is exact. Divisibility implies that any fractional number of shirts or dollars is possible.

(f) *Verbal Definition*: $D(q)$ = the price, in dollars, that you would have to charge in order for the average demand to be q shirts per week.
Symbol Definition: $D(q) = 0.5q^2 - 6.5q + 38$, for $0 \le q \le 6$.
Assumptions: Certainty and divisibility. Certainty implies that the relationship is exact. Divisibility implies that any fractional number of shirts or dollars is possible.

(g) $q^* = 4$ shirts and $p^* = \$20$. Yes, this is exactly the same as before.

(h) $CS = \$30.67$, and $PS = \$16$. Both values are higher than before, because the table calculations correspond to areas using *inscribed* rectangles (Right Rectangles for the demand and Left Rectangles for the supply).

(i) If the selling price is \$18, customers will demand 5 shirts, but you will only make 3. Thus 3 shirts will be sold at \$18 each. Your Producer Surplus will be \$6, and your customers' Consumer Surplus will be \$28. In other words, the customers have gained \$6 directly at your expense. The total, sometimes called the *social gain*, stays the same, however.

9. Answers may vary because of different model chosen. Possible answers:

(a) *Verbal Definition*: $D(q)$ = the price, in dollars, at which exactly q sodas will be sold.
Symbol Definition: $D(q) = -0.00002855\ldots q^2 + 0.005178\ldots q + 0.7606\ldots$, for $100 \le q \le 225$.
Assumptions: Certainty and divisibility. Certainty implies that the relationship is exact. Divisibility implies that any fractional number of sodas or dollars is possible.
The quadratic model is a good fit because the data are concave down.
Verbal Definition: $S(q)$ = the price, in dollars, at which the supplier will supply exactly q sodas.
Symbol Definition: $S(q) = 0.004183\ldots q + 0.08824\ldots$, for $q \ge 100$.
Assumptions: Certainty and divisibility. Certainty implies that the relationship is exact. Divisibility implies that any fractional number of sodas or dollars is possible.
The linear model fits the data very well.

(b) $q^* = 171.84\ldots$ rounded to 173. $p^* = 0.80721\ldots$ rounded to 0.81.

(c) $CS = 19.6722\ldots$ rounded to \$19.67. $PS = 62.25334\ldots$ rounded to \$62.25.

(d) \$139.32

(e) If price = \$0.85, according to the model, they would sell 162 sodas for a revenue of \$137.70. They would not want to do this because the revenue is less than that realized at the equilibrium price.

(f) Answers may vary. From the data, you could estimate sales at \$0.85 to be 159 sodas. Revenue would then be \$135.15. This suggests a margin of error of approximately 2%.

INDEX

A

accuracy, 10, 363
AIDS cases, 129
algebraic definition of derivative, 218, 219
aligning data, 57, 58, 91, 351
analysis, 351
annual percent change in Medical CPI, 196, 301
antiderivative, 472, 503
antidifferentiation, 472, 495
approximate, 439, 465
area between two curves, 523, 531
area under a curve, 419
area under constant derivative function, 440
ASCII code, 30
assumptions, 14, 44, 46–47
average cost, 387, 403, 408
average rate of change, 155, 157, 158, 159

B

bank failures, 129–30
best-fit curve, 6
beta value, 97
bias, 363
bounding, 51
Break-even point, 82, 83, 309

C

C, 476
central value, 363
certainty, 8, 42
chain rule, 244, 248
charitable contributions in U.S., 320, 346–47
Clarify the Problem, 3, 4–5, 11
composite functions, 35, 244, 246
concave down, 108, 292, 326
concave up, 292, 325
concavity, 106, 280, 322
concavity graphs, 332, 334
conclusion and solution, 3, 9–10, 13
conclusions, 372
constant, 80
constrained global maximization, 433
Consumer Price Index, 282
  percentage change in consumer price index for U.S., 138, 303
Consumer Price Index for Medical Care, 136, 172, 244
consumer surplus, 523, 526–27, 529–30
context, 405
context of problem, 355
continuous random variable, 422
converge, 176
cost, 34, 308–9
CPI, 348
crime index in the U.S., 134, 200, 301, 321, 329
critical point, 280, 287, 306, 325, 326
cubic functions, 131, 133
cumulative distribution function, 512
cumulative probability, 500

D

data point, 5
decimal places, 369

decision variables, 44
decreasing function, 73, 107, 285–86
definite integral, 469
degree of polynomial, 351
delta notation, 161
demand, price elasticity of, 387
demand function, 525, 527
denominator, 249
dependent variable, 23
derivative, 187–88, 200
  of constant, 219–20
  of constant function, 283
  of exponential function, 236
  of linear function, 223, 224
  of polynomial, 225, 233
  of quotient, 271
derivative rules, 235–36
dimensional analysis, 63
discrete data, 457
disposable personal income, 104
divisibility, 8, 42
domain, 19, 247

E

e (Euler's number), 114
elastic demand, 396
elasticity, 281, 381, 390, 391, 393
equations, 26
equilibrium price, 524, 528, 529
equilibrium quantity, 528, 529
estimating, 188
estimating area from data, 456
evaluating functions, 26
expected behavior, 116
exponent, 249
exponential distribution, 500
exponential models, 112
extrapolation, 50, 96, 356, 497
extrema, 280, 293
  global, 280, 307
  local, 306
  of slope graph, 337

F

first order conditions, 396
first variable, 19
Forecasting, 2
forecasting, 46, 356
Formulate a Model, 3, 5–6, 11–12
function, 6, 14, 15, 19–20
functional notation, 24
fundamental theorem of calculus, 419, 493, 501
fundamental theorem of calculus, part II, 512, 514

G

general antiderivative, 473, 482, 485, 486
global extrema, 280, 307
global maximization
  constrained, 433
  unconstrained, 431
global maximum, 280
global minimum, 280

H

heights of subintervals, 446
histogram, 33
histogram distributions, 426
horizontal line test, 32
households with cable TV, 175

I

improper integrals, 512, 514, 515
increasing function, 73–74, 107, 285
indefinite integral, 473, 478
independent variable, 23
inelastic demand, 395–96
inequalities, 398
infinity, 469
inflation, 236
inflection point, 133, 351

inherent probabilistic randomness or uncertainty, 366
inner function, 246–47
input, 23
input variable, 80
instantaneous rate of change, 179, 183, 185, 186
integration, 419, 469
  variable limits of, 512
Intel Corporation, 96
intercept, 73
interest, 236
interest paid on public debt, 173
International Paper, 105
interpolation, 53, 356, 357
interpretation, 204, 205
intersection of two functions, 529
inverse, 26, 31, 502

K

KISS, 54

L

least-squares regression, 6
left rectangle method, 449
limit, 179–80, 465, 516
limit of sums, 465, 469, 503
linear function, 79
local extrema, 306
local maximum, 280
local minimum, 280
logistic curve derivative, 342–43
logistic functions, 131, 142
lower limit, 142

M

magnitude, 394
marginal, 208–9
marginal analysis, 200
marginal average cost function, 273
marginal cost, 82, 207, 209, 310, 403, 408
marginal demand, 389
marginal profit, 210, 309
marginal revenue, 82, 83, 210, 310
margin of error, 7, 9, 281, 349
mathematical model, 1, 6, 42, 44
mathematical solution, 353
maxima, 293
maximum point, 287
  relative, 293
measurement error, 363
medians, 512, 517
Medical CPI
  annual percent change in, 196, 301
  percentage change in, 322
midpoint method, 449
minima, 293
minimum point, 287
  relative, 293
mode, 419, 428
model assumptions, 360

N

natural logarithm, 238–39
negative slope, 325
net change, 440, 443
net change in a function, 503
NHL, 16, 24–26
normalized scores, 245
notation, 200, 202, 205
numerator, 249
numerical format, 49

O

one-to-one correspondence, 38
optimal solution, 15
optimization, 2, 45, 280, 356, 357
ordered pairs, 5, 20
outer function, 246–47
output, 23

P

parabola, 107
parameters, 62
pdf, 422
percentage change in consumer price index for U.S., 138, 303
percentage change in Medical CPI, 322
percentage of households with cable television, 140, 197
percentage of households with computers, 138
percent change, 113
percent rate of change, 155–56, 164
percent rate of change at a point, 387, 389
personal income, 104
points of inflection, 280, 326, 330, 331–32
positive slope, 325
post-optimality analysis, 281, 353
power functions, 118
power rule, 250, 252
precision, 363
price elasticity of demand, 387
probability and statistics, 367
probability density function, 421, 500
producer surplus, 523, 525, 526, 529
product rule, 263, 267
profit, 34, 54, 308–9
public debt
  interest paid by U.S., 149
  per capita for U.S., 149
Public Storage, 104–5

Q

quadratic function, 107–9, 292
quartic functions, 131, 137, 351
quotient rule, 273

R

radical, 249
radius, 30
random, 367
randomness, 356
range, 20
rate of change, 155, 156
  average, 155, 156, 157, 158, 159
  as a function, 192
  at a point, 177
ratios, 113
regression analysis, 91
relation, 18, 20
relationship, 18
relative likelihood, 421, 422
relative maximum point, 293
relative minimum point, 293
reservation price, 427
revenue, 34, 54, 308–9
right rectangle method, 447, 456
rounding, 48
round-off, 6
rule, 21

S

sawtooth graph, 61
secant line, 161, 176, 181
second derivative, 322
second derivative test, 326, 327
second differences, 109–10
second order conditions, 396
second variable, 20
sensitivity analysis, 65, 281, 349, 362
sigma notation, 448–49
significant digits, 10, 48, 363–64, 369
Simpson's Rule, 455, 468
slope, 73, 74, 80
  of curve, 183
  negative, 325
  positive, 325
  of secant line, 162
slope graph, 280, 282, 332
smooth continuous distributions, 429
solution validation, 360–61
Solve/Analyze the Model, 3, 6–7, 12
Standard and Poor's, 96
straight lines, 72
strictly decreasing, 433
subintervals, 444

summation, 419
supply function, 525, 526
symbol definition, 44, 45

T

tangent line, 181
trapezoid method, 451–56
trapezoid rule, 451, 453, 456, 457
trend line, 91–92
true value, 363
truncating, 50
turning points, 351
12 steps for problem solving, 11–13

U

unconstrained global maximization, 431
undiversifiable risk, 97
unemployment in U.S., 173, 319, 346
uniform distribution, 422–23
unit elasticity, 393–94
upper limit, 142

V

validate, 65
  a model, 354–55
  a solution, 355
Validate the Model and Solution, 3, 7–9, 12–13
validation, 8, 48, 281, 349, 354
variable limits of integration, 512
velocity, 440
verbal definition, 44, 45
verification, 281, 349, 353, 460
verify, 7, 65
vertical line test, 21–22